PROCEEDINGS OF THE ELEVENTH ANNUAL SYMPOSIUM ON

COMPUTATIONAL GEOMETRY

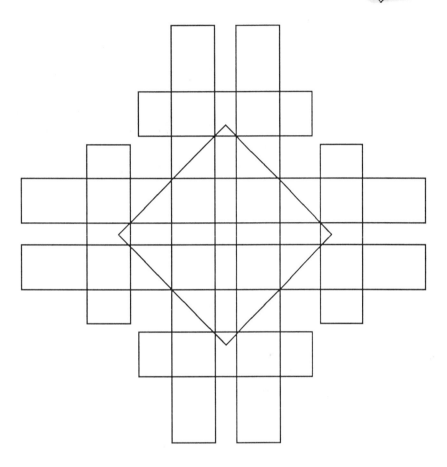

Vancouver, British Columbia, Canada
June 5-7, 1995

Sponsored by the ACM Special Interest Groups

for Graphics and Algorithms and Computation Theory

The Association for Computing Machinery
1515 Broadway
New York, N.Y. 10036

Copyright © 1995 by the Association for Computing Machinery (ACM). Copying without fee is permitted provided that the copies are not made or distributed for direct commercial advantage,and credit to the source is given. Abstracting with credit is permitted. For other copying of articles that carry a code at the bottom of the first or last page, copying is permitted provided that the per-copy fee indicated in the code is paid through the Copyright Clearance Center, 222 Rosewood Drive, Danvers, MA 01923. For permission to republish write to: Director of Publications, ACM. To copy otherwise or republish, requires a fee and/or specific permission.

ACM ISBN: 0-89791-724-3

Additional copies may be ordered prepaid from:

ACM Order Department
P.O. Box 12114
Church Street Station
New York, N.Y. 10257

Phone: 1-800-342-6626
(U.S.A. and Canada)
1-212-626-0500
(All other countries)
Fax: 1-212-944-1318
E-mail: acmhelp@acm.org
acm_europe@acm.org (in Europe)

ACM Order Number: 429950

Printed in the U.S.A.

Foreword

This volume contains papers presented at the *Eleventh Annual Symposium on Computational Geometry*, held June 5-7, 1995 in Vancouver, Canada. The symposium was sponsored by the Special Interest Groups for Graphics (SIGGRAPH) and for Algorithms and Computation Theory (SIGACT) of the Association for Computing Machinery and was partially funded with the generous support of the agencies shown below.

This proceedings contains the 41 papers and 18 communications that were selected from a total of 107 submitted extended abstracts and 22 short abstracts for communications. The program committee met in Berkeley, California, at the International Computer Science Institute. The selection was based on the quality and originality of the results and their relevance to computational geometry. The program committee wishes to thank all the people who helped us in refereeing the papers – see the list on the next page. The papers generally represent preliminary reports of ongoing research and it is expected that most of them will eventually appear in more polished form in refereed journals.

This year, for the first time, the program committee solicited communications: two-page reports, primarily on experimental results, experiences with real-world applications, and other practical issues that are relevant to computational geometry. It is the understanding of the program committee that appearance of a communication in this proceedings does *not* exclude submission of a corresponding paper to other conferences.

Descriptions of the accepted videos of the 4th Video Review of Computational Geometry appear in these proceedings. For details see the introduction on page Vl.

The program committee wishes to thank all the authors who submitted abstracts in response to the call for papers.

Program Committee

Mark de Berg
David Eppstein
Michael Goodrich
Dan Halperin
Hiroshi Imai
Raimund Seidel
Jack Snoeyink
Monique Teillaud
Seth Teller
Emo Welzl (Chair)

Conference Chair

Jack Snoeyink

Symposium Sponsored by

ACM Special Interest Groups for Graphics (SIGGRAPH) and for Algorithms & Computation Theory (SIGACT)

With Support From

Nat'l Sci. and Eng. Research Council (NSERC)
Media & Graphics Interdisciplinary Ctr (MAGIC)
University of British Columbia

Subreferees

David Alberts
Helmut Alt
Nancy Amato
Nina Amenta
Gill Barequet
Craig Becker
Avrim Blum
Jean-Daniel Boissonnat
Prosenjit Bose
Franz J. Brandenburg
Paul Callahan
Frédéric Cazals
Timothy M. Y. Chan
Bernard Chazelle
Robert Constable
Pascal Desnogues
Olivier Devillers
Katrin Dobrindt
Bruce Donald
Jacqueline Duquesne
Jeff Erickson
Andreas Fabri
Michael Facello
Stefan Felsner
Ulrich Fuchs
Bernd Gärtner
Bernhard Geiger
Michael Godau
Michael Goldwasser
Hirohisa Hirukawa
Frank Hoffmann
Christian Icking
Matthew Katz
Klara Kedem
Lutz Kettner
David Kirkpatrick
Rolf Klein

Klaus Kriegel
Sylvain Lazard
Ziming Li
Andrzej Lingas
Jim Little
Tomonari Masada
Jiří Matoušek
Michael McAllister
David Mount
Bernard Mourrain
Daniel Naiman
Franck Nielsen
Tsuyoshi Ono
Mark Overmars
Marco Pellegrini
Yuval Rabani
Kumar Ramaiyer
Edgar Ramos
Adi Rosén
Günter Rote
Jim Ruppert
Jeff Salowe
Otfried Schwarzkopf
Micha Sharir
Michiel Smid
Hisao Tamaki
Roberto Tamassia
Takeshi Tokuyama
Marc van Kreveld
Frank Wagner
Gerald Weber
Lorenz Wernisch
Peter Widmayer
Sue Whitesides
Mariette Yvinec
Günter Ziegler

Eleventh Annual ACM Symposium on Computational Geometry

June 5-7, 1995
Vancouver, B.C., Canada

Table of Contents

SESSION 1: CHAIRED BY RAIMUND SEIDEL

A Polynomial Time Algorithm for Minkowski Reconstruction
P. Gritzmann, A. Hufnagel 1

Output-Sensitive Results on Convex Hull, Extreme Points, and Related Problems
Timothy M. Chan 10

How Good are Convex Hull Algorithms?
D. Avis, D. Bremner 20

Invited Lecture. Computational Geometry
Michael Goodchild

SESSION 2: CHAIRED BY DAN HALPERIN

Immobilizing Polygons against a Wall
M. Overmars, A. Rao, O. Schwarzkopf, Ch. Wentink 29

Vertical Decomposition of Shallow Levels in 3-Dimensional Arrangements and its Applications
P. Agarwal, A. Efrat, M. Sharir 39

Efficient Collision Detection for Moving Polyhedra
E. Schömer, Chr. Thiel 51

SESSION 3: CHAIRED BY MARK DE BERG

A Comparison of Sequential Delaunay Triangulation Algorithms
P. Su, R. S. Drysdale 61

Voronoi Diagrams and Containment of Families of Convex Sets on the Plane
M. Abellanas, G. Hernandez, R. Klein, V. Neumann-Lara, Jorge Urrutia 71

Voronoi Diagrams in Higher Dimensions under Certain Polyhedral Distance Functions
J.-D. Boissonnat, M. Sharir, B. Tagansky 79

The Voronoi Diagram of Curved Objects
H. Alt, O. Schwarzkopf 89

SESSION 4: CHAIRED BY MICHAEL GOODRICH

A Combinatorial Approach to Cartograms
H. Edelsbrunner, R. Waupotitsch 98

Map Labeling Heuristics: Provably Good and Practically Useful
F. Wagner, A. Wolff 109

Overlaying Simply Connected Planar Subdivisions in Linear Time
U. Finke, K. Hinrichs 119

SESSION 5: CHAIRED BY EMO WELZL
New Lower Bounds for Hopcroft's Problem
J. Erickson 127

A Helly-Type Theorem for Unions of Convex Sets
J. Matoušek 138

On Conway's Thrackle Conjecture
L. Lovász, J. Pach, M. Szegedy 147

SESSION 6: CHAIRED BY HIROSHI IMAI
An Optimal Algorithm for Closest Pair Maintenance
S. N. Bespamyatnikh 152

The Rectangle Enclosure and Point-Dominance Problems Revisited
P. Gupta, R. Janardan, M. Smid, B. Dasgupta 162

Approximate Range Searching
S. Arya, D. Mount 172

Invited Lecture. Prediction, Data Compression, and Metric Dimension
D. Haussler

SESSION 7: CHAIRED BY DAVID EPPSTEIN
The Overlay of Lower Envelopes in Three Dimension and its Applications
P. K. Agarwal, O. Schwarzkopf, M. Sharir 182

Rounding Arrangements Dynamically
L. Guibas, D. Marimont 190

A New Technique for Analyzing Substructures in Arrangements
B. Tagansky 200

SESSION 8: CHAIRED BY MONIQUE TEILLAUD
An Optimal Algorithm for Finding Segments Intersections
I. Balaban 211

Triangulations Intersect Nicely
O. Aichholzer, F. Aurenhammer, M. Taschwer, G. Rote 220

How to Cut Pseudo-Parabolas into Segments
H. Tamaki, T. Tokuyama 230

New Algorithms and Empirical Findings on Minimum Weight Triangulation Heuristics
M. T. Dickerson, S. A. McElfresh, M. Montague 238

SESSION 9: CHAIRED BY DAN HALPERIN
Computing the Visibility Graph via Pseudo-triangulation
  M. Pocchiola, G. Vegter 248

Searching for the Kernel of a Polygon - A Competitive Strategy
  Chr. Icking, R. Klein 258

Stabbing Triangulations by Lines in 3D
  P. K. Agarwal, B. Aronov, S. Suri 267

A Complete and Practical Algorithm for Geometric Theorem Proving
  A. Rege 277

SESSION 10: CHAIRED BY MICHAEL GOODRICH
Monte Carlo Approximation of Form Factors with Error Bounded a Priori
  M. Pellegrini 287

Strategies for Polyhedral Surface Decomposition: An Experimental Study
  B. Chazelle, D. P. Dobkin, N. Shouraboura, A. Tal 297

An Experimental Comparison of Three Graph Drawing Algorithms
  G. Di Battista, A. Garg, G. Liotta, R. Tamassia, E. Tassinari, F. Vargiu 306

Visibility with Reflection
  B. Aronov, A. R. Davis, T. K. Dey, S. P. Pal, D. Chithra Prasad 316

Invited Lecture. Geometric Algorithms and the Rendering Equation
  P. Hanrahan

SESSION 11: CHAIRED BY SETH TELLER
Efficient Randomized Algorithms for Some Geometric Optimization Problems
  P. Agarwal, M. Sharir 326

Accounting for Boundary Effects in Nearest Neighbor Searching
  S. Arya, D. Mount, O. Narayan 336

A Parallel Algorithm for Linear Programming in Fixed Dimension
  M. Dyer 345

SESSION 12: CHAIRED BY MARK DE BERG
Precision-Sensitive Euclidean Shortest Path in 3-Space
  J. Choi, J. Sellen, C.-K. Yap 350

Approximation Algorithms for Geometric Tour and Network Design Problems
  Chr. Mata, J. B. Mitchell 360

Shortest Path Queries among Weighted Obstacles in the Rectilinear Plane
  D. Z. Chen, K. S. Klenk, H-Y. Tu 370

Rectilinear Geodesics in 3-Space
  J. Choi, C.-K. Yap 380

Communications

Union of Spheres (UoS) Model for Volumetric Data
  V. Ranjan, A. Fournier C2

Hexahedral Mesh Generation via the Dual
  S. Benzley, T. D. Blacker, S. A. Mitchell, P. Murdoch, T. J. Tautges C4

Fast and Robust Computation of Molecular Surfaces
  A. J. Olson, M. F. Sanner, J.-C. Spehner C6

Representation and Computation of Boolean Combinations of Sculptured Models
  S. Krishnan, D. Manocha, A. Narkhede C8

The Extensible Drawing Editor Ipe
  O. Schwarzkopf C10

Geomview: A System for Geometric Visualization
  A. Amenta, T. Munzner, S. Levy, M. Philips C12

VideHoc: A Visualizer for Homogeneous Coordinates
  R. Lewis C14

Evaluation of a New Method to Compute Signs of Determinants
  F. Avnaim, J.-D. Boissonnat, O. Devillers, F. P. Preparata, M. Yvinec C16

Exact Geometric Computation in LEDA
  Chr. Burnikel, J. Könnemann, K. Mehlhorn, S. Näher, S. Schirra, Chr. Uhrig ... C18

A Global Motion Planner for a Mobile Robot on a Terrain
  J.-D. Boissonnat, K. Dobrindt, B. Geiger, H. Michel C20

Complete Algorithms for Reorienting Polyhedral Parts Using a Pivoting Gripper
  A. Rao, D. Kriegman, K. Goldberg C22

Geometry in GIS is not Combinatorial: Segment Intersection for Polygon Overlay
  D. S. Andrews, J. Snoeyink C24

On Levels of Detail in Terrains
  M. de Berg, K. Dobrindt C26

An Animated Library of Combinatorial VLSI-Routing Algorithms
  D. Wagner, K. Weihe C28

Printed Circuit Board Simplification: Simplifying Subdivisions in Practice
  M. van Kreveld, B. van de Kraats, M. Overmars C30

An Implementation for Maintaining Arrangements of Polygons
  M. Goldwasser C32

Minimal Enclosing Parallelogram with Application
  B. L. Evans, Chr. Schwarz, J. Teich, A. Vainshtein, E. Welzl C34

Topologically Sweeping the Visibility Complex of Polygonal Scenes
  S. Rivière C36

Video Review

The Visibility Complex Made Visibly Simple
  F. Durand, C. Puech V2

An Animation of Euclid's Proposition 47: The Pythagorean Theorem
  S. Glassman, G. Nelson V3

HIPAIR: Interactive Mechanism Analysis, Design Using Configuration Spaces
  L. Joskowith, E. Sacks V5

Incremental Collision Detection for Polygonal Models
  M. K. Ponamgi, M. C. Lin, D. Manocha V7

Convex Surface Decomposition
  B. Chazelle, D. P. Dobkin, N. Shouraboura, A. Tal V9

3D Modeling Using the Delaunay Triangulation
  B. Geiger V11

Index of Authors

A Polynomial Time Algorithm for Minkowski Reconstruction

Peter Gritzmann and Alexander Hufnagel

Universität Trier, Fb IV, Mathematik,
D-54286 Trier, Germany
gritzman@dm1.uni-trier.de, hufnagel@dm5.uni-trier.de

Introduction

We deal with the following problem:

Given an $m \times n$ matrix A and positive reals $\mu_1, \ldots, \mu_m$, does there exist a polytope $P = \{x \colon Ax \leq b\}$ whose facets have volume $\mu_1, \ldots, \mu_m$, respectively, and if such a polytope exists, compute a corresponding right-hand side b of $\mathbb{R}^m$ efficiently.

This task is motivated by applications to problems in computer vision [13], [14], [9], [10] and other algorithmic questions in computational convexity (see [5]), and in view of the following classical theorem of Minkowski [15], [16] it seems fair to say that this problem is fundamental in computational convexity in that it essentially asks for the conversion of one data structure (facet normals & facet volumes) of polytopes into another (intersection of finitely many closed halfspaces).

Proposition 1. *[Minkowski 1897/1903] Let $u_1, \ldots, u_m$ be pairwise different unit vectors of Euclidean n-space $\mathbb{R}^n$ which span $\mathbb{R}^n$, and let $\mu_1, \ldots, \mu_m$ be positive reals such that $\sum_{i=1}^m \mu_i u_i = 0$. Then there exists a polytope P with outer facet normals $u_1, \ldots, u_m$ and facet volumes $\mu_1, \ldots, \mu_m$. Further, P is unique up to translation.*

Note, that the condition $\sum_{i=1}^m \mu_i u_i = 0$ simply reflects the fact that the orthogonal projections of a polytope in directions $\pm y$ coincide.

Research of the first author was supported in part by the *Alexander von Humboldt-Foundation* and by a *Max Planck Research Award*. Research of both authors was supported in part by the *Deutsche Forschungsgemeinschaft*.

Permission to copy without fee all or part of this material is granted provided that the copies are not made or distributed for direct commercial advantage, the ACM copyright notice and the title of the publication and its date appear, and notice is given that copying is by permission of the Association of Computing Machinery.To copy otherwise, or to republish, requires a fee and/or specific permission.
11th Computational Geometry, Vancouver, B.C. Canada
© 1995 ACM 0-89791-724-3/95/0006...$3.50

Proposition 1 solves the theoretical part of our problem completely. Even more, Minkowski's original proof is constructive. However, the question of efficiency is by no means clear: *Can a right hand side b be computed in polynomial time?*

We employ the *binary Turing machine model* of computation in which the input data is encoded in binary form, and the performance of an algorithm on a given input is measured in terms of the number of operations of the Turing machine [3]. This model imposes the condition on our problem that all input and all output be *rational*, hence A is a rational matrix and $b = (\beta_1, \ldots, \beta_m)^T$ is a rational vector of $\mathbb{R}^m$, whence

$$P = P_A(b) = \{x \in \mathbb{R}^n \colon Ax \leq b\}$$

is a *rational polytope*. Clearly, for rational polytopes P neither the facet volumes μ_i nor the outer facet *unit* normals need to be rational. But since the vertices of P are rational vectors it follows that for each facet $F_i = \{x \in P \colon \langle a_i, x \rangle = \beta_i\}$ of P, and each rational point $p \notin \text{aff}(F_i)$,

$$\mu_i = \text{Vol}_{n-1}(F_i) = \frac{n\|a_i\|}{|\beta_i - \langle a_i, p \rangle|} \text{Vol}_n(\text{conv}(F_i \cup \{p\})),$$

whence the product $\nu_i = \|a_i\|^{-1} \text{Vol}_{n-1}(F_i)$ is rational.

Therefore we replace the facet volumes $\mu_1, \ldots, \mu_m$ by

$$\nu_1 = \|a_1\|^{-1}\mu_1, \ldots, \nu_m = \|a_m\|^{-1}\mu_m,$$

and regard $\nu_1, \ldots, \nu_m$ as part of the input. Then

$$\sum_{i=1}^m \nu_i a_i = 0$$

and for each rationally presented polytope $P_A(b)$ the corresponding *Minkowski data* $a_1, \ldots, a_m; \nu$ is rational. This takes care of the rationality of the input.

The rationality requirement for the output b is a different problem: even if A and ν are rational, the Minkowski reconstruction problem may not admit a rational solution unless $n \leq 2$. Hence, for our algorithmic purposes we need to resort to the following problem

MINKAPP:
Given $m, n \in \mathbb{N}$, vectors $a_1, \ldots, a_m \in \mathbb{Q}^n \setminus \{0\}$, no two positively dependent, which span $\mathbb{R}^n$, positive rationals $\nu_1, \ldots, \nu_m$ such that $\sum_{i=1}^m \nu_i a_i = 0$, and a positive rational error bound ϵ; determine $\tilde{b} = (\tilde{\beta}_1, \ldots, \tilde{\beta}_m)^T \in \mathbb{Q}^m$ such that the facet volumes $\tilde{\mu}_1, \ldots, \tilde{\mu}_m$ of the polytope $P_A(\tilde{b})$ satisfy

$$\max_{i=1,\ldots,m} |\tilde{\mu}_i - \nu_i \|a_i\|| < \epsilon.$$

We will often "collect" the input data of an instance of MINKAPP in a string

$$(n, m; A, v; \epsilon),$$

where A is the matrix with rows $a_1^T, \ldots, a_m^T$, and $v = (\nu_1, \ldots, \nu_m)^T$. If the dimension n is not part of the input, but a constant that has been fixed a-priorily, we call the problem n-MINKAPP, while the "exact variant" of MINKAPP (where ϵ is a-priorily set to 0) will in the following be denoted by MINKRECON.

In his original proof [15], [16], MINKOWSKI models MINKRECON as the task to minimize a linear functional over a closed convex region $C \subset \mathbb{R}^m$ (see Subsection 2.1). LITTLE [14] (see also [13], [9]) used standard methods from convex optimization to develop an algorithm for MINKAPP in dimension 3, which was then applied to a practical problem of computer vision. However, even though [14]'s work is restricted to polytopes in $\mathbb{R}^3$, there is no polynomial bound on the running time of his algorithm.

Here we settle the related complexity question completely by showing:

Theorem 1. *For each fixed $n \in \mathbb{N}$, n-MINKAPP can be solved in polynomial time.*

Theorem 2. *MINKAPP is $\#\mathbb{P}$-equivalent, i.e. $\#\mathbb{P}$-hard and $\#\mathbb{P}$-easy.*

The proofs of our main Theorems 1 and 2 are indicated in Sections 3 and 4, while Sections 1 and 2 contain basic prerequisites and tools.

1. PRELIMINARIES

1.1. Presentations of General Convex Bodies. Unlike for polytopes, there is in general no finite manner to present arbitrary convex sets. To deal with convex bodies algorithmically, GRÖTSCHEL, LOVÁSZ & SCHRIJVER [8] augment the Turing-machine model by so-called "oracles" which provide information about the convex body in question. Let $\mathcal{C}^n$ denote the family of all convex subsets of $\mathbb{R}^n$ with nonempty interior, and let $\mathcal{K}^n$ denote the set of all proper convex bodies, i.e. of all compact members of $\mathcal{C}^n$. Further, define for $C \in \mathcal{C}^n$ and $\epsilon \geq 0$

$$C(\epsilon) = C + \epsilon \mathbb{B}^n = \{x + \epsilon y \colon x \in C \wedge y \in \mathbb{B}^n\},$$

and

$$C(-\epsilon) = C \setminus ((\mathbb{R}^n \setminus C) + \epsilon \operatorname{int} \mathbb{B}^n),$$

the *outer* and the *inner parallel sets* of C, respectively. Then the oracles for sets $C \in \mathcal{C}^n$ that are most relevant for our purposes solve one of the following problems.

WEAK MEMBERSHIP PROBLEM.
Given a vector $y \in \mathbb{Q}^n$ and a rational number $\epsilon > 0$, conclude with one of the following:
- *assert that $y \in C(\epsilon)$;*
- *assert that $y \notin C(-\epsilon)$.*

WEAK OPTIMIZATION PROBLEM.
Given $c \in \mathbb{Q}^n \setminus \{0\}$ and a rational number $\epsilon > 0$, conclude with one of the following:
- *compute a vector $y \in \mathbb{Q}^n \cap K(\epsilon)$ such that $\langle c, y \rangle \leq \langle c, x \rangle + \epsilon$ for every $x \in C(-\epsilon)$;*
- *assert that $C(-\epsilon)$ is empty;*
- *assert that $\langle c, x \rangle$ is not bounded below on C.*

If a set $C \in \mathcal{C}^n$ is given by an algorithm $\mathcal{O}$ that solves the weak membership problem or the weak optimization problem, we say that C is given by a *weak membership oracle* or a *weak optimization oracle* $\mathcal{O}$, respectively. C is called *well-guaranteed* if rational numbers r, R are given in advance such that C contains a ball of radius r and $C \subset R\mathbb{B}^n$. If a vector $a \in \mathbb{Q}^n$ and a positive rational r are given beforehand, such that $a + r\mathbb{B}^n \subset C$ we call C *centered*. Note that a well-guaranteed set $C \in \mathcal{C}^n$ is compact, whence contained in $\mathcal{K}^n$.

The *size* $\langle K \rangle$ of a centered well-guaranteed body $K \in \mathcal{K}^n$ with parameters r, R and a that is given by any of the above oracles is defined as

$$\langle K \rangle = n + \langle r \rangle + \langle R \rangle + \langle a \rangle,$$

where $\langle r \rangle$, $\langle R \rangle$, and $\langle a \rangle$ denote the usual *binary sizes* of the numbers r, R, and the vector a, respectively. Furthermore, the input size of the above three problems are $\langle y \rangle + \langle \epsilon \rangle$ and $\langle c \rangle + \langle \epsilon \rangle$, respectively.

Using the *ellipsoid-algorithm*, [8] prove, in particular, the following theorem.

Proposition 2. *The problem whose instance is a centered well-guaranteed body $K \in \mathcal{K}^n$ given by an oracle* *for one of the problems weak membership or weak optimization, and whose task it is to solve the other problem admits an oracle-polynomial time algorithm.*

A string $(n, m; A, b)$, where $n, m \in \mathbb{N}$, A is a rational $m \times n$-matrix and $b \in \mathbb{Q}^m$ such that $P = \{x \in \mathbb{R}^n : Ax \leq b\}$ is a polytope is called an $\mathcal{H}$-*polytope* in $\mathbb{R}^n$ (and is identified with the geometric object P). Note that the preceding assumptions "centered, well-guaranteed" can be satisfied for $\mathcal{H}$-polytopes in the following sense; see e.g. [8].

Proposition 3. *The following problems can be solved in polynomial time: Given an $\mathcal{H}$-polytope P,*

- *decide whether $P = \emptyset$, or*
- *find an (irredundant) rational $\mathcal{H}$-presentation of $\mathrm{aff}(P)$, and compute a point $a \in P$ and rational numbers r, R such that*

$$(a + r\mathbb{B}^n) \cap \mathrm{aff}(P) \subset P \subset R\,\mathbb{B}^n.$$

Clearly, $\langle a \rangle$, $\langle r \rangle$ and $\langle R \rangle$ are bounded by a polynomial in $\langle P \rangle$, specifically, $r \geq 2^{-4\langle P \rangle}$ and $R \leq 2^{4\langle P \rangle}$.

1.2. Polytope Volume Computation. Here we deal with the following problem VOLAPP (and its variant VOLCOMP for $\epsilon = 0$):

VOLAPP:
Given an $\mathcal{H}$-polytope $P = (n, m; A, b)$, and a positive rational ϵ; compute a rational number $\hat{V}$ such that

$$\left|\hat{V} - \mathrm{Vol}(P)\right| \leq \epsilon.$$

Let us point out that [12] has shown that $\langle \mathrm{Vol}(P) \rangle$ cannot be bounded by a polynomial in $\langle P \rangle$ if n is part of the input. This means that, in general, $\mathrm{Vol}(P)$ cannot be computed exactly in polynomial space. The nontrivial part of the following proposition is due to [1].

Proposition 4. *If the dimension n is fixed, VOLCOMP can be solved in polynomial time; otherwise, VOLAPP is $\#\mathbb{P}$-equivalent.*

For a survey on volume computation and its applications see [5].

2. A Convex Program Related to MINKRECON

2.1. Minkowski's Proof Revisited. Let A be a rational $m \times n$ matrix with rows $a_1^T, \ldots, a_m^T$, no two positively dependent, such that $\mathrm{pos}\{a_1, \ldots, a_m\} = \mathbb{R}^n$.

Then the input vectors $v = (\nu_1, \ldots, \nu_m)^T$ of MINKRECON are contained in the kernel N_A of the linear mapping $y \mapsto y^T A$. Observe that for $t \in \mathbb{R}^n$

$$t + P_A(b) = P_A(b + At).$$

This implies that for each vector $v = (\nu_1, \ldots, \nu_m)^T \in N_A$ the solutions of MINKRECON correspond to an equivalence class $b + N_A^\perp$, where $N_A^\perp = \{At : t \in \mathbb{R}^n\}$. Hence each instance of MINKRECON determines a point in the $(m - n)$-dimensional quotient space $X_A = \mathbb{R}^m / N_A^\perp$. Using the *Brunn-Minkowski-theorem*, and the *Minkowski-inequality* see [18] one can show the following result that is due to MINKOWSKI [15], [16].

Proposition 5. *Let $(n, m; A, v)$ denote an instance of MINKRECON, let b be a solution, (i.e. $P_A(b)$ has facet volumes $\nu_i \|a_i\|$). Further, let $b^\diamond$ be an optimizer of*

$$\min\{\langle v, z \rangle : \mathrm{Vol}(P_A(z)) = 1\}.$$

Then $P_A(b^\diamond)$ is homothetic to $P_A(b)$; specifically, $\lambda b^\diamond$ with

$$\lambda = \left(\tfrac{1}{n}\langle v, b^\diamond \rangle\right)^{1/(n-1)}$$

solves MINKRECON.

Clearly the optimum does not change when we replace the feasible region by

$$C = \{z \in \mathbb{R}^m : \mathrm{Vol}(P_A(z)) \geq 1\}.$$

Note that C is closed (since the volume is continuous) and, by Brunn-Minkowski's theorem, C is also convex. Hence MINKRECON is equivalent to the following convex program:

MINKCONVMIN:
Given an instance $(n, m; A, v)$ of MINKRECON; minimize the functional $z \mapsto \langle v, z \rangle$ on $C = \{z \in \mathbb{R}^m : \mathrm{Vol}(P_A(z)) \geq 1\}$.

Let us point out that though being constructive, the above characterization is not fully adequate for our algorithmic purposes. In particular, the convex set C is unbounded with n-dimensional lineality space, and it hence causes problems to apply results from the algorithmic theory of convex bodies. The main difficulty is, however, to handle the fact that only approximative solutions will be available: the oracles are weak. This means, we need strong *stability estimates* for MINKRECON between the error in the optimization problem and an appropriate distance measure for the corresponding polytopes.

2.2. Bounds for the Solutions of MinkConvMin.

In order to be able to apply Proposition 2, we replace the set C by a suitable centered well-guaranteed convex body $K \subset C$ without changing the optimal value of the objective function. Let L denote the size of the input $(n, m; A, v)$, let again $C = \{z : \text{Vol}(P_A(z)) \geq 1\}$, and set

$$S(v) = \{\tilde{v} \in N_A : \|\tilde{v} - v\|_\infty \leq \tfrac{1}{2}\min\{\nu_1, \ldots, \nu_m\}\}.$$

Further, for $K \in \mathcal{K}^n$, let $r(K)$ and $R(K)$ denote K's inradius, and circumradius, respectively, and set $\mathbf{1} = (1, \ldots, 1)^T \in \mathbb{R}^m$.

Then we can prove the following Lemma.

Lemma 1. *Let $(n, m; A, v)$ be an instance of MinkConvMin of size L, and let $b^\diamond$ denote a positive solution. Further, let $\vartheta = 2^{-16nL}$, $\Theta = \vartheta^{-1}$ and $I = [\vartheta, \Theta]$, and set*

$$K = \{z : 2^{-15nL}\mathbf{1} \leq z \leq 2^{2L}\mathbf{1} \wedge \text{Vol}(P_A(z)) \geq 1\}.$$

Then the following statements hold.

(i) *K is a well-bounded convex body of $\mathbb{R}^m$, $K \subset I^n$, $1 \leq r(K)$, and $\langle K \rangle$ is bounded by a polynomial in L.*
(ii) *$b^\diamond \in \tfrac{1}{3}2^{2L}[0, 1]^m$.*
(iii) *Whenever $\tilde{v} \in S(v)$, the body K contains an optimizer $\tilde{b}^\diamond$ for the instance $(n, m; A, \tilde{v})$ of MinkConvMin.*
(iv) *$R(P_A(z)) \leq \Theta$ for all $z \in K$.*
(v) *$\langle \tilde{v}, z\rangle \in nI$ for every $\tilde{v} \in S(v)$ and $z \in K$.*
(vi) *$\tilde{\lambda} = \left(\tfrac{1}{n}\langle \tilde{v}, \tilde{b}^\diamond\rangle\right)^{\frac{1}{n-1}} \in I$, for every $\tilde{v} \in S(v)$; ($\tilde{\lambda}$ is the scaling factor defined in Proposition 5).*
(vii) *$R(P_A(\tilde{b})) \leq \Theta^2$, where $\tilde{b}$ is any solution of MinkRecon on the input $(n, m; A, \tilde{v})$ with $\tilde{v} \in S(v)$.*

Proof. Omitted; see the full version [4] of this paper. □

3. Minkowski Reconstruction from a Volume Oracle

We will now show that MinkApp can be reduced to VolApp in polynomial time. Theorem 1 and the easyness part of Theorem 2 will then follow from Proposition 4.

The *Hausdorff-distance* of two bodies $K_1, K_2 \in \mathcal{K}^n$ is defined by

$$d(K_1, K_2) = \min\{\tau \geq 0 : K_1 \subset K_2 + \tau \mathbb{B}^n \wedge K_2 \subset K_1 + \tau \mathbb{B}^n\}.$$

The corresponding functional d on $\mathcal{K}^n \times \mathcal{K}^n$ is called *Hausdorff-metric*. Due to the translation invariance of our problem, it is more appropriate to utilize the *translative Hausdorff-metric* δ, which is defined by

$$\delta(K_1, K_2) = \min\{d(t + K_1, K_2) : t \in \mathbb{R}^n\},$$

for $K_1, K_2 \in \mathcal{K}^n$.

The reduction of MinkApp to VolApp is divided into three steps:

- Use an oracle for VolApp to construct a weak membership oracle for the well-bounded convex body $K \subset C$ of Lemma 1. Then Proposition 2 yields a weak optimization oracle for K.

This first reduction is rather straightforward, and is ommitted here; for a detailed proof see [11], Section 6.3.

- Approximate a solution of MinkRecon with respect to the *translative Hausdorff-metric*.

This reduction is the main part of the proof, it will be given in Subsection 3.2. Subsection 3.1 contains stability estimates which facilitate the transition from "weak solution" in the sense of weak optimization oracles to approximate solution of MinkRecon with respect to the translative Hausdorff-metric.

- Give the desired approximation with respect to the distance measure of MinkApp.

This last reduction is contained in Subsection 3.3.

3.1. Some Stability Estimates.
The following result is due to Groemer [6]; see also [18, p.318] and [7]. It gives a bound on the "stability term"

$$\Delta(K_1, K_2) = V(\overbrace{K_1, \ldots, K_1}^{n-1}, K_2) - \text{Vol}^{n-1}(K_1)\text{Vol}(K_2)$$

of Minkowski's inequality, where $V(\overbrace{K_1, \ldots, K_1}^{n-1}, K_2)$ is the first *mixed volume* of the bodies K_1, K_2; see [18], [2] and [17].

Proposition 6. *Let $K_1, K_2 \in \mathcal{K}^n$ with $\text{Vol}(K_1) = \text{Vol}(K_2) = 1$. Further define $R = \max\{R(K_1), R(K_2)\}$. Then*

$$\delta(K_1, K_2) \leq \left(\frac{1}{2n}\right)^{\frac{1}{n+1}}(4 \cdot 6.00025 \cdot n)$$
$$\cdot 2R \cdot \Delta(K_1, K_2)^{\frac{1}{n+1}} \leq 50Rn\Delta(K_1, K_2)^{\frac{1}{n+1}}.$$

While Proposition 6 is quite deep, the following Lemma is rather easy to show.

Lemma 2. Let $K_1, K_2 \in \mathcal{K}^n$, set again
$$R = \max\{R(K_1), R(K_2)\},$$
and let λ_1, λ_2 be nonnegative reals. Then
$$\delta(\lambda_1 K_1, \lambda_2 K_2) \leq |\lambda_1 - \lambda_2| R + \delta(K_1, K_2) \max\{\lambda_1, \lambda_2\}.$$

Proof. Omitted; see the full version [4] of this paper. □

3.2. Minkowski Reconstruction for the Translative Hausdorff-Distance.

We now give proofs of the tractability results for the variant of MINKAPP that is obtained by replacing its distance measure $\max_{i=1,\ldots,m}\{|\tilde{\mu}_i - \mu_i|\}$ by the translative Hausdorff-distance δ. This is in fact the main step in the proofs of our "positive" results.

Let $(n, m; A, v; \epsilon)$ be an instance of MINKAPP, let L denote its size, and suppose that $\epsilon < 1$. We want to construct a vector $\tilde{b} \in \mathbb{Q}^m$ such that
$$\delta(P_A(b), P_A(\tilde{b})) \leq \epsilon,$$
where b denotes again an (exact) solution of MINKRECON on the instance $(n, m; A, v)$. Without further notice, we use the notation of Lemma 1. The construction falls into two steps: first we use the membership oracle for the set K to compute a rational polytope $P = P_A(\tilde{b}^\diamond)$ with $\delta(P, P_A(b^\diamond)) \leq \epsilon_1$, where $b^\diamond$ is an exact solution to MINKCONVMIN on the input $(n, m; A, v)$, and $\epsilon_1 = \frac{1}{4}\vartheta\epsilon$. In the second step we compute $\tilde{b}$ by means of a suitable scaling.

Suppose we have a weak optimization oracle $\mathcal{O}$ for K. Let us call $\mathcal{O}$ with input v and error bound
$$\eta = \frac{1}{(200n)^{n+1} \cdot 8n} \vartheta^{n+2} \epsilon_1^{n+1}.$$

Since $K(-\eta) \neq \emptyset$, the oracle outputs a vector $\tilde{b}^\diamond = (\tilde{\beta}_1^\diamond, \ldots, \tilde{\beta}_m^\diamond)^T \in \mathbb{Q}^m$ satisfying
$$\tilde{b}^\diamond \in K(\eta),$$
$$\langle v, \tilde{b}^\diamond \rangle \leq \langle v, z \rangle + \eta \quad \text{for all } z \in K(-\eta).$$

Then $\tilde{b}^\diamond + \eta\mathbf{1} \in C$, and $b^\diamond + \eta\mathbf{1} \in K(-\eta)$, which implies
$$\text{Vol}(P_A(\tilde{b}^\diamond + \eta\mathbf{1})) \geq 1,$$
$$(\diamondsuit) \quad \langle v, \tilde{b}^\diamond \rangle \leq \langle v, b^\diamond \rangle + \eta\langle v, \mathbf{1}\rangle \leq \langle v, b^\diamond \rangle + \eta 2^L.$$

Since for $i = 1, \ldots, m$,
$$\tilde{\beta}_i^\diamond - (1 - 2\Theta\eta)(\tilde{\beta}^\diamond + \eta) \geq \eta(2\Theta\tilde{\beta}_i^\diamond - 1) \geq 0,$$
we have
$$P_A(\tilde{b}^\diamond) \supset P_A((1 - 2\Theta\eta)(\tilde{b}^\diamond + \eta\mathbf{1})),$$

whence
$$\text{Vol}(P_A(\tilde{b}^\diamond)) \geq (1 - 2\eta\Theta)^n.$$
Now let
$$\sigma = \left(\text{Vol}(P_A(\tilde{b}^\diamond))\right)^{-\frac{1}{n}}.$$
Then $\text{Vol}(P_A(b^\diamond)) = \text{Vol}(P_A(\sigma\tilde{b}^\diamond)) = 1$, and it follows that $\sigma \leq (1 - 2\eta\Theta)^{-1} < \frac{4}{3}$. Using the homogenity of mixed volumes, it follows from Proposition 5 that
$$V(\overbrace{P_A(b), \ldots, P_A(b)}^{n-1}, P_A(\sigma\tilde{b}^\diamond))$$
$$= \frac{1}{n}\langle v, b^\diamond\rangle \cdot V(\overbrace{P_A(b^\diamond), \ldots, P_A(b^\diamond)}^{n-1}, P_A(\sigma\tilde{b}^\diamond));$$
and since
$$V(\overbrace{P_A(b), \ldots, P_A(b)}^{n-1}, P_A(\sigma\tilde{b}^\diamond)) \leq \frac{\sigma}{n}\langle v, \tilde{b}^\diamond\rangle$$
and
$$\eta\langle v, \mathbf{1}\rangle \leq \eta\Theta \leq \frac{1}{8n},$$
we have
$$\Delta\left(P_A(b^\diamond), P_A(\sigma\tilde{b}^\diamond)\right) = V_1\left(P_A(b^\diamond), P_A(\sigma\tilde{b}^\diamond)\right)^n - 1$$
$$= \left(\frac{n}{\langle v, b^\diamond\rangle} V(\overbrace{P_A(b), \ldots, P_A(b)}^{n-1}, P_A(\sigma\tilde{b}^\diamond))\right)^n - 1$$
$$\leq \sigma^n \left(\frac{\langle v, \tilde{b}^\diamond\rangle}{\langle v, b^\diamond\rangle}\right)^n - 1$$
$$\leq \sigma^n \left(1 + \frac{\eta 2^L}{\langle v, b^\diamond\rangle}\right)^n - 1$$
$$\leq \left(\frac{1 + \eta\Theta}{1 - 2\eta\Theta}\right)^n - 1 \leq (1 + 4\eta\Theta)^n - 1$$
$$\leq 4n\eta\Theta + \sum_{j=2}^n (4n\eta\Theta)^j \leq 8n\eta\Theta.$$

To apply Proposition 6 we need an upper bound for $R(P_A(b^\diamond))$ and $R(P_A(\sigma\tilde{b}^\diamond))$. Since $\tilde{b}^\diamond \in K(\eta)$ there exists some $z = (\zeta_1, \ldots, \zeta_m)^T \in K$ such that
$$\tilde{\beta}_i^\diamond - \eta \leq \zeta_i \quad \text{for all } i = 1, \ldots, m;$$
using Lemma 1 (iv) this implies that
$$R(P_A(\tilde{b}^\diamond - \eta\mathbf{1})) \leq \Theta.$$
Since, further, $\tilde{\beta}_i^\diamond \leq \frac{3}{2}(\tilde{\beta}_i^\diamond - \eta)$ for $i = 1, \ldots, m$, by the choice of η, it follows that
$$R(P_A(\tilde{b}^\diamond)) \leq \frac{3}{2}\Theta.$$
Now $\sigma \leq \frac{4}{3}$, hence
$$\max\{R(P_A(b^\diamond)), R(P_A(\sigma\tilde{b}^\diamond))\} \leq 2\Theta.$$

Using Proposition 6, we obtain

$$\delta(P_A(b^\diamond), P_A(\sigma \tilde{b}^\diamond)) \leq 50n \cdot 2\Theta \cdot (8n\eta\Theta)^{\frac{1}{n+1}} \leq \frac{1}{2}\epsilon_1.$$

Now we use an oracle for VOLAPP to approximate $\mathrm{Vol}(P_A(\tilde{b}^\diamond))$ to absolute error $\vartheta^2\epsilon_1$. Using the fact that $\sigma \leq \frac{4}{3}$, and applying the mean value theorem of calculus to the function $x \mapsto x^{-\frac{1}{n}}$, we obtain an estimate $\tilde{\sigma}$ of σ such that

$$|\sigma - \tilde{\sigma}| \leq \vartheta^2 \epsilon_1 \frac{1}{n}\left(\frac{4}{3}\right)^{-\frac{n+1}{n}} \leq \frac{1}{4}\vartheta\epsilon_1.$$

By Lemma 2,

$$\delta\big(P_A(\tilde{\sigma}\tilde{b}^\diamond), P_A(\sigma\tilde{b}^\diamond)\big) \leq |\tilde{\sigma} - \sigma| \cdot \frac{3}{2}\Theta \leq \frac{\epsilon_1}{2},$$

whence

$$\delta\big(P_A(\tilde{\sigma}\tilde{b}^\diamond), P_A(b^\diamond)\big) \leq$$
$$\delta\big(P_A(\tilde{\sigma}\tilde{b}^\diamond), P_A(\sigma\tilde{b}^\diamond)\big) + \delta\big(P_A(\sigma\tilde{b}^\diamond), P_A(b^\diamond)\big)$$
$$\leq \epsilon_1.$$

This concludes the first step of the construction.

In order to obtain the desired vector $\tilde{b} \in \mathbb{Q}^m$ with $\delta\big(P_A(b), P_A(\tilde{b})\big) \leq \epsilon$, we will now apply a suitable scaling to $\tilde{b}^\diamond$. Let

$$\lambda = \left(\frac{1}{n}\langle v, b^\diamond\rangle\right)^{\frac{1}{n-1}} \quad \text{and} \quad \tilde{\lambda} = \left(\frac{1}{n}\langle v, \tilde{b}^\diamond\rangle\right)^{\frac{1}{n-1}};$$

see Proposition 5. With the aid of Lemma 1, we obtain

$$\frac{1}{n}\langle v, \tilde{b}^\diamond\rangle = \frac{1}{n}\langle v, \tilde{b}^\diamond + \eta\mathbf{1}\rangle - \frac{1}{n}\eta\langle v, \mathbf{1}\rangle \geq \vartheta - \eta\Theta \geq \frac{3}{4}\vartheta$$

and

$$\frac{1}{n}\langle v, b^\diamond\rangle \geq \frac{3}{4}\vartheta.$$

Further,

$$\langle v, b^\diamond\rangle \leq \sigma\langle v, \tilde{b}^\diamond\rangle \leq (1 + 4\eta\Theta)\langle v, \tilde{b}^\diamond\rangle \leq \langle v, \tilde{b}^\diamond\rangle + 4\eta\Theta^2.$$

With the aid of ($\Diamond$) and the mean value theorem applied to the function $x \mapsto x^{\frac{1}{n-1}}$, it follows that

$$|\tilde{\lambda} - \lambda| \leq \frac{1}{n-1}\left(\frac{3\vartheta}{4}\right)^{\frac{1}{n-1}-1} 4\eta\Theta^2 < 6\eta\Theta^3.$$

Now, we approximate $\tilde{\lambda}$ to absolute error $2\eta\Theta^3$; let $\hat{\lambda}$ denote the corresponding estimate. Then, using $8\eta \leq \vartheta^2$ and $\lambda \leq \Theta$, we have

$$|\hat{\lambda} - \lambda| \leq 8\eta\Theta^3, \quad \text{and} \quad \max\{\lambda, \hat{\lambda}\} \leq 2\Theta.$$

Setting $\tilde{b} = \hat{\lambda}\tilde{\sigma}\tilde{b}^\diamond$, and using $\eta \leq \frac{1}{8}\vartheta^3\epsilon_1$, Lemma 2 yields

$$\delta\big(P_A(b), P_A(\tilde{b})\big) = \delta\big(\lambda P_A(b^\diamond), \hat{\lambda} P_A(\tilde{\sigma}\tilde{b}^\diamond)\big)$$
$$\leq 8\eta\Theta^3 \cdot 2\Theta + 2\Theta \cdot \epsilon_1 \leq \epsilon,$$

and this is the desired estimate.

3.3. From Hausdorff to Minkowski. The following Lemma gives a relation between the Hausdorff-distances of two polytopes $P_A(z_1)$, $P_A(z_2)$ and the distances of corresponding facets.

Lemma 3. *Let $a_1, \ldots, a_m \in \mathbb{R}^n \setminus \{0\}$, mutually non-collinear, with $\mathrm{pos}\{a_1, \ldots, a_m\} = \mathbb{R}^n$, let A denote the matrix with rows $a_1^T, \ldots, a_m^T$, let $\epsilon > 0$ and let $z_1, z_2 \in \mathbb{R}^m$ such that*

$$d\big(P_A(z_1), P_A(z_2)\big) \leq \epsilon.$$

For $i = 1, \ldots, m$, let $F_i(z_1)$ and $F_i(z_2)$ denote the faces of $P_A(z_1)$ and $P_A(z_2)$, respectively, that correspond to a_i, and let $t_i \in \epsilon\mathbb{B}^n$ such that $F_i(c_1)$ and $t_i + F_i(c_2)$ lie in a common hyperplane H. Then we have for the Hausdorff-distance d_H relative to H,

$$d_H\big(F_i(z_1), t_i + F_i(z_2)\big) \leq \frac{2\epsilon}{1 - \alpha^2} \leq 2\epsilon\Theta,$$

where

$$\alpha = \max\left\{\left\langle \frac{a_k}{\|a_k\|}, \frac{a_l}{\|a_l\|}\right\rangle : 1 \leq k < l \leq m\right\}.$$

Proof. Omitted; see the full version [4] of this paper. □

Next we give a simple technical lemma.

Lemma 4. *Let $K_1, K_2 \in \mathcal{K}^n$, both contained in $R\mathbb{B}^n$. Then, with $\omega_n = \mathrm{Vol}(\mathbb{B}^n)$,*

$$|\mathrm{Vol}(K_1) - \mathrm{Vol}(K_2)| \leq$$
$$\omega_n R^n \left(\left(1 + \frac{d(K, L)}{R}\right)^n - 1\right).$$

Proof. Omitted; see the full version [4] of this paper. □

We are now able to finish the proof of our tractability results. Let $(n, m; A, b; \epsilon)$ be an instance of MINKAPP, let L denote its size, and assume that $\epsilon < 1$. Calling the algorithm of Subsection 3.2 with error bound

$$\epsilon_2 = \frac{1}{4n}\vartheta^{3n+1}\frac{1}{\lceil\omega_{n-1}\rceil}\epsilon,$$

we obtain a polytope $P_A(\tilde{b})$ satisfying $\delta\big(P_A(b), P_A(\tilde{b})\big) \leq \epsilon_2$, where b is a solution of MINKRECON on the input $(n, m; A, b)$. Now let $i \in \{1, \ldots, m\}$. If necessary, we apply a translation t_i to bring the faces $F_i(b)$ of $P_A(b)$ and $F_i(\tilde{b})$ of $P_A(\tilde{b})$ into the same hyperplane $H = \mathrm{aff}(F_i(b))$. Then by Lemma 3,

$$d = d(F_i(\tilde{b}), t_i + F_i(b)) \leq \frac{\vartheta^{3n}\epsilon}{2n\omega_{n-1}}.$$

Now $R(P_A(b)), R(P_A(\tilde{b})) \leq \Theta^3$, and hence, clearly,
$$R(F_i(b)), R(F_i(\tilde{b})) \leq \Theta^3.$$
So we can apply Lemma 4 and obtain
$$\begin{aligned}
|\nu_i\|a_i\| - \mu_i(\tilde{b})| &\leq \omega_{n-1} R^{n-1}\left(\left(1+\frac{d}{R}\right)^{n-1}-1\right) \\
&\leq \omega_{n-1} \sum_{j=1}^{n-1}(dn)^j R^{n-j-1} \\
&\leq \omega_{n-1} \Theta^{3n} \frac{dn}{1-dn} \\
&\leq 2\omega_{n-1}\Theta^{3n} dn \\
&\leq \epsilon.
\end{aligned}$$

This concludes the proof the #ℙ-easyness of MINKAPP and the proof of Theorem 1.

Note that we have, in fact, proved that MINKAPP can be solved in $\mathcal{A}$-oracle-polynomial time, where $\mathcal{A}$ is an oracle for VOLAPP; this statement persists when the problems are restricted to instances that correspond to subclasses of polytopes.

4. HARDNESS OF MINKOWSKI RECONSTRUCTION

In the last section we showed that any oracle for computing the volume of an $\mathcal{H}$-polytope can be used to devise an oracle-polynomial time algorithm for the weak optimization problem MINKCONVMIN, whence for MINKAPP. It it possible to reverse this argument and hence to show that any oracle for MINKAPP gives an oracle-polynomial time algorithm for computing the volume of $\mathcal{H}$-polytopes. Since the latter problem is #ℙ-hard if the dimension is part of the input, so is the former.

The proof of the #ℙ-hardness of MINKAPP consists of three reductions.

- Reduction of VOLAPP to the weak membership problem for the feasible regions C of MINKCONVMIN.

This step will be omitted; see the full version [4] of this paper.

- Reduction of the weak membership problem for such regions C to a the weak optimization problem for C.

Note that Proposition 2 is restricted to convex bodies, and hence does not directly apply to the unbounded set C. The special structure of C, however, allows us to handle this problem, see Subsection 4.1.

- Reduction of the weak optimization problem for the regions C to MINKAPP.

This last step will be outlined in Section 4.2.

4.1. Hardness of Minkowski-Optimization.

Now we show how a weak optimization oracle for MINK-CONVMIN can be used to devise an oracle-polynomial time algorithm for the weak membership problem for the underlying convex set C. Let, again, $(n, m; A)$ be given as before, let
$$C = \{z \in \mathbb{R}^m \colon \mathrm{Vol}(P_A(z)) \geq 1\},$$
and suppose that a weak optimization oracle for C is available for positive rational inputs (v, ϵ) with $v \in N_A$. Suppose, now, that (b, ϵ) is a given instance of the corresponding weak membership problem, i.e. $b \in \mathbb{Q}^m$ is the query point, and ϵ is the positive rational error bound. Since $C + A\mathbb{R}^n \subset C$, we may assume that $b > 0$.

Note that we are not really interested in points of C whose components are much greater than those of b. This can be used to show that for solving the weak membership problem on the input (b, ϵ) we need only calls to the optimization oracle whose set of optimizers contains a point not much greater than b. To make this observation precise, we will now cut off points of C by considering the polar of some translate of C.

It is quite easy to compute in poynomial time $p \in C \cap \mathbb{Q}^m$, and $\rho \in \mathbb{Q}$ such that $2p + \rho\mathbb{B}^m \subset C$, $b < p$, and $\langle p \rangle$ and $\langle \rho \rangle$ are bounded by a polynomial in the input size. Now, let $\hat{C} = -2p + C$ and let $\hat{b} = b - 2p$; note that $\hat{b} < 0$. To cut off all points of $\hat{C}$ that are not lying in the negative orthant, let
$$\hat{C}^\circ = \{y \in \mathbb{R}^m \colon \langle z, y \rangle \leq 1 + 2\langle p, y \rangle \text{ for all } z \in C\}.$$

Since C contains the ball $2p + \rho\mathbb{B}^m$, the set $\hat{C}^\circ$ is bounded. Further,
$$\hat{C} + [0, \infty[^m + N_A^\perp \subset \hat{C}$$
which implies that
$$\hat{C}^\circ \subset]-\infty, 0]^m \cap N_A.$$

It follows that $\hat{C}^\circ$ has nonempty interior relative to N_A. Clearly, the weak optimization oracle for C leads to a weak optimization oracle for $\hat{C}$. It is not hard to see, that this allows to solve the weak membership problem for $\hat{C}^\circ$ and some input $y \in N_A$ (relative to N_A); a detailed proof (following the lines of the proof of [8], Lemma 4.4.1) can be found in [11].

Now, $\hat{C}^\circ$ is a well-presented convex body (in aff $\hat{C}^\circ = N_A$), so using Proposition 2 (applied to $\hat{C}^\circ$ relative to

N_A) we obtain an oracle-polynomial time weak optimization algorithm for $\hat{C}^\circ$. Next we use linear programming to compute in polynomial time a point $\bar{v}$ of $]0,\infty[^m \cap N_A$, and set $S = \text{conv}(\hat{C}^\circ \cup \{\bar{v}\})$. Such a point $\bar{v}$ exists because $a_1,\ldots,a_m$ positively span $\mathbb{R}^n$. Note that S is again a centered well-bounded convex body in N_A, and $0 \in \text{relint}\,S$. Further, the weak optimization algorithm for $\hat{C}^\circ$ can easily be extended to S. But this means that we can solve the weak membership problem in oracle-polynomial time for the polar $\bar{C}$ of S relative to N_A. Note that

$$\bar{C} = \hat{C} \cap N_A \cap \{z \in \mathbb{R}^m : \langle \bar{v}, z \rangle \leq 1\}.$$

Further, $\langle \bar{v}, \hat{b} \rangle \leq 0 < 1$, thus $\hat{b} \in \hat{C}(\pm\epsilon)$ if and only if the orthogonal projection $\hat{b}'$ of $\hat{b}$ onto N_A is contained in $\hat{C}(\pm\epsilon)$.

4.2. From Minkowski to Weak Optimization.

We complete the #$\mathbb{P}$-hardness proof of Theorem 2 by showing that an oracle $\mathcal{A}$ for solving MINKAPP can be used to produce weak minima of linear functionals $z \mapsto \langle v, z \rangle$ over the sets C, where $v \in]0,\infty[^m \cap N_A$. We use the notation and results of Subsections 2.1 and 2.2.

Let $(n,m;A;v)$ be an instance of MINKCONVMIN of size L and $\epsilon \in \mathbb{Q}$, $0 < \epsilon \leq 1$. We have to compute a weak minimum of the linear functional $z \mapsto \langle v, z \rangle$ over $C = \{z : \text{Vol}(P_A(z)) \geq 1\}$. We call the oracle $\mathcal{A}$ on the input $(n,m;A;v;\eta)$, where $\eta = \vartheta^{10}\epsilon$, with ϑ as defined in Lemma 1, to produce a rational vector $\tilde{b} \in \mathbb{R}^m$ such that, for the facet volumes $\tilde{\mu}_i$ of $P_A(\tilde{b})$,

$$|\tilde{\mu}_i - \nu_i \|a_i\|| < \eta \quad \text{for } i = 1, 2, \ldots, m.$$

Let $\tilde{v} = (\tilde{\mu}_1 \|a_1\|^{-1}, \ldots, \tilde{\mu}_m \|a_m\|^{-1})^T$; then

$$\|v - \tilde{v}\| \leq \frac{\sqrt{m}\eta}{\min\{\|a_1\|,\ldots,\|a_m\|\}} \leq \eta 2^L,$$

whence, in particular, $\tilde{v} \in S(v)$. Now, let b° and $\tilde{b}^\circ$ be solutions of MINKCONVMIN in K for the inputs $(n,m;A,v)$ and $(n,m;A,\tilde{v})$, respectively. Then, as shown before,

$$b = \rho^{-1}b^\circ \quad \text{and} \quad \tilde{b} = \tilde{\rho}^{-1}\tilde{b}^\circ,$$

where

$$\rho = \text{Vol}(P_A(b))^{-\frac{1}{n}} = \left(\frac{n}{\langle v,b \rangle}\right)^{\frac{1}{n}},$$

and

$$\tilde{\rho} = \text{Vol}(P_A(\tilde{b}))^{-\frac{1}{n}} = \left(\frac{n}{\langle \tilde{v},\tilde{b} \rangle}\right)^{\frac{1}{n}}$$

are solutions of MINKRECON for the inputs $(n,m;A;v)$ and $(n,m;A;\tilde{v})$, respectively. Since $\tilde{v} \in S(v)$, Lemma 1 shows that all entries of $\tilde{b}^\circ$ and b° are bounded from above by Θ.

We obtain

$$|\langle v, \tilde{b}^\circ \rangle - \langle v, b^\circ \rangle| \leq |\langle v, \tilde{b}^\circ \rangle - \langle \tilde{v}, \tilde{b}^\circ \rangle| + |\langle \tilde{v}, \tilde{b}^\circ \rangle - \langle v, b^\circ \rangle|$$
$$\leq |\langle v - \tilde{v}, \tilde{b}^\circ \rangle| + \max\{\langle \tilde{v} - v, b^\circ \rangle, \langle v - \tilde{v}, \tilde{b}^\circ \rangle\}$$
$$\leq \|v - \tilde{v}\| (\|\tilde{b}^\circ\| + \max\{\|\tilde{b}^\circ\|, \|b^\circ\|\})$$
$$\leq \eta 2^L \cdot 2\sqrt{m}\Theta \leq \eta\Theta^2.$$

Note that $\hat{\rho} = \left(\frac{1}{n}\langle v, \tilde{b} \rangle\right)^{-\frac{1}{n}}$ is an estimate of $\tilde{\rho} = \left(\frac{1}{n}\langle \tilde{v}, \tilde{b} \rangle\right)^{-\frac{1}{n}}$ (the latter number cannot be computed directly since $\tilde{v}$ is not available), and we may use $\hat{\rho}\tilde{b}$ as an estimate for $\tilde{b}^\circ$.

With the aid of Lemma 1, it is easy to see that

$$\frac{1}{n}\langle v, \tilde{b} \rangle \geq \frac{1}{n}\tilde{\rho}^{-1}\langle v, b^\circ \rangle \geq \vartheta^2$$

and

$$\frac{1}{n}|\langle v, \tilde{b} \rangle - \langle \tilde{v}, \tilde{b} \rangle|$$
$$\leq \frac{1}{n}\|v - \tilde{v}\| \cdot \tilde{\rho}^{-1}\|\tilde{b}^\circ\| \leq \frac{\sqrt{m}}{n}\eta 2^L\Theta^2.$$

Using the mean value theorem for $x \mapsto x^{-\frac{1}{n}}$ we obtain

$$|\tilde{\rho} - \hat{\rho}| \leq \frac{1}{n}\eta\Theta^7.$$

Since $\hat{\rho}$ can be approximated to any absolute accuracy in polynomial time we can compute a rational number ρ' with

$$|\rho' - \tilde{\rho}| \leq \eta\Theta^7.$$

Setting $\tilde{b}' = \rho'\tilde{b}$ and combing the above estimates we have

$$|\langle v, b^\circ \rangle - \langle v, b' \rangle|$$
$$\leq |\langle v, b^\circ \rangle - \langle v, \tilde{b}^\circ \rangle| + |\langle v, \tilde{b}^\circ \rangle - \langle v, b' \rangle|$$
$$\leq \eta\Theta^2 + |\tilde{\rho} - \rho'| \|v\| \|\tilde{b}\|$$
$$\leq \eta\Theta^2 + \eta\Theta^7 m 2^L \Theta^2$$
$$\leq \eta\Theta^{10} = \epsilon.$$

So, all that remains to be shown is that $b' \in C(\epsilon)$. But this follows from the fact that $\tilde{b}^\circ \in C$, in conjunction with the inequality

$$\|\tilde{b}^\circ - b'\| \leq |\tilde{\rho} - \rho'| \|\tilde{b}\| \leq \eta\Theta^7\sqrt{m}\Theta^2 \leq \epsilon.$$

This completes the final reduction in the proof of the #$\mathbb{P}$-hardness of the problem MINKAPP.

REFERENCES

1. M.E. Dyer and A.M. Frieze, *The complexity of computing the volume of a polyhedron*, SIAM J. Comput. **17** (1988), 967–974.
2. M.E. Dyer, P. Gritzmann, and A. Hufnagel, *On the complexity of computing mixed volumes*, Manuscript, 1994.
3. M.R. Garey and D.S. Johnson, *Computers and intractability*, Freeman, San Francisco, 1979.
4. P. Gritzmann and A. Hufnagel, *On the algorithmic complexity of Minkowski's reconstruction theorem*, Manuscript, 1994.
5. P. Gritzmann and V. Klee, *On the complexity of some basic problems in computational convexity: II. Volume and mixed volumes*, Polytopes: Abstract, Convex and Computational (Boston) (T. Bisztriczky, P. McMullen, R. Schneider, and A. I. Weiss, eds.), Kluwer, 1994, pp. 373–466.
6. H. Groemer, *On an inequality of Minkowski for mixed volumes*, Geom. Dedicata **33** (1990), 117–122.
7. _____, *Stability of geometric inequalities*, Handbook of Convex Geometry Vol. A (Amsterdam) (P.M. Gruber and J.M. Wills, eds.), North-Holland, 1993, pp. 125–150.
8. M. Grötschel, L. Lovász, and A. Schrijver, *Geometric algorithms and combinatorial optimization*, Springer, Berlin, 1988.
9. B.K.P. Horn, *Extended Gaussian images*, Proceedings of the IEEE **72** (1984), 1671–1686.
10. _____, *Robot vision*, MIT Press, Mc Graw-Hill, Cambridge, Mass., 1986.
11. A. Hufnagel, *Algorithmic problems in Brunn-Minkowski theory*, Dissertation, Trier, 1995.
12. J. Lawrence, *Polytope volume computation*, Math. Comput. **57** (1991), 259–271.
13. J.J. Little, *An iterative method for reconstructing convex polyhedra from extended Gaussian images*, Proceedings of the AAAI, National Conference on Artificial Intelligence (Washington D.C.), 1983, pp. 247–250.
14. _____, *Extended Gaussian images, mixed volumes, and shape reconstruction*, Proc. 1st Conf. on Computational Geometry (Baltimore), 1985, pp. 15–23.
15. H. Minkowski, *Allgemeine Lehrsätze über konvexe Polyeder*, Nachr. Ges. Wiss. Göttingen (1897), 198–219.
16. _____, *Volumen und Oberfläche*, Math. Ann. **57** (1903), 447–495.
17. J.R. Sangwine-Yager, *Mixed volumes*, Handbook of Convex Geometry, Vol. A (Amsterdam) (P.M. Gruber and J.M. Wills, eds.), North-Holland, 1993, pp. 43–72.
18. R. Schneider, *Convex bodies: The Brunn-Minkowski theory*, Encyclopedia of Mathematics and its Applications, Vol. 44, Cambridge University Press, 1993.

Output-Sensitive Results on Convex Hulls, Extreme Points, and Related Problems

Timothy M. Chan*
Department of Computer Science
University of British Columbia

Abstract

We use known data structures for ray shooting and linear programming queries to derive new output-sensitive results on convex hulls, extreme points, and related problems. We show that the f-face convex hull of an n-point set P in a fixed dimension d can be constructed in $O(n \log f + (nf)^{1-1/(\lfloor d/2 \rfloor + 1)} \log^{O(1)} n)$ time. In particular, this yields new optimal output-sensitive convex hull algorithms in two and three dimensions. We also show that the h extreme points of P can be computed in $O(n \log^{d+2} h + (nh)^{1-1/(\lfloor d/2 \rfloor + 1)} \log^{O(1)} n)$ time. Our techniques are then applied to obtain improved time bounds for other problems including convex layers, levels in arrangements, and linear programming with few violated constraints.

1 Introduction

Let P be a set of n points in d-dimensional Euclidean space E^d, where $d \geq 2$ is a fixed constant. (We shall implicitly assume that the points are in general position.) The smallest convex set containing P is a polytope conv(P) called the *convex hull* of P. It is known that the number of faces, f, in this polytope is at worst $O(n^{\lfloor d/2 \rfloor})$ [23]. In the convex hull problem, we want to construct the facial structure of conv(P). This problem has been intensively studied in computational geometry [12, 26, 27, 30], and it has applications to other geometric problems such as computing intersections of halfspaces and computing Voronoi diagrams and Delaunay triangulations.

---

*Supported by a Killam Predoctoral Fellowship and an NSERC Postgraduate Scholarship.

Permission to copy without fee all or part of this material is granted provided that the copies are not made or distributed for direct commercial advantage, the ACM copyright notice and the title of the publication and its date appear, and notice is given that copying is by permission of the Association of Computing Machinery. To copy otherwise, or to republish, requires a fee and/or specific permission.
11th Computational Geometry, Vancouver, B.C. Canada
© 1995 ACM 0-89791-724-3/95/0006...$3.50

Chazelle [6] has solved the convex hull problem optimally in the worst case by giving an $O(n \log n + n^{\lfloor d/2 \rfloor})$-time algorithm. However, this bound depends only on the input size n and is insensitive to the output size f. An optimal $O(n \log f)$-time output-sensitive algorithm in two dimensions was given by Kirkpatrick and Seidel [19]. For dimension 3, Edelsbrunner and Shi [13] obtained an $O(n \log^2 f)$-time method, and Chazelle and Matoušek [8] demonstrated that optimal $O(n \log f)$ time is possible by derandomizing an earlier algorithm due to Clarkson and Shor [10]. In any fixed dimension, the "gift-wrapping" algorithm of Swart [36] and the "beneath/beyond" algorithm of Seidel [34] achieve $O(nf)$ and $O(n^2 + f \log n)$ time respectively. The latter is subsequently improved to $O(n^{2-2/(\lfloor d/2 \rfloor + 1) + \varepsilon} + f \log n)$ by Matoušek [20] using a data structuring technique that he has developed for linear programming queries. (Throughout this paper, $\varepsilon > 0$ denotes an arbitrarily small constant.) It is in fact possible to reduce the $O(n^{2-2/(\lfloor d/2 \rfloor + 1) + \varepsilon})$ term to $O(n^{2-2/(\lfloor d/2 \rfloor + 1)} \log^{O(1)} n)$ by using the static structures in [20]. Recently, Chan, Snoeyink, and Yap [4] have obtained an $O((n+f) \log^2 f)$-time algorithm in four dimensions; in higher dimensions, their method is less efficient, running in $O((n + f^{d-3}) \log^{d-2} f)$ time. Thus, there is still a large gap between the known upper bounds and the $\Omega(n \log f + f)$ lower bound for $d > 4$.

Here, we show that the gift-wrapping method can be further improved using the data structures for ray shooting queries in polytopes developed by Agarwal and Matoušek [1] and refined by Matoušek and Schwarzkopf [22]. Our convex hull algorithm runs in $O(n \log f + (nf)^{1-1/(\lfloor d/2 \rfloor + 1)} \log^{O(1)} n)$ time and is optimal when $f = O(n^{1/\lfloor d/2 \rfloor} / \log^K n)$ for a sufficiently large K. Furthermore, it is faster than all previous methods when $f = O(n/\log^K n)$ and $d > 4$. Note that in many cases, f can in fact be sublinear; for example, Raynaud [31] proved that the expected value of f is $O(n^{(d-1)/(d+1)})$ if the points of P are chosen uniformly at random from a d-dimensional ball. The expected number of hull vertices is even polylogarithmic in n if the points are chosen uniformly from a hypercube or from a normal distribution [2, 31].

Surprisingly, our method leads to new optimal output-sensitive algorithms in two and three dimensions, running in $O(n \log f)$ time. In the plane, our algorithm is as simple as Kirkpatrick and Seidel's algorithm [19] or its improvement by Chan, Snoeyink, and Yap [4], and has the added advantage that no median-finding subroutine is required; our method is essentially a refinement of Jarvis's march [18]. In three dimensions, our algorithm is simpler than Chazelle and Matoušek's [8] in that we do not use ε-approximations or other complex derandomization tools; the only data structure technique we use is the Dobkin-Kirkpatrick hierarchy [11].

Next, we turn to the problem of computing the *extreme points* of P, i.e., the vertices of $\text{conv}(P)$ (or equivalently, the set of points $p \in P$ with $\text{conv}(P - \{p\}) \neq \text{conv}(P)$). By Megiddo's linear programming algorithm [24], we can test whether a given point is an extreme point of P in linear time; this immediately yields an algorithm for the extreme point problem that runs in $O(n^2)$ time. Matoušek reduced the bound to $O(n^{2-2/(\lfloor d/2 \rfloor+1)+\varepsilon})$ using his data structures for linear programming queries [20]; again, the n^ε factor can be replaced by $\log^{O(1)} n$ if static structures are used. We further improve this to an output-sensitive bound. Let h denote the number of extreme points ($h \leq n$). By applying Matoušek's data structures to a simple $O(nh)$-time algorithm, we show that the extreme points can be computed in $O(n \log^{d+2} h + (nh)^{1-1/(\lfloor d/2 \rfloor+1)} \log^{O(1)} n)$ time. We have been informed that Clarkson [9] has independently obtained a similar result.

We then consider the problem of computing the *convex layers* of P, defined iteratively as follows: layer 1 is convex hull of P, and if layer i is nonempty, then layer $i + 1$ is defined as the convex hull of the points of P that are not vertices of the previous layers $1, \ldots, i$. It is known that this problem can be solved optimally in $O(n \log n)$ time by an algorithm of Chazelle [5] and quasi-optimally in $O(n^{1+\varepsilon})$ time by an algorithm of Agarwal and Matoušek [1].

For $d \geq 4$, Edelsbrunner [12, Problem 10.3(c)] asked whether the vertices of all layers can be identified in $o(n^3)$ time. This problem is equivalent to finding the *depth* of p, i.e., the index of the layer of which p is a vertex, for every $p \in P$. It is not difficult to get an $O(n^{3-3/(\lfloor d/2 \rfloor+1)+\varepsilon})$-time solution by applying Matoušek's technique [20]. Clarkson's $O(nh)$-time extreme point algorithm [9] yields a simple $O(n^2)$-time solution. Here, we show how the depth problem can be solved in $O(n^{2-\beta+\varepsilon})$ time with $\beta = 2/(\lfloor d/2 \rfloor^2 + 1)$;

for example, in four or five dimensions, the bound is $O(n^{8/5+\varepsilon})$. As a result, we can construct the convex layers in $O(n^{2-\beta+\varepsilon} + f \log n)$ time, where f is now the total number of faces in all layers (which is at least $\Omega(n)$ and at most $O(n^{\lfloor d/2 \rfloor})$).

Finally, we examine applications of our ideas to other related problems. The first application we consider is the construction of a level in an arrangement of hyperplanes. Given a set H of n hyperplanes in E^d, the *k-level* in the arrangement $\mathcal{A}(H)$ is defined as the set of all points in E^d that have at most k hyperplanes of H above it ($0 \leq k < n$). The 0-level is just the dual of a convex hull. In the plane, an output-sensitive algorithm for constructing the k-level was given by Edelsbrunner and Welzl [15]. We improve its running time from $O(n \log n + f \log^2 n)$ to $O(n \log f + f \log^2 n)$, where f denotes the size of the k-level. In higher dimensions, Agarwal and Matoušek [1] proposed a method based on ray shooting queries, which runs in $O(n \log n + f^{1+\varepsilon})$ time for $d = 3$ (actually they state a weaker $O((n + f)n^\varepsilon)$ bound). We improve this to $O(n \log f + f^{1+\varepsilon})$ and show that the time bound is $O(n \log f + (nf)^{1-1/(\lfloor d/2 \rfloor+1)+\varepsilon} + f n^{1-2/(\lfloor d/2 \rfloor+1)+\varepsilon})$ for $d \geq 4$.

Another related problem studied here is: given a set H of n hyperplanes in E^d, a direction ξ, and a small integer $0 \leq k < n$, find a point in the k-level of $\mathcal{A}(H)$ that is minimal w.r.t. ξ; in other words, find a minimal point that lies on or above all but at most k of the hyperplanes in H. This is the feasible case of the *linear programming problem with at most k violated constraints*. For this problem, Matoušek [21] has devised a method that runs in $O(n \log n + k^2 \log^2 n)$ time if $d = 2$ and $O(n \log n + k^{3+\varepsilon})$ time if $d = 3$; when $d \geq 4$ and k is sufficiently small (more precisely: $k^d \leq n^{1/\lfloor d/2 \rfloor+\varepsilon}$), the running time is $O(n \log n)$. The method actually enumerates all $O(k^d)$ local minima of the $(\leq k)$-levels. We show how the $O(n \log n)$ term can be reduced to $O(n \log k)$ in two dimensions or to $O(n \log \log n + n \log k)$ in higher dimensions.

As an aside, we point out that the Matoušek's results [21] can be used to improve an algorithm by Mulmuley [25] for constructing $(\leq k)$-levels of a nonredundant arrangement of n hyperplanes in E^d. The algorithm is an extension of Seidel's output-sensitive convex hull algorithm [34] and runs in $O(n^2 k^{d-1} + f \log n)$ time for an f-face output. We decrease the time bound to $O(n^{2-2/(\lfloor d/2 \rfloor+1)+\varepsilon} k^{d-1} + f \log n)$.

We remark that all our algorithms depend heavily on the assumption that the input points/hyperplanes are in general position. The justification for this assumption is provided by standard perturbation techniques [14, 16].

However, one should keep in mind that these perturbation methods may increase the output size when there are a large number of degeneracies.

The remainder of this paper is organized as follows. In Sections 2 and 3, we review some of the known data structures for ray shooting queries and linear programming queries, which serve as the basic tools of our approach. Our contribution is a (very simple) preprocessing time/query time tradeoff that allows us to obtain improved time bounds when there are only a small number of queries. We then apply these results to the output-sensitive construction of convex hulls in Section 4 and the output-sensitive computation of extreme points in Section 5. Applications to convex layers and depths are discussed in Section 6; further applications are given in Section 7. We then conclude with some final remarks in Section 8.

2 Ray Shooting Queries

We first investigate the problem of ray shooting in a convex polytope. Given a collection H of n (closed) halfspaces in E^d, where each halfspace contains a known point, say, the origin o, a *ray shooting query* is to determine the first bounding hyperplane h of $\bigcap H$ that is crossed by a query ray originating from $\bigcap H$ (a ray *crosses* a hyperplane h if it intersects h but is not contained in h).

In two dimensions, the ray shooting problem can be solved as follows: first compute the polygon $\bigcap H$ and store its vertices in an array in counterclockwise order; then a query can be done by a simple binary search. Observe that computing the intersection $\bigcap H$ is equivalent to computing a convex hull in the dual space, and thus takes $O(n \log n)$ time by Graham's scan for example [17]; and the binary search takes $O(\log n)$ time. Hence, this method requires $O(n \log n)$ preprocessing time, $O(n)$ space, and $O(\log n)$ query time.

The same preprocessing time, space, and query time can be obtained in three dimensions: in the preprocessing, compute the polytope $\bigcap H$ by the dual of Preparata and Hong's convex hull algorithm [29] and construct its Dobkin-Kirkpatrick hierarchical representation [11]; then use the query algorithm from [11].

Our first observation is that a preprocessing time/query time tradeoff is possible using a standard "grouping" technique. Using this observation, we can perform q queries in $O(n \log q)$ time rather than $O(n \log n)$ time for small q's.

Lemma 2.1 *There is a (static) data structure for ray shooting in a polytope defined by a set H of n halfspaces in E^2 or E^3 with $O(n \log m)$ preprocessing time, $O(n)$ space, and $O((n/m) \log m)$ query time, where m is a parameter between 1 and n.*

Proof: Partition H into $\lceil n/m \rceil$ subsets ("groups") $H_1, \ldots, H_{\lceil n/m \rceil}$, each of size at most m and build the above structures for each H_i. The total preprocessing time is $O(\frac{n}{m}(m \log m)) = O(n \log m)$, and the space complexity remains $O(n)$. Since ray shooting is a *decomposable* problem (i.e., the answer to a query on $H' \cup H''$ can be computed from the answers to the queries on H' and H'' in constant time), a query on H can be computed directly by querying on each H_i, taking $O((n/m) \log m)$ time. □

Corollary 2.2 *A sequence of q ray shooting queries in a polytope defined by a set H of n halfspaces in E^2 or E^3 can be performed in $O(n \log q + q \log n)$ time and $O(n)$ space.*

Proof: By Lemma 2.1, the total time needed to answer q queries is $O(n \log m + q(\frac{n}{m} \log m))$, where $1 \leq m \leq n$. Choose $m = q$ when $q \leq n$ and choose $m = n$ when $q > n$. □

For ray shooting queries in d-dimensional polytopes, Agarwal and Matoušek [1] have proposed a data structuring method that was subsequently improved by Matoušek and Schwarzkopf [22]. Table 1 shows their results. The "grouping" scheme may again be applied to obtain further preprocessing time/query time tradeoffs for Structure 1.

Lemma 2.3 *There is a (static) data structure for ray shooting in a polytope defined by a set H of n halfspaces in E^d with $O(n \log m)$ preprocessing time, $O(n)$ space, and $O((n/m^{1/\lfloor d/2 \rfloor}) \log^{O(1)} m)$ query time, where m is a parameter between 1 and n.*

Proof: By partitioning H into $\lceil n/m \rceil$ groups as in Lemma 2.1 and using Structure 1 to store each group, the preprocessing time becomes $O(\frac{n}{m}(m \log m)) = O(n \log m)$ and query time becomes $O(\frac{n}{m}(m^{1-1/\lfloor d/2 \rfloor} \log^{O(1)} m)) = O(\frac{n}{m^{1/\lfloor d/2 \rfloor}} \log^{O(1)} m)$. □

Corollary 2.4 *A sequence of q ray shooting queries in a polytope defined by a set H of n halfspaces in E^d can be performed in $O(n \log q + (nq)^{1-1/(\lfloor d/2 \rfloor+1)} \log^{O(1)} n + q \log n)$ time.*

Proof: CASE I. $q \leq n^{1/\lfloor d/2 \rfloor}/\log^K n$, where K is a sufficiently large constant. Use Lemma 2.3's modification of Structure 1 with $m = (q \log^K q)^{\lfloor d/2 \rfloor}$ ($1 \leq m \leq n$). Then the running time is

$$O\left(n \log m + q \frac{n}{m^{1/\lfloor d/2 \rfloor}} \log^{O(1)} m\right) = O(n \log q).$$

Structures	preprocessing time, space	update time (amortized)	ray shooting query time	linear programming query time
1	$n \log n$, n	N/A	$n^{1-1/\lfloor d/2 \rfloor} \log^{O(1)} n$	$n^{1-1/\lfloor d/2 \rfloor} \log^{O(1)} n$
2	$m \log^{O(1)} n$	N/A	$\frac{n}{m^{1/\lfloor d/2 \rfloor}} \log n$	$\frac{n}{m^{1/\lfloor d/2 \rfloor}} \log^{2d+1} n$
3	$n^{\lfloor d/2 \rfloor}/\log^{\lfloor d/2 \rfloor - \varepsilon} n$	N/A	$\log n$	$\log^{d+1} n$
1'	$n \log n$, n	$\log^2 n$	$n^{1-1/\lfloor d/2 \rfloor + \varepsilon}$	$n^{1-1/\lfloor d/2 \rfloor + \varepsilon}$
2'	$m^{1+\varepsilon}$	$m^{1+\varepsilon}/n$	$\frac{n}{m^{1/\lfloor d/2 \rfloor}} \log n$	$\frac{n}{m^{1/\lfloor d/2 \rfloor}} \log^{2d+1} n$
3'	$n^{\lfloor d/2 \rfloor + \varepsilon}$	$n^{\lfloor d/2 \rfloor - 1 + \varepsilon}$	$\log n$	$\log^{d+1} n$

Table 1: Known data structures for ray shooting queries in polytopes [22] and linear programming queries [20]. (For Structures 2 and 2', m is a parameter between n and $n^{\lfloor d/2 \rfloor}$.)

CASE II. $n^{1/\lfloor d/2 \rfloor} < q < n^{\lfloor d/2 \rfloor}$. Use Structure 2 with $m = (nq)^{1-1/(\lfloor d/2 \rfloor + 1)}$ ($n \leq m \leq n^{\lfloor d/2 \rfloor}$). Then the running time is

$$O\left(m \log^{O(1)} n + q \frac{n}{m^{1/\lfloor d/2 \rfloor}} \log n\right)$$
$$= O((nq)^{1-1/(\lfloor d/2 \rfloor + 1)} \log^{O(1)} n).$$

CASE III. $q \geq n^{\lfloor d/2 \rfloor}$. Use Structure 3. Then the running time is $O\left(n^{\lfloor d/2 \rfloor}/\log^{\lfloor d/2 \rfloor - \varepsilon} n + q \log n\right) = O(q \log n)$. □

Remark: In some applications, the number of queries q may not be known in advance. In that case, the parameter m cannot be set directly. This problem can be avoided by breaking the q queries into k clusters of $q_1, \ldots, q_k$ queries, where $q_1, q_2, \ldots$ is a known sequence and $q_1 + \ldots + q_{k-1} < q \leq q_1 + \ldots + q_k$. For example, in Case I of the proof of Corollary 2.4, if we choose the sequence $q_i = 2^{2^i}$ ($i = 1, 2, \ldots$), then the total running time is $O(\sum_{i=1}^{k} n \log q_i) = O(\sum_{i=1}^{\lceil \log \log q \rceil} n 2^i) = O(n \log q)$, as before. (Logarithms are in base 2.) Similarly, in Case II, we see that the complexity remains unchanged by choosing the sequence $q_i = n^{1/\lfloor d/2 \rfloor} 2^i$ ($i = 1, 2, \ldots$).

We now discuss dynamic ray shooting in polytopes, where halfspaces may be inserted or deleted. In two dimensions, a data structure by Overmars and van Leeuwen [28] has $O(n \log n)$ preprocessing time, $O(n)$ space, $O(\log^2 n)$ update time, and $O(\log n)$ query time. It is straightforward to extend Lemma 2.1 to get a method with $O(n \log m)$ preprocessing time, $O(n)$ space, $O((n/m) \log^2 m)$ update time, and $O((n/m) \log m)$ query time ($1 \leq m \leq n$). We can then obtain a dynamic planar version of Corollary 2.2:

Lemma 2.5 *A sequence of q ray shooting queries in a polygon defined by a dynamic set H of at most n halfplanes in E^2, and q insertions/deletions on H can be performed in $O(n \log q + q \log^2 n)$ time and $O(n)$ space.*

Proof: By the above method, the total time needed to perform q queries and updates is $O(n \log m + q(\frac{n}{m} \log^2 m))$, where $1 \leq m \leq n$. Choose $m = q \log q$ when $q \leq n/\log n$ and choose $m = n$ otherwise. □

In higher dimensions, Matoušek and Schwarzkopf [22] have provided dynamic versions of their data structures, as shown in the bottom half of Table 1. The grouping technique can be applied to get a preprocessing time/query time tradeoff for Structure 1'. This modification has $O(n \log m)$ preprocessing time, $O(n)$ space, $O((n/m) \log^2 m)$ amortized update time, and $O(n/m^{1/\lfloor d/2 \rfloor - \varepsilon})$ query time ($1 \leq m \leq n$). We then have:

Lemma 2.6 *A sequence of q ray shooting queries in a polytope defined by a dynamic set H of at most n halfspaces in E^d can be performed in*

i. $O(n \log q + (nq)^{1-1/(\lfloor d/2 \rfloor + 1) + \varepsilon} + q n^{1-2/(\lfloor d/2 \rfloor + 1) + \varepsilon})$ *time if the number of insertions/deletions is $O(q)$;*

ii. $O(n \log^2 n + (nq)^{1-1/(\lfloor d/2 \rfloor + 1) + \varepsilon} + q \log n)$ *time if the number of insertions/deletions is $O(n)$.*

Proof: For (i), use the above modification of Structure 1' with $m^{1/\lfloor d/2 \rfloor - \varepsilon} = q$ when $q \leq n^{1/\lfloor d/2 \rfloor - \varepsilon}$; use Structure 2' with $m = (nq)^{1-1/(\lfloor d/2 \rfloor + 1)}$ when $n^{1/\lfloor d/2 \rfloor} < q < n$ and Structure 2' with $m = n^{2-2/(\lfloor d/2 \rfloor + 1)}$ when $q \geq n$.

For (ii), use Structure 1' when $q \leq n^{1/\lfloor d/2 \rfloor - \varepsilon}$, Structure 2' with $m = (nq)^{1-1/(\lfloor d/2 \rfloor + 1)}$ when $n^{1/\lfloor d/2 \rfloor} < q < n^{\lfloor d/2 \rfloor}$, and Structure 3' when $q \geq n^{\lfloor d/2 \rfloor + \varepsilon}$. □

3 Linear Programming Queries

Given a collection H of n halfspaces in E^d, each containing the origin o, a *linear programming query* is to determine the vertex v of the polytope $\bigcap H$ that maximizes $\xi \cdot v$ for a query vector $\xi \in E^d$.

We begin by extending the grouping technique of Lemmas 2.1 and 2.3 to handle linear programming with a small number of queries. This is not trivial because linear programming, unlike ray shooting, is not a decomposable problem.

Lemma 3.1 *There is a dynamic data structure for linear programming queries on a set H of n halfplanes in E^2 with $O(n \log m)$ preprocessing time, $O(n)$ space, and $O((n/m) \log^2 m)$ update and query time, where m is a parameter between 1 and n.*

Proof: We consider the static case first. Partition H into $\lceil n/m \rceil$ groups $H_1, \ldots, H_{\lceil n/m \rceil}$, each of size at most m, compute the convex polygon $\Pi_i = \bigcap H_i$ for each i, and store each of them in an ordered array. The total preprocessing time is then $O(\frac{n}{m}(m \log m)) = O(n \log m)$, while space is linear. Reichling [32] showed that in $O(k \log^2 m)$ time, one can detect whether the intersection of k convex m-gons is empty, and if not, report the point in the intersection that is extreme in a given direction ξ; his method is based on Megiddo's prune-and-search technique. Using Reichling's algorithm on the $k = \lceil n/m \rceil$ polygons $\Pi_1, \ldots, \Pi_{\lceil n/m \rceil}$, we can answer a linear programming query in $O((n/m) \log^2 m)$ time.

The dynamic part can be proven using Overmars and van Leeuwen's data structure [28] to store each of the H_i's, which requires $O((n/m) \log^2 m)$ update time. □

As a result of this lemma, q linear programming queries in the plane can be answered in $O(n \log q)$ time for $q \leq n/\log n$.

Corollary 3.2 *A sequence of q linear programming queries and q insertions/deletions on a dynamic set H of at most n halfplanes in E^2 can be performed in $O(n \log q + q \log^2 n)$ time and $O(n)$ space.*

In higher dimensions, Matoušek [20] has obtained data structures for linear programming queries achieving the complexities shown in Table 1. His approach uses a multidimensional parametric search technique to reduce the problem of answering linear programming queries to that of answering halfspace-emptiness queries with witness. (In the dual setting, a *halfspace-emptiness query* on H is to determine whether a given query point p belongs to $\bigcap H$, and if not, provide a *witness* halfspace $h \in H$ that does not contain p.)

We now show how to obtain a preprocessing time/query time tradeoff for Structure 1. The bounds we get are similar to those in Lemma 2.3, except for an extra polylogarithmic factor in n in the query time; this causes an additional $O(n \log \log n)$ term in the overall time bound.

Lemma 3.3 *There is a (static) data structure for linear programming queries on a set H of n halfspaces in E^d with $O(n \log m)$ preprocessing time, $O(n)$ space, and $O((n/m^{1/\lfloor d/2 \rfloor}) \log^{O(1)} n)$ query time, where m is a parameter between 1 and n.*

Proof: In this proof, we assume that the reader is familiar with Matoušek's technique [20].

We consider the halfspace-emptiness queries first. Partition H into $\lceil n/m \rceil$ groups $H_1, \ldots, H_{\lceil n/m \rceil}$, each of size at most m. For each of the H_i's, we build a data structure [20] with $O(m \log m)$ preprocessing time and $O(m)$ space, so that each halfspace-emptiness query on H_i can be answered in $O(\log m)$ parallel steps using $O(m^{1-1/\lfloor d/2 \rfloor} \log^{O(1)} m)$ processors. The total preprocessing time is then $O(\frac{n}{m}(m \log m)) = O(n \log m)$ and the space requirement remains linear. Since the halfspace-emptiness problem is decomposable, a query on H can be performed in $\tau(n,m) = O(\log m + \log \frac{n}{m}) = O(\log n)$ parallel steps using $\pi(n,m) = O(\frac{n}{m}(m^{1-1/\lfloor d/2 \rfloor} \log^{O(1)} m)) = O(\frac{n}{m^{1/\lfloor d/2 \rfloor}} \log^{O(1)} m)$ processors; or sequentially, in $t(n,m) = O(\frac{n}{m}(m^{1-1/\lfloor d/2 \rfloor} \log^{O(1)} m)) = O(\frac{n}{m^{1/\lfloor d/2 \rfloor}} \log^{O(1)} m)$ time.

Matoušek has shown that any data structure for halfspace-emptiness queries (satisfying some reasonable conditions) can be used to answer linear programming queries by parametric search. The resulting query time is given by $O(t(n,m) \tau(n,m)^d \log^d \pi(n,m))$, which, in our case, is $O(\frac{n}{m^{1/\lfloor d/2 \rfloor}} \log^{O(1)} m \log^{2d} n)$. □

Corollary 3.4 *A sequence of q linear programming queries on a set H of n halfspaces in E^d can be performed in $O(n \log \log n + n \log q + (nq)^{1-1/(\lfloor d/2 \rfloor + 1)} \log^{O(1)} n + q \log^{d+1} n)$ time.*

Proof: The proof is as in Corollary 2.4, except that for Case I ($q \leq n^{1/\lfloor d/2 \rfloor}/\log^K n$) we use Lemma 3.3 with $m = (q \log^K n)^{\lfloor d/2 \rfloor}$ ($1 \leq m \leq n$). The running time for Case I now becomes $O\left(n \log m + q \frac{n}{m^{1/\lfloor d/2 \rfloor}} \log^{O(1)} n\right) = O(n \log \log n + n \log q)$. □

Remark: Again, the complexity remains the same even if the value of q is not known in advance. (Use the sequence $q_i = (\log n)^{2^i}$ ($i = 1, 2, \ldots$) for Case I.)

Since Matoušek's data structures can be made dynamic (see Table 1), the following analogue of Lemma 2.6 is straightforward:

Lemma 3.5 *A sequence of q linear programming queries on a dynamic set H of at most n halfspaces in E^d can be performed in*

i. $O(n \log \log n + n \log q + (nq)^{1-1/(\lfloor d/2 \rfloor+1)+\varepsilon} + qn^{1-2/(\lfloor d/2 \rfloor+1)+\varepsilon})$ time, if the number of insertions/deletions is $O(q)$;

ii. $O(n \log^2 n + (nq)^{1-1/(\lfloor d/2 \rfloor+1)+\varepsilon} + q \log^{d+1} n)$ time, if the number of insertions/deletions is $O(n)$.

Finally, we observe that for the semidynamic case, where there are no deletions, Lemma 3.5(ii) may be improved somewhat.

Lemma 3.6 *A sequence of q linear programming queries and n insertions on an initially empty set of halfspaces in E^d can be performed in $O(n \log^2 n + (nq)^{1-1/(\lfloor d/2 \rfloor+1)} \log^{O(1)} n + q \log^{d+2} n)$ time and $O(n + (nq)^{1-1/(\lfloor d/2 \rfloor+1)} \log^{O(1)} n)$ space.*

Proof: Again we consider the halfspace-emptiness problem first. Since, this problem is decomposable, the techniques by Bentley and Saxe [3] may be applied to convert a static structure to a semidynamic one (which increases building time and query time by a logarithmic factor). We then apply parametric search to use this structure for answering linear programming queries. The resulting time bound is only a logarithmic factor increase on the static bound in Corollary 3.4. □

4 Convex Hulls

We now show that the f-face convex hull of an n-point set can be constructed by performing $O(f)$ ray shooting queries in a polytope defined by n halfspaces. The algorithm we use is just the well-known gift-wrapping method [7, 30, 36] dualized, since a "gift-wrapping operation" corresponds to shooting a ray in the dual polytope. If the ray shooting queries are performed directly by scanning the halfspaces, then we get an $O(nf)$-time bound. We observe that this can be improved using the data structures from Section 2.

Theorem 4.1 *The convex hull of a set P of n points in E^d can be constructed in*

i. $O(n \log f)$ *time and* $O(n)$ *space, if $d = 2$ or 3;*

ii. $O(n \log f + (nf)^{1-1/(\lfloor d/2 \rfloor+1)} \log^{O(1)} n)$ *time, if $d \geq 4$;*

where f is the number of hull faces.

Proof: In the dual setting, our problem becomes computing an intersection of a set H of n halfspaces in E^d (assumed to be in general position), each containing the origin o. It suffices to compute the vertices of the intersection $\bigcap H$, from which one can easily generate the complete lattice structure of $\bigcap H$ in $O(f \log f)$ time by a dictionary.

First, an initial vertex v_0 can be found by performing d ray shooting queries in $\bigcap H$, since shooting a ray from o gives a point in a $(d-1)$-face, and shooting a ray from a point in a j-face inside its affine hull gives a point in a $(j-1)$-face ($1 \leq j < d$). Furthermore, given a vertex v, the vertices adjacent to v in the 1-skeleton (the graph formed by the vertices and edges of $\bigcap H$) can be found by performing d ray shooting queries: if $h_1, \ldots, h_d$ are the hyperplanes defining v, then shoot a ray from v along each of the d lines formed by intersecting $d-1$ hyperplanes from $\{h_1, \ldots, h_d\}$.

Since the 1-skeleton is connected, we can use a depth-first search (or a breadth-first search) to visit all vertices of $\bigcap H$; we can ensure that each vertex is visited only once by using a dictionary to detect replication. This shows that the vertices of $\bigcap H$ can be computed by performing $O(f)$ ray shooting queries in $\bigcap H$. The theorem then follows by applying Corollaries 2.2 and 2.4 (recall that $f = O(n^{\lfloor d/2 \rfloor})$). □

Notice that in the plane, the gift-wrapping method becomes Jarvis's march [18]. We now write out our two-dimensional output-sensitive convex hull algorithm, in primal space, to illustrate its simplicity. Given an n-point set $P \subset E^2$, the following procedure Hull2D(P) returns the f vertices of conv(P) in counterclockwise order. It calls a subroutine Hull2D(P, m, f_0) which computes the hull successfully if $f \leq f_0$. The subroutine uses the grouping scheme from Lemma 2.1 to improve the $O(nf)$ complexity of Jarvis's march.

Algorithm Hull2D(P)
1. for $i = 1, 2, \ldots$ do
2. $L \leftarrow$ Hull2D(P, f_0, f_0), where $f_0 = \min\{2^{2^i}, n\}$
3.   if $L \neq$ *incomplete* then return L

Subroutine Hull2D(P, m, f_0), where $1 \leq m, f_0 \leq n$
1. partition P into subsets $P_1, \ldots, P_{\lceil n/m \rceil}$ each of size at most m
2. for $i = 1, \ldots, \lceil n/m \rceil$ do
3.   compute conv(P_i) by Graham's scan and store its vertices in an ordered array
4. $p_0 \leftarrow (0, -\infty)$
5. $p_1 \leftarrow$ the rightmost point of P
6. for $k = 1, \ldots, f_0$ do
7.   for $i = 1, \ldots, \lceil n/m \rceil$ do
8.   compute a point $q_i \in P_i$ that maximizes $\angle p_{k-1} p_k q_i$ by a binary search on conv(P_i)
9. $p_{k+1} \leftarrow$ a point q from $\{q_1, \ldots, q_{\lceil n/m \rceil}\}$ that maximizes $\angle p_{k-1} p_k q$
10.   if $p_{k+1} = p_1$ then return $\langle p_1, \ldots, p_k \rangle$
11. return *incomplete*

Since line 3 takes $O(m \log m)$ time and line 8 takes $O(\log m)$ time, we see that the subroutine Hull2D(P, m, f_0) runs in $O(n(1 + f_0/m) \log m)$ time. (Jarvis's march is just the case $m = 1$; when $m = n$, the algorithm degenerates to just Graham's scan.) Setting $m = f_0$ gives an $O(n \log f_0)$-time algorithm. Since f is not known in advance, we have to use a sequence of f_0's (line 2 of Hull2D(P)) to "guess" its value in a way similar to what is suggested in the Remark after Corollary 2.4. (Guessing the output size has become quite a common technique; see Chazelle and Matoušek's algorithm [8], for instance.)

5 Extreme Points

We now consider the problem of computing the h extreme points of an n-point set. Since determining whether a point is extreme can be done by solving a certain linear program, it is not difficult to see that n linear programming queries on n halfspaces are sufficient. We show that we can do better if h is small: by a simple algorithm, the extreme points can be found using h queries on n halfspaces together with n queries on h halfspaces. With Megiddo's linear programming algorithm [24], this leads to a simple $O(nh)$-time extrema algorithm. The same $O(nh)$-time algorithm has recently been discovered by Clarkson [9]. We note that the time bound can be further improved using the results from Section 3.

Theorem 5.1 *The h extreme points of a set P of n points in E^d can be computed in $O(n \log^{d+1} h + (nh)^{1-1/(\lfloor d/2 \rfloor+1)+\varepsilon})$ time or in $O(n \log^{d+2} h + (nh)^{1-1/(\lfloor d/2 \rfloor+1)} \log^{O(1)} n)$ time.*

Proof: Without loss of generality, assume that the origin o is in the interior of conv(P). Consider the following incremental algorithm, which is essentially the same as Clarkson's algorithm:

Algorithm Extrema(P)
1. $Q \leftarrow \emptyset$
2. for each $p \in P$ (in any order) do
3.   if $p \notin Q$ and $p \notin$ conv$(Q \cup \{o\})$ then
4.   if p is an extreme point of P then
5. $Q \leftarrow Q \cup \{p\}$
6.   else find the facet f of conv(P) that intersects ray $\overrightarrow{op}$
7.   let v be a vertex of f that is not in Q
8. $Q \leftarrow Q \cup \{v\}$
9. return Q

Observe that v must exist in line 7, because otherwise all vertices of f would be in Q; since $p \in$ conv$(f \cup \{o\})$, this would imply that $p \in$ conv$(Q \cup \{o\})$: a contradiction with line 3. It is then clear that the algorithm correctly returns the set of extreme points of P.

We now analyze the cost of the algorithm. Note that line 3 can be done by solving a linear program on Q in the dual and lines 4 and 6 can be done by solving a linear program on P in the dual. (Line 7 takes constant time since each facet has d vertices by the general position assumption.) Observe that although line 3 is executed n times, lines 4–8 are executed only h times since each execution adds a new point to Q. Thus, the algorithm requires h linear programming queries on P, a static set of size n, and n linear programming queries on Q, a semidynamic set of size at most h.

By Corollary 3.4, the h queries on P can be done in $O(n \log \log n + n \log h + (nh)^{1-1/(\lfloor d/2 \rfloor+1)} \log^{O(1)} n)$ time. By Lemma 3.5(ii), the n queries and h insertions on Q can be done in $O((nh)^{1-1/(\lfloor d/2 \rfloor+1)+\varepsilon} + n \log^{d+1} h)$ time. The total running time is then $O(n \log \log n + n \log^{d+1} h + (nh)^{1-1/(\lfloor d/2 \rfloor+1)+\varepsilon})$.

Notice that when $h \leq n^\alpha$ for a constant $\alpha < (1/\lfloor d/2 \rfloor)^2$, the number of hull faces is $O(n^{1/\lfloor d/2 \rfloor - \varepsilon})$; so we can compute the entire convex hull in optimal $O(n \log h)$ time and $O(n)$ space by Theorem 4.1. This allows us to remove the $O(n \log \log n)$ term in the time bound.

The first part of the theorem is thus proven, and the second part follows similarly, using Lemma 3.6 instead of Lemma 3.5(ii) for Q. □

Theorem 5.1 has an interesting corollary. It implies a bound for the convex hull problem that is within a polylogarithmic factor of optimal in the worst case, if the complexity is measured in terms of n and the number of extreme points h. (Note that $\Omega(n \log h + h^{\lfloor d/2 \rfloor})$ is a lower bound.)

Corollary 5.2 *The convex hull of a set P of n points in E^d can be constructed in $O(n \log^{O(1)} h + h^{\lfloor d/2 \rfloor})$ time, where h is the number of hull vertices.*

Proof: Compute the extreme points by Theorem 5.1 and then construct the convex hull of these h points by Chazelle's algorithm [6] in $O(h^{\lfloor d/2 \rfloor})$ time (note that when $h = \Omega(n^{1/\lfloor d/2 \rfloor})$, we have $h^{\lfloor d/2 \rfloor} = \Omega((nh)^{1-1/(\lfloor d/2 \rfloor+1)})$). □

6 Convex Layers and Depths

We now consider the convex layers problem and the depth problem. For the depth problem, we use a hybrid of the methods of Sections 4 and 5 to obtain a sub-quadratic algorithm. Then we show how this leads to an output-sensitive convex layers algorithm using Seidel's convex hull algorithm [34].

Theorem 6.1 *The depth of all points in a set P of n points in E^d can be computed in $O(n^{2-\beta+\varepsilon})$ time, where $\beta = 2/(\lfloor d/2 \rfloor^2 + 1)$.*

Proof: We iteratively compute the vertices of the i-th layer ($i = 1, 2, \ldots$) as follows. We use the convex hull algorithm in Theorem 4.1 to construct the i-th layer, but as soon as more than n^β vertices are discovered in the layer, we stop the computation and switch to the extrema algorithm in Theorem 5.1 to compute the vertices of the layer. We then remove the vertices of the i-th layer from P and proceed to the $(i+1)$-th layer. At the end, we will have the depths of every point in P.

Let h_i denote the number of vertices of the i-th layer ($\sum_i h_i = n$). We first analyze the cost of the calls to the convex hull algorithm in Theorem 4.1, which involves a number of ray shooting queries and n deletions. Since we stop the computation in a layer when n^β vertices are found, we make at most $O(\min\{h_i^{\lfloor d/2 \rfloor}, n^{\beta \lfloor d/2 \rfloor}\}) = O(h_i n^{\beta(\lfloor d/2 \rfloor - 1)})$ queries for the i-th layer. The total number of queries is then $O(n^{1+\beta(\lfloor d/2 \rfloor - 1)})$. By Lemma 2.6(ii), we see that the cost of these queries is $O((n^{2+\beta(\lfloor d/2 \rfloor - 1)})^{1-1/(\lfloor d/2 \rfloor + 1)+\varepsilon}) = O(n^{2-\beta+c\varepsilon})$ by our choice of β (where c is an appropriate constant).

Next, we analyze the cost of the calls to the extrema algorithm in Theorem 5.1. Note that the extrema algorithm is called only for the layers i with $h_i > n^\beta$, and the number of h_i's with $h_i > n^\beta$ is at most $n^{1-\beta}$ (since $\sum_i h_i = n$). Ignoring logarithmic factors, the cost is then

$$\sum_{h_i > n^\beta} \left(n + (nh_i)^{1-1/(\lfloor d/2 \rfloor + 1)} \right)$$

$$\leq n^{2-\beta} + n^{1-1/(\lfloor d/2 \rfloor + 1)} \left(\sum_{h_i > n^\beta} h_i^{1-1/(\lfloor d/2 \rfloor + 1)} \right).$$

A simple application of Hölder's inequality shows that the first term dominates for any value of d. Therefore, the entire method runs in $O(n^{2-\beta+c\varepsilon})$ time. □

Corollary 6.2 *The convex layers of a set P of n points in E^d can be constructed in $O(n^{2-\beta+\varepsilon} + f \log n)$ time, where f is the total output size and $\beta = 2/(\lfloor d/2 \rfloor^2 + 1)$.*

Proof: Let P_i be the vertices of layer i (i.e., the points of depth i) and let h_i and f_i be the number of vertices and faces of the layer ($\sum_i h_i = n$, $\sum_i f_i = f$). We first compute P_i for all i in $O(n^{2-\beta+\varepsilon})$ time by Theorem 6.1 and then construct the convex hull of each P_i using Seidel's algorithm [34] with Matoušek's improvement [20]. The total time needed is $O(n^{2-\beta+\varepsilon} + \sum_i (h_i^{2-2/(\lfloor d/2 \rfloor + 1)+\varepsilon} + f_i \log h_i)) = O(n^{2-\beta+\varepsilon} + f \log n)$. □

Remarks: 1. A worst-case optimal convex layers algorithm for $d \geq 4$ is not difficult to get: just apply Clarkson's extrema algorithm [9] to compute the vertices of each layer and use Chazelle's convex hull algorithm [6] to construct the layers; then the running time is $O(n^2 + n^{\lfloor d/2 \rfloor})$.

2. For a more direct output-sensitive convex layers algorithm, we can simply do the following: iteratively use the convex hull algorithm in Theorem 4.1 to construct the i-th layer ($i = 1, 2, \ldots$) and delete points from P that are vertices of a layer after each iteration. This method is the same as the one by Agarwal and Matoušek [1] (proceedings version) for the three-dimensional case. It requires $O(f)$ ray shooting queries and n deletions, and by Lemma 2.6(ii), takes $O((nf)^{1-1/(\lfloor d/2 \rfloor + 1)+\varepsilon})$ time, which is superior to the bound in Corollary 6.2 only when f is near linear (recall $\Omega(n) = f = O(n^{\lfloor d/2 \rfloor})$).

7 Other Applications

We now consider applications of our techniques to the construction of a k-level in an arrangement and to linear programming with few violated constraints.

Theorem 7.1 *A k-level in an arrangement $\mathcal{A}(H)$ of n hyperplanes in E^d can be constructed in*

i. $O(n \log f + f \log^2 n)$ time, if $d = 2$;

ii. $O(n \log f + f^{1+\varepsilon})$ time, if $d = 3$;

iii. $O(n \log f + (nf)^{1-1/(\lfloor d/2 \rfloor + 1)+\varepsilon} + f n^{1-2/(\lfloor d/2 \rfloor + 1)+\varepsilon})$ time, if $d \geq 4$;

where f is the output size.

Proof: The depth-first search algorithm by Agarwal and Matoušek [1] (proceedings version) constructs the k-level using $O(f)$ polytope ray shooting queries and $O(f)$ insertions/deletions on two dynamic sets of at most n halfspaces. (In two dimensions, their algorithm is the same as Edelsbrunner and Welzl's [15].) Hence, the theorem follows from Lemmas 2.5 and 2.6(i). □

Theorem 7.2 *The linear programming problem on n constraints in E^d with at most k violations for the feasible case can be solved in*

i. $O(n \log k + k^2 \log^2 n)$ time, if $d = 2$;

ii. $O(n \log \log n + n \log k + k^{3+\varepsilon})$ time, if $d = 3$;

iii. $O(n \log \log n + n \log k)$ time, if $d \geq 4$ and $k^d \leq n^{1/\lfloor d/2 \rfloor - \varepsilon}$.

Proof: The depth-first search algorithm by Matoušek [21] solves this problem using $O(k^d)$ linear programming/membership queries and $O(k^d)$ insertions/deletions on two dynamic sets of at most n halfspaces. (A membership query is just a special case of a ray shooting query.) Hence, part (i) of the theorem follows from Lemma 2.5 and Corollary 3.2, and part (ii) and (iii) follow from Lemmas 2.6(i) and 3.5(i). □

The techniques here may also be applicable to the infeasible case of linear programming with k violated constraints, or to the smallest k-enclosing circle problem; see Matoušek's paper [21].

Finally, we mention an improvement to Mulmuley's output-sensitive algorithm [25] for constructing ($\leq k$)-levels. The algorithm assumes that the input hyperplanes H are *nonredundant*, i.e., every hyperplane in H supports the upper envelope of H. For applications to ($\leq k$)-order Voronoi diagrams, this assumption is automatically satisfied.

Theorem 7.3 *We can compute i-levels in an arrangement $\mathcal{A}(H)$ of n nonredundant hyperplanes in E^d for all $i = 0, 1, \ldots, k$ in $O(n^{2-2/(\lfloor d/2 \rfloor + 1) + \varepsilon} k^{d-1} + f \log n)$ time, where f is the output size.*

Proof: Let $L_i(H)$ denote the boundary of the i-level in $\mathcal{A}(H)$ and let f_i be its size ($\sum_{i=0}^{k} f_i = f$). For each $h \in H$, let $H_h = \{h \cap h' : h' \in H - \{h\}\}$, which is a set of $(d-1)$-dimensional hyperplanes.

Mulmuley [25] gave an algorithm which constructs the facial structure of $L_i(H)$ in $O((f_i + f_{i-1}) \log n)$ time, given the following information:

1. the local minima in $L_i(H) - L_{i-1}(H)$ (along some predefined direction),

2. the local minima in $L_i(H_h) - L_{i-1}(H_h)$ for each $h \in H$, and

3. the facial structure of $L_{i-1}(H)$.

Matoušek [21] has shown that the local minima of all i-levels in $\mathcal{A}(H)$ ($i = 0, 1, \ldots, k$) can be enumerated by performing $O(k^d)$ linear programming/membership queries and $O(k^d)$ insertions/deletions on two dynamic sets of at most n halfspaces. Similarly, the local minima of all i-levels in $\mathcal{A}(H_h)$ ($i = 0, 1, \ldots, k$) can be computed using $O(k^{d-1})$ linear programming/membership queries and $O(k^{d-1})$ insertions/deletions, for each $h \in H$. Observe that these $(d-1)$-dimensional queries do not have to done on separate dynamic structures for each H_h; they can be done on one structure in d dimensions. The total number of queries and updates is then $O(k^d + nk^{d-1}) = O(nk^{d-1})$. By Lemmas 2.6(i) and 3.5(i), this takes $O(n^{2-2/(\lfloor d/2 \rfloor + 1) + \varepsilon} k^{d-1})$ time.

Thus, items 1 and 2, for all $i = 0, 1, \ldots, k$, can be computed in $O(n^{2-2/(\lfloor d/2 \rfloor + 1) + \varepsilon} k^{d-1})$ time. Now, Mulmuley's algorithm can be used to construct the facial structure of $L_0(H), L_1(H), \ldots, L_k(H)$ incrementally, in additional $O(f \log n)$ time. □

8 Final Remarks

We remark that it is possible to remove the $O(n \log \log n)$ term in Lemma 3.5 (and thus in Theorem 7.2) using the known randomized algorithms for generalized linear programming ("LP-type") problems [35]; details will be included in the full paper.

Further applications of our ideas are also possible. For example, Theorem 4.1 can be extended to compute the intersection of a convex hull with a j-flat in an output-sensitive manner; in the dual, this corresponds to computing projections (shadows) of an intersection of halfspaces. More generally, we can obtain output-sensitive bounds for computing "skeletons" in a halfspace intersection, or with the known methods for general ray shooting [1], "skeletons" in a hyperplane arrangement; see Edelsbrunner [12, Chapter 9].

Many open questions remain, however. A major problem is to find an $O((n + f)^{1+\varepsilon})$-time convex hull algorithm in dimensions higher than four. Another question is: can the depth problem be solved in $O(n^{2-2/(\lfloor d/2 \rfloor + 1) + \varepsilon})$ time?

Acknowledgements

I am grateful to Jack Snoeyink for his guidance and encouragement as well as for many valuable suggestions and stimulating discussions. I would also like to thank the referees for pointing out Clarkson's result on extreme points.

References

[1] P. K. Agarwal and J. Matoušek. Ray shooting and parametric search. *SIAM J. Comput.*, 22:794-806, 1993. Also in *Proc. 24th ACM Sympos. Theory of Comput.*, 517-526, 1992.

[2] J. L. Bentley, H. T. Kung, M. Schkolnick and C. D. Thompson. On the average number of maxima in a set of vectors. *J. ACM*, 25:536-543, 1978.

[3] J. L. Bentley and J. Saxe. Decomposable searching problems I: static-to-dynamic transformations. *J. Algorithms*, 1:301-358, 1980.

[4] T. M. Chan, J. Snoeyink, and C.-K. Yap. Output-sensitive construction of polytopes in four dimensions and clipped Voronoi diagrams in three. In *Proc. 6th ACM-SIAM Sympos. Discrete Algorithms*, 282-291, 1995.

[5] B. Chazelle. An optimal algorithm for computing convex layers. *IEEE Trans. Information Theory*, IT-31:509-517, 1985.

[6] B. Chazelle. An optimal convex hull algorithm for point sets in any fixed dimension. *Discrete Comput. Geom.*, 9:145-158, 1993.

[7] D. R. Chand and S. S. Kapur. An algorithm for convex polytopes. *J. ACM*, 17:78-86, 1970.

[8] B. Chazelle and J. Matoušek. Derandomizing an output-sensitive convex hull algorithm in three dimensions. Manuscript, 1993.

[9] K. L. Clarkson. More output-sensitive geometric algorithms. In *Proc. 35th IEEE Sympos. Found. of Comput. Sci.*, 695-702, 1994.

[10] K. L. Clarkson and P. W. Shor. Applications of random sampling in computational geometry, II. *Discrete Comput. Geom.*, 4:387-421, 1989.

[11] D. P. Dobkin and D. G. Kirkpatrick. Fast detection of polyhedral intersection. *Theoretical Comput. Sci.*, 27:241-253, 1983.

[12] H. Edelsbrunner. *Algorithms in Combinatorial Geometry*. Springer-Verlag, Berlin, 1987.

[13] H. Edelsbrunner and W. Shi. An $O(n \log^2 h)$ time algorithm for the three-dimensional convex hull problem. *SIAM J. Comput.*, 20:259-277, 1991.

[14] H. Edelsbrunner and E. P. Mücke. Simulation of simplicity: A technique to cope with degenerate cases in geometric algorithms. *ACM Trans. Graphics*, 9:66-104, 1990.

[15] H. Edelsbrunner and E. Welzl. Constructing belts in two-dimensional arrangements with applications. *SIAM J. Comput.*, 15:271-284, 1986.

[16] I. Emiris and J. Canny. An efficient approach to removing geometric degeneracies. In *Proc. 8th ACM Sympos. Comput. Geom.*, 74-82, 1992.

[17] R. L. Graham. An efficient algorithm for determining the convex hull of a finite planar set. *Inform. Process. Lett.*, 1:132-133, 1972.

[18] R. A. Jarvis. On the identification of the convex hull of a finite set of points in the plane. *Inform. Process. Lett.*, 2:18-21, 1973.

[19] D. G. Kirkpatrick and R. Seidel. The ultimate planar convex hull algorithm? *SIAM J. Comput.*, 15:287-299, 1986.

[20] J. Matoušek. Linear optimization queries. *J. Algorithms*, 14:432-448, 1993. Also with O. Schwarzkopf in *Proc. 8th ACM Sympos. Comput. Geom.*, 16-25, 1992.

[21] J. Matoušek. On geometric optimization with few violated constraints. In *Proc. 10th ACM Sympos. Comput. Geom.*, 312-321, 1994.

[22] J. Matoušek and O. Schwarzkopf. On ray shooting in convex polytopes. *Discrete Comput. Geom.*, 10:215-232, 1993.

[23] P. McMullen. The maximal number of faces of a convex polytope. *Mathematica*, 17:179-184, 1970.

[24] N. Megiddo. Linear programming in linear time when the dimension is fixed. *J. ACM*, 31:114-127, 1984.

[25] K. Mulmuley. Output sensitive construction of levels and Voronoi diagrams in R^d of order 1 to k. In *Proc. 22nd ACM Sympos. Theory of Comput.*, 322-330, 1990.

[26] K. Mulmuley. *Computational Geometry: An Introduction Through Randomized Algorithms*. Prentice Hall, New York, 1993.

[27] J. O'Rourke. *Computational Geometry in C*, Cambridge University Press, 1994.

[28] M. H. Overmars and J. van Leeuwen. Maintenance of configurations in the plane. *J. Comput. Syst. Sci.*, 23:166-204, 1981.

[29] F. P. Preparata and S. J. Hong. Convex hulls of finite sets of points in two and three dimensions. *Commun. ACM*, 20:87-93, 1977.

[30] F. P. Preparata and M. I. Shamos. *Computational Geometry — An Introduction*. Springer-Verlag, New York, 1985.

[31] H. Raynaud. Sur l'enveloppe convexe des nuages des points aléatoires dans R^n. *J. Appl. Probab.*, 7:35-48, 1970.

[32] M. Reichling. On the detection of a common intersection of k convex objects in the plane. *Inform. Process. Lett.*, 29:25-29, 1988.

[33] R. Seidel. A convex hull algorithm optimal for point sets in even dimensions, Report 81-14, Depart. Comput. Sci., Univ. British Columbia, Vancouver, B.C., 1981.

[34] R. Seidel. Constructing higher-dimensional convex hulls at logarithmic cost per face. In *Proc. 18th ACM Sympos. Theory of Comput.*, 404-413, 1986.

[35] M. Sharir amd E. Welzl. A combinatorial bound for linear programming and related problems. In *Proc. 9th Sympos. Theoretical Aspects of Comput. Sci.*, Lecture Notes in Comput. Sci., vol. 577, 569-579, 1992.

[36] G. F. Swart. Finding the convex hull facet by facet. *J. Algorithms*, 6:17-48, 1985.

How Good are Convex Hull Algorithms?[*]

David Avis   David Bremner[†]

1 INTRODUCTION

A d dimensional *convex polyhedron* is the intersection of a finite number m of non-redundant halfspaces $\mathcal{H} = \{\,H_1, H_2, \ldots H_m\,\}$ of $\Re^d$. A bounded convex polyhedron is called a *polytope*. A classic theorem from convexity is that every polytope P can be expressed as the convex hull of its n extreme points (or vertices) $\mathcal{V}$. These descriptions of P will be referred to as the *halfspace* and *vertex* descriptions, respectively. The size of a polytope, denoted $size(P) = (m+n)d$, is the space required to store both descriptions of a polytope. Each vertex of P lies on d affinely independent hyperplanes each of which bounds a halfspace in $\mathcal{H}$. The set of indices of such a set of halfspaces is called a *basis*. A polytope is *simple* if each of its vertices has a unique basis. There are three closely related computational problems concerning the two descriptions of a polytope:

- The *vertex enumeration problem* is to compute $\mathcal{V}$ from $\mathcal{H}$.
- The *convex hull problem* it to compute $\mathcal{H}$ from $\mathcal{V}$.
- The *polytope verification problem* is to decide whether a given vertex description and halfspace description define the same polytope.

It is an open problem whether any of these problems can be solved in time polynomial in $size(P)$. Since the first two problems are essentially equivalent under point/hyperplane duality, we will mainly restrict ourself to the vertex enumeration problem, although all results apply to both problems.

There are two main classes of algorithms for solving these problems: pivoting algorithms and insertion algorithms. A pivoting algorithm for the vertex enumeration problem begins by finding a basis of some vertex of P. Two bases of P are adjacent if they differ in one index only. This adjacency relation defines a graph on the bases of P, where each edge corresponds to a *pivot*, the replacement of exactly one index in a basis. All vertices of this graph are generated, from which the extreme points of P are readily computed. Representatives of this class are the *gift wrapping* algorithm of Chand and Kapur [6], Seidel's algorithm [21] and the *reverse search* algorithm of Avis and Fukuda [1]. For a simple polytope P, pivoting algorithms can solve the vertex enumeration and polytope verification problems (and by duality the convex hull problem for simplicial polytopes) in time polynomial in $size(P)$. For non-simple polytopes the number of bases may be exponential in the size of P. For such polytopes pivoting algorithms rely on perturbation, either symbolic or numerical, to reduce the computation time. In Section 3 we give a fairly general model for pivoting algorithms and valid perturbation schemes for polytopes. We then give examples of polytopes for which no valid perturbation scheme can yield polynomial time pivoting algorithms in this model.

Insertion algorithms for the vertex enumeration problem compute the vertex description by intersecting the defining halfspaces sequentially. An initial simplex is constructed from a subset of $d+1$ halfspaces and its vertices are computed. Additional halfspaces are introduced sequentially and the vertex description is updated at each stage. Although the first explicit description of such an algorithm, now widely known as the *double description method*, appeared in the pioneering 1953 paper of Motzkin et al. [19], this paper seems to have been overlooked by the Computational Geometry community. Many of the same ideas were rediscovered and refined in the *beneath and beyond method* of Seidel [20], the *randomized algorithm* of Clarkson and Shor [9] and the *derandomized algorithm* of Chazelle [7]. In some sense the algorithms [20] and [7] can be considered optimal. The upper bound theorem of McMullen states that for any polytope P defined by m halfspaces, $size(P) \in O(m^{\lfloor d/2 \rfloor})$ and this bound is achieved (see e.g. [4]). The algorithms [20] and [7] solve the vertex enumeration problem in this time bound (for even and arbitrary d respectively). However, it is well-known that $size(P) \in \Omega(md)$ and this bound is also achieved. An efficient vertex enumeration algorithm for such polytopes should clearly be polynomial in md. It is not known whether such an algorithm exists. Examples that have arisen in practice (see e.g. [5]) suggest that the sizes of polytopes of interest are closer to the lower bound than the upper. Since there is such a wide variation in the output sizes of vertex enumeration problems, we suggest $size(P)$ as the appropriate measure of the problem size. The success of insertion algorithms is widely believed to depend on keeping the vertex complexity of the intermediate polytopes generated small by judicious choice of the order in which the halfspaces are inserted. While this is necessary for the success of such algorithms, it is not sufficient. The examples in Section 3 show that insertion algorithms which rely on triangulation (such as the the randomized and derandomized algorithms) can perform very badly indeed, even when the intermediate polytopes have small size and regardless of the order of insertion of the halfspaces.

For insertion algorithms that do not rely on triangulation, the choice of insertion order is indeed a critical factor. Dyer [10] uses a result of Klee [14] to give a family of polytopes for which there is an order of halfspace insertion that causes any insertion algorithm to build exponential sized intermediate polytopes. Dyer's

---

[*]This research supported by NSERC Canada, and FCAR Québec.

[†]The authors are with the McGill University School of Computer Science. email: bremner@cs.mcgill.ca

Permission to copy without fee all or part of this material is granted provided that the copies are not made or distributed for direct commercial advantage, the ACM copyright notice and the title of the publication and its date appear, and notice is given that copying is by permission of the Association of Computing Machinery.To copy otherwise, or to republish, requires a fee and/or specific permission.
11th Computational Geometry, Vancouver, B.C. Canada
© 1995 ACM 0-89791-724-3/95/0006...$3.50

construction is based on the dual of the cyclic polytope with two additional halfspaces that intersect every facet of the dual cyclic polytope. In Section 4 we give a set of sufficient combinatorial conditions for a class of hard polytopes for insertion algorithms and an example of a polytope family that meets these conditions. We show that any insertion algorithm that inserts halfspaces in a fixed order (independent of the input) or in lexicographic order must construct exponential sized intermediate polytopes in the worst case. In Section 5 we present experimental evidence that several popular heuristic insertion orders, including random order, are also exponential for our class of hard polytopes, but that there is a polynomial insertion order for the double description method for this class.

The examples of hard polytopes given in Sections 3 and 4 apply to computational models that include many published convex hull algorithms. A number of these algorithms have been implemented by various people, but an implementation is rarely completely faithful to the algorithm from which it is derived. For this reason in Section 5 of the paper we give actual computational experience obtained by trying to solve the vertex enumeration and convex hull problems for these hard polytopes.

2 PRELIMINARIES

Given a set of points $X = \{x_1, \ldots x_k\}$, y is called an *affine combination* of X if

$$y = \sum_{i=1}^{k} \lambda_i x_i \quad \text{and} \quad \sum_{i=1}^{k} \lambda_i = 1.$$

The *affine hull* of a set of points X (written aff(X)) is the set of all affine combinations of X. The set X is called *affinely independent* if no x_i is an affine combination of the other $k-1$ points in X. A *convex of combination* of X is an affine combination such each λ_i is non-negative. A *proper* convex combination is one where each λ_i is positive. A subset K of $\Re^d$ is called a convex set if every convex combination of points in K is also in K. The *convex hull* of X, denoted conv(X), is the set of all convex combinations of X. The relative interior of a convex set K, written relint(K), is the interior of K in the affine hull of K.

Lemma 1 (Theorem 3.3 [4]) *For any convex set C, any point x in* relint(C), *and any point $y \in C$, the open line segment from x to y is contained in* relint(C).

A d dimensional *convex polyhedron* is the intersection of a finite number of halfspaces of $\Re^d$. A bounded convex polyhedron is called a *convex polytope*. A halfspace H is called a *valid constraint* for convex polyhedron P if $P \subseteq H$. Given a set of halfspaces $\mathcal{H}$, $H \in \mathcal{H}$ is is called *non-redundant* if there is some point feasible for every constraint in $\mathcal{H} \setminus \{H\}$ but not feasible for $\mathcal{H}$. We use $\mathcal{H}(P)$ to denote the non-redundant halfspace description of P and $\mathcal{V}(P)$ to denote the vertex description of P. We abbreviate a d dimensional convex polyhedron (polytope) to a d-polyhedron (d-polytope). Let P be a d-polytope, containing the origin in its interior, with vertices $\{p_1, p_2, \ldots, p_n\}$ and defining halfspaces $\{a_i x \leq 1 \mid i = 1, \ldots, m\}$. The *dual* of P is the polytope with vertices $\{a_1, a_2, \ldots, a_m\}$ and is defined by the intersection of the halfspaces $\{p_j x \leq 1 \mid j = 1, \ldots, n\}$. It is well known and easy to show that the vertex enumeration problem for P is identical to the convex hull problem for the dual of P, and vice versa.

Given a convex polytope P, a point $x \in P$ is an *extreme* point of P if it is not a proper convex combination of any two points in P. The set of extreme points of a polyhedron P is written ext(P). A *supporting halfspace* of a convex polyhedron P is a halfspace H with bounding hyperplane h (called a *supporting hyperplane*) such that $P \subseteq H$ and $h \cap P \neq \emptyset$. A *face* of a convex polyhedron P is the intersection of P and one or more supporting hyperplanes of P. We abbreviate a k dimensional face to a k-face. The notation $f_j(P)$ denotes the number of j-faces of a polytope P. The $0, 1, d-2$ and $d-1$ dimensional faces of a d-polyhedron P are called the *vertices*, *edges*, *ridges*, and *facets* of P. Given a facet F of a polytope P, F^+ denotes the closed halfspace induced by aff(F) containing P and F^- the other closed halfspace induced by aff(F). We denote the interior of P as int(P).

Let the *centroid* of a set of points $\{x_1, \ldots x_k\}$ denote the point

$$\frac{1}{k} \sum_{i=1}^{k} x_i.$$

Define the centroid of a convex polytope as the centroid of its vertices. Observe that the centroid a set X of points is a proper convex combination, hence is contained in the relative interior of conv(X).

The convex hull of a set of $d+1$ affinely independent points in d dimensions is called a *simplex*.

Given a polyhedron P, a hyperplane h such that $h \cap \text{int}(P) = \emptyset$, and a point x, we say that x is *beneath* (respectively *beyond*) h if x is in the halfspace induced by h containing int(P) (respectively not containing int(P)). The point x is *strictly beneath* (respectively *strictly beyond*) h if it is beneath h (beyond h) but not in h. A hyperplane h *strictly separates* two points if one is in one open halfspace defined by h and one is in the other. We take beneath or beyond a facet F to mean beneath or beyond aff(F).

3 PERTURBATION, TRIANGULATION AND PIVOTING

Let P be a d-polytope with m facets and n vertices. A pivoting algorithm for the vertex enumeration problem starts at some basis for a vertex of P and generates all other feasible bases for P. From these bases the vertices of P are obtained, with repetitions if P is non-simple. This is efficient for simple polytopes, but can be extremely inefficient for polytopes with one or more highly degenerate vertices. In the worst case P may have a single vertex and as many as $\binom{m}{d}$ bases. Even though this is an extreme case, highly degenerate polytopes seem to arise in practice and generating all of the bases is impractical. To reduce the computation, P is perturbed to a simple polytope $\tilde{P}$ with m facets and $\tilde{n}$ vertices, where hopefully $\tilde{n}$ is much smaller than the number of bases of P. The vertices of the simple polytope $\tilde{P}$ are computed from which the vertices of the original polytope P are obtained. Let $\mathcal{B}$ and $\tilde{\mathcal{B}}$ denote the sets of bases of P and $\tilde{P}$ respectively. We call a perturbation *valid* if the following conditions are satisfied:

(i) $\tilde{\mathcal{B}} \subseteq \mathcal{B}$.

(ii) Each vertex of P has a basis B contained in $\tilde{\mathcal{B}}$.

It is clear that under the above conditions, all vertices of P can be generated as follows. Generate $\tilde{\mathcal{B}}$ and list all vertices in P corresponding to these bases, removing duplicates. If $\tilde{P}$ is simple then the number of bases is at most $O(m^{\lfloor d/2 \rfloor})$, by the upper bound theorem. A perturbation scheme can be either numerical or symbolic, and general framework for describing such schemes is contained in Yap [22]. A common numerical perturbation is obtained by perturbing the right hand side vector of the system of halfspaces defining P. This perturbed vector is used in calculations. A simple way of transforming the perturbed vertices back to vertices of P is simply to carry an extra column which is the original right hand side vector. This is pivoted along with the rest of the constraint matrix but does not play any role in the calculations. Instead of outputting the perturbed right hand side (corresponding to a vertex of $\tilde{P}$) this additional column is output (corresponding to a vertex of P); duplicates can be removed off-line.

In practice, perturbation has been extremely useful. In Ceder et al. [5], the 4862 vertices of a highly degenerate 8-dimensional polytope defined by 729 halfspaces were computed in exact arithmetic by the reverse search pivoting scheme. In this case, computing all the bases of the original polytope was infeasible, and even the perturbed polytope had 477,421 bases.

The effectiveness of a perturbation scheme obviously lies in the ability to generate a simple polytope $\tilde{P}$ with as few vertices as possible. The number of vertices in the perturbed polytope is therefore a natural measure of the quality of the perturbation. This measure of the quality of a perturbation does not seem to have been studied in the literature. The bound on the number of bases of the perturbed polytope given above assumes the worst case that P is perturbed onto the dual of the cyclic polytope. On the other hand, if a valid perturbation onto a lower bound polytope is possible, we would get very good results indeed. Here we answer a fundamental question: Is it always possible to find a perturbation such that the number of vertices in $\tilde{P}$ is polynomial in $size(P)$? A positive answer would lead to a polynomial time algorithm for the problems described in Section 1. Unfortunately, the answer is negative.

In order to see why a good perturbation is sometimes impossible, consider the following example suggested by Günter Rote. It is convenient here to consider the convex hull problem rather than the vertex enumeration problem. The input is the set of 2^d vertices of a d dimensional hypercube H_d, and the output is a list of its $2d$ facets. A basis of a facet F is a full dimensional (in $aff(F)$) simplex spanned by d of the 2^{d-1} vertices defining the facet. For the convex hull problem, a perturbation is applied to the input points to give a simplicial polytope, i.e. a polytope for which each facet has a unique basis. It can be checked that the criteria (i) and (ii) above for a valid perturbation (along with the condition that the perturbed polytope be simplicial) dualize into the condition that the facets of the convex hull of the perturbed point set induce a triangulation of the facets of the unperturbed hypercube. The duality between triangulations and perturbations was previously observed by Lee [18]. Note that the facets of the hypercube are $(d-1)$-cubes. It is well known (see for example [13]) that every triangulation of H_d requires at least $\Omega(\sqrt{d}!)$ simplices. Now $size(H_d) = d^2 2^{d+1}$, and hence the number of simplices required to triangulate the facets of a H_d is super polynomial in its size. Therefore there is no valid perturbation scheme that yields a perturbed point set with a polynomial sized convex hull. Dualizing the argument, we conclude that the halfspaces defining a d-cross polytope cannot be perturbed into a simple polytope with size polynomial in the size of the cross-polytope. Therefore pivoting algorithms for the vertex enumeration problem that rely on perturbation are super polynomial in the worst case.

Although the above argument gives a non-polynomial lower bound, the bound is only just super polynomial. If $N = size(H_d)$, then the number of pivots is

$$\Omega(e^{\log^2 N}).$$

The following polytopes were suggested by Bernd Sturmfels and give sharper lower bounds. Let w be a d-vector of all ones and for $1 \leq i \leq d$, let e_i denote the d-vector which is all zero except for $d+1$ in position i. We denote by T_d the simplex in $\Re^d$ spanned by the vertices $\{e_1 - w, e_2 - w, \ldots, e_d - w, -w\}$. Note that the centroid of T_d is the origin. In $\Re^{s+t}$ we construct the product $T_s \times T_t$ of the two simplices T_s and T_t in orthogonal $\Re^s$ and $\Re^t$ subspaces. The product contains the $(s+1)(t+1)$ vertices formed by concatenating each vertex of T_s with each vertex of T_t. The following results are known about these polytopes [12]:

(a) $T_d \times T_d$ has $n = (d+1)^2$ vertices and $m = 2d+2$ facets.

(b) Each triangulation of $T_s \times T_t$ requires exactly $\binom{s+t+2}{s+1}$ simplices.

(c) Each facet of $T_d \times T_d$ is combinatorially equivalent to $T_{d-1} \times T_d$.

From these facts we obtain the following result.

Theorem 1 *The polytope $T_d \times T_d$ has size $N = 2d(d^2+4d+3)$ and every valid perturbation yields a simple polytope of size $\Omega(2^{\sqrt[3]{N}})$.*

This theorem implies that solving the convex hull problem for $T_d \times T_d$ by a pivoting algorithm using perturbation requires the generation of at least $\Omega(2^{N^{1/3}})$ bases, where N is the size of the polytope. The bounds also apply to insertion algorithms that construct triangulations of the polytope, or of the facets of the polytope, such as the randomized [9] and derandomized [7] algorithms. The bound does not, however, apply to the double description method [19], which seems empirically to run in polynomial time on these polytopes. A similar result applies to the vertex enumeration problem for the dual of $T_d \times T_d$. In Section 5 we illustrate these results in practice on currently available implementations of these algorithms.

To conclude this section we observe that a similar argument can be applied to another technique which is frequently suggested for dealing with degeneracy: truncating degenerate vertices. Here the idea is to introduce an additional halfspace which contains in its interior all of the vertices of the polytope, except some known degenerate vertex. All vertices of this truncated polytope are enumerated, and those lying on the cutting hyperplane are associated with the original degenerate vertex. Such a technique can be applied on-line whenever a degenerate vertex is encountered. Arguments similar to those above again show this cannot lead to a polynomial time algorithm. As above, consider the dual convex hull problem. The cutting halfspace dualizes into the addition of a new point above a non-simplicial facet. The lower bounds on triangulation depend on volume arguments that still apply in this new setting, giving super polynomial lower bounds similar to those above.

4 INSERTION ALGORITHMS AND INTERMEDIATE SIZE

An insertion algorithm for vertex enumeration computes the vertices of $k \equiv |\mathcal{H}| - d$ intermediate polytopes, where $\mathcal{H}$ is the input set of halfspaces. Initially a simplex P_0 is constructed, using $d+1$ elements of $\mathcal{H}$. At each succeeding step, a halfspace $H \in \mathcal{H}$ is chosen according to some *insertion order* and the vertices of $P_i \equiv P_{i-1} \cap H$ are computed. The final output is the vertices of

$$P_k \equiv \bigcap_{H \in \mathcal{H}} H.$$

In order for such an algorithm to be polynomial in $size(P_k)$, the size of each of these intermediate polytopes must be polynomial in $size(P_k)$. In this section, we give examples of a polytope family and insertion orders for which this not the case.

The general outline of our construction is as follows. For input size n, we construct a hard polytope of size n as follows. We start with a polytope Q_n with vertex complexity exponential in n and facet complexity polynomial in n. We then construct a polytope P_n such that

1. $\mathcal{H}(P_n \cap Q_n) = \mathcal{H}(P_n) \cup \mathcal{H}(Q_n)$; i.e. no constraint of P_n or Q_n is redundant for $P_n \cap Q_n$, and

2. $P_n \cap Q_n$ has vertex complexity polynomial in n.

Our hard polytope will be $P_n \cap Q_n$. The lower bounds will result from forcing certain insertion orders to build the entire polytope Q_n before inserting any constraints of P_n. In Section 4.1 we derive general conditions on P_n and Q_n under which these properties hold. In Section 4.2 we give examples of polytope families P_n and Q_n for which these general conditions hold.

4.1 PAIRS OF PIERCING POLYTOPES

In this subsection we define a set of general conditions on a pair of polytopes P and Q such that for any pair of polytopes that satisfies these conditions, $size(P \cap Q)$ is polynomial in $|\mathcal{V}(P)|$ and $\mathcal{H}(P \cap Q) = \mathcal{H}(P) \cup \mathcal{H}(Q)$.

The edges and vertices of a polytope P define an undirected graph in an obvious way. Denote this graph $\mathcal{G}(P)$. Let V be a subset of the vertices of a polytope P: $\mathcal{G}(P \,|\, V)$ denotes the subgraph of $\mathcal{G}(P)$ induced by V.

Lemma 2 *Let P be a polytope with vertices $\mathcal{V}$ and halfspace description $\mathcal{H}$. Let H be a halfspace. Let $\mathcal{V}_H$ denote $\mathcal{V} \cap \text{int}(H)$. The graph $\mathcal{G}(P \,|\, \mathcal{V}_H)$ is connected.*

Proof: Define a linear program with the constraints $\mathcal{H} \cup \{H\}$ and an objective function of the inward normal of H. Every vertex of $P \cap H$ is either in $\mathcal{V}_H$ or contained in the bounding hyperplane of H. From the correctness of the simplex method with Bland's pivot rule (see e.g. [8]), there is a path in $\mathcal{G}(P \cap H)$ from any vertex of $\mathcal{V}_H$ to a unique optimum face F of $P \cap H$. Since the simplex method monotonically increases the value of the objective function (i.e. the distance from the bounding hyperplane of H), this path does not intersect the bounding hyperplane of H, hence is entirely contained in $\mathcal{G}(P \,|\, \mathcal{V}_H)$. By Balinski's Theorem (see e.g. [3]), $\mathcal{G}(F)$ is connected, hence $\mathcal{G}(P \,|\, \mathcal{V}_H)$ is connected. ∎

Given polytopes P and Q, a *connected component* of P with respect to Q is a maximal subset C of vertices of P such such that $\mathcal{G}(P \,|\, C)$ is connected and no edge of P between vertices of C intersects Q. A connected component C of P w.r.t. Q *properly pierces* a facet F of Q if $\text{aff}(F)$ strictly separates C from $\mathcal{V}(P) \setminus C$ and and $P \cap \text{aff}(F) \subset \text{relint}(F)$. The *proper intersection* of polytopes P and Q denotes $\text{relint}(P) \cap \text{relint}(Q)$.

Theorem 2 *Let P and Q be d-polytopes such that $P \cap Q \neq \emptyset$ and $\mathcal{V}(P) \cap Q = \emptyset$. The following conditions are equivalent.*

(a) Every connected component of P w.r.t. Q properly pierces some facet of Q.

(b) No ridge of Q intersects P.

(c) A point x is a vertex of $P \cap Q$ iff it is the proper intersection of an edge of P with a facet of Q.

Proof:

(a⇒b) Suppose (a) holds. Let F be a facet of Q. If F is properly pierced by a connected component of P w.r.t. Q, then by definition no ridge of Q contained in F intersects P. Otherwise let $\mathcal{V}_F$ be the subset of $\mathcal{V}(P)$ strictly beyond F. By Lemma 2, $\mathcal{G}(P \,|\, \mathcal{V}_F)$ is connected. By convexity, no edge with both endpoints in $\text{int}(F^-)$ can intersect Q. It follows that $\mathcal{V}_F$ is contained in some connected component C' of P w.r.t. Q. From (a), C' must properly pierces some facet F' of Q distinct from F.

Let E_F denote the intersection of the edges of P with the affine hull of F. We know the following:

$$\begin{aligned} P \cap F &= (P \cap \text{aff}(F)) \cap F \\ &= \text{conv}(\text{ext}(P \cap \text{aff}(F))) \cap F \\ &= \text{conv}(E_F) \cap F \end{aligned}$$

We argue that $\text{aff}(F')$ is a strict separating hyperplane for E_F and F.

The hyperplane $\text{aff}(F')$ strictly separates C' from Q; if it did not, there would be a vertex v of C' in F^+. By the connectedness of $\mathcal{G}(P \,|\, C')$, there would have to be some edge connecting the subset of C' strictly beyond $\text{aff}(F')$ with the subset of C' containing v. But such an edge would have to intersect $\text{relint}(F')$, contradicting the fact that v was in the connected component C'.

Each point in E_F is contained in the relative interior of a line segment with one endpoint in C' and one endpoint in $\text{relint}(F')$. By Lemma 1 any point in the relative interior of a line segment with one endpoint in $\text{aff}(F')$ and the other strictly beyond $\text{aff}(F')$ must be strictly beyond $\text{aff}(F')$. It follows that $\text{aff}(F')$ strictly separates E_F from F, hence $P \cap F = \emptyset$.

(a⇐b) Suppose (b) holds. Let C be a connected component of P w.r.t. Q. Let z be a point in $P \cap Q$. Let v be a vertex of C. Let F be the facet of Q intersected by segment $\overline{vz}$ (the facet F is unique since $\overline{vz} \cap \text{aff}(F) \subset \text{relint}(F)$). We argue that C properly pierces F.

First, we argue that $\text{aff}(F)$ strictly separates C from $\mathcal{V}(P) \setminus C$. By Lemma 2, any vertex of P strictly beyond $\text{aff}(F)$ must be in C. Suppose there were some vertex v_1 of C in F^+. Let x denote $\overline{vz} \cap F$. By convexity, $x \in P$. From (b), $x \in \text{relint}(F)$. From the connectedness of $\mathcal{G}(P \,|\, C)$, there must be some edge e of P that intersects $\text{aff}(F)$ but does not intersect Q. Let y denote $e \cap \text{aff}(F)$. By convexity, $y \in P$ and $\overline{xy} \subset P$. Since x and y are both in the affine hull of F, but one is in the relative interior of F, and one completely outside it, the segment $\overline{xy}$ must intersect some ridge of Q. Since this is a contradiction, every vertex of C is strictly beyond F. Since every vertex of P strictly beyond F is in C and every vertex of C is strictly beyond F, $\text{aff}(F)$ strictly separates C from $\mathcal{V}(P) \setminus C$.

Next we argue that $P \cap F \subset \text{relint}(F)$. Consider an edge e' of P such that exactly one endpoint of e' is in C and e' intersects Q. The edge e' must intersect $\text{aff}(F)$ since $\text{aff}(F)$ is a separating hyperplane for C and $\mathcal{V}(P) \setminus C$. If e' does not intersect $\text{relint}(F)$ then we again have a line segment $\overline{xy} \subset P$ with $x \in \text{relint}(F)$ and $y \in \text{aff}(F) \setminus \text{relint}(F)$, contradicting (b).

(b⇒c) Every vertex of $P \cap Q$ must be the intersection of at least d supporting hyperplanes of P or Q. Every intersection of 2 or more supporting hyperplanes of Q must be part of some ridge of Q, hence Q can contribute at most one hyperplane to a vertex of $P \cap Q$. Any intersection of d supporting hyperplanes of P is a vertex of P, hence Q must contribute at least one hyperplane to a vertex of $P \cap Q$. It follows that every vertex of $P \cap Q$ is the proper intersection of an edge of P and a facet of Q.

Suppose there were an edge e of P and a facet F of P such that $\text{relint}(e) \cap \text{relint}(F) \neq \emptyset$ but $x \equiv \text{relint}(e) \cap \text{relint}(F)$ is not a vertex of $P \cap Q$. In order for this to be the case x must be infeasible for $P \cap Q$; but this is impossible since $x \in e \subset P$ and $x \in F \subset Q$.

(b⇐c) We prove the contrapositive. Suppose some ridge r of Q has non-empty intersection with P. The vertices of $r \cap Q$ must be vertices of $P \cap Q$, but cannot be proper intersections of edges of P and facets of Q. ∎

If the conditions of Theorem 2 hold for polytopes P and Q, we say that P *pierces* Q.

Corollary 3 *If d-polytope P pierces d-polytope Q then*
$$f_0(P \cap Q) \leq 2f_1(P).$$

Corollary 4 *If d-polytope P pierces d-polytope Q and every connected component of P w.r.t. Q has precisely one vertex, then*
$$f_0(P \cap Q) = 2f_1(P).$$

The following technical lemma is useful in establishing that no constraint in $\mathcal{H}(P) \cup \mathcal{H}(Q)$ is redundant for the polytope $P \cap Q$.

Lemma 5 *Let P be a d-polytope and F a facet of P. Let K be a convex set. If $\mathrm{relint}(F) \cap \mathrm{relint}(K) \neq \emptyset$ and $K \not\subseteq \mathrm{aff}(F)$ then $\mathrm{relint}(K) \cap \mathrm{int}(P) \neq \emptyset$.*

If d-polytope P pierces d-polytope Q and every facet of Q is properly pierced by some connected component of P w.r.t. Q, we say that P *completely pierces* Q.

Theorem 3 *If d-polytope P completely pierces d-polytope Q and every connected component of P w.r.t. Q has exactly one vertex, then*
$$\mathcal{H}(P \cap Q) = \mathcal{H}(P) \cup \mathcal{H}(Q).$$

Proof: Let F_p be a facet of P. Let v be a vertex of F_p. By Theorem 2, there exists some facet F_q of Q such that $\mathrm{aff}(F_q)$ separates v from $\mathcal{V}(F_p) \setminus \{v\}$ and $F_p \cap \mathrm{aff}(F_q) \subset \mathrm{relint}(F_q)$. The facet F_p has at least two vertices so F_p must intersect $\mathrm{aff}(F_q)$. Let F' denote $F_p \cap \mathrm{aff}(F_q)$. Suppose $F' \cap \mathrm{relint}(F_p) = \emptyset$; it follows that $\mathrm{aff}(F_q)$ is a supporting hyperplane for F_p, but this is a contradiction. It follows that $\mathrm{relint}(F_p) \cap \mathrm{relint}(F_q) \neq \emptyset$. By Lemma 5, the relative interior of F_p intersects the interior of Q. This means that there is a point that strictly satisfies every constraint of $(\mathcal{H}(Q) \cup \mathcal{H}(P)) \setminus \{F_p^+\}$ and satisfies F_p^+ with equality, which implies that F_p^+ is not redundant.

Since every facet F_q of Q is properly pierced by some connected component of P w.r.t. Q, a symmetric argument implies that F_q^+ is not redundant either. ∎

4.2 TRUNCATED CUBES

We now give examples of polytope families that satisfy the conditions established in Section 4.1. We start with a scaled d-cube C_d. We then construct a simplicial polytope S_d with $2d$ vertices and $d^2 - d + 2$ facets that pierces each face of C_d. Our final construction is then the intersection of S_d and C_d, which we will show has $d^2 + d + 2$ facets and $3d^2 - d$ vertices ($d \geq 3$). The lower bound argument is based on showing that for some insertion orders, an insertion algorithm must compute all 2^d vertices of C_d.

We wish to construct a class of polytope whose facet and vertex complexity is small in terms of the dimension, but that cuts off all of the vertices of a scaled d-cube; our class will be *stacked* polytopes like those used in the proof of the Lower Bound Theorem. If P^* is the convex hull of $P \cup \{x\}$ where x is beyond exactly one facet of P, we say that P^* is a *stellation* of P. A polytope is called *stacked* if it is the stellation of a d-simplex, or the stellation of stacked polytope. Let $P_{n,d}$ be a stacked d-polytope with n vertices. The following is known about the face complexity of a stacked polytope (see e.g. [4]).

$$f_j(P_{n,d}) = \begin{cases} \binom{d}{d-j}n - \binom{d+1}{d-j}j, & \text{if } 0 \leq j \leq d-2; \\ (d-1)n - (d+1)(d-2), & \text{if } j = d-1 \end{cases}$$

Let S_d denote the convex hull of $d + 1$ *simplex vertices* t_i with coordinates:

$$t_{ij} = \begin{cases} 1 - d^{-1} & \text{if } 1 \leq i < d \text{ and } j = i; \\ 1 & \text{if } i = j = d; \\ -1 & \text{if } i = d+1 \text{ and } j = d; \\ -d^{-1} & \text{otherwise} \end{cases}$$

and $d - 1$ *pyramid vertices* p_i with coordinates

$$p_{ij} = \begin{cases} -\frac{3}{2d}, & \text{if } i = j; \\ 0, & \text{otherwise} \end{cases}$$

Lemma 6 *S_d is a stacked d-polytope with $2d$ vertices, $d^2 - d + 2$ facets and the following halfspace description:*

$$\begin{aligned} H_{kj} &= \{x \mid x_j + 2x_k \geq -\tfrac{3}{d}\} \\ & \quad 1 \leq k < d, \quad j \notin \{k, d\} \\ H_{kd} &= \{x \mid x_d - \sum_{i \notin \{k,d\}} x_i + 4x_k \geq -\tfrac{6}{d}\} \\ & \quad 1 \leq k < d \\ H_{k(d+1)} &= \{x \mid \sum_{i \neq k} x_i - 4x_k \leq \tfrac{6}{d}\} \\ & \quad 1 \leq k < d \\ H_d &= \{x \mid x_d - \sum_{i=1}^{d-1} x_i \geq -\tfrac{1}{d}\} \\ H_{d+1} &= \{x \mid x_d + \sum_{i=1}^{d-1} x_i \leq \tfrac{1}{d}\} \end{aligned}$$

We now consider the intersection of the stacked polytope S_d with a scaled d-cube. Let C_d denote the d dimensional parallelepiped defined by the following constraints.

$$\begin{aligned} L_i &= \{x \mid x_i \geq -\tfrac{5}{4d}\} \quad \text{for } 1 \leq i < d \\ U_i &= \{x \mid x_i \leq 1 - \tfrac{5}{4d}\} \quad \text{for } 1 \leq i < d \\ L_d &= \{x \mid x_d \geq -1 + \tfrac{1}{4d}\} \\ U_d &= \{x \mid x_d \leq 1 - \tfrac{1}{4d}\} \end{aligned}$$

Let K_d denote $S_d \cap C_d$. Figure 1 (courtesy of Geomview) shows K_3.

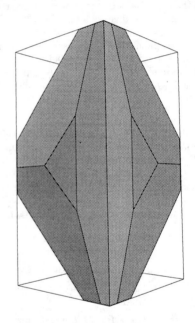

Figure 1: K_3

Theorem 4 *For $d \geq 3$,*

(a) $f_0(K_d) = 2f_1(S_d) = 3d^2 - d$

(b) $f_{d-1}(K_d) = f_{d-1}(S_d) + f_{d-1}(C_d) = d^2 + d + 2$

In order to prove this theorem, we establish two lemmas that will show that S_d completely pierces C_d, and that every connected component of S_d w.r.t. C_d consists of precisely one vertex of S_d. Part (a) of the theorem will then follow from Corollary 4 and part (b) will follow from Theorem 3.

Lemma 7 *For every vertex v of S_d, there exists a facet F of C_d, and for every facet F of C_d, there exists a vertex v of S_d such that: $\mathrm{aff}(F)$ strictly separates v from $\mathcal{V}(S_d) \setminus \{v\}$.*

Lemma 8 *For any facet F of C_d, $S_d \cap \text{aff}(F) \subset \text{relint}(F)$.*

There are several well known insertion orders for insertion algorithms.

minindex (maxindex) Insert the halfspaces in (the reverse of) the order given.

lexmin (lexmax) Insert the halfspaces in (reverse) lexicographic order of coefficient vector. There are several variations; one could reasonably reduce the constraint matrix to some canonical form, and there is the question of how to treat the right hand side vector. Here we assume that the input constraints are taken as is, and the right hand side entry is treated as the most significant entry of the coefficient vector (this is modelled on the program cdd [11]).

random Insert the halfspaces in random order.

maxcutoff (mincutoff) Insert the halfspace that causes the maximum (minimum) number of vertices and extreme rays to become infeasible.

Previous experiments (see [11]) have shown that there is no unique one of these orders that works well in practice on all problems.

Theorem 5 *Given $\mathcal{H}(K_d)$ as input, any insertion algorithm for vertex enumeration creates intermediate polytopes of size*

$$\Omega(2^{\sqrt[3]{size(K_d)}})$$

in the worst case when the halfspaces are inserted in an order independent of the input, or in lexicographic order.

Proof: In the case of an order such as minindex which ignores the input, it is trivial for an adversary to force an insertion algorithm to construct the entire d-cube. To fool lexicographic ordering, an adversary has merely to scale the constraints appropriately. ∎

Although the simplex method had long been regarded as a practical and efficient algorithm for linear programming, it was not until the seminal 1972 paper of Klee and Minty [15] that it was demonstrated to be an exponential time algorithm. The authors described a class of hard polytopes that cause the simplex method to make an exponential number of pivots using the greatest cost coefficient pivoting rule. Subsequent papers gave similar results for other pivot rules, and the search for a polynomial pivot rule (or the proof of its non-existence) still continues. Just as the Klee-Minty examples apply only to a particular pivoting order, this theorem applies only to particular insertion orders. Empirically, it would seem that there is indeed a polynomial insertion order for the truncated cubes (see Section 5.2). An interesting open question is whether every polytope has a polynomial insertion order.

It should be noted that the bound of Dyer [10] is somewhat stronger than that of Theorem 5. The polytope family of Klee that Dyer uses has several disadvantages as an empirical benchmark problem; it is not completely constructive, and it uses coordinates on the moment curve. These coordinates grow exponentially in the dimension of the polytope (i.e. the number of bits grows linearly) which quickly causes precision problems for programs working in floating point (both cdd and qhull — see Section 5 — run into numerical problems that prevent them from finding the correct number of vertices for the cyclic polytope at $d = 8$). Let $bits(P)$ denote the number of bits necessary to store both representations of polytope P in binary. By Lemma 6,

$$bits(K_d) \in O(d^3 \log d).$$

Corollary 9 *Given $\mathcal{H}(K_d)$ as input, for arbitrarily small positive ϵ, any insertion algorithm for vertex enumeration creates intermediate polytopes of size*

$$\omega(2^{bits(K_d)^{1/3-\epsilon}})$$

in the worst case when the halfspaces are inserted in an order independent of the input, or in lexicographic order.

5 EXPERIMENTAL RESULTS

The following convex hull/vertex enumeration programs were used: cdd, qhull and qrs. cdd is version 0.52b of Fukuda's implementation of the double description method [11] with local modifications to add timing facilities and optimize set operations. qhull is Barber and Huhdanpaa's implementation of "Quickhull" (a variant of the beneath and beyond algorithm), version 2.01 [2]. qrs is Avis' implementation of reverse search [1] using Edmonds Q-pivoting, version 2.4. We compiled qhull to use double precision. cdd uses double precision by default and qrs uses arbitrary precision rational arithmetic. The option C-0 was used to force qhull to merge the generated simplicial facets. Times are measured in CPU seconds.

5.1 PRODUCTS OF SIMPLICES

Table 1 shows the results for computing the convex hull of $T_{d/2} \times T_{d/2}$, the products of simplices described in Section 3. The column "#simplices" contains the number of simplices necessary to triangulate all facets of the polytope. For qhull we had to use the exhaustive search for a starting simplex. For cdd we used the default insertion order, lexmin. Figure 2 shows the times for cdd and Figure 3 shows the times for qrs and qhull.

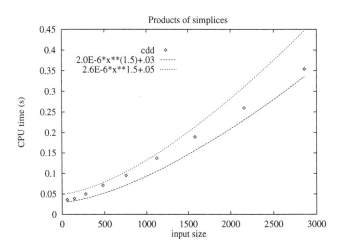

Figure 2: Products of simplices, cdd.

5.2 TRUNCATED CUBES

As well as qhull, we tested the following insertion orders for cdd: minindex, maxindex, maxcutoff, and random. To simulate random insertion order, we permuted the input 200 times using the *Combined Random Number Generator* of L'Ecuyer [16, 17] and reported the average time. In our data files, the cube constraints come last, so the results of Section 4 tell us to expect exponential performance for maxindex. The *scaled* order scales the cube constraints to be lexicographically larger than the stacked polytope constraints, then uses lexmax. Table 2 gives a summary of results. Figure 4 shows

d	vertices n	facets m	size $(m+n)d$	#simplices	times (CPU seconds) cdd	qhull	qrs	perturbed bases (qrs)
4	9	6	60	18	0.036	0.051	0.0	18
6	16	8	144	80	0.039	0.067	0.3	80
8	25	10	280	350	0.050	0.278	3.1	410
10	36	12	480	1512	0.071	2.06	27.5	1825
12	49	14	756	6468	0.095	20.1	253.5	7024
14	64	16	1120	27456	0.137	346.839	1740	29389
16	81	18	1584	115830	0.189	(a)	11060	117838
18	100	20	2160	486200	0.260		62553	653581
20	121	22	2860	2032316	0.354		432040	2803146

(a) Virtual memory usage exceeded available 342 megabytes.
All timings on a DEC3000/500 Alpha with 96M of memory, running OSF/1 1.3.

Table 1: Running times for convex hull on $T_{d/2} \times T_{d/2}$ (product of simplices in R^d).

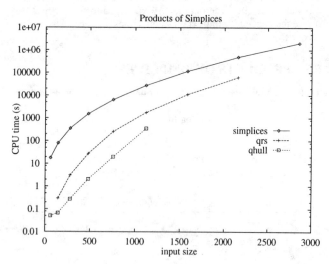

Figure 3: Products of simplices, qrs and qhull.

d	maxindex	scaled	maxcutoff	minindex
2	42	40	84	38
3	441	346	545	218
4	2953	3321	1576	647
5	15204	25825	4429	1515
6	54532	155263	12347	3049
7	169188	566793	18537	5495
8	551773	1968927	128807	9148
9	1366733	6966913	165586	14358
10	3153737	19860935	102658	21518
11	8433970	67547177	139569	31063
12	17531284	199965060	6698317	43472
13	35612382	476434767	657547	59268
14	89069240	1509324332	3189971	79018
15	224617847	3731767397	20328100	103333
16	656832668	8497874780	903479053	132868
17	2267294806	23782814756	123175667	168322
18	8296480578	in progress	321679986	210438

Table 3: Set intersection operations performed by cdd for truncated cubes

the timings for minindex, which builds the stacked polytope first, up to 35 dimensions.

Note that all insertion orders are monotonically increasing, except for maxcutoff, which is nonetheless bounded from below (in our experiments) by

$$3 \times 10^{-6} \times 2^{\sqrt[3]{size(P)}},$$

as are all of the insertion orders tested except minindex.

In order to abstract away some of the dependence of the timing experiments on a particular machine, we considered two machine independent measures of performance. The first is the number of basic (high level) operations performed by cdd; the second is the size of the intermediate polytopes generated. cdd represents a face f of a polytope by the set of facets containing f. Checking vertices for adjacency involves several intersections of facet sets representing vertices. In Table 3 we report the number of set intersections performed by various insertion orders. Another measure of the work done by the double description method is the maximum size of an intermediate polytope. Table 4 shows the maximum number of extreme rays in an intermediate polytope (cdd lifts each polytope to a cone in one higher dimension). This measure is only an approximation, as cdd reports it only after every 100 rays generated.

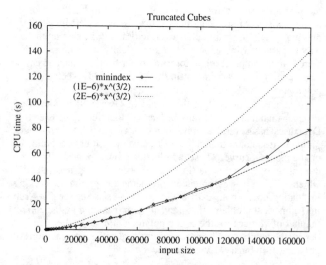

Figure 4: CPU time for truncated cubes, minindex insertion order.

| | | times (CPU seconds) | | | | | |
d	size	maxindex	scaled	random	maxcutoff	minindex	qhull
2	32	0.006	0.006	0.006	0.006	0.006	0.010
3	114	0.013	0.013	0.013	0.013	0.012	0.018
4	264	0.033	0.035	0.030	0.030	0.024	0.056
5	510	0.100	0.119	0.072	0.056	0.057	0.208
6	876	0.241	0.480	0.186	0.132	0.094	0.861
7	1386	0.616	1.42	0.422	0.186	0.121	2.20
8	2064	1.69	4.66	1.16	0.882	0.178	8.44
9	2934	3.88	15.1	3.35	1.18	0.284	37.9
10	4020	9.58	46.5	8.90	0.950	0.395	139
11	5346	25.1	171	28.8	1.24	0.541	1101
12	6936	50.7	535	76.3	28.4	0.765	(a)
13	8814	113	1343	267	6.23	1.038	
14	11004	302	4486	1001	17.17	1.362	
15	13530	1031	15077	4274	73.4	1.819	
16	16416	3817	41635	(b)	4825	2.390	
17	19686	14614	123326		664	2.977	
18	23364	58142	(b)		1337	3.871	

(a) Virtual memory usage exceeded available 120 megabytes.

(b) Experiment in progress.

All timings are on an SGI Indigo XS24 R4000 with 48M of memory and 72M swap, running IRIX 5.2.

Table 2: Timing results for truncated cubes.

What	Host	Internet Address	Directory	Files
cdd	ftp.epfl.ch	128.178.139.3	incoming/dma	cdd-*.tar.gz
qhull	geom.umn.edu	128.101.25.35	pub/software	qhull.tar.Z
Geomview	geom.umn.edu	128.101.25.35	pub/software/geomview	
qrs	mutt.cs.mcgill.ca	132.206.3.13	pub/C	qrs.c
products of simplices	mutt.cs.mcgill.ca	132.206.3.13	pub/C/ext/prod	t2t2.ver, t4t4.ver ...
truncated cubes	mutt.cs.mcgill.ca	132.206.3.13	pub/C/ine/trunc	trunc*.ine

Table 5: Availability of Software and Data by anonymous ftp.

d	maxindex	scaled	maxcutoff	minindex
4	40	52	(a)	(a)
5	82	152	63	70
6	148	401	92	59
7	214	653	109	137
8	392	1036	157	120
9	691	1433	180	194
10	1290	2063	240	266
11	2389	4090	272	335
12	4588	8189	720	402
13	8887	16288	468	467
14	17286	31487	831	528
15	33885	60186	2001	587
16	66984	113585	7776	740
17	132883	212384	2917	793
18	264282	(b)	3471	941

(a) No ray counts printed.
(b) Experiment in progress.

Table 4: Measured maximum intermediate sizes (number of extreme rays) for cdd on truncated cubes.

AVAILABILITY OF SOFTWARE AND DATA

All of the software and data files described in this paper are available by anonymous ftp. See Table 5 for details.

ACKNOWLEDGEMENTS

The authors would like to thank the Minnesota Geometry Center for the software packages qhull and Geomview, and Komei Fukuda for cdd. We would also like to thank Günter Rote and Bernd Sturmfels for suggesting the examples in Section 3, Luc Devroye for pointers to random number generators, Jesus deLoera for useful discussions, and Günter Ziegler for comments on a previous version of this paper.

REFERENCES

[1] D. Avis and K. Fukuda. A pivoting algorithm for convex hulls and vertex enumeration of arrangements and polyhedra. *Discrete Comput. Geom.*, 8:295–313, 1992.

[2] B. B. Barber, D. P. Dobkin, and H. Huhdanpaa. The quickhull algorithm for convex hull. Technical Report GCG53, The Geometry Center, University of Minnesota, July 1993.

[3] M. Bayer and C. Lee. Combinatorial aspects of convex polytopes. In P. Gruber and J. Wills, editors, *Handbook of Convex Geometry*, volume A, chapter 2.3, pages 485–534. North Holland, New York, NY, 1993.

[4] A. Brøndsted. *Introduction to Convex Polytopes*. Springer Verlag, 1981.

[5] G. Ceder, G. Garbulsky, D. Avis, and K. Fukuda. Ground states of a ternary lattice model with nearest and next-nearest neighbor interactions. *Physical Review B*, 49:1–7, 1994.

[6] D. Chand and S. Kapur. An algorithm for convex polytopes. *J. ACM*, 17:78–86, 1970.

[7] B. Chazelle. An optimal convex hull algorithm in any fixed dimension. *Discrete Comput. Geom.*, 10:377–409, 1993.

[8] V. Chvátal. *Linear Programming*. W. H. Freeman, New York, NY, 1983.

[9] K. L. Clarkson and P. W. Shor. Algorithms for diametral pairs and convex hulls that are optimal, randomized, and incremental. In *Proc. 4th Annu. ACM Sympos. Comput. Geom.*, pages 12–17, 1988.

[10] M. Dyer. The complexity of vertex enumeration methods. *Math. Oper. Res.*, 8(3):381–402, 1983.

[11] K. Fukuda. cdd Reference manual, version 0.52b. EPFL, Lausanne, Switzerland.

[12] I. Gel'fand, M. Kapranov, and A. Zelevinsky. *Discriminants, resultants, and multidimensional determinants*. Mathematics: theory & applications. Birkhauser, Boston, 1994.

[13] M. Haiman. A simple and relatively efficient triangulation of the n-cube. *Discrete Comput. Geom.*, pages 287–289, 1991.

[14] V. Klee. Polytope pairs and their relationship to linear programming. *Acta Math.*, 133:1–25, 1974.

[15] V. Klee and G. Minty. How good is the simplex method? In O. Shisha, editor, *Inequalities-III*, pages 159–175. Academic Press, 1972.

[16] L'Ecuyer. Efficient and portable combined random number generators. *Communications of the ACM*, 31, 1988.

[17] L'Ecuyer and Tezuka. Structural properties for two classes of combined random number generators. *Mathematics of Computation*, 57, 1991.

[18] C. W. Lee. Regular triangulations of convex polytopes. In *The Victor Klee Festschrift*, volume 4 of *DIMACS Series in Discrete Mathematics and Theoretical Computer Science*, pages 443–456. American Mathematical Society, 1991.

[19] T. Motzkin, H. Raiffa, G. Thompson, and R. M. Thrall. The double description method. In H. Kuhn and A. Tucker, editors, *Contributions to the Theory of Games II*, volume 8 of *Annals of Math. Studies*, pages 51–73. Princeton University Press, 1953.

[20] R. Seidel. A convex hull algorithm optimal for point sets in even dimensions. Technical report, University of British Columbia, Dept. of Computer Science, 1981.

[21] R. Seidel. Constructing higher-dimensional convex hulls at logarithmic cost per face. In *Proc. 18th Annu. ACM Sympos. Theory Comput.*, pages 404–413, 1986.

[22] C.-K. Yap. Symbolic treatment of geometric dependancies. *J. Symbolic Computation*, 10:349–370, 1990.

Immobilizing Polygons against a Wall

Mark Overmars*   Anil Rao*   Otfried Schwarzkopf*   Chantal Wentink*

Abstract

A familiar task in industrial applications is grasping an object to constrain its motions. When the external forces and torques acting on the object are uncertain or varying, form-closure grasps are preferred; these are grasps that constrain all infinitesimal and finite motion of the object. Much of previous work on computing form-closures has involved achieving it with point-contacts; for a planar object, four point-contacts were proven to be necessary and sufficient. Inspired by the intuitive habit of supporting an object against something flat to immobilize it, in this paper we propose a new class of contacts called edge-contacts; these offer a straight-line support against which the object rests. Our first result is that almost any polygonal part can be constrained in form-closure with an edge-contact and two point-contacts.

A related problem is that of immobilizing an object with modular fixtures. These typically comprise of a regular lattice of holes on which the object is placed and a set of precision contacting elements or fixels whose locations are constrained by the grid. Given the input object modeled as an n-gon, the grid size, and the shapes of the fixels available, we present an algorithm that returns all valid modular fixtures, each accomplished by an edge-fixel, a locator-fixel and a clamp. The algorithm uses sophisticated data-structures to achieve

---

Department of Computer Science, Utrecht University, P.O.Box 80.089, 3508 TB Utrecht, the Netherlands, {markov, anil, otfried, chantal}@cs.ruu.nl. This research was supported in part by ESPRIT Basic Research Action No. 6546 (project PROMotion) and the Dutch Organization for Scientific Research (NWO).

Permission to copy without fee all or part of this material is granted provided that the copies are not made or distributed for direct commercial advantage, the ACM copyright notice and the title of the publication and its date appear, and notice is given that copying is by permission of the Association of Computing Machinery.To copy otherwise, or to republish, requires a fee and/or specific permission.
11th Computational Geometry, Vancouver, B.C. Canada
© 1995 ACM 0-89791-724-3/95/0006...$3.50

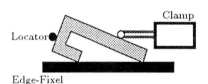

Figure 1: An example fixture with an edge-fixel, a locator and a clamp.

output-sensitivity; it runs in $O(n(n+p)^{4/3+\epsilon}+K)$ time, where K is the number of valid solutions (fixture configurations) returned, and p is the object's perimeter in grid units.

1 Introduction

Many manufacturing operations, such as machining, assembly, and inspection, require constraints on the motions of parts or subassemblies of parts [4, 6]. The concept of *form-closure* is over a century old [15] and refers to constraining, despite the application of an external wrench (force and moment), all motions of a rigid object (including infinitesimal motions) by a set of contacts on the object; any motion of an object in form-closure has to violate the rigidity of the contacts. Therefore, the problem becomes computing contact locations on a given part shape that achieve form-closure.

In this paper, we are interested in immobilizing planar objects and in particular, polygons. We refer to the set of contacts achieving form-closure as a *fixture*. We assume the contacts are frictionless; note that this is a conservative assumption since any fixture computed assuming zero friction also holds in presence of friction. By the *fixture model*, we imply the set of allowable contact types. The conceptually simplest model is that of point contacts. It has been known since Reuleaux that fixturing a planar object requires at least four frictionless contacts. Mishra, Schwartz, and Sharir [13] and Markenscoff, Ni and Papadimitriou [11] independently proved that four point-contacts are also sufficient.

While point-contacts are conceptually simple, they are not always easy to achieve in practice. The reason is that for form-closure, these point-contacts have to be capable of resisting arbitrary wrenches and therefore they have to be backed by bulky supports. This in turn implies difficulty of placing point-contacts at some points on the boundary of an object, in particular, at narrow concavaties. Therefore it becomes important to look for other possible practical fixture models in order to reduce the number of point-contacts. In everyday life, we frequently lean an object against a flat surface, such as a table or a wall, to constrain its motions. In the planar world, the analog of a wall is a supporting line. In this paper we extend the fixture model to include *edge-contacts* which offer straight-lines of support. Notice that an edge-contact can touch the object only along its convex hull. The object simply rests against it; there is no reaching into concavaties. See for an example Fig. 1.

How powerful are edge-contacts with respect to point-contacts? Although we cannot place point-contacts at convex vertices, it is possible to place edge-contacts touching a pair of convex vertices (a pair of non-adjacent vertices of a non-convex polygonal object but adjacent in the convex hull). On the other hand, in general, an edge-contact cannot replace more than two point-contacts. To see this, consider a convex polygonal object; an edge-contact is equivalent to two point-contacts at the two ends of contact. The question remains whether an edge-contact can always replace two point-contacts; in other words, can form-closure on polygons be achieved with one edge- and two additional point-contacts? It is easy to show that rectangular objects cannot be immobilized this way; there is always a direction to slide the rectangle out. However, we answer the question in the affirmative for any polygonal object which has no edge parallel to a different edge of its convex hull. Our proof is constructive and produces a fixture in $O(n \log n)$ time. Objects that have edges parallel to convex hull edges, a condition that can be prechecked in $O(n \log n)$ time, may or may not be fixturable.

The second part of our work relates to *modular fixtures* which is a subject of considerable popular interest in the manufacturing industry for the past ten years or so [2, 3, 8, 9]. Basically, this involves a regular square grid of lattice holes accompanied by fixture elements (or *fixels* [5]) that are constrained by the grid; the object then rests against these fixels which constrain its motions. Custom-built fixtures being expensive, the major benefit of modular fixtures stems from their reconfigurability; it is often necessary to fixture an object only for short periods of time after which the same set of fixels can be used to fixture different objects. Another advantage of modular fixtures is their easy assembly and disassembly.

Since research in computing form-closures generally involved point-contacts, it is not surprising that most fixels were designed to achieve point-contact. The simplest fixel is the *locator* which is a circular object centered at a lattice hole. Since achieving contact with four circles (constrained to a grid) is in general impossible ([17] shows this even for three circles), it is clear that we need a fixel that takes care of the slack. Such a fixel is called a *clamp* which has a fixed portion, the clamp *body*, attached to a movable rod, the clamp *plunger*, that can translate between certain limits along a grid line. The end of the plunger is the clamp *tip* which makes contact with the object. A clamp can be configured so that the motion of the plunger is parallel to either one of the axes; it is termed *horizontal* or *vertical* accordingly. See Fig. 1 for an example of a clamp.

Much of the research in modular fixtures to date has been on fixture analysis with detailed mechanical studies on part deformations and tolerencing; the fixture synthesis algorithms are generally incomplete or heuristic in nature. Among exceptions are the works by Mishra [12] and Brost and Goldberg [5] who investigated fixturing polygons with three locators and one clamp (the 3L1C model). However, these are either limited to simple object shapes ([12]: rectilinear objects) or are algorithmically inefficient and not output-sensitive because they involve some kind of a brute-force search for fixture locations among tuples of edges ([5, 16]).

Wallack and Canny [16] consider an interesting model of modular fixtures which uses four locators and no clamps; instead, the slack is countered by mounting the part on a split horizontal lattice and allowing the one half to slide horizontally relative to the other. They also give an algorithm to compute all possible fixtures which runs in time proportional to the number of part configurations contacting four locators. Notice that this is not output sensitive in the actual number of fixtures output as many contacting configurations may indeed not be in form closure.

Brost and Goldberg [5] and Wallack and Canny [16] do not identify a class of fixturable objects while [12, 17] identify a subset of rectilinear objects as fixturable under the respective model considered.

Corresponding to the edge-contact discussed before and to enhance our fixture model, we add *edge-fixels* to our repertoire. An edge-fixel is simply a bar-like object of appropriate dimensions fixed to the lattice offering a straight-edge of support. We assume that an edge-fixel is at least as long as the longest edge in the convex hull of the object. As before, we consider replacing two of the point-fixels (locators/clamps) by an edge-fixel and we describe an algorithm that generates the (possibly empty) list of all valid form-closure fixtures generated by one edge-fixel, one locator, and one clamp. We term this the ELC model.

An important distinguishing characteristic of our algorithm compared to previous algorithms is that it is output sensitive; it runs in $O(n(n+p)^{4/3+\epsilon} + K)$ time for any positive constant ϵ, where K is the number of fixtures output which can be as high as $n(n+p)^2$. By feeding these solutions through a given quality metric or Boolean comparator, an optimal fixture may be computed. Optimization criteria may include some function of the distance between the points of contact, or a function of the directions of contact (to prevent wear-and-tear, it might be necessary to orient fixels as far as possible towards or away from a particular direction), etc. The generated solutions can be sorted according to any given metric. The fixture that is the best under one metric may be bad under a different one; therefore, it is usually not sufficient to generate just one valid fixture but instead the whole set.

After introducing some preliminaries in Section 2, we consider fixturing polygons with an edge- and two point-contacts in Section 3. The results of this section are equivalent to considering locators and clamps of negligible size and locatable anywhere on the object's boundary. Section 4 considers the case of modular fixtures: a finitely sized grid and fixels. Finally, we close with some remarks and suggestions for future work in Section 5. Since there is a page limitation, we cannot afford to present all proofs here; these will be presented in a forthcoming paper.

2 Preliminaries and Notation

Recall that a fixture is said to provide *form-closure* if it precludes all (planar) motion, translations and rotations. Let us begin by examining what motions are ruled out by single point- or edge-contacts. From here onwards, we will consider rotations only with the understanding that translations in a direction are simply rotations about infinity along the perpendicular direction. Note that any infinitesimal motion of a polygon can be represented by a point in the plane, denoting the center of rotation (COR) of this motion. A point at infinity thus represents a infinitesimal translational motion.

Denote the input polygon by P and its boundary by ∂P; n is the number of edges forming ∂P. Let the polygon edge containing a point a on its boundary be denoted by $E(a)$; the directed line perpendicular to $E(a)$ through a pointing into the interior of P is $l(a)$. We distinguish between the following cases (See Fig. 2).

Point-contact at interior of an edge This is the fundamental contact and the motions allowed by other types of contacts can be deduced by composing those allowed by elementary point-contacts.

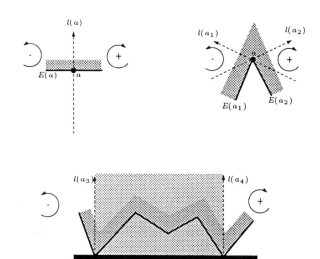

Figure 2: Motions possible under simple contacts.

Consider a point-contact at point a in the interior of edge $E(a)$ as shown. The allowed motions are defined by the line $l(a)$. If the object rotates (infinitesimally) in a clockwise (positive) direction with a point-contact at a, its COR will have to lie in the region to the right of (and including) $l(a)$. Furthermore, any point in this closed right-half-plane is a possible COR for a positive rotation. Similarly, the CORs for counter-clockwise or negative rotations lie in the closed half-plane to the left of $l(a)$. These are all and the only constraints imposed by the point-contact at a. For future reference, we term these as *half-plane constraints* imposed by the point contact.

Point-contact at concave vertex This is shown in the upper-right of Fig. 2; a is the concave vertex. Imagine two points a_1, a_2 infinitesimally close to a along the two edges defining it. The motions allowed by a can be determined by intersecting the motions allowed by the point-on-edge pairs $(a_1, E(a_1))$ and $(a_2, E(a_2))$ which may be individually analyzed as above. The result is shown in the figure. These may be termed the *wedge constraints* defined by a.

Consider rotating a line l through a from $l(a_1)$ to $l(a_2)$ in the clockwise direction. The wedge constraints defined by a is the intersection of each of the half-plane constraints obtained along the sweep. Therefore, the motions allowed by a are a subset of the motions allowed by any one of these half-plane constraints. However, the event that concave vertex is on a grid line is a degenerate one and in practice this will rarely happen. So, in future, for ease of analysis, we choose to approximate the wedge constraint by a half-plane constraint defined by a particular line along this sweep.

Edge-contacts An edge-contact can contact the polygon along an edge e of the polygon or along two vertices a_3, a_4 (adjacent on the convex hull). The latter case is shown in the figure; the former case can be similarly analyzed considering a_3, a_4 to be the end-vertices of e. Consider lines $l(a_3), l(a_4)$ perpendicular to the edge-contact and superimpose the two half-plane constraints to get the result shown. The closed left half-plane allows negative rotations while that on the right allows positive rotations. The open infinite "slab" in the middle, shown shaded in the figure and disallowing all motions is denoted $slab(a_3, a_4)$ and is a crucial entity in future analysis. The constraint imposed will be called a *slab constraint*.

3 A Class of Fixturable Objects

In this section we prove that any polygon having no edge parallel to a different edge in its convex hull can be immobilized with one edge- and two point-contacts. Recall that each contact gives a closed half-plane representing allowed positive rotations and another closed half-plane representing negative rotations. To constrain all motion and obtain form-closure, the set of allowable CORs must be empty; in other words, the intersection of the three closed half-planes for positive rotations must be empty and likewise the three closed half-planes for negative rotations. The following lemma gives a sufficient condition for the above which we exploit in our algorithms later. A set of vectors $v_1, v_2, \ldots, v_k$ is said to *positively span* a space Ξ if any vector $\xi \in \Xi$ can be written as $\sum_{i=1}^{k} \alpha_i v_i$ where the α_i are non-negative scalars. When $k = 3$ and $\Xi = \mathbf{R}^2$, it is easy to show that v_1, v_2, v_3 positively span $\mathbf{R}^2$ if and only if the angle between any two vectors not including the third is greater than zero and less than π.

Lemma 1 *An object P is in form-closure with point-contacts a_1, a_2 and edge-contact (a_3, a_4) if and only if*

1. *the three vectors along $l(a_1), l(a_2), l(a_3)$ positively span $\mathbf{R}^2$, and*

2. *the intersection point of $l(a_1)$ and $l(a_2)$ lies in the interior of $slab(a_3, a_4)$.*

See Fig. 3 for an example of an object that is in form closure with one edge-contact and two point-contacts. Notice that both conditions from Lemma 1 are satisfied. To compute one form-closure configuration, we look at maximal inscribed circles. Any maximal inscribed circle of a polygon without pairs of parallel edges touches the polygon in at least three points. The following result follows from Markenscoff et al. [11].

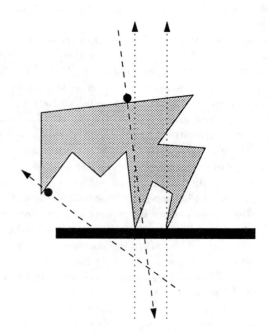

Figure 3: A polygon that is in form closure with one edge-contact and two point-contacts. The directed lines drawn indicate the constraints on the motion of the object by the individual contacts.

Lemma 2 [11] *For any maximal inscribed circle C of a polygon P without pairs of parallel edges, and c is the center of C, the three vectors $\overrightarrow{ca_i}$, $a_i \in (C(P) \cap \partial P)$, positively span $\mathbf{R}^2$.*

Theorem 3 *Let P be a convex polygon with no pairs of parallel edges. P can be held in form-closure with one edge-contact and two point-contacts.*

Proof: Let $C(P)$ be the maximal inscribed circle of P with center c. Since P does not have parallel edges, there are always at least three intersection points of P and $C(P)$; the circle is tangent to the polygon at these points. Consider three such points a_1, a_2 and b and their corresponding edges $E(a_1), E(a_2)$ and $E(b)$. Note that each of these points is strictly to the interior of the corresponding edge. Place one of these edges, say $E(b)$, against the edge-contact. The other two points, a_1, a_2, are the point-contacts. Let the extreme points of intersection between the edge-contact and P be b_1, b_2; since P is convex, these are simply the ends of $E(b)$.

From Lemma 2 it follows that the vectors along bc, a_1c, a_2c positively span $\mathbf{R}^2$. Since $l(a_1)$ is along a_1c, the same for a_2, and the vector along bc is parallel to $l(b_1)$, the first condition for form-closure in Lemma 1 is satisfied.

As for the second condition, note that $l(a_1)$ and $l(a_2)$ are along the diameters of C and therefore intersect at the center c. Since $E(b)$ and C are tangent, the center c of C, lies strictly within the open region $slab(b_1, b_2)$. $\square$

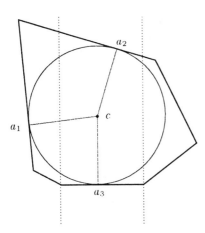

Figure 4: The tangent points of the largest inscribed circle of P with P give us the edge- and point-contacts for form-closure.

Theorem 4 *Let P be an arbitrary polygon. If no edge of P is parallel to a different edge of its convex hull then P can be held in form-closure with one edge- and two point-contacts.*

Proof: The proof of this theorem is similar to that given for the previous theorem. Let $C(P)$ be a maximal inscribed circle of P and let c be its center. Consider three intersection points of $C(P)$ with ∂P: a_1, a_2 and a_3. If none of these points is on one of the edges of the convex hull, grow $C(P)$ until it touches a edge e of the convex hull of P. Let this edge be bounded by points b_1, b_2. Fix the edge-contact against e. Since none of the edges $E(a_1), E(a_2)$ or $E(a_3)$ is parallel to e, the angles between the lines $l(a_1), l(a_2)$ and $l(a_3)$ are all smaller than π. (In case some a_i is a concave vertex, $l(a_i)$ can be considered as any from a range of lines.) Thus, at least one of these lines (assumed directed from a_i into the object) crosses $slab(b_1, b_2)$ from left to right, say $l(a_1)$, and another crosses it from right to left, say $l(a_2)$. Place the point-contacts at a_1, a_2. It is clear from the definitions of $l(a_1)$ and $l(a_2)$ that $l(a_1), l(a_2), l(b_1)$ positively span $\mathbf{R}^2$. Also, as before, c, the intersection point of $l(a_1)$ and $l(a_2)$ lies strictly within $slab(b_1, b_2)$. □

The above proofs immediately gives us an algorithm to compute one form-closure for a given polygon. First, we compute the maximal inscribed circle of the polygon. This can be done in time $O(n \log n)$ by computing the medial axis of P [10] and in time $O(n)$ when P is convex. Then, if necessary, we compute the edge of the convex hull that is closest to this inscribed circle; this gives us the edge-contact. The two point-contacts can be determined as detailed above.

4 Generating Modular Fixtures

In this section we consider fixturing of the object with modular fixtures. We are given a rectangular rigid and flat surface into which circular holes of some precise dimension have been drilled to form a regular square lattice. Grid points are hole centers while *grid lines* are the vertical and horizontal lines through the grid points. Unit distance is defined as the distance between two grid points. Let P denote the given polygonal object with perimeter p and diameter d. The edge-fixel is assumed to be attached to the lattice parallel to one of the grid lines (it can be prefabricated along with the lattice). The horizontal or x-axis of our reference frame lies along the edge-fixel with P in the positive y half-plane. In addition to the edge-fixel, we are given a locator and a clamp whose placements are constrained by the grid. The precise constraints will be detailed later.

We present algorithms that output all valid fixtures. For ease of exposition and to give the intuition to the general case, we first make the assumption that the locator is a point and so is the entire clamp. The constraints imposed by the grid are that the point locator should be placed at a grid-point and the point clamp can exist anywhere along a grid-line. A simple procedure to enumerate all fixtures (non-output sensitively) is presented in Section 4.1. An improved algorithm is presented in Section 4.2 in which we formulate the problem in terms of data structure queries and give an efficient realization of the data structure ensuring output sensitivity. Finally, we give modifications required to relax the simplifying assumptions made and consider circular locators and rectangular clamp bodies and plungers in Section 4.3.

4.1 Enumerating all fixtures

Consider the polygon P shown bold in Fig. 5 resting with some particular edge of its convex hull against the edge-fixel, the thick bar shown at the bottom, attached to the lattice. As P may not be touching a grid-point on its boundary, we slide it, say to the left, until it does. Now we have a possible position for the locator. To compute all possible locator positions, it is sufficient to slide the polygon one unit to the left and mark all grid-points encountered *during* the slide. Let $P = P(0)$ indicate the initial position of the polygon and $P(t)$, the polygon when shifted by t units to the left, $0 \leq t \leq 1$. Let t be termed the *shift variable*. A locator position L exists only at certain discrete values of the shift variable; at each one it contacts a different edge e of P. Let us term each such contact as a *locator-edge combination*. Let $t(L, e)$ denote the shift value when L touches e and $\alpha(L, e)$ the point of contact.

Clamp positions are those points on the boundary of P intersected by the grid lines in some $P(t)$. We

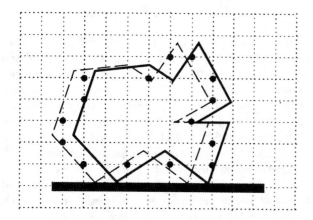

Figure 5: Sliding the polygon one unit along the edge-fixel to enumerate all feasible locator positions.

distinguish between two cases: those points intersected by the horizontal grid lines are called *horizontal clamps* while the others are *vertical clamps*. Notice that since the shift is horizontal, horizontal clamp positions remain constant for all shift values and can be computed from the unshifted polygon. On the other hand, vertical clamp positions vary with the shift value. A vertical grid line V might intersect several edges e of P during the shift resulting in *(vertical) clamp-edge* combinations. Let the interval $(t_s(V, e), t_f(V, e))$ denote the range of shift values for which the clamp-edge combination of (V, e) exists; for t from within this interval, let $\alpha(V, e, t)$, denote the point of intersection between edge e in $P(t)$ and V.

There are at most $O(p)$ grid points (locator positions) generated by this process (a better bound is $O(\min(d^2, p))$, where d is the diameter of P, but we'll assume $p < d^2$, without loss of generality). Further, the total number of locator-edge and clamp-edge combinations is $O(p+n)$ and can be computed in $O(n+p)$ time.

A naive procedure to generate all valid fixtures is therefore to consider each locator-edge combination and clamp-edge combination pair, check if they simultaneously exist (w.r.t. the shift variable; this is necessary only for locators with vertical clamps) and if so, test if they achieve form-closure. Iterating over all convex hull edges against the edge-fixel, this gives a $\Theta(n(p+n)^2)$ algorithm. In Section 4.2 we improve this to $O(n(p+n)^{4/3+\epsilon} + K)$ using efficient data structures.

Remark 1 In order for all the possible fixtures to be feasible, the lattice should have dimensions at least $\lceil d+1 \rceil \times \lceil d+1 \rceil$. A lattice of these dimensions also suffices. In the general case of locators and clamps of some non-zero diameters d_l, d_c, the sum $d + d_c + d_l$ should replace d.

Remark 2 The user may specify that certain locations on the boundary of P or that certain access-directions to the object be fixel-free. Edge-fixel, locator and clamp locations can be pruned out by checking each for violation. This will not affect the total time bound.

4.2 Efficient data structures

In the following we'll generate fixture configurations in which the locator is at the left side of the object. The procedure should be repeated for the other symmetrical case. The case of horizontal clamps is easier and gives some intuition to the vertical case and therefore is treated first.

Fixtures with a locator and horizontal clamp
As mentioned above, horizontal clamp positions can be determined by intersecting the horizontal grid lines with the object and they remain constant for all shifts. Thus the half-plane constraint corresponding to each horizontal clamp can be represented by a directed line at the point of contact and normal to the edge of contact, directed into the object (see Section 2). (If the point of contact is a concave vertex, we choose a suitable directed line that points "into" a cone of possible directions as discussed before. Convex vertices are not feasible clamp or locator positions.) Since we are interested only in clamps to the right of the object (because locators are to the left), only directed lines crossing the slab from right to left need to be included. There are $O(n+p)$ such directed lines. Call these the *clamp lines*.

Another set of directed lines comes from the locators. For each locator-edge combination (L, e), create a similar directed line at the contact point $\alpha(L, e)$. It is not necessary to take into consideration the shift value $t(L, e)$ at which the combination (L, e) exists because every horizontal clamp always exists and can exist simultaneously with any single locator irrespective of $t(L, e)$. Again, there are $O(n + p)$ *locator lines*. It is sufficient to consider directed lines crossing the slab from left to right.

Thus we obtain two sets of directed lines defining half-plane constraints on P; one set of lines associated with possible clamp positions and one set of lines for the possible locator positions. We want to find all combinations of lines, one from each set, such that Conditions 1 and 2 in Lemma 1 are satisfied. Intersecting the directed (infinite) lines with the slab corresponding to the fixed convex hull edge of P, we obtain two sets of directed line *segments*.

Given a locator line segment, we wish to detect, in an output-sensitive manner, all clamp line segments that (properly) intersect it and such that the vectors along the two segments and the upward vertical vector positively span $\mathbf{R}^2$. See Fig. 6. Formally, the basic query that we want to answer is:

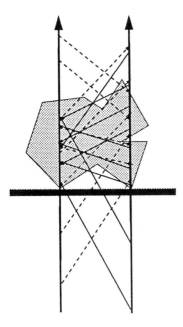

Figure 6: The directed line segments (directed lines representing half-plane constraints restricted to the slab formed by the edge-fixel at the bottom) resulting from the object in Figure 5. The dotted line segments (from left to right) correspond to the locators while the solid lines (from right to left) are from the clamps.

QUERY A Let S be a set of m directed line segments, each with an end-point along two vertical lines (the "slab"), and each with orientation from the open interval $(-\frac{1}{2}\pi, \frac{1}{2}\pi)$. Store S such that for a query segment q also similarly anchored on the slab but with orientation in the interval $(\frac{1}{2}\pi, \frac{3}{2}\pi)$ one can report all segments $s \in S$ such that

1. the vectors along s, q and the upward vertical vector positively span $\mathbf{R}^2$, and

2. s and q intersect strictly inside the slab.

Lemma 5 *There exists a data structure that solves QUERY A in time $O(\log^2 m + k)$ time, where k is the number of reported segments. Preprocessing takes $O(m \log^2 m)$ time.*

Proof: Since the orientation of all segments in S is in the interval $(-\frac{1}{2}\pi, \frac{1}{2}\pi)$, and the query segment q is oriented in direction $\alpha \in (\frac{1}{2}\pi, \frac{3}{2}\pi)$, it follows that all the line segments oriented in the subinterval $(\frac{3}{2}\pi, \alpha + \pi)$ satisfy the first condition in the query (this is because the angles between any two of the vectors not including the third are all smaller than π). The data structure that we need to compute this set of segments is a range tree on the angles of S; a query on this tree in order to find all segments in the desired interval takes $(\log m + k')$ time [14]. The k' angles can be reported in $O(\log m)$ canonical subsets in the nodes of the tree. In these subsets we have to search for intersections between the corresponding line segments and the query line segment. We do this by associating an intersection query structure with every node in the segment tree. This secondary structure is built on all the line segments that are associated with the angles in this node. Such an intersection query in a slab reduces to a dominance range query problem as will be described in the full paper. A query on this range query structure takes $O(\log m + k'')$ time, k'' being the number of answers. Combining these two structures gives us a query time of $O(\log^2 m + k)$, where k is the number of reported segments. The structure can easily be reported in time $O(m \log^2 m)$. □

Now we are ready to prove the following theorem.

Theorem 6 *Let P be a polygon with perimeter p and n edges. All configurations holding P in form-closure using one edge-fixel, one locator and one horizontal clamp can be enumerated in time $O(n(n+p)\log^2(n+p) + K)$, where K is the number of form-closure configurations.*

Proof: First compute the convex hull of P. For every edge of the convex hull along the edge-fixel do the following. Compute all the $O(n+p)$ locator lines and all the $O(n+p)$ clamp lines. This can be done in $O(n+p)$ time. For every locator line, we perform QUERY A. In all, this takes takes $O((n+p)\log^2(n+p) + k')$ time, since there are at most $O(n+p)$ of these lines; k' is the number of fixture configurations in this iteration of the algorithm. The total time therefore becomes $O(n(n+p)\log^2(n+p) + K)$. □

In contrast to querying the locator lines one at a time, we can "batch" together all the locator and horizontal clamp lines and by processing them together carefully, it is possible to improve the above time complexity to $O(n(n+p)\log(n+p) + K)$. However, we chose to retain the query algorithm since it gives some intuition towards handling vertical clamps.

Fixtures with a locator and vertical clamp As before, assume that one edge of the convex hull of P is flush against the edge-fixel. Shifting the polygon one unit to the right generates all locator-edge combinations along with a particular shift value at which each exists. However, in contrast to horizontal clamps, a vertical grid line V does not always touch an edge e of $P(t)$ at the same spot during the shift, i.e., the point of contact $\alpha(V, e, t)$ is a linear function of t. Also vertical clamp-edge combinations only exist for a subinterval of shift values $(t_s(V,e), t_f(V,e))$. For these reasons, the algorithm to compute form-closures using vertical clamps is different from the algorithm for horizontal clamps described above.

The major modification required is to explicitly handle shift by a third dimension. Firstly, the slab corresponding to the edge-fixel is extended into the t dimension by one unit to form a rectangular block. The directed line segment representing the half-plane constraint from a locator-edge combination (L, e) is lifted to the $t = t(L, e)$ plane. For vertical clamp-edge combinations (V, e), we need to represent the half-plane constraint at *every* shift instant in $(t_s(V, e), t_f(V, e))$. Fortunately, since e is linear, the corresponding directed line-segments are parallel to each other (perpendicular to e); the directed line at instant t has to be perpendicular to edge e in $P(t)$ and passing through $\alpha(V, e, t)$. This continuum of parallel directed line segments thus forms a parallel band between $t_s(V, e)$ and $t_f(V, e)$; the band is oriented in the direction of the lines comprising it.

In this representation, we look for proper intersections of oriented bands corresponding to vertical clamp-edge combinations with directed lines corresponding to locator-edge combinations that further satisfy the condition related to positively spanning $\mathbf{R}^2$. By restricting attention to the rectangular block, we get a parallelogram for every band and a directed line segment for every directed line.

As before we can consider right-to-left parallelograms to be intersected by left-to-right directed line-segments. The other symmetrical case can be treated in the same fashion.

QUERY B

Let S be a set of m directed line segments anchored between two parallel faces of a rectangular block as described above and each parallel to the xy plane. Store S such that for an anchored oriented parallelogram q between heights t_a and t_b as detailed above one can report all segments $s \in S$ such that

1. the vectors along s and q and the vertical vector in positive y-direction positively span $\mathbf{R}^2$ and

2. s and q intersect strictly inside the rectangular block.

Lemma 7 *There exists a data structure that solves QUERY B in time $O(m^{1/3+\epsilon} + k)$ time for any positive constant ϵ, where k is the number of reported segments. The preprocessing time is $O(m^{4/3+\epsilon})$.*

Proof: The data structure consists of a three-level tree. The first level is used to detect the segments of S that have the correct angle. The second level detects all segments that can possibly intersect the query parallelogram, because they exist at a time $t \in (t_a, t_b)$. The third layer is used to check for intersections between a query plane and a set of segments. First, we build a range tree on the orientations of the segments in S, which is the same as the one that we used for QUERY A, in order to satisfy Condition 1. Since the parallelogram only exists between shifts (t_a, t_b) which may not be $(0, 1)$. We build a range tree representing heights at which the segments occur. With every node in this tree, we associate a partition tree [1, 7] on the segments associated with the values that are present below this particular node. Such a partition tree can be queried with a plane in order to obtain all segments that intersect this query plane. Now we query the range tree with (t_a, t_b) in order to obtain all segments in between those shifts. Having done this, we can replace the parallelogram by the infinite plane containing it (because we only consider the segments extracted within the shifts of the parallelogram). In the corresponding partition trees associated with the $O(\log m)$ canonical subsets in the nodes bordering the search path in the range tree, we search for the segments that intersect the query plane. This takes $O(m^{1/3+\epsilon} + k)$ per query, where k is the number of answers. The logarithmic time that we need for the two queries on the range trees are subsumed by the m^ϵ factor. □

Theorem 8 *Let P be a polygon with perimeter p and n edges. All configurations holding P in form-closure using one edge-fixel, one locator and one vertical clamp can be enumerated in time $O(n(p+n)^{4/3+\epsilon} + K)$.*

Proof: We iterate over every edge of the convex hull of P against the edge-fixel. For a particular convex hull edge, the time to compute the vertical clamp-edge and locator-edge combinations as parallelograms and directed line-segments, respectively is $O(p + n)$. We query every parallelogram corresponding to a vertical clamp-edge combination by performing QUERY B; this takes $O((p+n)^{4/3+\epsilon} + k')$ time in all, since there are at most $(p + n)$ parallelograms; k' is the number of solutions for the current iteration. The total time complexity of the algorithm is $O(n(p+n)^{4/3+\epsilon} + K)$, since we have to perform this for every edge of the convex hull; K is the total number of form-closure configurations. □

4.3 Large locators and clamps

In the previous subsection, we assumed that the locator and clamp fixels were points and that clamps could exist anywhere along a grid line. In this section we consider the more realistic case of large locators and clamps. To consider a specific model, we assume the following. Locators are circular of some constant radius r and the clamp body is a rectangle. Perpendicular to one of its sides emanates the plunger which is also rectangular with a semicircular tip. The locations of the clamp body and the locator are constrained by grid points in that a reference point on each has to coincide with a grid point. Finally, the clamp can be configured in an axis-parallel fashion to allow either horizontal (horizontal clamp) or vertical (vertical clamp) motion of the plunger. We assume the maximal plunger extension is one grid unit.

To handle locators, we simply compute the Minkowski difference of the input polygon P with a circle of radius r to get a grown object P_r bounded by straight lines and circular arcs of radius r at convex vertices. This effectively shrinks the locators to points. Now we do as before: shift P_r by one unit to the left and identify the set of grid points crossed over during the shift. We can ignore the circular arcs in P_r as we perform the shift (this is because we cannot place locators at convex vertices). Although each grid point can be crossed by several edges, there are only $O(n+p)$ locator-edge combinations in all, each with an associate shift value. Further, these are computable in $O(p + n \log n)$ time (given the polygon edges in order). Now we go back to the original polygon P and radius r locators. For each locator-edge combination, we compute the directed line representing the half-plane constraint. In all, we obtain a set S_L of $O(n + p)$ such directed lines, each associated with a shift value.

For horizontal clamps, we first compute the Minkowski difference between the polygon and the horizontally configured clamp body. The resulting object is shifted by one unit as before and all grid-points that are at most one unit horizontally from the object are collected. These are locations where (the reference point on the) clamp body can be positioned to avoid collisions with P for certain intervals of the shift variable. However, unlike the case of point clamps treated before in which a clamp-edge combination was valid for all shifts, in the present case, because of the clamp body constraint, there can be several (disjoint) subintervals of valid shifts for a particular clamp-edge combination (See Fig. 7). By careful counting, it can be shown that the number of subintervals over all clamp-edge combinations totals $O(n + p)$. Each horizontal clamp-edge combination results in a directed line representing the corresponding half-plane constraint; this directed line is labelled with the appropriate subintervals for which the clamp-edge combination exists. To simplify matters, let us suppose that if there are β disjoint subintervals of $[0, 1]$ during which a clamp-edge combination exists, then we have β copies of the appropriate directed line, each copy representing a different disjoint subinterval. Thus, in all, we have a set S_H of $O(n+p)$ directed lines, each associated with a *single* interval of shift values.

Theorem 9 *Let P be a polygon with perimeter p and n edges. All configurations holding P in form closure*

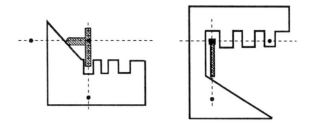

Figure 7: Several subintervals of valid shifts for a single clamp-edge combination. The figure on the left shows the case of a horizontal clamp while that on the right is a vertical clamp. Clamps are shown shaded.

using an edge-fixel, a round locator and a horizontal clamp can be enumerated in time $O(n(n + p) \log^3(n + p) + K)$, where K is the number of configurations.

Proof: From the above discussion, the polygon P results in a set S_L of $O(n + p)$ directed lines each associated with a single shift value and another set S_H of $O(n + p)$ directed lines associated with single intervals of shift values. These sets can be computed in time $O(p + n \log n)$. We can build a range tree on the shift values in S_L and query it with each subinterval (a, b) from S_H to extract all directed lines from S_L whose shift value lies in (a, b). With this tree we can associate a secondary structure as described in the algorithm in the previous subsection. This increases the query time for one element of S_H to $O(\log^3(n + p) + k)$, where k is the number of valid fixtures with this edge-clamp position combination. The total time complexity of the algorithm is therefore as claimed; K is the total number of valid fixtures. The preprocessing time is dominated by the total time of the algorithm. □

For vertical clamps, much of the initial analysis is the same. We compute the Minkowski difference between the polygon and the vertically configured clamp body. The resulting object is shifted by one unit and all grid-points that are at most one unit vertically from the object are locations where (the reference point of the) clamp body can be positioned freely without collisions for certain intervals of the shift variable. Fig. 7 shows that a single clamp-edge combination can exist over several (disjoint) subintervals. Also, the total number of such subintervals, taken over all clamp-edge combinations, is at most $O(n+p)$. However, unlike the case of horizontal clamps, a vertical clamp tip contacts different points of the edge for different shifts. Thus a single directed line does not suffice but a continuum of directed lines, or a parallelogram in the (x, y, t) space is required. Every subinterval of every clamp-edge combination results in a different parallelogram; thus there is a set S_V of $O(n + p)$ parallelograms, that can be computed in time $O(p + n \log n)$. From the above discussion and the proof of Theorem 8 we can conclude:

Theorem 10 *Let P be a polygon with perimeter p and n edges. All configurations holding P in form closure using an edge-fixel, a round locator and a vertical clamp can be enumerated in time $O(n(n+p)^{4/3+\epsilon} + K)$, where K is the number of configurations.*

Intersections between clamp and locator We have so far taken into account collisions between locators and the polygon as also clamps with the polygon. It could also happen that the locator and clamp intersect in certain configurations. However, recall that the clamp and the locator have a constant size. We can therefore check intersections between these two fixels in a generated solution in constant time, and therefore in $O(K)$ time over all the solutions. The number of wrong solutions can be bounded by $O(n+p)$, hence this does not change the time complexity of the algorithms.

Clamp plunger extension We have assumed so far that the maximum plunger extension $s = 1$. For general s, the clamp body positions are those within s units of a shifted polygon. The number of possible locations increases by a multiplicative factor of s. In other words, replace $n+p$ by $(n+p)s$ in the time complexity for the algorithm. Increased clamp extension capability can be useful if clamp bodies are large and have to be placed outside the convex hull of the polygon; a long plunger then might be able to reach into a deep cavity while a short one couldn't.

5 Discussion

In this paper we considered a new class of contacts called edge-contacts and showed that almost all polygonal objects can be fixtured by an edge- and two point-contacts. Based on the edge-contact, we proposed an edge-fixel for use in modular fixture design. In the full version of the paper, we will prove that all rectilinear polygons with edge length at least $2\sqrt{2}$ grid units that are not rectangular in convex hull can be modular fixtured in our model. We presented an output-sensitive algorithm that computes all valid edge-locator-clamp fixtures on polygonal objects. By feeding these solutions through a given quality metric or Boolean comparator, an *optimal* fixture may be computed.

We are currently investigating fixturing an object with two perpendicular edge-fixels and one point-fixel. We also wish to look ahead to non-convex curved objects. A promising direction of research appears to be considering T-slot clamps [17]. Here, clamping is done by a fixel traveling in a "T"-shaped slot. Such fixels can translate in both vertical and horizontal direction along the grid lines. We believe that similar techniques can be applied here.

References

[1] P. K. Agarwal and J. Matoušek. Ray shooting and parametric search. *SIAM J. Comput.*, 22(4):794–806, 1993.

[2] H. Asada and A. B. By. Kinematic analysis of workpart fixturing for flexible assembly with automatically reconfigurable fixtures. *IEEE Journal of Robotics and Automation*, RA-1(2), June 1985.

[3] J. J. Bausch and K. Youcef-Toumi. Kinematic methods for automated fixture reconfiguration planning. In *International Conference on Robotics and Automation*. IEEE, May 1990.

[4] W. Boyes, editor. *Handbook of Jig and Fixture Design, 2nd Edition*. Society of Manufacturing Engineers, 1989.

[5] R. C. Brost and K. Y. Goldberg. A complete algorithm for synthesizing modular fixtures for polygonal parts. In *International Conference on Robotics and Automation*. IEEE, May 1994.

[6] Y-C. Chou, V. Chandru, and M. M. Barash. A mathematical approach to automatic configuration of machining fixtures: Analysis and synthesis. *Journal of Engineering for Industry*, 111, November 1989.

[7] M. de Berg. *Ray Shooting, Depth Orders and Hidden Surface Removal*, volume 703 of *Lecture Notes in Computer Science*. Springer-Verlag, Berlin, 1993.

[8] M. V. Gandhi and B. S. Thompson. Automated design of modular fixtures for flexible manufacturing systems. *Journal of Manufacturing Systems*, 5(4), 1986.

[9] E. G. Hoffman. *Modular Fixturing*. Manufacturing Technology Press, Lake Geneva, Wisconsin, 1987.

[10] D. T. Lee. Medial axis transformation of a planar shape. *IEEE Trans. Pattern Anal. Mach. Intell.*, PAMI-4:363–369, 1982.

[11] X. Markenscoff, L. Ni, and C.H. Papadimitriou. The geometry of grasping. *IJRR*, 9(1), February 1990.

[12] B. Mishra. Workholding: Analysis and planning. In *International Conference on Intelligent Robots and Systems*. IEEE/RSJ, July 1991.

[13] B. Mishra, J. T. Schwartz, and M. Sharir. On the existence and synthesis of multifinger positive grips. *Algorithmica*, 2(4):641–558, 1987.

[14] F. P. Preparata and M. I. Shamos. *Computational Geometry: an Introduction*. Springer-Verlag, New York, NY, 1985.

[15] F. Reuleaux. *The Kinematics of Machinery*. Macmilly and Company, 1876. Republished by Dover in 1963.

[16] A. Wallack and J. Canny. Planning for modular and hybrid fixtures. In *International Conference on Robotics and Automation*. IEEE, May 1994.

[17] Y. Zhuang, K.Y. Goldberg, and Y-C. Wong. On the existance of modular fixtures. In *International Conference on Robotics and Automation*, pages 543–549. IEEE, May 1994.

Vertical Decomposition of Shallow Levels in 3-Dimensional Arrangements and Its Applications [*]

Pankaj K. Agarwal[†]   Alon Efrat[‡]   Micha Sharir[§]

Abstract

Let $\mathcal{F}$ be a collection of n bivariate algebraic functions of constant maximum degree. We show that the combinatorial complexity of the vertical decomposition of the $\leq k$-level of the arrangement $\mathcal{A}(\mathcal{F})$ is $O(k^{3+\varepsilon}\psi(n/k))$, for any $\varepsilon > 0$, where $\psi(r)$ is the maximum complexity of the lower envelope of a subset of at most r functions of $\mathcal{F}$. This result implies the existence of shallow cuttings, in the sense of [3, 31], of small size in arrangements of bivariate algebraic functions. We also present numerous applications of these results, including: (i) data structures for several generalized three-dimensional range searching problems; (ii) dynamic data structures for planar nearest and farthest neighbor searching under various fairly general distance functions; (iii) an improved (near-quadratic) algorithm for minimum-weight bipartite Euclidean matching in the plane; and (iv) efficient algorithms for certain geometric optimization problems in static and dynamic settings.

---

[*] Work on this paper by the first author has been supported by National Science Foundation Grant CCR-93-01259, an NYI award, and by matching funds from Xerox Corp. Work on this paper by the third author has been supported by NSF Grants CCR-91-22103 and CCR-93-11127, by a Max-Planck Research Award, and by grants from the U.S.-Israeli Binational Science Foundation, the Israel Science Fund administered by the Israeli Academy of Sciences, and the G.I.F., the German-Israeli Foundation for Scientific Research and Development.

[†] Department of Computer Science, Box 1029, Duke University, Durham, NC 27708-0129, USA

[‡] School of Mathematical Sciences, Tel Aviv University, Tel Aviv 69978, Israel

[§] School of Mathematical Sciences, Tel Aviv University, Tel Aviv 69978, Israel, and Courant Institute of Mathematical Sciences, New York University, New York, NY 10012, USA

Permission to copy without fee all or part of this material is granted provided that the copies are not made or distributed for direct commercial advantage, the ACM copyright notice and the title of the publication and its date appear, and notice is given that copying is by permission of the Association of Computing Machinery.To copy otherwise, or to republish, requires a fee and/or specific permission.
11th Computational Geometry, Vancouver, B.C. Canada
© 1995 ACM 0-89791-724-3/95/0006...$3.50

1 Introduction

In this paper we extend the recent range-searching techniques of Matoušek [31] and of Agarwal and Matoušek [3] to arrangements of bivariate algebraic functions, and derive many applications of the new techniques. As one motivation, consider the following *dynamic nearest-neighbor searching* problem: We maintain dynamically a set S of points in the plane. At any time, we wish to answer nearest-neighbor queries, in which we specify a point q and ask for the point of S nearest to q under some given metric δ. If δ is the Euclidean metric, a standard lifting transformation to 3-space (as in [15]) reduces the problem to that of dynamically maintaining the lower envelope of planes in $\mathbb{R}^3$, so that, at any time during the maintenance, we can answer efficiently point-location queries in the region below the envelope. The techniques of [3, 4, 31], which accomplish this goal, are based on the notion of *shallow cuttings*: Given an arrangement $\mathcal{A}$ of n planes in $\mathbb{R}^3$ and parameters $k, r \leq n$, there exists a $(1/r)$-*cutting* of the first k levels of $\mathcal{A}$ of size $O(r(1 + kr/n)^2)$, i.e., there exists a decomposition of the (union of the) cells constituting these levels into $O(r(1 + kr/n)^2)$ simplices, such that the interior of each simplex is intersected by at most n/r planes. Using the existence of such shallow cuttings, Matoušek [31] constructs an efficient static data structure for halfspace emptiness queries in $\mathbb{R}^3$: It requires $O(n \log \log n)$ storage and $O(n \log n)$ preprocessing time, and answers a halfspace emptiness query in $O(\log^2 n)$ time. This structure has later been dynamized in [3]. These techniques also apply to arrangements of hyperplanes in higher dimensions.

Although these techniques can be extended to arrangements of curves in the plane, they fail for arrangements of general surfaces in three and higher dimensions. Such an arrangement does arise, for example, in the nearest-neighbor searching problem if the underlying metric δ is not Euclidean. To tackle such cases efficiently, we need to extend the results of [3, 31] to the case of arrangements of more general functions, and this is one of the main goals of this paper.

The main technical result that we establish is the existence of shallow cuttings of small size in arrangements of low-degree, algebraic, bivariate functions. To obtain such cuttings, we need to use the technique of *vertical decomposition* of cells in arrangements. This technique, described in detail in [11], is the only known general-purpose technique for decomposing cells in such arrangements into a small number of subcells of constant description complexity. It is well known that the complexity of the vertical decomposition of a cell (2-face) C in a planar arrangement of low-degree, algebraic curves is proportional to the number of vertices of C. But this property does not hold in higher dimensions (even for an arrangement of planes in $\mathbb{R}^3$). In three dimensions, the complexity (number of subcells) of the vertical decomposition of the *entire* arrangement of n low-degree, algebraic, bivariate functions is shown in [11] to be close to $O(n^3)$, and is thus almost optimal (in higher dimensions, though, the bounds are considerably weaker). No similarly-sharp bounds were known for the vertical decomposition of the cells that lie only in the first k levels of the arrangement (even for arrangements of planes). Using a standard probabilistic analysis technique due to Clarkson and Shor ([13]; see also [36]), one can easily show that the combinatorial complexity of these (undecomposed) cells is $O(k^3 \psi(n/k))$, where $\psi(r)$ is the maximum combinatorial complexity of the lower envelope of any subset of at most r of the given surfaces. For low-degree, algebraic, bivariate functions, the results of [20, 35] imply that $\psi(n) = O(n^{2+\varepsilon})$, for any $\varepsilon > 0$, but in certain favorable cases, such as the case of planes, this complexity is smaller.

The first main result of the paper, derived in Section 2, is that the combinatorial complexity of the vertical decomposition of the cells in the first k levels of the arrangement is $O(k^{3+\varepsilon}\psi(n/k))$, for any $\varepsilon > 0$. This bound is close to optimal in the worst case. We then apply this bound to obtain a sharp bound on the size of shallow cuttings in arrangements of bivariate functions. Specifically, we show, in Theorem 3.1, that there exists a $(1/r)$-cutting of the first k levels in an arrangement of n low-degree, algebraic, bivariate functions in $\mathbb{R}^3$, as above, whose size is $O(q^{3+\varepsilon}\psi(r/q))$, for any $\varepsilon > 0$, where $q = k(r/n) + 1$. This bound is almost tight in the worst case, and almost coincides with the bound given in [31] for the case of planes. The proof adapts the analysis technique of [31]. If $r = O(1)$, a $(1/r)$-cutting of this size can be computed in $O(n)$ time.

An immediate consequence of Theorem 3.1 is an efficient algorithm for the following problem: Preprocess a set $\mathcal{F}$ of n bivariate functions into a data structure, so that, for a query point w, all functions whose graphs lie below w can be reported efficiently. We present a data structure of size $O(\psi(n)n^\varepsilon)$, for any $\varepsilon > 0$, which can be constructed in $O(\psi(n)n^\varepsilon)$ time, so that a query can be answered in $O(\log n + \xi)$ time, where ξ is the output size. If we are interested only in determining whether w lies below the graphs of all functions of $\mathcal{F}$, the query time is $O(\log n)$. (Here we are assuming a model of computation in which various operations involving a constant number of fixed-degree algebraic functions can be performed in $O(1)$ time; see below for details.) This data structure can be dynamized without increasing the asymptotic size and query-time, so that each update requires $O(\psi(n)/n^{1-\varepsilon})$ time.[1] We also obtain an efficient algorithm for constructing and searching in the first k levels of an arrangement of n bivariate functions, as above. These levels can be constructed in $O(k^{3+\varepsilon}n^\varepsilon\psi(n/k))$ time, for any $\varepsilon > 0$, in an appropriate model of computation, and can be stored into a data structure of similar size, so that, for any query point p, we can determine, in $O(\log n)$ time, whether the level of p in the arrangement is at most k, and, if so, return the level of p.

We next apply the new mechanism to a variety of geometric problems. First, in Section 4, we derive an efficient technique for dynamically maintaining a set S of points (or more general objects) in the plane, so that, given a query point w, we can efficiently compute the nearest neighbor (or the farthest neighbor) of w in the current set of points, under any 'reasonable' distance function δ (defined more precisely later), which can be fairly arbitrary, and does not have to satisfy any metric-like properties. Assume that the complexity of the lower (or upper) envelope of the functions $f_i(\mathbf{x}) = \delta(p_i, \mathbf{x})$, over all objects p_i in the current set S, is at most $g(|S|)$ (this is the same as the complexity of the nearest (or farthest) neighbor Voronoi diagram of S, if δ is a metric or a convex distance function). Then our solution requires $O(g(n) \cdot n^\varepsilon)$ storage, $O(g(n)/n^{1-\varepsilon})$ time for each update, and $O(\log^2 n)$ time for each nearest- (or farthest-) neighbor query.

Another interesting application of our technique is an improved algorithm for computing minimum-weight bipartite Euclidean matching for point sets in the plane (see Section 5). That is, we are given a set of n blue points and a set of n red points in the plane, and we wish to find a matching between the blue points and the red points, which minimizes the sum of the distances between the pairs of matched points. Our solution is based on the algorithm of Vaidya [38], and we improve the running time of Vaidya's algorithm from its original bound of $O(n^{2.5}\log n)$ to $O(n^{2+\varepsilon})$, for any $\varepsilon > 0$. We also obtain an $O(n^{7/3+\varepsilon})$ algorithm for an arbitrary 'reasonable' metric.

Another application of our results is to dynamic maintenance of the intersection of congruent balls in 3-space,

---

[1]Throughout this paper the update-time bounds are amortized. We believe that the same bounds can be achieved in the worst case, using known techniques. Nevertheless, for the sake of simplicity, we will stick to amortized bounds.

in the strong sense that we wish to determine, after each update, whether the current intersection is empty. For this, we combine our technique with the variant of parametric searching recently proposed in [37]. We obtain a data structure of size $O(n^{1+\varepsilon})$, for any $\varepsilon > 0$, which can be updated in $O(n^\varepsilon)$ time, and testing the current intersection for being empty takes $O(\log^4 n)$ time.

We next apply our technique, in Sections 7 and 8, to two problems in geometric optimization. First we study the following *dynamic smallest stabbing disk* problem: We maintain dynamically a set $\mathcal{C}$ of (possibly intersecting) simply-shaped, compact, convex sets in the plane, and we wish to compute, after each insertion or deletion of such a set, the smallest disk, or the smallest homothetic copy of any compact convex set of simple shape, that intersects all the sets in the current $\mathcal{C}$. The case where we have points instead of general convex sets was recently studied in [3]. Our solution requires $O(n^{1+\varepsilon})$ storage, and recomputes the smallest stabbing disk after each update in $O(n^\varepsilon)$ time, for any $\varepsilon > 0$. A byproduct of our analysis, which we believe to be of independent interest, is a near-linear bound on the complexity of the farthest-neighbor Voronoi diagram of (possibly intersecting) simply-shaped, compact, convex sets in the plane, under any simply-shaped, convex distance function. The bound is linear for (possibly intersecting) line segments and the Euclidean distance.

Another optimization problem that we study is the following variant of the *segment center* problem: We are given a set S of n points in the plane, and a parameter $k < n$, and we wish to compute the placement of a fixed-length segment that minimizes the k-th largest distance from the segment to the points of S. The special case of this problem, in which $k = 1$, was recently studied in [2, 17, 23]. Our solution for the generalized problem requires $O(n^{1+\varepsilon}k^2)$ time and storage, for any $\varepsilon > 0$, and is the first solution (that we are aware of) for this extended problem.

This collection of applications is by no means exhaustive, and the new techniques obtained in this paper have many additional applications. The main message of this paper, in our opinion, is that there is a real need to adapt and extend range-searching and related techniques, that were originally developed for arrangements of planes or hyperplanes, to arrangements of algebraic surfaces. This adaptation is by no means easy, and the present study achieves it only for 3-dimensional arrangements, but it enlarges significantly the range of range-searching applications.

2 Vertical Decomposition of Levels

Let $\mathcal{F}$ be a collection of n bivariate functions satisfying the following conditions:[2]

---

[2] Abusing the notation slightly, we will not distinguish between a function and its graph. We will use $\mathcal{F}$ to denote a collection

(F1) Each $f \in \mathcal{F}$ is a continuous, totally-defined, algebraic function of constant maximum degree b.

(F2) The functions in $\mathcal{F}$ are in *general position*. This excludes degenerate configurations where four function graphs meet at a point, a pair of graphs are tangent to each other, etc.

With some modifications of the analysis, we can also handle the case where the functions in $\mathcal{F}$ are only partially defined, and the boundary of the domain of each function is defined by a constant number of polynomial equalities and inequalities of constant maximum degree, say b too. Concerning the general position assumption, we refer the reader to the papers [20, 35] for more details. Following arguments given in these papers, we show (in the full version) that no real loss of generality is made by assuming general position.

The *arrangement* of $\mathcal{F}$, denoted as $\mathcal{A}(\mathcal{F})$, is the subdivision of $\mathbb{R}^3$ induced by the graphs of the functions in $\mathcal{F}$. We will refer to the 3-dimensional faces of $\mathcal{A}(\mathcal{F})$ as the *cells* of $\mathcal{A}(\mathcal{F})$. The complexity of a cell C, denoted as $|C|$, is the number of faces of all dimensions on the boundary of C. The *level* in $\mathcal{A}(\mathcal{F})$ of a point $p = (x_p, y_p, z_p) \in \mathbb{R}^3$, denoted as $\mu(p) = \mu_\mathcal{F}(p)$, is defined as the number of functions $f \in \mathcal{F}$ such that $z_p > f(x_p, y_p)$. The level of all points lying on a face ϕ (of any dimension) of $\mathcal{A}(\mathcal{F})$ is the same, which we denote by $\mu(\phi)$. The k-*level* of $\mathcal{A}(\mathcal{F})$, for any $0 \leq k \leq n-1$, is the closure of the union of all two-dimensional faces of $\mathcal{A}(\mathcal{F})$ whose level is k; the 0-level (resp. $(n-1)$-level) is the graph of the lower (resp. upper) envelope of $\mathcal{F}$. The $(\leq k)$-*level* of $\mathcal{A}(\mathcal{F})$, denoted $\mathcal{A}_{\leq k}(\mathcal{F})$, is the collection of all cells of $\mathcal{A}(\mathcal{F})$ whose levels are at most k. Let $\psi(\mathcal{F}, k) = \sum_{C \in \mathcal{A}_{\leq k}(\mathcal{F})} |C|$ denote the combinatorial complexity of $\mathcal{A}_{\leq k}(\mathcal{F})$. Let $\mathbf{F}$ denote a (possibly infinite) family of bivariate functions that satisfies (F1), and let $\psi(n, k) = \psi^{\mathbf{F}}(n, k) = \max_{\mathcal{F}} \psi(\mathcal{F}, k)$, where the maximum is taken over all collections $\mathcal{F} \subseteq \mathbf{F}$ of at most n functions (which satisfy (F1) and (F2)). For the sake of brevity, we will also use $\psi(n)$ to denote $\psi(n, 0)$, and will allow n in this notation to be any real non-negative number. A straightforward application of the probabilistic analysis technique of Clarkson and Shor [13] (see also [36]) implies that $\psi(n, k) = O(k^3 \psi(n/k))$. We also note that, by the results of [20, 35], we always have $\psi(n) = O(n^{2+\varepsilon})$, for any $\varepsilon > 0$, where the constant of proportionality depends on ε and on the maximum degree of the given surfaces. Hence, in the worst case, we have $\psi(n, k) = O(k^{1-\varepsilon} n^{2+\varepsilon})$, for any $\varepsilon > 0$. However, for certain favorable underlying families $\mathbf{F}$, these bounds can be much smaller.

Because of technical reasons (see the full version), we make the following assumption on $\mathbf{F}$.

---

of functions as well as the family of surfaces representing their graphs.

(F3) The upper bound $\psi(n)$ on the complexity of the lower envelope (and also of the upper envelope) of any collection $\mathcal{F} \subseteq \mathbf{F}$ of n functions is of the form $\psi(n) = O(n^\alpha \beta(n))$, where $\alpha \leq 2$ is a constant, and $\beta(n)$ is a function such that, for any $\delta > 0$, we have $\limsup_{n \to \infty} \beta(n)/n^\delta = 0$.

This assumption involves no real loss of generality. In fact, all known bounds for $\psi(n)$ have this form with either $\alpha = 1$ or $\alpha = 2$.

The *vertical decomposition* of a (3-dimensional) cell $C \in \mathcal{A}(\mathcal{F})$, denoted as $C^\star$, is defined in the following standard manner (see [11, 14, 36] for more details):

I. For each edge e of ∂C, we erect a z-vertical wall from e, which is the union of all maximal z-vertical segments passing through points of e and lying within (the closure of) C. The collection of these walls partitions C into subcells, each of which has a unique top facet (2-dimensional face) and a unique bottom facet, each contained in some facet of C. Every z-vertical line cuts such a subcell in a (possibly empty) interval.

II. We take each of the cells Δ generated in the first step, project it onto the xy-plane, and construct the two-dimensional vertical decomposition of the projection Δ^* of Δ, by erecting, from each vertex of Δ^* and from each locally x-extreme point of $\partial \Delta^*$, a maximal y-vertical segment contained in the closure of Δ^*. These segments partition Δ^* into trapezoidal-like subcells; each subcell $\tau^* \subseteq \Delta^*$ induces a subcell τ of Δ, obtained by intersecting Δ with the vertical cylinder $\tau^* \times \mathbb{R}$.

The cells obtained in this 2-step decomposition form the vertical decomposition of C. We define $\mathcal{A}^\star_{\leq k}(\mathcal{F})$ (resp. $\mathcal{A}^\star(\mathcal{F})$), the vertical decomposition of $\mathcal{A}_{\leq k}(\mathcal{F})$ (resp. of $\mathcal{A}(\mathcal{F})$), as the union of the vertical decompositions of all cells in $\mathcal{A}_{\leq k}(\mathcal{F})$ (resp. in $\mathcal{A}(\mathcal{F})$). The complexity of such a vertical decomposition is the total number of its cells. We will refer to the set $D(\tau)$ as the set of functions *defining* the cell τ. The main result of this section is:

Theorem 2.1 *Let $\mathcal{F}$ be a collection of n bivariate functions satisfying conditions (F1)–(F3). For any integer $k \leq n$, the combinatorial complexity of $\mathcal{A}^\star_{\leq k}(\mathcal{F})$ is $O(k^{3+\varepsilon}\psi(n/k))$, for any $\varepsilon > 0$, where the constant of proportionality depends on ε and on the maximum degree b of the functions in $\mathcal{F}$.*

Proof: It is a well-known property of vertical decompositions in $\mathbb{R}^3$ that the number of cells in $\mathcal{A}^\star_{\leq k}(\mathcal{F})$ is $O(k^3 \psi(n/k))$ plus the number of pairs of edges (e, e') such that (i) for some cell $C \in \mathcal{A}_{\leq k}(\mathcal{F})$, e and e' lie on the (top and bottom portions of the) boundary of C, and (ii) the xy-projections of e and e' intersect.

For a fixed pair of edges $e, e' \in \mathcal{A}(\mathcal{F})$ such that $\mu(e') < \mu(e)$, we define (e, e', σ) to be an *edge-crossing* if σ is an intersection point of the xy-projections of the (relative interiors of the) edges of e and e'. Let ℓ_σ denote the z-vertical line passing through σ. We define the *crossing number* of (e, e', σ) to be $\mu(e) - \mu(e') - 2$, which, by our general position assumption, is equal to the number of functions whose graphs intersect ℓ_σ strictly between e and e'. For an integer $\rho \leq \mu(e) - 2$, let $\mathcal{C}_\rho(e) = \mathcal{C}_\rho(e, \mathcal{F})$ denote the set of edge-crossings of the form (e, e', σ) whose crossing number is ρ; we put $\mathcal{C}_\rho(e, \mathcal{F}) = \emptyset$ for $\rho > \mu(e) - 2$. Set $\omega_\rho(e, \mathcal{F}) = |\mathcal{C}_\rho(e, \mathcal{F})|$ and $\omega_{\leq \rho}(e, \mathcal{F}) = \sum_{i=0}^{\rho} \omega_i(e, \mathcal{F})$. Define $\varphi_\rho(\mathcal{F}, k) = \sum_{e \in \mathcal{A}_{\leq k}(\mathcal{F})} \omega_\rho(e, \mathcal{F})$ and $\varphi_\rho(n, k) = \max \varphi_\rho(\mathcal{F}, k)$, where the maximum is taken over all subsets $\mathcal{F} \subseteq \mathbf{F}$ of size at most n. The corresponding quantities $\varphi_{\leq \rho}(\mathcal{F}, k)$ and $\varphi_{\leq \rho}(n, k)$ are defined in an analogous manner.

As follows from the above considerations, our goal is to show that $\varphi_0(n, k) \leq ck^{3+\varepsilon}\psi(n/k)$, for any $\varepsilon > 0$ and for some appropriate constant c. This will be achieved by deriving a recurrence relationship for $\varphi_0(n, k)$, whose solution will yield the desired bound. The recurrence is obtained using the following counting argument, which borrows ideas from [6].

Let e be an edge of $\mathcal{A}_{\leq k}(\mathcal{F})$, and let C be the cell of $\mathcal{A}(\mathcal{F})$ lying immediately below e. Let V_e be the vertical 2-manifold obtained as the union of all z-vertical rays emanating from the points of e in the negative z-direction. The intersection of the graph of each function $f \in \mathcal{F}$ with V_e is an algebraic arc $f^{(e)}$ of constant maximum degree, so each pair of these arcs intersect in at most a constant number of points, say, s, which depends only on the maximum degree b of the functions of $\mathcal{F}$. Let $\mathcal{A}^{(e)}(\mathcal{F})$ denote the cross-section of $\mathcal{A}(\mathcal{F})$ with V_e.

A simple but crucial observation is that (e, e', σ) is an edge-crossing with crossing number ξ if and only if the intersection point $e' \cap \ell_\sigma$ is a vertex at level $\mu(e) - \xi - 2$ in $\mathcal{A}^{(e)}(\mathcal{F})$. This implies that each edge-crossing of the form (e, e', σ) with crossing number 0 corresponds to a vertex of the cross-section $C^{(e)}$ of C with V_e (which lies on the lower portion of ∂C). Let $\mathcal{F}^{(e)} \subseteq \mathcal{F}$ be the set of functions that appear on the lower portion of $\partial C^{(e)}$, and let $t_e = |\mathcal{F}^{(e)}|$. For each function $f \in \mathcal{F}^{(e)}$, we consider the cross-section $f^{(e)}$ of f within V_e, as defined above. It is easily seen that the lower portion of $\partial C^{(e)}$ is the upper envelope of the functions in $\tilde{\mathcal{F}}^{(e)} \equiv \{f^{(e)} \mid f \in \mathcal{F}^{(e)}\}$. By the standard Davenport-Schinzel theory [7, 21], $\omega_0(e, \mathcal{F}) \leq \lambda_s(t_e)$, where s is as above, and where $\lambda_s(t)$ is the (near-linear) maximum length of a (t, s)-Davenport-Schinzel sequence.

Since the endpoints of all arcs in $\tilde{\mathcal{F}}^{(e)}$ lie on the vertical boundary of V_e and $\mu(e) \leq k$, it is easily seen that $t_e \leq k$, and thus $\omega_0(e, \mathcal{F}) \leq \lambda_s(k)$. Summing these bounds over all edges of $\mathcal{A}_{\leq k}(\mathcal{F})$, we get

$$\varphi_0(\mathcal{F},k) = \sum_{e \in \mathcal{A}_{\leq k}(\mathcal{F})} \omega_0(e,\mathcal{F}) \leq \lambda_s(k)\psi(\mathcal{F},k).$$

This yields the following weaker bound on $\varphi_0(n,k)$:

$$\varphi_0(n,k) \leq \lambda_s(k) \cdot \psi(n,k) \leq c_1 \lambda_s(k) k^3 \psi(n/k), \quad (1)$$

for some constant c_1. For $k = O(1)$, the bound in (1) is asymptotically what we want.

We proceed now to prove the sharper bound that we are after. We fix (a sufficiently small) $\varepsilon > 0$, and choose some threshold constant k_0, such that $4^\varepsilon k < c\lambda_s(k) < k^{1+\varepsilon^2/2}$, for all $k > k_0$ and for some absolute constant c, whose value will be determined later (the upper and lower bounds on $\lambda_s(k)$ in [7] imply that such a k_0 exists). If $k \leq k_0$, then the bound asserted in Theorem 2.1 follows from (1), with an appropriate choice of the constant of proportionality. Henceforth, we assume that $k > k_0$. Put $q = \lceil (c\lambda_s(k)/k)^{1/\varepsilon} \rceil$.

Now fix an edge e of $\mathcal{A}_{\leq k}(\mathcal{F})$, and continue to use the notations introduced above. If $t_e \leq q$ then $\omega_0(e,\mathcal{F})$, which is the same as the number of vertices of $C^{(e)}$, is at most $\lambda_s(q)$. Since the number of edges in $\mathcal{A}_{\leq k}(\mathcal{F})$ is at most $\psi(n,k)$, the overall number of edge-crossings with crossing number 0 and involving such edges e is at most $\lambda_s(q)\psi(n,k)$.

Next, assume that $t_e > q$. Let f, f' be a pair of distinct functions in $\mathcal{F}^{(e)}$. By continuity, f and f' must intersect within V_e at least once. Thus each function $f \in \mathcal{F}^{(e)}$ must cross at least $t_e - 1$ other functions of $\mathcal{F}$ within V_e, that is, each function $f \in \mathcal{F}^{(e)}$ is incident to at least $t_e - 1$ vertices of $\mathcal{A}^{(e)}(\mathcal{F})$. Since the graph of f contains points at level $\mu(e) - 2$ in this cross section, it follows that f is incident to at least q vertices of $\mathcal{A}^{(e)}(\mathcal{F})$ whose levels are between $\mu(e) - q - 2$ and $\mu(e) - 2$. The number of vertices of $\mathcal{A}^{(e)}(\mathcal{F})$ whose levels fall in this range is therefore $\Omega(t_e q)$, which, by the preceding analysis, implies that

$$\begin{aligned}\omega_{\leq q}(e,\mathcal{F}) &= \Omega(t_e q) = \Omega\left(qt_e \cdot \omega_0(e,\mathcal{F})/\lambda_s(t_e)\right) \\ &\geq (q/\beta(k)) \cdot \omega_0(e,\mathcal{F}),\end{aligned} \quad (2)$$

where $\beta(k) = \Theta(\lambda_s(k)/k)$ is an extremely slowly growing function of k [7, 21].

Summing (2) over all edges e of $\mathcal{A}_{\leq k}(\mathcal{F})$ for which $t_e > q$, adding the bound for the other edges of $\mathcal{A}_{\leq k}(\mathcal{F})$, and observing that each edge-crossing between any two edges of $\mathcal{A}_{\leq k}(\mathcal{F})$ with crossing number at most q is counted in this manner exactly once, we obtain:

$$\varphi_0(n,k) \leq \frac{\beta(k)}{q}\varphi_{\leq q}(n,k) + \lambda_s(q)\psi(n,k). \quad (3)$$

We obtain the following upper bound on $\varphi_{\leq q}(n,k)$ using a probabilistic argument similar to that in [13], whose proof is omitted from here.

$$\varphi_{\leq q}(n,k) \leq A(q+1)^4 \varphi_0(\lceil 2n/(q+1)\rceil, \lceil 2k/(q+1)\rceil) \quad (4)$$

for a constant $A > 0$. We thus obtain

$$\varphi_0(n,k) \leq B(q+1)^3 \beta(k)\varphi_0(\lceil 2n/(q+1)\rceil, \lceil 2k/(q+1)\rceil) \\ + B\lambda_s(q)k^3\psi(n/k), \quad (5)$$

where $B > 0$ is another constant. The solution of (5) is shown to be $\varphi_0(n,k) \leq Dk^{3+\varepsilon}\psi(n/k)$, where $D = D(\varepsilon)$ is a sufficiently large constant depending on ε. □

3 Shallow Cuttings and Range Searching

3.1 Shallow cuttings

Let $\mathcal{F}$ be a collection of bivariate functions satisfying conditions (F1) and (F3). We call a region (or cell) $\Delta \subseteq \mathbb{R}^3$ *primitive* if there exists a set $D = D(\Delta)$ of at most 6 functions in $\mathcal{F}$ such that $\Delta \in \mathcal{A}^\star(D)$. We call a cell Δ *weakly-primitive* if it is the intersection of two primitive cells. For a primitive or a weakly-primitive cell Δ, let $\mathcal{F}_\Delta \subseteq \mathcal{F}$ be the set of functions whose graphs intersect the interior of Δ, and put $n_\Delta = |\mathcal{F}_\Delta|$. A set Ξ of pairwise-disjoint weakly-primitive cells is called a $(1/r)$-*cutting* of $\mathcal{A}_{\leq k}(\mathcal{F})$ if the union of Ξ contains $\mathcal{A}_{\leq k}(\mathcal{F})$ and if $n_\Delta \leq n/r$ for every $\Delta \in \Xi$. We also refer to such a cutting as a *shallow cutting* in the arrangement $\mathcal{A}(\mathcal{F})$.

The ε-net theory [22] and the result of [11] easily imply that there exists a $(1/r)$-cutting of $\mathcal{A}(\mathcal{F})$ which consists of $O(r^3 \beta(r) \log^3 r)$ primitive cells, where $\beta(r)$ is the extremely slowly growing function defined in the previous section. Moreover, if r is a constant, such a cutting can be constructed in constant randomized time from a random sample of functions of $\mathcal{F}$, or in $O(n)$ deterministic time, using the technique of [30]. Using Theorem 2.1, we show

Theorem 3.1 *Let $\mathcal{F}$ be a collection of n bivariate functions, as above, and let $k, r < n$ be integers. Set $q = k(r/n) + 1$. Then there exists a $(1/r)$-cutting Ξ of $\mathcal{A}_{\leq k}(\mathcal{F})$ whose size is $O(q^{3+\varepsilon}\psi(r/q))$, for any $\varepsilon > 0$. Moreover, if $r = O(1)$, a $(1/r)$-cutting of this size can be computed in $O(n)$ time.*

The proof adapts the proof technique of Matoušek [31], where a similar result is obtained for arrangements of planes (or hyperplanes). Due to lack of space, we omit it altogether from this version.

3.2 Range searching

An immediate consequence of Theorem 3.1 is an efficient algorithm for the following problem: Given a set $\mathcal{F} \subseteq \mathbf{F}$ of n bivariate functions satisfying the conditions (F1) and (F3), preprocess it into a data structure so that, for a query point $p = (x_p, y_p, z_p)$, the subset $\{f \in \mathcal{F} \mid f(x_p, y_p) < z_p\}$ can be reported efficiently. We also want to update $\mathcal{F}$ dynamically by inserting a function $f \in \mathbf{F}$ into $\mathcal{F}$, or by deleting a function $f \in \mathbf{F}$ from $\mathcal{F}$.

Chazelle et al. [11] have shown that $\mathcal{F}$ can be preprocessed in time $O(n^{3+\varepsilon})$ into a data structure of size $O(n^{3+\varepsilon})$, so that a query can be answered in $O(\log n + \xi)$ time, where ξ is the output size. In this section we present a data structure that uses $O(\psi(n).n^{\varepsilon})$ storage and preprocessing time, answers a query in $O(\log n + \xi)$ time, and can be updated dynamically at a small (amortized) cost.

We construct a tree data structure T to answer these queries. Each node v of T is associated with a subset $\mathcal{F}_v \subseteq \mathcal{F}$ and with a weakly-primitive cell $\Delta \subseteq \mathbb{R}^3$. The root is associated with $\mathcal{F}$ and with the entire $\mathbb{R}^3$. Fix a sufficiently large constant r and set $k = \lfloor 2n/r \rfloor$. If $|\mathcal{F}|$ is less than some pre-specified constant n_0, then T consists of just the root. Otherwise, we compute a $(1/r)$-cutting Ξ of $\mathcal{A}_{\leq k}(\mathcal{F})$, whose size, by Theorem 3.1 and (F3), is $O(\psi(r))$ (where the constant of proportionality is independent of r). For each cell $\Delta \in \Xi$, let $\mathcal{F}_\Delta^- \subseteq \mathcal{F}$ be the set of functions whose graphs either intersect Δ or lie below Δ. If $|\mathcal{F}_\Delta^-| > 3n/r$, we discard Δ, because Δ lies completely above $\mathcal{A}_{\leq k}(\mathcal{F})$. Otherwise, we create a child v_Δ of the root, corresponding to Δ, associate Δ and $\mathcal{F}_\Delta^-$ with v_Δ, and recursively preprocess $\mathcal{F}_\Delta^-$ (in this recursive processing, we use the same parameter r, and set $k = \lfloor 2|\mathcal{F}_\Delta^-|/r \rfloor$). We show in the full version that the preprocessing time and storage for T is $O(\psi(n) \cdot n^\varepsilon)$, for any $\varepsilon > 0$.

Let p be a query point. To answer the query for p, we trace a path in T, starting from the root. Suppose we are at a node v. If v is a leaf, or if there is no child w_Δ of v for which $p \in \Delta$, we explicitly check for each $f \in \mathcal{F}_v$ (the set of functions associated with v) whether p lies above the graph of f, and report those functions that satisfy this condition. The time spent at v is then $O(|\mathcal{F}_v|)$, but either $|\mathcal{F}_v| \leq n_0$, or p lies above the $\lfloor 2|\mathcal{F}_v|/r \rfloor$-level in $\mathcal{A}(\mathcal{F}_v)$, so we will report at least $\lfloor 2|\mathcal{F}_v|/r \rfloor$ functions. If there is a child w_Δ of v such that $p \in \Delta$, we recursively visit that child. It thus follows that the overall query time is $O(\log n + \xi)$, where ξ is the output size of the query. This procedure can be modified so that it only determines whether p lies below the graphs of all the functions of $\mathcal{F}$. The modified procedure takes only $O(\log n)$ time, as is easily checked.

We can dynamize the above data structure, using the technique of Agarwal and Matoušek [3], so that a function f can be inserted into or deleted from $\mathcal{F}$ in time $O(\psi(n)/n^{1-\varepsilon})$, without increasing the asymptotic query time. Hence, we can conclude

Theorem 3.2 *Given a family $\mathbf{F}$ of bivariate functions satisfying (F1) and (F3), and a subset $\mathcal{F} \subseteq \mathbf{F}$ of size n, then $\mathcal{F}$ can be preprocessed (in an appropriate model of computation), in time $O(\psi(n) \cdot n^\varepsilon)$, into a data structure of size $O(\psi(n) \cdot n^\varepsilon)$, for any $\varepsilon > 0$, so that, for a query point p, all ξ functions whose graphs lie below p can be reported in time $O(\log n + \xi)$, and a function $f \in \mathbf{F}$ can be inserted into or deleted from $\mathcal{F}$ in $O(\psi(n)/n^{1-\varepsilon})$ time. The same dynamic data structure, with the same preprocessing, storage, and update costs, can be used to determine, in $O(\log n)$ time, whether a query point lies below the graphs of all the functions in the current set.*

3.3 Computing the ($\leq k$)-level

Next, consider the problem of computing $\mathcal{A}_{\leq k}(\mathcal{F})$, where $\mathcal{F} \subseteq \mathbf{F}$ is a set of n bivariate functions satisfying (F1) and (F3), for any $k < n$. If $\mathcal{F}$ is a set of linear functions in $\mathbb{R}^3$, this can be done by the randomized algorithm of [1] (see also [34]), whose expected time is $O(nk^2 + n\log^3 n)$, but this algorithm does not extend to nonlinear functions. We present a divide-and-conquer algorithm, based on the shallow cutting theorem, that computes all 0-, 1-, and 2-dimensional faces of $\mathcal{A}_{\leq k}(\mathcal{F})$ along with their incidence relations. A slight enhancement of the algorithm can store $\mathcal{A}_{\leq k}(\mathcal{F})$ into a data structure so that, for a query point p, we can determine, in $O(\log n)$ time, whether $\mu_\mathcal{F}(p) \leq k$, and, if so, return the value of $\mu_\mathcal{F}(p)$. A similar algorithm was developed by Clarkson [12] for computing the ($\leq k$)-level in an arrangement of hyperplanes. We omit all details from this version, and just state the result:

Theorem 3.3 *Let $\mathbf{F}$ be a family of bivariate functions satisfying (F1) and F(3). Given a set $\mathcal{F} \subseteq \mathbf{F}$ of size n and an integer parameter $k \leq n$, we can construct $\mathcal{A}_{\leq k}(\mathcal{F})$ in time $O(k^{3+\varepsilon} n^\varepsilon \psi(n/k))$, for any $\varepsilon > 0$, in an appropriate model of computation. Moreover, for any query point p, we can determine whether $\mu_\mathcal{F}(p) \leq k$, and, if so, compute the value of $\mu_\mathcal{F}(p)$, in $O(\log n)$ time.*

4 Nearest-Neighbor Searching

Let $S = \{p_1, \ldots, p_n\}$ be a set of n points in the plane, and let $\delta(\cdot, \cdot)$ be a given 'distance function' defined on $\mathbb{R}^2 \times \mathbb{R}^2$. Normally, we regard δ as a metric, or as a convex distance function, but many of the results obtained below apply for more general 'reasonable' functions δ. We want to store S into a data structure so that a point can be inserted into or deleted from S and, for a query point q, we can efficiently compute a *nearest neighbor* of q in S under the function δ (i.e., we want to return a point $p_i \in S$ such that $\delta(q, p_i) = \min_{1 \leq j \leq n} \delta(q, p_j)$). The function δ is called *reasonable* if the family $\mathbf{F} = \{\delta((x, y), (a_1, a_2)) \mid (a_1, a_2) \in \mathbb{R}^2\}$ satisfies conditions (F1) and (F3) of Section 2. For example, all L_p metrics, for integer $1 \leq p \leq \infty$, are reasonable. Here we do not assume any metric-like properties of δ, so δ can be fairly arbitrary. Let $\mathcal{F} = \{f_i(x, y) = \delta((x, y), (\xi_i, \eta_i)) \mid p_i = (\xi_i, \eta_i), 1 \leq i \leq n\}$. For a given point q, finding a nearest neighbor of q in S is equivalent to computing a function of $\mathcal{F}$ that attains the lower envelope $E_\mathcal{F}$ of $\mathcal{F}$ at q. Notice that the complexity of the lower envelope $E_\mathcal{F}$ is the same as the complexity of the Voronoi diagram of S [16] if δ is indeed a metric or a convex distance function.

The nearest-neighbor searching problem, also known as the post-office problem, has been widely studied because of its numerous applications [3, 4, 12, 32], but most of the work to date deals with the case where δ is the L_1, L_2, or L_∞ metric and where S does not change dynamically. If δ is a reasonable metric, then, using the technique by Clarkson [12] and the range searching structure of Chazelle et al. [11], a nearest-neighbor query can be answered in $O(\log^2 n)$ time using $O(n^{3+\varepsilon})$ storage, but this data structure is difficult to dynamize. On the other hand, using the range-searching data structures due to Agarwal and Matoušek [5], S can be preprocessed, in $O(n \log n)$ time, into a linear size data structure, so that a query can be answered in $O(n^{1/2+\varepsilon})$ time. In both cases, either the storage or the query time is too excessive for our purpose.

In this section we present a data structure, based on Theorem 3.2, that uses only $O(\psi(n) \cdot n^\varepsilon)$ space and preprocessing time, answers a query in $O(\log n)$ time, and inserts or deletes a point in $O(\psi(n)/n^{1-\varepsilon})$ time. These bounds are significantly better than the previously mentioned bounds. For example, we obtain a data structure of size $O(n^{1+\varepsilon})$ for L_p metrics, that can answer a nearest-neighbor query in $O(\log n)$ time, and that enables us to update S in $O(n^\varepsilon)$ time per insertion/deletion, thereby improving the query time significantly at a slight increase in the update time.

We preprocess S into the range searching data structure of Theorem 3.2. For a point $q \in \mathbb{R}^2$, we find a function of $\mathcal{F}$ that attains $E_\mathcal{F}$ at q, and thus answer a nearest neighbor of q in S, as follows. Let $\vec{\ell}$ be the vertical line in $\mathbb{R}^3$ passing through q, and oriented in the $(+z)$-direction. We trace a path in T, starting from the root. Suppose we are at a node $v \in T$. Recall that v is associated with a subset $\mathcal{F}_v \subseteq \mathcal{F}$, a $(1/r)$-cutting Ξ_v of $\mathcal{A}_{\leq k}(\mathcal{F}_v)$ (for appropriate parameters r and k), and a cell Δ_v. Inductively, assume that $E_\mathcal{F}(q) = E_{\mathcal{F}_v}(q)$. If v is a leaf, we explicitly search for a function of $\mathcal{F}_v$ that attains $E_{\mathcal{F}_v}$ at q, report that function and terminate the query. Otherwise, we compute the highest cell Δ of Ξ_v (in the $(+z)$-direction) that $\vec{\ell}$ intersects. Since Ξ_v covers $\mathcal{A}_{\leq k}(\mathcal{F}_v)$, Δ is the last cell in Ξ_v intersecting $\vec{\ell}$, and $\mathcal{F}_\Delta$ contains all functions of $\mathcal{F}_v$ whose graphs either intersect Δ or lie below Δ, it easily follows that $E_{\mathcal{F}_\Delta}(q) = E_{\mathcal{F}_v}(q) = E_\mathcal{F}(q)$. Hence, we can recursively search at the child corresponding to Δ. The overall query time is obviously $O(\log n)$. Again, the data structure can be dynamized without affecting its asymptotic query time, thereby obtaining the following theorem.

Theorem 4.1 *Let S be a set of n points in the plane and let δ be any reasonable distance function, as above. We can preprocess S in time $O(\psi(n)n^\varepsilon)$ into a data structure of size $O(\psi(n)n^\varepsilon)$, for any $\varepsilon > 0$, so that a nearest-neighbor query can be answered in $O(\log n)$ time. Moreover, we can insert and delete points in $O(\psi(n)/n^{1-\varepsilon})$ time per update operation. Here $\psi(n)$ is the maximum complexity of the envelope $E_\mathcal{F}$, as defined above.*

Remark 4.2: (i) For a point q, finding a farthest neighbor of q in S is equivalent to finding a function of $\mathcal{F}$ that attains the upper envelope of $\mathcal{F}$ at q. By reversing the direction of the z-axis, we can use a similar data structure, with similar performance bounds, for answering farthest-neighbor queries.
(ii) The above query procedures can be modified so that, for a query point q and an integer $\kappa \leq n$, we can compute the κ nearest (or farthest) neighbors of q in S in time $O(\log n + \kappa)$.

If δ is an L_p-metric, for any integer $1 \leq p \leq \infty$, then $\psi(n) = O(n)$ [27]. Similarly, if δ is an *(additive) weighted Euclidean* metric, i.e., each point $p_i \in S$ has a weight w_i and $\delta(q, p_i) = d(q, p_i) + w_i$, where $d(\cdot, \cdot)$ is the Euclidean distance, then also $\psi(n) = O(n)$ [28]. Hence, we obtain

Corollary 4.3 *Let S be a set of n points in the plane. We can store S into a data structure of size $O(n^{1+\varepsilon})$, for any $\varepsilon > 0$, so that points can be inserted into or deleted from S in time $O(n^\varepsilon)$ per update, and a nearest-neighbor query, under any L_p-metric or any weighted Euclidean metric, can be answered in $O(\log n)$ time.*

If the only known bound for $\psi(n)$ is the worst case $O(n^{2+\varepsilon})$, we can obtain a space/query-time tradeoff by combining Theorem 4.1 with the linear-size data structure of [5] for nearest-neighbor searching, mentioned in the beginning of this section, in a rather standard manner. In particular, if $\psi(n) = O(n^{2+\varepsilon})$ then, for any parameter $n \leq m \leq n^2$, we can store S into a data structure of size $O(m^{1+\varepsilon})$, for any $\varepsilon > 0$, so that a nearest-neighbor query can be answered in time $O(n^{1+\varepsilon}/\sqrt{m})$. Insertions and deletions can be performed in $O(m^{1+\varepsilon}/n)$ time, per update. Hence, we can conclude

Theorem 4.4 *Let S be a set of n points in the plane, $n \leq m \leq n^2$ a real parameter, and δ a reasonable distance function, as above. We can preprocess S in time $O(m^{1+\varepsilon})$ into a data structure of size $O(m^{1+\varepsilon})$, for any $\varepsilon > 0$, so that a nearest-neighbor query can be answered in $O(n^{1+\varepsilon}/\sqrt{m})$ time. Moreover, we can insert and delete points in $O(m^{1+\varepsilon}/n)$ time per update.*

Applying Corollary 4.3, Theorem 4.4, and some additional arguments to a result of Eppstein [18], we obtain

Theorem 4.5 (i) *Let R and B be two sets of points in the plane with a total of n points. We can store $R \cup B$ in a dynamic data structure of size $O(n^{1+\varepsilon})$, for any $\varepsilon > 0$, that maintains a closest pair in $R \times B$, under any L_p-metric or any additively-weighted Euclidean metric, in $O(n^\varepsilon)$ time per insertion or deletion. If we use an arbitrary, reasonable distance function, then the storage increases to $O(n^{4/3+\varepsilon})$, and the upate time to $O(n^{1/3+\varepsilon})$, for any $\varepsilon > 0$.*

(ii) *A minimum spanning tree of a set of n points in the plane, under any L_p metric, can be maintained in $O(\sqrt{n} \log^2 n)$ time for each insertion or deletion operation.*

5 Minimum-Weight Bipartite Euclidean Matching

Let $P = \{p_1, \ldots, p_n\}$ and $Q = \{q_1, \ldots, q_n\}$ be two sets of points in the plane. We wish to compute a *minimum-weight bipartite Euclidean matching* M of P and Q. A matching of P and Q is a set of n pairs $(p, q) \in P \times Q$ such that each point of $P \cup Q$ appears in exactly one pair. The weight of a pair is the Euclidean distance between its points, and the weight of a matching is the sum of the weights of its pairs. The standard *Hungarian method* [25, 26] yields an $O(n^3)$-time algorithm for computing M. Exploiting the fact that the weights are Euclidean distances, Vaidya obtained an $O(n^{2.5} \log n)$-time algorithm for computing M [38]. Recently, a number of efficient algorithms have been developed for computing a minimum-weight bipartite Euclidean matching when P and Q have some special structure [8, 10, 29], but no progress has been made when P and Q are arbitrary sets of points in the plane. In this section, we show that Vaidya's algorithm, for arbitrary sets A, B, can be modified so that its running time improves to $O(n^{2+\varepsilon})$, for any $\varepsilon > 0$. For the sake of completeness, we give a brief sketch of his algorithm.

The bipartite matching problem can be formulated as a linear program:

$$\min \sum_{i,j} d(p_i, q_j) x_{ij}$$
$$\sum_{j=1}^n x_{ij} = 1, \quad i = 1, \ldots, n,$$
$$\sum_{i=1}^n x_{ij} = 1, \quad j = 1, \ldots, n,$$
$$x_{ij} \geq 0, \quad i, j = 1, \ldots, n,$$

where (p_i, q_j) is an edge of M if and only if $x_{ij} = 1$. The dual linear program is

$$\max \sum_i \alpha_i + \sum_j \beta_j$$
$$\alpha_i + \beta_j \leq d(p_i, q_j), \quad \alpha_i, \beta_j \mathrel{?} 0, \; i, j = 1, \ldots, n,$$

where α_i (resp. β_j) is the dual variable associated with the point p_i (resp. q_j).

The Hungarian method computes a matching in n phases, each of which augments the matching by one edge and updates the dual variables. Let X be the current matching computed so far by the algorithm. Initially, $X = \emptyset$, $\alpha_i = 0$, and $\beta_j = \min_i\{d(p_i, q_j)\}$ for all i, j. An edge (p_i, q_j) is called *admissible* if $d(p_i, q_j) = \alpha_i + \beta_j$. A vertex of $P \cup Q$ is called *exposed* if it is not incident to any edge of X. An *alternating path* is one that alternately traverses edges of X, and edges not in X, starting with an edge not in X. An alternating path between two exposed vertices is called an *augmenting path*.

During each phase, we search for an augmenting path consisting only of admissible edges, as follows. For each exposed point $q \in Q$, we grow an 'alternating tree' whose paths are alternating paths starting at q. More precisely, each point of $P \cup Q$ in an alternating tree is reachable from its root by an alternating path that consists only of admissible edges. For a point w of P (resp. Q), the path leading to w ends at an edge not in X (resp. in X). (The alternating trees are maintained implicitly.) Let S (resp. T) denote the set of points of Q (resp. P) that lie in some alternating tree. In the beginning of each phase, S is the set of exposed vertices of Q and $T = \emptyset$. Let

$$\delta = \min_{p_i \in P-T, q_j \in S}\{d(p_i, q_j) - \alpha_i - \beta_j\}.$$

At each step, the algorithm takes one of the following actions, depending on whether $\delta = 0$ or $\delta > 0$:

Case 1: $\delta = 0$. Let (p_i, q_j), for $p_i \in P - T, q_j \in S$, be an admissible edge ($\delta = 0$ implies that such an edge must exist). If p_i is an exposed vertex, an augmenting path has been found, so the algorithm moves to the next phase. Otherwise, let q_k be the vertex such that $(p_i, q_k) \in X$. The algorithm adds the edges (p_i, q_j) and (p_i, q_k) to the appropriate alternating tree, the point p_i to T, and the point q_k to S.

Case 2: $\delta > 0$. The algorithm updates the dual variables, as follows. For each vertex $p_i \in T$, it sets $\alpha_i = \alpha_i - \delta$ and, for each $q_j \in S$, it sets $\beta_j = \beta_j + \delta$.

The algorithm repeats these steps until it reaches an exposed vertex of P, thereby obtaining an augmenting path. If an augmenting path Π is found, we delete the edge of $\Pi \cap X$ from the current matching X, add the other edges of Π (thereby increasing the size of the current matching by 1), and move to the next phase. This completes the description of the algorithm. Further details of the algorithm and the proof of its correctness can be found in [26, 38].

For arbitrary graphs, each step can be implemented in $O(n)$ time. Since there are at most n phases and each phase consists of $O(n)$ steps, the total running time of the above procedure is $O(n^3)$. Vaidya suggested the following approach to expedite the running time of each step. Maintain a variable Δ and associate a weight with each point in $P \cup Q$. In the beginning of each phase, $\Delta = 0$ and $w(p_i) = \alpha_i$, $w(q_j) = \beta_j$ for each $1 \leq i, j \leq n$. During each step, the weights and Δ are updated, as follows. If Case 1 occurs, then we set $w(p_i) = \alpha_i + \Delta$ and $w(q_k) = \beta_k - \Delta$, and do not change the value of Δ. (Note that $p_i \notin T$ and $q_k \notin S$, so the values of α_i and β_k are the same as in the beginning of the phase.) If Case 2 occurs, then we set $\Delta = \Delta + \delta$, and do not change the weights. Notice that $\beta_j = w(q_j) + \Delta$ for each $q_j \in S$, that $\alpha_i = w(p_i) - \Delta$ for each $p_i \in T$, and that the value of the dual variables for other points is equal to their values at the beginning of the phase. At the end of each phase, the value of the dual variables can be computed from Δ and from the weights of the corresponding points. The weight of each point changes only once during a phase, namely, when it is added either to S or to T. Moreover, at any time during a phase,

$$\delta = \min_{p_i \in P - T, q_j \in S} \{d(p_i, q_j) - w(p_i) - w(q_j)\} - \Delta.$$

Hence, δ can be computed in each step by maintaining the weighted closest pair between S and $P - T$. Since each step requires at most two update operations (inserting a point into S and deleting a point from $P - T$), each step can be performed in $O(n^\varepsilon)$ time, by Theorem 4.5. (This is where our solution differs from Vaidya's, where a considerably inferior updating mechanism was used.) If the distance between two points is measured by an arbitrary reasonable distance function, each step of the above algorithm will require $O(n^{1/3+\varepsilon})$ time, with an appropriately increased storage (the distance function continues to be reasonable, when weights are added to the points, as above). We can thus conclude

Theorem 5.1 *Given two sets P, Q, each consisting of n points in the plane, a minimum-weight bipartite Euclidean matching between P and Q can be computed in $O(n^{2+\varepsilon})$ time, for any $\varepsilon > 0$. If we use an arbitrary, reasonable distance function for computing distances between points of P and Q, a minimum weight bipartite matching between P and Q can be computed in time $O(n^{7/3+\varepsilon})$, for any $\varepsilon > 0$.*

6 Maintaining the Intersection of Congruent Balls in 3-Space

The next application of our techniques is an efficient algorithm for dynamically maintaining the intersection of congruent balls in 3-space, under insertions and deletions of balls. With no loss of generality, we assume that the balls have radius 1. For a point $p \in \mathbb{R}^3$, let $B(p)$ denote the ball of radius 1 centered at p. Let S be a set of n points in $\mathbb{R}^3$. Let $\mathcal{B} = \mathcal{B}(S) = \{B(p) \mid p \in S\}$, and let $K(S) = \bigcap_{p \in S} B(p)$ be the common intersection of $\mathcal{B}$. We wish to maintain $K(S)$ as we update S dynamically by inserting and deleting points. Although the complexity of $K(S)$ is $O(n)$ [19], a sequence of m updates in S may cause $\Omega(mn)$ changes in the structure of $K(S)$, even if we perform only insertions. Consequently, we cannot hope to maintain $K(S)$ explicitly. Instead, we store $K(S)$ into a data structure so that a number of different types of queries can be answered efficiently. The simplest query is whether a query point p lies in $K(S)$. A stronger type of query is to determine efficiently, after each update, whether $K(S)$ is empty. In this section, we present a data structure that answers these two queries efficiently.

Let $B^+(p)$ (resp. $B^-(p)$) denote the region consisting of all points that lie in or above (resp. in or below) $B(p)$. Let $K^+(S) = \bigcap_{p \in S} B^+(p)$ and $K^-(S) = \bigcap_{p \in S} B^-(p)$. If we regard the boundary of each region $B^+(p)$, for $p \in S$, as the graph of a partially defined bivariate function, then $\partial K^+(S)$ is the graph of the upper envelope of these functions; we denote this upper envelope by $z = F^+(\mathbf{x})$. The same argument as in [19] implies that the $K^+(S)$ also has linear complexity. Similar and symmetric properties hold for the region $K^-(S)$, and we use the notation $z = F^-(\mathbf{x})$ to denote the graph of its boundary.

A query point $p = (p_x, p_y, p_z)$ lies in $K(S)$ if and only if $F^+(p_x, p_y) \leq p_z \leq F^-(p_x, p_y)$. Therefore, the problem of determining whether $p \in K(S)$ reduces to evaluating $F^+(p_x, p_y)$ and $F^-(p_x, p_y)$. By Theorem 4.1, we can store $\{B^+(p) \mid p \in S\}$ and $\{B^-(p) \mid p \in S\}$ into two data structures, each of size $O(n^{1+\varepsilon})$, so that a point can be inserted into or deleted from S in $O(n^\varepsilon)$ time, and so that, for a query point $p = (p_x, p_y, p_z)$, $F^+(p_x, p_y), F^-(p_x, p_y)$ can be evaluated in $O(\log n)$ time.

The same data structure can be used to determine whether $K(S)$ is empty. Observe that the boundaries of the regions $K^+(S)$ and $K^-(S)$ are both (weakly) xy-monotone—one of them is a convex surface and the other is concave. These properties can be exploited to obtain a procedure, with $O(\log^4 n)$ query time, for detecting (after each update) whether $K^+(S)$ and $K^-(S)$ intersect, by applying a multi-dimensional variant of the parametric searching technique [33], recently proposed by Toledo [37]. Delegating the details of this procedure to the full version, we conclude

Theorem 6.1 *The intersection of a set of congruent balls in 3-space can be maintained dynamically, in a data structure of size $O(n^{1+\varepsilon})$, for any $\varepsilon > 0$, so that each*

insertion or deletion of a ball takes $O(n^\varepsilon)$ time, and the following queries can be answered: (a) For any query point p, we can determine in $O(\log n)$ time whether p lies in the current intersection, and (b) after performing each update, we can determine in $O(\log^4 n)$ time whether the current intersection is nonempty.

7 Smallest Stabbing Disk

Let **C** be a (possibly infinite) family of simply-shaped, compact, convex sets (called *objects*) in the plane. By 'simply-shaped' we mean that each object is described by a Boolean combination of a constant number of polynomial equalities and inequalities of constant maximum degree. Let $\mathcal{C} = \{c_1, \ldots, c_n\}$ be a finite subset of **C**. We wish to update $\mathcal{C}$ dynamically, by inserting an object $c \in \mathbf{C}$ into $\mathcal{C}$ or by deleting such a set, and maintain a smallest disk, or, more generally, a smallest homothetic copy of some simply-shaped, compact, convex set P, that intersects all sets of $\mathcal{C}$. This study extends recent work by Agarwal and Matoušek [3] who have obtained an algorithm that takes $O(n^\varepsilon)$ time per update operation, for the case where $\mathcal{C}$ is a set of points and P is a disk.

The set P induces a *convex distance function* defined by $d_P(x, y) = \min\{\lambda \mid y \in x + \lambda P\}$; (we assume here that P contains the origin in its interior). The function d_P is a metric if and only if P is centrally symmetric. We define the (nearest- or farthest-neighbor) Voronoi diagram $Vor_P(\mathcal{C})$ of $\mathcal{C}$, under the distance function d_P, in a standard manner (see [24] for details). In what follows we assume that P is strictly convex; the results can be extended to more general sets with some extra care. We first prove the following theorem, which is of independent interest. (We are not aware of any previous proof of this result.)

Theorem 7.1 *Given a set C of (possibly intersecting) simply-shaped, compact, convex objects in the plane, the complexity of the farthest-point Voronoi diagram of $\mathcal{C}$, under a convex distance function d_P induced by any compact, convex simply-shaped set P, is $O(\lambda_s(n))$, for some constant s depending on the shape of P and of the sets in $\mathcal{C}$. If $\mathcal{C}$ is a set of n line segments and d_P is the Euclidean distance function (i.e., when P is a disk), the complexity of the farthest-point Voronoi diagram is $O(n)$.*

Proof: Let R be one of the Voronoi cells of $Vor_P(\mathcal{C})$, and let $c \in \mathcal{C}$ be the farthest neighbor of all points of R. We show that R has the following 'anti-star-shape' property: Let $x \in R$, and let $q \in c$ be the nearest point to x (i.e., $d_P(x, q) = d_P(x, c)$); the convexity of c and the strict convexity of P imply that q is unique. Let ρ be the ray emanating from q towards x, and let y be any point on ρ past x (i.e., x lies in the segment qy). Then c is the farthest neighbor of y. This is a fairly standard argument, which holds even if the sets in $\mathcal{C}$ intersect; details are given in the full version.

This implies that R is unbounded, and so $Vor_P(\mathcal{C})$ is an outerplanar map. We next bound the number of faces of $Vor_P(\mathcal{C})$, as follows. Let Γ be the circle at infinity. We can identify Γ with the space of all orientations, and define, for each orientation θ, the set $c(\theta) \in \mathcal{C}$ to be the farthest neighbor in $\mathcal{C}$ of the points (r, θ) (in polar coordinates) for all sufficiently large r; it is easy to verify that $c(\theta)$ is well defined for all but a finite number of orientations θ. To bound the number of such orientations, we first note that, by a compactness argument, there exists a sufficiently large r_0 such that the circle σ_{r_0} of radius r_0 about the origin cuts the cells of $Vor_P(\mathcal{C})$ in the same sequence as does the circle at infinity, in the manner described above. For each $c \in \mathcal{C}$, let $f_c(\theta) = d_P((r_0, \theta), c)$, and let $F(\theta) = \max_{c \in \mathcal{C}} f_c(\theta)$. Clearly, the number of edges of $Vor_P(\mathcal{C})$ crossed by σ_{r_0} is equal to the number of breakpoints in the upper envelope F. Since we have assumed that P and each set $c \in \mathcal{C}$ has a simple shape, for each pair $c, c' \in \mathcal{C}$, the functions f_c and $f_{c'}$ have at most some constant number, s, of intersection points. Hence, the number of breakpoints of F is, by standard Davenport-Schinzel theory [21], at most $\lambda_s(n)$. This also bounds the number of faces of $Vor_P(\mathcal{C})$, and, by Euler's formula, also the total complexity of this diagram.

For the case of line segments and Euclidean distance, the bound can be improved to $O(n)$, using a more careful analysis which we omit from here. □

For each set c_i, define a bivariate function $f_i(\mathbf{x}) = -\min_{q \in c_i} d_P(q, \mathbf{x})$, for $\mathbf{x} \in \mathbb{R}^2$. Let $\mathcal{F} = \{f_i \mid 1 \leq i \leq n\}$. The minimization diagram of $\mathcal{F}$ is the same as the farthest point Voronoi diagram of $\mathcal{C}$ under the distance function d_P. Following the same argument as in Theorem 4.1, and using Theorem 7.1, we can show:

Corollary 7.2 *A set $\mathcal{C}$ of n (possibly intersecting) compact, convex, simply-shaped sets in the plane can be stored in a data structure of size $O(n^{1+\varepsilon})$, for any $\varepsilon > 0$, so that sets (of the same form) can be inserted into or deleted from $\mathcal{C}$ in $O(n^\varepsilon)$ time per update, and, for any query point p, a farthest neighbor of p in $\mathcal{C}$, under any convex distance function d_P as above, can be computed in $O(\log n)$ time.*

Returning to the smallest stabbing disk problem, we need to compute the highest point w^* on the lower envelope $E_\mathcal{F}$ of $\mathcal{F}$. In view of Corollary 7.2 and the easily-established fact that the cell of $\mathcal{A}(\mathcal{F})$ lying below $E_\mathcal{F}$ is convex, a natural approach to compute w^* is to use the multi-dimensional parametric searching technique of Toledo [37], already mentioned above.

Theorem 7.3 *A set $\mathcal{C}$ of n (possibly intersecting) simply-shaped, compact, convex sets in the plane can be* stored in a data structure of size $O(n^{1+\varepsilon})$, for any $\varepsilon > 0$, so that a smallest homothetic placement of some given simply-shaped, compact, strictly-convex set P, which intersects all sets of $\mathcal{C}$, can be computed in $O(n^\varepsilon)$ time, after each insertion or deletion of a set into/from $\mathcal{C}$.

8 Segment Center with $\leq k$ Violations

Let S be a set of n points in the plane, and let e be a fixed-length segment. A placement e^* of e is called a *segment center* of S if the maximum distance between the points of S and e^* is minimized. We call e^* a *segment center* of S *with k violations* if the $(k+1)$-st largest distance between e^* and the points of S is minimized. The problem of computing the segment center (without violations) has been studied in [2, 17, 23]; the best algorithm, given in [17], computes the segment center in $O(n^{1+\varepsilon})$ time, for any $\varepsilon > 0$.

In this section we study the problem of computing the segment center of S with k violations, for any integer $k < n$. Extending the result of [17], we show:

Theorem 8.1 *The segment center with k violations of a set S of n points in the plane can be computed in time $O(n^{1+\varepsilon}k^2)$, for any $\varepsilon > 0$.*

Proof: We adapt the analysis given in [17]. We first solve the *fixed size problem*: Given a real $d > 0$, we wish to determine whether there exists a placement of e for which the $(k+1)$-st largest distance to the points of S is $\leq d$. As follows from the analysis of [17], this is equivalent to the following problem: We need to determine whether there exists a (translated and rotated) placement Z of the *hippodrome* $H(e, d)$, defined as the Minkowski sum of e and a (closed) disk of radius d, which contains all but at most k points of S.

As in [17], for each point p_i of S we define two partially defined bivariate functions, F_i and G_i. Both functions are defined in the dual plane, where each point (ξ_1, ξ_2) is the dual of the line $\lambda(\xi_1, \xi_2) : y = \xi_1 x + \xi_2$. The function F_i (resp. G_i) measures the translation of a hippodrome, where its symmetry axis coincides with $\lambda(\xi_1, \xi_2)$, so that its left (resp. right) semicircle passes through p_i (these functions are only partially defined). Let $\mathcal{F}, \mathcal{G}$ denote the collections of the functions F_i, G_i, respectively. A placement (ξ_1, ξ_2, t) of the hippodrome (where t measures the amount of translation) contains all points of S if and only if it lies (in parametric 3-space) between the lower envelope of $\mathcal{F}$ and the upper envelope of $\mathcal{G}$. As shown in [17], the complexity of the region enclosed between these envelopes is $O(n \log n)$. Similarly, the placement (ξ_1, ξ_2) contains all but k points of S if and only if the sum of its levels in the arrangements $\mathcal{A}(\mathcal{F})$, $\mathcal{A}(\mathcal{G})$ is at most k (where the level is measured 'from the bottom' in $\mathcal{A}(\mathcal{F})$ and 'from the top' in $\mathcal{A}(\mathcal{G})$).

This suggests that we compute, using Theorem 3.3, the first k levels of each arrangement, superimpose these subarrangements, and test, for each vertex in the superimposed arrangement, whether the sum of its levels is at most k. There are several technical difficulties in a naive implementation of this approach, so we need to implement it using more sophisticated machinery (extending similar machinery developed in [17] and using Theorem 3.3). Omitting all further details, we show that the fixed-size problem can be solved in time $O(n^{1+\varepsilon}k^2)$, for any $\varepsilon > 0$.

We now apply the parametric searching paradigm to the algorithm that solves the fixed-size problem, so as to compute the desired center location of e. For this we first design a parallel version of this algorithm, in Valiant's comparison model [39], which is relatively easy to do. Again, omitting all details, we show that the resulting algorithm has the performance asserted in the theorem. □

References

[1] P. Agarwal, M. de Berg, J. Matoušek, and O. Schwarzkopf, Computing levels in arrangements and higher order Voronoi diagrams, *Proc. 10th Annual Symp. on Comp. Geom.*, 1994, 67–75.

[2] P. Agarwal, A. Efrat, M. Sharir, and S. Toledo, Computing a segment-center for a planar point set, *J. Algorithms* 15 (1993), 314–323.

[3] P. Agarwal and J. Matoušek, Dynamic half-space range searching and its applications, *Algorithmica* 14 (1995), in press.

[4] P. Agarwal and J. Matoušek, Ray shooting and parametric search, *SIAM J. Comput.* 22, 794–806.

[5] P. Agarwal and J. Matoušek, Range searching with semialgebraic sets, *Discrete Comput. Geom.* 11 (1994), 393–418.

[6] P. Agarwal, O. Schwarzkopf, and M. Sharir, Overlay of lower envelopes in three dimensions and its applications, *this proceedings*.

[7] P. Agarwal, M. Sharir, and P. Shor, Sharp upper and lower bounds for the length of general Davenport Schinzel sequences, *J. Combin. Theory, Ser. A.* 52 (1989), 228–274.

[8] A. Aggarwal, A. Bar-Noy, S. Khuller, D. Kravets, and B. Schieber, Efficient minimum cost matching using quadrangle inequality, *Proc. 33rd Symp. on Found. of Comp. Sci.*, 1992, 583–592.

[9] H. Brönnimann, B. Chazelle, and J. Matoušek, Product range spaces, sensitive sampling, and derandomization, *Proc. 34th Symp. on Found. of Comp. Sci.*, 1993, 400–409.

[10] S. Buss and P. Yianilos, Linear and $O(n \log n)$ time minimum-cost matching algorithms for quasi-convex tours, *Proc. 5th ACM-SIAM Symp. on Disc. Algo.*, 1994, 65–76.

[11] B. Chazelle, H. Edelsbrunner, L. Guibas, and M. Sharir, A singly exponential stratification scheme for real semi–algebraic varieties and its applications, *Proc. 16th Int. Colloq. on Auto., Lang. and Prog.*, 1989, 179–193. (See also *Theoretical Comp. Sci.* 84 (1991), 77–105.)

[12] K. Clarkson, New applications of random sampling in computational geometry, *Discrete Comput. Geom.* 2 (1987), 195–222.

[13] K. Clarkson and P. Shor, Applications of random sampling in computational geometry, II, *Discrete Comput. Geom.* 4 (1989), 387–421.

[14] M. de Berg, D. Halperin, and L. Guibas, Vertical decomposition for triangles in 3-space, *Proc. 10th Symp. on Comput. Geom.*, 1994, 1–10.

[15] H. Edelsbrunner, *Algorithms in Combinatorial Geometry*, Springer–Verlag, Heidelberg, 1987.

[16] H. Edelsbrunner and R. Seidel, Voronoi diagrams and arrangements, *Discrete Comput. Geom.* 1 (1986), 25–44.

[17] A. Efrat and M. Sharir, A near-linear algorithm for the planar segment center problem, to appear in *Discrete Comput. Geom.*

[18] D. Eppstein, Dynamic Euclidean minimum spanning trees and extrema of binary functions, *Discrete Comput. Geom.* 13 (1995), 111–122.

[19] B. Grünbaum, A proof of Vázsonyi's conjecture, *Bull. Research Council Israel, Sec. A* 6 (1956), 77–78.

[20] D. Halperin and M. Sharir, New bounds for lower envelopes in three dimensions, with applications to visibility of terrains, *Discrete Comput. Geom.* 12 (1994), 313–326.

[21] S. Hart and M. Sharir, Nonlinearity of Davenport-Schinzel sequences and of generalized path compression schemes, *Combinatorica* 6 (1986), 151–177.

[22] D. Haussler and E. Welzl, ϵ-nets and simplex range queries, *Discrete Comput. Geom.* 2 (1987), 127–151.

[23] H. Imai, D.T. Lee, and C. Yang, 1-Segment center covering problems, *ORSA J. of Comput.* 4 (1992), 426–434.

[24] D. Leven and M. Sharir, Planning a purely translational motion for a convex object in two-dimensional space using generalized Voronoi diagrams, *Discrete Comput. Geom.* 2 (1987), 9–31.

[25] H. Kuhn, The Hungarian method for the assignment problem, *Naval Research Logistics Quarterly* 2 (1955), 83–97.

[26] E. Lawler, *Combinatorial Optimization: Networks and Matroids*, Saunders College Publishing, Fort Worth, TX, 1976.

[27] D.T. Lee, Two-dimensional Voronoi diagrams in the L_p-metric, *J. ACM* 27 (1980), 604–618.

[28] D.T. Lee and S. Drysdale, Generalizations of Voronoi diagrams in the plane, *SIAM J. Comput.* 10 (1981), 73–87.

[29] O. Marcotte and S. Suri, Fast matching algorithms for points on a polygon, *SIAM J. Comput.* 20 (1991), 405–422.

[30] J. Matoušek, Approximations and optimal geometric divide-and-conquer, *Proc. 23. Annual ACM Symp. on Theory of Comput.*, 1991, 506–511.

[31] J. Matoušek, Reporting points in halfspaces, *Comput. Geom. Theory Appls.* 2 (1992), 169–186.

[32] J. Matoušek and O. Schwarzkopf, On ray shooting in convex polytopes, *Discrete Comp. Geom.* 10 (1993), 215–232.

[33] N. Megiddo, Applying parallel computation algorithms in the design of serial algorithms, *J. ACM* 30 (1983), 852–865.

[34] K. Mulmuley, *Computational Geometry: An Introduction Through Randomized Algorithms*, Prentice Hall, Englewood Cliffs, NJ, 1993.

[35] M. Sharir, Almost tight upper bounds for lower envelopes in higher dimensions, *Discrete Comput. Geom.* 12 (1994), 327–345.

[36] M. Sharir and P. Agarwal, *Davenport-Schinzel Sequences and Their Geometric Applications*, Cambridge University Press, New York, 1995.

[37] S. Toledo, Maximizing non-linear concave functions in fixed dimensions, *Proc. 33rd Symp. on Found. of Comp. Sci.*, 1992, 676–685.

[38] P. Vaidya, Geometry helps in matching, *SIAM J. Comput.*, 18 (1989), 1201–1225.

[39] L. Valiant, Parallelism in comparison problems, *SIAM J. Comput.* 4 (1975), 348–355.

Efficient collision detection for moving polyhedra

Elmar Schömer[*]   Christian Thiel[†]

Abstract

In this paper we consider the following problem: given two general polyhedra of complexity n, one of which is moving translationally or rotating about a fixed axis, determine the first collision (if any) between them. We present an algorithm with running time $O(n^{8/5+\epsilon})$ for the case of translational movements and running time $O(n^{5/3+\epsilon})$ for rotational movements, where ϵ is an arbitrary positive constant. This is the first known algorithm with sub-quadratic running time.

1 Introduction

The demands on quality, security and higher production capacity in manufacturing increase the need for planning during the phase of product design. To find potential faults in the design as soon as possible one uses simulation programs: these predict the physical properties and reactions of the product and check whether particular prefabricated parts can be easily assembled. For the latter purpose, the simulation of assemblies and robots, efficient methods for *collision detection* are needed. In general collision detection is an essential prerequisite of simulations of mechanical tools.

Regarding the significance of this problem we consider in our paper efficient algorithms for collision detection. It is known ([Bo79, CA84]) that in $\mathbb{R}^3$ a collision between a moving polyhedral object and a stationary obstacle is computable in time $O(n^2)$. Here, n denotes the complexity of the two objects, i.e. the number of vertices, edges and faces. We attempt to solve this problem in sub-quadratic time. Our results are justified by the following model: Objects are rigid bodies (polyhedra) in $\mathbb{R}^3$, their surfaces consist of planar faces with straight boundaries. An object may be moving translationally in an arbitrary direction or it may be rotating about an arbitrary axis. These restrictions are based on the fact that real objects can be easily modelled by polyhedra and every motion can be approximated by a sequence of translations and rotations. As our model of computation we take the standard Real-RAM-model ([PS88]).

1.1 Previous results

There are (up to now) only efficient solutions for some special cases of **translated** polyhedra. Dobkin and Kirkpatrick demonstrate in [DK85, DK90] the efficiency of the hierarchical representation for solving distance problems between convex polyhedra: using this data structure one can determine the collision between two convex polyhedra in time $O(\log^2 n)$. If one of the objects is not convex, an algorithm with running time $O(n \log n)$ is possible (for more details see [DHKS90, SCH94]). The collision between a translationally moving and a stationary c-iso-oriented polyhedron can be computed in time $O(c^2 n \log^2 n)$ (see [SCH94]). Also in this work the first sub-quadratic collision detection algorithm for two general polyhedra (one of which is translated and one is stationary) is developed.

For the case of **rotations** there are no sub-quadratic algorithms known. Even the special case of two convex polyhedra, one of which is rotating has not been solved up to now. This particular problem was posed as an open question by Jack Snoeyink during the Third Dagstuhl Seminar on Computational geometry in March 1993:

Given two convex polyhedra A, B, and an axis

---

[*]Universität des Saarlandes, Fachbereich 14, Informatik, Lehrstuhl Prof. Hotz, Im Stadtwald, D-66041 Saarbrücken, Germany. E-mail: schoemer@cs.uni-sb.de.

[†]Max-Planck-Institut für Informatik, Im Stadtwald, D-66123 Saarbrücken, Germany. E-mail: thiel@mpi-sb.mpg.de. This author was supported by the ESPRIT Basic Research Actions Program, under contract No. 7141 (project ALCOM II).

of rotation, compute the smallest angle by which B has to rotate to meet A. Can this be done in sub-quadratic time?

1.2 New results and outline

In this paper we give the first sub-quadratic algorithms, which solve the collision problem between two **general** polyhedra, one of which is moving translationally or rotating about a fixed axis, whereas the other is stationary. In particular we get an upper bound of $O(n^{8/5+\epsilon})$ for the translational movement and $O(n^{5/3+\epsilon})$ for the rotational movement*.

The first collision between two polyhedra can either be a collision between a vertex of one polyhedron and a facet of the other or a collision between two edges. The former case is the simpler one and will be treated in the last part of the paper by plane sweep techniques (see Section 6). The latter problem is the harder problem and we concentrate on it. We show how to preprocess the set of stationary segments, such that we can efficiently compute the first segment hit by a moving query segment. We proceed in three steps: In the first step we use the parametric search technique of Meggido (see [MEG83]) to reduce the problem of computing the first intersection during the motion to the problem of computing the total number of intersections during the motion. In the second step we show how to reduce the latter problem to a combination of halfspace and simplex range searching problems; the key technique here is linearization, which was for the first time suggested in [YY85]. In the third step we solve the range searching problems using known techniques of van Kreveld and Matoušek. After that description our general technique will be applied to the collision problem of line segments which move translationally or rotate about a fixed axis: in Section 4 we will deduce the needed appropriate linearizations and get an upper bound of $O(n^{5/3+\epsilon})$ for both kinds of motions. Applying a recent result of Pellegrini [PEL93] in Section 5 we sketch an improved solution for translationally moving edges with running time $O(n^{8/5+\epsilon})$. Section 6 considers the collision problem for facets and vertices.

2 General collision detection and parametric search

Let $\mathcal{T}$ be a class of (topologically closed) geometric objects, i.e. closed subsets of $\mathbb{R}^d$, and let $\mathcal{S}$ be some set of n objects in $\mathcal{T}$. Let $\mathcal{Q}$ be another class of (topologically closed) geometric objects in $\mathbb{R}^d$. Further let $\mathcal{M}$ be a set of *admissible motions* for the objects in $\mathcal{Q}$, i.e. in

---

*Throughout this paper, ϵ denotes an arbitrary small positive constant.

our case the set of all possible translations respectively rotations.

For an object S of $\mathcal{T}$ and an object $Q \in \mathcal{Q}$, which moves according to a formula ℓ, we denote the first time, such that S is hit by Q, with $\phi(S, Q, \ell)$. If there is no such collision we set $\phi(S, Q, \ell) = \infty$. Our goal is to build a data structure that, given a query object $Q \in \mathcal{Q}$ and the equation $\ell \in \mathcal{M}$ of a motion, computes quickly $\phi(\mathcal{S}, Q, \ell) := \min_{S \in \mathcal{S}} \phi(S, Q, \ell)$ together with a $S \in \mathcal{S}$ such that $\phi(S, Q, \ell) = \phi(\mathcal{S}, Q, \ell)$. We call this the *on-line collision problem* for $\mathcal{Q}$ with respect to $\mathcal{T}$.

Suppose we have an efficient algorithm A_s that, given a query object $Q \in \mathcal{Q}$ and a motion $\ell \in \mathcal{M}$, decides in T_s time, whether the moving object intersects some objects of $\mathcal{S}$ within a given time period $[0, t]$. In our case this time period is represented by the length of a translation or by the angle of a rotation. We also assume that the algorithm can detect the case when exactly one object of $\mathcal{S}$ is intersected and that it can identify this object. We call such a procedure an *emptiness algorithm*. Using this emptiness algorithm we can easily decide if a given time t is less, equal or greater than $\phi(\mathcal{S}, Q, \ell)$. Meggido's parametric search technique (see [MEG83]) replaces A_s by a parallel algorithm A_p that uses P processors and runs in T_p parallel time. Then it simulates A_p generically on the unknown value $t^* := \phi(\mathcal{S}, Q, \ell)$ and delivers an algorithm that computes t^* in time $O(PT_p + T_s T_p \log P)$.

3 The emptiness algorithm

Our strategy is to reduce the collision problem to a problem for other objects that do not move and then solve the latter by known techniques. We will proceed in two steps. Firstly we linearize the problem and construct a multilevel data structure for counting all collisions (respectively for testing, if there is any collision) within a given time interval. Then we modify this algorithm and get the emptiness algorithm needed for the parametric search technique.

In many applications one (complicated) query problem can be expressed as the combination of several other (easier) query problems. A general notion for the composition of general query problems was introduced in [KREV92]:

Let $\mathcal{P} = \{p_1, p_2, \ldots p_n\}$ be a set of n points in $\mathbb{R}^d$, let $\mathcal{R}$ denote the set of all simplices in $\mathbb{R}^d$, let $\mathcal{S} = \{s_1, \ldots, s_n\}$ be a set of n objects, and let $\mathcal{Q}$ denote a set of queries on $\mathcal{S}$. The *composed query problem* $(\mathcal{S}', \mathcal{Q}')$ is defined as follows: $\mathcal{S}' = \{(p_i, s_i) | 1 \leq i \leq n\}$, $\mathcal{Q}' = \mathcal{R} \times \mathcal{Q}$ and the answer set for a query $(R, Q) \in \mathcal{Q}'$ is given by $\{(p, s) | (p, s) \in \mathcal{S}'$ and $p \in R$ and $s \in Q\}$. We also say that $(\mathcal{S}', \mathcal{Q}')$ is obtained from $(\mathcal{S}, \mathcal{Q})$ by *simplex composition*.

Simplices in d-space are the intersection of at most $d+1$ many halfspaces. Therefore we can w.l.o.g. consider simplex compositions where the simplices are halfspaces. In this case we also use the term *halfspace composition*.

3.1 General form of linearization

In this section we introduce the concept of linearization. It allows to translate a complicated test in some low dimensional space into a test in some higher dimensional space but involving only linear tests. Here we want to test whether a moving object Q, whose location at time τ is described by $Q(\tau)$ intersects a stationary object S in some time interval $[0, t]$. To find a *linearization* of this problem means to establish the equivalence

$$[\exists \tau : 0 < \tau < t, Q(\tau) \cap S \neq \emptyset] \quad (1)$$
$$\iff \bigvee_{i=1}^{dis} \bigwedge_{j=1}^{con} \left[\sum_{k=1}^{dim} \delta_k^{ij}(Q,t)\, \zeta_k^{ij}(S) \bowtie 0 \right],$$

where $\bowtie \in \{<, >, =, \leq, \geq\}$, dis, con, dim are positive constants, and $\delta_k^{ij}(Q,t)$ respectively $\zeta_k^{ij}(S)$ are rational functions of constant degree depending on the kind of motion and the kind of objects.

Having such a linearization we map the objects $S \in \mathcal{S}$ into the points $p^{ij} := (\zeta_1^{ij}(S), \zeta_2^{ij}(S), \ldots, \zeta_{dim}^{ij}(S))$ in $\mathbb{R}^{dim}$ and the query object Q into the hyperplanes $h^{ij} := (\delta_1^{ij}(Q,t), \delta_2^{ij}(Q,t), \ldots, \delta_{dim}^{ij}(Q,t))$ in the same space. Then we can think of any $\sum_{k=1}^{dim} \delta_k^{ij}(Q,t)\,\zeta_k^{ij}(S) \bowtie 0$ as the condition, that (depending on $\bowtie$) the point p^{ij} lies on the hyperplane h^{ij} respectively in a halfspace bounded by h^{ij}. Because each conjunction of (1) can be interpreted as the composition of con halfspace range searching problems we can find the objects in $\mathcal{S}$ satisfying a particular conjunction by applying halfspace composition con times. The disjunctions of (1) correspond to the union of ranges. By rewriting the defining formula, we can assume that these are disjoint unions: a formula $A \vee B$ can be rewritten as $A \vee (B \wedge \neg A)$. Now for counting all objects hit by the moving query object we can just sum up the solutions of the dis composed problems (defined by the conjunctions).

In Section 4 we deduce the linearization for the collision problem between a set of moving line segments (all moving in the same direction or all rotating about the same axis) and a set of stationary segments in $\mathbb{R}^3$. There we get constant values dis, con and dim, especially $dim = 5$.

3.2 The data structure

In his Ph.D. thesis [KREV92] Marc van Kreveld investigated efficient solutions for simplex composition[†] of query problems:

Theorem 1 ([KREV92]) *Let $\mathcal{P}$ be a set of n points in dim-space, and let $\mathcal{S}$ be a set of n objects in correspondence with $\mathcal{P}$. Let T be a data structure on $\mathcal{S}$ having building time $p(n)$, size $s(n)$ and query time $q(n)$. For an arbitrary small constant $\epsilon > 0$, the application of simplex composition on $\mathcal{P}$ to T results in a data structure D of*

1. *size $O(n^\epsilon(n^{dim} + s(n)))$ and query time $O(q(n) + \log n)$, or*

2. *size $O(n + s(n))$ and query time $O(n^\epsilon(n^{1-1/dim} + q(n)))$, or*

3. *building time $O(m^\epsilon(m + p(n)))$, size $O(m^\epsilon(m + s(n))$ and query time $O(n^\epsilon(q(n) + n/m^{1/dim}))$ for every fixed m such that $n \leq m \leq n^{dim}$,*

assuming that $s(n)/n$ is non-decreasing and $q(n)/n$ is non-increasing. Reporting takes additional $O(k)$ time if there are k answers.

In our case we apply con halfspace compositions starting with a halfspace range searching problem. Therefore Theorem 1 leads to a data structure with building time and size $O(m^{1+\epsilon})$, which can count all objects in S satisfying a particular conjunction of (1) in query time $T_s := O(\frac{n^{1+\epsilon}}{m^{1/dim}})$, for every fixed m such that $n \leq m \leq n^{dim}$. We can parallelize that query algorithm such that it runs in $T_p := O(\text{polylog } n)$ parallel time with $P := O(\frac{n^{1+\epsilon}}{m^{1/dim}})$ processors. Using the parametric search technique this gives us the first time t^* of any collision in time $O(PT_p + T_s T_p \log P) = O(\frac{n^{1+\epsilon}}{m^{1/dim}})$. To get the first hit object we start the corresponding reporting algorithm satisfying the same resource bounds.

Theorem 2 *The on-line collision problem with linearization (1) can be solved with a data structure of size and preprocessing time $O(m^{1+\epsilon})$ and query time $O(\frac{n^{1+\epsilon}}{m^{1/dim}})$, for every fixed m such that $n \leq m \leq n^{dim}$.*

Assume we have n moving objects $Q \in \mathcal{Q}$ instead of only one, and we want to determine the first collision between any pair Q, S, for $Q \in \mathcal{Q}$ and $S \in \mathcal{S}$. We apply the solution to the on-line problem and query the data structure of Theorem 2 with each moving element. This gives us a list of n candidates in which we can find the first collision in time $O(n)$.

Using this approach we need $O(m^{1+\epsilon})$ preprocessing time and $n \times O(\frac{n^{1+\epsilon}}{m^{1/dim}})$ query time. To find the best

---
[†] Actually we use only halfspace composition

time bound we have to minimize the function

$$c_1 m^{1+\epsilon} + c_2 \frac{n^{2+\epsilon}}{m^{1/dim}},$$

where c_1, c_2 are the O-constants of the resource bounds. This function achieves its minimum for m satisfying $c_1 m^{1+\epsilon} = c_2 \frac{n^{2+\epsilon}}{m^{1/dim}}$ i. e.

$$m = \left(\frac{c_2}{c_1}\right)^{\frac{dim}{dim\epsilon+dim+1}} n^{\frac{(2+\epsilon)dim}{dim\epsilon+dim+1}} = O(n^{\frac{2dim}{dim+1}+\delta}).$$

This proves the following result.

Corollary 3 *Given a subset S of n objects from $\mathcal{S}$ and a set Q of n moving objects from $\mathcal{Q}$. Assume that there is a linearization of the collision problem for $\mathcal{Q}$ with respect to $\mathcal{T}$ in the form of (1). Then we can find in $O(n^{\frac{2dim}{dim+1}+\epsilon})$ time the first collision between any elements of Q and S.*

Corollary 4 *Given two polyhedra of complexity n, one of which is moving translationally respectively is rotating about a fixed axis. The first collision between any two edges of them can be computed in time $O(n^{5/3+\epsilon})$.*

4 Collision of translationally or rotationally moving line segments

Formulation of the problem:
Given: Two line segments l_{ab} and l_{cd} with endpoints $\mathbf{a}, \mathbf{b}$ and $\mathbf{c}, \mathbf{d}$. The line segment $l_{ab}(\tau)$ performs a translation in the direction of the positive x_3-axis or a counter-clockwise rotation about the x_3-axis, from time $\tau = 0$ to $\tau = t$.
Wanted: Linear conditions to describe the fact that there is a time τ, $0 < \tau < t$, such that $l_{ab}(\tau)$ and l_{cd} intersect.

In this section we show the following result: For a translational as well as for a rotational motion there exist natural numbers dis, con, dim, so that the following holds:

$$[\exists \tau : 0 < \tau < t, l_{ab}(\tau) \cap l_{cd} \neq \emptyset]$$
$$\iff \bigvee_{i=1}^{dis} \bigwedge_{j=1}^{con} \left[\sum_{k=1}^{dim} \zeta_k^{ij}(\mathbf{c}, \mathbf{d}) \, \delta_k^{ij}(\mathbf{a}, \mathbf{b}, t) \lessgtr 0 \right],$$

where $\zeta_k^{ij}(\mathbf{c}, \mathbf{d})$ is a polynomial in the coordinates of $\mathbf{c}$ and $\mathbf{d}$ and $\delta_k^{ij}(\mathbf{a}, \mathbf{b}, t)$ is a polynomial in t and the coordinates of $\mathbf{a}$ and $\mathbf{b}$. These polynomials depend on the kind of motion.

Let L_{ab} and L_{cd} be the lines that contain the segments l_{ab} and l_{cd} respectively. Let $T = \{\tau \mid L_{ab}(\tau) \cap L_{cd} \neq \emptyset\}$.

Then

$$[\exists \tau : 0 < \tau < t, l_{ab}(\tau) \cap l_{cd} \neq \emptyset]$$
$$\iff [\exists \tau \in T : 0 < \tau < t \land l_{ab}(\tau) \cap l_{cd} \neq \emptyset].$$

4.1 Plücker coordinates for lines in $\mathbb{R}^3$

If $L_{ab} \cap L_{cd} \neq \emptyset$, then all four points $\mathbf{a}, \mathbf{b}, \mathbf{c}, \mathbf{d}$ lie in a plane. In homogeneous coordinates this fact can be expressed by the equation:

$$\det \begin{bmatrix} a_0 & a_1 & a_2 & a_3 \\ b_0 & b_1 & b_2 & b_3 \\ c_0 & c_1 & c_2 & c_3 \\ d_0 & d_1 & d_2 & d_3 \end{bmatrix} = 0$$

Expansion of this 4×4 determinant according to the 2×2 minors of the submatrix formed by the coordinates of the points $\mathbf{a}$ and $\mathbf{b}$ and the minors of the submatrix formed by the points $\mathbf{c}$ and $\mathbf{d}$ yields the following homogeneous equation:

$$0 = \gamma_{23}\alpha_{01} + \gamma_{31}\alpha_{02} + \gamma_{12}\alpha_{03} \qquad (2)$$
$$+ \gamma_{03}\alpha_{12} + \gamma_{01}\alpha_{23} + \gamma_{02}\alpha_{31}$$

with $\alpha_{ij} = a_i b_j - a_j b_i$ and $\gamma_{ij} = c_i d_j - c_j d_i$. For the sequel it is convenient to assume that our lines are oriented from the lower to the higher end point, i.e. $a_3 \leq b_3$ and $c_3 \leq d_3$ and hence $\alpha_{03} \geq 0$ and $\gamma_{03} \geq 0$. Moreover we restrict ourselves to the case $\alpha_{03} > 0$ and $\gamma_{03} > 0$, the other cases being simpler.

The Plücker coordinates α_{ij} (and the Plücker coefficients γ_{ij}) are not independent. They fulfill the equations

$$\begin{aligned} \alpha_{01}\alpha_{23} + \alpha_{02}\alpha_{31} + \alpha_{03}\alpha_{12} &= 0, \\ \gamma_{01}\gamma_{23} + \gamma_{02}\gamma_{31} + \gamma_{03}\gamma_{12} &= 0. \end{aligned} \qquad (3)$$

With the help of the bilinear equation (2) one can interpret the collision of the two lines L_{ab} and L_{cd} in $\mathbb{R}^3$ as a collision of a point $\mathbf{p}_{ab}$ with a hyperplane H_{cd} in $\mathbb{R}^6$, where $\mathbf{p}_{ab}$ and H_{cd} are given by:

$$\begin{aligned} H_{cd} \; &: \; \gamma_{23}\xi_1 + \gamma_{31}\xi_2 + \gamma_{12}\xi_3 \\ &\quad + \gamma_{03}\xi_4 + \gamma_{01}\xi_5 + \gamma_{02}\xi_6 = 0 \\ \mathbf{p}_{ab} \; &: \; (\xi_1, \xi_2, \xi_3, \xi_4, \xi_5, \xi_6)^T = \\ &\quad (\alpha_{01}, \alpha_{02}, \alpha_{03}, \alpha_{12}, \alpha_{23}, \alpha_{31})^T \end{aligned}$$

4.2 Collision times for translationally moving lines

In this subsection we compute the possible times of a collision between a translationally moving line L_{ab} and a stationary line L_{cd}.

The translation of the line $L_{ab}(\tau)$ appears in Plücker space as a corresponding motion of the point $\mathbf{p}_{ab}(\tau)$. Its

Plücker coordinates are obtained as the 2×2 minors of the following matrix:

$$\begin{bmatrix} a_0 & a_1 & a_2 & a_3+\tau a_0 \\ b_0 & b_1 & b_2 & b_3+\tau b_0 \end{bmatrix},$$

$\mathbf{p}_{ab}(\tau) = (\alpha_{01}, \alpha_{02}, \alpha_{03}, \alpha_{12}, \alpha_{23}-\tau\,\alpha_{02}, \alpha_{31}+\tau\,\alpha_{01})$

Substituting these coordinates in the plane equation H_{cd} we obtain:

$$u_1 \tau + u_0 = 0 \quad \text{where}$$
$$u_1 = \gamma_{02}\alpha_{01} - \gamma_{01}\alpha_{02}$$
$$u_0 = \gamma_{23}\alpha_{01} + \gamma_{31}\alpha_{02} + \gamma_{12}\alpha_{03}$$
$$+\gamma_{03}\alpha_{12} + \gamma_{01}\alpha_{23} + \gamma_{02}\alpha_{31}.$$

In the general case, when the projections $\overline{L}_{ab}$ and $\overline{L}_{cd}$ of the lines onto the $x_1 x_2$-plane are not parallel, we get $u_1 \neq 0$ and therefore

$$\tau_0 = -\frac{u_0}{u_1}. \tag{4}$$

Otherwise, if $u_1 = 0$, a collision can only occur if $u_0 = 0$ and $\overline{L}_{ab} = \overline{L}_{cd}$. These conditions describe the following situation: the lines L_{ab} and L_{cd} have to be parallel or to intersect; additionally they must lie in the same plane perpendicular to the $x_1 x_2$-plane. In this case the collision detection of the line segments can be described as a two-dimensional problem.

It is easy to see that a collision between two segments in 2-space is always a collision between a vertex of one segment and the other segment. Appropriate case decompositions yield a linearization of dimension less than 5.

As far as the collision test for polyhedra is concerned these cases can be ignored, because they are detected during the collison test of vertices and facets (see Section 6).

4.3 Conditions for the collision of translationally moving lines

We want to derive linear expressions, which only depend on the coordinates of $\mathbf{a}$ and $\mathbf{b}$, for the predicate $[0 < \tau_0 < t]$. We have the equivalence

$[0 < \tau_0 < t] \iff$
$\quad [u_1 > 0] \;\wedge\; [u_0 < 0] \;\wedge\; [t\,u_1 + u_0 > 0]$
$\vee\; [u_1 < 0] \;\wedge\; [u_0 > 0] \;\wedge\; [t\,u_1 + u_0 < 0]$

where the term $tu_1 + u_0$ can be written in linearized form as follows:

$$tu_1 + u_0 = \gamma_{02}(t\alpha_{01} + \alpha_{31}) + \gamma_{01}(-t\alpha_{02} + \alpha_{23})$$
$$+\gamma_{03}\alpha_{12} + \gamma_{12}\alpha_{03} + \gamma_{23}\alpha_{01} + \gamma_{31}\alpha_{02}.$$

This gives a linearized form of dimension 6 for the predicate $[0 < \tau_0 < t]$.

4.4 Conditions for the collision of translationally moving line segments

We use the following relation in order to answer the question, whether the line segments really intersect, in case the corresponding lines collide:

$$[l_{ab}(\tau_0) \cap l_{cd} \neq \emptyset] \iff [\bar{l}_{ab}(\tau_0) \cap \bar{l}_{cd} \neq \emptyset]$$

With $\bar{l}_{ab}(\tau_0)$ and $\bar{l}_{cd}$ we denote the projection of the two line segments onto the $x_1 x_2$-plane. Note that $\bar{l}_{ab}(\tau_0) = \bar{l}_{ab}$ because l_{ab} is moving in the positive x_3-direction.

Projection of the line segments onto the $x_1 x_2$-plane

We project the line segments l_{ab} and l_{cd} onto the $x_1 x_2$-plane.

$\bar{l}_{ab}: \quad \overline{\mathbf{x}} = \overline{\mathbf{a}} + \lambda(\overline{\mathbf{b}} - \overline{\mathbf{a}}),$ where $0 < \lambda < 1$
$\bar{l}_{cd}: \quad \overline{\mathbf{x}} = \overline{\mathbf{c}} + \mu(\overline{\mathbf{d}} - \overline{\mathbf{c}}),$ where $0 < \mu < 1$

Then

$[\bar{l}_{ab} \cap \bar{l}_{cd} \neq \emptyset] \iff$

$\quad ([\overline{\mathbf{c}} \text{ left of } \overline{L}_{ab}] \wedge [\overline{\mathbf{d}} \text{ right of } \overline{L}_{ab}]$
$\quad \wedge [\overline{\mathbf{a}} \text{ right of } \overline{L}_{cd}] \wedge [\overline{\mathbf{b}} \text{ left of } \overline{L}_{cd}])$
$\vee$
$\quad ([\overline{\mathbf{c}} \text{ right of } \overline{L}_{ab}] \wedge [\overline{\mathbf{d}} \text{ left of } \overline{L}_{ab}]$
$\quad \wedge [\overline{\mathbf{a}} \text{ left of } \overline{L}_{cd}] \wedge [\overline{\mathbf{b}} \text{ right of } \overline{L}_{cd}]).$

The point $\overline{\mathbf{c}}$ lies to the left/right of the orientated line $\overline{L}_{ab}$ iff the following is true:

$((\mathbf{b} - \mathbf{a}) \times (\mathbf{c} - \mathbf{a}))_3 \lessgtr 0$
$\iff (\mathbf{a} \times \mathbf{b})_3 + (\mathbf{b} \times \mathbf{c})_3 + (\mathbf{c} \times \mathbf{a})_3 \lessgtr 0$
$\iff a_1 b_2 - a_2 b_1 + c_2 b_1 - c_1 b_2 + c_1 a_2 - c_2 a_1 \lessgtr 0.$

Therefore

$[\bar{l}_{ab} \cap \bar{l}_{cd} \neq \emptyset] \iff$

$(\; [a_1 b_2 - a_2 b_1 + c_2 b_1 - c_1 b_2 + c_1 a_2 - c_2 a_1 > 0]$
$\wedge [a_1 b_2 - a_2 b_1 + d_2 b_1 - d_1 b_2 + d_1 a_2 - d_2 a_1 < 0]$
$\wedge [c_1 d_2 - c_2 d_1 + d_1 a_2 - d_2 a_1 + c_2 a_1 - c_1 a_2 < 0]$
$\wedge [c_1 d_2 - c_2 d_1 + d_1 b_2 - d_2 b_1 + c_2 b_1 - c_1 b_2 > 0])$
$\vee$
$(\; [a_1 b_2 - a_2 b_1 + c_2 b_1 - c_1 b_2 + c_1 a_2 - c_2 a_1 < 0]$
$\wedge [a_1 b_2 - a_2 b_1 + d_2 b_1 - d_1 b_2 + d_1 a_2 - d_2 a_1 > 0]$
$\wedge [c_1 d_2 - c_2 d_1 + d_1 a_2 - d_2 a_1 + c_2 a_1 - c_1 a_2 > 0]$
$\wedge [c_1 d_2 - c_2 d_1 + d_1 b_2 - d_2 b_1 + c_2 b_1 - c_1 b_2 < 0]).$

4.5 Collision times for rotating lines

A counterclockwise rotation of the line $L_{ab}(\varphi)$ about the x_3-axis induces a corresponding motion of the point

$\mathbf{p}_{ab}(\varphi)$ in Plücker space. Its Plücker coordinates are given by the 2×2 minors of the following matrix:

$$\begin{bmatrix} a_0 & \cos\varphi\, a_1+\sin\varphi\, a_2 & -\sin\varphi\, a_1+\cos\varphi\, a_2 & a_3 \\ b_0 & \cos\varphi\, b_1+\sin\varphi\, b_2 & -\sin\varphi\, b_1+\cos\varphi\, b_2 & b_3 \end{bmatrix}$$

$$\mathbf{p}_{ab}(\varphi) = (\cos\varphi\,\alpha_{01}+\sin\varphi\,\alpha_{02}, -\sin\varphi\,\alpha_{01}+\cos\varphi\,\alpha_{02},$$
$$\alpha_{03}, \alpha_{12}, \cos\varphi\,\alpha_{23}+\sin\varphi\,\alpha_{31}, -\sin\varphi\,\alpha_{23}+\cos\varphi\,\alpha_{31})$$

Substituting these coordinates into the plane equation H_{cd} results in:

$$u_2\cos\varphi + u_1\sin\varphi + u_0 = 0 \quad \text{where} \quad (5)$$
$$u_2 = \gamma_{23}\alpha_{01} + \gamma_{31}\alpha_{02} + \gamma_{01}\alpha_{23} + \gamma_{02}\alpha_{31}$$
$$u_1 = -\gamma_{31}\alpha_{01} + \gamma_{23}\alpha_{02} - \gamma_{02}\alpha_{23} + \gamma_{01}\alpha_{31}$$
$$u_0 = \gamma_{12}\alpha_{03} + \gamma_{03}\alpha_{12}$$

The following parametric formulation

$$\sin\varphi = \frac{2\tau}{1+\tau^2}, \quad \cos\varphi = \frac{1-\tau^2}{1+\tau^2}, \quad \text{where } \tau = \tan\frac{\varphi}{2},$$

$0 < \varphi < \pi$, transforms equation (5) into a quadratic equation

$$u'_2\tau^2 + u'_1\tau + u'_0 = 0, \quad \text{where} \quad (6)$$
$$u'_2 = u_0 - u_2, \quad u'_1 = 2u_1, \quad u'_0 = u_0 + u_2$$

with the two roots:

$$\tau_1 = \frac{-u'_1 + \sqrt{u'^2_1 - 4u'_2 u'_0}}{2u'_2}, \quad \tau_2 = \frac{-u'_1 - \sqrt{u'^2_1 - 4u'_2 u'_0}}{2u'_2}$$

for $u'_2 \neq 0$. As expected there are in general two points in time where the two lines intersect. If $u'_2 = 0$ and $u'_1 \neq 0$ there is one collision at time $-\frac{u'_0}{u'_1}$ and an other at ∞, which corresponds to a rotation angle of $\varphi = \pi$. The degenerate case $u'_2 = u'_1 = 0$ means that a collision during the rotation can only occur if $u'_0 = 0$ and the lines lie on the same cone respectively cylinder for which the x_3-axis is the axis of symmetry. Therefore collision detection can be reduced to a 2-dimensional problem, so that we only need to test a collision between vertices and segments.

4.6 Conditions for the collision of rotating lines

τ_1 and τ_2 are real numbers only if $[u'^2_1 - 4u'_2 u'_0 \geq 0]$. Under this precondition the predicates $[0 < \tau_i < t]$ can be transformed as follows

$[0 < \tau_1 < t] \iff$

$\quad [u'_2 > 0] \wedge ([u'_1 < 0] \vee [u'_0 < 0]) \wedge [2tu'_2 + u'_1 > 0]$
$\quad \wedge [t^2 u'_2 + tu'_1 + u'_0 > 0]$
$\vee [u'_2 < 0] \wedge [u'_1 > 0] \wedge [u'_0 < 0] \wedge ([2tu'_2 + u'_1 < 0]$
$\quad \vee [t^2 u'_2 + tu'_1 + u'_0 > 0])$
$\vee [u'_2 = 0] \wedge [u'_1 > 0] \wedge [u'_0 < 0] \wedge [tu'_1 + u'_0 > 0].$

For $[0 < \tau_2 < t]$ we get similar conditions. We now derive the linear expressions for the various predicates. From equation (6) we obtain the following equations

$2tu'_2 + u'_1 =$
$\quad 2\gamma_{03}(t\alpha_{12}) + 2\gamma_{12}(t\alpha_{03}) + 2\gamma_{23}(-t\alpha_{01} + \alpha_{02})$
$\quad +2\gamma_{01}(-t\alpha_{23} + \alpha_{31}) + 2\gamma_{02}(-t\alpha_{31} - \alpha_{23})$
$\quad +2\gamma_{31}(-t\alpha_{02} - \alpha_{01})$

$t^2 u'_2 + tu'_1 + u'_0 =$
$\quad \gamma_{02}(-t^2\alpha_{31} - 2t\alpha_{23} + \alpha_{31}) + \gamma_{12}(t^2\alpha_{03} + \alpha_{03})$
$\quad +\gamma_{01}(-t^2\alpha_{23} + 2t\alpha_{31} + \alpha_{23}) + \gamma_{03}(t^2\alpha_{12} + \alpha_{12})$
$\quad +\gamma_{23}(-t^2\alpha_{01} + 2t\alpha_{02} - \alpha_{01})$
$\quad +\gamma_{31}(-t^2\alpha_{02} - 2t\alpha_{01} + \alpha_{02}),$

$$u'^2_1 - 4u'_2 u'_0 = \quad (7)$$
$\quad (\gamma^2_{23} + \gamma^2_{31})(\alpha^2_{01} + \alpha^2_{02}) + (\gamma^2_{01} + \gamma^2_{02})(\alpha^2_{23} + \alpha^2_{31})$
$\quad -2(\gamma_{02}\gamma_{23} - \gamma_{01}\gamma_{31})(\alpha_{02}\alpha_{23} - \alpha_{01}\alpha_{31})$
$\quad -\gamma^2_{12}(\alpha^2_{03}) - \gamma^2_{03}(\alpha^2_{12}).$

In each case the predicates $[t^2 u'_2 + tu'_1 + u'_0 \lessgtr 0]$, $[2tu'_2 + u'_1 \lessgtr 0]$ and $[u'^2_1 - 4u'_2 u'_0 \geq 0]$ are linear in at most 6 expressions $\delta^{i,j}_k(\mathbf{a}, \mathbf{b}, t)$, which are given as polynomials in the coordinates of $\mathbf{a}$ and $\mathbf{b}$ and the time parameter t. This means we have found a linearization of dimension 6 for the predicate $[0 < \tau_i < t]$.

4.7 Conditions for the collision of rotating line segments

We reduce the decision whether $[l_{ab}(\tau_i) \cap l_{cd} \neq \emptyset]$ (in case the corresponding lines collide) to the calculation of the x_3–coordinate z of the intersection points of the corresponding lines and a test whether $z \in [\min\{a_3, b_3\}, \max\{a_3, b_3\}] \cap [\min\{c_3, d_3\}, \max\{c_3, d_3\}]$. For the calculation of the x_3–coordinates of the possible intersection points we use cylindrical coordinates.

Representation of line segments in cylindrical coordinates

If we represent the line segments

$l_{ab}: \quad \mathbf{x} = \mathbf{a} + \lambda(\mathbf{b} - \mathbf{a}), \quad \text{where} \quad 0 \leq \lambda \leq 1,$
$l_{cd}: \quad \mathbf{x} = \mathbf{c} + \mu(\mathbf{d} - \mathbf{c}), \quad \text{where} \quad 0 \leq \mu \leq 1,$

in cylindrical coordinates (r, φ, z), we can easily check, whether the line segment l_{ab} can collide with the line segment l_{cd} during a full rotation about the x_3-axis. During its rotation the line segment l_{ab} generally describes a hyperboloid, whose projection onto the (r, z)–plane of the cylindrical coordinate system yields a hyperbolic segment. The rotating line segment l_{ab} can only collide with l_{cd}, if the two corresponding hyperbolic segments intersect in the (r,z)–plane. In order to compute this intersections we have to find the (r,z)–representation for each point in $\mathbb{R}^3$, which is given by its Cartesian coordinates (x_1, x_2, x_3). We get

$$z = x_3, \quad r = \sqrt{x_1^2 + x_2^2}.$$

Since $x_i = a_i + \lambda(b_i - a_i)$ and $\lambda = \dfrac{z - a_3}{b_3 - a_3}$, we have for $a_3 \neq b_3$:

$$r^2 = \frac{1}{(b_3-a_3)^2}((a_1 b_3 - a_3 b_1 + z(b_1 - a_1))^2 + (a_2 b_3 - a_3 b_2 + z(b_2 - a_2))^2).$$

We proceed with Plücker coordinates and get:

$$\begin{aligned} r^2 &= \frac{1}{\alpha_{03}^2}((\alpha_{31} - z\alpha_{01})^2 + (\alpha_{23} + z\alpha_{02})^2) \\ &= v_2 z^2 + v_1 z + v_0, \end{aligned}$$

where $v_2 = \dfrac{\alpha_{01}^2 + \alpha_{02}^2}{\alpha_{03}^2}$, $v_1 = 2\dfrac{\alpha_{02}\alpha_{23} - \alpha_{01}\alpha_{31}}{\alpha_{03}^2}$, and $v_0 = \dfrac{\alpha_{23}^2 + \alpha_{31}^2}{\alpha_{03}^2}$. The question whether the line segment l_{ab} collides with the stationary segment l_{cd} while it is rotating about the x_3–axis can be answered by calculating the intersection between the following two parabolic segments:

$$\begin{aligned} r_{ab}^2(z) &= v_2 z^2 + v_1 z + v_0 \text{ with } a_3 \leq z \leq b_3, \\ r_{cd}^2(z) &= w_2 z^2 + w_1 z + w_0 \text{ with } c_3 \leq z \leq d_3. \end{aligned}$$

W.l.o.g. let the line segments be given, such that $a_3 \leq b_3$ and $c_3 \leq d_3$. The intersection points of the two parabola can be found as the roots of a quadratic equation:

$$v_2' z^2 + v_1' z + v_0' = 0 \quad \text{where } v_i' = v_i - w_i, \qquad (8)$$

$$z_1 = \frac{-v_1' + \sqrt{v_1'^2 - 4v_2' v_0'}}{2v_2'}, \quad z_2 = \frac{-v_1' - \sqrt{v_1'^2 - 4v_2' v_0'}}{2v_2'},$$

for $v_2' \neq 0$.

But a collision of the rotating line segment l_{ab} with l_{cd} exists only if the quadratic equation has real roots ($[v_1'^2 - 4v_2' v_0' \geq 0]$) and these lie in the interval $[a_3, b_3] \cap [c_3, d_3]$. For $v_2' = 0$ and $v_1' \neq 0$ we get one solution at $-\dfrac{v_0'}{v_1'}$ and an other at ∞. The case $v_2' = v_1' = v_0' = 0$ occurs iff L_{cd} lies on the hyperboloid generated by the rotating line L_{ab}. In this case possible collisions can be ignored because they are detected when testing vertices against facets.

By using cylindrical coordinates we have succeeded in finding the x_3–coordinates of the possible intersection points of the rotating line $L_{ab}(\tau)$ with L_{cd}. But it remains open, to which x_3–coordinate the collision time τ_i corresponds. It can be shown that z_1 belongs to τ_1 and z_2 to τ_2, if we assume that $a_3 < b_3$ and $c_3 < d_3$. That is

$$\begin{aligned} &[l_{ab}(\tau_i) \cap l_{cd} \neq \emptyset] \Longleftrightarrow \qquad (9)\\ &[z_i > a_3] \wedge [z_i < b_3] \wedge [z_i > c_3] \wedge [z_i < d_3]. \end{aligned}$$

In the following we want to derive linearized conditions for the predicates $[z_i < Z]$ respectively $[z_i > Z]$. For example we get

$$\begin{aligned} &[z_1 < Z] \Longleftrightarrow \\ &[v_2' > 0] \wedge [2Zv_2' + v_1' > 0] \wedge [Z^2 v_2' + Z v_1' + v_0' > 0] \\ &\vee [v_2' < 0] \wedge ([2Zv_2' + v_1' < 0] \vee [Z^2 v_2' + Z v_1' + v_0' > 0]) \\ &\vee [v_2' = 0] \wedge \qquad [v_1' > 0] \wedge \qquad [Z v_1' + v_0' > 0] \end{aligned}$$

the other cases being similar. It holds:

$$\begin{aligned} v_2' &= \frac{\alpha_{01}^2 + \alpha_{02}^2}{\alpha_{03}^2} - \frac{\gamma_{01}^2 + \gamma_{02}^2}{\gamma_{03}^2}, \\ v_1' &= 2\frac{\alpha_{02}\alpha_{23} - \alpha_{01}\alpha_{31}}{\alpha_{03}^2} - 2\frac{\gamma_{02}\gamma_{23} - \gamma_{01}\gamma_{31}}{\gamma_{03}^2}, \quad (10)\\ v_0' &= \frac{\alpha_{23}^2 + \alpha_{31}^2}{\alpha_{03}^2} - \frac{\gamma_{23}^2 + \gamma_{31}^2}{\gamma_{03}^2}. \end{aligned}$$

The predicates $[v_1'^2 - 4v_2' v_0' \geq 0]$ and $[u_1'^2 - 4u_2' u_0' \geq 0]$ are equivalent, since both express the fact that the rotating line L_{ab} collides with the stationary line L_{cd} during a full rotation. On the basis of relations (10) and equation (3) it holds:

$$u_1'^2 - 4u_2' u_0' = \alpha_{03}^2 \gamma_{03}^2 \cdot (v_1'^2 - 4v_2' v_0'). \qquad (11)$$

According to equation (7) the predicate $[v_1'^2 - 4v_2' v_0' \geq 0]$ is linear in the expressions $\alpha_{01}^2 + \alpha_{02}^2$, $\alpha_{23}^2 + \alpha_{31}^2$, $\alpha_{01}\alpha_{31} - \alpha_{02}\alpha_{23}$, α_{03}^2 and α_{12}^2.

Now let us consider the predicates $[v_2' \lessgtr 0]$, $[2Zv_2' + v_1' \lessgtr 0]$ and $[Z^2 v_2' + Zv_1' + v_0' \lessgtr 0]$. A multiplication with $\alpha_{03}^2 \gamma_{03}^2$ yields:

$$\begin{aligned} &[v_2' \lessgtr 0] \Longleftrightarrow \\ &[\gamma_{03}^2(\alpha_{01}^2 + \alpha_{02}^2) - (\gamma_{01}^2 + \gamma_{02}^2)(\alpha_{03}^2) \lessgtr 0], \end{aligned}$$

$$\begin{aligned} &[2Zv_2' + v_1' \lessgtr 0] \Longleftrightarrow \\ &[2Z\gamma_{03}^2(\alpha_{01}^2 + \alpha_{02}^2) - 2Z(\gamma_{01}^2 + \gamma_{02}^2)(\alpha_{03}^2) \\ &+ 2\gamma_{03}^2(\alpha_{02}\alpha_{23} - \alpha_{01}\alpha_{31}) \\ &- 2(\gamma_{02}\gamma_{23} - \gamma_{01}\gamma_{31})(\alpha_{03}^2) \lessgtr 0], \end{aligned}$$

$$\begin{aligned} &[Z^2 v_2' + Zv_1' + v_0' \lessgtr 0] \Longleftrightarrow \\ &[Z^2 \gamma_{03}^2(\alpha_{01}^2 + \alpha_{02}^2) - Z^2(\gamma_{01}^2 + \gamma_{02}^2)(\alpha_{03}^2) \\ &+ 2Z\gamma_{03}^2(\alpha_{02}\alpha_{23} - \alpha_{01}\alpha_{31}) - (\gamma_{23}^2 + \gamma_{31}^2)(\alpha_{03}^2) \\ &- 2Z(\gamma_{02}\gamma_{23} - \gamma_{01}\gamma_{31})(\alpha_{03}^2) + \gamma_{03}^2(\alpha_{23}^2 + \alpha_{31}^2) \lessgtr 0]. \end{aligned}$$

If $Z = c_3, d_3$ (see condition 10), then these predicates are linear in the four expressions $\alpha_{01}^2 + \alpha_{02}^2$, α_{03}^2,

$\alpha_{02}\alpha_{23} - \alpha_{01}\alpha_{31}$, $\alpha_{23}^2 + \alpha_{31}^2$, and the corresponding coefficients only depend on the coordinates of the points **c** and **d**. However if $Z = a_3, b_3$, then one can define the expressions $Z^k(\alpha_{01}^2 + \alpha_{02}^2)$, $Z^k \alpha_{03}^2$, $Z^k(\alpha_{02}\alpha_{23} - \alpha_{01}\alpha_{31})$, $Z^k(\alpha_{23}^2 + \alpha_{31}^2)$ for $k = 0, 1, 2$, so that the former predicates are linear in at most six of these expressions, where again the corresponding coefficients only depend on the coordinates of the points **c** and **d**.

4.8 Linearization

To summarize we firstly computed the conditions for the fact that the moving line L_{ab} intersects the stationary line L_{cd} during a time interval $[0, t]$. In the next step (see Sections 4.4 and 4.7) we got the additional conditions for the intersection of the corresponding line segments. The combination of these two sets of conditions gives the wanted linearization, i.e.

$$[\exists \tau : 0 < \tau < t, l_{ab}(\tau) \cap l_{cd} \neq \emptyset]$$
$$\iff \bigvee_i ([0 < \tau_i < t] \wedge [l_{ab}(\tau_i) \cap l_{cd} \neq \emptyset])$$

Until now we have found linearizations with $dim = 6$ where $\delta_k^{ij}(\mathbf{c}, \mathbf{d})$ are polynomials in the coordinates of **c** and **d** and $\zeta_k^{ij}(\mathbf{a}, \mathbf{b}, t)$ are polynomials in t and the coordinates of **a** and **b**. In order to reduce the dimension we divide each inequality by a positive coefficient $\zeta_k^{ij}(\mathbf{a}, \mathbf{b}, t)$, which we can always find using the fact that $\alpha_{03} > 0$. So we get a linearization with dimension $dim = 5$. For example let us consider the condition $[2tu_2' + u_1' > 0]$ (see (7)).

We can divide the inequality by the term $2t\alpha_{03} > 0$ and replace the inequality $2tu_2' + u_1' > 0$ by

$$0 < \gamma_{01}\frac{-t\alpha_{23} + \alpha_{31}}{t\alpha_{03}} + \gamma_{02}\frac{-t\alpha_{31} - \alpha_{23}}{t\alpha_{03}}$$
$$+ \gamma_{03}\frac{t\alpha_{12}}{t\alpha_{03}} + \gamma_{23}\frac{-t\alpha_{01} + \alpha_{02}}{t\alpha_{03}}$$
$$+ \gamma_{31}\frac{-t\alpha_{02} - \alpha_{01}}{t\alpha_{03}} + \gamma_{12}.$$

Now we can apply the construction of Section 3.2, especially Corollary 3.

5 Improved solution for the translational case

Applying a recent result of Pellegrini we can construct a faster algorithm for translationally moving line segments. During its movement a segment moves over a quadrilateral which all segments, that are hit during the translation, have to intersect. We can triangulate the quadrilateral and determine all segments intersecting one of the resulting triangles. During this process it can happen that we count segments twice, if they intersect the common edge of the two triangles. But for our emptiness problem it is sufficient to work inaccurately. We can count the number of collisions and when there are less than three of them we compute them and check whether they are different. Therefore we have to consider the following subproblem:

> Given a set L of n line segments, construct a data structure such that for any query triangle one can efficiently count/compute the incidences with L.

In [PEL92] and [AM92] there is a solution based on Plücker coordinates of lines. In contrast to our method based on Theorem 1, the construction of the data structures for points and hyperplanes in Plücker space makes use of the result that the zone of the Plücker hypersurface among n hyperplanes has size $O(n^4 \log n)$ ([APS93]). This additional structure is essential to obtain a final bound of $O(n^{8/5+\epsilon})$.

Theorem 5 *Given n segments there is a data structure using $O(m)$ space, $n^{1+\epsilon} \leq m \leq n^{4+\epsilon}$, such that we can count the number of segments intersected by a query triangle in time $O(n^{1+\epsilon}/m^{1/4})$. For reporting these k segments we need $O(k)$ additional steps. The structure can be build in time $O(m)$.*

The query algorithm can be replaced by a parallel version running in $O(\text{polylog } n)$ parallel steps using $O(n^{1+\epsilon}/m^{1/4})$ processors, which allows us to apply the *parametric search technique* efficiently.

Corollary 6 *Given a set of n line segments, each of which is moving translationally in some fixed direction over distance t and a set of n stationary line segments in $\mathbb{R}^3$ we can compute the first collision between the two sets in time $O(n^{8/5+\epsilon})$.*

6 Collisions between facets and vertices

Recall that we consider two polyhedra one of which is moving (let us say P_1) whereas the other one, P_2, is stationary. Let V_i, E_i, F_i, $i = 1, 2$, denote the sets of vertices, edges and facets of the polyhedra. Only translations or rotations are permitted as motions. Until now we have shown how to compute the first collision between the edges of P_1 and P_2. We still have to determine the first facet of P_2 hit by a vertex of P_1 respectively the first facet of the moving polyhedron which collides with a vertex of P_2. A solution to this problem, which applies to both kinds of motions, is already presented in [SCH94] based on ideas from [NU85]. The facets and vertices are projected onto a 2-dimensional space and a plane sweep technique is applied. For completeness we present a sketch of this construction. We only determine the time of the first collision between the set of vertices of P_1 and the set of facets of P_2.

In case of a translation we project the facets and vertices onto a plane perpendicular to the direction of the motion. If the polyhedron P_1 is rotating about an axis (see Section 4.7) we can apply a similar method, but we have to work with cylindrical coordinates: the projection is done by removing the angle-component. In both cases we get in the projection-plane a point set $\overline{V_1}$ which is the image of the vertices of P_1 and 2-dimensional regions $\overline{F_2}$ bounded by line segments respectively hyperbola segments. Now we execute a plane sweep: stoppoints/halts are starting- and end-points, extremal- and intersection-points of the segments. Between two consecutive halts the ordering of the intersection-points between the segments and the plane sweep S is always the same. Therefore we can save the active segments in a balanced search tree which will be the primary structure for saving more informations.

Let R be a region between two segments, which are adjacent on S, and assume that S goes over R between two consecutive halts. Every vertex of P_1, whose projection $\overline{v}$ lies in R, can only collide with facets of P_2, the projections $\overline{f}$ of which contain R. Therefore for each region R we keep track of the set $F_R = \{f \in F_2 | R \subseteq \overline{f}\}$. Thus for every segment we save the set F_R of the region lying above it.

Dependent on the kind of motion we can define an ordering for each set F_R. During a rotation every point with projection in R will stab the regions in F_R with the same cyclic ordering; in the case of a translation we will get a linear order. As secondary structure which stores the elements in F_R we again use a balanced search tree which allows to find the first facet hit by a vertex projected onto R in logarithmic time.

The sweep line stops at every point $\overline{v} \in \overline{V}$. There we determine the region R containing $\overline{v}$ and search the facet of F_R which is hit by the corresponding vertex first. Both steps can be done in logarithmic time using the tree structure.

During the sweep we have to hold all regions intersected by the sweep plane S as well as the sets F_R. To save space we only store the changes of the sets F_R (see [Nu85]). Using this idea each set F_R can be stored with logarithmical costs.

The run time of the algorithm is $O((|V_1| + |E_2| + C_{\overline{E_2}}) \log |E_2|)$ where $C_{\overline{E_2}}$ denotes the number of intersections of the projected edges of P_2. Unfortunately this value could be quadratic in the complexity of the polyhedron. Therefore we divide the problem into several smaller subproblems. W.l.o.g. we assume that the facets of P_2 are triangles (a triangulation of the surface does not change the asymptotic complexity of P_2). We divide F_2 in $\sqrt{|F_2|}$ many subsets of size $\sqrt{|F_2|}$. For each subset we execute the above plane sweep algorithm.

Theorem 7 *Consider two polyhedra P_1, P_2, one of which is moving translationally or rotating about a fixed axis. The first collision between a vertex of one of them and a facet of the other can be computed in time $O((|P_1| + |P_2|)^{3/2} \log(|P_1| + |P_2|))$.*

As already in the case of edge/edge collisions there is an improved solution for the translational motion. This follows from the fact that vertical ray shooting into a set of disjoint triangles can be done with $O(n^{4/3+\epsilon})$ preprocessing and $O(n^{1/3})$ query time. The algorithm for this kind of ray shooting selects all triangles intersected by the line containing the query ray and determines the first triangle that is hit using binary search. We show how this can be done by simplex composition. We project all triangles and the query ray, respectively the supporting line, onto the plane perpendicular to the direction of the motion. A query line intersects a triangle if and only if the projection of the line lies within the projection of the triangle. Since the projection of the line results in a point in the plane the first part of the algorithm corresponds to planar point location which is dual to halfspace range searching in the plane; in the second step we have to find the first triangle hit by the query ray among all the triangles stabbed by the supporting vertical line. This can be done simply by sorting and searching. The projection of the triangles divides the plane into cells each of which corresponds to a certain stabbing order, i.e. all points in such a cell are the images of vertical lines intersecting the same triangles in the same order. The projection of the supporting line has to lie in the common intersection of all stabbed triangles. We compute this region, choose an arbitrary point in it and determine the stabbing order of the corresponding line, which is the same for all points in the region, in time $O(n \log n)$. This finishes the preprocessing. To find the first triangle hit by the query ray we do binary search on this order. Comparisons are made by determining the relative position of the ray's starting point with respect to a triangle. Applying Theorem 1 we get

Theorem 8 *Consider two polyhedra P_1, P_2, one of which is moving translationally. The first collision between a vertex of one of them and a facet of the other can be computed in time $O(n^{4/3+\epsilon})$.*

7 Conclusion

We have shown how to determine the collision between a stationary and a moving polyhedron in sub-quadratic time. For that we have computed the first collision between vertices of one polyhedron and facets of the other and the first collision between the edges of the polyhedra. We have reduced the latter task to the formulation of an appropriate linearization which is derived from an explicit computation of the collision times. We could do this because the equations of the motions have degree at most two. The natural question is how we can proceed if the motion of the polyhedron is more complicated, i.e. if the equations have degree greater than five (then no explicit formulation of the roots exists).

Actually we do not need an explicit representation of the collision times for the linearization. We only need to know whether two particular features of the polyhedra collide during a given time period. Formally we come upon the problem whether a system of (constant many) multivariate polynomial equations with constant degree has a common real solution in a box or not. Applying recent results on counting real zeros of a system of algebraic equations [P91, US92] one can decide this question using a finite number of algebraic operations on the coefficients of the polynomials and the vertices of the box. This extension of Sturm's theorem to arbitrary dimensions is purely symbolic so we can use it to get appropriate linearizations, which leads to subquadratic collision detection algorithms for a wide class of motions.

References

[AM92] P. K. Agarwal, J. Matoušek: *On range searching with semialgebraic sets*, Proc. 17th Symp. on Math. Foundation of CS: 1–13, (1992)

[AM92B] P. K. Agarwal, J. Matoušek: *Ray shooting and parametric search*, Proc. 24th STOC: 517–526, (1992)

[APS93] B. Aronov, M. Pellegrini, and M. Sharir: *On the zone of an algebraic surface in a hyperplane arrangement*, Discrete Comput. Geom 9: 177–188, (1993)

[Bo79] J. W. Boyse: *Interference detection among solids and surfaces*, CACM Vol. 22(1): 3–9, (1979)

[Ca84] J. Canny: *On detecting collision between polyhedra*, Proc. ECAI (1984), 533–542

[DK85] D. Dobkin, D. Kirkpatrick: *A linear algorithm for determining the seperation of convex polyhedra*, J. Algor. 6 (1985), 381–392

[DK90] D. Dobkin, D. Kirkpatrick: *Determining the seperation of preprocessed polyhedra – a unified approach*, Lecture Notes in Computer Science 443 (1990), 400–413

[DHKS90] D. Dobkin, J. Hershberger, D. Kirkpatrick, S. Suri: *Implicitily searching convolutions and computing depth of collisions*, Lecture Notes in Computer Science 450 (1990), 165–180

[Krev92] M. van Kreveld: *New Results on Data Structures in Computational Geometry*, Ph.D. Thesis, University of Utrecht, The Netherlands, (1992)

[Mat93] J. Matoušek: *Range searching with efficient hierarchical cuttings*, Discrete Comput. Geom 10: 159–182, (1993)

[Meg83] N. Megiddo: *Applying parallel computation algorithms in the design of serial algorithms*. Journal of the ACM **30**: 852–865, (1983).

[Nu85] O. Nurmi: *A fast algorithm for hidden-line elimination*, BIT, Vol. 25, 466–472, (1985)

[P91] P. Pedersen: *Counting real zeros*, New York University, NYU Technical Report 545-R243, (1991)

[Pel92] M. Pellegrini: *Incidence and nearest-neighbor problems for lines in 3-space*, Proc. 8th Annu. ACM Sympos. Comput. Geom.: 130–137, (1992)

[Pel93] M. Pellegrini: *Ray shooting on triangles in 3-space*, Algorithmica 9: 471–494, (1993)

[PS88] F. P. Preparata and M. I. Shamos: *Computational Geometry: an Introduction*, Springer-Verlag, New York, (1988)

[Sch94] E. Schömer: *Interaktive Montagesimulation mit Kollisionserkennung*, Ph.D. Thesis, Universität des Saarlandes, Germany, (1994)

[US92] A. Y. Uteshev and S. G. Shulyak: *Hermite's method of seperation of solutions of systems of algebraic equations and its applications*, Linear Algebra and its Applications 177: 49–88, (1992)

[YY85] A. C. Yao, F. F. Yao: *A general approach to D-dimensional geometric queries*, Proc. 17th Annu. Sympos. Theory Comput.: 163–168, (1985)

A Comparison of Sequential Delaunay Triangulation Algorithms

Peter Su
Transarc Corporation
The Gulf Tower, 707 Grant Street
Pittsburgh, PA 15219
psu+@transarc.com

Robert L. Scot Drysdale[*]
Department of Computer Science
6211 Sudikoff Laboratory
Dartmouth College, Hanover NH 03755-3510, USA
scot.drysdale@dartmouth.edu

1. Introduction

Sequential algorithms for constructing Delaunay triangulations come in three basic flavors: divide-and-conquer [6, 11], sweepline [8], and incremental [5, 9, 13, 15]. Which approach is best in practice? This paper presents an experimental comparison of a number of these algorithms. It considers both uniformly distributed points and several highly non-uniform distributions.

In addition, it describes a new version of the incremental algorithm that is simple to understand and implement, but is still competitive with the other, more sophisticated methods on a wide range of problems. The algorithm uses a combination of dynamic bucketing and randomization to achieve both simplicity and good performance.

The experiments in this paper go beyond measuring total run time, which is highly dependent on the computer system. We also analyze the major high-level primitives that each algorithm uses and do an experimental analysis of how often each algorithm performs each operation.

1.1. Divide-and-Conquer

Guibas and Stolfi [11] gave an $O(n \log n)$ Delaunay triangulation algorithm that used the quad-edge data structure and only two geometric primitives, a CCW orientation test and an in-circle test. This algorithm is asymptotically optimal in the worst case.

Dwyer [6] showed that a simple modification to this algorithm runs in $O(n \log \log n)$ expected time on uniformly distributed points. Dwyer's algorithm splits the point set into vertical strips of width $\sqrt{n/\log n}$, constructs the DT of each strip by merging along horizontal lines and then merges the strips together along vertical lines. His experiments indicate that in practice this algorithm runs in linear expected time. Another version of this algorithm, due to Katajainen and Koppinen [12] merges square buckets together in a "quad-tree" order. They show that this algorithm runs in linear expected time for uniform points. In fact, their experiments show that the performance of this algorithm is nearly identical to Dwyer's.

1.2. Sweepline Algorithms

Fortune [8] invented another $O(n \log n)$ scheme for constructing the Delaunay triangulation using a sweepline algorithm. The algorithm keeps track of two sets of state. The first is a list of edges called the *frontier* of the diagram. These edges are a subset of the Delaunay diagram, and form a tour around the outside of the incomplete triangulation. The algorithm also keeps track of a queue of events containing *site* events and *circle* events. Site events happen when the sweepline reaches a site, and circle events happen when it reaches the top of a circle formed by three adjacent vertices on the frontier. The algorithm sweeps a line up in the y direction, processing each event that it encounters.

In Fortune's implementation, the frontier is a simple linked list of half-edges, and point location is performed using a bucketing scheme. The x-coordinate of a point to be located is used as a hash value to get close to the correct frontier edge, and then the algorithm walks to the left or right until it reaches the correct edge. This edge is placed in the bucket that the point landed in, so future points that are nearby can be located quickly. This method works well as long as the query points and the edges are well distributed in the buckets. A bucketing scheme is also used to represent the priority queue. Members of the queue are bucketed according to their priorities, so finding the minimum involves searching for the first

---

[*]This study was supported in part by the funds of the National Science Foundation, DDM-9015851, and by a Fulbright Foundation fellowship.

non-empty bucket and pulling out the minimum element.

1.3. Incremental Algorithms

The third, and perhaps simplest class of algorithms for constructing the Delaunay triangulation consists of incremental algorithms. We will study two styles of incremental algorithms: incremental *construction* and incremental *search*. Incremental construction algorithms add sites to the diagram one by one and update the diagram after each site is added, while incremental search algorithms grow the diagram one valid triangle at a time.

In most incremental construction algorithms there are two basic steps. The first, Locate, finds the triangle containing the new point. (The algorithms are made simpler by enclosing all points within a very large triangle.) The second, Update, updates the diagram.

The algorithms perform Update by flipping edges until all edges invalidated by the new point have been removed, as in in Guibas and Stolfi [11]. The worst case time required for all updates is $O(n^2)$. However, if the points are inserted in a random order, Guibas, Knuth, and Sharir [10] show that the expected number of edge flips is linear.

Therefore the bottleneck in the algorithm is the Locate routine. Guibas and Stolfi start at a random edge in the current diagram and walk along a line in the direction of the new site until the right triangle is found, taking expected time $O(\sqrt{n})$ per Locate. Guibas, Knuth, and Sharir propose a tree-based data structure where internal nodes are triangles that have been deleted or subdivided at some point in the construction, and the current triangulation is stored at the leaves. The total expected cost of Locate is $O(n \log n)$ time. Sharir and Yaniv [17] prove a bound of about $12nH_n + O(n)$. Ohya, Iri and Murota [15] bucket the points and insert them using a breadth-first traversal of a quad-tree. They claim that their algorithm runs in expected linear time on sites that are uniformly distributed, and they provide experimental evidence for this fact.

1.4. A Faster Incremental Construction Algorithm

We present a Locate variant that leads to an easily implemented incremental algorithm that seems to perform better than those mentioned above when the input is uniformly distributed. We use a simple bucketing algorithm similar to the one that Bentley, Wiede and Yao used for finding the nearest neighbor of a query point [4]. This leads to an $O(n)$ time algorithm while maintaining the relative simplicity of the incremental algorithm.

The bucketing scheme places the sites into a uniform grid as it adds them to the diagram. To find a near neighbor, the point location algorithm first finds the bucket that the query point lies in and searches in an outward spiral for a non-empty bucket. It then uses any edge incident on the point in this bucket as a starting edge in the normal point location routine. If the spiral search fails to find a point after a certain number of layers, the point location routine continues from an arbitrary edge in the current triangulation.

The bucketing scheme uses a dynamic hash table to deal with the fact that sites are processed in an on-line fashion. The scheme also does not bother to store all the points that fall in a particular bucket, but just stores the last point seen. This is because the bucket structure does not have to provide the insertion algorithm with the true nearest neighbor of a query, it only has to find a point that is likely to be close the the query. Therefore, it makes sense not to use the extra space on information that we do not need. Let $c > 0$ be some small constant that we can choose later. The point location maintains the bucket table so that on average between c and $4c$ sites fall into each bucket. It does this on-line by quadrupling the size of the bucket grid and re-bucketing all points whenever the average goes above $4c$.

It is not hard to show that the expected cost of a Locate is $O(1)$ and the total cost of maintaining the buckets is $O(n)$ if the floor function is assumed to be constant time. Thus, the total expected run time of this algorithm is $O(n)$ time when the points are uniformly distributed in the unit square.

1.5. Incremental Search

Another class of incremental algorithms constructs the Delaunay triangulation by starting with a single Delaunay triangle and then incrementally discovering valid Delaunay triangles, one at a time. Each new triangle is grown from an edge of a previously discovered triangle by finding the site that joins with the endpoints of that edge to form a new triangle whose circumcircle is empty of sites. Several algorithms in the literature, including ones due to Dwyer [7], Maus [13], and Tanemura et. al. [18] are based on this basic idea. They all differ in the details, and only Dwyer's algorithm has been formally analyzed. Dwyer's analysis assumed that the input was uniformly distributed in the unit d-ball, and extending this analysis to the unit cube appears to be non-trivial. However, in the plane the difference is not great, and the experiments in the next section will show that the empirical runtime of the algorithm appears to be $O(n)$.

The data structure that is critical to the performance of this algorithm is the one that supports site searches. Dwyer's [7] uses a relatively sophisticated algorithm to implement site searching. First, the sites are placed in a bucket grid covering the unit circle. The search begins with the bucket containing the midpoint of (a, b). As each bucket is searched, its neighbors are placed into a priority queue that controls the order in which buckets are examined. Buckets are ordered in the queue according a measure of their distance from the initial bucket. Buckets are only placed in the queue if they intersect the half-plane to the right of (a, b) and if they intersect the unit disk from which the points are drawn. Dwyer's analysis shows that the total number of buckets that his algorithm will consider is $O(n)$.

The site search algorithm in our Algorithm IS is a simple variant on "spiral search" [4]. It differs from Dwyer's in several ways. The main difference is the lack of a priority queue to control the action of the search routine. Our spiral search approximates this order, but it is not exactly the same. Dwyer's algorithm is also careful not to inspect any buckets that are either to the left of (a, b) or outside of the unit circle. Our algorithm is somewhat sloppier about looking at extra buckets. The advantage of our scheme is that it is well tuned to the case when sites are distributed in the unit *square* and it avoids the extra overhead of managing a priority queue, especially avoiding duplicate insertions.

2. Empirical Results

In order to evaluate the effectiveness of the algorithms described above, we will study C implementations of each algorithm. Rex Dwyer provided code for his divide and conquer algorithm, and Steve Fortune provided code for his sweepline algorithm. We implemented the incremental search and construction algorithms. None of the implementations are tuned in any machine dependent way, and all were compiled using the GNU C compiler and timed using the standard UNIX^{tm} timers. The other performance data presented in this section was gathered by instrumenting the programs to count certain abstract costs. Each algorithm was tested for set sizes of between 1024 and 131072 sites. Ten trials with sites uniformly distributed in the unit square were run for each size, and the graphs either show the median of all the sample runs or a "box plot" summary of all ten samples at each size. In the box plots, a dot indicates the median value of the trials and vertical lines connect the 25th to the minimum and the 75th percentile to the maximum.

2.1. Performance of the Incremental Algorithms

The performance of the incremental algorithm is determined by the cost of point location and the number of circle tests the algorithm performs. While the standard incremental algorithm spends almost all of its time doing point location, the bucket-based point location routine effectively removes this bottleneck. Regression analysis on test runs indicates that the number of comparisons per point grows as $0.86\sqrt{n}$ for the standard incremental algorithm. By contrast the number of comparisons per site for our bucketing approach is bounded by a constant a bit less than 12. (About 10% of these comparisons are done in the spiral search routine and about 90% are done searching through triangles once a starting edge is found.) Other test runs show that this is 5-10% less than the number of comparisons needed by the quad-tree incremental approach.

For both the bucketing and the quad-tree methods the cost of point location fluctuates depending on whether $\log_2 n$ is even or odd. In the bucketing approach this is due to the fact that the algorithm re-buckets the sites at each power of four. Because of this, at each power of four, the average bucket density drops by a factor of 4. The quad-tree structure gains another level at each power of 4.

The number of comparisons needed for point location depends on the average density of the sites in the bucket grid. If the density of points in buckets is too high, then the point location routine will waste time examining useless edges. On the other hand, if it is too low, the algorithm will waste time examining empty buckets. An experiment over ten trials with $n = 8192$ and c ranging from 0.25 to 8 showed that $c = 2$ was the best choice, and we use this for the timing runs. Although the number of comparisons required in this test ranged from about 11.5 to 17, the actual effect on the runtime of the algorithm was less than 10%.

The runtime of the incremental algorithm depends mostly on how many circle tests it performs. Since the algorithm inserts the points in a random order, the analysis of Guibas, Knuth and Sharir [10] shows that the total number of circle tests is asymptotically $O(n)$. Sharir and Yaniv [17] tightened this bound to about $9n$. Figure 1 shows that this analysis is remarkably accurate. Surprisingly, the quad-tree insertion order actually performed about 5% more comparisons than the random insertion order.

Also shown on the plot is the cost of Dwyer's divide and conquer algorithm. Since this algorithm

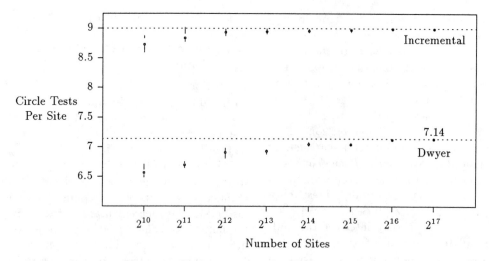

Figure 1: Circle tests per site for two algorithms.

is also based primarily on circle testing, it makes sense to compare them in this way. The plot shows that Dwyer's algorithm performs about 25% fewer circle tests than the incremental algorithm in this range. Profiling both programs shows that circle testing makes up roughly half the runtime of each, so Dwyer's algorithm should run roughly 10 to 15 percent faster than the incremental algorithm.

2.2. Performance of the Incremental Search Algorithm

The determining factor in the runtime of Algorithm IS is the cost of `Site-search`. This can be factored into the number of distance calculations performed, and the number of buckets examined. By *examined* we mean that a bucket is at least tested against an edge to see if the algorithm should search its contents for a new site.

We performed experiments for points uniformly distributed in the unit square and the unit circle. The number of distance calculations per site was similar in both cases, with just over 19 calculations for the unit square and about 21 for the unit circle. However, the number of boxes examined per site dropped from 23.3 for 2^{10} sites to 21.6 for 2^{17} sites for the unit square, while it increased from 25 for 2^{10} sites to 40 for 2^{17} sites for the unit circle.

The explanation for this behavior is that Algorithm IS is heavily influenced by the nature of the convex hull of the input. When dealing with an edge on or near the convex hull edge, the algorithm must search a lot of buckets before it can find a point lying outside of the edge or confirm that no such points exist. In the square distribution the expected number of convex hull edges is $O(\log n)$, while for points uniformly distributed in the unit circle the expected number of convex hull edges is $O(n^{1/3})$ [16]. This sensitivity to the distribution of the input is a major weakness of the algorithm, and indicates that a better approach is desirable for the `Site-search` routine.

Pre-computing the convex hull of the point set and searching from hull edges inward should help to minimize the effect of this problem, and we are working on a version that does this. But the bucket-based data structure will still be sensitive to clustering in the point set, and a non-uniform distribution of triangle sizes or shapes. The best way to fix these problems may be to replace the buckets with a nearest-neighbor search structure that is less sensitive to the distribution of sites, and can perform non-local searches more efficiently. Bentley's adaptive k-d tree [3] is a good example of such a data structure.

2.3. Performance of Fortune's Algorithm

The runtime of Fortune's algorithm is proportional to the cost of searching and updating the data structures representing the event queue and the state of the sweepline. Fortune's implementation uses hash tables for this purpose. We would expect that these data structures would perform well on uniform inputs. In fact, for small input sets, the algorithm seems to run in linear time.

For sites that are uniformly distributed in the x direction, our experiments show that the bucket structure representing the frontier performs exactly as we would expect. We find that the search procedure performs around 11.5 comparisons per point, on average, throughout our test range.

The main bottleneck in Fortune's algorithm ends up being the maintenance of the priority queue. The priority queue is represented using a uniform array of buckets in the y direction. Events are hashed according to their y-coordinate and placed in the appropriate bucket. We note that only circle events are explicitly placed in the event queue. The n site events are stored implicitly by initially sorting the sites.

The problem here is that while the sites are uniformly distributed, the resulting priorities are not. Circle events tend to cluster close to the current position of the sweepline. Animations of large runs of Fortune's algorithm clearly show that it has a tendency to create mostly small circles, and that large circles are soon removed and replaced by smaller ones. This clustering increases the cost of inserting or deleting events into Fortune's bucket structure. Each operation requires a linear time search through the list of events in a particular bucket. With heavily populated buckets, this becomes expensive. Regression analysis shows that the number of comparisons per point grows as $9.95 + .25\sqrt{n}$.

To see if an $O(\log n)$ priority queue would improve run times, we re-implemented Fortune's algorithm using an array-based heap to represent the priority queue. To test the effectiveness of the new implementation, we performed a more extensive experiment. Each algorithm was tested on uniform inputs with sizes ranging from 2^{10} to 2^{17} sites. Figure 2a shows the performance of the heap data structure in the experiment. The line $2\lg n + 11$ shows that the number of comparisons used to maintain the heap is growing logarithmically in n, rather than as $\sqrt{n}$.

In actual use, the heap does not significantly improve the runtime of the algorithm for the number of points that we are considering. Figure 2b compares the runtime of the two algorithms over the same range of inputs. In this graph, each data point is the ratio of the runtime of the bucketing algorithm to the runtime of the heap-base algorithm. The graph shows five trials for input sizes of between 2^{10} and 2^{17} sites at evenly spaced intervals.

The timings were taken using the machine and configuration described the next section. The plot shows that the bucketing algorithm has a slight edge up to 2^{16} points, when the heap version starts to dominate. At 2^{17} points, the heap version is roughly 10% better.

An interesting feature of the graph in Figure 2a is the fact that the number of comparisons periodically jumps to a new level, stays relatively constant, then jumps again. This is due to the fact that the heap is represented using an implicit binary tree. Thus, the number of comparisons jumps periodically when the size of the heap is near powers of two. On the graph, these jumps occur at powers of four, rather than two because the average size of the heap over one run of Fortune's algorithm is $O(\sqrt{n})$ rather than $O(n)$. We can prove this for uniformly distributed points using a lemma by Katajainen and Koppenin [12].

3. The Bottom Line

The point of all of this is to develop an algorithm that has the fastest overall runtime. In the following benchmarks, each algorithm was run on ten sets of sites generated at random from the uniform distribution in the unit square. Each trial used the same random number generator, and the same initial seed so all of the times are for identical point sets. Run times were measured on a Sparcstation 2 using the `getrusage()` mechanism in UNIX. The graphs show user time, not real time, and all of the inputs sets fit in main memory, and and were generated in main memory so I/O and paging would not affect the results. Finally, the graph for Fortune's algorithm shows the version using a heap rather than buckets, since this algorithm was better on the larger problems, and not much worse on smaller ones.

Figure 3 shows the results of the timing experiments. The graph shows that Dwyer's algorithm gives the best performance overall. The incremental algorithm is a bit faster than Fortune's algorithm, particularly for larger problems where the $O(\log n)$ growth of the heap data structure overhead is more of a factor. The quad-tree algorithm is a somewhat slower than the bucketing one. Algorithm IS is not competitive with any of the other four. These results are consistent with the analysis of primitives in the previous section.

4. Nonuniform Point Sets

Each of the algorithms that we have studied uses a uniform distribution of points to its advantage in a slightly different way. The incremental algorithm uses the fact that nearest neighbor search is fast on uniform point sets to speed up point location. Dwyer's algorithm uses the fact that only sites near the merge boundary tend to be effected by a merge step. Fortune's implementation uses bucketing to search the sweepline. Algorithm IS depends on a uniform bucket grid to support site searching.

In order to see how each algorithm adapts to its input, we will study further tests using inputs from very nonuniform distributions. In Table 1 the notation $N(s)$ refers to the normal distribution with mean 0 and standard deviation s, and $U(a,b)$ is the uniform distribution over the interval $[a,b]$.

The graphs show each algorithm running on five different inputs of 10K sites from each distribution. The uniform distribution serves as a benchmark. Figure 4 shows the effect of these point distributions on the incremental algorithms. As expected, point

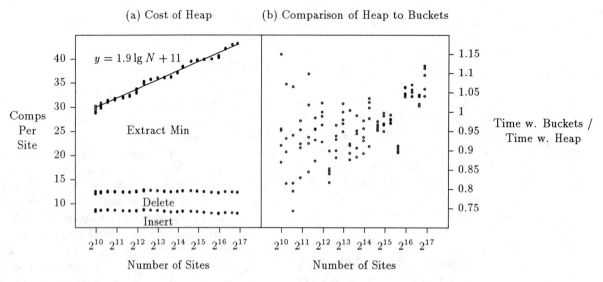

Figure 2: Cost of Fortune's algorithm using a heap. Part (a) shows a breakdown of the cost of using a heap to represent the event queue. Part (b) plots the ratio of the runtime of the old algorithm to the new.

Name	Description
unif	uniform in the unit square.
ball	uniform in a unit circle.
corners	$U(0, .01)$ at each corner of the unit square.
diam	$t = U(0,1)$, $x = t + U(0, .01) - .005$, $y = t + U(0, .01) - .005$
cross	$N/2$ points at $(U(0,1), .5 + U(-0.005, .005))$; $N/2$ at $(.5 + U(-0.005, .005), U(0,1))$.
norm	both dimensions chosen from $N(0, .01)$.
clus	$N(0, .01)$ at 10 points in the unit square.
arc	in a circular arc of width .01

Table 1: Nonuniform distributions.

distributions with heavy clustering, such as corners and normal, stress the point location data structures in each algorithm, increasing point location costs by up to a factor of ten. However, the comparison tests in the point location routine are substantially faster than circle tests, so even in the worst cases here, the whole runtime didn't increase by more than a factor of two. Surprisingly, the quad-edge algorithm degraded more than the bucketing one. The distribution of the input has little effect on the number of circle tests that each algorithm performs.

Figure 5 summarizes the performance of Fortune's algorithm in this experiment. The first graph shows that the bucket-based implementation of the event queue is very sensitive to site distributions that cause the distribution of priorities to become extremely nonuniform. In the cross distribution, this happens near the line $y = 0.5$. A substantial fraction of the $n/2$ points near the line will be on the frontier simultaneously, and the all of the circle events associated with them will cluster in the few buckets near this position. The corners distribution also causes a problem, but to a lesser degree. In both of these cases, the non-uniform distribution of sites in the x-direction also slows down site searches on the frontier, but this effect is less pronouced than the bad behavior of the event queue.

The second graph shows that the performance of the heap is much less erratic than the buckets. The small jumps that do appear are due to the fact that the event queue does become larger or smaller than its expected size on some distributions. However, since the cost of the heap is logarithmic in the size of the queue, this does not cause a large degradation in performance.

Figure 6 shows the performance of Dwyer's algorithm in the experiment. Dwyer's algorithm is slowed down by distributions that cause the algorithm to create many invalid edges in the subproblems, and then delete them later in the merge steps.

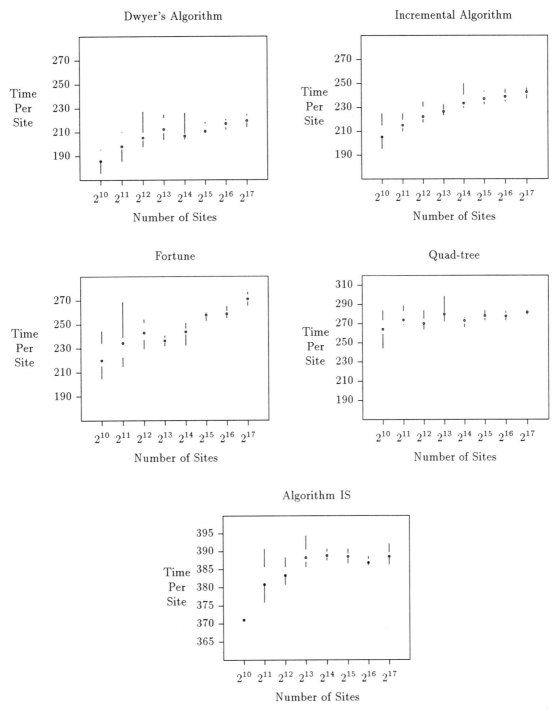

Figure 3: Comparison of the expected runtimes of different algorithms on sites chosen at random from a uniform distribution in the unit square. Times are in microseconds.

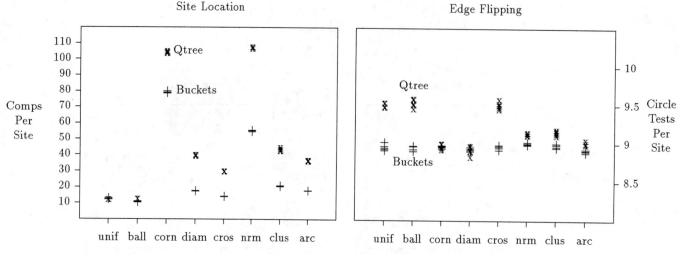

Figure 4: The incremental algorithm on non-uniform inputs, n is 10K.

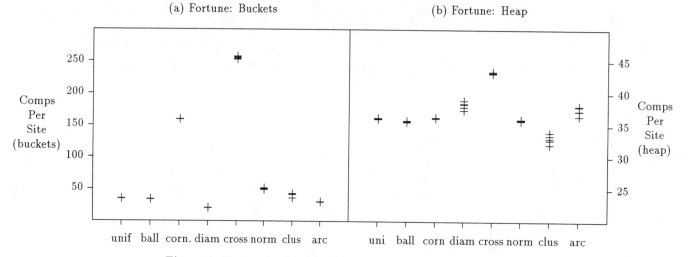

Figure 5: Fortune's algorithm on non-uniform inputs, n is 10K.

This effect is particularly pronounced in the run with the cross distribution because the group of sites near the line $x = 0.5$ is very tall and skinny, creating a worst case for the merge routine.

Figure 7 shows how the bad inputs affect Algorithm IS. These figures leave out the two worst inputs for this algorithm: corners and normal, because the algorithm would have taken several hours to finish the benchmark. Algorithm IS is easily the most sensitive to the distribution of its input. This is not surprising, since it depends on spiral search, which performed badly for the incremental routine on bad inputs. This did not handicap to the incremental algorithm to a large degree because the point location routine is not the major bottleneck in that algorithm. However, the performance of Site-search largely determines the runtime of Algorithm IS.

Finally, to understand how the abstract measures actually effect performance, Figure 8 shows the average runtime of the five trials with each algorithm except Algorithm IS. Since none of the runtimes in the graph are much greater than Algorithm IS's performance even in the uniform case, we eliminated it from this graph.

5. Notes and Discussion

The experiments in this paper led to several important observations about the performance of serial algorithms for constructing planar Delaunay triangulations. These observations are summarized below:

- Dwyer's algorithm is the strongest overall for this range of problem sizes. It is consistently faster than the incremental (and all other) algorithms, in part because it does fewer expensive in-circle tests. It is also the most resistant to bad data distributions, with an $O(n \log n)$ worst case. On the other hand, it not substantially

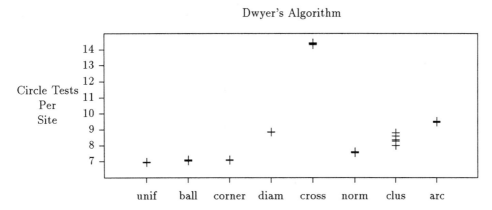

Figure 6: Dwyer's algorithm on non-uniform inputs, n is 10K.

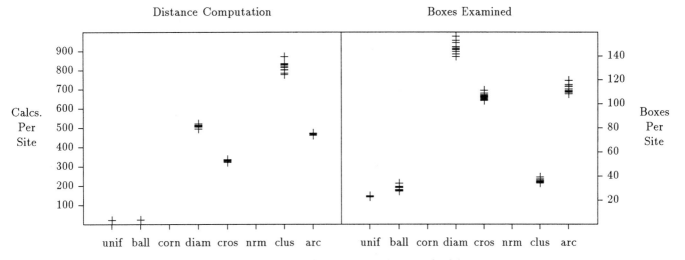

Figure 7: Algorithm IS is very sensitive to bad inputs.

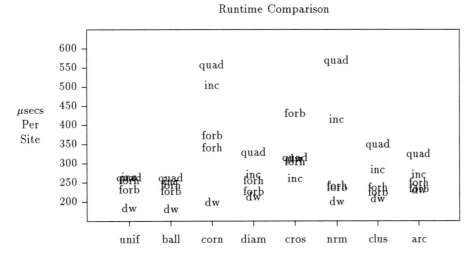

Figure 8: Runtimes on non-uniform inputs, n is 10K. inc and quad are the incremental algorithms, dw is Dwyer's algorithm, forb and forh is Fortune's algorithm with buckets and a heap respectively.

faster than the incremental algorithm and Fortune's algorithm. One of these algorithms could prove to be the fastest on some other system.

- A simple enhancement of the incremental algorithm results in an easy to implement on-line algorithm that is faster than some previous incremental variants and is competitive with other algorithms for all but very bad point set distributions.

- On uniformly distributed sites circle events in Fortune's algorithm cluster near the sweepline. This causes a hashing-based priority queue implementation to perform badly on large inputs. However, for up to 50,000 points this method is a bit faster than a heap-based priority queue.

5.1. Other Algorithms

While the experiments in this paper cover the most of the available algorithms for this problem, they were not quite comprehensive. We did not consider the divide and conquer algorithm discussed by Guibas and Stolfi because Dwyer's algorithm is always at least as fast. We also did not implement the incremental algorithm of Guibas, Knuth and Sharir [10], because it appeared that their more complex point location scheme would be slower than bucketing on all but very bad inputs. Finally, the convex hull algorithm developed by Barber [1] was not available soon enough for us to do a full analysis of it for this paper. However, the algorithm appears to be practical, efficient and robust and should be the subject of future studies.

References

[1] C. Brad Barber. *Computational geometry with imprecise data and arithmetic*. PhD thesis, Princeton, 1993.

[2] J. L. Bentley. Experiments on travel salesman heuristics. *ACM-SIAM Symposium on Discrete Algorithms*, pages 91–99, 1989.

[3] J. L. Bentley. K-d trees for semidynamic point sets. *Proc. 6th Annual ACM Symposium on Computational Geometry*, pages 187–197, 1990.

[4] J. L. Bentley, B. W. Weide, and A. C. Yao. Optimal expected time algorithms for closest point problems. *ACM Transactions on Mathematical Software*, 6(4):563–580, 1980.

[5] K. L. Clarkson and P. W. Shor. Applications of random sampling in computational geometry, ii. *Discrete and Computational Geometry*, 4:387–421, 1989.

[6] R. A. Dwyer. A faster divide-and-conquer algorithm for constructing delaunay triangulations. *Algorithmica*, 2:137–151, 1987.

[7] R. A. Dwyer. Higher-dimensional voronoi diagrams in linear expected time. *Discrete & Computational Geometry*, 6:343–367, 1991.

[8] S. Fortune. A sweepline algorithm for voronoi diagrams. *Algorithmica*, 2:153–174, 1987.

[9] P. Green and R. Sibson. Computing dirichlet tessellations in the plane. *Computing Journal*, 21:168–173, 1977.

[10] L. Guibas, D. Knuth, and M. Sharir. Randomized incremental construction of delaunay and voronoi diagrams. *Algorithmica*, 7:381–413, 1992.

[11] L. Guibas and J. Stolfi. Primitives for the manipulation of general subdivisions and the computation of voronoi diagrams. *ACM Transactions on Graphics*, 4(2):75–123, 1985.

[12] J. Katajainen and M. Koppinen. Constructing delaunay triangulations by merging buckets in quad-tree order. Unpublished manuscript, 1987.

[13] A. Maus. Delaunay triangulation and the convex hull of n points in expected linear time. *BIT*, 24:151–163, 1984.

[14] C. McGeoch. *Experimental Analysis of Algorithms*. PhD thesis, School of Computer Science, CMU, 1986.

[15] T. Ohya, M. Iri, and K. Murota. Improvements of the incremental method for the voronoi diagram with computational comparison of various algorithms. *Journal fo the Operations Research Society of Japan*, 27:306–337, 1984.

[16] L. A. Santaló. *Integral Geometry and Geometric Probability*. Addison-Wesley, Reading, MA, 1976.

[17] M. Sharir and E. Yaniv. Randomized incremental construction of delaunay diagrams: Theory and practice. *Annual ACM Symposium on Computational Geometry*, 1991.

[18] M. Tanemura, T. Ogawa, and N. Ogita. A new algorithm for three dimensional voronoi tessellation. *Journal of Computational Physics*, 51:191–207, 1983.

[19] B. Weide. *Statistical Methods for the Analysis of Algorithms*. PhD thesis, School of Computer Science, CMU, 1978.

Voronoi Diagrams and Containment of Families of Convex Sets on the Plane

M. Abellanas, G. Hernandez *Universidad Politécnica de Madrid, SPAIN*
R. Klein, *Fern Universität Hagen, GERMANY*
V. Neumann-Lara, *Universidad Nacional Autonoma de México, MEXICO*
J. Urrutia, *University of Ottawa, CANADA.*

Abstract

It is known that any family P_n of n points on the plane contains two elements such that any circle containing them contains $\frac{n}{4.7}$ elements of P_n. We prove: Let Φ be a family of n disjoint compact convex sets on the plane, S be a strictly convex compact set. Then there are two elements S_i, S_j of Φ such that any set S' homothetic to S that contains them contains $\frac{n}{c}$ elements of Φ, c a constant. Our proof method is based on a new type of Voronoi diagram, called the "closest covered set diagram" based on a convex distance function. We also prove that our result does not generalize to higher dimensions; we construct a set Ψ of n disjoint convex sets in $\Re^3$ such that for any subset H of Ψ there is a sphere containing all of the elements of H, and no other element of Ψ-H is contained in it. Finally, we show using closed covered Voronoi diagrams that the ordered set B_5 consisting of five bottom elements and ten top elements, one above each pair of bottoms, is not a circle order.

1. Introduction

It is known that any collection P_n of n points on the plane contains two elements u, v $\in P_n$ such that any circle containing them contains at least $\lceil \frac{n}{c} \rceil$ points of P_n. The first proved value for c was 60 [12], which was successively improved to 30 [2], then to $\frac{84}{5}$ [7] and at this point, the best known value for c is 4.7 (see [4]). Containment problems between families of points and circles originated from the study of circle orders, i.e. partial orders obtained from containment relations of families of circles on the plane; see [5, 13,14].

In this paper we prove the following generalization of the above mentioned result: Let S be a strictly compact convex set (i.e. it is closed and bounded and has no piece wise linear segments on its boundary) and $\Phi=\{S_1,...,S_n\}$ a family of n disjoint compact convex sets. Then there are two elements S_i, $S_j \in \Phi$ such that any set S' homothetic to S containing them contains $\lceil \frac{n-2}{30} \rceil$ elements of Φ. (S' is homothetic to S if S'=λS + v, where λ is a real value greater than 0 and v is a vector of $\Re^2$). Our proof relies heavily on a new type of *Voronoi diagram*, which we suggest calling the "closest covered set Voronoi diagram".

We study the case in which the sites form a disjoint family of closed convex sets in the plane, and where each point x of the plane belongs to the first site contained in a homothetically expanding compact and convex oval "centered" at x. The circle need not be Euclidean; we can take any strictly convex set S containing x in its interior, and consider growing homothetic copies of S. In other words, our diagram is based on the following distance function: With each site S_i and point x the distance, under a fixed convex distance function from x to the farthest point of S_i, is associated.

The original result for points and circles proved in [12] has been generalized to higher dimensions in [2] and in [1] to collections of points and ellipsoids in Euclidean spaces. Our result does not generalize to higher dimensions. We present a family Ψ of n disjoint convex sets in $\Re^3$ with the property that for *every* subset of H of Ψ there is a sphere that contains them and no other element of Ψ.

Using closest covered Voronoi diagrams, we show in Section 4 an easy proof that the bipartite partial order consisting of five bottom elements and ten top elements, one for each pair of bottom elements, is not a circle order. This is interesting since before this, it was known that for some n large enough, there is a bipartite order consisting of n bottoms and

$$\binom{n}{2}$$

top elements (one for each pair of bottoms) which is not a circle order. This proof was based on the use of Ramsey numbers [10,11].

2. Proof of Main Result

Our objective in this section is to prove:

Theorem 1: *Let $\Phi = \{S_1,...,S_n\}$ be a family of disjoint compact convex sets on the plane, and S any strictly convex compact set. Then there are two elements S_i, $S_j \in \Phi$ such that any convex set S' homothetic to S containing them contains at least $\lceil \frac{n-2}{30} \rceil$ elements of Φ.*

Our proof is based on the following theorem, whose proof is postponed to the next section:

Theorem 2: *Let H be any family of five compact convex sets, S a strictly convex compact convex set. Then there are two elements S_i, $S_j \in H$ such that any S' homothetic to S containing them contains another element of H.*

We now prove Theorem 1.

Proof of Theorem 1: Construct a bipartite graph G with $V(G) = X \cup Y$ where X consists of all subsets of pairs of elements of Φ and Y contains all of the subsets of Φ with exactly five elements. A vertex $T \in X$ is adjacent to a vertex $T' \in Y$ iff $T \subset T'$ and any homothetic copy of S containing the elements of T contains at least another element of T'. By Theorem 2, the degree of every element $T' \in Y$ is at least one. Then the sum of the degrees of all elements in X is at least

$$|Y| = \binom{n}{5}$$

and there is a vertex in X with degree at least

$$\frac{|Y|}{|X|} = \frac{\binom{n}{5}}{\binom{n}{2}} = \frac{(n-2)(n-3)(n-4)}{60}$$

In other words, there exists a pair T of Φ such that any set S' homothetic to S that contains them contains at least one other element of $\frac{(n-2)(n-3)(n-4)}{60}$ five-sets of Φ. Allowing for redundancies (each T, together with any other element, belongs to $\frac{(n-3)(n-4)}{2}$ different sets), we obtain that any S' homothetic to S containing both elements of T contains at least $\lceil \frac{n-2}{30} \rceil$ elements of Φ.

3. Proof of Theorem 2

In this section we prove Theorem 2. Some definitions and terminology will be needed before we can start our proof.

3.1 Closest Covered Set Voronoi Diagram

Suppose without loss of generality that S contains the origin $(0,0)$ in its interior. Let $\Psi = \{\lambda S = \{\lambda x : x \in S\} : \lambda \geq 0\}$. All elements $\lambda S \in \Psi$ are homothetic to S. Call $(0,0)$ the vortex of Ψ. In turn we call $(0,0)$ the vortex of $\lambda S \forall \lambda S \in \Psi$.

Let S' be homothetic to S. Then S' is a translation by a vector $t = (a,b)$ of some $\lambda S \in \Psi$, that is $S' = t + \lambda S$; $\lambda S \in \Psi$, $t = (a,b)$. The vortex of S' is now defined to be the image of $(0,0)$ under this translation, ie. vortex$(S') = t$.

Given any point p on the plane and a convex set Q, define the distance $d_S(p,Q)$ to be the smallest λ such that $S' = p + \lambda S$ and S' contains Q. Given two disjoint sets S_1 and S_2, we may now define the vortex bisector $b_S(S_1,S_2)$ to be the set of points p satisfying:

$$d_S(p,S_1) = d_S(p,S_2)$$

It is easy to see that under the restrictions imposed on S, the vortex bisector of S_1 and S_2 is well defined and is always a simple curve that partitions $\Re^2 - b_S(S_1,S_2)$ into two disjoint sets, one consisting of all points closer to S_1 than to S_2 and the other containing those closer to S_2 than to S_1.

Consider now a family $\Phi = \{S_1,...,S_n\}$ of disjoint compact convex sets. We may associate to each member of $S_i \in \Phi$ a Voronoi region $\text{Vor}_S(S_i)$ consisting of all points p on the plane such that $d_S(p,S_i) \leq d_S(p,S_j)$ for every $S_j \in \Phi$, $j \neq i$.

The set of regions $\text{Vor}_S(S_i)$ thus obtained will be called the *closest covered set Voronoi* diagram of Φ with respect to S and will be denoted by $\text{Vor}(S,\Phi)$.

It is easy to verify that if the sites of Φ are points, then we obtain precisely the well-known Voronoi diagram under a convex function; see [3] and [8]. There are, however, some important differences between $\text{Vor}(S,\Phi)$ and regular Voronoi diagrams. It might happen, for example, that there are elements of Φ such that $\text{Vor}_S(S_i) = \emptyset$. For example if $\Phi = \{S_1 = \{(-1,0)\}, S_2 = \{(x,y): x=0, -2 \leq y \leq 2\}, S_3 = \{(1,0)\}\}$ and C is a disk with center at the origin, then $\text{Vor}_C(S_2) = \emptyset$. See Figure 1.

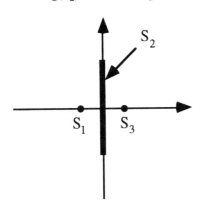

Figure 1

Given two convex sets, P, Q such that $P \subset Q$, we say that P splits Q if Q - P is not connected; see Figure 2.

We proceed now to characterize those sets $S_i \in \Phi$ for which $\text{Vor}_S(S_i) = \emptyset$. The following observation will be useful:

Observation 1: Let S' and S'' be two different strictly convex and homothetic sets. Then their boundaries intersect in at most two points.

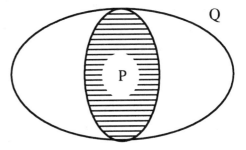

Figure 2

Lemma 1: $\text{Vor}_S(S_i) = \emptyset$ *iff there is S' homothetic to S such that S_i splits S' and there are two different components A, B of $S' - S_i$ and two elements S_j, S_k of Φ such that $S_j \subset A$, $S_k \subset B$.* See Figure 3.

Proof: Suppose that $\text{Vor}_S(S_i) \neq \emptyset$. Among all S' homothetic to S containing S_i, choose the one with smallest area; call it S''. It is easy to see that under the conditions assumed on S, that S_i splits S''. Since $\text{Vor}_S(S_i) = \emptyset$ the vortex v'' of S'' belongs to $\text{Vor}_S(S_j)$ for some $j \neq i$. This implies

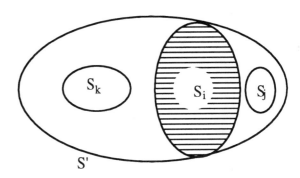

Figure 3

that S_j is contained in S'', and since S_i splits S'' it belongs to one component H of $S''-S_i$. Assume that no other element of Φ is contained in H (this restriction can be taken care of easily). Using tools of mathematical analysis, it is easy to show that there is a set S' homothetic to S that satisfies the following conditions:

a) S' contains S_i and S_j.

b) S_i splits S'.

c) S_j intersects the boundary of S'.

It is now easy to see using continuity arguments that if all elements S_k with $k \neq i,j$ are not contained in S' then we can find a homothetic copy Q of S that contains only S_i. That is, the vortex of Q belongs to $\text{Vor}_S(S_i)$, which is a contradiction.

Conversely, suppose that there is a set S' satisfying the conditions of our lemma. Then it is easy to see using Observation 1 that any homothetic copy S" of S containing S_i contains either S_j or S_k in its *interior*. In either case, the vortex of S" belongs to $\text{Vor}(S_i)$ or $\text{Vor}(S_j)$ respectively, and thus $\text{Vor}(S_i) = \emptyset$.

We shall also need the following result:

Lemma 2: *If* $\text{Vor}_S(S_i) \neq \emptyset$, *then it is also connected.*

The proof of this lemma is interesting on its own. To avoid breaking the flow of our paper we will postpone the proof of this lemma until the end of this section.

Next we prove:

Lemma 3: *Let* Φ *be a family with at least four convex sets, and let* $S_i \in \Phi$ *such that* $\text{Vor}_S(S_i) = \emptyset$. *Then there is* $S_j \in \Phi$, $i \neq j$ *such that any* S' *homothetic to* S *containing* S_i *and* S_j *contains other elements of* Φ.

Proof: By Lemma 1, since $\text{Vor}_S(S_i) = \emptyset$, there are two elements of Φ, say S_1 and S_2 different from S_i, and a homothetic copy S' of S such that S_i splits S' and S_1 and S_2 are in two different components, A and B, of $S' \setminus S_i$ respectively.

Since Φ has at least four elements, there is $S_j \in \Phi$ different from S_1, S_2 and S_i. We now prove that any S" homothetic to S containing S_i and S_j contains S_1 or S_2.

Call β_A the section of the boundary of S' that also belongs to the boundary of A. Suppose then that there is a homothetic copy of S" that contains S_i and S_j and does not contain S_1. Then S" intersects β_A in at least two points (see Figure 4). It now follows that B, and hence S_2, is totally contained in S". Similarly we can prove that if S_2 is not contained in S" then S_1 is, and our lemma is proved.

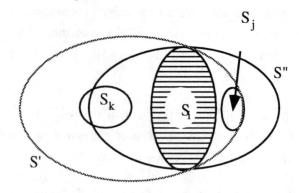

Figure 4

We are now ready to prove Theorem 2:

Proof of Theorem 2: Consider a family of five disjoint compact convex sets $\Phi = \{S_1, ..., S_5\}$ and a strictly convex compact set S. Consider the Voronoi diagram $\text{Vor}(S, \Phi)$. By Lemma 3 we can assume that $\text{Vor}(S_i) \neq \emptyset$ $i = 1, ..., 5$. Construct a graph G in which for every region of $\text{Vor}(S, \Phi)$ there is a vertex in G. Two vertices of G are adjacent if their corresponding Voronoi regions are adjacent. Then by planarity arguments there are two different values i and j such that $\text{Vor}(S_i)$ and $\text{Vor}(S_j)$ are not adjacent, otherwise a planar embedding of the complete graph on five vertices would be obtained, but this is impossible. It is now immediate that any S' homothetic to S containing both S_i and S_j contains at least another element S_k of Φ, that is the set S_k containing the vortex of S'.

3.2 Dimension 3

We now present an example of a family Ψ of n disjoint convex sets in $\Re^3$ for which for every subset Ψ_H of Ψ there is a sphere that contains all of the elements of Ψ_H and does not contain any other element of Ψ. First we construct a set $Q = \{P_1, ..., P_n\}$ of n polygons on the plane as follows: Consider a circle C on the x-y plane of $\Re^3$ together with a set of $2^n - 1$ disjoint intervals on it labeled C_H, where H is a non empty subset of $I_n = \{1,...,n\}$. For every H let C^-_H and C^+_H be the initial and final points of C_H in the counterclockwise direction. For every non empty subset $H=\{i(1),...,i(k)\}$ of I_n take n equidistant points in the interior of C_H and label the first k of them in the counterclockwise direction with the integers $i(1),...,i(k)$ and the remaining n-k points with the integers in $I_n - \{i(1),...,i(k)\}$. Let α_H be any point in C_H to the right of $i(k)$ appearing before any of the points on C_H with a label in $I_n - \{i(1),...,i(k)\}$. For every i let P_i be the convex closure of all of the $2^n - 1$ points in C labeled i.

It is easy to see that any circle that intersects C at the rightmost point C^-_H of C_H and α_H contains all of the polygons P_i, $i \in H$ and does not contain any of the remaining polygons. For every i, let P'_i be a translation of P_i in the direction of the z-axis by a small ε_i, $i=1,...,n$; $\varepsilon_i \neq \varepsilon_j$, $i \neq j$. Let $\Psi = \{P'_1,...,P'_n\}$. For every subset $H=\{i(1),...,i(k)\}$ of I_n let $\Psi_H = \{P'_{i(1)},...,P'_{i(k)}\}$. It is now easy to see that for any non empty subset Ψ_H of Ψ there is a sphere in R^3 that contains only the elements in Ψ_H and contains no other element of Ψ.

3.3 Proof of Lemma 2

We proceed now to prove that if a set S_i is such that $Vor_S(S_i) \neq \emptyset$ then $Vor_S(S_i)$ is connected. We will actually prove that if the interior $Int(Vor_S(S_i))$ of $Vor_S(S_i)$ is non-empty, then it is connected. This suffices to prove our result.

We tackle first the case when S_i is a *convex polygon*. To this end, let us assume that $Q_i \in \Phi$ is a convex polygon. We may assume that for this case, the boundary of $S(a')$ contains two vertices of Q_i. Otherwise, by first shrinking $S(a')$ while keeping its vortex fixed, we can assure that the boundary of $S(a')$ touches the boundary of Q_i in at least one vertex, say p. Next, consider the line segment L joining a' to p. For every point x of L let $S(x,p)$ be the homothetic copy of S with vortex x and containing p on its boundary. Clearly all $S(x,p)$ are contained in $S(a')$. If x is sufficiently close to p, $S(x,p)$ does not contain Q_i, and since $S(a')$ contains Q_i there is a point y in L such that $S(y,p)$ contains Q_i and intersects the boundary of Q_i at p and another vertex of Q_i different from p. Similarly, we can assume that $S(a'')$ intersects the boundary of Q_i in at least two vertices.

Let p and q denote the vertices of Q_i on the boundary of $S(a')$. Assume without loss of generality that the line L through p and q is horizontal. Note that one or both of p and q might be the intersection points of $\partial(S(a')) \cap \partial(S(a''))$.

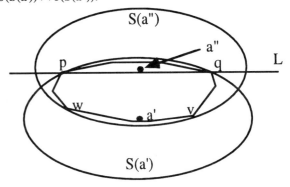

Figure 5

By assumption, a' belongs to the bisector $b_S(p, q)$ of p and q which is y-monotone since p and q have the same y-coordinate. This follows from a result in [8]. By definition, for each a' in $b_S(p, q)$ there is a unique homothetic copy $S(a')$ of S with vortex a' whose boundary contains p and q. Let $S^+(a')$ and $S^-(a')$ be the parts of $S(a')$ above (resp. below) the horizontal line L (see Figure 5).

Lemma 4: *As a moves along $b_S(p, q)$ in the upward direction, $S^+(a)$ is strictly growing while $S^-(a)$ is strictly shrinking.*

Proof: Let a and b be on $b_S(p, q)$. Suppose without loss of generality that $S^+(b)$ contains $S^+(a)$. We have to show that in this case, a lies below b. Let c be the intersection point of the common 'outer' tangents of $S(a)$ and $S(b)$. (See Figure 6.)

Clearly, c is the center of a homotheticity between $S(a)$ and $S(b)$. First, assume that c lies above L. From the point of view of c, the tangents intersect first S(b) and then S(a). Consequently, the ray from c through a and b hits b first. This implies that b has a higher coordinate than a. A similar argument applies if c is below L.

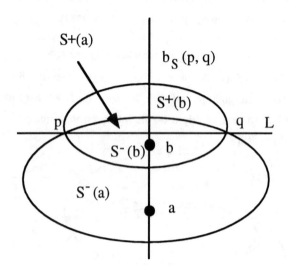

Figure 6

Now suppose that S(a') and S(a") are such that $S^+(a")$ contains $S^+(a')$ (see Figure 6). In order to transform S(a') into S(a") starting at a' we move a point a upwards along $b_S(p, q)$, and study how the set S(a) with vortex at a and whose boundary passes through p and q changes. By Lemma 4, $S^+(a)$ grows while $S^-(a)$ shrinks.

Lemma 5: *If S(a) contains Q_i, then the only set of Φ contained in S(a) is Q_i.*

Proof: Suppose that some $S_j \neq Q_i$ is also contained in S(a). Since Q_i and S_j are disjoint, S_j must be contained either in $S^-(a)$ which is contained in S(a') or $S^+(a)$ which is contained in S(a"), contradicting our assumptions.

We can now prove:

Lemma 6: *Let Q_i be a convex polygon such that $Int(Vor_S(Q_i)) \neq \emptyset$. Then $Int(Vor_S(Q_i))$ is connected.*

Proof: Let a' and a" be points in the interior of $Vor_S(Q_i)$ with S(a') and S(a") as defined before. We may assume that each of the boundaries of S(a') and S(a") contains two vertices of Q_i. Suppose that S(a') intersects Q_i at p and q and S(a") intersects Q_i at v and w. We have to prove that there is a path from a' to a" entirely contained in $Int(Vor_S(Q_i))$. This means that for each point a on this path, a is the vortex of a homothetic copy S(a) of S such that S(a) contains S_i and no other set $S_j \neq Q_i$. By assumption, S(a') and S(a") satisfy this condition.

There are two possible events that can prevent us from moving a upwards along $b_S(p, q)$. First, it could happen that the shrinking boundary of S(a) hits a third vertex, say r of Q_i (see Figure 7). Note that it cannot happen that r is a vertex on the boundary section of Q_i from v to w and does not contain p and q; this would cause S(a) to intersect S(a") in at least four points, which is impossible, since S(a') and S(a") are homothetic. At this point, we "update" a' to be a.

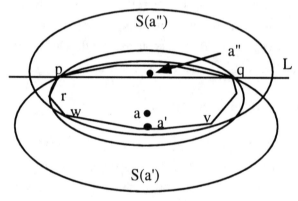

Figure 7

Now p and q are not the extreme points of Q_i contained on the part of $\partial(S(a'))$ contained in S(a"), these are now r and q. Now we continue moving a along the bisector $b_S(r,q)$ in the same way as before and update a' as we move along. Eventually, the boundary of $S^+(a)$ will hit the boundary of $S^+(a")$. At this point, S(a)=S(a") and our result follows.

We are now ready to prove Lemma 2.

Proof of Lemma 2. We will prove that $int(Vor(S_i))$ is connected. By Lemma 6 we can assume that S_i is not a convex polygon. Let a' and a" be points in the interior of Vor(S_i). This implies that there are sets S(a') and S(a") with vortices a' and a" such that S_1 is contained in the interior of S(a') and S(a"). Let Q_i be a convex polygon containing S_i such that Q_i is contained in S(a') $\cap$ S(a"). Since S_i is a subset of Q_i then Vor(Q_i) is contained in Vor(S_i). By Lemma 6, there is a path from a' to a" totally contained in int(Vor(Q_i)) which in turn is contained in Vor(S_i). Our result now follows.

4. Applications to Circle Orders

A partial order P(X, <) on a set X={$x_1,..., x_n$} is called a circle order if there is a family C={$C_1,...,C_n$} of circles such that $x_i < x_j$ in P(X,<) iff $C_i \subset C_j$, see [5,13,14].

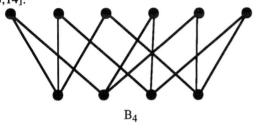

B_4

Figure 8

It can be shown using recent results that the bipartite order B_n consisting of n bottom elements {$b_1,...,b_n$} and $\binom{n}{2}$ top elements, {$t_{i,j}: 1 \le i < j \le n$} such that $t_{i,j}$ is greater that b_i and b_j and all b_i's and $t_{i,j}$'s are incomparable, is not a circle order, for n large enough [10,11].

In this section we use closest covered Voronoi diagrams to prove the following result:

Theorem 3: B_5 is not a circle order.

Proof: To prove this result, it is sufficient to show that given any family H={$C_1,...,C_5$} of five circles (not necessarily disjoint) representing $b_1,...,b_5$ there are two of them such that any circle that contains them contains a third element of H. First notice that we can assume that no circle C_i of H is contained in another element C_j of H; this would imply that $b_i < b_j$.

Given a point p of the plane and a circle C with center s and radius r let us define the distance d(p, C) of p to C to be the radius of the smallest circle centered at p that contains C.

Consider the Voronoi diagram induced on the plane by the elements of H under the metric defined above. It is easy to show in this case that the Voronoi partition is well defined and that the Voronoi cells are connected.

Our result now follows in the same manner as Theorem 2.

5. Conclusions and Open Problems

We have proved that for any Φ of disjoint convex sets, and S convex, there are two elements $S_i, S_j \in \Phi$ such that any homothetic S' containing S_i and S_j contains $\lceil \frac{n-2}{30} \rceil$ elements of Φ. We believe that the bound stated in Theorem 1 is far from optimal. At this point we are unable to give a good estimate for an optimal solution to our problem, but we believe that a bound of about $\frac{n}{4}$ or $\frac{n}{5}$ is achievable.

For points and circles, it has been proved that if the elements of P are vertices of a convex polygon, then there are two points of P such that any circle containing them contains $\lceil \frac{n-3}{3} \rceil$ points of P, and this is optimal [6]. This result does not extend, however, to the natural generalization to our problem. A family Φ of sets is called convexly independent if no element S_i of Φ is contained in the convex closure of $\Phi - S_i$. We have an example consisting of 4n elements such that for any pair of elements of Φ there is a circle containing them that contains at most $\frac{n}{4}$ elements of Φ (see Figure 9).

In this more restricted case we venture the next conjecture:

Conjecture: Let Φ be a family of n convexly independent sets, S a convex set. Then there are two elements of Φ such that any S' homothetic to S containing them contains at least $\frac{n}{4} \pm c$ elements of Φ, c a constant.

Finally, we mention that the construction of the new type of Voronoi diagrams introduced here can be carried out in kO(n log n) time, using techniques developed in [9], where k is the time it takes to calculate the vortex bisector of two elements of Φ.

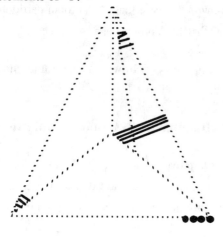

Figure 9

References

[1] I. Bárány and D.G. Larman, A combinatorial property of points and ellipsoids, *Discrete Comp. Geometry* **5** (1990) 375-382.

[2] I. Bárány. J. H. Schmerl, S. J. Sidney and J. Urrutia, A combinatorial result about points and balls in Euclidean space, *Discrete Comp. Geometry* **4** (1989) 259-262.

[3] P. Chew and S. Drysdale, Voronoi diagrams based on convex distance functions, In *Proc. 1st ACM Symposium on Computational Geometry*, pp. 235-244, Baltimore, 1985.

[4] H.Edelsbrunner, N. Hasan, R. Seidel and X. J. Shen, Circles through two points that always enclose many points, T.R. UIUCDCS-R-88-1400 Univ. of Illinois at Urbana-Champaign (1988).

[5] P. C. Fishburn, Interval orders and circle orders, *Order* **5** (1988) 225-234.

[6] R. Hyward, D. Rappaport and R. Wenger, Some extremal results on circles containing points, *Disc. Comp. Geom.* **4** (1989) 253-258.

[7] R. Hyward, A note on the circle containment problem, *Disc. Comp. Geom.* **4** (1989) 263-264.

[8] Ch. Ickling, R. Klein, M. Le, and L. Ma, Convex distance functions in 3-space are different, *Proc. 9th ACM Symposium on Computational Geometry*, pp. 116-123, 1993.

[9] R. Klein, K. Melhorn, and S. Meiser, Randomized incremental construction of abstract Voronoi diagrams, *Computational Geometry, Theory and Applications*, **3** (1993), pp.157-184.

[10] C. Lin, The crossing number of posets, *ORDER* **11** (1994) 1-25.

[11] C. Lin, An application of Ramsey Theory to the crossing number of posets, Preprint (1994).

[12] V. Neumann-Lara and J. Urrutia, A combinatorial result on points and circles in the plane, *Discrete Math.* **69** (1988) 173-178.

[13] J.B. Sydney, S.J. Sydney and J. Urrutia, Circle Orders, n-gon orders and the crossing number for partial orders, *ORDER* **5** (1988) 1-10.

[14] J. Urrutia, Partial orders and Euclidean geometry, in *Algorithms and Order*, ed. I. Rival 387-434 (1989) Kluwer Academic Publishers.

Voronoi Diagrams in Higher Dimensions under Certain Polyhedral Distance Functions

Jean-Daniel Boissonnat[*]   Micha Sharir[†]   Boaz Tagansky[‡]   Mariette Yvinec[§]

Abstract

The paper bounds the combinatorial complexity of the Voronoi diagram of a set of points under certain polyhedral distance functions. Specifically, if $\mathcal{S}$ is a set of n points in general position in $\mathbb{R}^d$, the complexity of its Voronoi diagram under the L_∞ metric, and also under a simplicial distance function, are both shown to be $\Theta(n^{\lceil d/2 \rceil})$. The upper bound for the case of the L_∞ metric follows from a new upper bound, also proved in this paper, on the complexity of the union of n axis-parallel hypercubes in $\mathbb{R}^d$. This complexity is $\Theta(n^{\lceil d/2 \rceil})$, for $d \geq 1$, and it improves to $\Theta(n^{\lfloor d/2 \rfloor})$, for $d \geq 2$, if all the hypercubes have the same size. Under the L_1 metric, the complexity of the Voronoi diagram of a set of n points in general position in $\mathbb{R}^3$ is shown to be $\Theta(n^2)$. We also show that the general position assumption is essential, and give examples where the complexity of the diagram increases significantly when the points are in degenerate configurations.

1 Introduction

Voronoi diagrams are among the most fundamental constructs in computational geometry, and, as such, have been studied a lot during the past two decades. Most of these studies, however, concentrated on Voronoi diagrams in the plane, with only few studies of diagrams in higher dimensions.

We assume in this paper familiarity of the reader with the standard definition and properties of Voronoi diagrams. They can be found in basic textbooks on computational geometry [4, 9, 10, 11], and in several survey papers [1, 8]. There are many variants of Voronoi diagrams. The three main parameters that can vary are (i) the type of sites defining the diagram (points, lines, etc.), (ii) the metric defining the distance to a site, and (iii) the dimension d. The 'classical' case is when the sites are points and the metric is euclidean. In this case, a standard lifting transform into $\mathbb{E}^{d+1}$ implies that the maximum combinatorial complexity of the diagram is $\Theta(n^{\lceil d/2 \rceil})$ (see [4]). However, for other metrics, or for other kinds of sites, this analysis does not apply. In this paper we will only consider Voronoi diagrams for point sites, so the only relevant parameters for us are the metric and the dimension.

As observed in [5], the Voronoi diagram of a set $\mathcal{S}$ of n sites in $\mathbb{R}^d$ can be interpreted as the lower envelope of a set of n d-variate functions, each measuring the distance from an arbitrary point of $\mathbb{R}^d$ to a site of $\mathcal{S}$. Under reasonable assumptions concerning the shape of the sites and the metric, these functions are (piecewise) algebraic of some fixed degree. Hence, applying the recent results of [12] concerning the complexity of the lower envelope of such a collection of functions, we immediately conclude that the complexity of the Voronoi diagram is $O(n^{d+\varepsilon})$, for any $\varepsilon > 0$, where the constant of proportionality depends on ε, d, and the maximum degree of the relevant functions. Since this is a much weaker bound than the one known for the euclidean case, it is tempting to conjecture that the actual complexity of the diagram is smaller, perhaps close to $O(n^{\lceil d/2 \rceil})$ for fairly general types of sites and metrics. This conjecture has been confirmed at least for $d = 2$, where linear bounds on the complexity of the diagram are known in fairly general settings.

Surprisingly, very little is known about generalized Voronoi diagrams in higher dimensions. Recently, Chew

---

[*] INRIA, 2004 Route des Lucioles, B.P.109, 06561 Valbonne cedex, France; E-mail: Jean-Daniel.Boissonnat@sophia.inria.fr.

[†] School of Mathematical Sciences, Tel Aviv University, Tel Aviv 69978, Israel and Courant Institute of Mathematical Sciences, New York University, New York NY 10012, USA; E-mail: sharir@math.tau.ac.il

[‡] School of Mathematical Sciences, Tel Aviv University, Tel Aviv 69978, Israel; E-mail: mia@math.tau.ac.il

[§] INRIA at Sophia Antipolis, and CNRS, URA 1376, Laboratoire I3S, Rue Albert Einstein, 06560 Valbonne, France; E-mail: Mariette.Yvinec@sophia.inria.fr.

Permission to copy without fee all or part of this material is granted provided that the copies are not made or distributed for direct commercial advantage, the ACM copyright notice and the title of the publication and its date appear, and notice is given that copying is by permission of the Association of Computing Machinery.To copy otherwise, or to republish, requires a fee and/or specific permission.
11th Computational Geometry, Vancouver, B.C. Canada
© 1995 ACM 0-89791-724-3/95/0006...$3.50

et al. [2] have shown that the complexity of the Voronoi diagram of a set of n lines in $\mathbb{R}^3$ under a convex polyhedral distance function (see below for a precise definition), induced by a polytope with a constant number of faces, is $O(n^2\alpha(n)\log n)$. Thus the conjecture holds in this case. The simpler case, of point sites under similar distance functions has not yet been investigated, and this paper initiates the study of such diagrams.

For certain technical reasons, the case of point sites is harder to analyze than the case of lines in 3-space. We have not been able to come up with a sharp bound for arbitrary polyhedral distance functions, even in $\mathbb{R}^3$. Nevertheless, we managed to substantiate the conjecture in the following special cases:

- We show that the maximum complexity of the Voronoi diagram of n points in $\mathbb{R}^3$ under the L_1 metric is $\Theta(n^2)$.

- We show that the maximum complexity of the Voronoi diagram of n points in $\mathbb{R}^d$ under the L_∞ metric is $\Theta(n^{\lceil d/2 \rceil})$.

- We show that the maximum complexity of the Voronoi diagram of n points in $\mathbb{R}^d$ under a simplicial distance function is also $\Theta(n^{\lceil d/2 \rceil})$.

In these bounds we assume that the given sites are in *general position* with respect to the relevant distance function (see below for a precise definition). It is interesting to note that this requirement is essential for the bounds to hold. We give examples of point sets in degenerate configurations for which the complexity of their L_1-Voronoi diagrams is much larger.

To obtain the bound concerning L_∞-Voronoi diagrams, we first derive a related new bound on the complexity of the union of n axis-parallel hypercubes in $\mathbb{R}^d$. We show that if the hypercubes have arbitrary sizes then the maximum complexity of their union is $\Theta(n^{\lceil d/2 \rceil})$, for $d \geq 1$. If all the hypercubes have the same size then the maximum complexity of their union is $\Theta(n^{\lfloor d/2 \rfloor})$, for $d \geq 2$. These results were known, and are easy to derive, for $d = 1, 2$. An alternative proof of a linear bound for equal-size cubes in $\mathbb{R}^3$ has been around for the past couple of years, but was not published.

The proofs of these bounds borrow ideas from the preceding paper [2]. The main ingredient of most of the proofs is a new technique for obtaining recurrence relationships for the number of vertices of the union. This technique is obtained by modifying and simplifying the proof technique developed in [6, 12] for the analysis of lower envelopes of multivariate functions. This improved technique has already been used in [2] and in [13].

2 Preliminaries

Let P be a convex polytope in $\mathbb{R}^d$ with a reference point o in its interior. A homothetic copy of P, having the form $p + \rho P$ for $p \in \mathbb{R}^d$ and $\rho \in \mathbb{R}$, is called a *placement* of P. The placement $p + \rho P$ is said to be centered at p and scaled by factor ρ. We define the distance induced by P from a point p to a point q as

$$d_P(p, q) = \min\{\rho : q \in p + \rho P\}.$$

We refer to d_P as a *(convex) polyhedral distance function* (induced by P). Note that $d_P(p, q)$ is not symmetric, and thus is not a metric, unless P is centrally symmetric with respect to o.

Let $\mathcal{S}$ be a set of n points in $\mathbb{R}^d$ and P be a convex polytope with m facets. The Voronoi diagram $\text{Vor}_P(\mathcal{S})$ of $\mathcal{S}$ for the distance d_P is defined as the decomposition of $\mathbb{R}^d$ into Voronoi cells, one cell $V(s_i)$ for each point $s_i \in \mathcal{S}$, given by

$$V(s_i) = \{p \in \mathbb{R}^d \mid d_P(p, s_i) \leq d_P(p, s_j) \ \forall s_j \neq s_i \in \mathcal{S}.\}$$

Each cell $V(s_i)$ is a star-shaped, generally nonconvex, d-polyhedron. More generally, for $1 \leq k \leq d + 1$, consider the locus of points p such that p is equidistant (under d_P) to the points of a subset $\mathcal{S}_k$ of cardinality k of $\mathcal{S}$, and such that p is strictly closer to the points of $\mathcal{S}_k$ than to any other point in $\mathcal{S} \setminus \mathcal{S}_k$. This locus is a $(d-k+1)$-dimensional piecewise linear surface, and each of its faces (of any dimension) is a face of the Voronoi diagram $\text{Vor}_P(\mathcal{S})$. (For this locus to have this dimension, the points of $\mathcal{S}$ must lie in *general position* with respect to P—see below for a precise definition and section 7 for further discussion.) The complexity of the Voronoi diagram $\text{Vor}_P(\mathcal{S})$ is defined as the total number of its faces of all dimensions. If we assume general position, then each face of $\text{Vor}_P(\mathcal{S})$ must have at least one vertex, and each vertex is incident to only a constant number of faces of any dimension. It follows that the complexity of the diagram is proportional to the number of its vertices, so we will concentrate in the foregoing analysis on bounding the number of vertices of the diagram.

We denote placements $p + \rho P$ of P by $\hat{P} = \hat{P}(p, \rho)$. A placement $\hat{P}$ is said to be *free* if it contains no points of $\mathcal{S}$ in its interior. If f is a face of P, $\hat{f}$ refers to the corresponding face of $\hat{P}$. If a point $p \in \mathcal{S}$ belongs to a facet $\hat{f}$ of $\hat{P}$, the pair (p, f) is said to be a *contact pair* of the placement $\hat{P}$. A point p is said to be a *simple contact point* of $\hat{P}$ if it belongs to the relative interior of facet $\hat{f}$. A point p of $\mathcal{S}$ which belongs to the relative interior of a face of $\hat{P}$ of codimension k, is said to be a *contact point with multiplicity k*. Thus, a contact point with multiplicity k is involved in at least k contact pairs (and exactly k contact pairs if the polytope P is simple).

The set $\mathbb{P}$ of all placements of a polytope P is a $(d+1)$-dimensional manifold. The set of placements such that a given point p belongs to the hyperplane which is the affine hull of a facet $\hat{f}$ is a hyperplane in $\mathbb{P}$ and the set of placements $\hat{P}$ such that p belongs to a specific facet $\hat{f}$ is a d-polytope. In the following, we shall assume that the set $\mathcal{S}$ is in *general position* with respect to the distance d_P, which means that the set of mn hyperplanes of $\mathbb{P}$, each representing contact of some point of $\mathcal{S}$ with the hyperplane in $\mathbb{R}^d$ spanned by some facet of P, is itself in general position, in the (weak) sense that the intersection of any k of these hyperplanes, if nonempty, is a $(d+1-k)$-flat, for $k = 1, \ldots, d+1$, and that the intersection of any $d+2$ of them is empty. This, in particular, implies that no two points of $\mathcal{S}$ belong to a common hyperplane parallel to the affine hull of a facet of P. This in particular implies that the multiplicities of the contact points of a placement sum up at most to $(d+1)$.

A placement whose contact point multiplicities sum up to $d+1$ is called a *rigid* placement. The free rigid placements of P are centered at the vertices of the Voronoi diagram $\text{Vor}_P(\mathcal{S})$, and each vertex is the center of such a placement, as follows easily from the definitions. The free rigid placements of P with $d+1$ distinct contact points are centered at what we call the *regular* vertices of the diagram. The center of such a placement is a point of $\mathbb{R}^d$ which is equidistant (under d_P) to $d+1$ points of $\mathcal{S}$ and closer to these points than to any other points of $\mathcal{S}$. Any other vertex of the diagram is called *singular*; it corresponds to a free rigid placement of P at which some points of $\mathcal{S}$ lie on lower-dimensional faces of $\hat{P}$. More generally, points in a k-face of the Voronoi diagram are centers of free placements whose contact points multiplicities sum up to $d+1-k$. The k-face is *regular* if all points in these contacts are distinct, and singular otherwise. The general position assumption implies that each (regular or singular) Voronoi vertex is incident to $d+1$ Voronoi edges, and, more generally, that each k-face, for $0 \leq k \leq d$, of $\text{Vor}_P(\mathcal{S})$ is incident to $d+1-k$ $(k+1)$-Voronoi faces. Thus the number of faces of the Voronoi diagram incident to each vertex is bounded by a constant depending on d. Hence, as already mentioned above, bounding the complexity of the Voronoi diagram reduces to bounding the number of Voronoi vertices and thus the number of free rigid placements.

3 Complexity of the Union of Axis-Parallel Hypercubes in Higher Dimensions

In this section we obtain a result that will be needed in our analysis of L_∞-Voronoi diagrams, but which is interesting in its own right.

Let $\mathcal{C}$ be a set of n axis-parallel hypercubes in $\mathbb{R}^d$. Let $\mathcal{A}(\mathcal{C})$ denote the arrangement of these hypercubes, and let $\mathcal{U}(\mathcal{C})$ denote their union. We want to bound the combinatorial complexity of $\mathcal{U}(\mathcal{C})$, which we measure by the number of vertices of the union (the number of all other faces of the union is clearly proportional to the number of vertices, where the constant of proportionality depends only on d). In this section we do not enforce the general position assumption. The main result of this section is:

Theorem 3.1 *The maximum number of vertices of the union of n axis-parallel hypercubes in $\mathbb{R}^d$ is $\Theta(n^{\lceil d/2 \rceil})$, for $d \geq 1$. If all the given hypercubes have the same size, then the maximum number of vertices of their union is $\Theta(n^{\lfloor d/2 \rfloor})$, for $d \geq 2$ (it remains $O(n)$ for $d = 1$). The constants of proportionality depend on d.*

Proof: Due to lack of space, we only prove the upper bounds, by induction on d. The bounds hold for $d = 1, 2$. This is trivial for $d = 1$ and follows for $d = 2$ e.g. from the results of [7]. Fix $d \geq 3$, assume that the theorem holds for all $d' \leq d-2$, and let $\mathcal{C}$ be a collection of n axis-parallel hypercubes in $\mathbb{R}^d$, as above.

For each hypercube $c \in \mathcal{C}$, define $x_j^+(c)$, $x_j^-(c)$ to be, respectively, the largest and smallest x_j-coordinate of the points in c, for $j = 1, \ldots, d$. Any hypercube $c \in \mathcal{C}$ has two facets normal to the x_j-axis, for each $j = 1, \ldots, d$, lying on the two respective hyperplanes $x_j = x_j^+(c)$, $x_j = x_j^-(c)$. The facet at $x_j^+(c)$ is said to be *positive* and the facet at $x_j^-(c)$ is said to be *negative*.

We use the following notational system for representing vertices of the arrangement of the given hypercubes. For a given ordered d-tuple, $(c_1, c_2, \ldots, c_d)$, of hypercubes in $\mathcal{C}$, let c_j^* be one of the symbols c_j, $\bar{c}_j$, for $j = 1, \ldots, d$. The tuple $(c_1^*, c_2^*, \ldots, c_d^*)$ represents the intersection point p of the facets $f_1, \ldots, f_d$, where f_j is a facet of c_j normal to the x_j-axis; it is the positive facet if $c_j^* = c_j$ and the negative facet if $c_j^* = \bar{c}_j$. Whenever we use this notation, we assume implicitly that the intersection point p exists (and is then unique).[1] The intersection point p is said to have *positive* representation if all the intersecting facets in such a representation are positive.

Such an intersection point (or, rather, a vertex of $\mathcal{A}(\mathcal{C})$) is said to be *outer* if it is contained in a $(d-2)$-face of some hypercube, and *inner* otherwise. If $(c_1^*, \ldots, c_d^*)$ is an inner vertex then the hypercubes $c_1, \ldots, c_d$ are distinct.

A vertex of $\mathcal{A}(\mathcal{C})$ is said to be a *k-level vertex* if it is contained in the interiors of exactly k of the hypercubes in $\mathcal{C}$. The vertices of the (boundary of the) union are 0-level vertices. Let $V_k(\mathcal{C})$ denote the number of inner k-level vertices of $\mathcal{A}(\mathcal{C})$, and let $D_k(\mathcal{C})$ denote the number of outer k-level vertices. We also denote by $V_k(n, d)$ the maximum of $V_k(\mathcal{C})$ over all possible collections of n axis-parallel hypercubes in $\mathbb{R}^d$, and, similarly, denote

---
[1] If the hypercubes are not in general position then p may have more than one representation.

by $D_k(n, d)$ the maximum of $D_k(\mathcal{C})$ over all possible such collections of hypercubes.

We first estimate the number of outer vertices of the union $\mathcal{U}(\mathcal{C})$. Such an outer vertex p belongs to at least one $(d-2)$-face of some hypercube $c \in \mathcal{C}$. Since every hypercube contains only $2d(d-1)$ such $(d-2)$-faces, we can reduce the problem to $2nd(d-1)$ 'smaller' problems, as follows. Fix a $(d-2)$-face f of some hypercube $c \in \mathcal{C}$, and let K be the affine hull of f. Form the intersections $K \cap c'$, for $c' \in \mathcal{C} - \{c\}$. These are $n-1$ axis-parallel hypercubes in the $(d-2)$-dimensional space K (and if the hypercubes of $\mathcal{C}$ are of equal size, so are these intersection hypercubes). Any outer vertex of $\mathcal{U}(\mathcal{C})$ that lies on f is clearly an (inner or outer) vertex of the union of these intersection hypercubes. It follows that

$$D_0(n, d) \le 2nd(d-1)\left(D_0^*(n-1, d-2) + V_0^*(n-1, d-2)\right)$$

where the functions D^* and V^* count, respectively, only outer and inner vertices of the union which lie inside some fixed $(d-2)$-dimensional hypercube. By the induction hypothesis, we have

$$D_0^*(n-1, d-2) + V_0^*(n-1, d-2) = O\left(n^{\lceil (d-2)/2 \rceil}\right).$$

If the hypercubes are of equal size, then we have

$$D_0^*(n-1, d-2) + V_0^*(n-1, d-2) = O\left(n^{\lfloor (d-2)/2 \rfloor}\right).$$

Indeed, this holds for $d = 3$, because the complexity of the union of equal intervals on a line, intersected with another interval of the same length, is $O(1)$. For $d > 3$, the bound follows by the induction hypothesis. Hence we obtain $D_0(n, d) = O(n^{\lceil d/2 \rceil})$, for hypercubes of arbitrary sizes, and $D_0(n, d) = O(n^{\lfloor d/2 \rfloor})$, for equal-size hypercubes.

In what follows we will also need a bound on $D_1(n, d)$. This is easy to obtain by a standard application of the Clarkson-Shor probabilistic technique [3] (using a random sample of, say, $n/2$ of the hypercubes). This yields, as is easily verified, $D_1(n, d) = O(n^{\lceil d/2 \rceil})$, for hypercubes of arbitrary sizes, and $D_1(n, d) = O(n^{\lfloor d/2 \rfloor})$, for equal-size hypercubes.

We next estimate the number of inner vertices of the union. Let p be a 0-level inner vertex, and assume, without loss of generality, that p has the positive representation $(c_1, \ldots, c_d)$. For each coordinate x_j, we will slide from p along an edge e_j in the negative x_j direction. This edge is contained in the intersection of the corresponding $d-1$ positive facets of the hypercubes c_k, for $k = 1, \ldots, d$ and $k \ne j$. As we start tracing e_j from p in the negative x_j-direction, we enter the hypercube c_j. We stop the sliding process as soon as we first encounter one of the following three types of events:

(i) We meet the negative facet of c_j at the 0-level vertex $(c_1, \ldots, c_{j-1}, \overline{c_j}, c_{j+1}, \ldots, c_d)$. This can happen only if c_j is smaller than the other $d-1$ hypercubes. For equal-size hypercubes, this cannot happen.

(ii) We meet another facet of one of the hypercubes c_k, for $k = 1, \ldots, d$ and $k \ne j$, at the 1-level outer vertex $(c_1, \ldots, c_{j-1}, c_k, c_{j+1}, \ldots, c_d)$, contained in the interior of c_j.

(iii) We meet a new hypercube c' at a 1-level inner vertex p', contained in the interior of c_j and having a positive representation of the form $(c_1, \ldots, c_{j-1}, c', c_{j+1}, \ldots, c_d)$. We say that p' and p are *neighbors* (in the arrangement $\mathcal{A}(\mathcal{C})$).

If we encounter an event of type (i), we simply ignore this edge, and do not use it in our charging scheme. As just noted, at most one such edge will be ignored.

If we encounter an event of type (ii), we charge the 1-level outer vertex by one unit. Since we can reach the outer vertex $(c_1, \ldots, c_{j-1}, c_k, c_{j+1}, \ldots, c_d)$ from an inner vertex only along one of the two corresponding facets of c_k (in a direction normal to the other facet), this outer vertex can be charged, by type (ii) events, at most twice, for a total of 2 units.

If we encounter an event of type (iii), we charge the 1-level inner vertex p' by one unit. The problem is that the vertex p' may be charged by up to d events of type (iii), and we need to account for such multiple charges. Suppose that p' is charged by w of its 0-level inner neighbors. If $w = 1$ (or $w = 0$) then p' pays one unit of charge for its unique charging neighbor (or does not pay at all). If $w > 1$, we will distribute $w - 1$ of the w units that p' is charged with to other outer vertices, so that p' still has to pay only one unit of charge.

Suppose that p' has the positive representation $(c_1, \ldots, c_d)$, and is contained in the interior of c_0. Suppose that $p_1 = (c_0, c_2, c_3, \ldots, c_d)$ and $p_2 = (c_1, c_0, c_3, \ldots, c_d)$ are two 0-level inner neighbors of p'. Let h be the 2-dimensional plane $x_i = x_i^+(c_i)$, for $i = 3, \ldots, d$, which contains the three vertices p', p_1, p_2. Let r be the axis-parallel rectangle in h having these points as three of its vertices (see Figure 1). For each hypercube $c \in \mathcal{C}$, let $s(c) = c \cap h$. The collection $\mathcal{S}$ of the nonempty intersections of this form is a set of at most n axis-parallel squares in h. By construction, the two edges $p_1 p', p_2 p'$ of r do not cross the boundary of any square in $\mathcal{S}$. Let q be the fourth corner of r. Clearly, q is an outer vertex of $\mathcal{A}(\mathcal{C})$ with the representation $q = (c_0, c_0, c_3, \ldots, c_d)$.

If r does not intersect the interior of any hypercube other than c_0, then q is a 0-level outer vertex of $\mathcal{A}(\mathcal{C})$, to which we pass 1 unit of charge from p'. The vertex q can be charged in this manner at most once. Indeed, given q, there is only one 2-dimensional plane in which q can be charged: this is the plane passing through q and spanned by the normal directions of the unique pair of facets in the representation of q that belong to the same hypercube (recall that $c_0, c_3, \ldots, c_d$ are all distinct, by construction). Moreover, r is the unique maximal

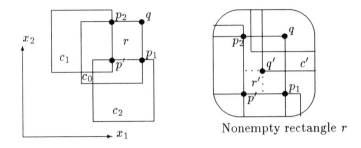

Figure 1: Charging outer vertices within the rectangle $p_1 p_2$

rectangle in $c_0 \cap h$ with corner q which is disjoint from (the interior of) any other square in $\mathcal{S}$. This implies that q can be charged at most once, namely, only by the opposite corner of r.

If the rectangle r meets some other square of $\mathcal{S}$, let q' be the point in $r \cap \mathcal{U}(\mathcal{C}')$ closest to p', where $\mathcal{C}' = \mathcal{C} - \{c_0, c_1, \ldots, c_d\}$. Note that q' cannot lie on the edges $p_1 p'$ or $p_2 p'$, since these edges do not intersect any hypercube in $\mathcal{C}'$, and that q' must be a 1-level outer vertex of $\mathcal{A}(\mathcal{C})$ having the representation $(\overline{c'}, \overline{c'}, c_3, \ldots, c_d)$, for some $c' \in \mathcal{C}'$; see Figure 1. Let r' be the axis-parallel rectangle in h having p' and q' as opposite corners. Again, (the interior of) r' is contained only in the interior of c_0, and meets no other hypercube of $\mathcal{C}$. We pass 1 unit charge from p' to q'. We claim that, in this case too, q' can be charged in this manner at most once. Indeed, given q', there is only one 2-dimensional plane h in which q' can be charged, which is shown by the same argument given above (since $c', c_3, \ldots, c_d$ are all distinct). Moreover, r' is the unique maximal rectangle in $c_0 \cap h$ with corner q', lying in the quadrant opposite to that occupied by the cube c', which is disjoint from (the interior of) any other square in $\mathcal{S}$. This implies, as above, that q' can be charged at most once, namely, only by the opposite corner of r'. Together with the previous charges in the case of type (ii) events, any 1-level outer vertex of $\mathcal{A}(\mathcal{C})$ can be charged a total of 3 units.

If the vertex p' has $w > 1$ 0-level inner neighbors, the number of pairs of these neighbors is always at least $w - 1$, so there is no problem in distributing $w - 1$ units of charge from p' to nearby outer vertices, in the manner described above.

Summing up the charges, each 0-level inner vertex p receives at least $d - 1$ units, by sliding in all directions parallel to the coordinate axes, with the possible exception of one direction in which we encounter a type (i) event (for equal-size hypercubes, p always receives d units). Each 0-level outer vertex pays at most 1 unit, each 1-level outer vertex pays at most 3 units, and each 1-level inner vertex pays at most 1 unit. We can thus conclude that

$$(d-1)V_0(\mathcal{C}) \leq V_1(\mathcal{C}) + 3D_1(\mathcal{C}) + D_0(\mathcal{C}). \quad (1)$$

We can now apply the following probabilistic argument, similar to that used in [2] and [13]. We have

$$\frac{n-1}{n} V_0(\mathcal{C}) = \frac{n-d}{n} V_0(\mathcal{C}) + \frac{d-1}{n} V_0(\mathcal{C})$$
$$\leq \frac{n-d}{n} V_0(\mathcal{C}) + \frac{1}{n} V_1(\mathcal{C}) + \frac{3}{n} D_1(\mathcal{C}) + \frac{1}{n} D_0(\mathcal{C})$$
$$= \mathbf{E}(V_0(\mathcal{R})) + O(n^\gamma),$$

where $\mathcal{R}$ is a random sample of $n - 1$ hypercubes of $\mathcal{C}$, where $\mathbf{E}$ denotes expectation with respect to the choice of $\mathcal{R}$, and where $\gamma = \lceil d/2 \rceil - 1$ (as follows from the bounds for D_0 and D_1 given above). This holds for the case of hypercubes of arbitrary sizes. For the case of equal-size hypercubes, we obtain, similarly, the improved recurrence

$$V_0(\mathcal{C}) \leq \mathbf{E}(V_0(\mathcal{R})) + O(n^{\gamma'}),$$

where $\gamma' = \lfloor d/2 \rfloor - 1$.

We can thus write, for the case of hypercubes of arbitrary sizes, the recurrence

$$\frac{n-1}{n} V_0(n, d) \leq V_0(n-1, d) + O(n^\gamma),$$

whose solution, for $d \geq 3$, is easily seen to be

$$V_0(n, d) = O(n^{\lceil d/2 \rceil}).$$

For the case of equal-size hypercubes, we obtain the recurrence

$$V_0(n, d) \leq V_0(n-1, d) + O(n^{\gamma'}),$$

whose solution, for $d \geq 2$, is easily seen to be

$$V_0(n, d) = O(n^{\lfloor d/2 \rfloor}).$$

This completes the proof of the upper bounds.

4 The L_∞ Voronoi Diagram of Points in Higher Dimensions

In this section, we address the complexity of the L_∞-Voronoi diagram of a set of n points in $\mathbb{R}^d$. The L_∞ distance function is the distance function associated with an axis parallel hypercube of $\mathbb{R}^d$ when the reference point is the center point of the hypercube. We show the following result:

Theorem 4.1 *The worst-case complexity of the L_∞-Voronoi diagram of a set of n points in $\mathbb{R}^d$ is $\Theta(n^{\lceil d/2 \rceil})$, provided that the set is in general position with respect to the L_∞ distance function.*

Proof: Again, we only prove the upper bound. Let $\mathcal{S}$ be a set of n points in general position in $\mathbb{R}^d$, with respect to an axis-parallel hypercube C. We denote by $\text{Vor}_\infty(\mathcal{S})$ the Voronoi diagram of $\mathcal{S}$ under the L_∞-distance. By the discussion in Section 2, a vertex of $\text{Vor}_\infty(\mathcal{S})$ corresponds to a free rigid placement of C with exactly $d+1$ contact pairs. The vertex is regular if all the contact points are distinct, and singular otherwise. By the general position assumption, no facet of C can contain more than one point of $\mathcal{S}$, and the $d+1$ contact pairs involve $d+1$ facets of C. Since C has d pair of parallel facets, a free rigid placement has at least two parallel contact pairs, that are contact pairs involving parallel facets. Moreover, by the general position assumption, there can be only one pair of parallel contact pairs. Indeed, a pair of parallel contact pairs fixes the scale factor of the placement, and thus two pairs of parallel contact pairs would imply a linear dependency between the four linear relations corresponding to these contact pairs. It follows that at this placement there is a vertex v of $\hat{C}$ incident to d facets of $\hat{C}$, each containing a point of $\mathcal{S}$. We can represent $\hat{C}$ as

$$\hat{C} = \{\mathbf{x} \mid x_j^-(\hat{C}) \le x_j \le x_j^+(\hat{C}), \text{ for } j = 1, \ldots, d\},$$

where $x_j^+(\hat{C}) - x_j^-(\hat{C}) = 2\rho(\hat{C})$ for all j, where ρ is the scaling factor of $\hat{C}$. With no loss of generality, assume that v is incident to the facets $x_j = x_j^-(\hat{C})$, for $j = 1, \ldots, d$, and that the facet $x_j = x_j^-(\hat{C})$ touches a point $p_j \in \mathcal{S}$, for $j = 1, \ldots, d$ (so v is the vertex of $\hat{C}$ all of whose coordinates are the smallest possible). As remarked above, these points do not have to be distinct.

We now shrink $\hat{C}$ towards v, keeping v fixed. We lose one contact of $\hat{C}$ with a point, but retain the d remaining contact pairs (between the points $p_1, \ldots, p_d$ and the corresponding facets of $\hat{C}$ incident to v). We stop the shrinking when one of these points comes to lie on another facet of $\hat{C}$. With no loss of generality, assume that this is the point p_1, and that the new facet it lies on is $x_2 = x_2^+(\hat{C})$ (clearly, the new facet must be a positive facet). The new placement that we have reached is free and rigid but singular. Let v' be the vertex of $\hat{C}$ incident to the facets $x_1 = x_1^-(\hat{C})$, $x_2 = x_2^+(\hat{C})$, and $x_j = x_j^-(\hat{C})$, for all $j = 3, \ldots, d$. These facets are incident to the points $p_1, p_3, \ldots, p_d$. We now shrink $\hat{C}$ towards v', losing the contact between p_2 and the facet $x_2 = x_2^-(\hat{C})$, but retaining the other d contact pairs, and stop when one of the contacting points comes to lie on another facet of $\hat{C}$.

We keep iterating this process. In the general step, just before starting a shrinking process, we have some number, k, of remaining points, call them $q_1, \ldots, q_k$, such that each q_i lies on some face f_i of codimension t_i, where $\sum_{i=1}^k t_i = d+1$. By the general position assumption, there is exactly one parallel pair of facets among the $d+1$ facets of $\hat{C}$ that are incident to the faces f_i.

Suppose first that in the present placement of $\hat{C}$ there is a face f_i of codimension 1 (that is, f_i is a facet), and that f_i is one of the pair of parallel facets. Then the remaining $k-1$ faces f_j, for $j \ne i$, have a common vertex w, and we can keep shrinking $\hat{C}$ towards w, losing only the one contact pair involving q_i and maintaining the other d contact pairs. We stop the shrinking, as above, when one of the other $k-1$ points comes to lie on another facet of $\hat{C}$.

Suppose next that the preceding subcase does not occur. Let q_1 and q_2 be the two (distinct) points incident to the (unique) pair of parallel facets. By assumption, the corresponding faces f_1, f_2 have each codimension at least 2. Hence there are at least 3 coordinates, say x_1, x_2, x_3, such that q_1 is incident to a pair of facets orthogonal to the x_1 and x_3 axes, and q_2 is incident to a pair of facets orthogonal to the x_2 and x_3 axes. Let us fix the points q_1 and q_2 (there are $O(n^2)$ choices for such a pair), and also fix the $\beta \ge 3$ coordinates such that the facets incident to q_1 and q_2 are orthogonal to these coordinates (there is a constant number of such choices). Then the scaling factor $\rho_0 = \rho(\hat{C})$ of $\hat{C}$ is fixed (under the above assumptions, it is equal to $\frac{1}{2}|x_3(q_1) - x_3(q_2)|$), and β coordinates of its center are also fixed. Hence the center of $\hat{C}$ must lie on an appropriate $(d-\beta)$-th dimensional flat K. For each point $p \in \mathcal{S} \setminus \{q_1, q_2\}$, let $\hat{C}_p$ be the intersection of K with the cube $p + \rho_0 C$. It is easily checked that the center of $\hat{C}$ in placements under consideration must be a vertex of the union of the equal-size axis-parallel hypercubes $\hat{C}_p$. By Theorem 3.1, the number of such vertices is $O(n^{\lfloor \frac{d-\beta}{2} \rfloor})$, so the number of placements under consideration is

$$O(n^2) \cdot O(n^{\lfloor \frac{d-\beta}{2} \rfloor}) = O(n^{\lceil d/2 \rceil}),$$

since $\beta \ge 3$.

To recap, the number of terminal placements of $\hat{C}$ that we can reach by our iterated shrinking process is $O(n^{\lceil d/2 \rceil})$. This also includes the case where the iterated shrinking process can continue all the way through, until the hypercube shrinks to a point; the number of such terminal placements is clearly only $O(n)$.

We claim that any such terminal placement $\hat{C}$ can be reached from only a constant number of initial placements of C. To see this, suppose first that the shrinking process has not terminated at a singleton hypercube. Pick a terminal placement $\hat{C}$, and reverse the shrinking process: choose a vertex v of $\hat{C}$ incident to all but one of the $d+1$ facets touched by points of $\mathcal{S}$. By construction, there is always at least one such vertex at the end of a shrinking step, and the discarded facet is necessarily one of the pair of parallel facets. When we expand $\hat{C}$ from v, none of the points touching $\hat{C}$ can enter into the interior of $\hat{C}$ (the point touching the discarded facet also touches another facet incident to v, so it remains on the boundary of $\hat{C}$ while we expand). We stop the expanding process when $\hat{C}$ hits another point, and then continue to expand from some (possibly different) vertex of $\hat{C}$. There are at most d expanding steps, and in each of them we have a constant number of choices for the vertex from which we expand, implying that only a constant number of initial placements (where the constant depends on d) can reach the same terminal placement $\hat{C}$. A similar (and actually simpler) argument also applies to the case where the terminal placement is a singleton hypercube.

So far we have only counted vertices of the diagram, but the arguments in Section 2 imply that the overall complexity of the diagram is proportional to the number of its vertices, which completes the proof of the upper bound in Theorem 4.1. □

5 Voronoi Diagrams for Simplicial Distance Functions

In this section, we consider the Voronoi diagram $\text{Vor}_T(\mathcal{S})$ of a point set $\mathcal{S}$ in $\mathbb{R}^d$ for a distance function d_T induced by a d-simplex T, and prove the following:

Theorem 5.1 *The worst-case complexity of the Voronoi diagram of a set of n points in $\mathbb{R}^d$, under the distance function induced by a d-simplex, is $\Theta(n^{\lceil d/2 \rceil})$, provided that the points are in general position with respect to the simplex.*

Proof: For lack of space, we omit the entire proof in this version. □

6 Voronoi Diagrams under the L_1 Metric in Three Dimensions

This section studies the complexity of Voronoi diagrams of point sets under the L_1 norm. The L_1-distance between two points p and q of $\mathbb{R}^d$ is

$$d_{L_1}(p,q) = \sum_{i=1}^{d} |p_i - q_i|.$$

This distance function is a metric, and is polyhedral, induced by the d-polytope which is the dual of the d-cube. This polytope is the convex hull of the $2d$ unit vectors $\pm\mathbf{e}_i$, where $\mathbf{e}_i$ is the unit vector in the positive x_i-direction, for $i=1,\ldots,d$. We will call this polytope a d-co-cube. In the case of the L_1-metric, we have only been able to prove the conjectured bound for $d=3$:

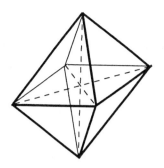

Figure 2: The regular octahedron

Theorem 6.1 *If $\mathcal{S}$ is a set of n points in $\mathbb{R}^3$ in general position with respect to the L_1-metric, then the worst-case complexity of the Voronoi diagram, $\text{Vor}_{L_1}(\mathcal{S})$, of $\mathcal{S}$ under the L_1-metric, is $\Theta(n^2)$.*

Proof: Again, we only prove the upper bound. In the three-dimensional case, the co-cube is just the regular octahedron O which is the convex hull of the six vertices $u_1(+1,0,0)$, $\overline{u_1}(-1,0,0)$, $u_2(0,1,0)$, $\overline{u_2}(0,-1,0)$, $u_3(0,0,1)$ and $\overline{u_3}(0,0,-1)$. The octahedron has twelve edges and eight faces and is shown in figure 6.2.

Let $\mathcal{S}$ be a set of n points in $\mathbb{R}^3$ in general position with respect to the octahedron O. Our goal is to bound the number of free rigid placements of O among the points of $\mathcal{S}$. For this, we bound, in succession, (i) the number of free rigid placements with at least one contact point of multiplicity at least two (which we call hereafter a *double contact point*); (ii) the number of free rigid placements with three contact pairs involving three faces of O sharing a common vertex; and (iii) the number of all other free rigid placements.

Let $\hat{O}$ be a placement with a double contact point. That is, there is a point $p \in \mathcal{S}$ that lies on an edge $\hat{e}$ of $\hat{O}$; we will denote this double contact by the pair (p,e). For each pair (p,e) of a point p of $\mathcal{S}$ and an edge e of O, the subset of placements attaining the double contact pair (p,e) is contained in a two-dimensional linear subspace $\mathbb{P}(p,e)$ of the set $\mathbb{P}$ of placements. The subspace $\mathbb{P}(p,e)$ can be parametrized by the position t of one of the endpoints of $\hat{e}$ on the line parallel to e through p, and by the scaling factor ρ. In this subspace, any other contact pair (p',f) at a placement $\hat{O}$ appears as a (possibly empty) segment $s(p',f)$. Then a rigid placement with the double contact pair (p,e) corresponds to a vertex of the planar arrangement of those segments, and a free placement with the double contact pair (p,e) corresponds to a point which lies on or below the lower envelope of those segments, relative to the ρ-direction. This follows from the observation that if we fix t and increase ρ then $\hat{O}$ expands, so that, once a point enters the expanding octahedron, it will never leave it again (see also [2] for a similar argument). Hence, the number of free rigid placements with the double contact pair (p, e) is exactly the number of vertices of the lower envelope of the at most $8n$ segments representing the contact pairs (p', f) in $\mathbb{P}(p, e)$. (Note that the vertices formed by the intersection of two segments represent placements with three contact points (one of which is p), whereas segment endpoints represent placements with two contact points (one of which is p), each being a double-contact point.) Moreover, it is easy to verify that the segments that represent contacts with a fixed face f of O are all parallel, thus we have eight families of at most n parallel segments each, so the complexity of their overall lower envelope is linear. Summing over the $12n$ possible pairs (p, e), we conclude:

Lemma 6.2 *Given a set S of n points in $\mathbb{R}^3$ in general position with respect to the regular octahedron O, the number of free rigid placements of O amidst the points of S with a double contact point is $O(n^2)$.*

Let us next bound the number of free rigid placements $\hat{O}$ with contact pairs involving three facets sharing a vertex. If such a placement has no double contact point, we apply a shrinking process to $\hat{O}$, in which the vertex of $\hat{O}$ incident to three contact faces is fixed. This process maintains at least three contact pairs and does not encounter any new contact point, since at any time during the shrinking the octahedron is contained in the initial placement. The shrinking process stops as soon as one of the contact points reaches an edge of the octahedron. Then the reached placement is a free rigid placement with a double contact point. Moreover, each free placement with a double contact point can be reached from at most two rigid free placements without double contact points. Indeed, from a terminal placement with a double contact point on edge e we can recover an initial free rigid placement without a double contact point by expanding the octahedron from one of the two vertices opposite to edge e in the two facets incident to e. Thus the number of placements without double contact point but with contact pairs involving at least 3 facets sharing a vertex is no more than twice the number of free rigid placements with a double contact point, which proves the following lemma.

Lemma 6.3 *Given a point set S as above, the number of free rigid placements of O with contact pairs involving three facets sharing a common vertex is $O(n^2)$.*

Finally, we bound the number of all other rigid free placements. Let us consider a rigid placement $\hat{O}$ with no double contact point and with no vertex common to three facets involved in contact pairs. Only two cases are then possible:

(a) The 4 contact facets f_1, f_2, f'_1, f'_2 of O form two pairs $(f_1, f_2), (f'_1, f'_2)$ of adjacent facets (i.e., with a common edge) and two complementary pairs $(f_1, f'_1), (f_2, f'_2)$ of parallel facets.

(b) The 4 contact facets have no pair of adjacent facets. This case can be realized only by one of the two following complementary subsets of four facets of O: the first set is $\{u_1 u_2 u_3, \overline{u_1 u_2} u_3, \overline{u_1} u_2 \overline{u_3}, u_1 \overline{u_2 u_3}\}$, and the second set is $\{\overline{u_1 u_2 u_3}, u_1 \overline{u_2} u_3, u_1 u_2 \overline{u_3}, \overline{u_1} u_2 \overline{u_3}\}$; see figure 2.

The first case does not occur for sets of points in general position with respect to O. Indeed, it is easy to see that four contact pairs involving two pairs of parallel facets yield a system of four linearly-dependent hyperplanes in the 4-dimensional placement manifold $\mathbb{P}$. Thus it remains to bound the number of free rigid placements of type (b). For this we apply the following scheme. Let $\hat{O}$ be a rigid free placement of type (b) with the four contact pairs $(p_1, f_1), (p_2, f_2), (p_3, f_3)$ and (p_4, f_4). We choose three of these four contact pairs, say $(p_1, f_1), (p_2, f_2), (p_3, f_3)$, and slide $\hat{O}$ while maintaining these three contact pairs, and having the fourth point p_4 penetrate the octahedron. The three contact pairs $(p_1, f_1), (p_2, f_2)$ and (p_3, f_3) determine a line in the space $\mathbb{P}$ of placements, and we just have to follow this line in the (unique) direction where p_4 penetrates into the octahedron. Let us add to O the three internal facets $u_1 u_2 \overline{u_1 u_2}$, $u_1 u_3 \overline{u_1 u_3}$, and $u_2 u_3 \overline{u_2 u_3}$ (see Figure 6.2); these are the intersections of the octahedron with its three symmetry planes, each containing four vertices of O. In the following, we refer to the octahedron augmented with these three internal facets as the *augmented octahedron*. The sliding process is stopped as soon as one of the following events occurs:

1. Point p_4 reaches one of the three internal facets.
2. One of the points p_1, p_2, p_3 reaches an edge of O and thus becomes a double contact point.
3. A contact with a new point is encountered on a face other than f_4.
4. A contact with a new point on face f_4 is encountered.

In the first case, we reach a rigid placement of the *augmented octahedron*. This rigid placement will be called quasi-free because it has no point of S inside the octahedron, except for one point on an internal facet. Consider the number of pairs (v, f), where f is one of the contact facets and v is a vertex of O incident to f. Since we have three triangular contact facets and one quadrangular contact facet, the number of these pairs is 13, which implies that one of the six vertices of O has to be shared by three of those contact facets. Thus we can apply to this placement the above shrinking scheme, retaining the three contact pairs whose facets share the common vertex, and stopping when we reach a quasi-free rigid placement of O with a double contact point. As argued above, such a terminal placement can be reached by at most two initial quasi-free rigid placements of O.

To bound the number of these terminal placements, we proceed as above. With each double contact (p, e) we associate a 2-dimensional subspace $I\!P(p, e)$ of $I\!P$, where the locus of all placements with an additional contact pair (with an external or an internal facet) is represented in this subspace by a line segment. The quasi-free placements that we are interested in appear as vertices of the arrangement of these segments lying at level at most 4 (i.e., with at most 4 segments lying below the vertex in the ρ direction). Indeed, if we fix the double contact point on e and increase the scale factor ρ from zero until we reach a terminal placement, the point on the internal face has crossed at most two external facets of O (when it gets into the octahedron through an edge, this point cannot get into the octahedron through a vertex because the set of point would not be in general position) and two internal facets (because it cannot cross the internal facet incident to edge e). Using standard arguments (based on the Clarkson-Shor analysis technique [3]), it is easily seen that the number of such vertices is $O(n)$. Hence the number of stopping events of the first case is $O(n^2)$.

In the second case, the reached placement is a rigid placement of the octahedron with a double contact pair and at most one point inside the octahedron. As previously argued, such a placement corresponds to a vertex of level at most 6 in the planar arrangement of segments representing the contact pairs in the two dimensional subspace of $I\!P$ representing the reached double contact pair. Thus the number of placements reached in this case is also bounded by $O(n^2)$.

In the third case, the reached placement is a rigid placement of the octahedron with three contact facets sharing a vertex and at most one point inside the octahedron. Again, arguing as above and applying the Clarkson-Shor technique, it follows that the number of terminal placements that we reach in this case is proportional to the number of free rigid placements with three contact facets sharing a vertex (and with no point inside the octahedron). Thus the number of placements reached in the third case is also $O(n^2)$.

In the last case, the reached placement is a rigid placement of the octahedron with four contact pairs involving four nonadjacent facets and one point inside the octahedron. In the following, we denote by $c_j(\mathcal{S})$ the number of rigid placements $\hat{O}$ with four contact pairs involving four nonadjacent facets and with j points of $\mathcal{S}$ inside the octahedron.

Before continuing, it is important to observe that each terminal placement reached in case 4 of the above sliding process is reached from a unique initial rigid free placement. Indeed, since the single point p_4 inside the octahedron never crosses any internal facet of the augmented octahedron, it lies in one of the eight octants into which the three internal facets partition O, and the external facet bounding that octant must be the contact facet f_4, so the initial placement $\hat{O}$ is uniquely determined.

On the other hand, we have four choices of the triple of the contact pairs that are preserved in the sliding process. If in one of these choices we reach a terminal placement of one of the first three types, we charge the initial free rigid placement to this terminal placement, observe that any such terminal placement can be charged in this manner only a constant number of times, and thus conclude that the number of initial free rigid placements of this kind is $O(n^2)$. If each of the four sliding processes terminate in a placement of type 4, then the initial free rigid placement can be charged to four terminal placements. Moreover, every initial and terminal placement in this case involves four contact pairs with four nonadjacent contact facets, except that the initial placements are free and the terminal placements contain a point inside the octahedron.

Thus, the preceding case analysis leads to the following recurrence relationship:

$$4c_0(\mathcal{S}) \leq c_1(\mathcal{S}) + O(n^2),$$

from which we obtain

$$c_0(\mathcal{S}) \leq \frac{n-4}{n} c_0(\mathcal{S}) + \frac{1}{n} c_1(\mathcal{S}) + O(n).$$

Now, $\frac{n-4}{n} c_0(\mathcal{S}) + \frac{1}{n} c_1(\mathcal{S})$ is just the expected number of free rigid placements with four contact pairs involving four nonadjacent facets for a random sample $\mathcal{R}$ of $n-1$ points of $\mathcal{S}$; see [2, 13] for a related argument. Thus, if we denote by $c_0(n)$ the maximum of $c_0(\mathcal{S})$ over all sets $\mathcal{S}$ of n points in general position with respect to O, we obtain the recurrence

$$c_0(n) \leq c_0(n-1) + O(n),$$

whose solution is

$$c_0(n) \leq O(n^2).$$

Thus, the number of free rigid placements with four contact pairs involving four nonadjacent facets is also bounded by $O(n^2)$, which completes the proof of the upper bound in Theorem 6.1 □

Remark: An obvious open problem is to extend this result to higher dimensions. Informally, the reason we have failed in doing so is that the corresponding polytope that generates the L_1 norm in $I\!R^d$, the d-co-cube, has a large number of facets (that is, 2^d facets). Consequently, there are too many combinatorially-different types of free rigid placements of the d-co-cube, which so far impeded a successful analysis of their number. A first goal in this direction would be to obtain a sharp bound on the complexity of the L_1-Voronoi diagram in four dimensions.

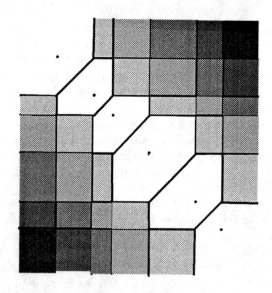

Figure 3: A degenerate configuration for the L_1 metric

7 Degenerate Configurations

Next, we show that the general position assumption is essential for the upper bounds of Theorems 4.1, 5.1 and 6.1 to hold. Specifically, we show:

Theorem 7.1 *For any polyhedral convex distance function d_P and any dimension $d \geq 2$, there exist sets of points, not in general position with respect to distance d_P, whose Voronoi diagram has complexity $\Omega(n^d)$.*

Figure 3 shows the L_1-Voronoi diagram of a degenerate set of points in $\mathbb{R}^2$. Due to lack of space, we omit the proof in this version.

8 Conclusion

In this paper we have studied the complexity of Voronoi diagrams of point sets in higher dimensions under certain special polyhedral convex distance functions, including simplicial distance functions and the L_1 and L_∞ norms. We have obtained tight worst-case bounds for all the cases that we studied. Some of these bounds match the known maximum complexity of euclidean Voronoi diagrams, namely $\Theta(n^{\lceil d/2 \rceil})$, lending support to the conjecture that this is the correct 'order of magnitude' of the complexity of fairly general types of Voronoi diagrams in higher dimensions.

There are quite a few open problems that this paper raises. The first problem is to extend the bound obtained for L_1-Voronoi diagrams to four and higher dimensions. Another, more challenging problem is to extend our analysis to Voronoi diagrams under arbitrary polyhedral convex distance functions, even just in three dimensions. An even more challenging problem is to extend our analysis further to cases where the sites are general convex polytopes, rather than points. For this, one would probably need an appropriate combination of the techniques used here and in the previous papers [2, 13].

References

[1] F. Aurenhammer, Voronoi diagrams: a survey of a fundamental geometric data structure, *ACM Comput. Surv.*, 23 (1991), 345–405.

[2] L.P. Chew, K. Kedem, M. Sharir, B. Tagansky and E. Welzl, Voronoi diagrams of lines in three dimensions under a polyhedral convex distance function, *Proc. 6th ACM-SIAM Symp. on Discrete Algorithms* (1995), 197–204.

[3] K. Clarkson and P. Shor, Applications of random sampling in computational geometry II, *Discrete Comput. Geom.* 4 (1989), 387–421.

[4] H. Edelsbrunner, *Algorithms in Combinatorial Geometry*, Springer-Verlag, Heidelberg 1987.

[5] H. Edelsbrunner and R. Seidel, Voronoi diagrams and arrangements, *Discrete Comput. Geom.* 1 (1986), 25–44.

[6] D. Halperin and M. Sharir, New bounds for lower envelopes in three dimensions with applications to visibility of terrains, *Discrete Comput. Geom.* 12 (1994), 313–326.

[7] K. Kedem, R. Livne, J. Pach and M. Sharir, On the union of Jordan regions and collision-free translational motion amidst polygonal obstacles, *Discrete Comput. Geom.* 1 (1986), 59–71.

[8] D. Leven and M. Sharir, Intersection and proximity problems and Voronoi diagrams, in *Advances in Robotics*, Vol. I, (J. Schwartz and C. Yap, Eds.), 1987, 187–228.

[9] K. Mulmuley, *Computational Geometry: An Introduction through Randomized Algorithms*, Prentice Hall, 1993.

[10] J. O'Rourke, *Computational Geometry in C*, Cambridge University Press, Cambridge 1994.

[11] F. Preparata and M. Shamos, *Computational Geometry: An Introduction*, Springer-Verlag, New York 1985.

[12] M. Sharir, Almost tight upper bounds for lower envelopes in higher dimensions, *Discrete Comput. Geom.* 12 (1994), 327–345.

[13] B. Tagansky, A new technique for analyzing substructures in arrangements, *this proceedings*.

The Voronoi Diagram of Curved Objects*

Helmut Alt
Freie Universität Berlin, Institut für Informatik
Takustr. 9, D-14195 Berlin

Otfried Schwarzkopf
Universiteit Utrecht, Vakgroep Informatica
Postbus 80.089, NL-3508 TB Utrecht

Abstract

Voronoi diagrams of curved objects can show certain phenomena that are often considered artifacts: The Voronoi diagram is not connected; there are pairs of objects whose bisector is a closed curve or even a two-dimensional object; there are Voronoi edges between different parts of the same site, (so-called self-Voronoi-edges); these self-Voronoi-edges may end at seemingly arbitrary points, and, in the case of a circular site, even degenerate to a single isolated point.

We give a systematic study of these phenomena, characterizing their differential geometric and topological properties. We show how a given set of curves can be refined such that the resulting curves define a "well-behaved" Voronoi diagram. We also give a randomized incremental algorithm to compute this diagram. The expected running time of this algorithm is $O(n \log n)$.

1 Introduction

Voronoi diagrams are among the most extensively studied objects in computational geometry (see for instance Aurenhammer's survey [1] or the book by Okabe, Boots, and Sugihara [13]). Naturally the first type of Voronoi diagrams being considered was the one for point sites and the Euclidean metric in two dimensions. Subsequent research was concerned with generalizations of all of these features. In the two-dimensional case these generalizations were particularly motivated by applications in motion planning which lead to the so-called *retraction method* [12].

This method makes use of the fact that if there is a collision-free motion of a disk-shaped object within a collection of obstacles, from a source to a target position then there is also one that essentially follows the edges of the Euclidean Voronoi diagram of the obstacles. Since in general the obstacles are not single points, Voronoi diagrams for other types of sites were investigated, mostly for line segments. Also, if the object to be moved is not a disk but some other convex body B the retraction approach can be applied to translational motions [9]. In this case the Euclidean distance has to be generalized to a *convex distance function* that has B, with some fixed reference point inside, as its "unit circle".

The Voronoi diagram under this distance function is usually referred to as the *B-Voronoi diagram*. Klein [5, 6] gave a unified approach for many of the different variants of two-dimensional Voronoi diagrams, the so-called *abstract Voronoi diagrams*. They are not specified by distance functions but by certain topological conditions which the vertices and edges have to satisfy. Klein, Mehlhorn, and Meiser [7] gave a general paradigm for a randomized $O(n \log n)$ algorithm for constructing an abstract Voronoi diagram for a set of n sites.

Concerning the construction of B-Voronoi diagrams, so far it was mostly assumed that the obstacles are bounded by line segments and B is a convex polygon. In fact, more complex shapes can be approximated by polygons to arbitrary precision. However, in general a good approximation requires very many line segments leading to large running times of the construction algorithms. Therefore it should be interesting to consider the construction of B-Voronoi diagrams where the sites and B are bounded by more general curves. Yap [17] solves the problem for the Euclidean metric and second degree curves. Further steps in a more general direction

---

*This research was supported by the Netherlands' Organization for Scientific Research (NWO), by the ESPRIT Basic Research Action No. 7141, Project ALCOM II, and partially by the ESPRIT III Basic Research Action No.6546 (PROMotion). Part of this research was done while O.S. visited Freie Universität Berlin.

Permission to copy without fee all or part of this material is granted provided that the copies are not made or distributed for direct commercial advantage, the ACM copyright notice and the title of the publication and its date appear, and notice is given that copying is by permission of the Association of Computing Machinery.To copy otherwise, or to republish, requires a fee and/or specific permission.
11th Computational Geometry, Vancouver, B.C. Canada
© 1995 ACM 0-89791-724-3/95/0006...$3.50

are made in [18], where an idea of an algorithm is given for the case that the bounding curves are circular arcs or line segments.

Unfortunately, Voronoi diagrams of curved objects do not satisfy the conditions of abstract Voronoi diagrams. Figure 1 shows the particularities that can occur even in the Euclidean case. Here we simply define the Voronoi

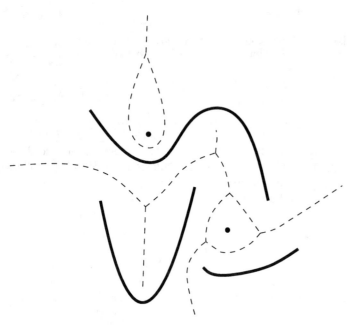

Figure 1: A Voronoi diagram of curves

diagram as the set of all points having more than one closest point on the union of all sites: The Voronoi diagram is not connected and there are Voronoi edges between different parts of the same site, which we will call *self-Voronoi-edges*; these self-Voronoi-edges may end at seemingly arbitrary points, and, in the case of a circular site, even degenerate to a single isolated point. Furthermore the bisector between two objects may be a closed curve. In the case of convex distance functions additionally there may be even pairs of objects whose bisector is a two-dimensional region.

The aim of this paper is twofold:

1. To investigate all the mentioned phenomena of Voronoi diagrams of curves and characterize their differential geometric and topological properties.

2. To show how these difficulties can be overcome (under certain preconditions) by breaking up the curves into so-called "harmless" pieces so that the idea of randomized incremental construction can be applied.

The algorithmic result of this paper is an efficient randomized algorithm for constructing the B-Voronoi diagram of a set of curves. B is assumed to be a convex body bounded by finitely many harmless curves. The algorithm will consider curves as abstract objects and assume that certain elementary operations are available as black boxes. These include finding the points having the same distance from three given sites, finding all points of a given slope, finding points where the curvature has a local maximum, given the representations of two curves finding the representation of a bisector, and finding intersection points of given curves. The details of these operations including numerical problems involved will depend on the particular application of our paradigm. For example, if it is applied to algebraic curves of some fixed degree, the elementary operations would consist of solving systems of algebraic equations of constant degree and constantly many variables.

As far as we know, this is the first systematic treatment of the phenomenon of self-Voronoi-edges. Self-Voronoi-edges may long have been considered as artifacts. We argue that self-Voronoi-edges play an essential role in Voronoi diagrams of curved objects. If the Voronoi diagram is used to do motion planning using the retraction method, for instance, then the self-Voronoi edges are necessary to capture the connectivity of the workspace. Without them, the robot may not be able to reach concavities formed by a single curve.

In this extended abstract, we can show our results in more detail for the case of the Euclidean metric only. We will give an idea as to how our results can be generalized to convex distance functions.

We will first characterize sets of curved sites that induce "well-behaved" Voronoi diagrams. As the conditions we will impose cannot be expected to hold for a given set of curves that may arise in an application, we then describe how such a given set of curves can—under some mild conditions—be refined by cutting up the curves and adding point sites on curves. We will then describe a randomized incremental algorithm of running time $O(n \log n)$ to compute the Voronoi diagram of a set of "harmless" curves. Combined with our technique to refine curves that are not yet sufficiently well-behaved, this will give us an algorithm to compute the Voronoi diagram of a set of arbitrary curves in time $O(n \log n)$.

2 The Voronoi diagram of harmless curves

A *curve* is given by a function $\gamma : I \to \mathbf{R}^2$ where $I \subset \mathbf{R}$ is some closed interval. Unless stated otherwise we will assume that curves are *regular* in the differential geometric sense, that is γ is twice continuously differentiable and $\gamma'(t) \neq 0$ for all $t \in I$. We say that two curves *touch* each other in some point $p \in \mathbf{R}^2$ iff they both pass through p without properly intersecting there. The radius of *curvature* at some point $\gamma(t)$ is (informally) the radius of the largest circle touching (and not intersecting) γ at $\gamma(t)$ on the "concave side" of γ; it is positive if that circle lies left of the direction of γ given by the parametrization and negative otherwise. The *curvature* $\gamma(t)$ is the reciprocal of the radius of curvature (for more details see Stoker [16]).

We will assume that curves are *simple*, that is $\gamma(t) \neq \gamma(t')$ for $t \neq t'$. Furthermore, each curve will not be considered as one, but as three sites: the two endpoints and the interior of the curve. We call an open curve (a curve without its endpoints) *harmless* if there is no circle that touches it in more than one point. A *harmless site* is either a point, an open circular arc, or a harmless curve. A *harmless site collection* is a finite set S of pairwise disjoint harmless sites with the condition that for every circular arc and harmless curve $\gamma \in S$ its endpoints are also members of S. For example, the parabolic arc $\{(t, t^2) \mid -3 < t < 3\}$ is not a harmless curve. We can cut it as its apex by adding the point site $(0, 0)$, and can obtain a harmless site collection of two parabolic arcs and three point sites. Observe that it is possible that several curves share one endpoint, so we allow arbitrary planar subdivisions by regular curves. Curves may not intersect but this case can be handled by making the intersection points additional point sites.

Throughout this paper, we will denote by $d(x, y)$ the Euclidean distance of points $x, y \in \mathbf{R}^2$ and for $A \subset \mathbf{R}^2$, $x \in \mathbf{R}^2$ we define $d(x, A) := \inf_{y \in A} d(x, y)$. Also, let $\psi_A(x)$ be that point of A with $d(x, \psi_A(x)) = d(x, A)$. ψ_A is not defined whenever there is no such point or when there is more than one. The following lemma is concerned with the set of points closest to some given point on a harmless curve.

Lemma 1 *Let γ be an open circular arc or a harmless curve and let p, q be its endpoints. Then:*

- *a) ψ_γ is defined for all x where $d(x, p) > d(x, \gamma)$ and $d(x, q) > d(x, \gamma)$, and ψ_γ is continuous wherever it is defined.*

- *b) Let a be a point on γ, c the center of the circle of curvature at a. Only points on the ray $\vec{cp}$ are mapped onto a by ψ_γ. Other points on the same straight line, which is the normal of γ in a, that are not mapped to a are closer to one of the endpoints than to γ.*

Proof: a) By definition, γ, together with its endpoints, is a compact subset of $\mathbf{R}^2$. Consequently, by continuity $a \mapsto d(x, a)$ assumes its minimum on $\gamma \cup \{p, q\}$ for some $a_0 \in \gamma$. By the harmlessness of γ there is only one such a_0, so ψ_γ is defined. Let now $x \in \mathbf{R}^2$ be some point where $a := \psi_\gamma(x)$ is defined and let $r := d(x, p)$. For some arbitrary $\varepsilon > 0$ consider the ε-neighborhood $U_\varepsilon(a)$ and the rest of the curve $A := \gamma \setminus U_\varepsilon(a)$. Let ε be sufficiently small so that $A \neq \emptyset$. Since for any $b \in A$, $d(x, b) > r$ and A is compact, there is some $\delta > 0$ such that

$$d(x, b) > r + 2\delta \qquad (1)$$

for all $b \in A$. Let y be any point in $U_\delta(x)$ so that $\psi_\gamma(y)$ is defined. Since by the triangle inequality $d(y, \gamma) \leq r + \delta$, and by (1) $\delta(y, b) > r + \delta$ for any $b \in A$, it must be $\psi_\gamma(y) \in U_\varepsilon(a)$. This shows the continuity of ψ_γ.

b) Consider $x \in \mathbf{R}^2$ with $\psi_s(x) = p$. The circle around x with radius $d(x, p)$ touches s in p. Therefore the line segment $\overline{xp}$ is part of the normal through p and x does not lie beyond c, since then the circle would contain parts of s in its interior. □

Figure 2 shows the partition of the plane into 3 regions depending on whether the closest point is p, q or in the interior of γ. By Lemma 1 b) the Voronoi region of γ cannot go beyond the curve η which is the locus of all centers of curvature, the so-called *evolute* of γ.

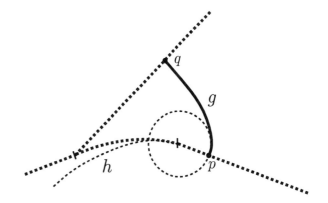

Figure 2: The Voronoi regions of p, q, and γ.

Given a harmless site collection S and a site $s \in S$, we define the *Voronoi region* of s in S, $VR(s, S)$ as follows:

$$VR(s, S) := \{x \in \mathbf{R}^2 \mid \exists p \in s \text{ with } d(x, p) = \min_{s' \in S} d(x, s')\}.$$

Note the slight twist in this definition. Had we chosen the simpler condition "$d(x, s) = \min_{s' \in S} d(x, s')$", then the Voronoi region of a curve would contain the Voronoi region of its endpoints. Our definition avoids that.

We will find it helpful to restrict our attention to the "finite" part of the Voronoi diagram. Therefore, we will add a large circle ω with center in the origin to our site collection. The radius of this circle will be assumed so large that it contains all *Voronoi vertices* and furthermore the circle is not their nearest site. The Voronoi vertices are the points whose shortest distance to any site is assumed to at least three different sites.[1] This means that nothing from the topological structure of the Voronoi diagram gets lost by inserting ω. So we

---

[1]Such a radius exists, in fact, it can be shown by an easy geometric argument that any triple of sites has at most two such points.

only have to consider the Voronoi diagram inside ω and, consequently, all Voronoi regions are bounded.

The *Voronoi diagram $V(S)$* is defined as the set of all Voronoi regions $VR(s, S)$, for $s \in S$. Next we investigate the shape of Voronoi regions:

Lemma 2 *Let S be a harmless site collection, $s \in S$.*

a) If s is a point then the intersection of any straight line through s with $VR(s, S)$ is a line segment.

b) If s is a curve and p a point on S, then the normal through p intersects $\partial VR(s, S)$ in exactly two points $h_l(p)$ and $h_r(p)$ lying on either side of p. The line segment ℓ in between belongs completely to $VR(s, S)$, in fact, these are exactly the points having p as their closest point on any site. The interior points of ℓ do not belong to any other Voronoi region.

c) h_l and $h_r : s \to \mathbf{R}^2$ are continuous.

Proof: [Proof of Part b), a) follows by the same argument]
Consider some intersection point q of the normal n with $\partial VR(s, S)$. Then the circle C through p with center q contains no point of any site in its interior but another one $r \neq p$ on its boundary. Suppose there is some point $t \in n$ beyond q (seen from p) that belongs to $\partial VR(s, S)$. Then by Lemma 1b) p must be the point on s closest to t as well. But the circle with center t through p contains r, a contradiction. On the other hand, for any point $u \in \overline{pq}$ the circle around u through p is inside C, so it is empty, therefore $u \in VR(s, S)$.

c) Suppose that h_l is not continuous. So there exists a sequence of points in s converging to p whose h_l-values do not converge to $h_l(p)$. Since the Voronoi region of s is compact there exists a subsequence converging to some $q \neq h_l(p)$. Because of the continuity of ψ_s q lies on the straight line through p and $h_l(p)$. Since ∂V is closed $q \in \partial V$. This is a contradiction to part b). □

We summarize important topological properties of Voronoi diagrams in the following theorem:

Theorem 3 *Let S be a harmless site collection of n sites.*

(i) The union of the Voronoi regions covers ω, and no Voronoi region is empty.

(ii) For $R \subset S$ and $s \in R$ we have $VR(s, S) \subseteq VR(s, R)$.

(iii) The intersection of two Voronoi regions lies on the boundary of both.

(iv) A Voronoi region $VR(s, S)$ is simply connected, for point sites it is even star-shaped.

(v) The boundary of each Voronoi-region $VR(s, S)$ is a Jordan-curve except if s is an endpoint where several curves meet. In this case the Voronoi-region might be a line segment or the point itself.

Proof: (i) Since the union of all sites is a compact set, for any point $x \in \omega$ the minimum distance to that union is assumed, so x lies in some Voronoi region. Any Voronoi region contains the site itself and, therefore is not empty.

(ii) Let $s \in R \subset S$, and let $x \in VR(s, S)$. This means that there is a point $p \in s$ with $d(x, p) \leq d(x, s)$ for all $x \in S$. Clearly this implies $x \in VR(s, R)$.

(iii) Let $x \in VR(s, S) \cap VR(t, S)$. So there must be points $p \in s$, $q \in t$ lying on the maximal empty circle around x. By Lemma 2b) any point in the interior of the line segments $\overline{xq}$ and $\overline{xp}$ lies in $VR(s, S) \setminus VR(t, S)$ and $VR(t, S) \setminus VR(s, S)$, respectively. This shows that x lies on the boundary of both.

(iv) Let α be some closed curve within $VR(s, S)$. Let q be some arbitrary point surrounded by α and n the normal to s through q which is unique by Lemma 1b) (the straight line sq if s is a point). n intersects α in at least two points $a, b \in VR(s, S)$ surrounding q. Then by Lemma 2b) $q \in VR(s, S)$. Since this holds for any point q in the region encircled by α, α is within $VR(s, S)$ contractible to a point. Since this holds for any α, $VR(s, S)$ is simply connected.

(v) Let s be a harmless curve and consider it oriented by its parametrization. Then to any $x \in S$ there exist unique points $h_\ell(x), h_r(x) \in \partial VR(x, S)$ to the left and right of s, respectively with $\psi_s(h_\ell(x)) = \psi_s(h_r(x)) = x$ (see Lemma 2b)). Since h_ℓ and h_r are continuous (Lemma 2c)) and because of the harmlessness of s, they are also one-to-one the images $A := h_\ell(s), B := h_r(s)$ are homeomorphic to s, so they are simple curves. By continuity reasons their endpoints have the endpoints p, q of s as closest points. Altogether, the boundary of $\partial VR(s, S)$ consists of A, B and two segments I, J (that may degenerate to a single point) of the normals through p, q, respectively (see Figure 3). Since I, A, J, B are pairwise non-intersecting, their

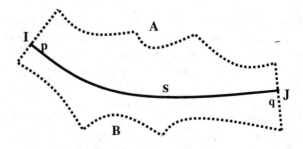

Figure 3: Voronoi region $VR(s, S)$

concatenation forms a Jordan curve. If s is a circular segment the argument is similar only that one of A, B may degenerate to a single point.

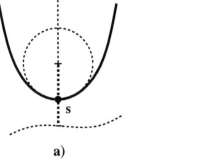
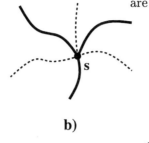

Figure 4: Degenerate Voronoi-cells

Similar techniques prove the statement, if s is an isolated point or an endpoint of a curve. If several curves share s as an endpoint $VR(s, S)$ may degenerate to a line segment (Figure 4a)) or just s itself (Figure 4b)). The former phenomenon occurs when the closed half-spaces beyond the normals of those curves in s intersect only in a straight line, the latter when they intersect only in s itself. □

The Voronoi diagram can also be represented as a graph as follows: The vertices of the graphs of are the points of $V(S)$ which belong to the boundary of three or more Voronoi regions. As was mentioned before, it can be shown that there are at most two such points per triple of sites. The edges of the graph correspond to the maximal connected subsets belonging to the boundary of exactly two Voronoi regions. They are curves by Theorem 3(v). The faces of the graph correspond to the Voronoi regions. We will use $V(S)$ to denote the graph as well. Using Theorem 3(iv) and Euler's formula we can prove:

Theorem 4 *Given a harmless site collection S of n sites. The Voronoi graph $V(S)$ is a planar connected graph with at most $n + 1$ faces, at most $3n - 3$ edges, and at most $2n - 2$ vertices.*

Here the outer circle ω is not counted as a site, but the edges and vertices where it is involved are counted.

3 Partitioning curves into harmless pieces

Here we will see that the points responsible for self-Voronoi-edges or, in other words, the non-harmlessness of curves are the points where the absolute value of the curvature has a local maximum. Figure 5 shows this situation. There is a circle around a touching γ in two points. When decreasing the radius the center of the circle traces a self-Voronoi-edge that ends at c where the two tangent points fall together. c is the center of the circle of curvature C of γ at p which is a local maximum of curvature. Points on the line segment $\overline{cp}$ are centers of circles that touch γ in p.

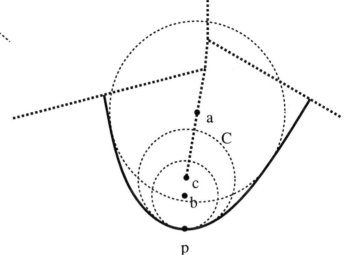

Figure 5: A local maximum of curvature

If we remove these local maxima by cutting the curve there and making them separate sites, the Voronoi diagram will have the nice properties described in Theorem 3. In fact, it holds

Theorem 5 *A regular curve that does not contain its endpoints is harmless if it has no two parallel tangents, it contains no circular segments, and the absolute value of its curvature has no local maximum.*

Proof: Let γ be such a curve. Since it has no two parallel tangents we can assume wlog. that it has no vertical tangents. So it can be parametrized as the graph of a function $f : I \to \mathbf{R}$, where $I \subset \mathbf{R}$ is some interval, and $\gamma(t) = \bigl(t, f(t)\bigr)$. Suppose some circle C is not intersecting γ but touching it in two points p and q. Since there is no vertical tangent C lies in p and q on the same side of γ. Without loss of generality we can assume that this is the left side, so the curvature of γ is nonnegative, and that both p and q lie in the lower semi-circle of C. Now the curvature of γ is at most the curvature of C in p and q, the latter is constant, and the former has no maximum in between. It follows that the curvature of γ is not larger than the curvature of C between p and q. Because of the following lemma this is only possible if γ coincides with C between p and q, a contradiction. □

Lemma 6 *Let γ, δ be two regular curves that are graphs of functions $f, g : I \to \mathbf{R}$ where $I \subset \mathbf{R}$ is some*

finite interval, so $\gamma(t) = \bigl(t, f(t)\bigr)$, $\delta(t) = \bigl(t, g(t)\bigr)$ for all $t \in I$. If γ touches δ in two points $t_1, t_2 \in I$ and its curvature is not greater than δ's for all $t \in [t_1, t_2]$ then γ and δ must coincide in $[t_1, t_2]$.

This lemma follows from the fact, that the curvature of the graph of a function f is given by

$$\kappa(t) = \frac{f''(t)}{\left(1 + \bigl(f'(t)\bigr)^2\right)^{3/2}}$$

by standard analytical considerations.

In Section 5 we will see how to compute the Voronoi diagram of a harmless site collection in time $O(n \log n)$. Theorem 5 allows us to apply that algorithm to more general sets of curves: In fact, it allows us to partition the given set of curved objects into harmless pieces and circular arcs. If, for instance, the given curves are algebraic of constant degree, each can have at most a constant number of points of vertical tangency or maxima of the curvature. By cutting the n original curves at these points we obtain a collection of $O(n)$ harmless curves. We can then compute the Voronoi diagram of these harmless pieces, and obtain a Voronoi diagram of complexity $O(n)$. If that is desired, we can then merge the Voronoi cells of curves that are pieces of the same original curve. In most applications, however, that is probably not what is needed: If the Voronoi diagram is used for motion planning with the retraction method, for instance, the additional self-Voronoi-edges are essential to guarantee that the resulting road map captures the connectivity of the workspace.

4 Convex distance functions

Let $B \subset \mathbf{R}^2$ be some convex body with some fixed reference point o inside. The B-*distance* from a point $p \in \mathbf{R}^2$ to a point q is defined as the factor by which B, when placed at p with o, has to be stretched around o until it touches q. The B-distance is in general not symmetric, but it satisfies the triangle inequality.

Here we will assume that B is bounded by a constant number k of harmless curve segments. So the boundary of B consists of $2k$ *features*: The k harmless curves and k vertices where they meet. We further partition the curve sites into pieces so that one piece can be touched only by one feature of B. This can be done as follows: To each feature of B there is an interval of possible tangent slopes. We partition the sites at points where the slope equals one of the interval boundaries (see Figure 6). Notice that we have to distinguish between the two sides of a curve and partition both accordingly. The number of sites produced this way equals the original number of sites plus the number of places on sites where one of the tangent slopes occurs. This number cannot be bounded in general, but for example for algebraic curves of constant degree the increase is by a constant factor.

In order to derive our results from the previous section to arbitrary B-distance we have to generalize our definitions: The B-*curvature* of a regular curve γ in a point p is the largest copy of B that touches γ in p without intersecting it. If the feature of B touching p is a curve δ, the B-curvature is the ratio of the curvatures of γ and δ at that point. If the feature is a non-smooth vertex, the B-curvature is 0, no matter what γ looks like.

A B-*harmless* site is a curve that cannot be touched by any homothet of B in more than one place. A B-*harmless site collection* consists of open B-harmless sites, their endpoints, other points, and curves that are homothets of segments of the boundary of B.

The B-*normal* of a curve γ in a point p is the orbit of the reference point o considering all homothets of B that touch γ in p (see Figure 7). Any B-normal consists of a ray on either side of γ.

With these definitions, all results and proofs of the previous section go through, if the boundary of B contains no straight segments.

Otherwise, the following problems will occur: Let e be a straight line segment on the boundary of B. Then for any point p on some site where the tangent is parallel to e the B-curvature is not defined since an arbitrarily large copy of B can be placed there. (With the exception that p is an inflection point of γ, i.e. has curvature 0.) p is a site by itself, as was mentioned before. There could be more point sites all lying on a straight line parallel to e. In this case there are "two-dimensional Voronoi-edges" and vertices, i.e. two-dimensional areas of points with the same distance to several of the sites. The exact partition may have quadratic complexity.

However, as we shall see, our algorithm will first construct the Voronoi diagram of the point sites and then insert the curves. For point sites however, there is an efficient algorithm by Klein and Wood [8] managing the mentioned degeneracies which can be applied. For the curve sites these situations cause no problems any more and the properties of Theorem 3 hold.

5 Randomized incremental construction

The Voronoi diagram of curved objects does not fit into the framework of abstract Voronoi diagrams by Klein et al. [7] and cannot be computed with their randomized incremental algorithm. The reason for this is that they assume that the bisector of any pair of sites is an unbounded simple curve. Even for a point and a circular arc, this is no longer true under the Euclidean metric: The bisector can be a closed curve.

If, however, the objects form a harmless site collec-

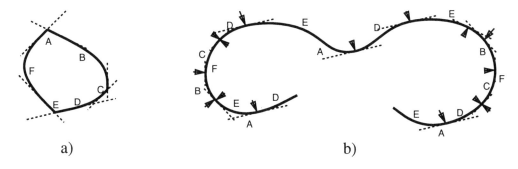

Figure 6: a) convex body B with its features b) partitioning of a curve γ

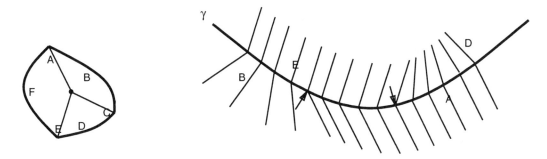

Figure 7: B-normals at γ

tion, this situation can be remedied. In the following we demonstrate how to compute the Voronoi diagram of a harmless site collection in time $O(n \log n)$ using a kind of randomized incremental algorithm, based on the framework set by Clarkson and Shor [4], Mulmuley [11], and Boissonnat et al. [2]. We have to make sure that during the algorithm which constructs the Voronoi diagram by inserting the sites one by one, the intermediate set of sites is always a harmless site collection, that is that no curve is inserted before both of its endpoints are.

Therefore, we compute the Voronoi diagram of a harmless site collection S in two stages. In the first stage, we compute a Voronoi diagram of n points $V(P \cup Q)$. Here $P \subset S$ is the set of all *point sites*. Q is obtained by selecting for each curve site $s \in S$ a point q_s in the relative interior of s.

The points in Q serve as "placeholders" for the curves they stem from which, in the second stage are added one by one in random order. This replacement of the placeholders by the actual curves is made easy by the fact that we already know where to insert a new site s. We will restrict ourselves to the description of the Euclidean case.

We will need to represent the Voronoi diagram $V(R)$ of a subset $R \subset S \cup Q$. This can be done using any standard structure for planar subdivisions, such as the doubly-connected edge list [10, 14]. In the first stage of the algorithm, we simply compute the Voronoi diagram $V(P \cup Q)$. This can be done using any efficient algorithm for the construction of Voronoi diagrams and takes time $O(n \log n)$.

As discussed in Section 2, we add a large circle ω to our site collection S. If necessary, this circle has to be handled symbolically. After computing $V(P \cup Q)$, we add ω to obtain $V(P \cup Q \cup \{\omega\})$. This can be done in time $O(n)$.

Let now $s_1, s_2, \ldots, s_m$, be a random permutation of the curve sites in S. Let q_r denote q_{s_r}, the point in the interior of s_r we had chosen. We consider the sites s_r in this order. In every step of the algorithm, we have to replace a point site q_r in the current Voronoi diagram $V(P \cup \{\omega, s_1, \ldots, s_{r-1}, q_r, \ldots, q_m\})$ by a curve site s_r to obtain $V(P \cup \{\omega, s_1, \ldots, s_r, q_{r+1}, \ldots, q_m\})$. Let's look at this in more detail.

For brevity, let $s := s_r$, let $q := q_s = q_r$, let $R := P \cup \{s_1, \ldots, s_{r-1}, q_r, q_{r+1}, \ldots, q_m\}$, and let $R' := P \cup \{s_1, \ldots, s_{r-1}, s_r, q_{r+1}, \ldots, q_m\}$. By Theorem 3, the Voronoi region $VR(s, R')$ of s is a simply connected region whose boundary is a closed Jordan-curve (recall that s is not a point site). To obtain $V(R')$ from $V(R)$ means to remove the portion $\mathcal{I}$ of $V(R)$ that lies in $VR(s, R')$, and to add the boundary of $VR(s, R')$. We first prove a lemma.

Lemma 7 *The "skeleton" $\mathcal{I}$ contains the boundary of $VR(q, R)$, is connected, and contains no cycle except for the boundary of $VR(q, R)$. All its leaves lie on the boundary of $VR(s, R')$, and its complexity is linear in* *the number of these leaves.*

Proof: Since q lies on s, clearly $VR(q,R)$ is completely contained in $VR(s,R')$, and hence $\mathcal{I}$ contains the boundary of $VR(q,R)$. If $\mathcal{I}$ contained any other cycle, some Voronoi region $VR(r,R)$, $r \in R$, $r \neq q$, must lie inside $VR(s,R')$ which would imply that r lies in $VR(s,R')$, which is impossible.

Assume now that $\mathcal{I}$ is not connected. That means that there are at least two connected components $\mathcal{I}_1$ and $\mathcal{I}_2$ of $\mathcal{I}$. By Theorem 3(iv), none of these can be contained in the interior of $VR(s,R')$, they both must have some connection with $\partial VR(s,R')$. So there is a path $\gamma \subset VR(s,R') \setminus \mathcal{I}$ connecting two points x and y on the boundary of $VR(s,R')$ and separating $\mathcal{I}_1$ from $\mathcal{I}_2$. Since $\gamma \cap V(R) = \emptyset$, γ lies in the interior of some $VR(r,R)$, for $r \in R$. This implies that there must be points x' and y' arbitrarily close to x and y that lie in $VR(r,R')$. That means that x and y can be connected by a path γ' in $VR(r,R')$. The combination $\gamma \cup \gamma'$ is a closed loop in $VR(r,R)$ containing either $\mathcal{I}_1$ or $\mathcal{I}_2$, a contradiction. It follows that $\mathcal{I}$ is connected.

Leaves of $\mathcal{I}$ must clearly lie on $\partial VR(s,R')$. By removing one edge on $\partial VR(q,R)$ from $\mathcal{I}$, it becomes a tree all of whose interior vertices have degree at least three. Consequently, its complexity is linear in the number of its leaves. □

Consider now the boundary of $VR(s,R')$. As discussed in the proof of Theorem 3(v), it consists of two segments I and J—possibly degenerated to a point—through the endpoints p and q of s and two simple curves A and B. The curves A and B consist of a sequence of

- edges that lie in the interior of some $VR(r,R)$, $r \in R$, and are hence part of the bisector of r and s,
- crossings between edges of $V(R)$ and $\partial VR(s,R')$, and
- vertices of $V(R)$.

We first identify I, J, and the skeleton $\mathcal{I}$. This can be done using a graph search starting at any edge on the boundary of $VR(q,R)$ and takes time linear in the complexity of $\mathcal{I}$. The leaves of $\mathcal{I}$ are the vertices of $VR(s,R')$, and from that information we can then construct the curves A and B to obtain $VR(s,R')$ in time linear in the complexity of $\mathcal{I}$ and $VR(s,R')$.

It remains to bound the running time of the algorithm. As we observed, the first stage of the algorithm takes time $O(n \log n)$. Inserting curve site s_r takes time linear in the complexity of $\mathcal{I}$ and $VR(s_r,R')$ and thus by Lemma 7 linear in the complexity of the new Voronoi region $VR(s_r,R')$.

It remains to bound the expected size of $VR(s_r,R')$. We use a standard backwards-analysis argument [3, 15]:

Fix R', and let s be a random curve site in R'. The total complexity of $V(R')$ is $O(n)$ by Theorem 4, and there are r possible choices for s. Consequently, the expected complexity of $VR(s,R')$ is $O(n/r)$. Summing this over all curve sites, we find that the second stage of the algorithm takes expected time $O(n \log n)$ as well.

Theorem 8 *The two-stage randomized incremental algorithm constructs the Voronoi diagram of a harmless site collection of n sites in time $O(n \log n)$.*

Acknowledgments. We would like to thank Stefan Felsner and Lorenz Wernisch for fruitful discussions and valuable hints.

References

[1] F. Aurenhammer. Voronoi diagrams: a survey of a fundamental geometric data structure. *ACM Comput. Surv.*, 23:345–405, 1991.

[2] J.-D. Boissonnat, O. Devillers, R. Schott, M. Teillaud, and M. Yvinec. Applications of random sampling to on-line algorithms in computational geometry. *Discrete Comput. Geom.*, 8:51–71, 1992.

[3] L. P. Chew. Building Voronoi diagrams for convex polygons in linear expected time. Technical Report PCS-TR90-147, Dept. Math. Comput. Sci., Dartmouth College, Hanover, NH, 1986.

[4] K. L. Clarkson and P. W. Shor. Applications of random sampling in computational geometry, II. *Discrete Comput. Geom.*, 4:387–421, 1989.

[5] R. Klein. Abstract Voronoi diagrams and their applications. In *Computational Geometry and its Applications*, volume 333 of *Lecture Notes in Computer Science*, pages 148–157. Springer-Verlag, 1988.

[6] R. Klein. *Concrete and Abstract Voronoi Diagrams*, volume 400 of *Lecture Notes in Computer Science*. Springer-Verlag, 1989.

[7] R. Klein, K. Mehlhorn, and S. Meiser. Randomized incremental construction of abstract Voronoi diagrams. *Comput. Geom. Theory Appl.*, 3(3):157–184, 1993.

[8] R. Klein and D. Wood. Voronoi Diagrams and Mixed Metrics. Proc. 5th Sympos. Theoret. Aspects Comput. Sci. (STACS), Lecture Notes in Computer Science 294, Springer-Verlag 1988, 281–291.

[9] D. Leven and M. Sharir. Planning a purely translational motion for a convex object in two-dimensional space using generalized Voronoi diagrams. *Discrete Comput. Geom.*, 2:9–31, 1987.

[10] D. E. Muller and F. P. Preparata. Finding the intersection of two convex polyhedra. *Theoret. Comput. Sci.*, 7:217–236, 1978.

[11] K. Mulmuley. A fast planar partition algorithm, I. In *Proc. 29th Annu. IEEE Sympos. Found. Comput. Sci.*, pages 580–589, 1988.

[12] C. Ó'Dúnlaing and C. K. Yap. A "retraction" method for planning the motion of a disk. *J. Algorithms*, 6:104–111, 1985.

[13] A.Okabe, B. Boots, and K. Sughihara. *Spatial Tesselations: Concepts and Applications of Voronoi diagrams.* J. Wiley & Sons, 1992.

[14] F. P. Preparata and M. I. Shamos. *Computational Geometry: an Introduction.* Springer-Verlag, New York, NY, 1985.

[15] R. Seidel. Backwards analysis of randomized geometric algorithms. In J. Pach, editor, *New Trends in Discrete and Computational Geometry*, volume 10 of *Algorithms and Combinatorics*, pages 37–68. Springer-Verlag, 1993.

[16] J. J. Stoker. *Differential Geometry.* Wiley-Interscience, 1969.

[17] C. K. Yap. An $O(n \log n)$ algorithm for the Voronoi diagram of a set of simple curve segments. *Discrete Comput. Geom.*, 2:365–393, 1987.

[18] Chee K. Yap and Helmut Alt. Motion planning in the CL-environment. In *Proc. 1st Workshop Algorithms Data Struct.*, volume 382 of *Lecture Notes in Computer Science*, pages 373–380, 1989.

A Combinatorial Approach to Cartograms*

Herbert Edelsbrunner and Roman Waupotitsch
Department of Computer Science, University of Illinois at Urbana-Champaign, Urbana, Illinois 61801, USA.

Abstract

A homeomorphism from $\mathbb{R}^2$ to itself distorts metric quantities, such as distance and area. We describe an algorithm that constructs homeomorphisms with prescribed area distortion. Such homeomorphisms can be used to generate cartograms, which are geographic maps purposely distorted so its area distribution reflects a variable different from area, such as for example population density. The algorithm generates the homeomorphism through a sequence of local piecewise linear homeomorphic changes. Sample results produced by the preliminary implementation of the method are included.

1 Introduction

Cartograms and homeomorphisms. The starting point for the research reported in this paper is a problem in cartography that has to do with deformations of geographic maps [2]. A *cartogram* is a geographic map purposely distorted so its spatial properties represent quantities not directly associated with position on the globe [3]. Of particular interest are distortions that do not sacrifice continuity, that is, cartograms generated by applying a homeomorphism to the plane carrying the original geographic map.

---

*This research is partially supported by the National Science Foundation, under grant ASC-9200301. Herbert Edelsbrunner is also supported through the Alan T. Waterman award, grant CCR-9118874.

Permission to copy without fee all or part of this material is granted provided that the copies are not made or distributed for direct commercial advantage, the ACM copyright notice and the title of the publication and its date appear, and notice is given that copying is by permission of the Association of Computing Machinery.To copy otherwise, or to republish, requires a fee and/or specific permission.
11th Computational Geometry, Vancouver, B.C. Canada
© 1995 ACM 0-89791-724-3/95/0006...$3.50

A more general and more mathematical view of the problem studied in this paper is the generation of a homeomorphism $\mathbb{R}^2 \to \mathbb{R}^2$ whose effect on area distortion can be prescribed. The generalization to 3 and higher dimensions will be described elsewhere.

Continuous versus combinatorial methodology. Previous approaches to producing cartograms are based on continuous methods dealing with force fields and differential equations. The algorithmic ideas are usually limited to iterations approximating a continuous solution [3, 10]. The approach in this paper is different and based on combinatorial methods in geometry [6], topology [8], and algorithms [1]. We use simplicial complexes tiling $\mathbb{R}^2$ in the construction of piecewise linear homeomorphisms. The final deformation is the composition of a sequence of local homeomorphic deformations. Each local deformation maintains area except in a small region specified within the tiling.

We also study the question of quality of the homeomorphism expressed in terms of local distance distortions. The assessment of quality opens the way to an objective comparison of different homeomorphisms or cartograms. Possibly more importantly, it can be used to improve quality through area invariant deformations.

The approach to constructing homeomorphic deformations described in this paper has been implemented. Applications of the resulting software to specific problem instances can be found in section 7. We describe two data structures representing the homeomorphism, one based on arrays and the other on trees. They differ in storage and time requirements. The final picture, the image under the homeomorphism, is constructed by searching through either one of these data structures.

Outline of the paper. Section 2 presents definitions of the topology and geometry concepts needed for our method. Section 3 describes the main steps of the homeomorphism construction. Section 4 develops a data structure for constructing and evaluating a homeomorphism based on arrays. Section 5 studies an alternative data structure based on trees. Section 6 discusses the assessment of quality of the thus constructed homeomorphisms. Section 7 mentions a few applications and exhibits sample results produced with our software. Finally, section 8 offers a few concluding remarks.

2 Definitions and Concepts

We use periodic tilings to impose a discrete structure on the real plane, $\mathbb{R}^2$. Finite portions of these infinite tilings are algorithmically manipulated. It is convenient to tile with triangles so barycentric coordinates can be used to extend vertex maps to continuous piecewise linear maps in an unambiguous manner. For most of the definitions we follow the notation in [8], but see also [4, 9].

Simplicial complexes. A *k-simplex*, σ, is the convex hull of $k+1$ affinely independent points. The *dimension* of σ is $\dim \sigma = k$. In $\mathbb{R}^2$, at most 3 points can be affinely independent, so we have only 0-simplices (points or vertices), 1-simplices (line segments or edges), and 2-simplices (triangles). Each subset of the $k+1$ points defines also a simplex called a *face* of σ. The face is *proper* if the subset is proper. It is convenient to call the empty set a (-1)-simplex and to consider it an (improper) face of every simplex.

For a vector e, the simplex $\sigma + e = \{p + e \mid p \in \sigma\}$ is a *translate* of σ, and for a scalar λ, $\lambda \sigma = \{\lambda p \mid p \in \sigma\}$ is a *scaled copy* of σ. An edge with endpoints u and v is usually denoted uv. The *cone* of u over A, $uA = \bigcup_{a \in A} ua$, is defined if $ua \cap ub = u$ for all $a \neq b$ in A. For example, if u is a point not in the affine hull of a k-simplex σ then $u\sigma$ is a $(k+1)$-simplex. By convention, $u\emptyset = u$.

A *simplicial complex* is a collection K of simplices for which $\sigma \in K$ and τ a face of σ implies $\tau \in K$, and $\sigma_1, \sigma_2 \in K$ implies $\sigma_1 \cap \sigma_2$ is a face of both. We allow countably infinite collections but require each vertex belong to only finitely many simplices. A simplicial complex $L \subseteq K$ is a *subcomplex* of K. Not every subset $M \subseteq K$ is also a subcomplex, but its *closure*, $\operatorname{Cl} M = \{\tau \in K \mid \tau \text{ a face of } \sigma \in M\}$ is a subcomplex of K. The *open star* of $\tau \in K$ is $\operatorname{St} \tau = \{\sigma \in K \mid \tau \text{ a face of } \sigma\}$. In contrast to the *closed star*, $\overline{\operatorname{St}} \tau = \operatorname{Cl} \operatorname{St} \tau$, $\operatorname{St} \tau$ is in general not a subcomplex of K. The *frontier* of a subset $L \subseteq K$ is $\operatorname{Fr} L = \{\tau \in L \mid \operatorname{St} \tau \not\subseteq L\}$.

The *vertex set* of K is $\operatorname{vert} K = \{\sigma \in K \mid \dim \sigma = 0\}$. The *underlying space* of K is $|K| = \bigcup_{\sigma \in K} \sigma$. It is often cumbersome to distinguish a complex from its underlying space, and occasionally we will prefer ambiguous over awkward language. A *triangulation* of $|K|$ is a simplicial complex L with $|L| = |K|$. A *subdivision* of K is a triangulation L of $|K|$ so each simplex of L is contained in a simplex of K. A subdivision of K can be generated by *starring* from a new vertex, u, added to K: each simplex $\sigma \in K$ that contains u is replaced by cones of u over all faces of σ disjoint from u. A *stellar subdivision* is the result of repeated starring.

Simplicial maps and homeomorphisms. Each point $x \in |K|$ belongs to the (relative) interior of exactly one simplex $\sigma \in K$. If σ is the convex hull of vertices $a_0, a_1, \ldots, a_k$, then there are unique positive real numbers $\alpha_0, \alpha_1, \ldots, \alpha_k$ with

$$\sum_{i=0}^{k} \alpha_i a_i = x \quad \text{and} \quad \sum_{i=0}^{k} \alpha_i = 1.$$

The *barycentric coordinate* of x with respect to a vertex u of K is $q_u(x) = \alpha_i$ if $u = a_i$ and $q_u(x) = 0$ if u is not a vertex of σ. It is important to note that $q_u : |K| \to [0, 1]$ is continuous for each u; q_u is non-zero only for $x \in |\overline{\operatorname{St}} u|$. We remark that barycentric coordinates have been used before in cartography to construct so-called rubber-sheet maps solving a problem different from the construction of cartograms [12].

Let K and L be two simplicial complexes. A *vertex map* $f : \operatorname{vert} K \to \operatorname{vert} L$ has the property that the vertices of a simplex in K are mapped to (not necessarily distinct) vertices of a simplex in L. f can be extended in an unambiguous way to a *simplicial map* $g : |K| \to |L|$ defined by

$$g(x) = \sum_{u \in \operatorname{vert} K} q_u(x) \cdot f(u).$$

g is continuous because each q_u is continuous. Furthermore, if f is a bijection, then g is a *homeomorphism*, that is, g is bijective and g and g^{-1} are both continuous. In this case, we call g a *simplicial homeomorphism*. K and L are *isomorphic* if such a homeomorphism exists.

The composition of two simplicial maps is again a simplicial map, and the composition of two homeomorphisms is again a homeomorphism. We will make extensive use of this composition property and generate homeomorphisms from possibly many components, each a simple piecewise linear homeomorphism.

Tilings. A simplicial complex K *tiles* $\mathbb{R}^2$ if $|K| = \mathbb{R}^2$. In the specialized literature, K would be referred to as a simplicial face-to-face tiling [9], but since we consider only this type we call K simply a *tiling*. A *symmetry* of K is a Euclidean motion (translation, rotation, reflection) that maps each simplex of K to a simplex of

K. It is a particularly simple homeomorphism from $|K|$ to itself. The set of all symmetries forms the *symmetry group*. The set of translations in this group forms the *translation subgroup*. K is *periodic* if this group contains translations in 2 linearly independent directions.

As an example, consider a tiling T of $\mathbb{R}^2$ consisting of equilateral triangles and their faces, see figure 1. T is unique up to similarity, and for specificity we define T so its vertices have coordinates $(i-\frac{j}{2}, \frac{j\sqrt{3}}{2})$, $i, j \in \mathbb{Z}$. There

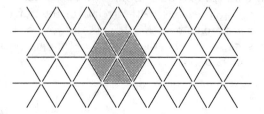

Figure 1: Every vertex in T belongs to 6 triangles defining its star.

is a one-to-one correspondence between the symmetries of T and the injective maps from the 3 vertices of a particular triangle to the vertices of any triangle in T. The symmetry group is therefore isomorphic to $S_3 \times \mathbb{Z} \times \mathbb{Z}$, where S_3 is the group of permutations of 3 elements. The translation subgroup is isomorphic to $\mathbb{Z} \times \mathbb{Z}$, so T is periodic.

3 Construction

Homeomorphisms in this paper are composed of instances of simple units. The most important such units are the inflations and their inverses, the deflations. They can be defined using closed stars, domes, and corridors, which are subcomplexes of the periodic tiling T defined above. Stellar subdivisions are used in the creation of inflations. We begin by defining and discussing inflations on the lowest level of resolution.

Domes and corridors. The closed stars of any two vertices in T are isomorphic; each is defined by 6 triangles, see figure 1. A *dome* is the closure of a set of 4 triangles in contiguous positions in a star $\bar{\text{St}}\, u$. u is the *center vertex* of the dome. The middle 2 triangles form the *diamond* of the dome. A closed star contains 6 different domes. The frontier of a closed star and a dome are isomorphic, each being a cycle of 6 edges and vertices. This will be important for our construction, which maps the interior of a dome to the interior of a star.

A dome has one of 6 orientations specified by the 2 rows of triangles that need to be altered when the dome expands to a star. To formalize this observation, consider a dome D with center vertex u, and let uv be the edge decomposing D into 2 triangles on each side. v is the *start vertex*, $e = u - v$ the *direction vector*, and $C = \{\sigma + ie \mid \sigma \in D, i \geq 1\}$ is the *corridor* of D, see figure 2.

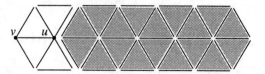

Figure 2: The corridor of a dome is an infinite subcomplex of T. It shares 2 edges and 3 vertices with the dome.

Inflations and deflations. Let $D \subseteq T$ be a dome, u the center vertex, and C the corridor. We construct 2 stellar subdivisions T' and T'' of T and a simplicial homeomorphism $\iota : |T'| \to |T''|$ equal to the identity except inside D and C. The 2 subdivisions are obtained through sequences of starring operations inside D and C, see figures 3 and 4. Observe that T' and T'' are

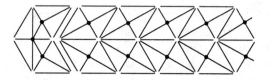

Figure 3: To obtain T' from T subdivide the 4 triangles of the dome D into 10 triangles using starring operations adding the midpoints of the 3 interior edges of D. The corridor is decomposed into convex quadrilaterals. Each quadrilateral consists of 2 triangles and is further subdivided by a starring operation adding the midpoint of the dividing edge.

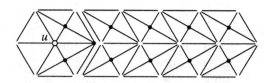

Figure 4: To obtain T'' from T subdivide the 6 triangles of $\bar{\text{St}}\, u$ into 10 triangles using starring operations adding the midpoints of 2 interior edges. The part of the corridor outside $\bar{\text{St}}\, u$ is decomposed into convex quadrilaterals similar to but different from before. Each quadrilateral consists of 2 triangles and is further subdivided by a starring operation adding the midpoint of the dividing edge.

isomorphic, so there is a unique bijective vertex map $f : \text{vert}\, T' \to \text{vert}\, T''$ equal to the identity everywhere except inside D and C. Its extension $\iota : |T'| \to |T''|$ is a simplicial homeomorphism. Since $|T'| = |T''| = |T| = \mathbb{R}^2$, $\iota : |T| \to |T|$ is a piecewise linear homeomorphism albeit not defined by a vertex map for T.

We call ι an *inflation*; its inverse, $\delta = \iota^{-1}$, is called a *deflation*. ι is completely determined by the choice of the dome, D. Intuitively, it moves every vertex in the interior of $D \cup C$ by one position along e. The first vertex in this sequence is freed up and used to expand the dome to a star. The triangles in C are deformed but area remains the same. Triangles outside D and C are left untouched. ι increases area only inside D. More precisely, the diamond of D is increased to twice the area, and the area of the 2 end triangles remains unchanged. The corresponding deflation halves the area of the quadrilateral inside St u whose image is the diamond of D. The center vertex, u, is removed by the deflation.

Composition. The stellar subdivision T' of T is really the superposition of T and the preimage of T under ι. Similarly, T'' is the superposition of T and the image of T under ι. It is interesting to look at the preimage and image of T under the composition of several inflations. Let us consider an example of 3 inflations which distribute the impact of local growth in 3 different directions. Figure 5 shows the 3 domes that define the inflations ι_1, ι_2, ι_3 with center vertices 1, 2, 3. We execute the inflations in this sequence and thus consider $h = \iota_3 \circ \iota_2 \circ \iota_1$. The expansion effect is strongest for the central triangle shared by all 3 domes, see figure 5. Composition is not commutative, which can be seen from the asymmetry of the image. Figure

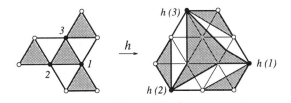

Figure 5: The 3 inflations are defined by overlapping domes covering 7 triangles of the tiling. Any 2 domes overlap in 2 triangles, and all 3 domes overlap in 1 triangle. The image of the union of domes is a hexagon decomposed into 7 regions each corresponding to one of the 7 triangles.

6 shows the hexagon decomposed by triangles of T and their preimages under h. Symmetrically, figures 5 and 6 show the preimage and image of the homeomorphism $\delta_1 \circ \delta_2 \circ \delta_3 = h^{-1}$ composed of 3 deflations.

Instead of defining h as the composition of 3 inflations, we can also design our own image of the 7 triangles decomposing the union of domes. For example,

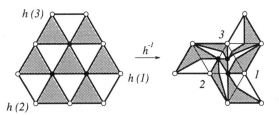

Figure 6: In T, the image of the union of 3 domes is decomposed into $13 = 7 + 3 \cdot 2$ triangles. The preimages of the 13 triangles form a pattern of 13 regions inside the union of domes.

the central region in the right part of figure 5 could be drawn with straight edges thus producing a uniform expansion to 7 times the original area. There are indeed plenty of possibilities resulting in different homeomorphic deformations. Similarly, we can create other operations, or *aggregates*, for special deformation tasks.

Knees and forks. Two particular aggregates are now described. They can be used to redirect the directional effect of an inflation or a deflation. The *knee* bends a corridor by an angle of 60 degrees, see figure 7. A knee,

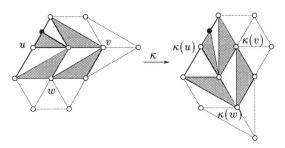

Figure 7: The knee bends the corridor of an inflation by 60 degrees.

k, is similar to a deflation, δ, followed by an inflation, ι, with dome inside the corridor of δ and direction vector 60 degrees different from that of δ. The difference between k and $\iota \circ \delta$ is the local deformation in the knee area, which is less for k.

The *fork* is shown in figure 8. It splits a corridor into two. If applied together with an inflation, it halts the deformation along the corridor and propagates half the amount to each of the new corridors. A fork, y, is similar to a deflation, δ, followed by two half inflations, ι_0 and ι_1. Half inflations will be defined in the next paragraph considering coarse resolution levels. The local deformation in the forking region is again less for y than it is for $\iota_1 \circ \iota_0 \circ \delta$. The part of the tiling affected by a knee or a fork can be described as the union of 2 or 3 corridors. This is important in the discussion of data structures used in the construction of homeomorphisms.

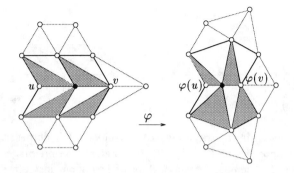

Figure 8: The fork diverts the impact of an inflation into 2 directions.

Coarse resolution levels. For $\varrho \in \mathbb{R}$ define $\varrho T = \{\varrho\sigma \mid \sigma \in T\}$, where $\varrho\sigma = \{\varrho p \mid p \in \sigma\}$ is a scaled copy of σ. Clearly, $1T = T$, and $\varrho T = (-\varrho)T$ because central reflection through the origin belongs to the symmetry group of T. For every positive integer k, T is a subdivision of kT. Consider some fixed $k \geq 2$ and an inflation ι defined within kT, see figure 9. ι doubles the area of

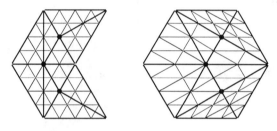

Figure 9: An inflation in $4T$. The triangles in the diamond of the dome are expanded uniformly.

$2k^2$ triangles in T, which is the same amount of growth achieved by k^2 inflations within T. The number of rows in the corridor of ι is $2k$, which is considerably less than the total of $2k^2$ rows in the corridors of the k^2 inflations on the lowest level of resolution. This suggests a hierarchical approach to constructing homeomorphisms as the composition of simple deformations on different levels of resolution. This idea is further developed in section 4 where only powers of 2 are allowed as values of k. In a sense, we write the homeomorphism in binary notation.

A possibility suggested by revisiting inflations on a coarse resolution level is the idea of a *fractional* inflation: the center vertex, u, is moved by less than one position in the direction $e = u - v$. Indeed, in kT there are $k - 1$ vertices of T strictly between u and $u + e$. Any one of these can be chosen as the image of u. The effect of a fractional inflation is a milder expansion of the diamond region and a milder shift along the corridor. Of course, in principle the image of u can be chosen anywhere along the axis of the inflation and does not

even have to be a vertex of T. The only limitation to the freedom of design is the increase in complexity and difficulty of an efficient algorithmic realization.

Matching strategy. The basic idea for finding simple deformations that compose to a desired homeomorphism is to transport area from regions of surplus to regions of need. An inflation-deflation pair with matching corridors transports area from the deflation to the inflation diamond. If no common corridor is available, we can use knees and forks to design more complicated transportation channels.

The way surplus regions are matched with regions of need has a profound influence on the quality of the generated homeomorphisms and the running time of the algorithm. At the same time, this is currently the least developed part of our method. The matching algorithm we use is heuristic, and at this time we merely have a list of recommendations supported by experimental evidence.

(i) Use fractional operations (inflations, deflations, knees, etc.) to produce homeomorphisms without large distance distortion.

(ii) Plan non-crossing transportation channels, if possible. Indeed, preliminary results suggest short transportation channels more or less evenly distributed over the geographic map work well.

(iii) To save execution time, choose and execute inflation-deflation pairs in large groups before re-computing the deformed map.

4 Array Based Data Structure

We describe a hierarchical data structure keeping track of the homeomorphism as it evolves. Each level of the hierarchy is a 2-dimensional array representing $T_\ell = kT$, with $k = 2^\ell$ and ℓ a non-negative integer. The array also represents the inflations defined within T_ℓ. See also section 5 where trees are used to represent the hierarchy. Levels are indexed from bottom up, so the lowest level is $T_0 = T$. We begin by describing a single level and then discuss how they are connected.

Array of queues. The data structure for the tiling T_ℓ is a 2-dimensional array $\mathtt{A}_\ell[i, j]$, $i, j \in \mathbb{Z}$. For now we assume T_ℓ and $\mathtt{A}_\ell$ are infinite in all directions. Each element $\mathtt{A}_\ell[i, j]$ is a vertex of T_ℓ, and the neighboring vertices are given by index pairs $[i-1, j-1]$, $[i-1, j]$, $[i, j-1]$, $[i, j+1]$, $[i+1, j]$, $[i+1, j+1]$, see figure 10. The coordinates of the vertex $\mathtt{A}_\ell[i, j]$ in T_ℓ are

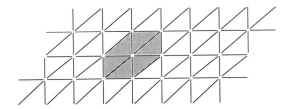

Figure 10: The regular tiling with squares decomposed into 2 triangles each by mutually parallel diagonals is isomorphic to T.

$$\begin{pmatrix} 2^\ell & -\frac{2^\ell}{2} \\ 0 & \frac{2^\ell \sqrt{3}}{2} \end{pmatrix} \cdot \begin{pmatrix} i \\ j \end{pmatrix}. \quad (1)$$

It is convenient to implicitly assume the transformation and to call i and j the coordinates of the vertex.

An inflation, ι, is specified by a dome, D, determined by its center vertex, $u = u_\iota$, and its direction vector, $e = e_\iota$. There are only 6 possibilities for e, namely $(-1,-1)$, $(-1,0)$, $(0,-1)$, $(0,1)$, $(1,0)$, and $(1,1)$. To record the changes caused by ι, each vertex $u_\lambda = u + \lambda e$, $\lambda \geq 0$ and $\lambda \in \mathbb{Z}$, stores e to indicate its motion under ι. We have $\iota(u_\lambda) = u_{\lambda+1}$ for all $\lambda \geq 0$ and $\iota^{-1}(u_\lambda) = u_{\lambda-1}$ for all $\lambda \geq 1$. The preimage of u_0 is $u_0 - \frac{1}{2}e$, which is not a vertex in A_ℓ. To indicate that ι *creates* u_0 we mark the label e recorded with u_0.

In the general situation we deal with a sequence of inflations on level ℓ. Let $h = \iota_n \circ \ldots \circ \iota_2 \circ \iota_1$ be their composition, and call the index the *time stamp*. A single vertex $v = \mathsf{A}_\ell[i,j]$ may be affected by several of these inflations, and it stores them in chronological order in a queue of its own, Q_v. The entry $e_t = e_{\iota_t}$ in Q_v indicates the image and preimage of v under ι_t. To speed up path traversals, we equip e_t with a pointer $\mathsf{n}_{v,t}$ to the entry $e_{t'}$ in $\mathsf{Q}_{\iota_t(v)}$, with minimal $t' > t$. $\mathsf{n}_{v,t}$ is undefined if there is no such t'. Similarly, we equip e_t with a pointer $\mathsf{p}_{v,t}$ to the entry $e_{t''}$ in $\mathsf{Q}_{\iota_t^{-1}(v)}$ with maximal $t'' < t$. $\mathsf{p}_{v,t}$ is undefined if there is no such t'' or $\iota_t^{-1}(v)$ is not a vertex in A_ℓ. In the latter case, e_t is marked indicating v is created by ι_t.

Tracing paths. The queue entries provide enough information to trace the path a point takes under the sequence of inflations. Consider first a vertex v in A_ℓ. The path leading from v to its image under h can be traced by following n pointers. The path starts with the first entry in Q_v and ends when n is not defined. Similarly, the path from v to its preimage under h can be traced by following p pointers. The path starts with the last entry in Q_v and ends when p is undefined.

Every point $x \in \mathbb{R}^2$ can be specified by its non-zero barycentric coordinates with respect to vertices of T_ℓ.

In all cases, the image of x is obtained with the same coordinates from the images of the vertices: if $x = q_u u + q_v v + q_w w$ then
$$\iota(x) = q_u \iota(u) + q_v \iota(v) + q_w \iota(w).$$
In many cases, the triangle in T_ℓ that contains $\iota(x)$ is $\iota(uvw)$. Otherwise, it is one of at most 3 triangles overlapping $\iota(uvw)$. Similarly,
$$\iota^{-1}(x) = q_u \iota^{-1}(u) + q_v \iota^{-1}(v) + q_w \iota^{-1}(w),$$
except in cases where the preimage of uvw is not a geometric triangle. For each inflation there are 4 such triangles, namely the ones that partially overlap the image of the dome diamond, see figure 4. Such a triangle is cut into half by adding the midpoint of an edge. After computing the new barycentric coordinates of x, considering this midpoint a new vertex, we can find $\iota^{-1}(x)$ as before.

Only minor modifications are necessary to adapt the data structure and path tracing strategy to handle deflations and aggregates. We need pointers along all corridors making up an aggregate. The other changes have to do with the fact that deflations and aggregates do not necessarily map vertices to vertices. This is the case for one vertex in a deflation and can be the case for most of the vertices affected by an aggregate. As a result, an undefined n pointer for v does not imply no deformation applies to v, but merely suggests v needs to be treated like an arbitrary point of $\mathbb{R}^2$ from then on. To distinguish these two possibilities we add a field to the queue entries identifying the corresponding deformations.

Hierarchy of tilings and arrays. We call the infinite sequence $T_0, T_1, \ldots$ a *tiling hierarchy* and we use the corresponding sequence of arrays $\mathsf{A}_0, \mathsf{A}_1, \ldots$ to represent the homeomorphism defined by inflations within the T_i. T_ℓ is a subdivision of $T_{\ell+1}$. More precisely, each triangle in $T_{\ell+1}$ is subdivided into 4 triangles in T_ℓ. It follows the vertex $\mathsf{A}_{\ell+1}[i,j]$ is the same as $\mathsf{A}_\ell[2i,2j]$. Only vertices with even coordinates in A_ℓ exist also in $\mathsf{A}_{\ell+1}$.

By simple index manipulation (multiplication with and division by 2) we can go from one array to the next. This does not mean it is easy to compose inflations on different levels in any arbitrary sequence. Indeed, we need to avoid pointers from lower to higher levels. An inflation on a high level can otherwise require a pointer each from a large number of triangles on lower levels. The effort to establish these pointers would defeat the whole purpose of inflations on high levels speeding up the computations. For this reason we require the inflations on level $\ell+1$ precede the inflations on level ℓ, for all ℓ. The gain in efficiency seems to justify the associated loss in flexibility. After all, it is natural to first get the rough features of the homeomorphism right before working on its details.

5 Tree Based Data Structure

As an alternative to arrays, we can base the representation of the tiling hierarchy on trees. The storage improves to constant per inflation, but tracing a path slows down and is no longer constant time per step. The basic idea is to store corridors with a constant number of descriptors. It is thus suitable for inflations, deflations, knees, and forks alike.

Two Key Search. Suppose we are given a sequence of integer items, $\phi_1, \phi_2, \ldots, \phi_n$. We seek a data structure for the ϕ_i so the following kind of query can be answered quickly: given a pair of integers (t, x), find the smallest $i > t$ with $\phi_i < x$, see figure 11. We store the items

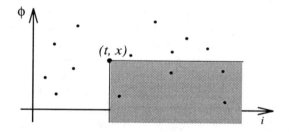

Figure 11: Plot each item ϕ_i as a point with coordinates (i, ϕ_i) and convert the query into the open orthant below and to the right of the point (t, x). The answer is the leftmost point (i, ϕ_i) inside the orthant.

in the leaves of a binary tree T [1, chapter III]. T is balanced and the leaves are sorted by item index from left to right. Each interior node $\nu \in$ T stores the range of indices in its subtree, $[a_\nu, b_\nu]$, and the index, $m(\nu) \in [a_\nu, b_\nu]$, of the smallest item.

To answer a query (t, x), traverse T starting at its root. At an interior node ν we distinguish several cases:

Case 1. $\phi_{m(\nu)} \geq x$. Return the empty answer.

Case 2. $\phi_{m(\nu)} < x$.

    Case 2.1. $b_\nu \leq t$. Return the empty answer.

    Case 2.2. $t < b_\nu$. Compute the answer for the left subtree and return it if non-empty. Otherwise, compute and return the answer for the right subtree.

The time required to find the answer is proportional to the height of the tree. To see this note there are 3 types of nodes ν characterized by $a_\nu \leq b_\nu \leq t$, $a_\nu \leq t < b_\nu$, and $t < a_\nu \leq b_\nu$. The nodes of second type lie on a single path traversed in case 2.2 of the algorithm. A node of first type takes only constant time, case 2.1, and is visited only if it is the root or the child of a node of second type. A node of third type either takes only constant time for one test, case 1, or it for sure returns a non-empty answer. Also, it is visited only if it is the right child of a node whose left child returns the empty answer. After the first time a node of third type returns a non-empty answer, this case cannot happen any more. It follows a query is answered by traversing at most 2 paths in T.

Storing corridors. Consider an inflation, ι, on the finest level of resolution, in the tiling $T = T_0$. There are 6 possible direction vectors, and we have a data structure for each possibility. The *axis* of ι is the oriented line $l : x = u + \lambda e$, $\lambda \in \mathbb{R}$, determined by the center vertex, $u = u_\iota$, and the direction vector, $e = e_\iota$. Observe the axis splits the dome and its corridor along the middle. The two key item associated with ι consists of its time stamp or index, i, and the position ϕ_i of the start vertex, $v = u - e$, along l. All inflations with the same axis l are stored in a tree T_l as described above. Similarly, all other deformation types are represented by corridors, and the ones with axis l are stored in T_l, together with the inflation corridors.

For a vertex $x \in T$ on l and a time t, the next deformation affecting x can be computed by searching T_l as explained. Since x belongs to 6 axes, one per direction vector, we really need to search 6 trees and take the earliest deformation among the 6 answers.

For each direction vector, there is a countably infinite sequence of parallel axes. Only finitely many are associated with deformations performed by the algorithm, and these are stored in a dictionary, maybe a sorted tree [1]. Each entry in the dictionary corresponds to an oriented line, l, and stores the corresponding tree T_l. To find the next deformation for a vertex $x \in T$ we thus search all 6 dictionaries and one tree inside each dictionary.

The same idea also works for tilings T_ℓ on higher levels of resolution. Note, however, that an axis l in T_ℓ affects vertices on $2^\ell - 1$ lines in T parallel to l. Only the middle line, l itself, also belongs to T_ℓ. If a vertex v lies on a line in T but not in T_ℓ, we need to search 2 axes for this direction vector, namely the 2 axes in T_ℓ next to v on both sides.

The tree based data structure consists of 6 dictionaries per level of the tiling hierarchy. Each entry stores a non-empty tree T. Each deformation is stored in a leaf each of 6 trees T. The total amount of storage is therefore proportional to n, the number of unit deformations. The cost per step in tracing a path is $O(\log n)$ time per level of resolution. There are at most $1 + \log_2(N+1)$ levels if $N \geq 2$ is an upper bound on the absolute size of the coordinates of all relevant vertices in T.

Intervals of axes. It is interesting to observe that in the tree based data structure deformations on different levels of resolution can be mixed freely. Indeed, with a minor modification of the data structure it is possible to perform deformations in tilings kT for any positive integer k. The modification substitutes two segment trees [7] for the list of dictionaries per direction vector. They are used to discriminate between deformations affecting an axis in T and deformations that have no effect on this axis. More specifically, the two segment trees store two intervals (of parallel axes) for each corridor. If the deformation is defined in kT, the two intervals consist of the $k-1$ lines on one side of the axis and the k lines on the other side, including the axis itself. The reason the $2k-1$ lines are stored in two intervals, rather than one, is that the starting points at which the deformation affects a line are aligned differently on the two sides of the axis.

A vertex $x \in T$ determines an axis and the two segment trees are used to identify the set of intervals containing this axis. This set is given as the disjoint union of roughly $2\log_2 n$ smaller sets, each represented by a tree T as described above. To answer a query we first search the two segment trees and then search logarithmically many trees T, which in total takes time $O(\log^2 n)$. The entire data structure consists of 12 segment trees, two per direction vector, and the amount of storage it requires is $O(n \log n)$.

6 Measuring Quality

The homeomorphism solving a cartogram problem is by no means unique. It is therefore essential to develop a computable notion of quality that permits the comparison of different homeomorphisms. We consider two such notions: the worst and the average distance distortion. The discussion focuses on inflations, and all results can be extended to include other types of deformations.

Linear pieces. An inflation consists of 5 linear (or rather affine) pieces. To describe them, let ι be an inflation with dome D, center vertex u, direction vector e, and corridor C. The diamond of D consists of 2 triangles, σ_ℓ and σ_r, to the left and right of the axis $l : x = u + \lambda e$. The two sides of C (together with the 2 non-diamond triangles of D) to the left and the right of l are denoted C_ℓ and C_r. Outside $D \cup C$, ι is the identity map. Inside $D \cup C$ it consists of a linear map each for σ_ℓ, σ_r, C_ℓ, and C_r. In each case, the map is of the form $\iota(x) = Ax + b$, and A and b are readily computed from the vertex coordinates of a single triangle and its image. For example, assuming $e = (1,1)$, the 4 matrices A for σ_ℓ, σ_r, C_ℓ, C_r are

$$\begin{pmatrix} 2 & 0 \\ 1 & 1 \end{pmatrix}, \begin{pmatrix} 1 & 1 \\ 0 & 2 \end{pmatrix}, \begin{pmatrix} 2 & -1 \\ 1 & 0 \end{pmatrix}, \begin{pmatrix} 0 & 1 \\ -1 & 2 \end{pmatrix}; \quad (2)$$

the corresponding translation vectors b are not important for the discussion in this section. The growth factor for area is the determinant of A, and indeed it is 2 for σ_ℓ and σ_r, and 1 for C_ℓ and C_r.

Singular values. The effect of A on distance between points is measured by the two *Lipschitz numbers*,

$$L_A = \max\left\{\frac{\|Ax\|}{\|x\|} \mid 0 \neq x \in \mathbb{R}^2\right\} \text{ and}$$

$$\ell_A = \min\left\{\frac{\|Ax\|}{\|x\|} \mid 0 \neq x \in \mathbb{R}^2\right\},$$

where $\|y\| = |y0|$ is the Euclidean distance of y from the origin $0 \in \mathbb{R}^2$. It can be studied by considering the image of the unit circle, $\mathbb{S}^1 = \{x \in \mathbb{R}^2 \mid \|x\| = 1\}$. $A(\mathbb{S}^1)$ is an ellipse determined by the singular value decomposition of A, see [5, chapter 2.5]. This is a decomposition of the form

$$A = U \cdot \begin{pmatrix} s_1 & 0 \\ 0 & s_2 \end{pmatrix} \cdot V, \quad (3)$$

where U and V are orthogonal matrices (rotations in $\mathbb{R}^2$). The diagonal entries, $s_1 \geq s_2$, are the *singular values* of A. s_1 and s_2 are the lengths of the ellipse axes, see figure 12. It follows that $L_A = s_1$ and $\ell_A = s_2$. To

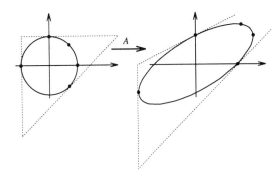

Figure 12: The matrix expanding the upper triangle of a diamond maps the unit circle to the ellipse with axes of length approximately 2.288 and 0.874.

compute singular values we can use their relationship

with the eigenvalues of $A^T A$. Specifically, the singular values are square-roots of the roots of the characteristic polynomial

$$\det\left(A^T A - \begin{pmatrix} \lambda & 0 \\ 0 & \lambda \end{pmatrix}\right),$$

see [5, chapter 8.5]. As an example, consider the characteristic polynomial for the first matrix A in (2), which is

$$(5-\lambda)(1-\lambda) + 1.$$

Its roots are $\lambda = 3 \pm \sqrt{5}$, and the Lipschitz numbers are $L_A = s_1 = \sqrt{3+\sqrt{5}} \approx 2.288$ and $\ell_A = s_2 = \sqrt{3-\sqrt{5}} \approx 0.874$. Note, however, that this only measures the distortion in the decomposed square tiling, and the distortion in the corresponding regular tiling T is less. It is given by the Lipschitz numbers

$$L_C = \sqrt{\frac{8+\sqrt{28}}{3}} \approx 2.105$$

and

$$\ell_C = \sqrt{\frac{8-\sqrt{28}}{3}} \approx 0.950$$

of $C = B \cdot A \cdot B^{-1}$, where B is the matrix in (1).

Dag of matrices. In general, we study a piecewise linear homeomorphism $h = \iota_n \circ \ldots \circ \iota_2 \circ \iota_1$. The deformation of each linear piece is given by a product of matrices as in (2). To determine the singular values of such a product there is little else we can do than first compute the product matrix and then its singular values. All matrix entries are small integers, $-2, -1, 0, 1, 2$, so we can compute products exactly. We can improve the overall running time by exploiting that the matrix chains overlap in possibly large portions.

Specifically, we define a directed acyclic graph, G, whose nodes are triangles in T with time stamps in $\{0, 1, \ldots, n\}$. The source nodes are triangles with time stamp 0 and the sink nodes have time stamp n. There is an edge from (σ, i) to $(\sigma', i+1)$ if $\iota_i(\sigma)$ and σ' overlap. Each edge is labeled with the corresponding matrix, A. Paths from source to sink nodes are chains of matrices. Most edges are labeled with the identity matrix, and we may choose to ignore them by identifying their 2 triangles, which are the same anyway only with different time stamps. By definition of inflation, each node has at most 2 incoming and at most 3 outgoing edges, see figures 13 and 3. G is implicit in the data structures described in sections 4 and 5, which provide images of triangles indirectly through the images of their vertices. The composition of the n inflations can be computed by

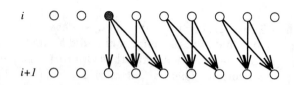

Figure 13: A row of triangles in T at time i and $i+1$. The shaded node corresponds to the triangle in the dome diamond of ι_i. It expands to twice its original area and it is the only triangle whose image overlaps 3 triangles in T.

traversing the entire G from sources to sinks. A breadth-first traversal avoids recomputing products of partial matrix chains [1, chapter VI.23]. A triangle (σ, i) is the last node of a number of initial matrix chains, each associated with a piece of σ. The pieces subdivide σ. If (σ, i) has 2 (or 3) outgoing edges then this does not imply each of its pieces is split into 2 (or 3). Indeed, many paths do not correspond to any non-empty linear piece of h. Computing matrix products along such paths can be avoided by keeping track of the pieces subdividing each σ.

Distance distortion. The extreme distance distortions are captured by the *Lipschitz numbers* L_h and ℓ_h, which are the maximum and minimum of

$$\left\{\frac{|h(x)h(y)|}{|xy|} \mid x \neq y \in \mathbb{R}^2\right\}.$$

Both can be computed from the matrix products at the sink nodes of G. Another relevant quality measure is the *condition number*,

$$C_h = \max_\zeta \left\{\frac{L_{h|\zeta}}{\ell_{h|\zeta}} - 1\right\},$$

defined over all maximal $\zeta \subseteq \mathbb{R}^2$ so $h|\zeta$, the restriction of h to ζ, is linear. Large ratios occur in transition regions between small and large area distortions. Small condition numbers indicate good quality homeomorphisms. More tolerant to large condition numbers in small regions than C_h is the *average condition number*,

$$\bar{C}_h = \sum_\zeta \text{area}(\zeta) \cdot C_{h|\zeta}.$$

By definition, the condition number $C_{h|\zeta}$ vanishes in undeformed regions of $\mathbb{R}^2$, so their contribution to $\bar{C}_h$ is zero.

7 Implementation and Results

The performance of our approach in terms of quality of generated cartograms is illustrated with examples redrawing the states of the USA. Area for each state represents the number of electoral votes in the 1992 presidential election.

Actual implementation. The algorithm outlined in the earlier sectons is occasionally vague and allows implementations that differ in possibly significant details. We list a few design heuristics used in the current version of our software, which is based on trees representing the tiling hierarchy.

Following recommendation (i) in section 3, we use deflation factors ranging from 0.5 though $1 - \frac{1}{\varrho+1}$ and inflation factors ranging from $1 + \frac{1}{\varrho}$ through 2.0 on resolution level $\varrho \geq 1$.

To simplify the matching problem, we replace the deflation operation described in section 3 by a similar but different aggregate. It restricts the area reduced by the deformation to within a diamond of two triangles, just as in the case of an inflation.

To limit the number of crossings among the transportation channels, we first pick a direction and attempt to choose all initial inflation-deflation pairs with corridors along this direction. Remaining surplus is propagated to the closest rows with deficiency. This is accomplished by a combination of left- and right-knees. With this strategy, it is possibly to inflate or deflate different portions of the same district or state in a uniform manner.

It was found that visually unpleasant side effects of deformations decrease with finer tilings. Experience seems to suggest that the triangle edges should be shorter by a factor between 10 and 20 than the "smallest feature size" of the geographic map. Finer tilings do not imply a direct increase in complexity because many operations are executed on course resolution levels.

Sample deformations. Figure 14 shows the geo-

Figure 14: The states of the USA approximating actual shape and size.

graphically accurate decomposition of the USA into states, not including Alaska and Hawaii. Figure 15

Figure 15: The deformed United States at an intermediate stage of the computations.

shows the deformation of the USA at an intermediate stage of the algorithm, with average cartographic error of about 50%. In [3], the *error* for a district or state ζ is defined as

$$\max\{\frac{a(\zeta)}{t(\zeta)}, \frac{t(\zeta)}{a(\zeta)}\} - 1,$$

where $a(\zeta)$ is the actual area (after deformation) and $t(\zeta)$ is the targeted area in the cartogram. The home-

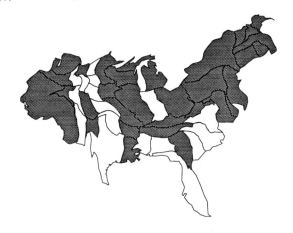

Figure 16: A cartogram of the United States based on the electoral votes in the 1992 presidential elections.

omorphism whose image is shown in figure 16 deforms the USA so the area of the states reflects the number of electoral votes in the 1992 presidential elections. Shading indicates states with a majority of votes for the elected president, Bill Clinton. The cartogram has average cartographic error of less than 1% and the maximum error for any state is 2.33%. In addition to the techniques described in this paper, the computation leading to this cartogram iterated the deformation by slowly deforming states one by one, see also [11]. The

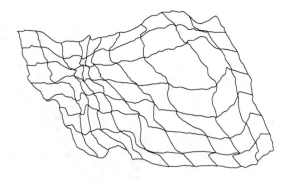

Figure 17: The image of a regular square grid under the homeomorphism of figure 16.

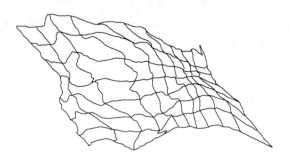

Figure 18: The preimage of a regular square grid under the homeomorphism of figure 16.

deformation resulting from the application of a homeomorphism is intuitively visualized through the image and the preimage of the regular square grid, see figures 17 and 18.

8 Remarks

This paper describes a combinatorial approach to homeomorphisms generating cartograms. Even though combinatorial methods are introduced to describe homeomorphisms and to manipulate them, the numerical nature of the problem is still reflected by the overall iterative strategy and the resolution of the tiling underlying the deformation process.

It is premature to judge the direct practical relevance of the algorithm described in this paper. Several steps still have to be refined and tuned to produce accurate as well as visually pleasing cartograms. For example, none of the quality measuring and improving mechanisms suggested in section 6 have been implemented. The matching strategy moving area by pairing inflations with deflations is currently rather heuristic. A more rational choice of pairs has the potential to vastly improve the quality of the generated homeomorphisms, and indirectly the running time of the algorithm. It would be interesting to analyze convergence rates for different matching strategies.

Acknowledgements

The authors thank Jack Snoeyink for bringing the cartogram problem to their attention, and Michael McAllister for providing pointers to the literature on cartograms.

References

[1] T. H. CORMEN, CH. E. LEISERSON AND R. L. RIVEST. *Introduction to Algorithms.* MIT Press, Cambridge, Massachusetts, 1990.

[2] D. DORLING. Map design for census mapping. *The Geographic J.* **30** (1993), 167–179.

[3] J. A. DOUGENIK, N. R. CHRISMAN AND D. R. NIEMEYER. An algorithm to construct continuous area cartograms. *Professional Geographer* **37** (1985), 75–81.

[4] P. J. GIBLIN. *Graphs, Surfaces and Homology.* 2nd edition, Chapman and Hall, London, 1981.

[5] G. H. GOLUB AND CH. F. VAN LOAN. *Matrix Computations.* The Johns Hopkins University Press, 2nd edition, Baltimore, 1989.

[6] P. M. GRUBER AND J. M. WILLS (EDS.) *Handbook of Convex Geometry, volumes A and B.* Elsevier, North-Holland, Amsterdam, 1993.

[7] K. MEHLHORN. *Data Structures and Algorithms 3: Multi-dimensional Searching and Computational Geometry.* Springer-Verlag, Heidelberg, Germany, 1984.

[8] J. R. MUNKRES. *Elements of Algebraic Topology.* Addison-Wesley, Redwood City, California, 1984.

[9] E. SCHULTE. Tilings. In *Handbook of Convex Geometry*, eds. P. M. Gruber and J. M. Wills, Elsevier, North-Holland, Amsterdam, 1993.

[10] W. R. TOBLER. A continuous transformation useful for districting. *Annals, New York Acad. Sci.* **219** (1973), 215–220.

[11] V. S. TIKUNOV. Anamorphated cartographic images: historical outline and construction technique. *Cartography (Austral.)* **17** (1988), 1–8.

[12] M. S. WHITE, JR. AND P. GRIFFIN. Piecewise linear rubber-sheet map transformation. *The American Cartographer* **12** (1985), 123–131.

Map Labeling Heuristics:
Provably Good and Practically Useful *

Frank Wagner †   Alexander Wolff ‡

Abstract

The lettering of maps is a classical problem of cartography that consists of placing names, symbols, or other data near to specified sites on a map. Certain design rules have to be obeyed. A practically interesting special case, the Map Labeling Problem, *consists of placing axis parallel rectangular labels of common size so that one of its corners is the site, no two labels overlap, and the labels are of maximum size in order to have legible inscriptions.*

The problem is $\mathcal{NP}$-hard; it is even $\mathcal{NP}$-hard to approximate the solution with quality guaranty better than 50 percent. There is an approximation algorithm A with a quality guaranty of 50 percent and running time $\mathcal{O}(n \log n)$. So A is the best possible algorithm from a theoretical point of view. This is even true for the running time, since there is a lower bound on the running time of any such approximation algorithm of $\Omega(n \log n)$.

Unfortunately A is useless in practice as it typically produces results that are intolerably far off the maximum size.

The main contribution of this paper is the presentation of a heuristical approach that has A's advantages while avoiding its disadvantages:
1. It uses A's result in order to guaranty the same optimal running time efficiency; a method which is new as far as we know.
2. Its practical results are close to the optimum.

The practical quality is analysed by comparing our results to the exact optimum, where this is known; and to lower and upper bounds on the optimum otherwise.

The sample data consists of three different classes of random problems and a selection of problems arising in the production of groundwater quality maps by the authorities of the City of München.

1 Introduction

Map lettering is one of the classical key problems that has to be solved in the process of map production. Usually the map producer does not only want to show the exact geographic positions of the features depicted but also explain properties of these features. She has to arrange this information on the map so that:
— for every piece of information it is intuitively clear which feature is described;
— the information is of legible size;
— different texts do not overlap.
These and in addition a lot of esthetic criteria are described by Imhof [5] in an attempt to characterize good quality map lettering having mostly manual map making in mind. Nowadays there is an increasing need for large, especially technical maps, for which legibility is much more important than beauty.

The application which brought the problem to our attention is the design of groundwater quality maps by the municipal authorities of the City of München. They have a net of drillholes spread over the city. The map has to contain the location of these holes and for every hole a block of measuring results such as the concentration of certain chemicals.

The growing importance of such technical maps induces a need for the computerization of map making,

---

*This work was done at the Institut für Informatik, Fachbereich Mathematik und Informatik, Freie Universität Berlin, Takustraße 9, 14195 Berlin-Dahlem, Germany. It was supported by the ESPRIT BRA Project ALCOM II.

† wagner@math.fu-berlin.de
‡ awolff@inf.fu-berlin.de

Permission to copy without fee all or part of this material is granted provided that the copies are not made or distributed for direct commercial advantage, the ACM copyright notice and the title of the publication and its date appear, and notice is given that copying is by permission of the Association of Computing Machinery.To copy otherwise, or to republish, requires a fee and/or specific permission.
11th Computational Geometry, Vancouver, B.C. Canada
© 1995 ACM 0-89791-724-3/95/0006...$3.50

Figure 1: A valid labeling

Figure 2: An optimal labeling for the example of Figure 1

the need for fully automated algorithms. Typically, labels in technical maps are axis-parallel rectangles of identical sizes. By rescaling one of the axes we can assume that the rectangles are squares. An adequate formalization is as follows:

Problem MAP LABELING

Given n distinct points in the plane. Find the supremum σ_{opt} of all reals σ such that there is a set of n closed squares with side length σ, satisfying the following two properties.

1. Every point is a corner of exactly one square.

2. All squares are pairwise disjoint.

We call σ_{opt} the *optimal size*. A set of non-intersecting squares fulfilling (1) and (2) is called a *valid labeling*, see Figure 1 and 2.

Previously [4], we showed by reduction from 3-SAT that the corresponding decision problem is $\mathcal{NP}$-complete. The main result of that paper is an approximation algorithm A that finds a valid labeling of at least half the optimal size. In addition, it is shown that, provided that $\mathcal{P} \neq \mathcal{NP}$, no polynomial time approximation algorithm with a quality guaranty better than 50 percent exists. Related results were reported in [1] and [8]. The running time of A is in $\mathcal{O}(n \log n)$. In [10] we showed that there is a matching lower bound on the running time of $\Omega(n \log n)$.

A conceptually works as follows: We start with infinitesimal equally sized squares attached to each point in all four possible positions. Then all squares are expanded uniformly. In order to resolve conflicts between them, we eliminate all those which would contain another point if they were twice as big. It is easy to show that after this process, a point p can not have more than two squares left which overlap other squares. If we consider p a boolean variable and associate its squares with the values p and $\bar{p}$, we can generate a boolean formula consisting of clauses which encode all conflicts. Suppose the square p was overlapping the square $\bar{q}$ of a point q,

this would give us the clause $\overline{(p \wedge \bar{q})} = (\bar{p} \vee q)$ meaning that we do *not* want p and $\bar{q}$ to be simultanously in the solution. If we join all such clauses with the $\wedge$-operator, the satisfiability of the formula tells us exactly whether there is a solution of the current size. Since all clauses consist of two laterals, the formula is of 2-SAT type, and can be evaluated in time proportional to its length [2].

This works only because we make sure that no point has more than two squares left after the elimination phase. On the other hand, we often eliminate both of two conflict partners, where it would have sufficed to delete one to resolve the conflict. This seems to be the reason for the practically very bad behaviour of A. In fact, A usually produces solutions not much better than 50 percent of the optimum, which makes it nearly useless for practical problems. So we developed a heuristical approach that uses strongly the ideas of A, maintains its quality and running time guaranty, and yields very convincing results. Instead of eliminating the squares as early as possible, it eliminates a square just when it is clear that it cannot be in any solution of the current size. The bad side effect of this is, that some points might have three or four squares left after the elimination phase. In order to handle this, we suggest three different heuristics to bring their number down to two.

The simplest of these heuristics is used by the City of München for the application mentioned above, by the PTT Research Labs of the Netherlands to produce online maps for mobile radio networks, and in a computer system for the automated search for matching constellations in a star catalogue [11] as a tool to label the output on the screen. With a very similar algorithmic approach we were able to solve the so-called METAFONT labeling problem posed by Knuth and Raghunathan [6].

2 Description of the Heuristics

2.1 A Theoretical Foundation

Definition 1 *For a point p in the plane denote by p_i, $i \in \{1,2,3,4\}$, an axis-parallel unit square with p in its southwest, southeast, northeast respectively northwest corner. The enumeration is chosen like that of quadrants. We will call p_i a candidate of the site p.*

For a real $\sigma \geq 0$ denote by σp_i analogously a candidate with edge length σ. Where this edge length is omitted, we refer to a candidate of the current label size. A solution of size σ is a valid labeling with candidates of side length σ.

For technical reasons, we will from now on consider a candidate an open square, plus the open edges incident to the site. Note that this excludes all corner points, especially the site itself. The idea is that we shrink the squares by a tiny bit, so that an optimal labeling is a valid labeling, too.

Definition 2 *of some special label sizes:*

σ_{dead} = *largest label size at which all sites still have a candidate which does not contain a site.*

σ_{opt} = *size of the maximum valid solution. This is equivalent to the previous definition of σ_{opt}.*

σ_{lower} = *size of the solution of the Approximation Algorithm A*

σ_{upper} = $2\sigma_{lower}$

Corollary 3 $\sigma_{lower} \leq \sigma_{opt} \leq \sigma_{upper} \leq \sigma_{dead}$

Proof. $\sigma_{opt} \leq \sigma_{upper}$ is of course due to A's approximation guaranty, see [4].
$\sigma_{upper} \leq \sigma_{dead}$: A stops at the latest at size $\sigma_{dead}/2$, because then there is a point all of whose candidates are eliminated. Therefore $\sigma_{lower} \leq \sigma_{dead}/2$. □

We say that two candidates *overlap* or have a *conflict* if they intersect. Analogously, two points are in conflict if any of their candidates are. One of the key words in the description of the heuristics is that of a *conflict size*. For a pair of candidates we define its conflict size as the largest edge length at which do not intersect. We call a conflict size *interesting*, if both candidates do not contain a site, and if it is not larger than σ_{upper}.

Lemma 4 *The number of interesting conflict sizes is linear.*

Proof. Let s be the vector $(\sigma_{upper}, \sigma_{upper})$, and $\prec$ the lexicographical order on $\mathbb{R}^2$. Given

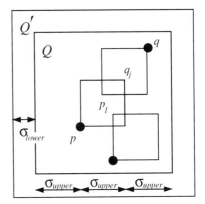

Figure 3:

a candidate p_i, say p_1, we define two squares as in Figure 3, $Q := \{z \in \mathbb{R}^2 \mid p - s \preceq z \preceq p + 2s\}$ and $Q' := \{z \mid p - \frac{3}{2}s \preceq z \preceq p + \frac{5}{2}s\}$, such that $\sigma_{upper}p_1 \subset Q \subset Q'$. Then clearly all sites q with candidates q_j, which might have a conflict with p_1 of size not greater than σ_{upper}, must lie within Q, because its border runs around p_1 at a distance of σ_{upper}. We know that there must be a partial solution of size σ_{lower} for the sites in Q. All candidates of such a solution must lie in Q', so Q cannot contain more than 64 sites. Therefore the number of conflicts of interesting size per candidate is constant. □

2.2 Structure

All three heuristics use a common framework. We first need to run the Approximation Algorithm A to get σ_{upper} and a solution of size σ_{lower}. This takes $\mathcal{O}(n \log n)$ time. What they do then, can be split up into the following parts:

1. Find all interesting conflict sizes.

2. Do a binary search on the interesting conflict sizes between σ_{lower} and σ_{upper}, and check for each size you look at, whether there is a solution or not, by going through the following three phases:

   Phase I: Preprocessing.
   Phase II: Make all decisions, which do not destroy a possible solution.
   Phase III: For those points which still have two or more "active" candidates left, choose exactly two, and check whether this remaining problem is solvable by 2-SAT, as described in the introduction.

The heuristics differ in the way in which they choose those two candidates in Phase III.

2.3 Finding interesting conflict sizes

Since A supplies us with σ_{upper} which is an upper bound for σ_{opt}, we know that during the search for an optimal solution, only conflicts between sites at a distance of at most $2\sigma_{upper}$ in the L_∞-metric, have to be considered. Therefore, we can use a sweep line — or rather, sweep window, approach to determine these conflicts of interest. As usual, we need two data structures: firstly, an event point queue as horizontal structure. This is a queue which holds pointers to the lexicographically ordered sites in the window, that is to all sites of distance at most $2\sigma_{upper}$ left of the sweep line which moves to the right. Further, we need a vertical structure, the sweep window status, which allows us to look up efficiently neighbours of new sites entering the window according to the y-coordinate.

The result of the sweep is a list of all conflict sizes between σ_{lower} and σ_{upper}. We do not have to consider any other label size, since the conflict graph does not change inbetween two consecutive interesting conflict sizes. We use this list afterwards to do a binary search for the best possible solution. In addition to this long list, for every candidate p_i we create a short list consisting of pointers to other candidates q_j, which are overlapping p_i before p_i touches the first site which we call $\delta(p_i)$, or reaches the size σ_{upper}. So for every p_i we need to know $\delta(p_i)$ and $d(p_i) := \|p - \delta(p_i)\|_\infty$ or ∞ if there is no site in the ith quadrant relative to p. This information can be obtained by eight plane sweeps — one for the closest site in every $45°$ octant — in $\mathcal{O}(n \log n)$ time according to [3].

What happens when the right border of the window moves to the lexicographically next site? We want to keep the invariant that we have computed all interesting conflict sizes between the candidates of all sites left of the right border of the window.

OUT: Since there cannot be any such conflict between the new site p entering the window on its right, and sites q leaving it on the left side, we first of all remove them from both the event point queue and the sweep window status. This can be done in constant time per site.

IN: Then we look at all successors (and predecessors) r of p in the vertical structure and compute all conflicts between r's and p's four candidates. With similar arguments as in the proof of Lemma 4 we show that there can only be a constant number of other sites r with $\|p - r\|_\infty \le 2\sigma_{upper}$ in the window, and only the conflicts between those sites r and p are interesting.

We use $(2, 4)$-trees to implement the sweep window status, so inserting p costs $\mathcal{O}(\log n)$ time (see [7]), but accessing a successor or predecessor of p, or deleting p can then be done in constant time, computing the conflicts between its and p's candidates of course, too.

This sums up to a running time of $\mathcal{O}(n \log n)$ for sorting the sites and for the sweep. As a consequence of Lemma 4, it requires only linear space — for the list of all conflict sizes and the short lists stored with every candidate, which have constant length.

2.4 Check whether there is a solution for a fixed label size σ

2.4.1 Phase I: Preprocessing

We run through all the candidates p_i. If $d(p_i) < \sigma$ we *eliminate* p_i, i. e. we will not consider it any more because then σp_i contains $\delta(p_i)$. Otherwise we create a new list of overlap information which is an excerpt from p_i's conflict list. Its elements consist of pointers to the overlap information of those candidates which actually overlap p_i for the given label size σ, the area of the intersection (needed for Heuristic J), and a pointer back to the candidate it belongs to. This can be done in linear time since the sum of the lengths of all conflict lists is linear, confer Section 2.3.

2.4.2 Phase II: Making Decisions

We run once through all sites p. There are three cases:

- If all candidates of p have been eliminated, we stop and return "no solution" to the program which does the binary search on the conflict list.

- If p has candidates free of intersections with other candidates, we choose an arbitrary one of them (say p_i), and eliminate all other candidates p_j of p. Before their deletion, we have to do some updates for each of them: we delete its list of overlap information and the symmetric entries stored with those candidates which overlap it.

- If p has only one candidate p_i left, do the same updates with all candidates q_j which overlap p_i, and then delete them.

While we do this we maintain a stack. On this stack we put all those candidates which now fulfill the same properties as p_i did before, i. e. do not intersect any other squares, or are the last candidates of their sites. Before we look at the next site p, we do all the decisions waiting for us on the stack. Since there is just a linear number of conflicts, and we can detect and delete each of them in constant time, Phase II takes us linear time.

Corollary 5 *If there is a solution of the current label size σ, then there is still one after Phase II.*

Proof. Suppose to the contrary that p_i is the first candidate after whose elimination the remaining problem becomes unsolvable. Then the following statement is true:

($\star$) Every solution π of the problem just before this elimination must contain p_i.

Consider the circumstances under which p_i could have been eliminated:

1. p_i contains a site q. This contradicts ($\star$).

2. p_i does not overlap other candidates, but the same holds for some p_j, and the algorithm decides to eliminate p_i.
   In this case we could replace p_i in π by p_j, contradicting ($\star$).

3. p_i overlaps q_j which is the last candidate of q. Then also q_j must be part of π, which again contradicts ($\star$). □

At the end of Phase II we are done if all sites have exactly one candidate left. Otherwise we know that candidates of sites with several candidates — call them *active* — never intersect with those that are "the last of their breed", i. e. belong to sites with exactly one square left, because then the former ones would have been eliminated. So it is enough to focus on active candidates from now on. The others are already chosen as part of the solution, and do not interfere with the active ones any more.

As a consequence of Corollary 5 we also know that we have not yet returned "no solution" if there is one of size σ. So we could still find a solution with the help of 2-SAT as described before if no site had more than two candidates left. If some do, our heuristics try to get rid of the additional candidates in different ways until they all hand over the remaining problem to 2-SAT. Eliminating candidates, is of course, where we might lose a possible solution of the current size.

2.4.3 Phase III: The Heuristics Come into Play

Heuristic H We randomly choose two of the possible four candidates left per point, before we hand them over to 2-SAT. To increase the probability of a choice which enables a solution, this process can be repeated in case of a negative answer. Three repetitions yield good results without prolonging the running time too much.

Since we look at a (hopefully small) part of the linear number of conflicts, we will only get a linear number of clauses, resulting in a running time of $\mathcal{O}(n)$ for 2-SAT, and for this part of Heuristic H as well.

Heuristic I Here we run through all points with active candidates twice. In the first run, we only look at those with four candidates left, eliminate the one with most conflicts, and make all decisions of the type we did in Phase II. During the second run, we do the same for points which still have three active candidates. Then the remaining problem (consisting only of points with exactly two active candidates) is handed over to 2-SAT.

This takes linear time.

Heuristic J For the third variant, we put all active candidates left into a priority queue according to the sum of all intersection areas of a candidate p_i. We then delete the minimum p_i from the queue, and eliminate all candidates q_j which overlap it, and the other active candidates p_k belonging to p. If any of these decisions induces new ones according to the pattern used in Phase II, then these are made as well, before the next minimum is deleted from the queue. Naturally the sizes of the intersection areas, and the data structure, have to be updated accordingly. This process is repeated until either one point runs out of candidates ("no solution"), or no point has more than two of them left, so the remaining problem can be handed over to 2-SAT.

Using Fibonacci heaps to realize a priority queue that allows inserting and minimum deletions in $\mathcal{O}(\log n)$, and decreasing a key in constant time, this part of Heuristic J can be implemented to run in time $\mathcal{O}(n \log n)$, since there is just a constant number of conflicts to be resolved per candidate we look at.

Since we have to look at $\mathcal{O}(\log n)$ conflict sizes during the binary search for the best solution, these running times sum up to a total of $\mathcal{O}(n \log n)$ for Heuristic H and I, while J takes $\mathcal{O}(n \log^2 n)$ time.

3 Experiments

3.1 The Exact Solver

The exact solver we used was implemented by Erik Schwarzenecker from Saarbrücken in C++. It uses some ideas of our Heuristic H but solves the problem in Phase III exactly. Thanks to its fine tuning it handles examples of up to 300 points even slightly faster than the heuristics, but we were forced to introduce a time limit of 5 minutes for larger *hard* and *dense* problem sets (see Section 3.2) to be able to perform any test row in reasonable time. This exact algorithm X shows exponential behaviour. For small examples it is very fast, for larger ones it is unreliable. Only few of the largest *hard* and *dense* examples took less than five minutes, and we have observed that the solution of examples beyond that bound then easily takes half an hour or much more. The CPU times of X are not comparable to those of the heuristics, since the latter are implemented in a very different way.

Still X is much better in practice than the exact solver with a subexponential time bound suggested in [9]. It normally runs out of memory for more than 60–80 points, which we could improve to 120–150, when we made it solve only the problem remaining in Phase III. Even splitting this up into its connected regions, and dealing with those seperately, did not help a great deal.

3.2 Example Generators

Random. We just choose a given number of points uniformly distributed in a rectangle of given size.

Dense. Here we try to place as many squares as possible of a given size σ on a rectangle. We do this by randomly choosing points p and then checking whether σp_1 intersects with any of the σq_1 chosen before. We stop when we have unsuccessfully tried to place a new square 200 times. In a last step we assign a random corner point to each of the squares we were able to place without intersection, and return its coordinates. This method gives us a lower bound for the label size of the optimal solution.

Hard. In principle we use the same method as for Dense, that is, trying to place as many squares as possible into a given rectangle. In order to do so, we put a grid of cell size σ on it. In a random order, we try to place a square of edge length σ into each of the cells. This is done by randomly choosing a point within the cell and putting a fixed corner of the square on it. If it overlaps any of those chosen before, we try to place it into the same cell a constant number of times.

Real World. The municipal authorities of Munich provided us with the coordinates of roughly 1200 ground water drill holes within a 10 by 10 kilometer square centred approximately on the city centre. From this list we extract a given number of points being closest to some centre point according to the L_∞-norm, thus getting all those lying in a square around this extraction centre, where the size of the square depends on the number of points asked for. For our tests we chose five different centres; that of the map and those of its four quadrants in order to get results from different areas of the city with strongly varying point density. This is due to the fact that many of the holes were drilled during the construction of subway lines which are concentrated in the city centre, see Figure 5.

The choice of these four example generators might be justified by the following considerations. The need for real world data for testing is obvious. Random and Dense are intuitively the first things one would come up with, and differ enough in their behaviour to make them worth looking at. Hard examples might serve as a reminder that we are looking at an $\mathcal{NP}$-complete problem, and that no heuristic can be proved to do better than 50 percent of the optimal solution [4].

3.3 Experimental Set-up

Since the problem generators Dense and Hard ask for a label size σ, while Random and Real World directly use the number of points as input, the problem sizes differ. We run the exact solver, the Approximation Algorithm A, and the heuristics on each of the examples. For every size we averaged the approximation quality and running time over 50 tests.

Actually we do not use σ_{upper} (that is twice the result of A) as an upper bound for the conflicts we have to look at in the heuristics, because then we would have to add the computation time of A to that of the heuristics. Though losing the theoretical bounds, it turned out to be much faster and to yield results of the same quality if we compute σ_{dead} and work with a longer list of conflict sizes (between 0 and σ_{dead} instead of σ_{lower} and σ_{upper}) on which we do the binary search. Even the longer conflict lists of each candidate did not play a great roll, because σ_{upper} and σ_{dead} normally do not differ a lot in any case, especially not for large Hard or Dense examples where we have the highest number of conflicts per candidate.

3.4 Results

We show the two classical kinds of plots; time and quality. Quality here means the quotient of the solutions of a heuristic and the exact solver. Time is measured in CPU time, which is sufficient since it is closely related to the number of square–square conflicts. This on the other hand determines the number of crucial steps, namely finding all interesting conflicts once, and then extracting those valid for a certain σ in every step of the binary search.

The results both for time and quality are averaged only over those tests the exact solver managed within the time bound.

The standard deviation is represented by the length of the vertical bars in each point of the result plots.

3.4.1 Running Time

In Figure 4 we plot the running times of the slowest of the three heuristics, namely J, on the different example

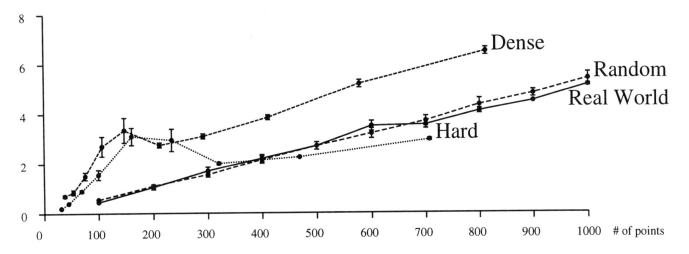

Figure 4: Running time of Heuristic J on different example classes

sets. H and I are slightly faster. Above 300 points the plot shows a rather stable $\mathcal{O}(n)$-behaviour with very small standard deviation. So far we are neither able to analyse the running time for small dense and hard examples nor to support the empirically linear running time by a theoretical analysis.

3.4.2 Approximation Quality

In Figures 8, 9, 10, and 11, the approximation quality of the three heuristics on the different example sets is plotted. On random and real world problems all three heuristics yield extremely good results. For an example, see Figure 6 and 7. On dense examples the differences between the heuristics become more clearly visible. Heuristic I is the best, yielding results of very high average quality with a slightly larger standard deviation. The behaviour on hard examples is still quite good but clearly becoming worse with an increasing number of points.

The quality of Algorithm A is extremely bad on Hard and Dense, and still useless from a practical point of view on random and real world examples.

A remark on the examples for which X did not give a result within the time bound: As mentioned above we did not include those in the calculation of the quality plots. But using the bound σ_{upper} resulting from the approximation algorithm A, and taking into consideration the typical quality of A, we found out that the behaviour of the heuristics on those examples does not differ significantly from that on the other examples.

4 Implementation

The implementation of the heuristics follows the structure listed in 2.2. The code was written in C++, and we strongly took advantage of data structures and algorithms provided by LEDA [7]. The commands LEDA offers, helped a great deal to shorten and simplify the code. It was not optimized with respect to running time but rather kept "legible". All heuristics and problem generators can be tested on the WWW under
http://www.inf.fu-berlin.de/~awolff
/html/labeling.html.

5 Conclusion and Acknowledgements

Our experiences with the Map Labeling Problem and its solution can be summed up as follows: We started with the purely mathematical formulation of the problem which was communicated to us by Kurt Mehlhorn from Saarbrücken, who received the problem from Rudi Krämer of the Amt für Informations- und Datenverarbeitung in München. Quickly we showed the $\mathcal{NP}$-hardness, were surprised to hear of the practical relevance, and started developing an approximation algorithm. We found one, analysed it, and showed its theoretical optimality. The problem was solved perfectly—in theory!

Applied to real world data, the algorithm proved useless. We used the insight into the problem structure gained during the design of A and our insight into the reasons for its practical failure, to develop Heuristic

H which produced satisfiably good results. Meanwhile Bettina Preis et. al. developed an exact algorithm which could solve small problems up to about 80 points, which enabled us to estimate the quality of our heuristic. We improved H to I, and to the even more sophisticated Heuristic J which turned out to be a little worse than our champion I. Erik Schwarzenecker used our heuristical concept to enable X to solve larger problems in reasonable time. He also suggested the class of hard examples. Thus we were able to do a thorough experimental analysis of the quality of our heuristics. We also owe thanks to Stefan Lohrum who helped us to make our heuristics accessible on the WWW.

Our intense contacts with the practitioners were successful in two respects: We could solve their problems, and they gave us the opportunity to get to know interesting related problems that come up in this context. We are now adapting our heuristics to these variants of the original problem and hope to be able to solve them with similar success.

References

[1] H. AONUMA, H. IMAI, Y. KAMBAYASHI, *A visual system of placing characters appropiatly in multimedia map databases*, Proceedings of the IFIP TC 2/WG 2.6 Working Conference on Visual Database Systems, North Holland (1989) 525-546.

[2] S. EVEN, A. ITAI, A. SHAMIR, *On the complexity of Timetable and Multicommodity Flow Problems*, SIAM J. Comput. **5** (1976) 691-703

[3] M. FORMANN, *Algorithms for Geometric Packing and Scaling Problems*, Dissertation, Fachbereich Mathematik, Freie Universität Berlin (1992)

[4] M. FORMANN, F. WAGNER, *A Packing Problem with Applications to Lettering of Maps*, Proceedings of the 7th ACM Symposium on Computational Geometry (1991) 281-288

[5] E. IMHOF, *Positioning Names on Maps*, The American Cartographer **2** (1975) 128-144

[6] D. E. KNUTH AND A. RAGHUNATHAN, *The Problem of Compatible Representatives*, SIAM Journal on Discrete Mathematics **5** (1992) 422-427

[7] K. MEHLHORN, S. NÄHER, *LEDA, a Library of Efficient Data Types and Algorithms*, TR A 04/89, FB10, Universität des Saarlandes, Saarbrücken, 1989

[8] H. IMAI, T. ASANO, *Efficient Algorithms for Geometric Graph Search Problems*, SIAM J. Comput. **15** (1986) 478-494

[9] L. KUČERA, K. MEHLHORN, B. PREIS, E. SCHWARZENECKER, *Exact Algorithms for a Geometric Packing Problem (Extended Abstract)*, Proceedings of the 10th Annual Symposium on Theoretical Aspects of Computer Science (STACS 93), Lecture Notes in Computer Science **665** (1993) 317-322

[10] F. WAGNER *Approximate Map Labeling is in $\Omega(n \log n)$*, Information Processing Letters **52** (1994) 161-165

[11] G. WEBER, L. KNIPPING, H. ALT, *An Application of Point Pattern Matching in Astronautics*, Journal of Symbolic Computation **17** (1994) 321-340

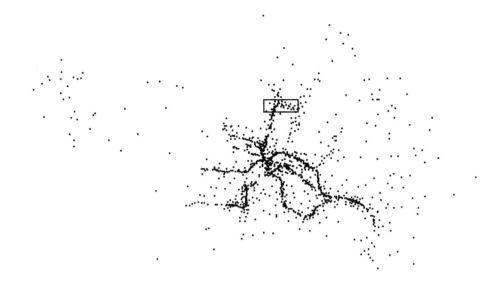

Figure 5: Map showing our sample data from Munich, and the section tested below. There are no conflicts between this section and the rest. The subway lines can be detected easily.

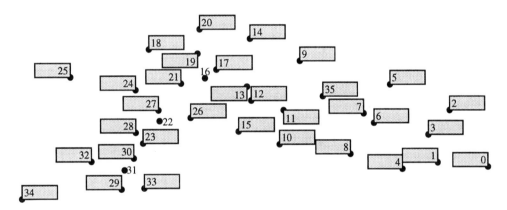

Figure 6: Solution of the program used by the authorities of the City of München before (label height 5000, 3 sites not labelled). It tries to maximize the number of sites labelled for a given size.

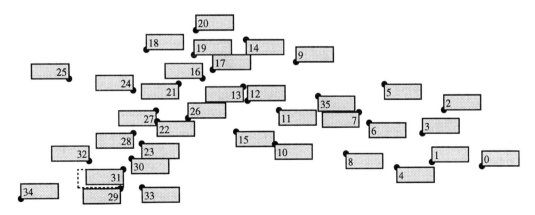

Figure 7: Solution produced by all of our heuristics (label height 5400, optimal). The dashed rectangle shows the candidate with label height $\sigma_{dead} = 6650$.

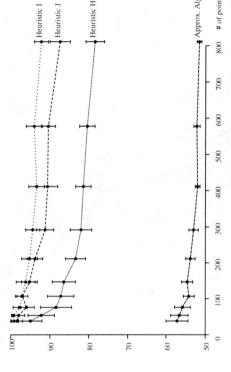

Figure 8: Quality of the heuristics on real world examples

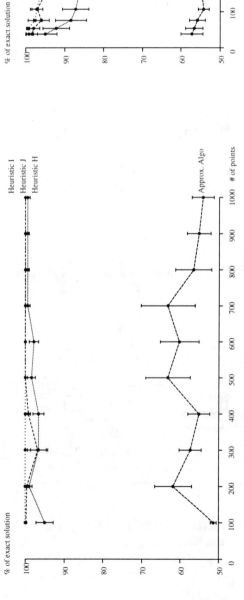

Figure 9: Quality of the heuristics on random examples

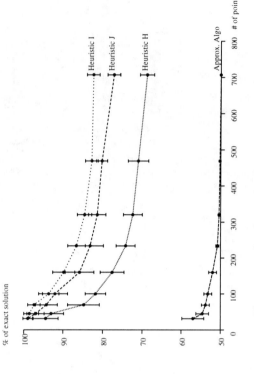

Figure 10: Quality of the heuristics on dense examples

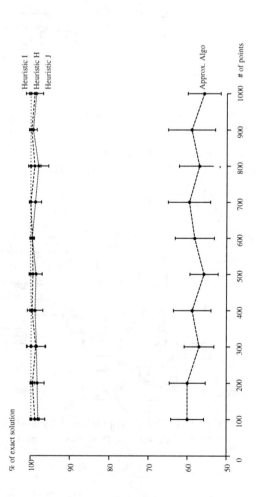

Figure 11: Quality of the heuristics on hard examples

Overlaying simply connected planar subdivisions in linear time

Ulrich Finke   Klaus H. Hinrichs

FB 15, Informatik, Westfälische Wilhelms-Universität
Einsteinstr. 62, D - 48149 Münster, Germany
E-Mail: finke, khh @ math.uni-muenster.de

Abstract

We present an algorithm which computes the overlay $\Pi_b \cap^\Pi \Pi_g$ of two simply connected planar subdivisions Π_b and Π_g; we assume that Π_b (resp. Π_g) and all its components are colored in blue (resp. green). The algorithm runs in $O(n + k)$ time and space, where n denotes the total number of edges of Π_b and Π_g and k the number of intersections between blue and green edges.

1 Introduction

Algorithms to compute the overlay of planar subdivisions are of theoretical and practical significance. They can be applied to solve the map overlay problem in spatial information systems, the hidden surface problem in computer graphics or design rule checking in VLSI layout design.

A *planar subdivision* is a partition Π of the 2-dimensional plane $\mathbb{R}^2$ into three finite collections of disjoint parts: the set of *vertices* V (each vertex $v \in V$ is a singleton $\{p\}$ where $p \in \mathbb{R}^2$), the set of (straigt-line) *edges* E and the set of *faces* F, i.e. $\Pi = V \cup E \cup F$. The exact definitions can be found in [GuSt 85].

Planar subdivisions can be further classified by imposing additional restrictions. Each closed path lying completely in one region of a *simply connected subdivision* can be topologically contracted to a point. Therefore a region in a simply-connected subdivision cannot contain any other region.

In this paper we present an algorithm which computes the overlay $\Pi_b \cap^\Pi \Pi_g$ of two simply connected planar subdivisions Π_b and Π_g; we assume that Π_b (resp. Π_g) and all its components are colored in blue (resp. green). The overlay $\Pi_b \cap^\Pi \Pi_g$ consists of all non-empty sets $b \cap g$ with $b \in \Pi_b$ and $g \in \Pi_g$.

Input and output of our algorithm are stored in the quad view data structure - a data structure which represents a trapezoidal decomposition of a subdivision. The algorithm is based on a graph exploration technique and computes from the trapezoidations of Π_b and Π_g the trapezoidation of the overlay $\Pi_b \cap^\Pi \Pi_g$. However, the algorithm does not compute the overlay of the trapezoidations, i.e. it does not compute the intersections between the verticals obtained by the trapezoidations and the edges of Π_b and Π_g. The algorithm operates destructively on its input subdivisions, i.e. the algorithm interweaves Π_b and Π_g into the overlay $\Pi_b \cap^\Pi \Pi_g$. Those components of Π_b and Π_g which are also contained in $\Pi_b \cap^\Pi \Pi_g$ are taken over implicitly and are not touched by the algorithm.

The overlay problem is related to the red/blue segment intersection problem: Given a set of non-intersecting red line segments and a set of non-intersecting blue line segments in the plane, with a total of n segments, report all k intersections of red segments with blue segments. This problem can be solved in $O(n \cdot \log n + k)$ time and $O(n)$ space [Chan 94]. If the line segments in the red and blue set each form a convex planar subdivision of the plane the intersection of these subdivisions can be determined in $O(n + k)$ time and $O(n)$ space [GuSe 86].

2 Preliminaries

In order to simplify the subsequent discussion we make the following assumptions:

- The plane $\mathbb{R}^2$ is topologically equivalent to a cylinder by identifying the points at "infinity", i.e. $(-\infty, y)$ and (∞, y), $y \in \mathbb{R}$, are identical. The plane $\mathbb{R}^2$ is bounded "above" and "below" by two horizontal edges e_∞ and

e_∞, i.e. each point of $\mathbb{R}^2$ lies above (resp. below) the edge $e_{-\infty}$ (resp. e_∞).

- In the overlay $\Pi_b \cap^\Pi \Pi_g$ the edge e_∞ (resp. $e_{-\infty}$) is assumed to be blue (resp. green).
- The blue and the green partition contain one common vertex $v_\infty = \{(\infty, 0)\}$.

[Fin 94] shows how to overcome the following restrictions:

- Except for v_∞ no other vertex of one partition is coincident with an edge or vertex of the other partition.
- A blue and a green edge intersect in at most one point, i.e. they may not overlap.
- No edge is vertical.

3 Views of a partition

Consider a trapezoidal decomposition of a subdivision Π obtained by drawing through each vertex v of the partition a vertical which ends at the edges lying directly above and below v (Fig. 1).

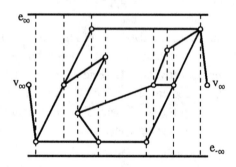

Fig. 1: Trapezoidal decomposition

The trapezoids and triangles obtained by this partitioning are called *views*. The left (resp. right) boundary of a view t is formed by a vertical which passes through the *left* (resp. *right*) *bounding vertex* $v_0(t)$ (resp. $v_1(t)$) of t. The lower (resp. upper) boundary of t is formed by the *lower* (resp. *upper*) *bounding edge* $e_0(t)$ (resp. $e_1(t)$) (Fig. 2). In the following we denote by $T(\Pi)$ the set of views of a planar partition Π.

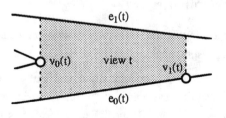

Fig. 2: View

In order to express neighborhood among views we define the neighborhood relation ‡ on the set of views $T(\Pi)$:

Definition: Two views $v_1 \neq v_2$ are neighbors, denoted by $v_1 ‡ v_2$, if they have a common bounding vertex and a common bounding edge.

It is easy to show that a view can have at most four neighbors. Since ‡ is a symmetric relation on the set of views the *view graph* with vertex set $T(\Pi)$ and edge set $‡(T(\Pi))$ is an undirected graph.

4 Intersection graph

We can distinguish three different types of views in the overlay $\Pi_b \cap^\Pi \Pi_g$ (Fig. 3):

- The blue (resp. green) views for which both the lower and the upper bounding edge are blue (resp. green).
- The *dichromatic views* for which the lower and upper bounding edge have different colors.

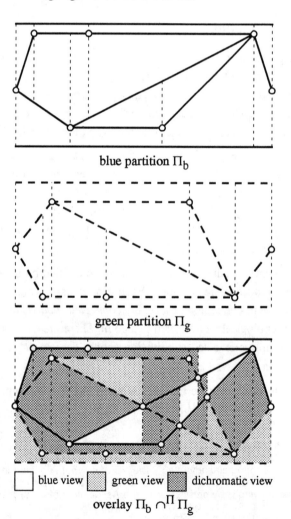

Fig. 3: Overlay of two subdivisions

Definition: The intersection graph Ig(Π) of an overlay partition Π = Π$_b$ ∩^{Π} Π$_g$ contains the left and right bounding vertices $v_0(t)$ and $v_1(t)$ of the dichromatic views t ∈ T(Π) as nodes. These vertices are called *dichromatic vertices*. Each dichromatic view represents an edge between its left and right bounding vertex (Fig. 4).

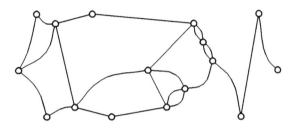

Fig. 4: Intersection graph

The intersection graph Ig(Π$_b$ ∩^{Π} Π$_g$) has two important properties: It contains all the intersection points between edges of Π$_b$ and edges of Π$_g$ as nodes. Furthermore the intersection graph consists of a single connected component if Π$_b$ and Π$_g$ are simply connected.

A blue (resp. green) view t in the overlay is identical to or contained in a view of the blue (resp. green) input subdivision. In the first case t is called an *original view*, in the latter case at least one of the bounding vertices of t is a dichromatic vertex and t is called a *clipped view*. Since the clipped and dichromatic views are bounded by dichromatic vertices the overlay is obtained by 1) computing for each dichromatic vertex v all views which have v as a bounding vertex, and 2) taking over the original views from the input subdivisions which are not bounded by at least one dichromatic vertex.

Our overlay algorithm takes advantage of the connectivity of the intersection graph to determine the overlay. It uses the graph exploration technique to build the intersection graph.

5 Graph exploration

A connected graph can be constructed successively from its already known components by *graph exploration*. During the graph exploration vertices and half edges have to perform different tasks.

When a vertex v is created it has to generate all half edges emanating from v. A half edge only knows the vertex by which it was created. Each half edge h has to perform three tasks:

1. First, h has to determine the vertex v_p of its partner half edge p with which it forms an edge of the graph.
2. If v_p does not yet exist then h generates this vertex.
3. Finally, h and p merge to form an edge of the graph.

If the graph is connected and all vertices and half edges perform their tasks the complete graph can be constructed by starting at any known vertex (Fig. 5).

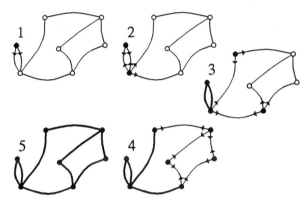

Fig. 5: Graph exploration

Graph exploration can be applied to intersection graphs as follows: Each vertex of an intersection graph corresponds to a bounding vertex of a dichromatic view. Each half edge corresponds to one "half" of a dichromatic view and is therefore called a *half view*. Each half view knows one of its bounding vertices and both bounding edges.

A vertex of the intersection graph has to generate all half views which have this vertex as a bounding vertex.

A half edge h has to determine the vertex v_p of its partner half edge p. The vertex v_p is either an intersection point of a blue and a green edge, or it is a blue or green vertex. If v_p has not yet been generated, i.e. if the partner half view p does not yet exist, then h generates the vertex v_p. Then the new vertex v_p generates all half views emanating from v_p, and therefore also the partner half view p. The half view h can now merge with its partner p to form an edge of the intersection graph, i.e. a dichromatic view of the overlay. The implementation of this graph exploration for intersection graphs requires

- a data structure for the representation of views and half views,
- an algorithm to perform the task of a vertex (vertex algorithm),

- an algorithm to perform the tasks of a half edge (half edge algorithm), and
- the implementation of a control mechanism which initializes and controls the process of graph exploration.

6 Quad view data structure

View graphs are represented by the *quad view data structure*: A view t is represented by arrays data and next each consisting of four elements (Fig. 6):

- data[i] points for $i \in \{0, 2\}$ to $e_{1-i/2}(t)$ and for $i \in \{1, 3\}$ to $v_{(i-1)/2}(t)$.
- next[i] points to that neighbor of t which has $v_{\lfloor((i+3) \bmod 4)/2\rfloor}(t)$ as bounding vertex and $e_{\lfloor((i+2) \bmod 4)/2\rfloor}(t)$ as bounding edge.

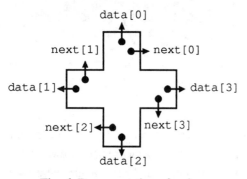

Fig. 6: Representation of a view

The next-pointers ensure that the views form doubly-linked rings around the vertices and edges of a partition. The ring of views around a vertex is called a *vertex ring*, the ring of views around an edge is called an *edge ring* (Fig. 7).

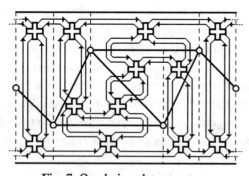

Fig. 7: Quad view data structure

The basic building blocks of the quad edge data structure [GuSt 85] represent the edges of a planar subdivision. Each edge references its two end points and the two regions bounded by the edge. Furthermore each edge references its neighbor edges at the end points in clockwise and counterclockwise direction. In contrast the basic building blocks of the quad view data structure are the views. Each view references its two bounding vertices and its two bounding edges. Furthermore each view references its four neighbor views defined by the neighborhood relation ‡.

On a quad view data structure a number of operations can be defined which allow

- the traversal of vertex rings and edge rings;
- the splitting of views; this operation is needed to add points as new vertices to a view graph; and
- the splitting of edges; this operation is needed to add intersection points as new vertices to a view graph.

Because of space restrictions we omit a detailed description of these operations. All operations except the one for splitting edges have constant time and storage requirements.

7 Representation of half views

Now we consider how to represent half views in the quad view data structure. The elements of the array data of a half view h point to its bounding vertex v_h and its two bounding edges e_0 and e_1, the element of data which corresponds to the unknown bounding vertex v_p of the partner half view p of h is initialized with NULL. In the following we assume that the index of this array element is i. This index determines the direction in which one has to search for the vertex v_p.

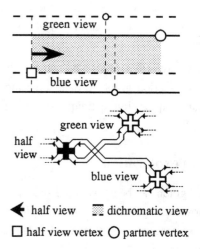

Fig. 8: Half view

Topologically a half view h is part of a blue and a green edge ring around the bounding edges e_0 and e_1 (Fig. 8). The next[i] and next[(i+1) mod 4] array elements of a half view h point to a blue view t_b and a green view t_g which are determined by v_h, e_0 and e_1. The views t_b and t_g are not contained in the overlay and determine the position of h in the blue partition Π_b and the green partition Π_g. The next[(i+2) mod 4] and next[(i+3) mod 4] array elements point to other half views, dichromatic views, green views or blue views which are part of the overlay.

8 Vertex algorithm

During the graph exploration a dichromatic vertex v has to generate all half views emanating from v. We assume that the vertex v for which the vertex algorithm has to be executed is contained in the blue and green partition, i.e. there exists a blue and a green vertex ring around v. To ensure that this condition is satisfied we extend the second task to be performed by an half edge: Before the vertex algorithm is executed for a dichromatic vertex v, the half edge has to ensure that v is contained as a vertex in both the blue and the green partition; eventually views and edges must be split to insert a vertex which is not yet contained in a partition.

The vertex algorithm traverses the blue vertex ring T_b and the green vertex ring T_g around v synchronously in counterclockwise direction. The traversal is initialized with two views $(t_b, t_g) \in T_b \times T_g$ having a non-empty intersection. These views are obtained from the half view which triggers the execution of the vertex algorithm. We explain in the next section how the half edge determines these two views. The synchronous traversal is performed in such a way that the current pair of views $(t_b, t_g) \in T_b \times T_g$ always has a non-empty intersection. During the traversal a new vertex ring for v in $\Pi_b \cap^\Pi \Pi_g$ is constructed as follows. If the intersection $t_b \cap t_g$ is bounded by two edges having

- the same color then the view having these two edges as bounding edges is taken over in the new vertex ring.
- different colors a half view is added to the new vertex ring. This half view is linked to t_b and t_g.

After the synchronous traversal is finished the new vertex ring contains all half views emanating from v. Furthermore it contains all blue (resp. green) views of the overlay $\Pi_b \cap^\Pi \Pi_g$ having v as a bounding vertex. After the graph exploration has been completed the vertex algorithm has been executed for all dichromatic vertices. Those blue (resp. green) views in the overlay not having a dichromatic bounding vertex are implicitly taken over into the overlay since these views are connected directly or indirectly to blue (resp. green) views which have a dichromatic bounding vertex and are therefore explicitly taken over by the vertex algorithm. Hence we do not only construct the intersection graph but also the overlay.

This synchronous traversal and construction of a new vertex ring is similar to a merge of two linear lists. The time and storage requirements are bounded linearly by the number of views in the two vertex rings around v. Fig. 9 shows the effect of the vertex algorithm on the quad view data structure.

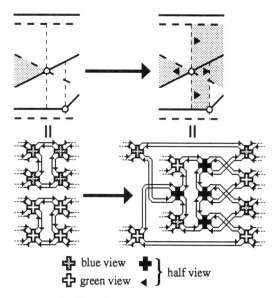

➕ blue view
➕ green view ◄ } half view

Fig. 9: Building a new vertex ring

9 Half edge algorithm

First, a half view h has to determine the vertex v_p of its partner half edge p with which it forms an edge of the intersection graph. h starts this search with the two views $t_0 = $ next[i] and $t_1 = $ next[(i+1) mod 4] (see section 7 for the definition of i). Let T_0 and T_1 be the edge rings of the bounding edges of h such that $t_0 \in T_0$ and $t_1 \in T_1$.

t_0 and t_1 determine v_p if the intersection of their bounding edges or one of the bounding vertices $v_{(i-1)/2}(t_0)$ or $v_{(i-1)/2}(t_1)$ lies on the boundary $\partial(t_0 \cap t_1)$ of their intersection.

If t_0 and t_1 do not determine v_p we exchange that view $t \in \{t_0, t_1\}$ for which the bounding vertex data[i] is closer in horizontal direction to the bounding vertex data[(i+2) mod 4] of h. The view t is replaced by that neighbor in its edge ring which has not yet been considered, i.e. which lies in the search direction determined by i (Fig. 10).

If t_0 and t_1 determine v_p the half edge algorithm has to check in its second step whether v_p has already been generated by the vertex algorithm before. This is the case if t_0 and t_1 are linked to the partner half view p, and the vertex algorithm must not be executed. If this is not the case the half view calls the vertex algorithm for v_p. As mentioned in section 8, eventually the half edge algorithm has to split views and edges before calling the vertex algorithm to ensure that v_p is contained as a vertex in both Π_b and Π_g. The vertex algorithm starts its traversal with the views t_0 and t_1.

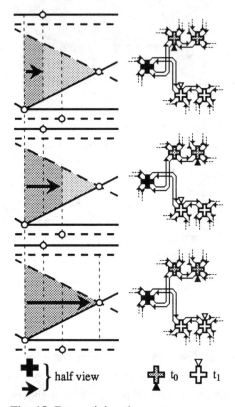

Fig. 10: Determining the partner vertex v_p

The last step guarantees the existence of the partner half view p of h. In the third step h and p are merged to form a new dichromatic view of $\Pi_b \cap^\Pi \Pi_g$ (Fig. 11). This merge operation disconnects those views of Π_b and Π_g which are not contained in the overlay.

10 Control of graph exploration

The control mechanism which initializes and controls the process of graph exploration is implemented by a waiting system W which stores references to the currently existing half views. Since there is no distinguished sequence in which the half views should be processed it does not matter in what order the half views are fetched from W, i.e. W can be implemented by a stack or a first-in first-out queue.

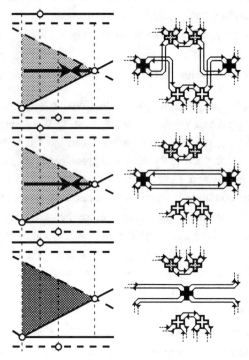

Fig. 11: Merging two half views

For starting the graph exploration it is necessary to know at least one half view. The initial half views are obtained by executing the vertex algorithm for the vertex v_∞ which is common to both Π_b and Π_g. The references to these initial half views are passed to the waiting system W. Then we iterate a loop until W becomes empty. In each iteration we fetch a reference to a half view h from W. If h has not yet merged with another half view to form a dichromatic view in the overlay partition we execute the half edge algorithm for h. References to the half views generated by the half edge algorithm are passed to W. If h has already been merged with its partner half view before, the storage occupied by h can be released.

Since the algorithm works destructively on its input the quad view data structures of the input subdivisions no longer exist after the loop has terminated. The overlay $\Pi_b \cap^\Pi \Pi_g$ is represented in a quad view data structure and can therefore be used as input for further overlay operations or be converted to another representation, e.g. quad edge data structure [GuSt 85] or doubly-connected-edge-list.

11 Analysis

In the following we denote by n the total number of edges in the blue partition Π_b and the green partition Π_g, and by k the number of intersections between blue and green edges. The number of views in Π_b and Π_g is bounded linearly by n, since in an incremental construction of a partition the number of views increases by at most three with each additional edge. For the same reason the number of views in the overlay partition $\Pi_b \cap^\Pi \Pi_g$ is bounded linearly by n + k. Hence the storage requirement of this algorithm is O(n + k).

The time needed for all executions of the vertex algorithm is O(n + k), since the time to process a view is constant and the total number of views in $\Pi_b \cap^\Pi \Pi_g$ is bounded linearly by n + k. The same is true for the control mechanism. The time needed for all executions of the half edge algorithm is O(n + k) if we use the following algorithm for splitting edges:

If an edge e has to be split in a vertex v we first have to determine those views t_0 und t_1 in the edge ring T around e which enclose v horizontally between their left and right bounding vertices. The view t_0 is determined by the half edge algorithm. To find the other view t_1, the edge ring T is traversed synchronously in clockwise and counterclockwise direction to determine t_1 and the shortest path P between t_0 and t_1. The edge e is not only split in v, but also along P such that we obtain subedges of e with edge rings consisting of at most a constant number c > 4 views. [Fin 94] shows that the time needed for all edge splits is O(n + k). Therefore we have

Theorem: If Π_b und Π_g are two simply connected subdivisions their overlay $\Pi_b \cap^\Pi \Pi_g$ can be computed in O(n + k) time and storage space.

12 Concluding remarks

We have presented an algorithm for overlaying simply connected planar subdivisions in linear time and space measured in the size of the resulting partition. Both the input subdivisions and the output subdivision are represented by the quad view data structure. The quad view representation for a simply connected planar subdivision which is given by another representation (e.g. quad edge data structure [GuSt 85] or doubly-connected-edge-list) can be obtained in O(n) time by applying the trapezoidal decomposition algorithm described in [Cha 91], or in $O(n \cdot \log^* n)$ expected time by the more practical randomized algorithm described in [Sei 91].

By using the trapezoidal decomposition algorithm in [Cha 91] we can determine the visibility map of a simple polygon in O(n) time and storage space where n denotes the number of edges of the polygon. The quad view data structure of a simple polygon can be extracted from the visibility map in O(n) time. By modifying the overlay algorithm presented above the intersection of two simple polygons can be determined in optimal O(n + k) time and storage space.

[Fin 94] shows a generalization of the overlay algorithm to path connected subdivisions, i.e. the faces of subdivisions can contain holes. For path connected subdivisions the intersection graph is no longer connected. If z is the total number of connected components of the input subdivisions the time complexity of the algorithm is $O(z \cdot n + k)$. The problem can be solved in $O(n \cdot \log^* n + k + z \cdot \log n)$ expected time by using the point location structure created as a by-product by the randomized algorithm [Sei 91]. It remains an open question whether the overlay of two path connected subdivisions can be determined in O(n + k) time and storage space.

The overlay algorithm presented in this paper and a number of further algorithms operating on quad view data structures are embedded in a test environment which has been implemented in C++ [Wen 94]. This test environment allows to create planar subdivisions interactively; an interface allows to import maps from a commercial geographic information system (GIS). All implemented algorithms operating on quad view data structures can be animated. It is possible to observe the half views and the waiting system during execution of the overlay algorithm. Furthermore the actual time and storage requirements of the algorithm can be measured.

References

[Chan 94] T. Chan: A simple trapezoid sweep algorithm for reporting red/blue segment intersections, 6th Canadian Conference on Computational Geometry (1994), 263 - 268.

[Cha 91]   B. M. Chazelle: Triangulating a simple polygon in linear time, Discrete & Computational Geometry 6 (1991), 485 - 524.

[Fin 94]   U. Finke: Algorithmen für Verschneidungs-operationen, Ph. D. Thesis, University of Siegen, 1994.

[GuSe 86] L. Guibas, R. Seidel: Computing convolutions by reciprocal search, 2nd ACM Symposium on Computational Geometry (1986), 90 - 99.

[GuSt 85] L. Guibas, J. Stolfi: Primitives for the manipulation of general subdivisions and the computation of Voronoi diagrams, ACM Transactions on Graphics 4 (1985), 74 - 123.

[Sei 91]   R. Seidel: A simple and fast incremental randomized algorithm for computing trape-zoidal decompositions and for traingulating polygons, Computational Geometry: Theory and Applications 1 (1991), 51 - 64.

[Wen 94]   M. Wenzel: Verschneidungsoperationen für Geo-Informationsysteme, Diploma Thesis, University of Münster, 1994.

New Lower Bounds for Hopcroft's Problem*

(Extended Abstract)

Jeff Erickson

Computer Science Division
University of California
Berkeley, CA 94720 USA
jeffe@cs.berkeley.edu

Fachbereich Informatik
Universität des Saarlandes
D-66123 Saarbrücken, Germany

Abstract

We establish new lower bounds on the complexity of the following basic geometric problem, attributed to John Hopcroft: Given a set of n points and m hyperplanes in $\mathbb{R}^d$, is any point contained in any hyperplane? We define a general class of *partitioning algorithms*, and show that in the worst case, for all m and n, any such algorithm requires time $\Omega(n \log m + n^{2/3} m^{2/3} + m \log n)$ in two dimensions, or $\Omega(n \log m + n^{5/6} m^{1/2} + n^{1/2} m^{5/6} + m \log n)$ in three or more dimensions. We obtain slightly higher bounds for the counting version of Hopcroft's problem in four or more dimensions. Our planar lower bound is within a factor of $2^{O(\log^*(n+m))}$ of the best known upper bound, due to Matoušek. Previously, the best known lower bound, in any dimension, was $\Omega(n \log m + m \log n)$. We develop our lower bounds in two stages. First we define a combinatorial representation of the relative order type of a set of points and hyperplanes, called a *monochromatic cover*, and derive lower bounds on the complexity of this representation. We then show that the running time of any partitioning algorithm is bounded below by the size of some monochromatic cover.

1 Introduction

In the early 1980's, John Hopcroft posed the following problem to several members of the computer science community.

> Given a set of n points and n lines in the plane, does any point lie on a line?

Hopcroft's problem arises as a special case of many other geometric problems, including point location,

range searching, motion planning, collision detection, ray shooting, and hidden surface removal.

The earliest sub-quadratic algorithm for Hopcroft's problem, due to Chazelle [6], runs in time $O(n^{1.695})$. A very simple algorithm, attributed to Hopcroft and Seidel [12], runs in time $O(n^{3/2} \log^{1/2} n)$. (See [13, p. 350].) Cole et al. [12] combined these two algorithms, achieving a running time of $O(n^{1.412})$. Edelsbrunner et al. [15] developed a randomized algorithm with expected running time $O(n^{4/3+\varepsilon})$.[1] Further research replaced the n^ε term in this upper bound with a succession of smaller and smaller polylogarithmic factors [10, 14, 1, 8]. The fastest known algorithm, due to Matoušek [22], runs in time $n^{4/3} 2^{O(\log^* n)}$.[2] Matoušek's algorithm can be tuned to detect incidences among n points and m lines in the plane in time $O(n \log m + n^{2/3} m^{2/3} 2^{O(\log^*(n+m))} + m \log n)$ [5], or more generally among n points and m hyperplanes in $\mathbb{R}^d$ in time $O(n \log m + n^{d/(d+1)} m^{d/(d+1)} 2^{O(\log^*(n+m))} + m \log n)$.

The lower bound history is much shorter. The only previously known lower bound is $\Omega(n \log m + m \log n)$, in the algebraic decision tree and algebraic computation tree models, by reduction from the problem of detecting an intersection between two sets of real numbers [23, 3].

In this paper, we establish new lower bounds on the complexity of Hopcroft's problem. We formally define a general class of *partitioning algorithms*, which includes most (if not all) of the algorithms mentioned above, and show that any such algorithm can be forced to take time $\Omega(n \log m + n^{2/3} m^{2/3} + m \log n)$ in two dimensions, or $\Omega(n \log m + n^{5/6} m^{1/2} + n^{1/2} m^{5/6} + m \log n)$ in three or more dimensions. We improve this lower bound slightly in dimensions four and higher for the *counting* version of Hopcroft's problem, where we want to know the number of incident point-hyperplane pairs.

---

*This research was partially supported by NSF grant CCR-9058440. An earlier version of this paper was published as Technical Report A/04/94, Fachbereich Informatik, Universität des Saarlandes, Saarbrücken, Germany, November 1994.

[1] In time bounds of this form, ε refers to an arbitrary positive constant. The multiplicative constants hidden in the big-Oh notation depend on ε, and tend to infinity as ε approaches zero.

[2] The iterated logarithm $\log^* n$ is defined to be 1 for all $n \leq 2$ and $1 + \log^*(\log_2 n)$ for all $n \geq 2$.

Some related results deserve to be mentioned here. Erdős constructed a set of n points and n lines in the plane with $\Omega(n^{4/3})$ incident point-line pairs [13]. It follows immediately that any algorithm that reports all incident pairs requires time $\Omega(n^{4/3})$ in the worst case. Of course, we cannot apply this argument to either the decision version or the counting version of Hopcroft's problem, since the output size for these problems is constant. We use Erdős' construction to establish our planar lower bounds.

Chazelle has established lower bounds for the closely related *simplex range counting* problem: Given a set of points and a set of simplices, how many points are in each simplex? For example, in the online case, any data structure of size s that supports arbitrary triangular range queries among n points in the plane requires $\Omega(n/\sqrt{s})$ time per query [7]. It follows that answering n queries over n points requires $\Omega(n^{4/3})$ time in the worst case. For the offline version of the same problem, where all the triangles are known in advance, Chazelle establishes a slightly weaker bound of $\Omega(n^{4/3}/\log^{4/3} n)$ [9], although an $\Omega(n^{4/3})$ lower bound follows easily from the Erdős construction using Chazelle's methods. Both lower bounds hold in the Fredman/Yao semigroup arithmetic model [19], in which we give the points arbitrary weights from a semigroup, and count the number of arithmetic operations required to calculate the answer. Unfortunately, this model is inappropriate for studying Hopcroft's problem. If there are no incidences, the we perform no additions; conversely, if we perform a single addition, there must be an incidence. In a later section of this paper, we extend Chazelle's offline lower bounds to a counting version of Hopcroft's problem.

The paper is organized as follows. In Section 2, we derive a *quadratic* lower bound for Hopcroft's problem in a restricted model of computation. In Section 3, we define a combinatorial representation of the relative order type of a set of points and hyperplanes, called a *monochromatic cover*, and derive lower bounds on the complexity of this representation. In Section 4, we formally define the class of partitioning algorithms, and prove that the running time of such an algorithm that decides Hopcroft's problem is bounded below by the complexity of some monochromatic cover. In Section 5, we discuss a number of related geometric problems for which our techniques give new lower bounds. Finally, in Section 6, we offer our conclusions and suggest directions for further research.

2 A Simple Quadratic Lower Bound

Erickson and Seidel [18] have proven a number of lower bounds on other geometric degeneracy-detection problems, under a model of computation in which only a limited number of geometric primitives are allowed. For example, $\Omega(n^2)$ sidedness queries are required to decide whether a set of n points in the plane contain a collinear triple.

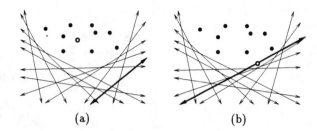

Figure 1. A quadratic lower bound for Hopcroft's problem. (a) The original adversary input. (b) The "collapsed" input.

The corresponding simple primitive for Hopcroft's problem is the *relative orientation query*: Given a point and a hyperplane, does the point lie above, on, or below the hyperplane? Surprisingly, we can easily establish a *quadratic* lower bound for Hopcroft's problem if this is the only primitive we are allowed.

Theorem 2.1. *In the worst case, $\Omega(mn)$ relative orientation queries are required to decide Hopcroft's problem in $\mathbb{R}^d$, for any $d \geq 1$.*

Proof: The lower bound follows from a simple adversary argument. The adversary presents the algorithm with a set of n points and m hyperplanes in which every point is above every hyperplane. If the algorithm does not perform a relative orientation query for some point/hyperplane pair, the adversary can move that point onto that hyperplane without changing the relative orientation of any other pair. See Figure 1. The algorithm cannot tell the two sets apart, even though one has an incidence and the other does not. □

In fact, this argument applies to the much easier problem "Is every point above every hyperplane?", for which there is a linear-time algorithm in one dimension and $O(n \log m + m \log n)$-time algorithms in two and three dimensions, all of which are optimal. In two and higher dimensions, we can strengthen the model of computation to include several other primitives, such as sidedness queries and coordinate comparisons, and still maintain the quadratic lower bound. It appears that higher-order primitives such as "Is this point to the left or right of the intersection of these two lines?" are necessary to achieve nontrivial upper bounds. If we allow either of these two primitives, however, it seems unlikely that the techniques developed in [18] can be used to derive nontrivial lower bounds.

We omit further details from this extended abstract.

3 Monochromatic Covers

Let $P = \{p_1, p_2, \ldots, p_n\}$ be a set of points and $H = \{h_1, h_2, \ldots, h_m\}$ a set of hyperplanes in $\mathbb{R}^d$. These two sets induce a *relative orientation matrix* $M(P, H) \in \{+, 0, -\}^{n \times m}$ whose (i, j)'th entry denotes whether the point p_i is above, on, or below the hyperplane h_j. Any minor of the matrix $M(P, H)$ is itself a relative orientation matrix $M(P', H')$, for some $P' \subseteq P$ and $H' \subseteq H$. Hopcroft's problem is to decide, given P and H, whether the matrix $M(P, H)$ contains a zero.

We call a sign matrix *monochromatic* if all its entries are equal. A *minor cover* of a matrix is a set of minors whose union is the entire matrix. If every minor in the cover is monochromatic, we call it a *monochromatic cover*. The *size* of a minor is the number of rows plus the number of columns; the size of a minor cover is the sum of the sizes of the minors in the cover.

Monochromatic covers for 0-1 matrices have been previously used to prove lower bounds for various communication complexity problems [21]. Typically, however, these results make use of the number of minors in the cover, not the size of the cover as we define it here.[3] A similar concept was introduced by Tarján [26] in the context of switching theory. He considers (in our terminology) sets of monochromatic minors that cover the ones in a given 0-1 matrix, or equivalently, sets of bipartite cliques that cover a given bipartite graph. Tuza [27] showed that every $n \times m$ 0-1 matrix has such a cover of size $O(nm/\log(\max(m, n)))$, and that this bound is tight in the worst case, up to constant factors. These results apply immediately to monochromatic covers of arbitrary sign matrices. See also [2] for a geometric application of bipartite clique covers.

Given a set of points and hyperplanes, a monochromatic cover of its relative orientation matrix provides a succinct combinatorial representation of the relative order type of the set. In particular, if no point lies on any hyperplane, a monochromatic cover provides a *proof* of this fact. If we consider a model of computation in which algorithms are allowed to ask questions of the form "Is this minor monochromatic?" at a cost equal to the size of the minor[4], then the size of the smallest monochromatic cover is a lower bound for the running time of any algorithm that decides Hopcroft's problem, given a set of points and hyperplanes with no incidences.

Let $\mu(P, H)$ denote the minimum size of any monochromatic cover of the relative orientation matrix $M(P, H)$. Let $\mu_d(n, m)$ denote the maximum of $\mu(P, H)$, where P ranges over all sets of n points in $\mathbb{R}^d$, and H ranges over all sets of m hyperplanes in $\mathbb{R}^d$. Let $\mu_d^*(n, m)$ denote the maximum of $\mu(P, H)$ over all sets of points and hyperplanes with no incidences. Clearly, $\mu_d(n, m) \geq \mu_d^*(n, m)$.

We are also interested in the following related quantity. Call any collection of monochromatic minors that covers all (and only) the zero entries in a sign matrix a *zero cover*. Let $\zeta(P, H)$ denote the minimum size of any zero cover of the relative orientation matrix $M(P, H)$, and let $\zeta_d(n, m)$ be the maximum of $\zeta(P, H)$ over all sets of n points and m hyperplanes in $\mathbb{R}^d$. If P and H have no incidences, $\zeta(P, H) = 0$. Since every monochromatic cover must include a zero cover, we have $\mu_d(n, m) \geq \zeta_d(n, m)$.

In the remainder of this section, we develop asymptotic lower bounds for $\mu_d^*(n, m)$ and $\zeta_d(n, m)$, which in turn imply lower bounds for $\mu_d(n, m)$.

3.1 Simple Covers

Relative orientation matrices are defined in terms of a fixed (projective) coordinate system, which determines what it means for a point to be "above" or "below" a hyperplane. More generally, we can assign an absolute orientation to each point and hyperplane, and define relative orientations with respect to these assigned orientations. (See [24].) The orientations of the points and hyperplanes determine which minors of the relative orientation matrix are monochromatic, and therefore determine the minimum monochromatic cover size.

Surprisingly, however, the minimum monochromatic cover size is independent of any choice of absolute orientations, up to a factor of two. We prove an even stronger result. Call a sign matrix *simple* if it can be changed in to a monochromatic matrix by inverting some set of rows and columns. A simple cover is a minor cover in which every minor is simple. A set of points and hyperplanes has a simple relative orientation matrix if and only if every hyperplane partitions the points into the same two subsets, or equivalently, if and only if the points all lie in the same open full-dimensional (projective) cell of the arrangement of hyperplanes. See Figure 2.

Let $\sigma(P, H)$ denote the minimum size of any simple cover of the relative orientation matrix $M(P, H)$. Note that changing the orientations of the points P and hyperplanes H preserves all simple minors. Thus, $\sigma(P, H)$ is strictly independent of the orientations assigned to P and H. We bound $\mu(P, H)$ in terms of $\sigma(P, H)$ as follows.

---

[3] Any sign matrix can be covered by $3\min(m, n)$ monochromatic minors. Furthermore, there are sets of n points and m lines in the plane whose relative orientation matrices require $3\min(m, n)$ monochromatic minors to cover them.

[4] This cost assumption is rather unrealistic. Deciding whether a set of n points and m hyperplanes has a monochromatic relative orientation matrix requires $\Omega(n \log m + m \log n)$ time, and almost certainly more in dimensions four and higher. Fortunately for us, the cost assumption is unrealistic in the right direction.

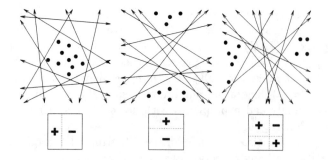

Figure 2. Three collections of points and lines with simple relative orientation matrices.

Theorem 3.1. $\sigma(P,H) \leq \mu(P,H) \leq 2\sigma(P,H)$

Proof: Every monochromatic minor is also a simple minor. Every simple minor can be partitioned into four monochromatic minors, whose total size is twice that of the original minor. □

3.2 Two Dimensions

Let $I(P,H)$ denote the number of incident point-hyperplane pairs between a set of points P and a set of hyperplanes H. To derive lower bounds for $\mu_2^*(n,m)$ and $\zeta_2(n,m)$, we use the following combinatorial result of Erdős. (See [19] or [13, p.112] for proofs.)

Lemma 3.2 (Erdős). *For all n and m, there is a set of n points and m lines in the plane with $\Omega(n + n^{2/3}m^{2/3} + m)$ incident pairs.*

Fredman [19] uses Erdős' construction to prove lower bounds for dynamic range query data structures in the plane.[5] The Erdős lower bound is asymptotically tight. The corresponding upper bound was first proven by Szemerédi and Trotter [25]. A much simpler proof, with better constants, was later given by Clarkson et al. [11]

Theorem 3.3. *The two-dimensional zero cover size $\zeta_2(n,m) = \Omega(n + n^{2/3}m^{2/3} + m)$.*

Proof: It is not possible for two distinct points to both be adjacent to two distinct lines; any mutually incident set of points and lines has either exactly one point or exactly one line. It follows that for any set P of points and H of lines in the plane, $\zeta(P,H) \geq I(P,H)$. The theorem now follows from Lemma 3.2. □

Theorem 3.4. *The two-dimensional monochromatic cover size $\mu_2^*(n,m) = \Omega(n + n^{2/3}m^{2/3} + m)$.*

---
[5]Perhaps it is more interesting that Chazelle's static lower bounds [7] do *not* use this construction.

Proof: Consider any configuration of n points and $m/2$ lines with $\Omega(n + n^{2/3}m^{2/3} + m)$ point-line incidences, as given by Lemma 3.2. Replace each line ℓ in this configuration with a pair of lines, parallel to ℓ and at distance ε on either side, where ε is a constant, sufficiently small that all point-line distances in the new configuration are at least ε. The resulting configuration of n points and m lines clearly has no point-line incidences. We call a point-line pair in this configuration *close* if the distance between the point and the line is ε. There are $\Omega(n + n^{2/3}m^{2/3} + m)$ such pairs.

Now consider a single monochromatic minor in the relative orientation matrix of these points and lines. Let P' denote the set of points and H' the set of lines represented in this minor. We claim that the number of close pairs between P' and H' is small.

Without loss of generality, we can assume that all the points are above all the lines. If a point is close to a line, the point must be on the convex hull of P', and the line must support the upper envelope of H'. Thus, we can assume that both P' and H' are in convex position. In particular, we can order both the points and lines from left to right.

Either the leftmost point is close to at most one line, or the leftmost line is close to at most one point. It follows inductively that the number of close pairs is at most $|P'| + |H'|$, which is exactly the size of the minor. The theorem follows immediately. □

3.3 Three Dimensions

The technique we used in the plane does not generalize immediately to higher dimensions. Even in three dimensions, there are collections of points and planes where every point is incident to every plane. See Figure 3. Of course, we can cover the relative orientation matrix of such a configuration with a single zero minor. If we apply the plane-doubling trick to such a set to eliminate all incidences, the relative orientation matrix of the resulting set is always simple. Thus, in order to derive a lower bound for either $\zeta_3(m,n)$ or $\mu_3^*(n,m)$, we need a construction of points and planes with many incidences, but without large sets of mutually incident points and planes.

We use the notation $[n]$ to denote the set of integers $\{1, 2, \ldots, n\}$, and $i \perp j$ to mean that i and j are relatively prime. We also use (without proof) a number of simple number-theoretic results concerning the Euler totient function $\phi(n)$, the number of positive integers less than n that are relatively prime to n. We refer the reader to [20] for relevant background.

Lemma 3.5. *For all n and m such that $\lfloor n^{1/3} \rfloor < m$, there exists a set P of n points and a set H of m planes,*

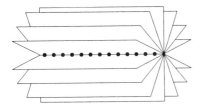

Figure 3. Every point is incident to every plane.

such that $I(P,H) = \Omega(n^{5/6}m^{1/2})$ and any three planes in H intersect in at most one point.

Proof: Fix sufficiently large n and m such that $\lfloor n^{1/3} \rfloor < m$. Let $h(a,b,c;i,j)$ denote the plane passing through the points (a,b,c), $(a+i,b+j,c)$ and $(a+i,b,c+i-j)$. Let $p = \lfloor n^{1/3} \rfloor$ and $q = \lfloor \alpha(m/p)^{1/4} \rfloor$ for some suitable constant $\alpha > 0$. (Note that with n sufficiently large and m in the indicated range, p and q are both positive integers.)

Now consider the points $P = [p]^3 = \{(x,y,z) \mid x,y,z \in [p]\}$ and the hyperplanes

$$H = \{h(a,b,c;i,j) \mid \\ i \in [q], j \in [i], i \perp j, a \in [i], b \in [j], c \in [\lfloor p/2 \rfloor]\}$$

The number of planes in H is

$$\left\lfloor \frac{p}{2} \right\rfloor \sum_{i=1}^{q} \sum_{\substack{j=1 \\ j \perp i}}^{i} j = \left\lfloor \frac{p}{2} \right\rfloor \sum_{i=1}^{q} \frac{i^2 \phi(i)}{2} = O(pq^4) = O(m).$$

By choosing the constant α appropriately and possibly adding in $o(m)$ extra planes, we can ensure that H contains exactly m planes. We claim that this collection of points and planes satisfies the lemma.

Consider a single plane $h = h(a,b,c;i,j) \in H$. Since i, j, and $i-j$ are pairwise relatively prime, h intersects exactly one point (x,y,z) such that $x \in [i]$ and $y \in [j]$, namely, the point (a,b,c). Thus, for each fixed i and j we use, the planes $h(a,b,c;i,j) \in H$ are distinct. Since planes with different "slopes" are clearly different, it follows that the planes in H are distinct.

For all $k \in [\lfloor p/2i \rfloor]$, the intersection of $h(a,b,c;i,j) \in H$ with the plane $x = a + ki$ contains at least k points of P. It follows that

$$|P \cap h(a,b,c;i,j)| \geq \sum_{k=1}^{\lfloor p/2i \rfloor} k > \frac{1}{2} \left\lfloor \frac{p}{2i} \right\rfloor^2.$$

Thus, the total number of incidences between P and H can be calculated as follows.

$$I(P,H) \geq \left\lfloor \frac{p}{2} \right\rfloor \sum_{i=1}^{q} i \sum_{\substack{j=1 \\ i \perp j}}^{i} \frac{j}{2} \left\lfloor \frac{p}{2i} \right\rfloor^2$$

$$\geq \left\lfloor \frac{p}{2} \right\rfloor^3 \sum_{i=1}^{q} \sum_{\substack{j=1 \\ i \perp j}}^{i} \frac{j}{2i}$$

$$= \left\lfloor \frac{p}{2} \right\rfloor^3 \sum_{i=1}^{q} \frac{\phi(i)}{4}$$

$$= \Omega(p^3 q^2)$$

$$= \Omega(n^{5/6} m^{1/2})$$

Finally, If H contains three planes that intersect in a line, the intersection of those planes with the plane $x = 0$ must consist of three concurrent lines. It suffices to consider only the planes passing through the point $(1,1,1)$, since for any other triple of planes in H there is a parallel triple passing through that point. The intersection of $h(1,1,1;i,j)$ with the plane $x = 0$ is the line through $(0, 1-j/i, 1)$ and $(0, 1, j/i)$. Since $i \perp j$, each such plane determines a unique line. Furthermore, since all these lines are tangent to a parabola, no three of them are concurrent. It follows that the intersection of any three planes in H consists of at most one point. □

Edelsbrunner *et al.* [15] prove an upper bound of $O(n \log m + n^{4/5 + 2\epsilon} m^{3/5 - \epsilon} + m)$ on the maximum number of incidences between n points and m planes, where no three planes contain a common line. Using the probabilistic counting techniques of Clarkson *et al.* [11], we can improve this upper bound to $O(n + n^{4/5} m^{3/5} + m)$.

Theorem 3.6. *The three-dimensional zero cover size* $\zeta_3(n,m) = \Omega(n + n^{5/6} m^{1/2} + n^{1/2} m^{5/6} + m)$.

Proof: Consider the case $n^{1/3} < m \leq n$. Fix a set P of n points and a set H of m hyperplanes satisfying Lemma 3.5. Any mutually incident subsets of P and H contain either at most one point or at most two planes. Thus, the number of entries in any zero minor of $M(P,H)$ is at most twice the size of the minor. It follows that any zero cover of $M(P,H)$ must have size $\Omega(I(P,H)) = \Omega(n^{5/6} m^{1/2})$. The dual[6] construction gives us a lower bound of $\Omega(n^{1/2} m^{5/6})$ for all m in the range $n \leq m < n^3$, and the trivial lower bound $\Omega(n+m)$ applies for other values of m. □

Lemma 3.7. *Let P be a set of n points and H a set of m planes in $\mathbb{R}^3$, such that every point in P is either on or above every plane in H, and any three planes in H intersect in at most one point. Then $I(P,H) \leq 3(m+n)$.*

Proof: Call any point (resp. plane) *lonely* if it is incident to less than three planes (resp. points). Without loss of generality, we can assume that none of the points in P or planes in H is lonely, since each lonely point and plane contributes at most three incidences.

---
[6]We assume the reader is familiar with the concept of point-hyperplane duality. Otherwise, see [13] or [24].

No point in the interior of of the convex hull of P can be incident to a plane in H. Any point in the interior of a facet of the convex hull can be on at most one plane in H. Consider any point $p \in P$ in the interior of an edge of the convex hull. Any plane containing p also contains the two endpoints of the edge. There cannot be more than two such planes in H, so p must be lonely. It follows that every point in P is a vertex of the convex hull of P.

No plane can contain a point unless it touches the upper envelope of H. Any plane that only contains a vertex of the upper envelope must be lonely. For any plane h that contains only an edge of the envelope, two other planes also contain that edge, and any points on h must also be on the other two planes. Then h must be lonely, since any three planes in H intersect in at most one point. It follows that every plane in H spans a facet of the upper envelope of H. Furthermore, every point in P is a vertex of this upper envelope.

Construct a bipartite graph with vertices P and H and edges corresponding to incident pairs. This graph is clearly planar, and thus has at most $3(m+n)$ edges. $\square$

Note that this lemma still holds if we weaken the general position requirement to rule out only mutually incident sets of s points and t planes, where s and t are any fixed constants. We reiterate here that without some sort of general position assumption, we can easily achieve mn incidences.

Theorem 3.8. *The three-dimensional monochromatic cover size $\mu_3^*(n,m) = \Omega(n + n^{5/6}m^{1/2} + n^{1/2}m^{5/6} + m)$.*

Proof: Consider the case $2n^{1/3} < m \leq n$. Fix a set P of n points and a set H of $m/2$ hyperplanes satisfying Lemma 3.5.

Call a sign matrix *loosely monochromatic* if either none of its entries is $+$ or none of its entries is $-$. For all subsets $P' \subseteq P$ and $H' \subseteq H$, Lemma 3.7 implies that if $M(P', H')$ is loosely monochromatic, then $I(P', H') = O(|P'| + |H'|)$.

Now replace each plane $h \in H$ with a pair of parallel planes at distance ε on either side of h, for some suitably small constant $\varepsilon > 0$. Call the resulting set of m hyperplanes H_ε. We say that a point is *close* to a plane if the distance between them is exactly ε. There are $\Omega(n^{5/6}m^{1/2})$ close pairs between P and H_ε, and no incidences.

For every monochromatic minor of the matrix $M(P, H_\varepsilon)$, there is a corresponding loosely monochromatic minor of $M(P, H)$. Furthermore, there is a one-to-one correspondence between the close pairs in the first minor and the incident pairs in the second. It follows that any monochromatic minor of $M(P, H_\varepsilon)$ orients only a linear number of close pairs. Thus, any

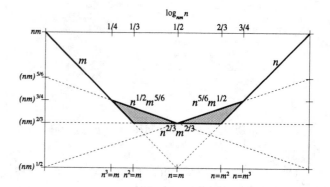

Figure 4. Comparison of lower bounds for $\mu_2^*(n,m)$ and $\mu_3^*(n,m)$.

monochromatic cover for P and H_ε must have size $\Omega(n^{5/6}m^{1/2})$.

Similar arguments apply to other values of m. $\square$

Unfortunately, for the special case $m = \Theta(n)$, this does not improve the $\Omega(n^{4/3})$ bound we derived earlier for the planar case. For all other values of m between $\Omega(n^{1/3})$ and $O(n^3)$, however, the new bound is an improvement. See Figure 4.

3.4 Higher Dimensions

In the full version of the paper, we prove the following results.

Lemma 3.9. *For any $\lfloor n^{1/d} \rfloor < m$, there exists a set P of n points and a set H of m hyperplanes in $\mathbb{R}^d$, such that $I(P,H) = \Omega(n^{1-2/d(d+1)}m^{2/(d+1)})$ and any d hyperplanes in H intersect in at most one point.*

Theorem 3.10. *The d-dimensional zero cover size $\zeta_d(n,m) =*

$$\Omega\left(\sum_{i=1}^{d}\left(n^{1-2/i(i+1)}m^{2/(i+1)} + n^{2/(i+1)}m^{1-2/i(i+1)}\right)\right).$$

Unfortunately, the best lower bounds we can derive for $\mu_d^*(m,n)$ in higher dimensions derive trivially from Theorem 3.8. In particular, we are unable to generalize Lemma 3.7 even to four dimensions. The best upper bound we can prove on the number of incidences between n points and m hyperplanes in $\mathbb{R}^4$, where every point is above or on every hyperplane and no four hyperplanes contain a line, is $O(n + n^{2/3}m^{2/3} + m)$. (See [16] for the derivation of a similar upper bound.) No superlinear lower bounds are known in any dimension, so there is some hope for a linear upper bound.

However, we can achieve a super-linear number of incidences in five dimensions, under a weaker combinatorial general position requirement. Unlike in lower dimensions, therefore some sort of *geometric* general position requirement is necessary to keep the number of incidences small. (We do not know of any such requirement that is *sufficient*, except for the trivial requirement that at most $d+1$ hyperplanes contain any point.)

Lemma 3.11. *For all n and m, there exists a set P of n points and a set H of m hyperplanes in $\mathbb{R}^5$, such that every point is on or above every hyperplane, no two hyperplanes in H contain more than one point of P in their intersection, and $I(P, H) = \Omega(n + n^{2/3} m^{2/3} + m)$.*

Proof: Define the function $\sigma : \mathbb{R}^3 \to \mathbb{R}^6$ as follows.

$$\sigma(x, y, z) = (x^2, y^2, z^2, \sqrt{2}\, xy, \sqrt{2}\, yz, \sqrt{2}\, xz)$$

For any $v, w \in \mathbb{R}^3$, we have $\langle \sigma(v), \sigma(w) \rangle = \langle v, w \rangle^2$, where $\langle \cdot, \cdot \rangle$ denotes the usual inner product of vectors. In a more geometric setting, σ maps points and lines in the plane, represented in homogeneous coordinates, to points and hyperplanes in $\mathbb{R}^5$, also represented in homogeneous coordinates [24]. For any point p and line ℓ in the plane, the point $\sigma(p)$ is incident to the hyperplane $\sigma(\ell)$ if and only if p is incident to ℓ; otherwise, $\sigma(p)$ lies above $\sigma(\ell)$. Thus, we can take P and H to be the images under σ of any sets of n points and m lines with $\Omega(n + n^{2/3} m^{2/3} + m)$ incidences, as given by Lemma 3.2. □

4 Partitioning Algorithms

A *partition graph* is a directed acyclic graph, with one source, called the *root*, and several sinks, or *leaves*. Every non-leaf is either a *primal node* or a *dual node*. Associated with each primal or dual node v is a set $\mathcal{R}_v$ of query regions, satisfying three conditions.

1. The cardinality of $\mathcal{R}_v$ is at most some constant Δ.
2. Each region in $\mathcal{R}_v$ is connected.
3. The union of the regions in $\mathcal{R}_v$ is $\mathbb{R}^d$.

In particular, we do not require the query regions to be disjoint, convex, simply connected, or even semialgebraic, nor do we require that each query region have constant descriptional complexity.[7] Each query region in $\mathcal{R}_v$ is associated with an outgoing edge of v. Thus, the out-degree of the graph is at most Δ.

Every point $p \in \mathbb{R}^d$ induces a subgraph of a partition graph as follows. We say that a point *reaches* every node in its subgraph and *traverses* every edge in its subgraph. The point p reaches a node v if either v is the root, or

[7] In fact, all three of the conditions we list are stronger than necessary to prove our results.

p traverses some edge into v. If p reaches a primal node v, then it also traverses every edge corresponding to a query region $R \in \mathcal{R}_v$ that contains p. If p reaches a dual node v, then it also traverses every edge corresponding to a query region that intersects the dual hyperplane p^*. The subgraph induced by p contains all the nodes that p reaches and all the edges that p traverses. Similarly, every hyperplane h induces a subgraph, according to the primal query regions that intersect h and the dual query regions that contain the dual point h^*.

Given a set of points and hyperplanes, a *partitioning algorithm* constructs a partition graph and determines the subgraphs induced by each point and hyperplane. For the purpose of proving lower bounds, we charge unit time whenever a point or hyperplane traverses an edge. In particular, we do not charge for the construction of the partition graph itself, nor for constructing its query regions. We emphasize that partitioning algorithms are *nondeterministic*, since the partition graph and its query regions depend on the input.

A partitioning algorithm decides Hopcroft's problem by reporting an incidence if and only if some leaf in its partition graph is reached by both a point and a hyperplane. It is easy to see that if a point and hyperplane are incident, then there is at least one leaf in every partition graph that is reached by both the point and the hyperplane. Thus, given a set P of points and a set H of hyperplanes, a partition graph in which no leaf is reached by both a point and a hyperplane provides a *proof* that there are no incidences between P and H.

In the remainder of this section, we derive lower bounds for the worst-case running time of partitioning algorithms that solve Hopcroft's problem. With the exception of the basic lower bound of $\Omega(n \log m + m \log n)$, which we prove directly, all of our lower bounds are derived from the cover size bounds in Section 3.

4.1 The Basic Lower Bound

Theorem 4.1. *Any partitioning algorithm that decides Hopcroft's problem in any dimension must take time $\Omega(n \log m + m \log n)$ in the worst case.*

Proof: It suffices to consider the following configuration, where n is a multiple of m. P consists of n points on some vertical line in $\mathbb{R}^d$, say the x_d-axis, and H consists of m hyperplanes normal to that line, placed so that n/m points lie between each hyperplane and the next higher hyperplane, or above the top hyperplane. For each point, call the hyperplane below it its *partner*. Each hyperplane is the partner of n/m points.

Let G be the partition graph generated by some partitioning algorithm. The out-degree of any node in G is bounded by some constant Δ. The *level* of any node in G is the length of the shortest path from the root to that node. There are at most Δ^k nodes at level k. We say that a node v *separates* a point-hyperplane pair if both the point and the hyperplane reach v, but none of the outgoing edges of v is traversed by both the point and the hyperplane. Finally, we say that a hyperplane h is *active at level k* if none of the nodes in the first k levels separates h from any of its partners.

Suppose v is a primal node. For each hyperplane h that v separates from one of its partner points p, mark some query region in $\mathcal{R}_v$ that contains p, but misses h. The marked region lies completely above h, but not completely above any hyperplane higher than h. It follows that the same region cannot be marked more than once. Since there are at most Δ regions, at most Δ hyperplanes become inactive. By similar arguments, if v is a dual node, then v separates at most Δ points from their partners.

Thus, the number of hyperplanes that are inactive at level k is less than Δ^{k+2}. In particular, at level $\lfloor \log_\Delta m \rfloor - 3$, at least $m(1 - 1/\Delta)$ hyperplanes are still active. It follows that at least $n(1 - 1/\Delta)$ points each traverse at least $\lfloor \log_\Delta m \rfloor - 3$ edges, so the total running time of the algorithm is at least

$$n(1 - 1/\Delta)(\lfloor \log_\Delta m \rfloor - 3) = \Omega(n \log m).$$

Similar arguments establish a lower bound of $\Omega(m \log n)$ when $n < m$. □

4.2 The Decision Problem Lower Bound

Let $T_\mathcal{A}(P, H)$ denote the running time of an algorithm $\mathcal{A}$ that decides Hopcroft's problem in $\mathbb{R}^d$ for some d, given points P and hyperplanes H as input.

Theorem 4.2. *Let $\mathcal{A}$ be a partitioning algorithm that decides Hopcroft's problem, and let P be a set of points and H a set of hyperplanes such that $I(P, H) = 0$. Then $T_\mathcal{A}(P, H) = \Omega(\mu(P, H))$.*

Proof: Recall that the running time $T_\mathcal{A}(P, H)$ is defined in terms of the edges of the partition graph as follows.

$$T_\mathcal{A}(P, H) \triangleq \sum_{p \in P} \#\text{edges } p \text{ traverses} + \sum_{h \in H} \#\text{edges } h \text{ traverses}$$

We say that a point or hyperplane *misses* an edge from v to w if it reaches v but does not traverse the edge. (It might still reach w by traversing some other edge.) For every edge that a point or hyperplane traverses, it misses at most $\Delta - 1$ other edges.

$$\Delta \cdot T_\mathcal{A}(P, H) \geq$$
$$\sum_{p \in P} (\#\text{edges } p \text{ traverses} + \#\text{edges } p \text{ misses}) +$$
$$\sum_{h \in H} (\#\text{edges } h \text{ traverses} + \#\text{edges } h \text{ misses})$$

Call any edge that leaves a primal node a primal edge, and any edge that leaves a dual node a dual edge.

$$\Delta \cdot T_\mathcal{A}(P, H) \geq$$
$$\sum_{p \in P} (\#\text{primal edges } p \text{ traverses} + \#\text{dual edges } p \text{ misses}) +$$
$$\sum_{h \in H} (\#\text{dual edges } h \text{ traverses} + \#\text{primal edges } h \text{ misses})$$
$$= \sum_{\substack{\text{primal} \\ \text{edges } e}} (\#\text{points transversing } e + \#\text{hyperplanes missing } e) +$$
$$\sum_{\substack{\text{dual} \\ \text{edges } e}} (\#\text{hyperplanes traversing } e + \#\text{points missing } e)$$

Consider, for some primal edge e, the set P_e of points that traverse e and the set H_e of hyperplanes that miss e. The edge e is associated with some query region R, such that every point in P_e is contained in R, and every hyperplane in H_e is disjoint from R. It follows immediately that the relative orientation matrix $M(P_e, H_e)$ is simple. Similarly, for any dual edge e, the relative orientation matrix of the set of points that miss e and hyperplanes that traverse e is also simple.

Now consider any point $p \in P$ and hyperplane $h \in H$. Since $\mathcal{A}$ correctly decides Hopcroft's problem, no leaf is reached by both p and h. It follows that some node v separates p and h. If v is a primal node, then h misses the outgoing primal edges that p traverses. If v is a dual node, then p misses the outgoing dual edges that h traverses.

Thus, for each edge in the partition graph, we can associate a simple minor, and this collection of minors covers the relative orientation matrix $M(P, H)$. Furthermore, the size of this simple cover is exactly the lower bound we have for $\Delta \cdot T_\mathcal{A}(P, H)$ above. Splitting each simple minor into monochromatic minors at most doubles the size of the cover. Thus, the algorithm induces a monochromatic cover of size at most $2\Delta \cdot T_\mathcal{A}(P, H)$. Since this must be at least $\mu(P, H)$, we have the lower bound $T_\mathcal{A}(P, H) \geq \mu(P, H)/2\Delta$. □

Corollary 4.3. *The worst-case running time of any partitioning algorithm that solves Hopcroft's problem in $\mathbb{R}^d$ is $\Omega(n \log m + n^{2/3}m^{2/3} + m \log n)$ for $d = 2$ and $\Omega(n \log m + n^{5/6}m^{1/2} + n^{1/2}m^{5/6} + m \log n)$ for all $d \geq 3$.*

Proof: Theorems 4.1 and 4.2 together imply that the worst case running time is $\Omega(n \log m + \mu_d^*(n,m) + n \log m)$. Theorem 3.4 gives us the planar lower bound, and Theorem 3.8 gives us the lower bound in higher dimensions. □

We emphasize here that the condition $I(P, H) = 0$ is necessary for the lower bound to hold. If there is an incidence, then the trivial algorithm correctly "detects" it. The partition graph contains one leaf, and since it is reached by every point and hyperplane, the algorithm reports an incidence. This is consitent with the intuition that it is trivial to prove the existence of an incidence, but much harder to prove the nonexistence of incidences.

4.3 The Counting Problem Lower Bound

The counting version of Hopcroft's problem is to determine, given a set of points and hyperplanes, the number of incident pairs. A partitioning algorithm solves the counting version of Hopcroft's problem as follows. The number of incidences associated with a leaf in its partition graph is the number of points that reach it times the number of hyperplanes that reach it. The algorithm returns as its output the sum of these products over all leaves in its partition graph.

Since every incident point-hyperplane pair is guaranteed to reach at least one leaf, it is not possible for a partitioning algorithm to count too few incidences. The only ways the algorithm can go wrong are counting the same incidence more than once and counting incidences that don't exist. Thus, in order to be correct, the algorithm must ensure that every non-incident point-hyperplane pair is separated, and that every incident pair reaches exactly one leaf.

Theorem 4.4. *Let $\mathcal{A}$ be a partitioning algorithm that solves the counting version of Hopcroft's problem, and let P be a set of points and H a set of hyperplanes. Then $T_{\mathcal{A}}(P, H) = \Omega(\mu(P, H))$.*

Proof: We follow the proof for the decision lower bound almost exactly. We associate a simple minor with every edge just as before. We also associate a monochromatic minor with every leaf, consisting of all points and hyperplanes that reach the leaf. Every non-incident point-hyperplane pair is represented in some edge minor, and every incident pair in exactly one leaf minor. Thus, the minors form a simple cover. The total size of the leaf minors is certainly less than $T_{\mathcal{A}}(P, H)$, since every point and hyperplane that reaches a leaf must traverse one of the leaf's incoming edges. The total size of the edge minors is at most $\Delta \cdot T_{\mathcal{A}}(P, H)$, as established previously. Splitting each edge minor into monochromatic minors at most doubles their size. Thus, we get a monochromatic cover of size at most $(2\Delta + 1)T_{\mathcal{A}}(P, H)$, which implies $T_{\mathcal{A}}(P, H) \geq \mu(P, H)/(2\Delta + 1)$. □

Corollary 4.5. *Any partitioning algorithm that solves the counting version of Hopcroft's problem in $\mathbb{R}^d$ requires time*

$$\Omega\left(n \log m + \sum_{i=2}^{d}\left(n^{1-\frac{2}{i(i+1)}}m^{\frac{2}{i+1}} + n^{\frac{2}{i+1}}m^{1-\frac{2}{i(i+1)}}\right) + m \log n\right)$$

in the worst case.

We can prove the following much stronger bound by only paying attention to the minors induced at the leaves. We define an *unbounded partition graph* to be just like a partition graph except that we place no restrictions on the number of query regions associated with each node. Call the resulting class of algorithms *unbounded partitioning algorithms*.

Theorem 4.6. *Let $\mathcal{A}$ be an unbounded partitioning algorithm that solves the counting version of Hopcroft's problem, and let P be a set of points and H a set of hyperplanes. Then $T_{\mathcal{A}}(P, H) = \Omega(\zeta(P, H))$.*

Proof: We associate a zero minor with every leaf, and these minors form a zero cover. The total size of the leaf minors is certainly less than $T_{\mathcal{A}}(P, H)$, since every point and hyperplane that reaches a leaf must traverse one of the leaf's incoming edges. □

Corollary 4.7. *Any unbounded partitioning algorithm that solves the counting version of Hopcroft's problem in $\mathbb{R}^d$ requires time*

$$\Omega\left(\sum_{i=1}^{d}\left(n^{1-2/i(i+1)}m^{2/(i+1)} + n^{2/(i+1)}m^{1-2/i(i+1)}\right)\right)$$

in the worst case.

Our results also imply a lower bound for a variant of the counting version of Hopcroft's problem, in the Fredman/Yao semigroup arithmetic model. The lower bound follows from the following result of Chazelle [9, Lemma 3.3]. (Chazelle's lemma only deals with the case $n = m$, but his proof generalizes immediately to the more general case.)

Lemma 4.8. *If A is an $n \times m$ incidence matrix with I ones and no $p \times q$ minor of ones, then the complexity of computing Ax over a semigroup is $\Omega(I/pq - n/p)$.*

Theorem 4.9. *Given n weighted points and m hyperplanes in $\mathbb{R}^d$,*

$$\Omega\left(\sum_{i=1}^{d}\left(n^{1-2/i(i+1)}m^{2/(i+1)} + n^{2/(i+1)}m^{1-2/i(i+1)}\right)\right)$$

semigroup operations are required to determine the sum of the weights of the points on each hyperplane, in the worst case.

Proof: The lower bound follows immediately from Lemma 3.9. □

5 Related Problems

In the full version of the paper, we prove lower bounds for a number of other problems, either by reduction to Hopcroft's problem, or from direct application of our earlier proof techniques. No lower bound bigger than $\Omega(n \log m + m \log n)$ was previously known for any of these problems.

Extreme caution must be taken when applying reduction arguments to partitioning algorithms. It is quite easy to apply a "standard" reduction argument, only to find that the reduction also changes the model. A simple example illustrates the difficulty. Consider the problem of detecting incidences between a set of points and a set of *lines* in *three* dimensions. This problem is clearly harder than Hopcroft's problem in the plane. Nevertheless, there is an extremely simple partitioning algorithm that solves this problem in linear time! The partition graph consists of a single primal node with two query regions, one of which contains all the points but does not intersect any of the lines. Even in this case, however, Theorem 4.6 implies an $\Omega(n^{4/3})$ lower bound for the *counting* version of this problem.

We list here some of our results. We omit the proofs from this extended abstract.

Theorem 5.1. *Any partitioning algorithm that decides, given n red and m blue line segments in the plane, whether any red segment intersects a blue segment, requires time $\Omega(n \log m + n^{2/3} m^{2/3} + m \log n)$ in the worst case.*

Theorem 5.2. *Any (unbounded) partitioning algorithm that counts, given n lines in $\mathbb{R}^3$, the number of intersecting pairs, requires time $\Omega(n^{4/3})$ in the worst case.*

Theorem 5.3. *Any partitioning algorithm that computes, given n points and m halfplanes, the sum over all halfplanes of the number of points contained in each halfplane, requires time $\Omega(n \log m + n^{2/3} m^{2/3} + m \log n)$ in the worst case.*

Theorem 5.4. *Any partitioning algorithm that determines, given n points and m triangles in the plane, whether any triangle contains a point, requires time $\Omega(n \log m + n^{2/3} m^{2/3} + m \log n)$ in the worst case.*

Theorem 5.5. *Any partitioning algorithm that detects unit distances among n points in the plane requires time $\Omega(n^{4/3})$ in the worst case.*

Theorem 5.6. *Any (unbounded) partitioning algorithm that counts incidences between n points and m hyperplanes in $\mathbb{R}^5$, where every point lies on or above every hyperplane, requires time $\Omega(n + n^{2/3} m^{2/3} + m)$ in the worst case.*

6 Open Problems

A number of open problems remain to be solved. The most obvious problem is to improve our lower bounds, in particular for the case $n = m$. The true complexity of Hopcroft's problem almost certainly increases with the dimension, but the best lower bound we can achieve in higher dimensions comes trivially from the two-dimensional case. The most obvious approach is to improve our cover size bounds. Is there a set of n points and n planes in $\mathbb{R}^3$ whose minimum monochromatic cover size is $\omega(n^{4/3})$?

Another possible approach is to consider restrictions of the partitioning model. Can we achieve better bounds if we only consider algorithms whose query regions are convex? What if the query regions at every node must be distinct? What if the running time depends on the complexity of the query regions, or the number of nodes in the partition graph?

The class of partitioning algorithms is general enough to directly include many, but not all, existing algorithms for deciding Hopcroft's problem. The model requires that a single data structure be used to determine which points and hyperplanes intersect each query region, but many algorithms use a tree-like structure to locate the points and an iterative procedure to locate the hyperplanes. We can usually modify such algorithms so that they do fit our model, at the cost of only a constant factor in their running time, but this is a rather ad hoc solution. Any extension of our lower bounds to a more general model, which would explicitly allow different strategies for locating points and hyperplanes, would be interesting.

The partitioning algorithm model is specifically tailored to detect intersections or containments between pairs of objects. There are a number of similar geometric problems for which the partitioning algorithm model simply does not apply. We mention one specific example, the *cyclic overlap problem*. Given a set of non-intersecting line segments in $\mathbb{R}^3$, does any subset form a cycle with respect to the "above" relation? The fastest known algorithm for this problem, due to de Berg *et al.* [4], runs in time $O(n^{4/3+\varepsilon})$, using a divide-and-conquer strategy very similar to algorithms for Hopcroft's problem. In fact, in the algebraic decision tree model, the cyclic overlap problem is at least as hard as Hopcroft's problem [17]. However, it is not clear that this problem can even be solved by a partitioning algorithm, since the answer might depend on arbitrarily large tuples of segments, arbitrarily far apart. Extending our lower bounds into more traditional models of computation is an important and very difficult open problem.

Acknowledgements. The author gratefully thanks Kurt Mehlhorn and the Max-Planck-Institut für Informatik in Saarbrücken for their generous hospitality, and Raimund Seidel for his continuing support, encouragement, suggestions, and patience.

References

[1] P. K. Agarwal. Partitioning arrangements of lines: II. Applications. *Discrete Comput. Geom.*, 5:533–573, 1990.

[2] P. K. Agarwal, N. Alon, B. Aronov, and S. Suri. Can visibility graphs be represented compactly? In *Proc. 9th Annu. ACM Sympos. Comput. Geom.*, pages 338–347, 1993.

[3] M. Ben-Or. Lower bounds for algebraic computation trees. In *Proc. 15th Annu. ACM Sympos. Theory Comput.*, pages 80–86, 1983.

[4] M. de Berg, M. Overmars, and O. Schwarzkopf. Computing and verifying depth orders. In *Proc. 8th Annu. ACM Sympos. Comput. Geom.*, pages 138–145, 1992.

[5] M. de Berg and O. Schwarzkopf. Cuttings and applications. Report RUU-CS-92-26, Dept. Comput. Sci., Utrecht Univ., Utrecht, Netherlands, Aug. 1992.

[6] B. Chazelle. Reporting and counting segment intersections. *J. Comput. Syst. Sci.*, 32:156–182, 1986.

[7] B. Chazelle. Lower bounds on the complexity of polytope range searching. *J. Amer. Math. Soc.*, 2:637–666, 1989.

[8] B. Chazelle. Cutting hyperplanes for divide-and-conquer. *Discrete Comput. Geom.*, 9(2):145–158, 1993.

[9] B. Chazelle. Lower bounds for off-line range searching. To appear in *Proc. 27th Annu. ACM Sympos. Theory Comput.*, 1995.

[10] B. Chazelle, M. Sharir, and E. Welzl. Quasi-optimal upper bounds for simplex range searching and new zone theorems. *Algorithmica*, 8:407–429, 1992.

[11] K. Clarkson, H. Edelsbrunner, L. Guibas, M. Sharir, and E. Welzl. Combinatorial complexity bounds for arrangements of curves and spheres. *Discrete Comput. Geom.*, 5:99–160, 1990.

[12] R. Cole, M. Sharir, and C. K. Yap. On k-hulls and related problems. *SIAM J. Comput.*, 16:61–77, 1987.

[13] H. Edelsbrunner. *Algorithms in Combinatorial Geometry*, volume 10 of *EATCS Monographs on Theoretical Computer Science*. Springer-Verlag, Heidelberg, West Germany, 1987.

[14] H. Edelsbrunner, L. Guibas, J. Hershberger, R. Seidel, M. Sharir, J. Snoeyink, and E. Welzl. Implicitly representing arrangements of lines or segments. *Discrete Comput. Geom.*, 4:433–466, 1989.

[15] H. Edelsbrunner, L. Guibas, and M. Sharir. The complexity of many cells in arrangements of planes and related problems. *Discrete Comput. Geom.*, 5:197–216, 1990.

[16] H. Edelsbrunner and M. Sharir. A hyperplane incidence problem with applications to counting distances. In P. Gritzman and B. Sturmfels, editors, *Applied Geometry and Discrete Mathematics: The Victor Klee Festschrift*, volume 4 of *DIMACS Series in Discrete Mathematics and Theoretical Computer Science*, pages 253–263. AMS Press, 1991.

[17] J. Erickson. On the relative complexities of some geometric problems. Unpublished manuscript, 1995.

[18] J. Erickson and R. Seidel. Better lower bounds on detecting affine and spherical degeneracies. *Discrete Comput. Geom.*, 13(1):41–57, 1995.

[19] M. L. Fredman. Lower bounds on the complexity of some optimal data structures. *SIAM J. Comput.*, 10:1–10, 1981.

[20] G. Hardy and E. Wright. *The Theory of Numbers*. Oxford University Press, London, England, 4th edition, 1965.

[21] L. Lovàsz. Communication complexity: A survey. In *Paths, Flows, and VLSI Layout*, volume 9 of *Algorithms and Combinatorics*, pages 235–265. Springer-Verlag, 1990.

[22] J. Matoušek. Range searching with efficient hierarchical cuttings. *Discrete Comput. Geom.*, 10(2):157–182, 1993.

[23] J. M. Steele and A. C. Yao. Lower bounds for algebraic decision trees. *J. Algorithms*, 3:1–8, 1982.

[24] J. Stolfi. *Oriented Projective Geometry: A Framework for Geometric Computations*. Academic Press, 1991.

[25] E. Szemerédi and W. T. Trotter, Jr. Extremal problems in discrete geometry. *Combinatorica*, 3:381–392, 1983.

[26] T. G. Tarján. Complexity of lattice-configurations. *Studia Sci. Math. Hungar.*, 10:203–211, 1975.

[27] Z. Tuza. Covering of graphs by complete bipartite subgraphs; complexity of 0-1 matrices. *Combinatorica*, 4:111–116, 1984.

A Helly-type theorem for unions of convex sets*

JIŘÍ MATOUŠEK
Department of Applied Mathematics
Charles University
Malostranské nám. 25, 118 00 Praha 1
Czech Republic

Abstract

We prove that for any $d, k \geq 1$ there exist numbers $q = q(d, k)$ and $h = h(d, k)$ such that the following holds: Let $\mathcal{K}$ be a family of subsets of the d-dimensional Euclidean space, such that the intersection of any at most q sets of $\mathcal{K}$ can be expressed as a union of at most k convex sets. Then the Helly number of $\mathcal{K}$ is at most h. We also obtain topological generalizations of some cases of this result. The main result was independently obtained by Alon and Kalai, by a different method.

1 Introduction

Let $\mathbb{R}^d$ denote the d-dimensional Euclidean space. A famous theorem of Helly, discovered in 1913, asserts that if $\mathcal{F}$ is a finite family of convex sets in $\mathbb{R}^d$ such that the intersection of any $d+1$ of these sets is nonempty, then also the intersection of all sets of $\mathcal{F}$ is nonempty. Over the years, a vast body of analogs and generalizations of this result has been accumulated in the literature, see [Eck93] for

---

*This research was supported by Czech Republic Grant GAČR 201/94/2167, Charles University grants No. 351 and 361 and by EC Cooperative Action IC-1000 (project AL-TEC: *Algorithms for Future Technologies*).

Permission to copy without fee all or part of this material is granted provided that the copies are not made or distributed for direct commercial advantage, the ACM copyright notice and the title of the publication and its date appear, and notice is given that copying is by permission of the Association of Computing Machinery.To copy otherwise, or to republish, requires a fee and/or specific permission.
11th Computational Geometry, Vancouver, B.C. Canada
© 1995 ACM 0-89791-724-3/95/0006...$3.50

a recent survey. Here we give a brief overview of previous work related to our result.

A general scheme of a Helly-type theorem is captured by the following definition. Let $\mathcal{K}$ be an arbitrary family of sets. We say that $\mathcal{K}$ has *Helly number h* (h a natural number), if the following holds for any finite subfamily $\mathcal{F} \subseteq \mathcal{K}$: If the intersection of any h sets of $\mathcal{F}$ is nonempty, then $\bigcap \mathcal{F}$ (the intersection of all sets of $\mathcal{F}$) is nonempty. Thus, the family $\mathcal{C}$ of all convex sets in $\mathbb{R}^d$ has Helly number $d + 1$.

Recently, a new motivation for studying Helly-type results came from geometric optimization algorithms. Sharir and Welzl [SW92] defined a class of optimization problems, the so-called *LP-type problems*, which encompasses linear programming, convex programming and other natural geometric optimization problems. They gave an efficient algorithm for solving problems in this class (provided that certain primitive operations can be implemented efficiently for the problem in question), and also several other algorithms can be applied, see [Gär92], [Cla88], [MSW92], [Mat94]. A crucial parameter in these problems is their *dimension*, and this is closely related to the Helly number of suitable set systems in $\mathbb{R}^d$. Roughly speaking, if one wants to show that the above mentioned algorithms work fast for an optimization problem with some set of constraints, one needs to bound the Helly number of certain derived set systems by a (possibly small) number. These relations have been investigated by Amenta [Ame94], [Ame93]. We believe that studying Helly-type properties is important for understanding the structure of LP-type problems and potentially also for developing and analyzing yet more efficient algorithms for these problems, or perhaps proving lower bounds for such algorithms.

Disjoint unions. Grünbaum and Motzkin [GM61] considered Helly-type theorems for disjoint unions of convex sets. They conjectured that if $\mathcal{K}$ is a family such that the intersection of any its subfamily of size at most k can be expressed as a disjoint union of k closed convex sets, then the Helly number of $\mathcal{K}$ is at most $k(d+1)$ (examples show that this bound is best possible in general). In fact, they conjecture a more general result in an abstract setting. Let $\mathcal{B}$ be a family of sets with the following properties:

(i) $\mathcal{B}$ is *intersectional*, that is, $B_1 \cap B_2 \in \mathcal{B}$ for any $B_1, B_2 \in \mathcal{B}$.

(ii) $\mathcal{B}$ is *nonadditive*, that is, no finite disjoint union of at least 2 nonempty sets of $\mathcal{B}$ belongs to $\mathcal{B}$.

(iii) $\mathcal{B}$ has Helly number at most h, h a natural number.

Let $[\mathcal{B}]_k$ denote the family of all disjoint unions of at most k members of $\mathcal{B}$. The conjecture of [GM61] can be formulated as follows:

> If $\mathcal{K}$ is a family of sets such that the intersection of any at most k sets of $\mathcal{K}$ belongs to $[\mathcal{B}]_k$, with $\mathcal{B}$ satisfying (i)-(iii), then the Helly number of $\mathcal{K}$ is at most kh.

The assumptions (i)-(iii) hold for the family of all closed (resp. open) convex sets in $\mathbb{R}^d$. On the other hand, the nonadditivity fails for the family of all convex sets.

Grünbaum and Motzkin [GM61] proved their conjecture for $k=2$, the $k=3$ case was established by Larman [Lar68], and the general case was proved by Morris [Mor73][1]. A short elegant proof of the result with $\mathcal{B}$ being the family of all closed convex sets was given by Amenta [Ame94]. Her result can be also stated in an abstract framework, but different from the one described above.

Let us remark that the following holds (and it is not too difficult to prove):

Proposition 1 [GM61] *Let $\mathcal{B}$ satisfy (i) and (ii), and let $\mathcal{K}$ be a family such that the intersection of any $\leq k$ sets of $\mathcal{K}$ is in $[\mathcal{B}]_k$. Then the intersection of any number of sets of $\mathcal{K}$ belongs to $[\mathcal{B}]_k$.*

On the other hand, it is not sufficient to assume only that the members of $\mathcal{K}$ are in $[\mathcal{B}]_k$. For instance, the family of all unions of disjoint pairs of closed convex sets has no finite Helly number.

Grünbaum and Motzkin [GM61] consider a family $\mathcal{K}$ in the plane such that each set as well as the intersection of each 2 sets is a disjoint union of 2 arbitrary convex sets. They give an example of such a family with Helly number 9 (instead of 6 as one would get with convex sets replaced by closed convex sets), and they conjecture that under these assumptions on $\mathcal{K}$ the Helly number is always at most 9. Our results imply, in particular, that if one assumes that the intersection of any at most 3 sets of $\mathcal{K}$ is a union of at most 2 convex sets, then the Helly number is bounded by a constant.

Main result. We prove the following:

Theorem 2 *For any $k \geq 1$, $d \geq 1$ there exist numbers $q = q(d,k)$ and $h = h(d,k)$ with the following property. Let $\mathcal{K}$ be a family of subsets of $\mathbb{R}^d$, such that the intersection of any q or fewer sets of $\mathcal{K}$ can be expressed as a union (not necessarily disjoint) of at most k convex sets. Then $\mathcal{K}$ has Helly number at most h.*

This result was independently obtained by Alon and Kalai [AK95]. Their method is more complicated than the one presented here (and quite different), on the other hand, they obtain this result as a consequence of a powerful theorem concerning piercing numbers of certain families, which they prove by a method developed by Alon and Kleitman [AK92].

Topological generalizations. In the plane, Theorem 2 can be generalized as follows:

---

[1] To which Eckhoff [Eck93] remarks "However, Morris' proof ... is extremely involved and the validity of some of his arguments is, at best, doubtful".

Theorem 3 *For any $k \geq 1$ there exists a number $h = h(k)$ with the following property. Let $\mathcal{K}$ be a family of sets in the plane, such that the intersection of any finite subfamily of $\mathcal{K}$ has at most k path-connected components. Then $\mathcal{K}$ has Helly number at most h.*

Currently we do not know whether it is sufficient to restrict the assumption to at most q-wise intersections for some bounded $q = q(k)$ (similarly as one can do in Theorem 2). For dimensions higher than 2, we have no general topological analogue of Theorem 2. With $k = 1$, however, the following topological analogue can be established easily.

Theorem 4 *For any $d \geq 2$ there exists a number $h = h(d)$ (where $h(d) \leq d+2$ for d even and $h(d) \leq d+3$ for d odd) with the following property. Let $\mathcal{K}$ be a family of subsets of $\mathbb{R}^d$, such that the intersection of any nonempty finite subfamily of $\mathcal{K}$ is either empty or $(\lceil \frac{d}{2} \rceil - 1)$-connected[2]. Then $\mathcal{K}$ has Helly number at most h.*

The author learned the argument for the 2-dimensional case of Theorem 4 (in a slightly different context) from Nina Amenta. We suspect it was also observed by others long time ago, but currently we have no explicit reference.

A number of topological Helly-type theorems is known. Helly gave a topological version of his theorem in [Hel30]. We recall that a topological space[3] X is called a *homology cell* if it is nonempty and its (singular) homology groups of all dimensions vanish; in particular, convex sets are homology cells. Helly's result can be rephrased as follows (see [Deb70]): Let $\mathcal{F}$ be a finite family of open subsets of $\mathbb{R}^d$ such that the intersection of any at most d members of $\mathcal{F}$ is a homology cell, and the intersection of any $d+1$ members is nonempty. Then $\bigcap \mathcal{F}$ is a homology cell, in particular, it is

nonempty. As a consequence, assuming that the intersection of any at most d members of a family $\mathcal{K}$ of open sets in $\mathbb{R}^d$ is a homology cell, we get that the Helly number of $\mathcal{K}$ is $d+1$. Debrunner [Deb70] shows that it is enough to assume that the j-wise intersections of sets of $\mathcal{F}$ are $(d-j)$-acyclic[4] ($j = 1, 2, \ldots, d$), that is, all their reduced homology groups up to dimension at most $d-j$ vanish. He also generalizes the results to families of open subsets of an arbitrary d-manifold.

Thus, the topological theorems by [Hel30], [Deb70] require that the sets in the considered family are homology cells, while our result allows them to have 'holes', more precisely, a nonzero homology in dimensions $\lceil \frac{d}{2} \rceil$ thru $d-1$. On the other hand, we assume the $(\lceil \frac{d}{2} \rceil - 1)$-connectedness for all intersections, which is a very strong requirement.

Remarks and open problems. The estimates of the numerical values of the Helly numbers and of the numbers $q(d, k)$ following from our proof of Theorem 2 are quite large. For instance, for unions of convex sets in the plane, the current proof gives estimates $h(2,2) \leq 20$, $h(2,3) \leq 90$, $h(2,4) \leq 231$ etc. Small improvements are possible by refining the counting argument, but still we are probably quite far from the best values. It would be interesting to get better upper bounds and/or some nontrivial lower bound examples.

Another interesting question is to what extent an analogue of Proposition 1 holds, that is, for which d and k there exist numbers $c = c(d, k)$, $K = K(d, k)$ such that the following statement (*) holds: *Suppose that $\mathcal{K}$ is a family in $\mathbb{R}^d$ such that the intersection of any at most c sets of $\mathcal{K}$ can be expressed as a union of at most k convex sets. Then the intersection of any finite subfamily of $\mathcal{K}$ is a union of at most K convex sets.*

Example 9 below (noticed by Valtr) shows that with $K = k$ (which would be the strongest form), (*) holds for no c, even in the simplest case $d = k =$

---

[2] Let us recall the definition of j-connectivity. Let B^d denote the d-dimensional unit ball in $\mathbb{R}^d$ and S^{d-1} its boundary, the $(d-1)$-dimensional sphere. A topological space X is said to be (-1)-connected if it is nonempty. For $j \in \{0, 1, 2, \ldots\}$, X is j-connected if it is $(j-1)$-connected and moreover any continuous mapping $f : S^j \to X$ can be extended to a continuous mapping $\bar{f} : B^{j+1} \to X$.

[3] All spaces we consider are subspaces of $\mathbb{R}^d$ with the usual topology.

[4] Let us remark that a j-connected topological space is j-acyclic, and for $j \geq 2$ a j-acyclic 1-connected space is j-connected. There exist acyclic spaces which are not 1-connected, since the fundamental group is in general not commutative and it may be nontrivial even if the homology in dimension 1 vanishes. Such a situation doesn't occur for sets in the plane, however.

2, and Example 10 implies that for $d \geq 4, k \geq 2$, (*) holds with no c, K at all. Later the author found out that a related question has been investigated in several previous papers ([PS90], [BK76] and references therein), and the examples were essentially known (in particular, the 4-dimensional example is, most likely, similar to an unpublished example due to Perles mentioned in [BK76]). For completeness, we sketch the examples in section 5. For $d = 2$ and perhaps for $d = 3$, (*) might be true with large enough c and K. For $d = 2$, the results of Perles and Shelah [PS90] easily imply that (*) holds, after replacing 'convex sets' by 'closed convex sets', with $c(2,k) \leq \binom{k+1}{2}$, $K(2,k) \leq k^6$. It appears possible to get rid of the closedness assumption (this has not been worked out in detail yet). The case $d = 3$ seems to be wide open.

A number of other questions can be asked, for instance ones concerning the topological generalization. Does the claim of Theorem 2 hold (for all d) with 'convex' replaced by 'contractible' or perhaps even '($\lceil \frac{d}{2} \rceil - 1$)-connected'?

2 Proof for unions of convex sets

In this section we prove Theorem 2.

Let $[n]$ denote the set $\{1, 2, \ldots, n\}$. For a set X and a natural number t, let $\binom{X}{t}$ denote the set of all t-element subsets of X (we will sometimes call them t-sets).

We need a suitable d-dimensional generalization of the well-known fact that a planar graph with n vertices has only $O(n)$ edges. Let us set $b = \lceil d/2 \rceil + 1$.

Lemma 5 *Let d and $\alpha > 0$ be fixed, and let $n = n(d, \alpha)$ be sufficiently large. Let P be an n-point set in $\mathbb{R}^d$, and let $\mathcal{S} \subseteq \binom{P}{b}$ be a family of b-sets of P with $|\mathcal{S}| \geq \alpha \binom{n}{b}$. Then there exist disjoint sets $S, S' \in \mathcal{S}$ such that $\text{conv}(S) \cap \text{conv}(S') \neq \emptyset$.*

Let us remark that for our proof, it would be sufficient to have this lemma e.g., with $d+1$ instead of b, only we get somewhat worse values for $q(d, k)$ and $h(d, k)$. Such a statement is explicitly proved by Alon et al. [ABFK92] (see also Bárány et al. [BFL90]).

For $d = 3$, Lemma 5 is a special case of results of Dey and Edelsbrunner [DE93]. Dey and Pach [DP94] consider extensions of a similar method into higher dimensions; they explicitly prove the statement of the Lemma with d instead of b, and it seems that their method can be extended to yield the Lemma itself.

Another proof for a general dimension can be given along the lines of Alon et al. [ABFK92]; we only sketch the method here. Consider the hypergraph $\mathcal{H}$ with vertex set $V(\mathcal{H}) = \{(i, j); i = 1, 2, 3, j = 1, 2, \ldots, b\}$ and edge set $E(\mathcal{H}) = \{\{(i_1, 1), (i_2, 2), \ldots, (i_b, b); i_1, i_2, \ldots, i_b \in \{1, 2, 3\}\}$ (a complete b-partite b-uniform hypergraph). By a result of Van Kampen [vK32] (mentioned in [Sar92]), the simplicial complex whose maximal simplices are the edges of $\mathcal{H}$ cannot be embedded into $\mathbb{R}^d$ so that no two vertex-disjoint simplices intersect. By the Erdős-Stone theorem, the hypergraph with vertex set P and edge set $\mathcal{S}$ contains a copy of $\mathcal{H}$, see [ABFK92], and the Lemma follows. □

Definition 6 *Let t, j be natural numbers, $t \geq j$, and let X be a set in $\mathbb{R}^d$. We say that X has property $P(t, j)$ if for any t-set $T \subseteq X$ there exists a j-set $J \subseteq T$ such that the convex hull of J is contained in X.*

We note that if a set X is a union of at most k convex sets and $t \geq k(j - 1) + 1$, then X has property $P(t, j)$ (by the pigeonhole principle). Property $P(t, j)$ more convenient to work with for our purposes than the property "being a union of $\leq k$ convex sets", because of the following easy

Lemma 7 *Let $\mathcal{K}$ be a family of sets in $\mathbb{R}^d$ such that the intersection of any at most $\binom{t}{j}$ sets of $\mathcal{K}$ has property $P(t, j)$. Then the intersection of any subfamily of $\mathcal{K}$ has property $P(t, j)$.*

Proof. Let X be the intersection of a subfamily $\mathcal{F} \subseteq \mathcal{K}$, and suppose that property $P(t, j)$ fails for X, that is, one can find a t-set $T \subseteq X$ such that none of the convex hulls of its j-sets is fully contained in X. For each $J \in \binom{T}{j}$, fix a set $F_J \in \mathcal{F}$ with $\text{conv}(J) \not\subseteq F_J$. Then property $P(t, j)$ fails also for the intersection

$$\bigcap_{J \in \binom{T}{j}} F_J.$$

□

Let us fix $t = k(b-1)+1$, $q = q(d,k) = \binom{t}{b}$. Then the assumption of Theorem 2 with this value of q implies that the intersection of any at most q sets of $\mathcal{K}$ has property $P(t,b)$, and by Lemma 7 any intersection of sets of $\mathcal{K}$ has property $P(t,b)$. Hence Theorem 2 is proved by establishing the following:

Proposition 8 *For any $d \geq 1$, $t \geq b = \lceil d/2 \rceil + 1$ there exists a number $h = h(d,t)$ with the following property. Let $\mathcal{K}$ be a family of subsets of $\mathbb{R}^d$, such that the intersection of any its finite subfamily has property $P(t,b)$. Then $\mathcal{K}$ has Helly number at most h.*

Proof. Throughout the proof, d and t are treated as constants (in the O, Ω notation). Let h be sufficiently large so that all estimates below are valid (by tracking down all the constants, one can find a specific value for $h = h(d,t)$).

For proving that $\mathcal{K}$ has Helly number h, it is enough to show that if $\mathcal{F} \subseteq \mathcal{K}$ is any subfamily of $h+1$ sets of $\mathcal{K}$ such that the intersection of each h sets of $\mathcal{F}$ is nonempty, then $\bigcap \mathcal{F} \neq \emptyset$ (the case of a larger $\mathcal{F}$ is then handled by induction on the number of elements of $\mathcal{F}$).

Let the sets of $\mathcal{F}$ be numbered $F_1, F_2, \ldots, F_n$, $n = h + 1$. For each i, choose a point $p_i \in \bigcap_{j \in [n] \setminus \{i\}} F_j$.

Consider a t-set $J \in \binom{[n]}{t}$. All the points p_j, $j \in J$, belong to the intersection $X_J = \bigcap_{j \in [n] \setminus J} F_j$. Since X_J has property $P(t,b)$, we can fix a b-set $K = K(J) \subseteq J$ such that $\Delta_K := \text{conv}\{p_j; j \in K\}$ is contained in X_J (if there are more choices for $K(J)$ fix one arbitrarily). In this situation, we assign the set $J \setminus K$ to the b-set $K = K(J)$ as a label.

As we fix the $K(J)$ for each t-set J, some b-sets K receive various labels. If L is one of the labels of K, then we know that the $(b-1)$-simplex Δ_K is contained in all sets F_j with $j \notin L \cup K$. Let us call a b-set K *good* if it has at least one label but the intersection of all its labels is empty, and call K *bad* otherwise.

Each bad b-set is assigned at most $\binom{n-b-1}{t-b-1}$ labels, and there are $\binom{n}{b}$ b-sets, thus at most $O(n^{t-1})$ labels are assigned to bad b-sets. Since $\binom{n}{t}$ labels are assigned altogether, at least $\Omega(n^t)$ labels are assigned to good b-sets. A good b-set is assigned $O(n^{t-b})$ labels, therefore there are $\Omega(n^b)$ good b-sets.

By Lemma 5, there are two disjoint good b-sets K, K' with $\Delta_K \cap \Delta_{K'} \neq \emptyset$. By the definition of a good b-set, Δ_K is contained in all sets F_j with $j \notin K$, and similarly for $\Delta_{K'}$. Thus, $\emptyset \neq \Delta_K \cap \Delta_{K'} \subseteq \bigcap_{j=1}^n F_j$. □

3 Proof for the planar topological case

The proof begins similarly as for Proposition 8. We consider a family $\mathcal{F} = \{F_1, \ldots, F_n\}$, with $n = h+1$ sufficiently large and points $p_1, \ldots, p_n$, with p_i belonging to all sets of $\mathcal{F}$ but possibly F_i. We set $t = k+1$, and consider a t-set $J \in \binom{[n]}{t}$. We know that all points p_j, $j \in J$ belong to the intersection $X_J = \bigcap_{i \in [n] \setminus J} F_i$, and this is a union of at most k path-connected sets. Therefore, there is a pair $P(J) = \{j_1, j_2\} \subset J$ of indices such that p_{j_1} and p_{j_2} belong to a common path-connected subset of X_J (in fact, there may be many pairs, so we fix one arbitrarily for each t-set J).

The considered t-set is linearly ordered, and the pair $P(J)$ can be uniquely encoded by specifying a pair of elements of the set $\{1, 2, \ldots, t\}$ (let us call it the *type* of the pair $P(J)$). The number of possible pairs is $\binom{t}{2}$, the important fact is that it is bounded by a function of t. In this way, each $J \in \binom{[n]}{t}$ is assigned one of a bounded number of types, which can be viewed as coloring all t-sets on $[n]$. If m is a prescribed parameter and $n = n(m,t)$ is chosen sufficiently large, by Ramsey's theorem (see e.g., [GRS90]) there exists an m-element subset $M \subseteq [n]$ such that all t-sets $J \in \binom{M}{t}$ have the same color. In our case this means that they all have the same type of the pair $P(J)$.

Choose a fixed non-planar graph G, say $G = K_5$, and let V be its vertex set and E its edge set. We assume that a set $M \subseteq [n]$ has been selected as above, so that all t-sets have the same type of $P(J)$, and $|M|$ is sufficiently large (as we will see, $|M| \geq 5 + 10(t-2)$ suffices). We illustrate the method for $k = 2$ (then $t = 3$) and assuming that the type of the pair is (• • ○), the other cases are quite similar. To each vertex $v \in V$ we assign a point $\varphi(v) \in M$ and to each edge $e \in E$ we assign a t-set $\Phi(e)$ of points of M, in such a way that

(C1) For any two disjoint $e, e' \in E$, we have $\Phi(e) \cap \Phi(e') = \emptyset$, and if $e \neq e'$ share a vertex v, then $\Phi(e) \cap \Phi(e') = \{\varphi(v)\}$.

(C2) For each $e = \{u, v\} \in E$, the points $\varphi(u)$ and $\varphi(v)$ are just the points of the pair $P(\Phi(e))$.

Such an assignment, for the particular type of the pair $P(J)$ of 3-sets, may be chosen as indicated in fig. 1 (the full circles represent the points $\varphi(v_1), \ldots, \varphi(v_5)$, the empty circles the other points of M, and the triples are marked by the segments at various levels connected to their points).

Now by (C2), for each edge $e = \{u, v\} \in E$, the points $p_{\varphi(u)}$ and $p_{\varphi(v)}$ lie in the same path-connected piece of the intersection $X_{\Phi(e)}$, so we can choose a path $\Delta_e \subseteq X_{\Phi(e)}$ connecting $p_{\varphi(u)}$ and $p_{\varphi(v)}$. These paths together yield a drawing of K_5, so some two paths belonging to vertex-disjoint edges e, e' cross. The intersection of these paths belongs to all sets F_i with $i \notin \Phi(e)$, and also to all F_i with $i \notin \Phi(e')$. But $\Phi(e) \cap \Phi(e') = \emptyset$ by (C1) and we are done. □

4 Proof for the topological case with $k = 1$

We need a finite $\lceil \frac{d}{2} \rceil$-dimensional simplicial complex $\mathcal{S}$ with the following property: Whenever $f : \|\mathcal{S}\| \to \mathbb{R}^d$ is a continuous mapping, there exist vertex-disjoint simplices $s, s' \in \mathcal{S}$ such that $f(\|s\|) \cap f(\|s'\|) \neq \emptyset$ (here $\|\mathcal{S}\|$ denotes the *polyhedron* of $\mathcal{S}$, which is the topological space of some geometric realization of $\mathcal{S}$, and for a simplex $s \in \mathcal{S}$, $\|s\|$ means the closed subset of $\|\mathcal{S}\|$ corresponding to the simplex s; see e.g., [Mun84] for an introduction to simplicial complexes). We may use a classical result of Van Kampen [vK32] and Flores [Flo34], which says that if $\mathcal{S}$ is the j-skeleton of the $(2j + 2)$-dimensional simplex, then it has the required property for $d \leq 2j$.

To prove Theorem 4, we fix the simplicial complex $\mathcal{S}$ as above (for a given d) and we let $n = h + 1$ be the number of its vertices. As in the previous proofs, let $F_1, \ldots, F_n$ be sets from $\mathcal{K}$, and for each i, we fix $p_i \in \bigcap_{j \in [n] \setminus \{i\}} F_j$.

We construct a continuous mapping $f : \|\mathcal{S}\| \to \mathbb{R}^d$. This mapping is constructed inductively; as a first step, we let $f(v_i) = p_i$, where v_i denotes the ith vertex of $\mathcal{S}$ (in some arbitrary numbering of vertices). Further, the construction proceeds by induction on j, which takes the values $0, 1, \ldots, \lceil \frac{d}{2} \rceil$.

Suppose that the mapping f has already been defined at all points of each $\|s\|$, where $s \in \mathcal{S}$ is a simplex of dimension less than j, so that the following holds:

$$f(\|s\|) \subseteq X_{J(s)} = \bigcap_{i \in [n] \setminus J(s)} F_i, \qquad (1)$$

where $J(s) = \{i \in [n]; v_i \in s\}$. We are going to define f on j-dimensional simplices. Consider a j-dimensional simplex $s \in \mathcal{S}$. The portion of $\|s\|$ corresponding to proper faces of s (on which f has already been defined) is homeomorphic to S^{j-1}, and by (1), the images of all such faces are contained in the set $X_{J(s)}$. Since this set is assumed to be $(\lceil \frac{d}{2} \rceil - 1)$-connected, we can extend f continuously also to the relative interior of $\|s\|$, in such a way that the image is still contained in $X_{J(s)}$. This finishes the induction step.

Having defined f on the whole $\|\mathcal{S}\|$, by the choice of $\mathcal{S}$ we know there are two vertex-disjoint faces $s, s' \in \mathcal{S}$ with $f(\|s\|) \cap f(\|s'\|) \neq \emptyset$. By (1), we have $f(\|s\|) \cap f(\|s'\|) \subseteq X_{J(s)} \cap X_{J(s')} = \bigcap_{i \in [n]} F_i$. This concludes the proof. □

5 Examples

Example 9 *For any odd integer n, there exist sets $F_1, \ldots, F_n \subseteq \mathbb{R}^2$, such that the intersection of any at most $n - 1$ of them is a union of 2 convex sets, while $\bigcap_{i=1}^n F_i$ cannot be expressed as a union of fewer than 3 convex sets.*

Proof. First, let C be a regular convex n-gon, let $v_1, \ldots, v_n$ denote its vertices numbered along the circumference. We let F_i be C minus the relative interior of the edge $v_i v_{i+1}$ (v_{n+1} meaning v_1). If the intersection $\bigcap_{i=1}^n F_i$ were a union of 2 convex sets, there would be 2 consecutive vertices v_i, v_{i+1} belonging to the same convex sets, but this is impossible, since the interior of the edge $v_i v_{i+1}$ is missing from the intersection. On the other hand,

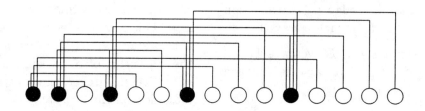

Figure 1: The embedding Φ for the graph K_5.

for any intersection X of fewer than n of the F_i, there is one edge, say $v_n v_1$, which is contained in X. In this situation, if we let A_1 be the set X minus all vertices v_i with i odd, and A_2 is X minus all vertices v_i with i even, we have $X = A_1 \cup A_2$ and it is easily checked that A_1 and A_2 are convex.

The example can be easily modified so that sets F_i are closed (or open). Namely, consider the midpoint m_i of the edge $v_i v_{i+1}$, move it a little bit towards the center of C, obtaining a point $\bar{m}_i$, and let F_i be C minus the interior of the triangle $v_i \bar{m}_i v_{i+1}$. $\square$

Example 10 *For any given integers c, K, there exist n and sets $F_1, \ldots, F_n \subseteq \mathbb{R}^4$, such that the intersection of any at most c of them is a union of 2 convex sets, while $\bigcap_{i=1}^n F_i$ cannot be expressed as a union of fewer than K convex sets.*

Proof. Let G be a K-chromatic graph such that any its subgraph with at most c edges is 2-colorable (the existence of such a graph follows from [Erd59], say). Let C be a cyclic polytope in $\mathbb{R}^4$ with $|V(G)|$ vertices, and suppose that its vertices are identified with the vertex set V of G. Any two vertices of C are connected by an edge (1-dimensional face) of C. Let $e_1, \ldots, e_n$ be the edges of G. For $e_i = \{u, v\}$, we let F_i be C minus the relative interior of the edge uv of the polytope C. The set $\bigcap_{i=1}^n F_i$ cannot be expressed as a union of fewer than K convex sets, since this would induce a coloring of G by less than K colors. On the other hand, if $X = \bigcap_{e_i \in E'} F_i$, where E' is a set of at most c edges of G, we let $\chi : V \to \{1, 2\}$ be a 2-coloring of the graph with edge set E'. For $j = 1, 2$ we set $A_j = X \setminus \{v \in V; \chi(v) \neq j\}$. As in the previous example, one may check that the A_j are convex and cover X. A modification with all F_i closed or all F_i open is again possible. $\square$

Acknowledgment. I would like to acknowledge two significant contributions to this paper by other people. Nina Amenta brought the problems to my attention and she gave an argument which can be interpreted as a proof of Theorem 4 for $d = 2$; this inspired the basic approach of all the proofs. Pavel Valtr pointed out Example 9, showing that a direct generalization of Proposition 1 doesn't hold with general unions of at most k convex sets, which directed me to using property $P(t, j)$ instead.

References

[ABFK92] N. Alon, I. Bárány, Z. Füredi, and D. Kleitman. Point selections and weak ε-nets for convex hulls. *Combin., Probab. Comput.*, 1(3):189–200, 1992.

[AK92] N. Alon and D. Kleitman. Piercing convex sets. *Adv. Math.*, 96(1):103, 1992.

[AK95] N. Alon and G. Kalai. Bounding the piercing number. *Discr. Comput. Geom.*, 1995. Special volume (L. Féjes Tóth Festschrift). To appear.

[Ame93] N. Amenta. Helly theorems and generalized linear programming. In *Proc. 9th Annu. ACM Sympos. Comput. Geom.*, pages 63–72, 1993.

[Ame94] N. Amenta. Bounded boxes, Hausdorff distance, and a new proof of an interesting Helly theorem. In *Proc. 10th Annu. ACM Sympos. Comput. Geom.*, pages 340–347, 1994.

[BFL90] I. Bárány, Z. Füredi, and L. Lovász. On the number of halving planes. *Combinatorica*, 10:175–183, 1990.

[BK76] M. Breen and D. C. Kay. General decomposition theorems for m-convex sets in the plane. *Israel J. Math.*, 24:217–233, 1976.

[Cla88] K. L. Clarkson. A Las Vegas algorithm for linear programming when the dimension is small. In *Proc. 29th Annu. IEEE Sympos. Found. Comput. Sci.*, pages 452–456, 1988.

[DE93] T. K. Dey and H. Edelsbrunner. Counting triangle crossings and halving planes. In *Proc. 9th Annu. ACM Sympos. Comput. Geom.*, pages 270–273, 1993.

[Deb70] H. Debrunner. Helly type theorems derived from basic singular homology. *Amer. Math. Monthly*, 77:375–380, 1970.

[DP94] T. Dey and J. Pach. Extremal problems for geometric hypergraphs. Manuscript in preparation, 1994.

[Eck93] J. Eckhoff. Helly, Radon and Carathéodory type theorems. In *Handbook of Convex Geometry*. North-Holland, 1993.

[Erd59] P. Erdős. Graph theory and probability. *Canad. J. Math.*, 11:34–38, 1959.

[Flo34] A. Flores. Über n-dimensionale Komplexe die im R_{2n+1} absolut selbstverschlungen sind. *Ergeb. Math. Kolloq.*, 6:4–7, 1932/1934.

[Gär92] B. Gärtner. A subexponential algorithm for abstract optimization problems. In *Proc. 33rd Annu. IEEE Sympos. Found. Comput. Sci.*, pages 464–472, 1992.

[GM61] B. Grünbaum and T. Motzkin. On components in some families of sets. *Proc. Amer. Math. Soc.*, 12:607–613, 1961.

[GRS90] R. L. Graham, B. L. Rothschild, and J. Spencer. *Ramsey Theory*. J. Wiley & Sons, 1990.

[Hel30] E. Helly. Über Systeme von abgeschlossenen Mengen mit gemeinschaftlichen Punkten. *Monaths. Math. und Physik*, 37:281–302, 1930.

[Lar68] D. Larman. Helly type properties of unions of convex sets. *Mathematika*, 15:53–59, 1968.

[Mat94] J. Matoušek. On geometric optimization with few violated constraints. In *Proc. 10th Annu. ACM Sympos. Comput. Geom.*, pages 312–321, 1994.

[Mor73] H. Morris. Two pigeon hole principles and unions of convexly disjoint sets. Ph. D. thesis, California Institute of Technology, Pasadena, 1973.

[MSW92] J. Matoušek, M. Sharir, and E. Welzl. A subexponential bound for linear programming. In *Proc. 8th Annu. ACM Sympos. Comput. Geom.*, pages 1–8, 1992.

[Mun84] J. R. Munkres. *Elements of algebraic topology*. Addison-Wesley, 1984.

[PS90] M. Perles and S. Shelah. A closed $(n+1)$-convex set in R^2 is a union of n^6 convex sets. *Israel J. Math.*, 70:305–312, 1990.

[Sar92] K. Sarkaria. Shifting and embeddability of simplicial complexes. Tech. Report 91/51, Max-Planck Institut für Mathematik, Bonn, 1992.

[SW92] M. Sharir and E. Welzl. A combinatorial bound for linear programming and related problems. In *Proc. 9th Sympos. Theoret. Aspects Comput. Sci.*, volume 577 of *Lecture Notes in Computer Science*, pages 569–579. Springer-Verlag, 1992.

[vK32]   R. E. van Kampen. Komplexe in euklidischen Räumen. *Abh. Math. Sem. Hamburg*, 9:72–78, 1932. Berichtigung dazu, *ibid.* (1932) 152–153.

On Conway's Thrackle Conjecture

*László Lovász**
Dept. Computer Science, Yale University

János Pach[†]
City College, CUNY and Courant Institute, NYU

Mario Szegedy
AT&T Bell Laboratories

Abstract

A *thrackle* is a graph that can be drawn in the plane so that its edges are represented by Jordan arcs and any two distinct arcs either meet at exactly one common vertex or cross at exactly one point interior to both arcs. About thirty years ago, J. H. Conway conjectured that the number of edges of a thrackle cannot exceed the number of its vertices. We show that a thrackle has at most twice as many edges as vertices. Some related problems and generalizations are also considered.

1 Introduction

Let G be a graph with vertex set $V(G)$ and edge set $E(G)$, and assume that it has no loops or multiple edges. A *drawing* of G is a representation of G in the plane such that every vertex corresponds to a point, and every edge is represented by a Jordan arc connecting the corresponding two points without passing through any other vertex. Two edges (arcs) are said to *cross* each other if they have an interior point p in common. For simplicity, we always assume that no three edges cross at the same point. A crossing p is called *proper* if in a small neighborhood of p one edge passes from one side of the other edge to the other side. Due to its aesthetic appeal and wide range of applications in VLSI layout, computer-aided-design, software visualization, etc., the area of graph drawings has received a lot of attention in the past two decades. For a recent bibliography of graph drawing algorithms, see [DETT94].

There are many interesting results in topological graph theory characterizing all graphs embeddable on a given surface without crossing (see [WB78]). However, we know very little about the possible intersection patterns determined by the edges of a graph drawn on a surface. In particular, even for some very simple graphs we do not how to find the *crossing number* of G, i.e., the minimum number of crossing pairs of edges in a planar drawing of G. In the case when G is a complete bipartite graph, this is Turán's brick factory problem [T77, G72]. The determination of the crossing number is known to be NP–complete [GJ83].

Another well-known open problem that illustrates our ignorance about graph drawings was raised by Conway more than thirty years ago. He defined a *thrackle* as a drawing of a graph G with the property that any two distinct edges either

(i) share an endpoint, and then they do not have any other point in common; or

(ii) do not share an endpoint, in which case they meet exactly once and determine a proper crossing.

Thrackle Conjecture: *The number of edges of a thrackle cannot exceed the number of its vertices.*

A graph that can be drawn as a thrackle is said to be *thrackleable*. Assuming that the above conjecture is true, Woodall [W69] characterized all thrackleable graphs. With this assumption, a finite graph is thrackleable if and only if it has at most one odd cycle, it has no cycle of length four, and each of its connected components contains at most one cycle. Note that it is quite straightforward to check the necessity of these conditions (see Lemma 2.1). Using a construction suggested by Conway, the thrackle conjecture can be reduced to the following statement: If a graph G consists of two even cycles meeting in a single vertex then G

---

*Supported by NSF grant CCR-94-02916.

[†]Supported by NSF grant CCR-91-22103, PSC-CUNY Research Award 663472 and OTKA-4269.

is not thrackleable (cf. [W69, PRS94]). It is worth mentioning that the thrackle conjecture is true for *straight-line thrackles*, i.e., for drawings where every edge is represented by a segment [HP34, FS35, PA95]. See [LST94] for a surprising relation between straight-line thrackles and triangulations of certain polytopes, and [G75] for another geometric application.

Any two edges of a thrackle intersect in exactly one point, including the endpoints. For finite set-systems satisfying a similar condition we have the following well-known result [F40, BE48].

Fisher Inequality: *Let F be a family of subsets of a finite set X such that any two members of F have exactly one element in common. Then F has at most as many members as the number of elements of X.*

An interesting modular version of this inequality was discovered by Berlekamp [B69]. Suppose that every member of F has an *odd* number of elements and that the intersection of any two members is *even*. Then $|F| \leq |X|$. These results and their generalizations originate in linear algebra and play a central role in finite geometries and in the theory of combinatorial designs (see [BF88]).

Since thrackles do not contain cycles of length four, it follows from [KST54] that the maximum number of edges a thrackle of n vertices can have is $O(n^{3/2})$. Our next theorem represents a substantial improvement on this bound.

Theorem 1.1 *Every thrackle of n vertices has at most $2n - 3$ edges.*

The proof is based on the following result.

Theorem 1.2 *Every thrackleable bipartite graph is planar.*

Just like the Fisher inequality, the thrackle conjecture has some modular versions, too. For example, call a graph drawing a *generalized* (or *modulo 2-*) *thrackle* if any two edges meet an *odd* number of times, where "meet" means either "meet at a common vertex" or "meet at a proper crossing".

Theorem 1.3 *Every generalized thrackle of n vertices has at most $3n - 4$ edges.*

Theorem 1.4 *A bipartite graph can be drawn as a generalized thrackle if and only if it is planar.*

Woodall [W72] asked whether the thrackle conjecture remains true for generalized thrackles. Our last theorem implies that the answer to this question is in the negative, because a bipartite planar graph of n vertices can have as many as $2n - 4$ edges.

2 Three lemmas

In the sequel, a thrackle and its underlying "abstract" graph are both denoted by G. If there is no danger of confusion, we make no notational distinction between a vertex (edge) of the graph and the corresponding point (arc).

Lemma 2.1 *Let G be a thrackleable graph. Then G contains (i) no cycle of length four; (ii) no two vertex disjoint odd cycles.*

Proof: To show (ii), notice that a pair of vertex disjoint odd cycles would be represented in a thrackle by two closed curves that properly cross each other an odd number of times. □

Lemma 2.2 *Let C_1 and C_2 be two cycles in a graph G that have precisely one vertex v in common. Suppose that G can be drawn as a thrackle.*

Then the two closed curves representing C_1 and C_2 cross each other in a small neighborhood of v if and only if both cycles are odd.

Proof: Let k_i denote the length of C_i, $i = 1, 2$. The closed curve representing C_1 divides the plane into $k_1(k_1 - 3)/2 + 2$ connected cells. Color these cells with black and white so that no two cells that share a boundary arc have the same color. The curve representing C_2 intersects C_1 exactly $2(k_1 - 2) + (k_2 - 2)k_1 \equiv k_1 k_2$ times (mod 2), not counting v. Every time C_2 intersects C_1, it passes from one cell to another whose color is different. Assume that in a small neighborhood of v the initial segment of an edge of C_2 incident to v lies in a white region. Then the initial segment of the other edge of C_2 incident to v lies in a black region if and only if $k_1 k_2$ is odd. □

A graph consisting of three vertex disjoint paths P_i, $i = 1, 2, 3$ between u and v is called a Θ-*graph*. A drawing of this Θ-graph is said to be a *preserver* if in a small neighborhood of u the initial pieces of the paths P_i follow each other in the same circular order as the final pieces do around v. Otherwise, the drawing is called a *converter*. Note that, using this terminology, if G is a planar graph drawn in the plane without crossing then any Θ-subgraph of this drawing is a converter.

The proof of the next lemma is very similar to that of the previous one.

Lemma 2.3 *A Θ-subgraph of a thrackle is a converter if and only if at most one of its three paths has odd length.*

Remark: With the exception of Lemma 2.1(i), all statements and proofs in this section remain valid for generalized thrackles.

3 Bipartite thrackles

Proof of Theorem 1.2: By Kuratowski's theorem, it is sufficient to show that a thrackleable bipartite graph G does not contain a subdivision of K_5 or a subdivision of $K_{3,3}$.

Suppose that G contains a subdivision of K_5, whose vertices are $v_0, \ldots, v_4$. Assume without loss of generality that in a thrackle-drawing of G the initial pieces of the edges incident to v_0 follow each other in the clockwise order $v_0v_1, \ldots, v_0v_4$. Then there are two (even) cycles through v_0, v_1, v_3 and v_0, v_2, v_4 that have no vertex in common other than v_0. The corresponding two curves cross each other in a small neighborhood of v_0, contradicting Lemma 2.2.

Suppose next that G contains a subdivision of $K_{3,3}$ with vertex classes $\{u_1, u_2, u_3\}$ and $\{v_1, v_2, v_3\}$. Denote this subdivision by K. Assume first that the lengths of all nine paths in K connecting the u_i's and the v_j's have the same parity. Deleting from K the point u_3 together with the three paths connecting it to the v_j's, we obtain a Θ-graph. In view of Lemma 2.3, it is a converter between u_1 and u_2. Similarly, deleting u_2 (u_1) we obtain a converter between u_1 and u_3 (u_2 and u_3, respectively). We say that the type of u_i is *clockwise* or *counterclockwise* according to the circular order of the initial segments of the paths u_iv_1, u_iv_2, u_iv_3 around u_i. It follows from the definition of a converter that any two u_i's must have opposite types, which is impossible.

There are two other essentially different cases according to the parities of the nine paths forming K. It turns out that in both cases one can arrive at a contradiction by showing that there is exactly one pair of points among u_1, u_2, u_3 having opposite types. □

Proof of Theorem 1.4: In view of the remark at the end of the previous section, the above argument also proves that every bipartite graph that can be drawn as a *generalized thrackle* is planar. To establish the theorem, we have to show that the reverse of this statement is also true, i.e., every bipartite planar graph G can be drawn as a generalized thrackle. To see this, consider a crossing-free embedding of G in the plane such that

(i) $V(G) = V_1 \cup V_2$, where all points of V_1 are mapped into the upper half-plane and all points of V_2 below the line $y = -1$;

(ii) every edge $e \in E(G)$ connects a vertex of V_1 to a vertex of V_2, and each piece of e belonging to the strip $-1 \leq y \leq 0$ is a vertical segment.

Now erase the part of the drawing in the strip $-1 \leq y \leq 0$, and replace the part in the upper half-plane by its reflection about the y-axis. Reconnecting the corresponding pairs of points on the lines $y = -1$ and $y = 0$ by straight-line segments, we obtain a drawing of G such that any pair of independent edges meet an odd number of times. This can be turned into a generalized thrackle by slightly modifying the edges in a small neighborhood of their endpoints so as to reverse the circular order of edges around each vertex of G. □

We could have completed our proof without using Lemma 2.2. The fact that a thrackle contains no subdivision of K_5 can also be deduced from Lemma 2.3 in a slightly more complicated way.

Corollary 3.1 *A graph is planar if and only if it has a drawing whose every Θ-subgraph is a converter.*

For a related result, see [T70].

4 Reduction to the bipartite case

Every graph can be made bipartite by the removal of fewer than half of its edges. It follows from Euler's polyhedral formula that any bipartite planar graph of n vertices has at most $2n - 4$ edges ($n > 2$). If in addition the graph has no cycles of length four then this bound can be replaced by $\lfloor 3n/2 \rfloor - 3$ ($n > 3$). Thus, Theorem 1.4 and Lemma 2.1(i) immediately imply the following.

Corollary 4.1 *Let $n > 3$. Then*
(i) every thrackle of n vertices has at most $3n-7$ edges;
(ii) every generalized thrackle of n vertices has at most $4n - 9$ edges.

In the rest of this section we sketch how to reduce the bound in Corollary 4.1(i) roughly by n.

Let G be a thrackle of n vertices, $n > 3$. One can assume that G is not bipartite, otherwise its number of edges cannot exceed $\lfloor 3n/2 \rfloor - 3$. Let C denote a shortest odd cycle of G with length c. By Lemma 2.1(i) and by the minimality of C, any vertex of G has at most one neighbor belonging to C. Hence, there are at most n edges of G incident to some vertex of C. It follows from Lemma 2.1(ii) that the graph $G - C$ obtained from G by the removal of all points of C is bipartite. Thus,

$$|E(G)| \leq |E(G-C)| + n \leq 3(n-c)/2 + n = 5n/2 - 3c/2.$$

One can refine this argument, as follows. The closed curve representing C cuts the plane into a number of cells that can be colored with black and white so that no two cells with a common boundary arc have the same color. Let b and w denote the number of vertices of $G-C$ lying in black and in white cells, respectively.

Clearly, $c + b + w = n$, and one can assume without loss of generality that $b \leq w$. Observe that if an edge e connects a point of C to (say) a black vertex, then in a small neighborhood of this point the initial piece of e must be white. There are at most b such edges, and if remove all of them together with all edges of C, the resulting graph (thrackle) becomes bipartite. This yields the inequality

$$|E(G)| \leq \lfloor 3n/2 \rfloor - 3 + b + c \leq 2n + c/2 - 3.$$

Comparing the last two inequalities, we obtain that $|E(G)| < (2 + 1/8)n$.

One can further reduce this bound by utilizing an idea of Conway (see [W69, G93, PRS91, PRS94]). Now we replace each vertex and edge of C by two nearby vertices and edges, respectively. More precisely, we split each vertex v of C into two vertices, v_b and v_w, and connect all black and white neighbors of v to v_b and v_w, respectively. Furthermore, if v and v' are two consecutive vertices of C, we connect v_b to v'_w and v_w to v'_b. It is not hard to see that this construction can be carried out in such a way that the resulting drawing G' is a thrackle, which becomes bipartite after the removal of all edges between v_b's and black vertices. Thus,

$$|E(G')| - b = |E(G)| + c - b \leq \lfloor 3(n+c)/2 \rfloor - 3,$$

which implies that

$$|E(G)| \leq 2n - 3,$$

as stated in Theorem 1.1.

5 Small forbidden configurations

All of the results in the previous sections were based on parity arguments. Theorem 1.4 shows that if we want to settle Conway's original conjecture, we have to go beyond these methods. In the proof of Theorem 1.1 we were able to explore a property of thrackles that does not hold for generalized thrackles. Namely, we used the fact that a thrackleable graph has no cycle of length four (Lemma 2.1(i)). By excluding some other small configurations that would contradict the thrackle conjecture, one can easily improve the bound in Theorem 1.1. The trouble is that it is quite difficult to find any new non-trivial forbidden subgraph, because even a relatively small graph may have an enormous number of topologically different drawings such that no two edges meet more than once. In this section, we illustrate these difficulties by an example.

Let Θ_3 denote a graph consisting of two vertices connected by three vertex disjoint paths of length three.

Theorem 5.1 *A thrackleable graph cannot contain Θ_3 as a subgraph.*

For the proof we need some preparation. Let G be a fixed thrackle whose edges are smooth curves. Given two directed edges e and f that do not share an endpoint, we say that e meets f *clockwise* if at their intersection point a tangent vector to e can be carried into a tangent vector of f by a clockwise turn with angle less than π.

Let $P = e_1 e_2 e_3 e_4$ be a directed path in G with length four, directed towards e_4. Associate P with a 4×4 matrix M such that $M_{ij} = 0$ if $i = j$ or if e_i and e_j do not have an interior point in common. Otherwise, let $M_{ij} = 1$ or -1 depending on whether e_i meets e_j clockwise or counterclockwise. Clearly, M is antisymmetric and it is determined by the triple (M_{13}, M_{14}, M_{24}). This triple is called the *type* of P. It turns out that their are only six possible types: $a=(1,1,-1); b=(1,-,1,-1); c=(1,-1,1);$ $A=(-1,-1,1); B=(-1,1,1); C=(-1,1,-1).$

Lemma 5.2 *Let $e_1, e_2, \ldots, e_6$ be six directed edges of a thrackle that form a simple directed cycle, and let $P_i = e_i e_{i+1} e_{i+2} e_{i+3}$, where the indices are taken mod 6.*

Then $type(P_1) type(P_2) \ldots type(P_6)$ must be one of the following sequences: $AaAaAa$, $aAaAaA$, $BbBbBb$, $bBbBbB$.

Given a directed path $P = e_1 e_2 e_3 e_4$, let the *reverse* of P be defined as $P^{-1} = e_4^{-1} e_3^{-1} e_2^{-1} e_1^{-1}$, where e_i^{-1} denotes the same edge as e_i but with reversed orientation. If $e_1 \ldots e_5$ is a simple directed path, we say that $P' = e_2 e_3 e_4 e_5$ can be obtained from $P = e_1 e_2 e_3 e_4$ by a *shift*.

Lemma 5.3 *Let P be a path of length four in a thrackle, and assume that $type(P) \in \{a, b, A, B\}$.*
(i) $type(P^{-1}) = b, a, B$ or A according to whether $type(P) = a, b, A$ or B.
(ii) If P' can be obtained from P by a shift and $type(P') \in \{a, b, A, B\}$, then $type(P) type(P')$ must be one of the following six pairs: aA, aB, bB, Aa, Ab, Bb.

Proof of Theorem 5.1: Assume that there is a thrackle containing Θ_3 as a subgraph. By Lemma 5.2, the type of every directed path of Θ_3 belongs to the set $\{a, b, A, B\}$. Consider a path P whose type belongs $\{b, B\}$. (If P does not satisfy this condition then its reverse does.) Observe that the topology of Θ_3 allows us to transform P into its reverse by a series of shifts. It follows from Lemma 5.3(ii) that the types of all paths obtained during this process belong to $\{b, B\}$. However, by Lemma 5.3(i), type$(P^{-1}) \in \{a, A\}$, which is a contradiction. □

References

[BF88] L. Babai and P. Frankl, *Linear Algebra Methods in Combinatorics, Part I*, Tech. Report, University of Chicago.

[B69] E.R. Berlekamp, On subsets with intersections of even cardinality, *Canadian Math. Bull.* **12**, 363–366.

[BE48] N.G. de Bruijn and P. Erdős, On a combinatorial problem, *Proc. Konink. Nederl. Akad. Wetensch., Ser. A* **51**, 1277–1279 (=*Indagationes Math.* **10**, 421–423.)

[DETT94] G. Di Battista, P. Eades, R. Tamassia, and I.G. Tollis, Algorithms for drawing graphs: an annotated bibliography, *Computational Geometry: Theory & Applications* **4**, 235–282.

[FS35] W. Fenchel and J. Sutherland, Lösung der Aufgabe 167, *Jahresbericht der Deutschen Mathematiker-Vereinigung* **45**, 33–35.

[F40] R. A. Fisher, An examination of the different possible solutions of a problem in incomplete blocks, *Annals of Eugenics (London)* **10**, 52–75.

[GJ83] M.R. Garey and D.S. Johnson, Crossing number is NP-complete, *SIAM J. Algebraic Discrete Methods* **4**, 312–316.

[G75] R.L. Graham, The largest small hexagon, *J. Combinatorial Theory* **18**, 165–170.

[G93] J. Green-Cottingham, *Thrackles, Surfaces and Drawings of Graphs*, Doctoral Dissertation, Clemson University.

[G72] R. K. Guy, Crossing numbers of graphs, in: *Graph Theory and Applications, Lecture Notes in Mathematics, Vol. 303*, Springer–Verlag, 111–124.

[HP34] H. Hopf and E. Pannwitz, Aufgabe Nr. 167, *Jahresbericht der Deutschen Mathematiker-Vereinigung* **43**, 114.

[KST54] T. Kővári, V. T. Sós, and P. Turán, On a problem of K. Zarankiewicz, *Colloqium Mathematicum* **3**, 50–57.

[LST94] J. A. de Loera, B. Sturmfels, and R. R. Thomas, Gröbner bases and triangulations of the second hypersimplex, *Combinatorica*, to appear.

[PA95] J. Pach and P.K. Agarwal, *Combinatorial Geometry*, J. Wiley, New York.

[PRS94] B. Piazza, R. Ringeisen, and S. Stueckle, On Conway's reduction of the thrackle conjecture and some associated drawings,, to appear.

[PRS91] B. Piazza, R. Ringeisen, and S. Stueckle, Properties of non-minimum crossings for some classes of graphs, in: *Graph Theory, Combinatorics, and Applications, Vol. 2* (Y. Alavi et al., eds.), J. Wiley, New York, 975–989.

[RSP91] R. Ringeisen, S. Stueckle, and B. Piazza, Subgraphs and bounds on maximum crossings, *Bulletin ICA* **2**, 33–46.

[T77] P. Turán, A note of welcome, *J. Graph Theory* **1**, 7–9.

[T70] W.T. Tutte, Toward a theory of crossing number, *J. Combinatorial Theory* **8**, 45–53.

[WB78] A.T. White and L.W. Beineke, Topological graph theory, in: *Selected Topics in Graph Theory* (L. Beineke, R. Wilson, eds.), Academic Press, London, 15–49.

[W69] D. R. Woodall, Thrackles and deadlock, in: *Combinatorial Mathematics and Its Applications* (D.J.A. Welsh, ed.), Academic Press, London, 335–348.

[W72] D. R. Woodall, Open problems, in: *Combinatorics, Proc. Conference on Combinatorial Mathematics* (D.J.A. Welsh, D.R. Woodall, eds.), Institute of Mathematics and Its Applications, Southend-on-Sea, Essex, England, 341–350.

An optimal algorithm for closest pair maintenance

(Extended abstract)

Sergei N. Bespamyatnikh

Department of Mathematics and Mechanics,
Ural State University,
51 Lenin St., Ekaterinburg 620083, Russia.
e-mail: `bsn@ubd.e-burg.su`.

Abstract

Given a set S of n points in k-dimensional space, and an L_t metric, the dynamic closest pair problem is defined as follows: find a closest pair of S after each update of S (the insertion or the deletion of a point). For fixed dimension k and fixed metric L_t, we give a data structure of size $O(n)$ that maintains a closest pair of S in $O(\log n)$ time per insertion and deletion. The running time of algorithm is optimal up to constant factor because $\Omega(\log n)$ is lower bound, in algebraic decision-tree model of computation, on the time complexity of any algorithm that maintains the closest pair (for $k = 1$).

1 Introduction

The dynamic closest pair problem is one of the very well-studied proximity problem in computational geometry [8, 13, 14, 15, 16, 19, 21, 22, 23, 25, 26, 27, 28]. We are given a set S of n points in k-dimensional space, $k \geq 1$, and a distance metric $L_t, 1 \leq t \leq \infty$. The point set is modified by insertions and deletions of points. Each point p is given as a k-tuple of real numbers $(p_1, \ldots, p_k)$.

The closest pair of S is a pair (p, q) of distinct points $p, q \in S$ such that the distance between p and q is minimal. The dynamic closest pair problem is defined as follows: find a closest pair (any) of S after each update of S.

We assume that the dimension k and the distance metric L_t are fixed. We use $d(p, q)$ to denote the distance between p and q.

A survey can be found in Schwarz's Ph.D.Thesis [21]. For the static closest pair problem and dimension $k = 2$, Shamos and Hoey [20] gave an algorithm with running time of $O(n \log n)$. Shortly after that, Bentley and Shamos [6] obtained this result for general dimension $k \geq 2$. In the *on-line* closest pair problem only insertions are allowed. For this problem Smid [25] obtained a data structure of size $O(n)$ that supports insertions in $O(\log^{k-1} n)$ amortized time. Schwarz, Smid and Snoeyink [23] presented a data structure of size $O(n)$ that maintains the closest pair in $O(\log n)$ amortized time per insertion.

Several algorithms are obtained for the dynamic closest pair problem [15, 16, 19, 21, 26, 27, 28]. In [16, 19, 26] the problem is solved with $O(\sqrt{n}\log n)$ update time using $O(n)$ space. In [15] Kapoor and Smid gave data structures of size $S(n)$ that maintain the closest pair in $U(n)$ amortized time per update, where for $k \geq 3$, $S(n) = O(n)$ and $U(n) = O(\log^{k-1} n \log\log n)$; for $k = 2$, $S(n) = O(n \log n/(\log\log n)^m)$ and $U(n) = O(\log n \log\log n)$; for $k = 2$, $S(n) = O(n)$ and $U(n) = O(\log^2 n/(\log\log n)^m)$ (m is an arbitrary non-negative integer constant). In [8] the author obtained an algorithm with $O(\log^{k+1} n \log\log n)$ update time and $O(n \log^{k-2} n)$ space. Callahan and Kosaraju [10] developed a tree-maintenance technique to solve a general class of dynamic algorithms. This technique can be used to maintain the closest pair in $O(\log^2 n)$ time and $O(n)$ space.

We give a linear size data structure that maintains the closest pair in $O(\log n)$ time per update. The algorithm is deterministic and the update time is worst-case. The algorithm fits in the algebraic computation tree model. In the algebraic computation tree model, there is a lower bound of $\Omega(n \log n)$ on the time complexity of any algorithm that solves the static closest pair problem for dimension $k = 1$ [4, 17, 18]. So the running time of our algorithm is optimal up to a constant factor.

Our algorithm is based on the following idea. We use a hierarchical subdivision of space into boxes. Several proximity algorithms build hierarchical subdivisions of space [30, 12, 11, 25, 22, 21, 3, 10]. These subdivisions differ by the shape of boxes, the overlap allowance, the manner of box splitting, the number of points in a box stored at a leaf. Our algorithm maintains almost cubical boxes. The boxes are split by almost middle cutting. Any smallest boxe contains exactly one of the given points. For each point we store some neighbor points. The closest pair is one of these pairs. To maintain efficiently these pairs we apply the dynamic tree of Sleator and Tarjan [24]. To insert a point we implement the point location. The point location also uses the dynamic tree. The idea to use dynamic trees for point location in hierarchical subdivisions is due to Cohen and Tamassia [12] and Chiang, Preparata and Tamassia [11]. Schwarz [21] applied the dynamic trees for the on-line closest pair problem and obtained an algorithm with worst-case $O(\log n)$ time per insertion and $O(n)$ space. Our hierarchical subdivision is similar to the *box decomposition* of [2] and the *fair split tree* of [10]. Arya et al. [3] presented a dynamic algorithm for approximate nearest neighbor queries. The tree they maintain is closely related to the quadtree of [7] and, as a result, their algorithm requires non-algebraic bit operations.

In Section 2 we describe a hierarchical subdivision. In Section 3 we show how to maintain the hierarchical subdivision (without point location). Section 4 explains how to maintain a neighbor information of points and the closest pair. In Section 5 we show how to emplement the search on the dynamic tree. Finally, in Section 6 we give some concluding remarks.

2 The hierarchical subdivision

The essential component of the data structure is a hierarchical subdivision of space into boxes. We define a box to be the product $[a_1, a_1') \times \ldots \times [a_k, a_k')$ of k semiclosed intervals. The boxes of the subdivision are almost cubical. We shall call such boxes as c-boxes.

Definition 2.1 Let B be a box with sides $s_1, \ldots, s_k$. The box B is said to be a *c-box* if, for any $i, j \in \{1, \ldots, k\}$, $s_i/s_j \in [1/3, 3]$.

The boxes are split by almost middle cutting.

Definition 2.2 Let $[a, a')$ be an interval in $\mathbf{R}$. Divide the interval into 3 intervals of equal length. We shall call the interval $[\frac{2a+a'}{3}, \frac{a+2a'}{3}]$ as *middle interval* of $[a, a')$.

Definition 2.3 Let $B = [a_1, a_1') \times \ldots \times [a_k, a_k')$ be a box and $c_i \in (a_i, a_i')$ be a real number for some i. The cut of B by the hyperplane $x_i = c_i$ into the boxes $B \cap \{x | x_i < c_i\}$ and $B \cap \{x | x_i \geq c_i\}$ is *almost middle cut* of B if c_i belongs to middle interval of $[a_i, a_i')$, i.e. $c_i \in [\frac{2a_i+a_i'}{3}, \frac{a_i+2a_i'}{3}]$.

Definition 2.4 Let A and B be k-dimensional boxes. The box A is said to be a *s-sub-box* of B if there exists a sequence of boxes $B_0, B_1, B_1', \ldots, B_l, B_l'$ such that $B = B_0$, $A = B_l$ and the boxes B_i, B_i' are obtained by almost middle cut of the box B_{i-1} for $i = 1, \ldots, l$.

The notion of s-sub-box is important and we give another definition.

Definition 2.5 Let $[a, a')$ and $[b, b')$ be intervals in $\mathbf{R}$. Let $[a, a')$ is the sub-interval of $[b, b')$, i.e. $b \leq a \leq a' \leq b'$. The interval $[a, a')$ is called *s-sub-interval* of the interval $[b, b')$ if one of the following conditions holds

1. $[a, a') = [b, b')$, or
2. $a = b$ and $|a' - a| \leq \frac{2}{3}|b' - b|$, or
3. $a' = b'$ and $|a' - a| \leq \frac{2}{3}|b' - b|$, or
4. $|a' - b| \leq \frac{2}{3}|b' - b|$ and $|a' - a| \leq \frac{2}{3}|a' - b|$, or
5. $|b' - a| \leq \frac{2}{3}|b' - b|$ and $|a' - a| \leq \frac{2}{3}|b' - a|$.

Definition 2.6 Let $A = [a_1, a_1') \times \ldots \times [a_k, a_k')$ and $B = [b_1, b_1') \times \ldots \times [b_k, b_k')$ be k-dimensional boxes. The box A is said to be a *s-sub-box* of B if, for $i = 1, \ldots, k$, $[a_i, a_i')$ is s-sub-interval of $[b_i, b_i')$.

We leave the proof of the equivalence of definitions 2.4 and 2.6 to the reader. The definition 2.4 implies the following Lemma.

Lemma 2.7 (Transitivity of s-sub-box relation) *Let $A, B,$ and C be k-dimensional boxes. If A is s-sub-box of B and B is s-sub-box of C, then A is s-sub-box of C.*

The hierarchical subdivision is represented by the binary tree T. With each node v of the tree T, we store a box $B(v)$ and a shrunken box $SB(v)$. The boxes satisfy the following conditions.

1. For any node v, the boxes $B(v)$ and $SB(v)$ are c-boxes.
2. For any node v, the box $SB(v)$ is a s-sub-box of $B(v)$.
3. For any node v, $SB(v) \cap S = B(v) \cap S$.
4. If w has two children u and v, then boxes $B(u)$ and $B(v)$ are the results of an almost middle cut of the box $SB(w)$.
5. If v is a leaf, then $|S \cap SB(v)| = 1$ and $B(v) = SB(v)$.

For a point $p \in S$ corresponding to the leaf v, let $B(p)$ denotes the box $B(v)$.

Let $parent(v)$, $lson(v)$, and $rson(v)$ denote parent, left son, and right son of the node v of T. Two nodes with the same parent are *siblings*.

3 The maintenance of the subdivision

In this Section we shall show how to maintain the tree T under insertions and deletions of points. The deletion is simpler than insertion and we consider the deletion first.

Let p be a point to be deleted. Let us w be a leaf corresponding p, i.e. $p \in B(w)$, v be the parent of w and $u \neq w$ be the sibling of v. We consider 2 cases.

1) u is a leaf (see Fig. 1 a)). Then set $SB(v) = B(v)$ and delete the leaves u and w.

2) u is an internal node (see Fig. 1 b)). Then delete the node w, set $B(u) = B(v)$, and collaps the edge (u, v), i.e. set $parent(u) = parent(v)$, delete the node v, and rename the node u as v.

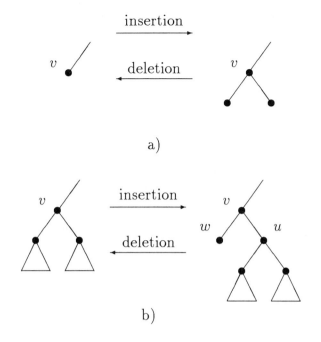

Figure 1: Updating the dynamic tree. a) The inserted point p belongs to $B(v)$. The deleted point p belongs to $B(u)$ where u is a son of v. b) The inserted point p belongs to $B(v) \setminus SB(v)$. The deleted point p belongs to $B(w)$.

Now consider the insertion. Let p be a point to be inserted. The insertion algorithm has two steps. At first we find the smallest box contained the point p. Then we update a finite set of nodes and boxes of the tree T. The first step uses the point location algorithm which is described in Section 5. After point location there are 3 cases.

1. The point p does not belong to $B(v_{\text{root}})$, where v_{root} is the root of T.

2. The point p belongs to the box $B(v)$, where v is a leaf (see Fig. 1 a)).

3. The point p belongs to the set $B(v) \setminus SB(v)$ for some node v (see Fig. 1 b)).

The cases 1 and 2 can be handled similar the case 3. Consider the case 3. We want to construct the box D and the almost middle cut of D into the boxes D_1 and D_2 that satisfy the following conditions

1) the box D is a s-sub-box of $B(v)$,

2) the box $SB(v)$ is a s-sub-box of D_1, and

3) the point $p \in D_2$.

After finding D we remove the edges from v to children v' and v'', create two nodes u and w below v, add edges joining u and v', v'', and set $SB(u) = SB(v), B(u) = D_1, SB(v) = D, B(w) = D_2$ (see Fig. 1 b)).

Denote $SB(v) = [a_1, b_1) \times \ldots \times [a_k, b_k)$. The algorithm uses a box D and repeatedly shrinks the box D until the almost middle cut of D is found. Initially $D = B(v)$. Denote $D = [d_1, e_1) \times \ldots \times [d_k, e_k)$. After each iteration of the algorithm

1) the box D is a c-box and s-sub-box of $B(v)$,

2) the box $SB(v)$ is a s-sub-box of D, and

3) the box D contains the point p.

The algorithm has $O(k)$ iterations because after each iteration the number of coordinates a_i, b_i coincided with endpoints of $[d_i, e_i)$ is increased, i.e. $\sum_{i=1}^{k} |\{a_i, b_i\} \cap \{d_i, e_i\}|$ is increased. We omit details but describe the basic procedure.

procedure middle-cut(D)   (* almost middle cut of the box $D = [d_1, e_1) \times \ldots \times [d_k, e_k)$ *)

1) Find i such that $e_i - d_i$ is maximal. In Step 2 we shall choose $const \in [d_i, e_i)$ to partition D by the hyperplane $x_i = const$.

2) If a_i or b_i lies on middle interval of $[d_i, e_i)$, then $const = a_i$ or $const = b_i$, respectively. Otherwise the interval $[a_i, b_i]$ does not intersect the middle interval of $[d_i, e_i)$. Without loss of generality $b_i < \frac{2d_i + e_i}{3}$. Choose $const = \max(\frac{2d_i + e_i}{3}, 2b_i - d_i)$.

3) Partition the box D by the hyperplane $x_i = const$. If this hyperplane partition the box $SB(v)$ and the point p, the cut of D into the boxes $D \cap \{x, x_i < const\}$ and $D \cap \{x, x_i \geq const\}$ is almost middle cut which satisfies above conditions (1), (2), and (3). In this case we stop the iterations. Otherwise one of these boxes contains both the box $SB(v)$ and the point p. Choose this box as D.

4) End of procedure.

Theorem 3.1 *Let the dimension k is fixed and the point location takes $COST$ time. The hierarchical subdivision can be maintained in $O(1) + COST$ time per insertion and $O(1)$ time per deletion.*

4 The maintenance of the closest pair

To maintain the closest pair we shall store the set E of some pairs of points of S.

Definition 4.1 A point $p \in S$ is a *nearest neighbor of* q if, for any $r \in S \setminus \{q\}$, $d(p,q) \leq d(q,r)$. For points $p, q \in S$, we call the pair (p,q) a *neighbor pair* if p is the nearest neighbor of q and vice versa.

The set E contains the neighbor pairs. It is clear that the closest pair of S is a neighbor pair of S and the closest pair belongs to E.

Let a heap H stores the distances of the pairs of E. The heap item is the pair of the points. The key of the item (p,q) is the L_t-distance $d(p,q)$. The pair of points with minimal key is a closest pair of S.

With each point $p \in S$, we store a list $E_p = \{q \mid (p,q) \in E\}$. With each point q in E_p, we store a pointer to the item (p,q) of the heap H.

The set E satisfies the following property.

Invariant. For any distinct points $a, b \in S$, if the pair (a,b) does not belong to the set E then there exists a node v such that
1. $B(a) \cap B(v) = \emptyset$, and
2. $d(B(v)) \leq 2d(B(a))$, and
3. $d_{\min}(a, B(v)) \leq 3d(B(v))$, and
4. $d_{\max}(a, B(v)) < d(a,b)$.

Lemma 4.2 *Let the invariant holds for the set E. Then the set E contains the neighbor pairs of S.*

Let N_k be a constant independent of n. (N_k depends on the dimension k and the metric L_t. One can prove that $N_k \leq (27k+1)^k$.)

Theorem 4.3 *Let the invariant holds for the set E. If, for a point $p \in S$, the set E_p contains at least N_k+1 pairs, then there exists the point $q \in E_p$ such that the invariant holds for $E \setminus \{(p,q)\}$.*

Theorem 4.3 ensures the existence of a set E such that, for any $p \in S$, the cardinality of E_p is at most $O(1)$. Theorem 4.3 follows from Theorem 4.4 which is useful in the insertion algorithm. To find a set E_p, for an inserted point p, we use a search on the dynamic tree. We need to limit the number of nodes that are used in search at the same time. Let $V = \{v_1, \ldots, v_N\}$ be a set of these nodes. We associate the set $S_i = B(v_i) \setminus \bigcup_{B(v_j) \subseteq B(v_i)} B(v_j)$ with every node $v_i \in V$.

Theorem 4.4 *Let p be a point of S, $V = \{v_1, \ldots, v_N\}$ be a set of nodes of T, and E be a set of some point pairs. Let E satisfies the conditions of the invariant and $E_p \cap S_i \neq \emptyset$ for $i = 1, \ldots, N_k$. If $N > N_k$, there exists i such that the invariant holds for $E \setminus \{(p,q) \mid q \in E_p \cap S_i\}$. (Choosing of i does not depend on layout of the points of E_p in the associated sets). The cost of finding i is $O(kN_k^2)$.*

The insertion of the point p causes insertions of some pairs into E and deletions of some pairs from E. Let us look at the updates of boxes. Note that the boxes, corresponding to the nodes, are only insert and, in the case $B(v_{\text{root}})$, are enlarge. Hence to provide that the invariant holds for E we need not to insert a pairs that are not incident to inserted point. Using the dynamic tree we find at most N_k pairs that are adjacent to p. Add these pairs into E. Now in fact the invariant holds for E. However, for some points, the number of incidented pairs may exceed N_k. These points are adjacent to p and can be determined when we add a pairs into E. For these points, we remove some pairs from E using Theorem 4.4.

Now we consider the deletion of the point p. The deletion causes insertions of some pairs into E and deletions of some pairs from E. Delete the pairs adjacent to p, i.e. $\{(p,q) \mid q \in$

$S, (p,q) \in E\}$. Note that always two boxes are deleted. These boxes are the results of an almost middle cut of the box $SB(parent(w))$ where the node w corresponds to p.

We consider the deletion of the box $B(v)$. Let the pair (a,b) was deleted by conditions of the invariant for node v. Then $d(B(a)) \geq d(B(v))/2$ and $d_{\min}(a, B(v)) \leq 3d(B(v))$. One can show that the number of such points is at most $O(1)$. The argument is similar to the proof of Theorem 4.4. Let $A(v)$ denote this set, i.e.

$$A(v) = \{a \in S \mid d(B(a)) \geq d(B(v))/2 \text{ and } d_{\min}(a, B(v)) \leq 3d(B(v))\}.$$

To find a set $A(v)$ we use a search on the dynamic tree. As in finding of E_p we bound the number of nodes that are used in search at the same time. For each $a \in A(v)$, we find the set E_a. This gives the set E such that the invariant holds for E.

5 Dynamic tree

In this Section we shall describe the dynamic tree. Using the dynamic tree, we shall implement the point location and the search for the sets E_p and $A(v)$.

A dynamic tree $\Delta(T)$ based on the binary tree T has the same nodes and the same edges as T. The dynamic tree is a partition of edges into two kinds, *solid* and *dashed*, with property that each node has at most one child linked to it by a solid edge. Thus the solid edges define a collection of *solid path* that partition the vertices. (A vertex with no incident solid edge is a one-vertex solid path). The head of the path is its bottommost node; the tail is its topmost node.

For a node v of T, let $size(v)$ be the number of nodes in the subtree of T rooted at v. Let (v, w) be an edge of T from v to its parent w. The edge is *heavy* if $size(v) > size(w)/2$ and *light* otherwise. A node v of $\Delta(T)$ fulfills the *size invariant* if, for each edge e to one of its children, e is solid if it is heavy and light if it is dashed. We say that the size invariant holds for the dynamic tree $\Delta(T)$ if it holds for each node of T.

A solid path is represented by a *path tree*. We use *globally biased binary trees* [5] to implement path trees. A biased binary tree stores an ordered sequence of *weighted* items in its leaves. The weight of a node v of T (and corresponding leaf of biased binary tree) is defined as

$$weight(v) = \begin{cases} size(v), \text{ if no solid edge} \\ \qquad \text{enters } v \\ size(v) - size(w), \text{ if the} \\ \qquad \text{solid edge } (w,v) \\ \qquad \text{enters } v \end{cases}$$

The weight of an internal node of biased binary tree is inductively defined as the sum of the weight of its children.

Each node v of biased binary tree has an integer rank denoted $rank(v)$ that satisfies the following properties:

(i) If v is a leaf, $rank(v) = \lfloor \log weight(v) \rfloor$. If v is an internal node, $rank(v) \leq 1 + \lfloor \log weight(v) \rfloor$.

(ii) If node w has parent v, $rank(w) \leq rank(v)$, with the inequality strict if w is external. If w has grandparent u, $rank(w) < rank(u)$.

Each internal node v of biased binary tree contains four pointers [24]: $bleft(v)$ and $bright(v)$, which point to the left and right child of v, and $bhead(v)$ and $btail(v)$, which point to the head and tail of the subpath corresponding to v (the leftmost and rightmost external descendants of v). For a topmost point v of a solid path P, there is the pointer $pt\_root(v)$ to the root of the path tree for P.

Lemma 5.1 ([24]) *If v is a leaf of a biased binary tree with root u, the depth of v is at most $2(rank(u) - rank(v)) \leq 2\log(weight(u)/weight(v)) + 4$.*

The updates of T can be performed using the following operations [5] on rooted trees.

$link(v, w)$: If v is the root of one tree and w is a node in another tree, combine the trees containing v and w by adding an edge joining v and w.

$cut(v, w)$: If there is an edge joining v and w, delete it, thereby breaking the tree containing v and w into two trees, one containing v and one containing w.

The time bound of these operations is $O(\log n)$. This gives the following result.

Lemma 5.3 *The dynamic tree can be maintained under insertions and deletions of points in $O(\log n)$ time per update.*

In the rest of this Section we discuss the search algorithms. We have to implement the point location and the search for the sets E_p and $A(v)$. Our point location algorithm is similar to the algorithm of Schwarz [21].

function point_location(p)
  $v := root(T)$
  while v is an internal node of T **do**
    (* Note that $p \in B(v)$ and v is the
      topmost node of some path P *)
    $u := pt\_root(v)$ (* u is the root of
      the path tree for P *)
    while u is an internal node of the path
      tree **do**
      if $p \in B(btail(bleft(u)))$ **then**
        $u := bright(u)$
      else $u := bleft(u)$
      fi
    od
    (* u is the bottommost node of the path
      P such that $p \in B(u)$ *)
    $v := u$
    if the edge $(v, rson(v))$ is dashed and $p \in B(rson(v))$ **then**
      $v := rson(v)$
    else if the edge $(v, lson(v))$ is dashed
        and $p \in B(lson(v))$ **then**
      $v := lson(v)$
      else return v
      fi
    fi
  od
  return v
end (* of the function *)

It is clear that the point location algorithm is correct. Let us analyze the running time of the algorithm. Let $P_1, \ldots, P_l$ be the solid paths that are searched during the algorithm. Let $u_1, \ldots, u_l$ be the roots of path trees and $v_1, \ldots, v_l$ be the bottommost nodes on path trees that are searched. Note that v_i is the parent of u_{i+1} in T for $i = 1, \ldots, l-1$. The number l of paths is at most $\log n$ by the size invariant. The depth of v_i in the path tree for P_i is at most $2(rank(u_i) - rank(v_i))$ by Lemma 5.1. For $i = 1, \ldots, l-1$, $rank(v_i) \geq rank(u_{i+1})$ by definition of $rank$. The total running time of the point location algorithm is

$$O(\log n + \sum_{i=1}^{l} 2(rank(u_i) - rank(v_i))) =$$
$$O(\log n + rank(u_1) - rank(v_l)) =$$
$$O(\log n).$$

Now we shall describe the searching for the sets E_p and $A(v)$. We consider this problem as the point location problem for at most $O(1)$ points (N_k points for E_p). In fact we can build the search tree such that

1. the external nodes corresponds the points S, and

2. the path from the root of the search tree to an external node v corresponds to the nodes of the path trees that are searched during the point location for the point corresponding to v.

The search for the sets E_p and $A(v)$ applies the *breadth-first search* on the search tree. *node_set* denotes a set of nodes that is stored in the breadth-first search. We use the pointer $depth(v)$ that is a depth of the node v in search tree. For simplicity we extend the pointers *btail* to the external nodes of any path trees. (It is not necessary to store these pointers). Using Theorem 4.4., the procedure $refine()$ finds at most N_k nodes among the nodes $\{btail(v) \mid v \in node\_set\}$ and removes another nodes from *node_set*.

```
function search()        (* the search
                           for E_p or A(v) *)
  v := pt_root(root(T))
  node_set := {v}
  depth(v) := 0
  while there is a node v in node_set such
      that btail(v) is an internal node of T do
    v is a node in node_set with minimal
        depth such that btail(v) is an internal
        node of T
    if v is an internal node of some path tree
      then
        node_set := node_set ∪
          {bleft(v), bright(v)}
        depth(bleft(v)) := depth(v) + 1
        depth(bright(v)) := depth(v) + 1
      else (* v is an external node of some path
             tree *)
        u := btail(v) (* u is the corresponding
                node of v in T *)
        if the edge (u, rson(u)) is dashed then
          w := pt_root(rson(u))
          node_set := node_set ∪ {w}
          depth(w) := depth(v) + 1
        fi
        if the edge (u, lson(u)) is dashed then
          w := pt_root(lson(u))
          node_set := node_set ∪ {w}
          depth(w) := depth(v) + 1
        fi
    fi
    node_set := node_set \ {v}
    if |node_set| > N_k then
      refine({btail(v) | v ∈ node_set})
      (* by Theorem 4.4 for E_p and
         analogous result for A(v) *)
    fi
  od
  return the points corresponding the nodes
      btail(v) for v ∈ node_set
end (* of the function *)
```

Finally, we formulate the main result.

Theorem 5.4 *There is a data structure of size $O(n)$ that maintains the closest pair of S in $O(\log n)$ time per update.*

6 Conclusion

We have presented an algorithm for maintaining the closest pair in $O(\log n)$ time per update, using $O(n)$ space. The running time of the algorithm is optimal up to a constant factor in the algebraic decision-tree model of computation. The algorithm can be adapted (by changing some constants, including N_k) for another metric such that $d(p,q) = O(d_\infty(p,q))$. In fact algorithm may give the list of the closest pairs (if any) in the time proportional to its number.

The algorithm maintains a set E of point pairs that contains the neighbor pairs.

Unfortunately our hierarchical subdivision does not allow efficiently maintain the set of the neighbor pairs. It would be interesting to solve the problem of the neighbor pairs maintenance with $O(\log n)$ update time and $O(n)$ space.

Acknowledgment. The author thanks Emo Welzl for comments and pointing out to the paper of Arya et al. [3] and the paper of Callahan and Kosaraju [10]. (Unfortunately, when I was writing this paper I did not know about these results.)

References

[1] S. Arya and D. M. Mount. *Approximate Nearest-Neighbor Queries in Fixed Dimensions.* Proceedings 4th Annual Symposium on Discrete Algorithms, 1993, pp. 271–280.

[2] S. Arya, D. M. Mount, N. S. Netanyahu, R. Silverman, and A. Wu. *An Optimal Algorithm for Approximate Nearest-Neighbor Searching.* Proceedings 5th Annual Symposium on Discrete Algorithms, 1994, pp. 573–582.

[3] S. Arya, D. M. Mount, N. S. Netanyahu, R. Silverman, and A. Wu. *An Opti-*

mal Algorithm for Approximate Nearest-Neighbor Searching. (revised version), 1994.

[4] M. Ben-Or. Lower Bounds for Algebraic Computation Trees. Proc. 15th Annual ACM Symp. Theory Comput., 1983, pp. 80-86.

[5] S. W. Bent, D. D. Sleator and R.E. Tarjan Biased Search Trees. SIAM Journal of Computing, 1985, 14, pp. 545-568

[6] J. L. Bentley and M. I. Shamos. Divide-and-Conquer in Multidimensional Space. Proc. 8th Annual ACM Symp. Theory of Computing, 1976, pp. 220-230.

[7] M. Bern, D. Eppstein, and S.-H. Teng. Parallel Construction of Quadtrees and Quality Triangulations. Algorithms and Data Structures, Third Workshop, WADS '93, Lecture Notes in Computer Science 709, Springer-Verlag, 1993, pp. 188–199.

[8] S. N. Bespamyatnikh. The Region Approach for Some Dynamic Closest-Point Problems. Proc. 6th Canadian Conf. Comput. Geom., 1994.

[9] P. B. Callahan and S. R. Kosaraju. A Decomposition of Multi-Dimensional Point-Sets with Applications to k-Nearest-Neighbors and n-Body Potential Fields. Proceedings 24th Annual AMC Symposium on the Theory of Computing, 1992, pp. 546–556.

[10] P. B. Callahan and S. R. Kosaraju. Algorithms for Dynamic Closest Pair and n-Body Potential Fields. Proc. 6th Annual ACM-SIAM Symposium on Discrete Algorithms, 1995.

[11] Y.-J. Chiang, F. T. Preparata, and R. Tamassia. A Unified Approach to Dynamic Point Location, Ray Shooting, and Shortest Paths in Planar Maps. Proc. 4th ACM-SIAM Symp. on Discrete Algorithms, 1993, pp. 44-53.

[12] R. F. Cohen and R. Tamassia. Combine and Conquer: a General Technique for Dynamic Algorithms. Proc. First Europ. Symp. on Algorithms, Lecture Notes in Computer Science. Springer-Verlag, 1993.

[13] M. J. Golin, R. Raman, C. Schwarz, and M. Smid. Randomized Data Structures for the Dynamic Closest-Pair Problem. Proc. 4th Annual ACM-SIAM Symp. on Discrete Algorithms, 1993, pp. 301-310.

[14] M. J. Golin, R. Raman, C. Schwarz, and M. Smid. Simple Randomized Algorithms for Closest Pair Problems. Proc. 5th Canadian Conf. Comput. Geom., 1993, pp. 246-251.

[15] S. Kapoor and M. Smid. New Techniques for Exact and Approximate Dynamic Closest-Point Problems. Proc. 10th Annual ACM Symp. Comput. Geom., 1994, pp. 165–174.

[16] H.-P. Lenhof and M. Smid. Enumerating the k Closest Pair Optimally. Proc. 33rd Ann. IEEE Symp. Found. Comput. Sci., 1992, pp. 380-386.

[17] F. P. Preparata and M. I. Shamos. Computational Geometry: An Introduction. Springer-Verlag, New York Berlin Heidelberg Tokyo, 1985.

[18] F. P. Preparata and M. I. Shamos. Computational Geometry: An Introduction. Springer-Verlag, New York Berlin Heidelberg Tokyo, second edition, 1988.

[19] J. S. Salowe. Shallow Interdistance Selection and Interdistance Enumeration. International Journal of Computational

Geometry & Applications, 2, 1992, pp. 49–59.

[20] M. I. Shamos and D. Hoey. *Closest-Point Problem.* Proc. 16th Annual IEEE Symp. Found. Comput. Sci., 1975, pp. 151-162.

[21] C. Schwarz. *Data Structures and Algorithms for the Dynamic Closest Pair Problem.* Ph.D. Thesis, Universität des Saarbrücken, 1993.

[22] C. Schwarz and M. Smid. *An $O(n \log n \log \log n)$ Algorithm for the On-Line Closest Pair Problem.* Proc. 3th ACM-SIAM Symp. Discrete Algorithms, 1992, pp. 280–285.

[23] C. Schwarz, M. Smid, and J. Snoeyink. *An Optimal Algorithm for the On-Line Closest-Pair Problem.* Proc. 8th Annual ACM Symp. Comput. Geom., 1992, pp. 330–336.

[24] D. D. Sleator and R. E. Tarjan. *A Data Structure for Dynamic Trees.* Journal of Computer and System Sciences, 26, 1983.

[25] M. Smid. *Dynamic Rectangular Point Location, with an Application to the Closest Pair Problem.* Technical Report MPI-I-91-101, Max-Plank-Institut fürInformatik, Saarbrücken, Germany, 1991

[26] M. Smid. *Maintaining the Minimal Distance of a Point Set in Less Than Linear Time.* Algorithms Rev., 2, 1991, pp. 33-44.

[27] M. Smid. *Maintaining the Minimal Distance of a Point Set in Polylogarithmic Time.* Discrete & Computational Geometry, 7, 1992, pp. 415–431.

[28] K. L. Supowit. *New Techniques for Some Dynamic Closest-Point and Farthest-Point Problems.* Proc. 1-st Annual ACM-SIAM Symposium on Discrete Algorithms, 1990, pp. 84–90.

[29] P. M. Vaidya. *An Optimal Algorithm for All-Nearest-Neighbors Problem.* Proceedings 27th Annual Symposium Found. Comp. Sc., 1986, pp. 117–122.

[30] P. M. Vaidya. *An $O(n \log n)$ Algorithm for All-Nearest-Neighbors Problem.* Discrete Comput. Geom., 1989, pp. 101–115

The rectangle enclosure and point–dominance problems revisited

Prosenjit Gupta[*†]   Ravi Janardan[*†]   Michiel Smid[‡]   Bhaskar Dasgupta[§†]

Abstract

We consider the problem of reporting the pairwise enclosures in a set of n axes-parallel rectangles in $\mathbb{R}^2$, which is equivalent to reporting dominance pairs in a set of n points in $\mathbb{R}^4$. Over a decade ago, Lee and Preparata [LP82] gave an $O(n \log^2 n + k)$-time and $O(n)$-space algorithm for these problems, where k is the number of reported pairs. Since that time, the question of whether there is a faster algorithm has remained an intriguing open problem.

In this paper, we give an algorithm which runs in $O(n \log n \log \log n + k \log \log n)$ time and uses $O(n)$ space. Thus, although our result is not a strict improvement over the Lee-Preparata algorithm for the full range of k, it is, nevertheless, the first result since [LP82] to make *any* progress on this long-standing open problem. Our algorithm is based on the divide–and–conquer paradigm. The heart of the algorithm is the solution to a red–blue dominance reporting problem (the "merge" step). We give a novel solution for this problem which is based on the iterative application of a sequence of non-trivial sweep routines. This solution technique should be of independent interest.

We also present another algorithm whose bounds match the bounds given in [LP82], but which is simpler. Finally, we consider the special case where the rectangles have at most a constant number, α, of different aspect ratios, which is often the case in practice. For this problem, we give an algorithm which runs in $O(\alpha n \log n + k)$ time and uses $O(n)$ space.

1 Introduction

Problems involving sets of rectangles have been studied widely in computational geometry since they are central to many diverse applications, including VLSI layout design, image processing, computer graphics, and databases. (See, for instance, Chapter 8 in each of the books [PS88, PL88].) For most of these problems, efficient (indeed, optimal) algorithms are known. In this paper, we investigate the following rectangle problem, whose complexity has not yet been resolved satisfactorily.

Problem 1.1 *Given a set $\mathcal{R}$ of n axes-parallel rectangles in the plane, report all pairs (R', R) of rectangles such that R encloses R'.*

By mapping each rectangle $R = [l, r] \times [b, t]$ to the point $(-l, -b, r, t)$ in $\mathbb{R}^4$, we can formulate this problem as a dominance problem: If $p = (p_1, p_2, p_3, p_4)$ and $q = (q_1, q_2, q_3, q_4)$ are points in $\mathbb{R}^4$, then we say that p *dominates* q if $p_i \geq q_i$

for all i, $1 \leq i \leq 4$. We call the pair (q,p) a *dominance pair*. Using this terminology, Problem 1.1 is transformed—in linear time—into the following one:

Problem 1.2 *Given a set V of n points in $\mathbb{R}^4$, report all dominance pairs in V.*

In fact, a result of Edelsbrunner and Overmars [EO82] implies that Problems 1.1 and 1.2 are equivalent, i.e., in linear time, Problem 1.2 can also be transformed into Problem 1.1.

Here is a brief history of the problem. Let k denote the number of pairs (R', R) of rectangles such that R encloses R', or, equivalently, the number of dominance pairs in V. The rectangle problem was first considered by Vaishnavi and Wood [VW80], who gave an $O(n \log^2 n + k)$-time and $O(n \log^2 n)$-space algorithm. This result was also obtained independently by Lee and Wong [LW81]. In 1982, Lee and Preparata [LP82] gave an algorithm which ran in $O(n \log^2 n + k)$ time and used only $O(n)$ space. Ever since, the question of whether there is a faster algorithm has remained intriguingly open [PS88, page 371].

1.1 Summary of contributions

Our main result is an algorithm for Problems 1.1 and 1.2 which runs in $O(n \log n \log \log n + k \log \log n)$ time and uses $O(n)$ space. While our result is not a strict improvement over [LP82] for the full range of k (it is an improvement for $k = o(n \log^2 n / \log \log n)$), it is, nevertheless, the first result since [LP82] to make *any* progress on this long-standing open problem, and we hope that our approach will spur further research on finally putting this problem to rest.

Our approach is to transform Problem 1.2 to a grid and then apply divide-and-conquer. The heart of the algorithm is the solution to a red-blue dominance reporting problem (the "merge" step). We give a novel solution to this problem which is based on the iterative application of a sequence of non-trivial sweep routines. We regard this solution technique as the second contribution of the paper since it could find applications in other grid-based problems.

We also present a second algorithm whose bounds match the bounds in [LP82], but which is simpler. In particular, our algorithm employs just one level of divide-and-conquer (as opposed to two levels in [LP82]) and uses simple data structures.

Finally, we consider the special case, where the rectangles have only a constant number, α, of different aspect ratios, where the *aspect ratio* of a rectangle is its height divided by its width. This is a reasonable assumption in VLSI design. For this problem, we give an algorithm which runs in $O(\alpha n \log n + k)$ time and uses $O(n)$ space. (Previously, no results were known for this special case.)

The full version of this paper appears as [GJSD94].

1.2 Overview of the main result

Throughout the paper (except in Section 4) we consider Problem 1.2. Our first observation is that we can afford to (in $O(n \log n)$ time) normalize the problem to a grid. This allows us to bring into play efficient structures such as van Emde Boas trees [vEB77a, vEB77b]. Specifically, we map the n points in V to a set S of n points in U^4, where $U = \{0, 1, \ldots, n-1\}$, such that dominance pairs in V are in one-to-one correspondence with dominance pairs in S. We divide S along the fourth coordinate into two equal halves and recurse on these sets. In the merge step, we (effectively) have a set of red and blue points in U^3 and need to report all red–blue dominances. We solve this problem by an iterative sequence of sweeps, as follows:

We first "clean" the red set so as to remove those red points that are not dominated by any blue points. We also "clean" the blue set to eliminate those blue points that do not dominate any red point. Intuitively, this *cleaning step* gets rid of points that do not contribute to any dominance pair and this allows us to bound the running time of the next step—the reporting step.

In the *reporting step*, we report all red–blue dominances in the cleaned sets. Assume, wlog, that there are more red points than blue points in the cleaned sets. To do the reporting correctly and efficiently, we do not consider the blue points all at once. Instead, we report red–blue dominances involving only those blue points that are maximal (in three dimensions) in the blue set. We find these maximal points by a single sweep. Then we sweep in the opposite direction and incrementally reconstruct the blue contour using information computed in the first sweep. During this second sweep, we report red–blue dominances involving blue maximal points. In both the reporting step and the cleaning step, we need to dynamically maintain certain two–dimensional contours of maximal points. For this we use the van Emde Boas trees mentioned earlier.

Because of the cleaning step, we are guaranteed to find a number of dominance pairs which is at least proportional to the number of red and blue points that remain after the cleaning step. Hence, we can charge the time for this reporting step to the number of reported dominance pairs. Since we have found all dominance pairs in which the maximal elements of the blue points occur, we can remove them. Then, we perform the cleaning step on the remaining red and blue points and, afterwards, we perform a reporting step again. We repeat this until either there are no red points left or there are no blue points left. At the end of the algorithm, we will have reported all red-blue dominance pairs.

2 A divide-and-conquer algorithm

Let V be a set of n points in $\mathbb{R}^d$, where $d \geq 2$. Point p *dominates* point q if $p_i \geq q_i$ for all i, $1 \leq i \leq d$. A point of V is called *maximal in V* if it is not dominated by any other point of V. The *maximal layer* of V is defined as the subset of all points that are maximal in V.

If V is a set of points in the plane, i.e., if $d = 2$, then the maximal points, when sorted by their x-coordinates, form a *staircase*, also called a *contour*. The ordering of the maximal points by x-coordinate is the same as the ordering by y-coordinate. Consider the contour of V. Let p be any point in the plane. We say that p is *inside* the contour if it is dominated by some point of the contour. Otherwise, we say that p is *outside* the contour.

2.1 The normalization step

Let V be a set of n points in $\mathbb{R}^4$. For each i, $1 \leq i \leq 4$, we sort the vectors in the set $\{(p_i, p_1, \ldots, p_{i-1}, p_{i+1}, \ldots, p_4) : (p_1, p_2, p_3, p_4) \in V\}$ lexicographically. Then we replace the i-th coordinate of each point of V by its rank in this ordering. We denote the resulting set of points by S. The following lemma can easily be proved.

Lemma 2.1 *The above normalization step takes $O(n \log n)$ time. This produces a set $S \subseteq \{0, 1, 2, \ldots, n-1\}^4$ of n points such that (q, p) is a dominance pair in V iff the corresponding pair in S is also a dominance pair. Moreover, for each i, $1 \leq i \leq 4$, no two points of S have the same i-th coordinate.*

2.2 The algorithm

Let S be the set of n points from Lemma 2.1. Note that during the normalization we can obtain the points of S sorted by their third coordinates. Our algorithm for finding all dominance pairs in S follows the divide-and-conquer paradigm. Since in each recursive call the number of points decreases, but the size of the universe remains the same, we introduce the latter as a separate variable u. Note that in our case $u = n$—the initial number of points. However, to keep our discussion general, we will derive our bounds in terms of both u and n and finally substitute n for u to get our main result. The algorithm is as follows:

1. Compute the median m of the fourth coordinates of the points of S. By walking along the points of S in their order according to the third coordinate, compute the sets $S_1 = \{p \in S : p_4 \leq m\}$ and $S_2 = \{p \in S : p_4 > m\}$. Both these sets are sorted by their third coordinates.

2. Using the same algorithm recursively, solve the problem for S_1 and S_2.

3. Let R (resp. B) be the set of "red" (resp. "blue") points in U^3 obtained by removing the fourth coordinate from each point of S_1 (resp. S_2). Compute all dominance pairs (r, b), where $r \in R$ and $b \in B$.

2.3 The merge step

We give two solutions for the merge step (step 3 above). Let x, y and z denote the coordinate axes in U^3. In the first solution, we sweep a plane parallel to the xy-plane downward along the z-direction.

During the sweep, we maintain a radix priority search tree (PST), see [McC85], for the projections onto the sweep plane of all points of B that have been visited already. If the sweep plane visits a point (b_x, b_y, b_z) of B, then we insert (b_x, b_y) into the PST. If a point (r_x, r_y, r_z) of R is encountered, we query the PST and find all points (b_x, b_y) such that $b_x > r_x$ and $b_y > r_y$. For each such point, we report the corresponding pair in $R \times B$, or, in fact, in $S_1 \times S_2$.

If k_{RB} denotes the number of red-blue dominance pairs in $R \times B$, then the merge step takes time $O(n \log u + k_{RB})$. This implies that the algorithm for Problem 1.2 takes $O(n \log n \log u + k) = O(n \log^2 n + k)$ time and $O(n)$ space. Thus this algorithm matches the bounds in [LP82], but is simpler since it uses only one level of divide and conquer. Moreover, because of the normalization step, we can use a radix PST, which is a simple data structure not requiring any rebalancing.

In the next section, we give an alternative algorithm for the three-dimensional red-blue dominance problem, taking $O(n \log \log u + k_{RB} \log \log u)$ time. This will lead to an $O(n \log n \log \log n + k \log \log n)$ time algorithm for Problem 1.2.

3 Red-blue dominance reporting in three dimensions

In the final algorithm (Section 3.3), we first construct an empty van Emde Boas tree (vEB-tree) on the universe U. (See [vEB77a, vEB77b].) During the entire algorithm, elements will be inserted and deleted in this tree and we will perform queries on it. Its construction time is $O(u)$, its query and update times are $O(\log \log u)$ and it uses $O(u)$ space. In the rest of this section, we assume that we have this tree available.

3.1 The cleaning step

One of the essential steps in our algorithm is to remove all red points that are not dominated by any blue point, and all blue points that do not dominate any red point. We denote this as the "cleaning" of the red (resp. blue) set w.r.t. the blue (resp. red) set.

In [KO88], Karlsson and Overmars give an $O(n \log \log u)$ time and $O(u)$ space algorithm, which given n points in U^3, computes the maximal elements. We modify this algorithm to find all red points that are not dominated by any blue point, within the same time and space bounds:

We sweep a plane parallel to the xy-plane downward along the z-direction, stopping at each point. During the sweep, we maintain the contour of the two-dimensional maximal elements of the projections (onto the sweep plane) of the blue points already seen. We store these maximal elements in the initially empty vEB-tree, sorted by their x-coordinates. When the sweep plane visits a blue point b, we update the contour and the vEB-tree, as follows: We search in the vEB-tree with the x-coordinate of b and determine if b's projection is inside or outside the blue contour. If it is outside, then we delete from the vEB-tree all blue points on the contour whose projections are dominated by b's projection and we insert b as a new contour point. Note that the points to be deleted can easily be found since they are contiguous in the vEB-tree.

On visiting a red point r, we query the vEB-tree with the x-coordinate of r and determine if r's projection lies inside or outside the blue contour. If it is inside, we insert r into an initially empty set R_1.

At the end of the sweep, we delete all elements from the vEB-tree. The empty tree will be used later on in the algorithm.

Lemma 3.1 We have $R_1 = \{r \in R : \exists b \in B \text{ such that } r \text{ is dominated by } b\}$ at the end of the algorithm. Moreover, the algorithm takes $O(n \log \log u)$ time and uses $O(u)$ space.

The given algorithm cleans the red set R w.r.t. the blue set B. To clean B w.r.t. R, we use the mapping $\mathcal{F}$ that maps the point (a, b, c) in U^3 to the point $(u-1-a, u-1-b, u-1-c)$ in U^3. This mapping reverses all dominance relationships. Also, the mapping $\mathcal{F}$ is equal to its inverse. We run our sweep algorithm on the sets $\mathcal{F}(R)$ and $\mathcal{F}(B)$, maintaining a red contour and querying with the blue points. As a result, we get a set $B_0 \subseteq \mathcal{F}(B)$, where each point in B_0 is dominated by some point in $\mathcal{F}(R)$. Then the set $B_1 = \mathcal{F}(B_0)$ satisfies $B_1 = \{b \in B : \exists r \in R \text{ such that } b \text{ dominates } r\}$.

Lemma 3.2 *Let R and B be sets of points in U^3 that are sorted by their third coordinates, and let $u = |U|$ and $n = |R| + |B|$. Assume that $n \leq u$. Also, assume we are given an empty vEB-tree on the universe U. In $O(n \log \log u)$ time and using $O(u)$ space, we can compute sets $R_1 \subseteq R$ and $B_1 \subseteq B$ such that $R_1 = \{r \in R : \exists b \in B \text{ such that } r \text{ is dominated by } b\}$ and $B_1 = \{b \in B : \exists r \in R \text{ such that } b \text{ dominates } r\}$.*

The procedure that cleans the red set R w.r.t. the blue set B and returns the set R_1 will be denoted by $Clean(R, B)$. The set B_1 is obtained as $\mathcal{F}(Clean(\mathcal{F}(B), \mathcal{F}(R)))$.

Remark 3.1 Observe that it does not matter whether we first clean R w.r.t. B and then clean B w.r.t. R, or vice versa. In either case, we get the same clean sets R_1 and B_1.

3.2 The sweep and report step

Let R_1 and B_1 be the sets of Lemma 3.2. Wlog, let us assume that $|R_1| \geq |B_1|$. Let B_1' denote the three-dimensional maxima of B_1. The procedure $Sweep(R_1, B_1)$, which will be described in this section, reports all red-blue dominance pairs (r, b), where $r \in R_1$ and $b \in B_1'$. Note that because of the cleaning step, there are at least $|R_1|$ such pairs. (If $|R_1| < |B_1|$, then we invoke the procedure $Sweep(\mathcal{F}(B_1), \mathcal{F}(R_1))$.)

Step 1: We sweep along the points of B_1 *downwards* in the z-direction and determine the set B_1': During the sweep, we maintain the contour of the 2-dimensional maximal elements of the projections of the points of B_1 already seen. These maximal elements are stored in the initially empty vEB-tree, sorted by their x-coordinates. We also maintain a list M, in which we store all updates that we make in the vEB-tree.

When the sweep plane visits a point b of B_1, we add b to an initially empty list L iff b's projection lies outside the current contour. In this case, we also update the contour by updating the vEB-tree, and we add the sequence of updates made to the list M.

Lemma 3.3 *After Step 1, list L contains the set B_1' of three-dimensional maxima of B_1.*

Remark 3.2 $B_1' \subseteq B_1$ is the set of points whose projections are added to the two-dimensional contour during Step 1. Note that once a point has been added, it may be removed again from the contour later on during the sweep in Step 1 itself.

Step 2: We now sweep along the points of $R_1 \cup B_1'$ *upwards* in the z-direction. Using the list M, we reconstruct the contour of the projections of the points of B_1' that are above the sweep plane. With each blue point b on the contour, we store a list $C_b \subseteq R_1$ of candidate red points.

Initially, the sweep plane is at the point having minimal z-coordinate and the vEB-tree stores the final contour from Step 1. For each blue point b on this contour, we initialize an empty list C_b.

When the sweep plane reaches a blue point b of B_1', we do the following:

2.1. Using M, we undo in the vEB-tree the changes we made to the two-dimensional blue contour when we visited b during the sweep of Step 1. Call each blue point which now appears on the contour a *new* point; call all remaining blue points on the contour *old*. Note that the new points form a single continuous staircase.

2.2. For each $r \in C_b$, we report (r, b) as a dominance pair.

2.3. For each new blue point q on the contour, we have to create a list C_q: We look at all points of C_b. For each such point r, we search with its x-coordinate in the vEB-tree. If r's projection is inside the new contour, then we find the leftmost blue point p of the new contour that is to the right of r. Starting at p we walk right along the contour. For each blue point q encountered such that q is new, we insert r into the list C_q. We stop walking as soon as we find a blue point q whose projection does not dominate r's projection or we reach the end of the contour. (See Figure 1.)

When the sweep plane reaches a red point r, we search with its x-coordinate in the vEB-tree and determine if its projection is inside or outside the current contour. If it is inside, we start walking along the contour from the point immediately to the right of r and insert r into the list C_q for each

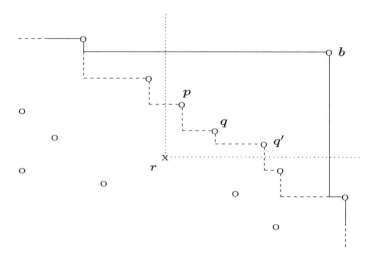

Figure 1: *Illustrating Step 2.3. Point r is inserted into the lists C_p, C_q and $C_{q'}$.*

blue point q on the contour, until we reach a blue point whose projection does not dominate r's projection or we reach the end of the contour. Note that at the end of Step 2, the vEB-tree is empty.

Lemma 3.4 *In Step 2 of the algorithm, all dominance pairs (r, b), where $r \in R_1$ and $b \in B'_1$, are reported. Moreover, only such pairs are reported.*

Proof. Suppose that (r, b) is reported. Then, $r \in R_1$ and $b \in B'_1$. Also, $r \in C_b$ when b is reached and so $r_z < b_z$. Also, by construction of C_b, b's projection dominates r's projection. Thus b dominates r.

Now let $r \in R_1$ and $b \in B'_1$ such that b dominates r. We prove that the pair (r, b) is reported. At the moment when the sweep plane reaches b, this point is removed from the contour. We have to show that r is contained in C_b at this moment.

When the sweep plane reaches r, we insert this point into the lists C_q for all points q that are on the contour at that moment and whose projection onto the xy-plane dominate r's projection. If b is one of these points, then we are done, because r stays in C_b until the sweep plane reaches b. Otherwise, b's projection lies inside the contour. Let q be the point with smallest z-coordinate that is on the contour at the moment when the sweep plane reaches r and whose projection onto the xy-plane dominates b's projection. Then, r is inserted into C_q. Note that $b_z > q_z$, because otherwise b would be dominated by q, contradicting the fact that b belongs to B'_1.

When the sweep plane reaches q, point r is inserted into the lists C_p for all points p that appear on the contour at that moment and whose projections dominate r's projection. If b is one of these points, then we are done. Otherwise, let p be the point with smallest z-coordinate that is on the contour at the moment when the sweep plane reaches q and whose projection onto the xy-plane dominates b's projection. Point r is inserted into C_p. We have $b_z > p_z$. Now we consider the moment when the sweep plane reaches point p, and repeat the same argument. Continuing in this way, and observing that point b must appear on the contour, it follows that r will be inserted into C_b. □

Lemma 3.5 *Let k_{RB} be the number of dominance pairs (r, b) such that $r \in R_1$ and $b \in B'_1$. Algorithm Sweep(R_1, B_1) takes $O(k_{RB} \log \log u)$ time and uses $O(u)$ space.*

Proof. Let $n = |R_1| + |B_1|$. Step 1 of the algorithm takes $O(n \log \log u)$ time. The total time for updating the contour in Step 2.1 is upper-bounded by the time for Step 1. The total time for Step 2.2 is obviously $\Theta(k_{RB})$. It remains to estimate the time for updating the C-lists in Step 2.3. Let $r \in C_b$ be a red point to be added to the C-lists of the new contour points that appear as a result of undoing the changes at b in Step 2.1. Deciding whether r's projection lies inside or outside the two-dimensional contour takes $O(\log \log u)$ time. If it lies outside, then we charge this cost to the pair (r, b) just reported in Step 2.2. The total number of such charges due to all red points is $O(k_{RB} \log \log u)$. If r lies inside the contour and if it is inserted into m C-lists (m will be at least one), then the time taken is $O(m + \log \log u) = O(m \log \log u)$. We charge $O(\log \log u)$ to each of the m instances of r thus inserted. Likewise, when we encounter a red point r in the upward sweep, we use a similar charging scheme.

Thus the algorithm takes $O((n + k_{RB}) \log \log u)$ time. We know that $k_{RB} \geq |R_1|$ because of the cleaning step. Also, since $|R_1| \geq |B_1|$, we have $n \leq 2|R_1| \leq 2k_{RB}$. This proves the bound on the running time. It is clear that the algorithm uses $O(u)$ space. □

3.3 The overall three-dimensional red-blue dominance algorithm

The algorithm for reporting all red-blue dominance pairs in $R \times B$ is given in Figure 2. This algorithm uses the procedures *Clean* and *Sweep* that were given in Sections 3.1 and 3.2, respectively. Also recall the mapping $\mathcal{F}$ that was defined in Section 3.1. We assume that we have constructed already the empty vEB-tree on the universe U.

Lemma 3.6 *Algorithm 3Ddom(R, B) terminates and reports all dominance pairs (r, b), where $r \in R$ and $b \in B$. Moreover, if a pair (r, b) is reported, then it is a red-blue dominance pair.*

Proof. The algorithm terminates because after each iteration of the while-loop either $|B_{i+1}| = |B_i \setminus B'_i| < |B_i|$ (since $|B'_i| > 0$) or $|R_{i+1}| = |\mathcal{F}(\mathcal{F}(R_i) \setminus R'_i)| < |R_i|$ (since $|R'_i| > 0$). We now prove that (r, b) is reported iff b dominates r.

Suppose that (r, b) is reported. Since a report happens only during one of the calls to *Sweep*, it follows from the correctness of this procedure (see Lemma 3.4) that b dominates r.

Conversely, suppose that b dominates r. Note that a point is discarded in algorithm $3Ddom(R, B)$ either during a call to *Clean* or right after that call to *Sweep* during which it becomes a three-dimensional maximal element. Since b dominates r, it follows that if neither r nor b has been discarded just before one of the calls to *Clean* within the while-loop then neither will be discarded during that call. (Similarly, if neither r nor b has been discarded just before the two calls to *Clean* outside the while-loop, then neither will be discarded during those two calls.) Moreover, at least one of r and b will be discarded sometime during the algorithm since the algorithm terminates. Wlog, assume that b is discarded. Then it follows that b becomes a three-dimensional maximal element before r becomes one (if ever). Let b become a three-dimensional maximal element in Step 1 of $Sweep(R_i, B_i)$ for some i. Thus, when Step 2 of $Sweep(R_i, B_i)$ commences, $r \in R_i$. By the correctness of the *Sweep* routine, (r, b) is reported as a dominance pair. □

Theorem 3.1 *Let R and B be two sets of points in U^3 that are sorted by their third coordinates. Assume we are given an empty vEB-tree on the universe U. Let $u = |U|$ and $n = |R| + |B|$, and let k' be the number of dominances (r, b), where $r \in R$ and $b \in B$. Assume that $n \leq u$. Algorithm 3Ddom finds all these dominance pairs in $O((n + k') \log \log u)$ time and $O(u)$ space.*

Proof. Let $n_i = |R_i| + |B_i|$ and let k_i be the number of dominance pairs that are reported during the i-th iteration. Because of the cleaning step and because we distinguish between the cases where $|R_i| \geq |B_i|$ and $|R_i| < |B_i|$, we have $n_i \leq 2k_i$. Also, since during each iteration, we output different dominance pairs, we have $\sum_i k_i = k'$. The initial cleaning of R and B takes $O(n \log \log u)$ time. By Lemmas 3.2 and 3.5, the i-th iteration takes time $O((n_i + k_i) \log \log u)$, which is bounded by $O(k_i \log \log u)$. It follows that the entire algorithm takes time

$$O(n \log \log u + \sum_i k_i \log \log u) = O((n+k') \log \log u).$$

The algorithm uses space $O(n + u)$, which is bounded by $O(u)$. □

Algorithm $3Ddom(R, B)$
(* R and B are sets of points in U^3; the algorithm reports all pairs (r, b) such that $r \in R$, $b \in B$ and b dominates r *)
begin
$R_1 := Clean(R, B);$
$B_1 := \mathcal{F}(Clean(\mathcal{F}(B), \mathcal{F}(R)));$
$i := 1;$
while $R_i \neq \emptyset$ **and** $B_i \neq \emptyset$
do if $|R_i| \geq |B_i|$
   then $Sweep(R_i, B_i);$
       (* this procedure computes the set B'_i of three-dimensional maxima
         of B_i and reports all dominances (r, b) where $r \in R_i$ and $b \in B'_i$ *)
       $H := B_i \setminus B'_i;$
       $R_{i+1} := Clean(R_i, H);$
       $B_{i+1} := H$
       (* B_{i+1} is clean w.r.t. R_{i+1}; *)
   else $Sweep(\mathcal{F}(B_i), \mathcal{F}(R_i));$
       (* this procedure computes the set R'_i of three-dimensional maxima
         of $\mathcal{F}(R_i)$ and reports all dominances (r, b) where $r \in \mathcal{F}(R'_i)$
         and $b \in B_i$ *)
       $H := \mathcal{F}(R_i) \setminus R'_i;$
       $B_{i+1} := \mathcal{F}(Clean(\mathcal{F}(B_i), H));$
       $R_{i+1} := \mathcal{F}(H)$
       (* R_{i+1} is clean w.r.t. B_{i+1}; *)
   fi;
   $i := i + 1$
od
end

Figure 2: The three-dimensional red-blue dominance reporting algorithm.

3.4 Analysis of the four-dimensional dominance reporting algorithm

Consider again our divide-and-conquer algorithm of Section 2.2 for solving the 4-dimensional dominance reporting problem on the normalized set $S \subseteq U^4$. We implement Step 3—the merge step—using algorithm 3Ddom.

Let $T(n, u)$ denote the total running time on a set of n points in U^4, that are sorted by their third coordinates. Recall that it is assumed that $n = u$ (however, the sizes of the sets in the recursive calls will be smaller than u). We do not include in $T(n, u)$ the time that is charged to the output.

Step 1 of the algorithm takes $O(n)$ time, and Step 2 takes $2T(n/2, u)$ time. By Theorem 3.1, Step 3—except for the reporting—takes $O(n \log \log u)$ time. Hence, $T(n, u) = O(n \log \log u) + 2T(n/2, u)$, which solves to $T(n, u) = O(n \log n \log \log u)$. For each dominance pair, we spend an additional amount of $O(\log \log u)$ time. Since each such pair is reported exactly once, the total running time of the divide-and-conquer algorithm is bounded by $O(n \log n \log \log u + k \log \log u)$, where k denotes the number of dominance pairs in S. Moreover, the algorithm uses $O(u)$ space.

Our original problem was to solve the dominance reporting problem on a set V of n points in $\mathbb{R}^4$. In $O(n \log n)$ time, we normalize the points, giving a set S of n points in $U^4 = \{0, 1, \ldots, n-1\}^4$. Then, in $O(n)$ time, we construct an empty vEB-tree on the universe U. Finally, in $T(n, n) + O(k \log \log n)$ time we find all k dominance pairs in S. This gives all k dominance pairs in V. The entire algorithm takes $O(n \log n \log \log n + k \log \log n)$ time and it uses $O(n)$ space. This proves our main result:

Theorem 3.2 *Problems 1.1 and 1.2 can be solved in $O(n \log n \log \log n + k \log \log n)$ time and $O(n)$ space, where k is the number of pairs of enclosing rectangles or, equivalently, the number of dominance pairs.*

4 A faster algorithm for a special case

Assume that there are only $\alpha = O(1)$ different aspect ratios in the set $\mathcal{R}$ of rectangles. By a *diagonal* of a rectangle we mean the line-segment joining its SW and NE corners. Clearly, there are α different slopes among the diagonals in $\mathcal{R}$. For some such slope ρ, let $\mathcal{R}' \subseteq \mathcal{R}$ consist of the rectangles whose diagonals have slope ρ. Let $R = [l, r] \times [b, t]$ and $R' = [l', r'] \times [b', t']$ be rectangles in $\mathcal{R}$ and $\mathcal{R}'$, respectively. (Throughout, we view rectangle sides as closed line segments, i.e., endpoints are included.)

Lemma 4.1 *Let L be a line with slope ρ which moves over the plane from the northwest to the southeast. Consider the moment at which L coincides with the diagonal of R'. If L intersects R, then one of the following holds:*

1. *L meets the left and top sides of R. In this case, we have $R' \subseteq R$ iff $l' \geq l$ and $t' \leq t$.*

2. *L meets the left and right sides of R. In this case, we have $R' \subseteq R$ iff $l' \geq l$ and $r' \leq r$.*

3. *L meets the bottom and top sides of R. In this case, we have $R' \subseteq R$ iff $b' \geq b$ and $t' \leq t$.*

4. *L meets the bottom and right sides of R. In this case, we have $R' \subseteq R$ iff $b' \geq b$ and $r' \leq r$.*

Note that L meets the corners of R in a specific order, namely, NW, NE, SW, SE (resp. NW, SW, NE, SE), depending on whether R's diagonal has slope less (resp. greater) than ρ. The NE and SW corners will be met simultaneously if R's diagonal has slope ρ; this case is covered by Lemma 4.1 since rectangle sides are closed line segments.

4.1 The algorithm

For each diagonal-slope ρ we do the following: We project all the rectangle corners in $\mathcal{R}$ onto a line $\hat{L}$ normal to L and sort them in non-decreasing order. Note that the SW and NE corners of each rectangle in $\mathcal{R}'$ projects to the same point on $\hat{L}$. We treat these two points as a composite point.

Using L, we sweep over $\hat{L}$ from $-\infty$ to $+\infty$, maintaining four priority search trees, PST_i, $1 \leq i \leq 4$. (PST_i will handle condition i of Lemma 4.1.) Let v be the current event point. The following actions are taken:

1. v corresponds to the NW corner of $R = [l, r] \times [b, t]$. We insert (l, t) into PST_1.

2. v corresponds to the NE corner of $R = [l, r] \times [b, t]$. If the SW corner of R has not been seen so far then we delete (l, t) from PST_1 and insert (l, r) into PST_2. Otherwise, we delete (b, t) from PST_3 and insert (b, r) into PST_4.

3. v corresponds to the SW corner of $R = [l, r] \times [b, t]$. If the NE corner of R has not been seen so far then we delete (l, t) from PST_1 and insert (b, t) into PST_3. Otherwise, we delete (l, r) from PST_2 and insert (b, r) into PST_4.

4. v corresponds to the SE corner of $R = [l, r] \times [b, t]$. We delete (b, r) from PST_4.

5. v corresponds to the SW and NE corner of $R' = [l', r'] \times [b', t'] \in S'$. We query PST_1 with (l', t') and report all points (l, t) in it such that $l' \geq l$ and $t' \leq t$. Similarly, we query PST_2 with (l', r'), PST_3 with (b', t'), and PST_4 with (b', r'). Then we delete (l', t') from PST_1 and insert (b', r') into PST_4.

Theorem 4.1 *Given a set $\mathcal{R}$ of n axes-parallel rectangles in $\mathbb{R}^2$ with at most α different aspect ratios, where α is a constant, all k pairs of rectangles (R', R) such that R encloses R' can be reported in $O(\alpha n \log n + k)$ time and $O(n)$ space.*

5 Concluding remarks

We have given an algorithm for solving the rectangle enclosure reporting problem, or, equivalently, the four-dimensional dominance reporting problem, that runs in $O(n \log n \log \log n + k \log \log n)$ time, where k is the number of reported pairs. Previously, the problem had been solved in $O(n \log^2 n + k)$ time by Lee and Preparata [LP82].

We leave open the question of whether the problem can be solved in $O(n \log n + k)$ time. It seems very difficult to remove the $\log \log n$ term that occurs in the "reporting" part of our running time.

We have given a new technique to solve the three-dimensional red-blue dominance reporting problem. Using the same approach we can solve the two-dimensional version of this problem, where the red and blue points are sorted by their x-coordinates, optimally, i.e., in $O(n + k)$ time.

References

[EO82] H. Edelsbrunner and M.H. Overmars. On the equivalence of some rectangle problems. *Information Processing Letters*, 14:124–127, 1982.

[GJSD94] P. Gupta, R. Janardan, M. Smid, and B. Dasgupta. The rectangle enclosure and point–dominance problems revisited. Rept. MPI–I–94–142, Max–Planck–Institut für Informatik, Saarbrücken, Germany, 1994.

[KO88] R.G. Karlsson and M.H. Overmars. Scanline algorithms on a grid. *BIT*, 28:227–241, 1988.

[LP82] D.T. Lee and F.P. Preparata. An improved algorithm for the rectangle enclosure problem. *Journal of Algorithms*, 3:218–224, 1982.

[LW81] D.T. Lee and C.K. Wong. Finding intersection of rectangles by range search. *Journal of Algorithms*, 2:337–347, 1981.

[McC85] E.M. McCreight. Priority search trees. *SIAM Journal on Computing*, 14:257–276, 1985.

[PL88] B. Preas and M. Lorenzetti, Eds. *Physical design automation of VLSI systems*. Benjamin/Cummings, Menlo Park, CA, 1988.

[PS88] F.P. Preparata and M.I. Shamos. *Computational Geometry – An Introduction*. Springer–Verlag, Berlin, 1988.

[vEB77a] P. van Emde Boas, R. Kaas and E. Zijlstra. Design and implementation of an efficient priority queue. *Mathematical Systems Theory*, 10:99–127, 1977.

[vEB77b] P. van Emde Boas. Preserving order in a forest in less than logarithmic time and linear space. *Information Processing Letters*, 6:80–82, 1977.

[VW80] V. Vaishnavi and D. Wood. Data structures for the rectangle containment and enclosure problems. *Computer Graphics and Image Processing*, 13:372–384, 1980.

Approximate Range Searching

Sunil Arya[*] David M. Mount[†]

Abstract

The range searching problem is a fundamental problem in computational geometry, with numerous important applications. Most research has focused on solving this problem exactly, but lower bounds show that if linear space is assumed, the problem cannot be solved in polylogarithmic time, except for the case of orthogonal ranges. In this paper we show that if one is willing to allow approximate ranges, then it is possible to do much better. In particular, given a bounded range Q of diameter s and $\epsilon > 0$, an approximate range query treats the range as a fuzzy object, meaning that points lying within distance ϵs of the boundary of Q either may or not be counted. We show that in any fixed dimension d, a set of n points in R^d can be preprocessed in $O(n \log n)$ time and $O(n)$ space, such that approximate queries can be answered in $O(\log n + (1/\epsilon)^d)$ time. The only assumption we make about ranges is that the intersection of a range and a d-dimensional cube can be answered in constant time (depending on dimension). For convex ranges, we tighten this to $O(\log n + (1/\epsilon)^{d-1})$ time. We also present a lower bound argument for approximate range searching based on partition trees of $\Omega(\log n + (1/\epsilon)^{d-1})$, which implies optimality for convex ranges. Finally we give empirical evidence showing that allowing small relative errors can significantly improve query execution times.

1 Introduction.

The range searching problem is among the fundamental problems in computational geometry. A set P of n data points is given in d-dimensional real space, R^d, and a space of possible *ranges* is considered (e.g. d-dimensional rectangles, spheres, halfspaces, or simplices). The goal is to preprocess the points so that, given any query range Q, the points in $P \cap Q$ can be counted or reported efficiently. More generally, one may assume that the points have been assigned weights, and the problem is to compute the accumulated weight of the points in $P \cap Q$, $weight(P \cap Q)$, under some commutative semigroup.

There is a rich literature on this problem. In this paper we consider the weighted counting version of the problem. We are interested in applications in which the number of data points is sufficiently large that one is limited to using only linear or roughly linear space in solving the problem. For orthogonal ranges, it is well known that range trees can be applied to solve the problem in $O(\log^{d-1} n)$ time with $O(n \log^{d-1} n)$ space (see e.g. [12]). Chazelle and Welzl [7] showed that triangular range queries can be solved in the plane in $O(\sqrt{n} \log n)$ time using $O(n)$ space. Matoušek [10] has shown how to achieve $O(n^{1-1/d})$ query

---

[*]Max-Planck-Institut für Informatik, D-66123 Saarbrücken, Germany. Email: arya@mpi-sb.mpg.de. This author was supported by the ESPRIT Basic Research Actions Program, under contract No. 7141 (project ALCOM II).

[†]Department of Computer Science and Institute for Advanced Computer Studies, University of Maryland, College Park, MD 20742. Email: mount@cs.umd.edu. This author was supported by the National Science Foundation under grant CCR-9310705.

time for simplex range searching with nearly linear space. This is close to Chazelle's lower bound of $\Omega(n^{1-1/d}/\log n)$ [6] for linear space. For halfspace range queries, Brönnimann, et al. [3] give a lower bound of $\Omega(n^{1-2/(d+1)})$ (ignoring logarithmic factors) assuming linear space. This lower bound applies to the more general case of spherical range queries as well.

Unfortunately, the lower bound arguments defeat any reasonable hope of achieving polylogarithmic performance for arbitrary (nonorthogonal) ranges. This suggests that it may be worthwhile considering variations of the problem, which may achieve these better running times. In this paper we consider an approximate version of range searching. Rather than approximating the count, we consider the range to be a *fuzzy* range, and that data points that are "close" to the boundary of the range (relative to the range's diameter) may or may not be included in the count.

To make this idea precise, we assume that ranges are bounded sets of bounded complexity. (Thus our results will not be applicable to halfspace range searching). Given a range Q of diameter s, and given $\epsilon > 0$, define Q^- to be the locus of points whose distance from a point exterior to Q is at least $s\epsilon$, and Q^+ to be the locus of points whose distance from a point interior to Q is at most $s\epsilon$. (Equivalently, Q^+ and Q^- can be defined in terms of the Minkowski sum of a ball of radius s and either Q or its complement.) Define a *legal answer* to an ϵ-approximate range query to be $weight(P')$ for any subset P' such that

$$P \cap Q^- \subseteq P' \subseteq P \cap Q^+.$$

This definition allows for two-sided errors, by failing to count points that are barely inside the range, and counting points barely outside the query range. It is trivial to modify the algorithm so that it produces one-sided errors, forbidding either sins of omission or sins of commission (but obviously not both).

Approximate range searching is probably interesting only for fat ranges. Overmars [11] defines an object Q to be *k-fat* if for any point p in Q, and any ball B with p as center that does not fully contain Q in its interior, the portion of B covered by Q is at least $1/k$. For ranges that are not k-fat, the diameter of the range may be arbitrarily large compared to the thickness of the range at any point. However, there are many applications of range searching that involve fat ranges.

There are a number of reasons that this formulation of the problem is worth considering. It is well known that what seems to make range queries "hard" to solve are the points that are near the boundary of the range. However, there are many applications where data are imprecise, and ranges themselves are imprecise. For example, the user of a geographic information system that wants to know how many single family dwellings lie within a 60 mile radius of Manhattan, may be quite happy with an answer that is only accurate to within a few miles. Also range queries are often used as part of an initial filtering process to very large data sets, after which some more complex test will be applied to the points within the range. In these applications, a user may be quite happy to accept a coarse filter that runs faster. The user is free to adjust the value of ϵ to whatever precision is desired (without the need to apply preprocessing again), with the understanding that a tradeoff in running times is involved.

In this paper we show that by allowing approximate ranges, it is possible to achieve significant improvements in running times, both from a theoretical as well as practical perspective. We show that (for fixed dimension) after $O(n \log n)$ preprocessing, and with $O(n)$ space, ϵ-approximate range queries can be answered in time $O(\log n + 1/\epsilon^d)$. Under the assumption that ranges are convex, this can be strengthened to $O(\log n + 1/\epsilon^{d-1})$. Some of the features of our method are

- The data structure and preprocessing time are independent of the space of possible ranges and ϵ. We only assume that in constant time (depending on dimension) it is possible to determine whether there is a nonempty intersection between the inner and outer ranges (Q^- and Q^+) and a cube.

- Space and preprocessing time are free of exponential factors in dimension. Space is $O(dn)$ and preprocessing time is $O(dn \log n)$. (Assuming the binarized version of the data structure, discussed in Section 2.)

- The algorithms are quite simple. The data

structure is a variant of the well-known quadtree data structure.

- Our experimental results show that even for uniformly distributed points in dimension 2, there is a significant improvement in the running time if a small approximation error is allowed. Furthermore, on average the *effective error* (defined in Section 5) committed by the algorithm is much smaller than the allowed error, ϵ.

We also present lower bound of $\Omega(\log n + 1/\epsilon^{d-1})$, for the complexity of answering ϵ-approximate range queries assuming a partition tree approach for cubical range in fixed dimension. Thus our approach is optimal under these assumptions for convex ranges.

2 The BBD Tree.

In this section we describe the data structure from which queries will be answered. We call this structure a *balanced box-decomposition tree* (or BBD tree) for the point set. The BBD tree is a balanced variant of a number of well-known data structures based on hierarchical subdivision of space into rectilinear regions. Examples of this class of structure include point quadtrees [13], *k-d* trees [2], or (unbalanced) box-decomposition tree (also called a fair-split tree) [1, 4, 8, 14].

Except for the *k-d* tree, none of these data structures need be balanced, in the sense that their depth is not bounded by $O(\log n)$. However, all of these data structures, except the *k-d* tree subdivide space into regions of bounded aspect ratio. The BBD tree achieves both of these properties. This is not the first example of such a tree. Arya et al. [1] showed that balance could be imposed on a box-decomposition by computing a centroid decomposition tree or topology tree for the box-decomposition tree. This idea was also used by Callahan and Kosaraju to modify the fair-split tree [5]. The BBD tree is a somewhat more direct implementation of this same idea. We present a brief description of this structure for the sake of completeness.

The BBD tree is a balanced 2^d-ary tree associated with a hierarchical subdivision of space into cells each of $O(d)$ complexity. For our purposes, a *box* in R^d is d-dimensional cube with faces orthogonal to the coordinate axes. We assume that all the data points have been scaled to lie within a d-dimensional unit hypercube H. This can be done in linear time before preprocessing. Define a *quadtree box* to be any box obtained by a finite number of applications of the following recursive rules.

- H is a quadtree box.
- If b is a quadtree box, then so are each of the 2^d boxes of half the side length formed by splitting b using d hyperplanes, each orthogonal to one of the coordinate axes.

Observe that each quadtree box has sides of length $(1/2)^i$ for some $i \geq 0$. All the boxes appearing in our BBD tree are quadtree boxes. Data points can be thought of as degenerate quadtree boxes of side length 0.

Each node v in the BBD tree is associated with a region denoted, $cell(v)$. Each cell is the set theoretic difference of two quadtree boxes, a bounding box, $BB(v)$, and a (possibly empty) inner box, $IB(v)$, which is properly contained within $BB(v)$. The set of data points associated with v, $P(v)$ is the intersection of P with v's cell, that is,

$$P(v) = P \cap (BB(v) - IB(v)).$$

The *size* of a node $size(v)$ is the side length of its bounding box $BB(v)$. Observe that the Euclidean distance between any two points in a cell of size s is at most $s\sqrt{d}$. The *weight* of a node, $weight(v)$, is the cardinality of $P(v)$ (or more generally, the sums of weights of the points in the cell). We assume that each node v contains the quantities, $BB(v)$, $IB(v)$ (and a flag indicating whether $IB(v)$ is empty), $size(v)$, $weight(v)$.

The root of the tree is associated with the bounding hypercube H, and its inner box is empty. Leaf nodes are each associated with a single data point (or generally a small constant number of data points), and have no inner box. Internal nodes are two types, *splitting nodes* and *shrinking nodes*, according to the way in which they subdivide the associated cell. Each splitting node v subdivides the associated cell by splitting its bounding quadtree box into 2^d quadtree boxes each of half the size by the process described earlier. These smaller boxes are associated with each of the 2^d children of v. By the nature of quadtree boxes, the inner box of v will either be entirely contained within one of v's children, or it will be equal to one of these boxes. In the former case, the inner box is made the inner box of the appropriate child cell, and all the other children have no inner box. In the latter case, the corresponding child is null (because the associated region is empty).

Each shrinking node v subdivides the associated cell by splitting it about a box b which contains the inner box of v (if any). The two children of v consist of one cell whose bounding box is $BB(v)$ and inner box b, and the other cell whose bounding box is b and inner box is $IB(v)$. (Note that shrinking is different from the shrinking operation defined in [1], because the region between the outer and inner box need not be free of data points.) These operations are illustrated in Figure 1(a) and (b).

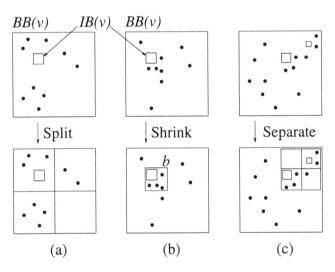

Figure 1: Splitting and shrinking.

For the purposes of implementation in higher dimensions, nodes with 2^d children are quite impractical. This can be overcome by implementing each split by a series of d consecutive binary splits, each cutting along a plane orthogonal to a different coordinate axis. The resulting boxes are not necessarily cubes, but they have bounded aspect ratio. However, for the purposes of describing our algorithms, it is somewhat simpler to think of splitting as an atomic operation, rather than this alternative *binarized* tree.

We claim that for every cell it is possible to apply a shrinking operation so that each of the resulting subcells contains at most a fraction of $2^d/(2^d+1)$ of the points in the original cell. To see this, consider a cell with m points, and consider a sequence of consecutive splits, starting with the original cell. After each split, apply the next split to the subcell with the largest number of points. Repeat until the largest subcell c has at most $m2^d/(2^d+1)$ points. It is easy to show that at most $m2^d/(2^d+1)$ points can lie outside c. Thus, it is always possible to find a shrinking cell containing a constant fraction of the points in a given cell. Such a shrinking operation is called a *centroid shrink*. (The term arises from the fact that the box c corresponds to finding a centroid node in the unbalanced box-decomposition tree, as described in [1].)

However, performing a centroid shrink may result in two inner boxes within the same bounding box. Repeated application could result in cells of arbitrary complexity. To remedy this, whenever a centroid shrink results in a box which is disjoint of the existing inner box, we separate these two inner boxes as follows. First, we find the smallest quadtree box which contains both of these boxes. We shrink to this box, and then we split this box. By the minimality of the containing box, the two inner boxes will become inner boxes of two different boxes in the split (or they will be equal to one of these boxes, in which case the box can be ignored as a child). This process of separating inner boxes increases the size and depth of the tree by a constant factor. However, it has the nice feature that each cell in the tree consists of a single bounding box and a (possibly empty) inner box, and so has complexity $O(d)$.

By alternating splitting rules, first applying a split followed by a centroid shrink (and box separation), and repeating until each cell contains at most a single point, we produce the balanced box-decomposition tree (BBD tree). The following lemma summarizes the main elements of the BBD tree.

Lemma 1 *Given a set of n points in a d-dimensional unit hypercube:*

(i) *the BBD tree has height $O(d \log n)$,*

(ii) *each node is associated with a cell of complexity $O(d)$,*

(iii) with every 2 levels of descent in the tree the size of the associated cells decreases by at least a factor of $1/2$.

The following lemma, which follows from the fact that cells cannot be too "skinny" will also be of importance. This type of *packing lemma* is common to many analyses of box-decomposition trees. The proof is omitted from this version.

Lemma 2 *The number of cells in a BBD tree of size at least s with pairwise disjoint interiors, and which intersect a range of diameter $2r$ is at most $(1 + \lceil \frac{2r}{s} \rceil)^d$.*

Lemma 3 *The number of cells in a BBD tree of size at least s with pairwise disjoint interiors, and which intersect the boundary of a convex range of diameter $2r$ is at most $O(1 + \lceil \frac{2r}{s} \rceil)^{d-1}$.*

Lemma 4 *The BBD tree can be built in $O(n \log n)$ time and $O(n)$ space.*

Proof: (Sketch) The construction of the BBD tree follows largely from the construction described by Callahan and Kosaraju for the fair-split tree [4]. The only nontrivial step not described there is the implementation of the centroid shrink. The problem is that it is not generally possible to bound the number of split operations needed until the number of points falls below some constant (for example, if the points are densely clustered). To do this, it suffices to determine the smallest quadtree box enclosing the current set of points at each step. This can be done in constant time, under the assumption that the model of computation supports the operations of exclusive-or, integer logarithm, and integer division on the coordinates of the data points [1]. Callahan and Kosaraju have shown that this assumption can be overcome with a somewhat more careful choice of splitting rules [5]. □

3 Range Searching Algorithm.

In this section we present the algorithm for answering range queries using the BBD tree. For simplicity, we consider the case where the query range is a ball of radius r centered at a query point q, but the generalization to other types of ranges is straightforward. We use the terms *inner range* and *outer range* to refer to the balls of radius $r^- = r/(1 + \epsilon)$ and radius $r^+ = r(1 + \epsilon)$ centered at q. Although we assume that $\epsilon > 0$ for the purposes of analysis, the algorithm runs correctly even if $\epsilon = 0$.

Generalizing range search algorithms for partition trees, the main idea of the algorithm is to simply descend the tree and classify nodes as lying completely inside the outer range or completely outside the inner range. If a node cannot be classified we recursively explore its children. The algorithm starts with the root of the tree. Let v denote the current node being visited.

(1) If $cell(v)$ lies completely inside outer range, then return($weight(v)$).

(2) If $cell(v)$ lies completely outside inner range then return(0).

(3) If v is a leaf node, check for each associated point whether it lies inside the true range. Return(m), where m is the number of points that do lie inside.

(4) Otherwise, recursively call the procedure with the left and right child and return the sum of the two weights obtained.

The correctness of this simple algorithm is quite straightforward, and has been omitted from this version.

The main result of this section is the following theorem, which establishes the running time of the range counting algorithm.

Theorem 1 *After $O(n \log n)$ preprocessing time, and data structure of size $O(n)$ can be built, so that given a spherical query range and $\epsilon > 0$, a $(1+\epsilon)$-approximate range count can be computed in $O((\log n) + 1/\epsilon^d)$ time. (Constant factors in preprocessing time and space are linear in d, and constant factors in query time are on the order of $d^2 2^d$.)*

Proof: Preprocessing has already been discussed. We start with two definitions. A node v is said to be *visited* if the algorithm is called with node v as argument. A node v is said to be *expanded* if the the algorithm visits the children of node v. We distinguish between two kinds of expanded nodes depending on size. An expanded node v for which $size(v) \geq 2r$ is *large* and otherwise it is *small*. We will show that the number of large expanded nodes is bounded by $O(2^d \log n)$ and the number of small expanded nodes is bounded by $O((2\sqrt{d}/\epsilon)^d)$. Because each node can be expanded in $O(2^d)$ time, it will follow that the total running time of the algorithm is $O(2^d(2^d \log n + (2\sqrt{d}/\epsilon)^d))$. A factor of 2^d can be saved in expansion time if the 2^d-ary tree is replaced by a binarized tree described in Section 2, while adding a factor of d in the depth of the tree and a factor of d to the expansion time. This yields a better total time of $O(d^2 2^d \log n + d(2\sqrt{d}/\epsilon)^d)$. In either case, the running time is $O(\log n + 1/\epsilon^d)$ for fixed d.

We first show the bound on the number of large expanded nodes. In the descent through the BBD tree, the sizes of nodes decrease monotonically. Consider the set of all expanded nodes of size greater than $2r$. These nodes induce a subtree in the BBD tree. Let V denote the leaves of this tree. The cells associated with the elements of V are pairwise disjoint from one another, and furthermore they intersect the range (for otherwise they would not be expanded). It follows from Lemma 2 (applied to the cells associated with V) that there are at most $(1 + \lceil 2r/(2r) \rceil)^d = 2^d$, such boxes. Because the depth of the tree is $O(\log n)$, the total number of expanded large nodes is $O(2^d \log n)$, as desired.

Next we bound the number of small expanded nodes. First we claim that any node of size less than $r\epsilon/\sqrt{d}$ cannot be expanded. For a node to be expanded its cell must intersect the inner range of radius $r^- = r/(1+\epsilon)$ and the complement of the outer range of radius $r^+ = r(1+\epsilon)$. Hence the cell must have diameter of at least $r^+ - r^- \geq r\epsilon$. Since the diameter of a cell of size s is at most $s\sqrt{d}$, a cell of size less than $r\epsilon/\sqrt{d}$ is too small to be expanded.

To complete the analysis of the number of small expanded nodes, it suffices to count the number of expanded nodes of sizes from $2r$ down to $r\epsilon/\sqrt{d}$. Because sizes are powers of $1/2$, it suffices to count the number of expanded nodes of *size group* $(1/2)^i$, where i varies over an appropriate range. Consider the expanded nodes in the i-th size group. Because these nodes have the same size, the corresponding cells have pairwise disjoint interiors, and they overlap the query range. Applying Lemma 2, it follows that the number of maximal nodes in the i-th group is $(1 + \lceil 2^{i+1}r \rceil)^d$. Thus the total number of expanded nodes in all the size groups is at most

$$\sum_{i=a}^{b} \left(1 + \lceil 2^{i+1}r \rceil\right)^d,$$

where $a = -\lg 2r$ and $b = -\lg(r\epsilon/\sqrt{d})$. This is a geometric series, which is dominated asymptotically by its largest term,

$$\left(1 + \left\lceil \frac{2r\sqrt{d}}{r\epsilon} \right\rceil\right)^d = O\left(\left(\frac{2\sqrt{d}}{\epsilon}\right)^d\right),$$

as desired. □

For convex ranges, we can easily show a tighter bound of $O((\log n) + 1/\epsilon^{d-1})$ by using Lemma 3 in place of Lemma 2.

4 Lower Bounds

The method we use in this paper to solve the approximate range counting problem falls under the partition tree paradigm. This paradigm is also commonly used for solving the exact version of this problem. In the context of exact range counting, Chazelle and Welzl [7] have developed an interesting lower bound argument for any algorithm that uses partition trees. In this section we develop a similar argument for the approximate problem, which will establish the optimality of our algorithm in this paradigm.

We start by reviewing the notion of a partition tree. We are given a set P of n data points. A partition tree is a rooted tree of bounded degree in which each node v of the tree is associated with a set of points $P(v)$, according to the following rules. (For simplicity we will assume that the degree is at least two; it will be easy to see that the argument we develop here also holds without this assumption.)

(a) The leaves of the tree have a one-to-one correspondence with the data points.

(b) The set associated with an internal node v is formed by taking the union of all the points in the leaves of the subtree rooted at v.

With each node v we also store its $weight(v)$ defined as the cardinality of the set $P(v)$. Given a range Q we can recursively search the partition tree to count the number of points inside Q as follows. We start at the root of the tree and initialize a global variable *count* to 0. At a node v we do the following.

(a) if $P(v) \subset Q$, we add *weight(v)* to the count.

(b) if $P(v) \cap Q = \emptyset$, we do nothing.

(c) Otherwise, we recursively search its children.

It is easy to see that the algorithm correctly solves the range counting query. Assuming that the conditions in Step (a) and (b) can be carried out in $O(1)$ time, the number of nodes visited by the algorithm accurately reflects its running time. Chazelle and Welzl [7] have shown that in the worst case the number of nodes visited is $\Omega(n^{1-1/d})$ for *any* partition tree. Using similar techniques, we show a lower bound on the number of nodes visited for the approximate version of the problem.

First we modify the above algorithm to solve the approximate range counting query. Let Q be the query range and let Q^+ and Q^- be the ϵ expansion and contraction of this range, respectively, as defined in the introduction. For the approximate range counting problem, we make a few straightforward modifications to the search algorithm given above.

(a) if $P(v) \subset Q^+$, we add *weight(v)* to the count.

(b) if $P(v) \cap Q^- = \emptyset$, we do nothing.

(c) Otherwise, we recursively search its children.

Define the *visiting number* of a partition tree as the maximum number of nodes visited by the above algorithm over all query ranges. We show a lower bound of $\log n + (1/\epsilon)^{d-1}$ on the visiting number of any partition tree. First, we modify some of the definitions of Chazelle and Welzl [7] to apply to the approximate problem. We say that a set $P(v)$ is *stabbed* if neither $P(v) \subset Q^+$ nor $P(v) \cap Q^- = \emptyset$ is true. In other words, $P(v)$ contains both a point inside Q^- and a point outside Q^+. We define the stabbing number of a spanning path as the maximum number of edges on the path (each edge is a set of its two end points) that are stabbed. Here the maximum is computed over all query ranges. Along the lines of Lemma 3.1 in [7], we can easily establish the following.

Lemma 5 *If T is any partition tree for P, then there exists a spanning path whose stabbing number does not exceed the visiting number of T.*

We will now exhibit a point set in d dimensions and a set of query ranges such that any spanning path will have a stabbing number of at least $1/\epsilon^{d-1}$ with respect to one of the query ranges. We assume that the dimension d is fixed. Consider a unit hypercube $[0,1]^d$ divided into a regular grid consisting of k^d cells of equal size. We choose $k = \lceil 1/(4\epsilon) \rceil$ and use $\epsilon' = 1/k$ to denote the grid spacing. The data set consists of one point located at each of the vertices of the grid, which gives a total of $\Omega(1/\epsilon^d)$ points. The query ranges in our set are balls in the L_∞ metric of radius unity. The centers of these balls are located along the d principal axis at distances from the origin of $-1 + \epsilon'/2, -1 + 3\epsilon'/2, \ldots, -\epsilon'/2$, respectively. This gives a total of $O(d/\epsilon)$ query ranges. Now consider any spanning path on the set of points P. From the construction it is straightforward to show that every edge on this spanning path is stabbed by at least one query range. Thus the number of stabbed edges on the spanning path equals the total number of edges on the path which is $\Omega(1/\epsilon^d)$. Dividing this by the total number of query ranges implies that a query range stabs on average $\Omega(1/(d\epsilon^{d-1}))$ edges of the spanning path. Therefore there must exist a query range which stabs $\Omega(1/(d\epsilon^{d-1}))$ edges. Thus the stabbing number of any spanning path exceeds this quantity. Combining this with Lemma 5 gives us a lower bound on the visiting number of partition trees.

We next show a $\Omega(\log n)$ lower bound on the visiting number of any partition tree. Consider a set of n distinct data points and a set of n query ranges consisting of L_∞ ball centered at each of these points. The radius of these balls is chosen to be sufficiently small so that the $(1+\epsilon)$ expansion of the ball contains no other point. It is easy to see that the ball centered at p stabs the point sets corresponding to every proper ancestor of p. Since there must be a leaf at depth $\Omega(\log n)$, this gives a lower bound on the number of nodes whose point sets are stabbed by the ball centered at the point associated with the leaf. All such nodes are visited by the algorithm, hence this is also a bound on the visiting number of the partition tree. Combining this with the results of the last paragraph, we have the following lower bound on the visiting number of any partition tree. In fact, the lower bound holds even under the restriction to L_∞ balls.

Theorem 2 *For the set of query ranges consisting of balls in the L_∞ metric, the visiting number of any partition tree exceeds $\Omega(\log n + 1/\epsilon^{d-1})$.*

The theorem implies the optimality of our algorithm in the partition tree paradigm for convex ranges, and near optimality for the more general class of query ranges discussed in the introduction.

5 Experimental Results

To establish the validity of our claims empirically, we implemented our algorithm and tested it on a number of data sets of various sizes, various distributions, and with various sizes and types of ranges. To enhance performance, we implemented a variation of the data structure described in Section 2. First, we implemented the *binarized* version of the tree mentioned in this section (each node has two children rather than 2^d). Second, we did not always split boxes through the midpoint, but used a somewhat more sophisticated decomposition method, called the *fair-split rule* [1]. Intuitively, this splitting rule attempts to partition the point set of each box as evenly as possible, subject to maintaining boxes with bounded aspect ratio. Finally, our decomposition process attempted to *avoid* centroid shrinking whenever it was not warranted. The reason is that there are optimizations that can be performed at splitting nodes that are not possible at shrinking nodes. As long as the fair-split rule produced trees whose depth was within a constant factor of $\log_2 n$, we did not introduce centroid shrinking, and for the data sets we tested it was never neeed.

We ran our program for approximate range counting for ϵ ranging from 0 (exact searches) to 0.5. Our experiments were conducted for data points drawn from a number of distributions. Due to space limitations, only the following two are presented in this paper.

Uniform: Each coordinate was chosen uniformly from the interval $[0, 1]$.

ClusNorm: Ten points were chosen from the uniform distribution over the unit hypercube and a Gaussian distribution with standard deviation 0.05 centered at each.

For each distribution we generated data sets ranging in size from $2^6 = 64$ to $2^{16} = 65,536$. Experiments were run in dimensions 2 and 3, and the query ranges were either L_2 balls (circles) or L_∞ balls (squares). Due to space limitations, we only show the results for dimension 2 and for circular ranges. We tested radii, ranging in size from 1/256 to 1/2. For each experiment, we fixed ϵ and the radius of the query balls and measured a number of statistics, averaged over 1,000 queries. The center of the query ball was chosen from the same distribution as the query points.

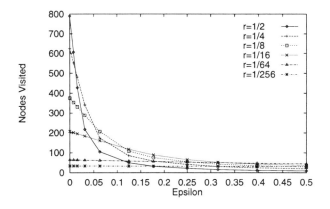

Figure 2: Number of nodes visited vs. ϵ. Uniform distribution.

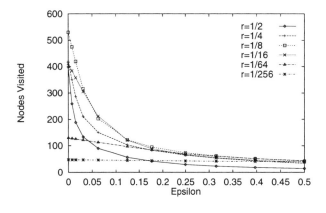

Figure 3: Number of nodes visited vs. ϵ. Clustered normal distribution.

In Figures 2 and 3, for each of the distributions, we show the number of nodes visited as a function of the accuracy of approximation, ϵ, for 65,536 data point. Since the algorithm does a constant amount of work for each node visited, the average number of nodes visited accurately reflects its running time. (We also measured floating point operations, and found that in dimension 2 on the average there were from 10 to 20 floating point operations for each node visited.) The key observation is that as ϵ increases (even to relatively small values in the range from 0.05 to 0.1), there are significant improvements in running time (factors as high as 10 to 1, and often around 4 to 1) for larger ranges. As ϵ grows, the running times tend to converge, irrespective of radius. Improvements for smaller ranges were not as significant, because the running times on small ranges are uniformly small. Results for square ranges were similar, and results in 3-space were similar, although the improvements were not quite as dramatic.

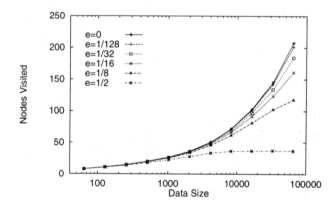

Figure 4: Number of nodes visited vs. number of data points. Uniform distribution.

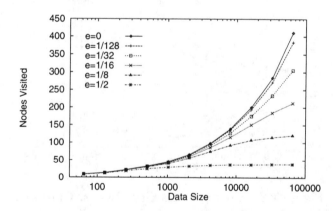

Figure 5: Number of nodes visited vs. number of data points. Clustered normal distribution.

In Figures 4 and 5 we show the number of nodes visited as a function of the number of data points. Note that the x-axis is given on a log scale. The key observation is that, as ϵ increases, the running times show a decreasing dependence on the number of data points. We also ran the experiments for smaller values of radius, and observed that the decrease of dependence occurs, but for larger values of ϵ or n. Results with square ranges and in 3-space showed a similar behavior.

We measured one interesting statistic, called *effective error* or *effective epsilon*. Consider a range of radius r and a point at distance r'. If $r' < r$ but the point was classified as being outside the range, the associated *misclassification error* is defined to be the relative error, $(r - r')/r'$; and if $r' > r$ but the point was classified as being inside the range, the associated *misclassification error* is $(r' - r)/r$. By definition, there can be no classification error greater than ϵ. But the algorithm may be doing better than this. To see how much better it is doing, we measured this relative error for every misclassified point, and averaged this over all the points which were eligible for misclassification (that is, points lying in the difference of the outer and inner ranges). This quantity is the *effective error* of the query. If no points were eligible for misclassification, then this quantity is zero.

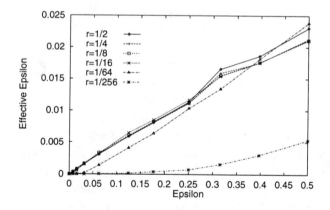

Figure 6: Effective error vs. ϵ. Uniform distribution.

In Figures 6 and 7 we show the effective errors as a function of ϵ, for 65,536 data points. The key observation is that effective error appears to vary almost linearly with ϵ (depending on distribution, dimension, and other factors). In dimension 2, effective errors were frequently less than 0.06ϵ, and in all distributions effective error was never greater

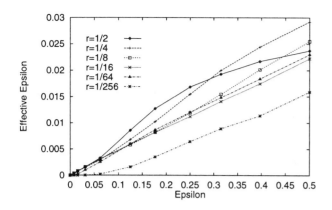

Figure 7: Effective error vs. ϵ. Clustered normal distribution.

than 0.1ϵ. These bounds were observed across all distributions tested, for both circular and square ranges, and in both dimensions 2 and 3. This explains in part, one of the reasons that we ran experiments with such large values of ϵ. Even with ϵ as large as 0.5 (allowing a maximum 50% error), we were often observing much smaller effective errors in the range of 1.5% to 3%.

References

[1] S. Arya, D. M. Mount, N. S. Netanyahu, R. Silverman, and A. Wu. An optimal algorithm for approximate nearest neighbor searching. In *Proc. 5th ACM-SIAM Sympos. Discrete Algorithms*, pages 573–582, 1994.

[2] J. L. Bentley. K-d trees for semidynamic point sets. In *Proc. 6th Ann. ACM Sympos. Comput. Geom.*, pages 187–197, 1990.

[3] H. Brönnimann, B. Chazelle, and J. Pach. How hard is halfspace range searching. *Discrete Comput. Geom.*, 10:143–155, 1993.

[4] P. B. Callahan and S. R. Kosaraju. A decomposition of multi-dimensional point-sets with applications to k-nearest-neighbors and n-body potential fields. In *Proc. 24th Annu. ACM Sympos. Theory Comput.*, pages 546–556, 1992.

[5] P. B. Callahan and S. R. Kosaraju. Algorithms for dynamic closest pair and n-body potential fields. In *Proc. 6th ACM-SIAM Sympos. Discrete Algorithms*, 1995.

[6] B. Chazelle. Lower bounds on the complexity of polytope range searching. *J. Amer. Math. Soc.*, 2:637–666, 1989.

[7] B. Chazelle and E. Welzl. Quasi-optimal range searching in spaces of finite VC-dimension. *Discrete Comput. Geom.*, 4:467–489, 1989.

[8] K. L. Clarkson. Fast algorithms for the all nearest neighbors problem. In *Proc. 24th Ann. IEEE Sympos. on the Found. Comput. Sci.*, pages 226–232, 1983.

[9] N. Farvardin and J. W. Modestino. Rate-distortion performance of DPCM schemes for autoregressive sources. *IEEE Transactions on Information Theory*, 31:402–418, 1985.

[10] J. Matoušek. Range searching with efficient hierarchical cuttings. *Discrete Comput. Geom.*, 10(2):157–182, 1993.

[11] M. H. Overmars. Point location in fat subdivisions. *Inform. Process. Lett.*, 44:261–265, 1992.

[12] F. P. Preparata and M. I. Shamos. *Computational Geometry: an Introduction*. Springer-Verlag, New York, NY, 1985.

[13] H. Samet. *The Design and Analysis of Spatial Data Structures*. Addison Wesley, Reading, MA, 1990.

[14] P. M. Vaidya. An $O(n \log n)$ algorithm for the all-nearest-neighbors problem. *Discrete Comput. Geom.*, 4:101–115, 1989.

The Overlay of Lower Envelopes in Three Dimensions and Its Applications*

Pankaj K. Agarwal[†] Otfried Schwarzkopf[‡] Micha Sharir[§]

Abstract

Let $\mathcal{F}$ and $\mathcal{G}$ be two collections of a total of n bivariate (possibly partially-defined) algebraic functions of constant maximum degree. The minimization diagrams of $\mathcal{F}, \mathcal{G}$ are the planar subdivisions obtained by the projections of the lower envelopes of $\mathcal{F}, \mathcal{G}$, respectively, onto the xy-plane. We show that the combinatorial complexity of the overlay of the minimization diagrams of $\mathcal{F}$ and $\mathcal{G}$ is $O(n^{2+\varepsilon})$, for any $\varepsilon > 0$ (the actual bound that we prove is somewhat stronger). This result has several applications: (i) an $O(n^{2+\varepsilon})$ upper bound on the complexity of the region in $\mathbb{R}^3$ enclosed between the lower envelope of one such collection of functions and the upper envelope of another collection; (ii) an efficient and simple divide-and-conquer algorithm for constructing lower envelopes in three dimensions; and (iii) a near-quadratic upper bound on the combinatorial complexity of the space of plane transversals of n compact convex simply-shaped sets in $\mathbb{R}^3$.

---

*Work on this paper by the first author has been supported by National Science Foundation Grant CCR-93-01259, an NYI award, and matching funds from Xerox Corporation. Work on this paper by the second author has been supported by the Netherlands' Organization for Scientific Research (NWO) and partially supported by ESPRIT Basic Research Action No. 6546 (project PROMotion). Work on this paper by the third author has been supported by NSF Grant CCR-91-22103, by a Max-Planck Research Award, and by grants from the U.S.-Israeli Binational Science Foundation, the Fund for Basic Research administered by the Israeli Academy of Sciences, and the G.I.F., the German-Israeli Foundation for Scientific Research and Development.

[†] Department of Computer Science, Box 1029, Duke University, Durham, NC 27708-0129, USA

[‡] Department of Computer Science, Utrecht University, P.O. Box 80.089, 3508 TB Utrecht, the Netherlands

[§] School of Mathematical Sciences, Tel Aviv University, Tel Aviv 69978, Israel, and Courant Institute of Mathematical Sciences, New York University, New York, NY 10012, USA

Permission to copy without fee all or part of this material is granted provided that the copies are not made or distributed for direct commercial advantage, the ACM copyright notice and the title of the publication and its date appear, and notice is given that copying is by permission of the Association of Computing Machinery.To copy otherwise, or to republish, requires a fee and/or specific permission.
11th Computational Geometry, Vancouver, B.C. Canada
© 1995 ACM 0-89791-724-3/95/0006...$3.50

1 Introduction

Let $\mathcal{F} = \{f_1, \ldots, f_n\}$ be a collection of n bivariate, possibly partially-defined, algebraic functions of some constant maximum degree b (and if they are partially defined, the domain of definition of each f_i is also bounded by a constant number of algebraic arcs of maximum degree b). Abusing the notation slightly, we will not distinguish between a function and its graph. The *lower envelope* $E_\mathcal{F}$ of $\mathcal{F}$ is defined as

$$E_\mathcal{F}(\mathbf{x}) = \min_i f_i(\mathbf{x}),$$

where the minimum is taken over all functions of $\mathcal{F}$ that are defined at $\mathbf{x}$. Similarly, we define the *upper envelope* $E'_\mathcal{F}$ of $\mathcal{F}$ as

$$E'_\mathcal{F}(\mathbf{x}) = \max_i f_i(\mathbf{x}).$$

The *minimization diagram* $M_\mathcal{F}$ of $\mathcal{F}$ is the planar subdivision into maximal connected relatively open cells, of dimensions 0,1, and 2 (called, respectively, vertices, edges and faces), so that within each cell the same subset of functions (and/or function boundaries) appear on the envelope $E_\mathcal{F}$. The *combinatorial complexity* of $M_\mathcal{F}$ and of $E_\mathcal{F}$ is the number of vertices, edges, and faces in $M_\mathcal{F}$. The *maximization diagram* and its combinatorial complexity are defined in an analogous manner for the upper envelope.

Recently there has been significant progress in the analysis of the combinatorial complexity of lower envelopes of multivariate functions [14, 18]. In particular, it was shown that the maximum complexity of $M_\mathcal{F}$ is $O(n^{2+\varepsilon})$, for any $\varepsilon > 0$, where the constant of proportionality depends on ε and b [18]. This result almost settles a major open problem and has already led to many applications [14, 18, 19]. The bound has also been extended to envelopes of d-variate functions, for any $d > 2$ [18].

In some applications, however, one has to consider the interaction between the lower envelope of one collection of functions and the upper envelope of another collection. A major application of this type, which was one of the motivations of the present paper, is the analysis of the combinatorial complexity of the space of *k-transversals* of a collection $\mathcal{C}$ of n compact convex sets in d dimensions (see e.g. [10, 11, 12]); a k-transversal is a k-flat that intersects all the sets of $\mathcal{C}$. Using an appropriate coordinate system for representing the space of k-flats in $\mathbb{R}^d$ (as is well known, the dimension of that space is $N = (k+1)(d-k)$), one can show that the space of k-transversals of $\mathcal{C}$ can be represented as the region enclosed between the upper envelope of one collection of functions and the lower envelope of another collection, where each function in the first (resp. second) collection represents all k-flats that are tangent to one of the given sets from below (resp. from above). Hence, the study of spaces of transversals calls for combinatorial (as well as algorithmic) analysis of the region enclosed between two envelopes in higher dimensions [10, 12]. Following this approach, Edelsbrunner et al. [10] proved an $O(n^{d-1}\alpha(n))$ upper bound on the complexity of the space of hyperplane transversals of a set of convex polytopes in $\mathbb{R}^d$; here n is the total number of vertices of the polytopes and $\alpha(n)$ is the inverse Ackermann function. Their approach, however, does not extend to arbitrary convex sets.

In this paper we provide an analysis of the region enclosed between two envelopes of bivariate functions, which results in improved complexity bounds for spaces of plane transversals of arbitrary convex sets in $\mathbb{R}^3$. The analysis depends on the following main result of the paper, which we consider to be interesting in its own right. Let M denote the *overlay* (or superposition) of the minimization diagrams $M_{\mathcal{F}}$ and $M_{\mathcal{G}}$, for any two collections $\mathcal{F}, \mathcal{G}$ of bivariate functions with the above properties. Put $m = |\mathcal{F}|$ and $n = |\mathcal{G}|$. We show that the combinatorial complexity of M is $O((m+n)^{2+\varepsilon})$, for any $\varepsilon > 0$, where the constant of proportionality depends, as in the case of a single envelope, on ε and on the maximum degree of the given functions and, in case of partial functions, of their domain boundaries. In other words, the worst-case complexity of the overlay of two (totally unrelated) minimization diagrams is asymptotically no worse than that of a single diagram.

Notice that the overlay problem is easy for the case of univariate functions, because the complexity of the overlay of the x-projections of two envelopes of univariate functions is proportional to the sum of the complexities of the individual envelopes. This is, however, not true for envelopes of bivariate functions; see Figure 1. This highlights the significance of our result.

As a matter of fact, we prove a stronger result, which can be relevant in cases where the sizes $m = |\mathcal{F}|$

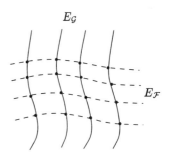

Figure 1: Overlay of two minimization diagrams with quadratic complexity

and $n = |\mathcal{G}|$ differ significantly. Specifically, we show that the number of intersection points between edges of $M_{\mathcal{F}}$ and $M_{\mathcal{G}}$ is $O(m^{1+\varepsilon}n^{1+\varepsilon})$, for any $\varepsilon > 0$.

We next show that these bounds on the complexity of overlays of minimization diagrams lead directly to a similar bound on the complexity of the region enclosed between a lower envelope and an upper envelope in $\mathbb{R}^3$. This step follows fairly routinely from the preceding analysis.

We also apply our bound to derive a very simple, deterministic, divide-and-conquer algorithm for constructing the lower envelope of n bivariate functions, as above, whose running time is $O(n^{2+\varepsilon})$, for any $\varepsilon > 0$. (This assumes an appropriate model of computation, where various standard operations on a constant number of the given functions can be performed in constant time—see below for details.) We believe that our algorithm is conceptually much simpler than competing techniques (see [4, 8, 18]). These techniques either use randomization, or are based on complicated constructions of ε-nets and cuttings.

Finally, we apply this latter bound to obtain an $O(n^{2+\varepsilon})$ bound, for any $\varepsilon > 0$, on the complexity of the space of plane transversals of a collection $\mathcal{C}$ of n compact convex sets in $\mathbb{R}^3$, where we assume that each of the given sets has 'constant description complexity', meaning that it is defined by a constant number of polynomial equalities and inequalities of constant maximum degree. If the sets in $\mathcal{C}$ do not have constant description complexity, the complexity of the space of their plane transversals can be arbitrarily large. However, if one assumes, in addition, that the sets are *separated*, in the sense that no 3 of them are crossed by a common line, then a recent result of Cappell et al. [5] gives an upper bound of $O(n^2)$ for the complexity of this space. Thus the separation assumption implies a slightly better bound, and does not require the sets to be simply shaped.

2 Complexity of the Overlay of Two Minimization Diagrams

Let $\mathcal{F}$ and $\mathcal{G}$ be two given families of m and n bivariate functions, respectively, satisfying the following condition:

($\star$) Each $f \in \mathcal{F} \cup \mathcal{G}$ is a continuous, totally or partially defined, bivariate, algebraic function of constant maximum degree b; if f is only partially defined, the domain of definition of f is bounded by a constant number of algebraic arcs of constant maximum degree, say b too.

Let M denote the planar map obtained by superimposing $M_\mathcal{F}$ and $M_\mathcal{G}$. We refer to M as the *overlay* of $M_\mathcal{F}$ and $M_\mathcal{G}$.

Theorem 2.1 *Let $\mathcal{F}$ and $\mathcal{G}$ be two collections of totally or partially defined bivariate functions, satisfying the condition ($\star$). Let $m = |\mathcal{F}|$ and $n = |\mathcal{G}|$. Then the number of intersections between edges of $M_\mathcal{F}$ and edges of $M_\mathcal{G}$ is $O(m^{1+\varepsilon} n^{1+\varepsilon})$, for any $\varepsilon > 0$. Hence, the combinatorial complexity of the overlay M is $O((m+n)^{2+\varepsilon})$, for any $\varepsilon > 0$ (where the constant of proportionality depends on ε and on b).*

Proof: For the sake of simplicity, we assume that the functions in $\mathcal{F} \cup \mathcal{G}$ are in *general position*. This excludes degenerate configurations where four function graphs meet at a point, a pair of graphs are tangent to each other, a singular or boundary point on one graph lies on an intersection curve between two other graphs, etc.) Similar conditions were assumed in the papers [14, 18]. We refer the reader to these papers for more details, and for an argument that no real loss of generality is made by assuming general position. An appropriate variant of this argument shows that our proof can also be extended to collections $\mathcal{F}$, $\mathcal{G}$ not in general position.

Our general position assumption implies that over each face of $M_\mathcal{F}$ the envelope is attained by a single function (or by no function at all), that over each edge the envelope is attained by two functions simultaneously or by the boundary of a single function graph, and that over each vertex of $M_\mathcal{F}$ the envelope is attained by three functions simultaneously, by the intersection of the boundary of one function graph with another function, by a point on the boundary of one function graph that lies directly below the boundary of another function graph or below an intersection curve of two other functions (so that this higher point is vertically visible from the point on the lower boundary), or by a vertex of a function graph boundary (a point where two arcs forming this boundary meet).

By Euler's formula for planar maps, the complexity of the overlay M is proportional to the number of vertices of M. Each vertex of M is a vertex of $M_\mathcal{F}$, a vertex of $M_\mathcal{G}$, or an intersection point of an edge of $M_\mathcal{F}$ and an edge of $M_\mathcal{G}$. Since the total number of vertices in $M_\mathcal{F}$ and $M_\mathcal{G}$ is $O(m^{2+\varepsilon} + n^{2+\varepsilon})$, as proved in [14, 18], the second assertion of the theorem follows from the first one. It thus suffices to bound the number of intersection points between the edges of $M_\mathcal{F}$ and the edges of $M_\mathcal{G}$.

We call an intersection between an edge of $M_\mathcal{F}$ and an edge of $M_\mathcal{G}$ an *edge-crossing* in M. For the purpose of analysis, we also generalize this notion, as follows. Let $\mathcal{A}(\mathcal{F})$ denote the arrangement of $\mathcal{F}$, namely the three-dimensional space decomposition induced by the graphs of the functions of $\mathcal{F}$ (see [9] for a more detailed definition). The *level* of a point w in $\mathcal{A}(\mathcal{F})$ is defined as the number of (the relative interiors of) function graphs of $\mathcal{F}$ that lie vertically below w (note that 0-level points are precisely those that lie on or below the lower envelope $E_\mathcal{F}$). We define the arrangement $\mathcal{A}(\mathcal{G})$, and the level of a point in this arrangement, in an analogous manner for the collection $\mathcal{G}$. Let e be an edge of $\mathcal{A}(\mathcal{F})$, and let e' be an edge of $\mathcal{A}(\mathcal{G})$, such that the xy-projections of e and e' cross each other at a point σ. Let ξ, ξ' be the levels of the respective points on e, e' that project onto σ. Then we say that (e, e', σ) is an *edge-crossing* in $(\mathcal{A}(\mathcal{F}), \mathcal{A}(\mathcal{G}))$ at *level* (ξ, ξ'). (By our assumptions, for any pair of edges (e, e'), as above, there is only a constant number of points σ that appear in edge-crossings of the form (e, e', σ).) Note that the original edge-crossings in M correspond to edge-crossings in $(\mathcal{A}(\mathcal{F}), \mathcal{A}(\mathcal{G}))$ at level $(0, 0)$.

We define the *index* of an edge-crossing (e, e', σ) in $(\mathcal{A}(\mathcal{F}), \mathcal{A}(\mathcal{G}))$, borrowing a similar notation from [18], to be the number of edge-crossings (e, e', ζ) such that the x-coordinate of ζ is greater than that of σ (by the general position assumption, we may exclude cases where two such crossings have the same x-coordinate).

Let $C_{p,q}(\mathcal{F}, \mathcal{G})$ denote the number of edge-crossings in $(\mathcal{A}(\mathcal{F}), \mathcal{A}(\mathcal{G}))$ whose level is (p', q') for some $p' \leq p$, $q' \leq q$, and let

$$C_{p,q}(m,n) = \max_{\substack{|\mathcal{F}|=m \\ |\mathcal{G}|=n}} C_{p,q}(\mathcal{F}, \mathcal{G}),$$

where the maximum is taken over all collections $\mathcal{F}$ and $\mathcal{G}$ satisfying ($\star$) and in general position. The goal is thus to obtain an upper bound for $C_{0,0}(m,n)$. We also denote by $C_{p,q}^{(t)}(\mathcal{F}, \mathcal{G})$ the number of edge-crossings in $(\mathcal{A}(\mathcal{F}), \mathcal{A}(\mathcal{G}))$ whose index is at most t, and whose level is (p', q') for some $p' \leq p$, $q' \leq q$, and let

$$C_{p,q}^{(t)}(m,n) = \max C_{p,q}^{(t)}(\mathcal{F}, \mathcal{G}),$$

where the maximum is taken over all collections $\mathcal{F}$ and $\mathcal{G}$, as above.

Let e be an edge of $M_{\mathcal{F}}$, let $\bar{e}$ be the edge of $E_{\mathcal{F}}$ projecting onto e, and let V_e be the vertical 2-manifold obtained as the union of all z-vertical lines passing through points of e. We assume that e is x-monotone; otherwise, we partition e into a constant number of x-monotone pieces, and apply the following analysis to each piece separately. The intersection of the graph of each function $g \in \mathcal{G}$ within V_e is an algebraic arc of constant maximum degree, so each pair of these arcs intersect in at most some constant number, s, of points (where s depends only on the maximum degree of the functions of $\mathcal{F} \cup \mathcal{G}$ and of their graph boundaries, but not on e). Let $\mathcal{A}^{(e)}(\mathcal{G})$ denote the cross-section of $\mathcal{A}(\mathcal{G})$ with V_e, and let $C_{0,q}(e, \mathcal{G})$ denote the number of edge-crossings of the form $(\bar{e}, e')$ whose level in $(\mathcal{A}(\mathcal{F}), \mathcal{A}(\mathcal{G}))$ is $(0, q')$, for any $q' \leq q$. See Figure 2 for an illustration.

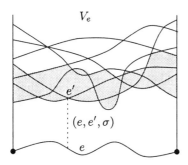

Figure 2: The arrangement $\mathcal{A}^{(e)}(\mathcal{G})$; the shaded region consists of points at level ≤ 3.

The following property is straightforward, but is crucial to our analysis: Let e' be an edge of $\mathcal{A}(\mathcal{G})$. Then $(\bar{e}, e', \sigma)$ is an edge-crossing in $(\mathcal{A}(\mathcal{F}), \mathcal{A}(\mathcal{G}))$ at level $(0, \xi)$, if and only if the point of $e' \cap V_e$ that lies on the z-vertical line through σ is a vertex at level ξ in $\mathcal{A}^{(e)}(\mathcal{G})$. This property implies that each edge-crossing in M incident to e corresponds to a vertex in the cross-section $E_{\mathcal{G}}^{(e)}$ of the lower envelope $E_{\mathcal{G}}$ within V_e.

Let (e, e', σ) be an edge-crossing of M, as above, that corresponds to a vertex v of $E_{\mathcal{G}}^{(e)}$. Suppose that the index of (e, e', σ) is at most t, and that v is incident to the graphs of $g, g' \in \mathcal{G}$. Assume that g' lies below g at points on V_e slightly to the right of v.

Let $k = k_t$ be some threshold parameter, depending on t. We will choose $k_t = k_0^{(\varepsilon^t)}$, for $t = 0, 1, \ldots, s-1$, where k_0 is a sufficiently large constant.

We trace $g \cap V_e$ from v to the right, and look for vertices of $\mathcal{A}^{(e)}(\mathcal{G})$ along that curve. The tracing stops as soon as we encounter one of the following events:

(a) We reach an endpoint w of $g \cap V_e$. We consider separately the following two subcases:

 (a.i) w lies on the vertical line passing through an endpoint of e.
 (a.ii) w is a point on the boundary of g.

(b) We reach a point w that lies vertically above an endpoint of $g'' \cap V_e$, for some $g'' \in \mathcal{G}$.

(c) We reach another point w of intersection of g and g' (within V_e).

(d) We have encountered k_t vertices of $\mathcal{A}^{(e)}(\mathcal{G})$.

If the tracing stops at an event of type (a), we charge v to w; it is easily checked that each endpoint w is charged at most a constant number of times. If this is an event of type (a.i), then there can be at most k_t such points over each endpoint of e, and the number of such endpoints is proportional to the complexity of $M_{\mathcal{F}}$, which is $O(m^{2+\varepsilon})$, for any $\varepsilon > 0$ [18]. Thus the number of vertices v of this type, over all choices of e, is $O(k_t m^{2+\varepsilon})$.

In case (a.ii), we estimate the total number of points w of this form as follows. Let γ denote (one of the constant number of arcs forming) the boundary of g. Define V_γ in an analogous manner to the definition of V_e. Note that the points w, over all choices of e, correspond to the breakpoints in the lower envelope $E_{\mathcal{F}}^{(\gamma)}$, defined in complete analogy to the above definitions. The total number of such breakpoints, over all $g \in \mathcal{G}$, is $O(\lambda_{s+2}(m)n)$, for some appropriate constant s, as defined above, where $\lambda_{s+2}(m)$ is the near-linear maximum length of $(m, s+2)$-Davenport-Schinzel sequences [2, 16]. Hence, the number of vertices v in this case, over all choices of e, is $O(\lambda_{s+2}(m)n) = O(m^{1+\varepsilon} n^{1+\varepsilon})$.

If the tracing stops at an event of type (b), we charge v to w. Again, each such point w is charged by a constant number of vertices. To estimate the total number of points w of this form, fix a function $g'' \in \mathcal{G}$, and let γ denote (one of the algebraic arcs forming) its boundary. Define V_γ as above. Note that each of the above points w that lie over γ is co-vertical with a breakpoint in the lower envelope $E_{\mathcal{F}}^{(\gamma)}$ and lies at level at most k_t in $\mathcal{A}(\mathcal{G})$. Arguing as in case (a.ii) above, the total number of such points w is thus $O(k_t m^{1+\varepsilon} n^{1+\varepsilon})$.

If the tracing stops at an event of type (d), then none of the k_t encountered vertices lies on $E_{\mathcal{G}}^{(e)}$, and each of these vertices corresponds to an edge-crossing at level $(0, k')$ in $(\mathcal{A}(\mathcal{F}), \mathcal{A}(\mathcal{G}))$, for some $k' \leq k_t$. In this case we proceed as in [18], charging v to the block of the first k_t encountered vertices. Each vertex of the $(\leq k_t)$-level is charged a constant number of times, so there are $O(C_{0,k_t}(e, \mathcal{G})/k_t)$ vertices in this case, and the total number of these vertices, over all edges e, is thus $O(C_{0,k_t}(\mathcal{F},\mathcal{G})/k_t)$.

If the tracing stops at an event of type *(c)*, then we charge v to w. As above w, is charged by a constant number of vertices. We note that w corresponds to an edge-crossing of index at most $t-1$ in $(\mathcal{A}(\mathcal{F}), \mathcal{A}(\mathcal{G}))$, whose level is $(0, k')$, for some $k' \leq k_t$, and that no such w is charged in this way more than a constant number of times. The number of vertices v in this case is thus bounded by $O(C_{0,k_t}^{(t-1)}(\mathcal{F},\mathcal{G}))$.

In conclusion, we obtain the recurrence:

$$C_{0,0}^{(t)}(\mathcal{F},\mathcal{G}) = O\left(\frac{1}{k_t}C_{0,k_t}(\mathcal{F},\mathcal{G}) + C_{0,k_t}^{(t-1)}(\mathcal{F},\mathcal{G}) + k_t(m^{2+\varepsilon} + m^{1+\varepsilon}n^{1+\varepsilon})\right). \quad (1)$$

We next transform (1) using the probabilistic technique of Clarkson and Shor [7]. We keep $\mathcal{F}$ fixed and choose a random sample $\mathcal{R}$ of $r = n/k_t$ functions of $\mathcal{G}$. Let (e, e', σ) be an edge-crossing of level $(0, \xi)$ in $(\mathcal{A}(\mathcal{F}), \mathcal{A}(\mathcal{G}))$, for $\xi \leq k_t$, and let g, g' be the two functions of $\mathcal{G}$ that define the edge e' of this crossing. Then (e, e', σ) appears as an edge-crossing in the overlay of $M_\mathcal{F}$ and $M_\mathcal{R}$ if an only if g and g' are chosen in $\mathcal{R}$ and none of the ξ functions of $\mathcal{G}$ whose graphs lie over σ below e' are chosen in $\mathcal{R}$. Hence, a standard application of the technique of [7] (as in [18]) implies that

$$C_{0,k_t}(\mathcal{F},\mathcal{G}) = O(k_t^2 \mathbf{E}(C_{0,0}(\mathcal{F},\mathcal{R})))$$
$$C_{0,k_t}^{(t-1)}(\mathcal{F},\mathcal{G}) = O(k_t^2 \mathbf{E}(C_{0,0}^{(t-1)}(\mathcal{F},\mathcal{R}))),$$

where $\mathbf{E}$ denotes expectation (with respect to the choice of $\mathcal{R}$). Passing to the maximum quantities $C_{0,0}(m,n)$, etc., we obtain the recurrence

$$C_{0,0}^{(t)}(m,n) = O\left(k_t C_{0,0}\left(m, \frac{n}{k_t}\right) + k_t^2 C_{0,0}^{(t-1)}\left(m, \frac{n}{k_t}\right) + k_t(m^{2+\varepsilon} + m^{1+\varepsilon}n^{1+\varepsilon})\right).$$

We will use this recurrence only when $m \leq n$, in which case it can be simplified to:

$$C_{0,0}^{(t)}(m,n) = O\left(k_t C_{0,0}\left(m, \frac{n}{k_t}\right) + k_t^2 C_{0,0}^{(t-1)}\left(m, \frac{n}{k_t}\right) + k_t m^{1+\varepsilon}n^{1+\varepsilon}\right), \quad \text{for } m \leq n. \quad (2)$$

If $m \geq n$, we interchange the roles of $\mathcal{F}$ and $\mathcal{G}$, and apply a fully symmetric analysis, to obtain the recurrence

$$C_{0,0}^{(t)}(m,n) = O\left(k_t C_{0,0}\left(\frac{m}{k_t},n\right) + k_t^2 C_{0,0}^{(t-1)}\left(\frac{m}{k_t},n\right) + k_t m^{1+\varepsilon}n^{1+\varepsilon}\right), \quad \text{for } m \geq n. \quad (3)$$

Using the same analysis as in [18] (where the specific choices of the values of k_t play an important role) we can now verify that the solution to (2) and (3) is

$$C_{0,0}(m,n) = O(m^{1+\varepsilon'}n^{1+\varepsilon'}),$$

for a slightly larger, but still arbitrarily small $\varepsilon' > \varepsilon > 0$. The constant of proportionality depends on $\varepsilon, \varepsilon'$, and on the maximum degree of the given functions and of their graph boundaries. We have thus established Theorem 2.1. □

3 Applications

In this section we present some applications of Theorem 2.1.

3.1 Computing lower envelopes in 3-space

Let $\mathcal{F}$ be a collection of n bivariate functions satisfying the condition $(\star)$. Our goal is to construct the lower envelope $E_\mathcal{F}$ of $\mathcal{F}$. This is equivalent to constructing the minimization diagram $M_\mathcal{F}$, as defined above, so that each face ϕ of $M_\mathcal{F}$ is labeled with the unique function of $\mathcal{F}$ (if exists) attaining $E_\mathcal{F}$ over ϕ. Several algorithms for this construction have recently been designed (see [4, 8, 18]), but they are either rather complicated or require the use of randomization. Here we present a simple deterministic algorithm based on the divide-and-conquer approach, which is similar to Atallah's algorithm [3] for computing the minimization diagram of univariate functions.

The algorithm partitions $\mathcal{F}$ into two subcollections, $\mathcal{F}_1, \mathcal{F}_2$, of roughly $n/2$ functions each, constructs recursively the minimization diagrams $M_{\mathcal{F}_1}$, $M_{\mathcal{F}_2}$, and then merges these diagrams to obtain the final minimization diagram $M_\mathcal{F}$.

The merging step is done as follows. We first compute the overlay M of $M_{\mathcal{F}_1}$ and $M_{\mathcal{F}_2}$. This can be done, e.g., by applying a standard sweep-line procedure, whose running time is $O((|M_{\mathcal{F}_1}| + |M_{\mathcal{F}_2}| + |M|)\log n)$; by Theorem 2.1, this is $O(n^{2+\varepsilon})$, for any $\varepsilon > 0$. We can implement the sweep so that it also constructs the *vertical decomposition* of M; this is a refinement of M, obtained by drawing a vertical segment upwards and downwards (in the y-direction) from each vertex of M and from each point on any edge of M that has y-vertical tangency, and by extending each segment until it hits another edge of M (or else goes all the way to $\pm\infty$). The number of resulting cells, usually referred to as 'pseudo-trapezoids', is proportional to the complexity of M, namely it is also $O(n^{2+\varepsilon})$.

Let c be a pseudo-trapezoid in this vertical decomposition. Note that, over c, the envelope $E_{\mathcal{F}_1}$ is attained by a single function $f_1 \in \mathcal{F}_1$ (or by no function at all), and $E_{\mathcal{F}_2}$ is attained by a single function $f_2 \in \mathcal{F}_2$ (or by no function at all). Hence, the envelope $E_{\mathcal{F}}$ is equal to $\min\{f_1, f_2\}$ if both functions are defined in c, $E_{\mathcal{F}}$ is equal to one of these functions if the other is not defined in c, or $E_{\mathcal{F}}$ is undefined if both functions do not exist in c. In any case, we can compute $E_{\mathcal{F}}$ over c in constant time.[1] We repeat this computation over all pseudo-trapezoids of M, in overall $O(n^{2+\varepsilon})$ time, and thus obtain the entire envelope $E_{\mathcal{F}}$. We still need to apply a final clean-up stage, in which the computed portions of $E_{\mathcal{F}}$ are properly glued together, removing, as appropriate, any redundant data concerning the behavior of $E_{\mathcal{F}}$ over edges of the pseudo-trapezoids of M. This stage also produces the final minimization diagram $M_{\mathcal{F}}$, with its faces labeled in the required manner. We omit the details of this step, and note that it also takes only $O(n^{2+\varepsilon})$ time. It follows that the cost of the entire divide-and-conquer process is also $O(n^{2+\varepsilon})$. In conclusion, we thus have:

Theorem 3.1 *The lower envelope of a collection of n bivariate functions satisfying the condition $(\star)$, can be computed, in an appropriate model of computation, by a deterministic divide-and-conquer algorithm, in time $O(n^{2+\varepsilon})$, for any $\varepsilon > 0$, where the constant of proportionality depends on ε and on the maximum algebraic degree of the given functions (and of their domain boundaries).*

3.2 Complexity of the region enclosed between two envelopes in 3-space

Let $\mathcal{T}$ and $\mathcal{B}$ be two given families of a total of n, possibly partially-defined, bivariate functions satisfying the condition $(\star)$. We denote by $\mathcal{L}_{\mathcal{T}}$ the lower envelope of the 'top' family $\mathcal{T}$, and by $\mathcal{U}_{\mathcal{B}}$ the upper envelope of the 'bottom' family $\mathcal{B}$. We consider the region $K = \{(x, y, z) \mid \mathcal{U}_{\mathcal{B}}(x, y) \leq z \leq \mathcal{L}_{\mathcal{T}}(x, y)\}$ of points lying between the two envelopes, and our goal is to derive an $O(n^{2+\varepsilon})$ bound on the combinatorial complexity of K. As mentioned in the introduction, Edelsbrunner et al. [10] proved $O(n^2 \alpha(n))$ upper bound on the size of K for the special case when $\mathcal{T}$ and $\mathcal{B}$ were partially-defined linear functions.

We establish the desired bound as follows. Let $M_{\mathcal{T}}$, $M_{\mathcal{B}}$ denote the minimization and maximization diagrams of the envelopes $\mathcal{L}_{\mathcal{T}}$, $\mathcal{U}_{\mathcal{B}}$, respectively. By Theorem 2.1, the combinatorial complexity of the overlay M of these two planar maps is $O(n^{2+\varepsilon})$, for any $\varepsilon > 0$. Construct the vertical decomposition of M, as defined above. As noted, the number of pseudo-trapezoids of this decomposition is proportional to the complexity of M, i.e., it is $O(n^{2+\varepsilon})$. Observe that, for each resulting pseudo-trapezoid τ, there is a single function $f \in \mathcal{T}$ and a single function $g \in \mathcal{B}$ such that $\mathcal{L}_{\mathcal{T}} \equiv f$ and $\mathcal{U}_{\mathcal{B}} \equiv g$ over τ (if the given functions are only partially defined then either f or g or both may not exist at all, in which case the corresponding envelope(s) are undefined over τ). This implies that the portion of K that projects onto τ has constant complexity—it is defined by the interaction between f, g, and the functions defining the (at most 4) edges of τ. Since the number of pseudo-trapezoids is $O(n^{2+\varepsilon})$, we immediately obtain:

Theorem 3.2 *The combinatorial complexity of the region enclosed between a lower envelope and an upper envelope of two respective collections of n bivariate functions satisfying the condition $(\star)$, is $O(n^{2+\varepsilon})$, for any $\varepsilon > 0$, where the constant of proportionality depends on ε and on the maximum algebraic degree of the given functions (and of their domain boundaries).*

It is also easy to construct the desired region K, in a manner that resembles the divide-and-conquer algorithm presented above. That is, we compute $\mathcal{L}_{\mathcal{T}}$ and $\mathcal{U}_{\mathcal{B}}$ separately, in time $O(n^{2+\varepsilon})$, using the algorithm of the preceding subsection. Next we compute the overlay of the minimization diagram $M_{\mathcal{T}}$ and of the maximization diagram $M_{\mathcal{B}}$, using the same sweep technique described above, and decompose the resulting map into pseudo-trapezoids. Finally we compute the portions of K over each pseudo-trapezoid separately, and 'glue' together the resulting pieces to obtain the whole K. It is easily verified that the overall complexity of the algorithm is $O(n^{2+\varepsilon})$, so we have:

Theorem 3.3 *The region enclosed between two envelopes in 3-space, as above, can be computed in (deterministic) time $O(n^{2+\varepsilon})$, for any $\varepsilon > 0$.*

Remarks: (1) The proof of Theorem 3.2 actually implies the stronger result that the complexity of the *vertical decomposition* of the region K enclosed between two envelopes, as above, is $O(n^{2+\varepsilon})$, for any $\varepsilon > 0$. That is, K can be decomposed into $O(n^{2+\varepsilon})$ subcells of constant description complexity (see [6] for more details). This result has already been used in a recent companion paper [1] to obtain bounds on the complexity of certain vertical decompositions in 4-space.
(2) It is instructive to compare Theorem 3.2 with the recent results of [15], that the complexity of a single cell in an arrangement of n low-degree algebraic

---

[1] We are implicitly assuming an appropriate model of computation, in which computing the pointwise minimum of two given functions, as well as various primitive operations involving edges of the minimization diagrams, can be performed in constant time. For example, we can use precise rational arithmetic to perform each of these operations in constant time, using standard techniques from computational real algebraic geometry; see, e.g., [17].

surface patches in $\mathbb{R}^3$ is also $O(n^{2+\varepsilon})$, for any $\varepsilon > 0$. On one hand, the result of [15] is stronger, because it deals with arbitrary cells, of potentially rather complex topologies. On the other hand, Theorem 3.2 is stronger, because the region K may consist of more than one cell (up to about quadratically many cells in the worst case). In fact, if the given surfaces are graphs of totally-defined continuous functions, then Theorem 3.2 is stronger, because a single cell in an arrangement of such surfaces must be contained in some region enclosed between a lower envelope and an upper envelope of appropriate subcollections of the given surfaces.

3.3 Complexity of the space of plane transversals

In this subsection we obtain new bounds on the combinatorial complexity of the space of plane transversals of a collection of simply-shaped convex sets in 3-space. Let $\mathcal{C} = \{C_1, \ldots, C_n\}$ be a collection of n compact convex sets in 3-space. A plane π is a *transversal* of $\mathcal{C}$ if it intersects every set in $\mathcal{C}$. The space of all plane transversals of $\mathcal{C}$ is denoted by $T(\mathcal{C})$.

It is more convenient to represent $T(\mathcal{C})$ in the dual space, where each nonvertical plane $z = \xi x + \eta y + \zeta$ is mapped to a point (ξ, η, ζ), and each point (u, v, w) is mapped to a plane $z = -ux - vy + w$. Note that a plane $z = \xi x + \eta y + \zeta$ intersects a compact convex set C if and only if $\phi_C(\xi, \eta) \leq \zeta \leq \psi_C(\xi, \eta)$, where $\phi_C(\xi, \eta)$, $\psi_C(\xi, \eta)$ are defined so that the plane $z = \xi x + \eta y + \phi_C(\xi, \eta)$ (resp. $z = \xi x + \eta y + \psi_C(\xi, \eta)$) is tangent to C from below (resp. from above). Thus, in the dual space, the set of all (nonvertical) plane transversals of $\mathcal{C}$ is the set

$$\left\{ (\xi, \eta, \zeta) \mid \max_{C \in \mathcal{C}} \phi_C(\xi, \eta) \leq \zeta \leq \min_{C \in \mathcal{C}} \psi_C(\xi, \eta) \right\}.$$

That is, $T(\mathcal{C})$ is, in the dual space, the region enclosed between a lower envelope and an upper envelope of two respective collections of bivariate functions.

We can therefore apply Theorem 3.2 to this case, but we first have to ensure that the functions ϕ_C and ψ_C satisfy the assumptions of that theorem. This will be the case if we assume that each $C \in \mathcal{C}$ has *constant description complexity*, that is, it is defined by a constant number of algebraic equalities and inequalities of constant maximum degree. In this case one can show that the functions ϕ_C and ψ_C do indeed satisfy the required conditions.[2] (The general position assumption can be enforced by imposing a similar general position assumption on the sets in $\mathcal{C}$.) We thus have:

Theorem 3.4 *The complexity of the space of plane transversals of a collection of n compact convex sets in 3-space, each of constant description complexity, is $O(n^{2+\varepsilon})$, for any $\varepsilon > 0$.*

Remarks: (1) Convexity is not essential here, since we can replace each set in $\mathcal{C}$ by its convex hull, without affecting the transversality of any plane.
(2) As already mentioned in the introduction, if the sets in $\mathcal{C}$ are not each of constant description complexity, the complexity of $T(\mathcal{C})$ can be arbitrarily large. However, if one assumes, in addition, that the sets are *separated*, in the sense that no three of these sets is crossed by a common line, then it is shown by Cappell et al. [5] that, for such a collection $\mathcal{C}$, the complexity of $T(\mathcal{C})$ is $O(n^2)$. This bound, in this restricted case, is slightly better than the bound derived above. The result of [5] applies in higher dimensions too: Under an appropriate assumption of separation of the sets in $\mathcal{C}$, the complexity of the space $T(\mathcal{C})$ of hyperplane transversals of $\mathcal{C}$ is $O(n^{d-1})$. See also [12, 13] for related results on transversals.

Acknowledgments

We wish to thank Boris Aronov, Leo Guibas, and Olivier Devillers for useful discussions concerning the problems studied in this paper. Part of the work on the paper has been carried out in the Mathematisches Forschungsinstitut in Oberwolfach, and we would like to thank the institute for its hospitality.

References

[1] P. Agarwal and M. Sharir, Efficient randomized algorithms for some geometric optimization problems, *this proceedings*.

[2] P. Agarwal, M. Sharir and P. Shor, Sharp upper and lower bounds for the length of general Davenport Schinzel sequences, *J. Combin. Theory, Ser. A.* 52 (1989), 228-274.

[3] M. Atallah, Some dynamic computational geometry problems, *Computers and Mathematics with Applications* 11 (1985), 1171–1181.

[4] J.D. Boissonnat and K. Dobrindt, On-line randomized construction of the upper envelope of triangles and surface patches in $\mathbb{R}^3$, Tech. Rept. 1878, INRIA, Sophia-Antipolis, France, 1993.

[5] S. Cappell, J.E. Goodman, J. Pach, R. Pollack, M. Sharir and R. Wenger, Common tangents and common transversals, *Advances in Math.* 106 (1994), 198-215.

---
[2] Actually, ϕ_C and ψ_C may only be piecewise-algebraic, so we may have to replace them by a constant number of appropriate partially defined algebraic functions, and apply Theorem 3.2 to the resulting new collections of functions.

[6] B. Chazelle, H. Edelsbrunner, L. Guibas and M. Sharir, A singly exponential stratification scheme for real semi–algebraic varieties and its applications, *Proc. 16th Int. Colloq. on Automata, Languages and Programming* (1989) pp. 179–193. (Also in *Theoretical Computer Science* 84 (1991), 77–105.)

[7] K. Clarkson and P. Shor, Applications of random sampling in computational geometry, II, *Discrete Comput. Geom.* 4 (1989), 387–421.

[8] M. de Berg, K. Dobrindt and O. Schwarzkopf, On lazy randomized incremental construction, *Proc. 26th Annual ACM Symp. Theory of Computing*, 1994, 105–114.

[9] H. Edelsbrunner, *Algorithms in Combinatorial Geometry*, Springer-Verlag, Berlin, 1987.

[10] H. Edelsbrunner, L. Guibas, and M. Sharir, The upper envelope of piecewise linear functions: Algorithms and applications, *Discrete Comput. Geom.* 4 (1989), 311–336.

[11] H. Edelsbrunner and M. Sharir, The maximum number of ways to stab n convex non-intersecting objects in the plane is $2n-2$, *Discrete Comput. Geom.* 5 (1990), 35–42.

[12] J. Goodman, R. Pollack, and R. Wenger, Geometric transversal theory, in: *New Trends in Discrete and Computational Geometry* (J. Pach, ed.), Springer-Verlag, New York–Berlin–Heidelberg, 1993, pp. 163–198.

[13] J. Goodman, R. Pollack, and R. Wenger, Bounding the number of geometric permutations induced by k-transversals, *Proc. 10th Annual Symp. Computational Geometry*, 1994, pp. 192–197.

[14] D. Halperin and M. Sharir, New bounds for lower envelopes in three dimensions, with applications to visibility in terrains, *Discrete Comput. Geom.* 12 (1994), 313–326.

[15] D. Halperin and M. Sharir, Almost tight upper bounds for the single cell and zone problems in three dimensions, to appear in *Discrete Comput. Geom.*

[16] S. Hart and M. Sharir, Nonlinearity of Davenport-Schinzel sequences and of generalized path compression schemes, *Combinatorica* 6 (1986), 151–177.

[17] J. Heintz, T. Recio and M.-F. Roy, Algorithms in real algebraic geometry and applications to computational geometry, in *Discrete and Computational Geometry: Papers from the DIMACS Special Year* (J.E. Goodman, R. Pollack and W. Steiger, Eds.), AMS Press, Providence, RI 1991, pp. 137–163.

[18] M. Sharir, Almost tight upper bounds for lower envelopes in higher dimensions, *Discrete Comput. Geom.* 12 (1994), 327–345.

[19] M. Sharir and P. Agarwal, *Davenport-Schinzel Sequences and Their Geometric Applications*, Cambridge University Press, Cambridge-New York-Melbourne, 1995.

Rounding Arrangements Dynamically

Leonidas J. Guibas†‡ and David H. Marimont†

†Xerox Palo Alto Research Center, 3333 Coyote Hill Rd., Palo Alto, Calif. 94304
‡Department of Computer Science, Stanford University, Stanford, Calif. 94305

Abstract

We describe a robust, dynamic algorithm to compute the arrangement of a set of line segments in the plane, and its implementation. The algorithm is robust because, following Greene [6] and Hobby [8], it rounds the endpoints and intersections of all line segments to representable points, but in a way that is globally topologically consistent. The algorithm is dynamic because, following Mulmuley [13], it uses a randomized hierarchy of vertical cell decompositions to make locating points, and inserting and deleting line segments, efficient. Our algorithm is novel because it marries the robustness of the Greene and Hobby algorithms with Mulmuley's dynamic algorithm in a way that preserves the desirable properties of each.

1 Introduction

The goal of this paper is to describe a new, robust, and dynamic algorithm for constructing arrangements of line segments in the plane, and its implementation. The problem of constructing line-segment arrangements has been well studied in Computational Geometry and is the focus of some of the most famous algorithms in the field. Let n denote the number of segments we are given and A the complexity of their arrangement (say the number of its vertices). An early solution for this problem was provided by the Bentley-Ottmann sweep which ran in time $O((n + A) \log n)$ [1]; a long sequence of improvements followed, culminating in the optimal but complex $O(n \log n + A)$ algorithm of Chazelle and Edelsbrunner [2]. The introduction of randomization into Computational Geometry revitalized the problem, and several new randomized incremental algorithms for the problem were invented, all with optimal randomized complexity $O(n \log n + A)$ and notable for their simplicity [3, 12, 13]. Among these authors, Mulmuley [13] was especially successful at providing an algorithm that was dynamic, allowing efficient segment insertion and deletion.

The key data structure in Mulmuley's algorithm is the vertical cell decomposition (VCD) of a set of line segments. The VCD of a set of line segments is a refinement of their arrangement into cells that are all trapezoids (possibly degenerate) with two vertical sides. As is well established by now, there is a large class of applications in Computational Geometry for which this further refinement of the arrangement into cells each with only a bounded number of sides (four in our case) is very useful. The Mulmuley algorithm is randomized and dynamic: that is, it is possible to insert and delete line segments from the VCD. The output of the algorithm is a hierarchy of vertical cell decompositions (VCDs) of subsets of the line segments. The lowest, most detailed level contains all the line segments. Each higher level contains a randomly chosen subset of the line segments present one level down. The hierarchy makes it possible to locate points, insert and delete line segments, and otherwise navigate efficiently around the arrangement.

This, as well as all previously mentioned algorithms, were developed assuming an infinite-precision model of computer arithmetic. In any practical implementation of a line segment arrangement computation, however, the implementors have to consider the effects of finite precision arithmetic on the above techniques. This issue of robust implementation of geometric algorithms has been addressed in several papers [9, 11, 15, 4, 14, 5], but with mixed success. We do not attempt to survey this extensive literature here. In our work, we will address the robustness problem by perturbing all vertices of our arrangement to lie on an integer grid. However, in order to ensure that the perturbed arrangement has a topology consistent with the original, we will need to perturb to the grid additional features of our arrangement as well. In the end, all vertices, edges, and faces of our perturbed arrangement will have exact representations with finite arithmetic. We call this operation *rounding* the arrangement. Exactly how to accomplish such a perturbation of the arrangement to the grid was first studied by Greene and Yao [7]. As we explain below, the Greene-Yao rounding has a number of undesirable properties, which were overcome in another rounding scheme proposed by Greene, and independently by Hobby, in as yet unpublished manuscripts [6, 8].

The key contribution of our paper is to show how to combine the ideas of Mulmuley's dynamic segment arrangement algorithm, while maintaining (and producing) only the rounding of Greene and Hobby of the arrangement of the current segments. The algorithms proposed by Greene and

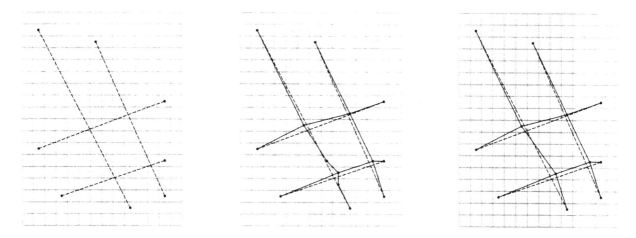

Figure 1: Left, a small line segment arrangement; center, its Greene-Yao perturbation; right, its snap-rounded form.

Hobby are robust but not dynamic; they are based on finite-precision arithmetic but operate in a batch mode that assumes all the segments are given at once. The Mulmuley algorithm is dynamic but not robust; it provides for inserting and deleting line segments but is based on infinite-precision arithmetic. The technical challenge we have to overcome is how, using only the VCD of the rounded arrangement of the present segments (and its hierarchy of sampled counterparts), to simulate the effect of doing an insertion or deletion in the ideal Mulmuley structure and then rounding the result. We guarantee that the rounded arrangement we compute dynamically is exactly the same as what would have been produced by the batch algorithms of Greene and Hobby. We have implemented and extensively tested this new algorithm.

In the paper we begin by describing the rounding of Greene and Hobby and a new and elementary way to derive its desirable topological properties (sections 2 and 3). We then provide a succinct summary of our data structures and discuss the key algorithmic issues in adding a new segment to the arrangement (sections 4 and 5). Section 6 discusses deletions and the effect of the hierarchy on the algorithm. In section 7 we provide a brief analysis of our method. In section 8 we talk about some of the experiences from our implementation. We end by presenting some conclusions in section 9.

2 Snap rounding

In this section we briefly discuss a way to round an arrangement of line segments that is especially economical in terms of the number of "kinks" introduced in the segments. This method, as already mentioned, was introduced by Greene [6] and Hobby [8] — we shall refer to it from here on as *snap rounding* for reasons that become apparent below.

The setting is as follows: the Euclidean plane is tiled into unit squares in the obvious way; we refer to these tiles as *pixels*. We coordinatize the plane so that pixel centers have integral coordinates and refer to these pixel centers as *integral points*. When we round an arrangement of line segments, we require that all its vertices (endpoints of segments, as well as intersections of pairs of segments) be perturbed to integral points — in other words, the only points we allow as vertices in our rounded representation are the integral points.

Any rounding scheme must have at least two goals: (1) to keep the perturbed segments near the originals, and (2) to preserve as much as possible the topology of the original arrangement. Requirement (1) suggests that each vertex of the ideal arrangement be perturbed to its nearest integral point. However, it is well known that just doing this can cause topological inconsistencies between the ideal and the rounded arrangements. In order to avoid this problem, about eight years ago, Greene and Yao [7] suggested that we treat each representable point as an "obstacle" and do not allow our segments to go over these obstacles while vertices move to their nearest integral point. These obstacles can create more "kinks" in the perturbed segments. Greene and Yao showed that, with this additional fragmentation, no topological inconsistencies arise. They also gave an algorithm for computing their perturbation efficiently.

The difficulty with the Greene-Yao method is that the requirement of not going over the obstacles adds a large number of additional breaks to each rounded segment. See, for example, figure 1(left) showing an ideal line arrangement (dashed lines), and figure 1(center) its Greene-Yao perturbation (solid lines). In that figure the grid lines (gray lines) correspond to pixel boundaries.

To get around the excess fragmentation, snap rounding proceeds as described below. To fix the terminology, we call the original unrounded segments *ursegments*. After the perturbation, each ursegment becomes a polygonal line that we call a *polysegment*. A polysegment consists of smaller line segments which themselves are called *fragments*. If s denotes an ursegment, we denote the corresponding polysegment by the corresponding greek letter σ. For brevity in this abstract, we also assume that every vertex of our arrangement has a unique nearest integral point — degeneracies do not introduce substantial difficulties.

We declare all pixels containing an ursegment endpoint, or the intersection point of two ursegments, to be *hot*. In other words, any pixel containing a vertex of the ideal arrangement becomes hot — note that a pixel may become hot for multiple reasons. Snap rounding is then this: if an ursegment terminates in a hot pixel, it is perturbed to terminate at the that pixel's center; and if an ursegment passes through a hot pixel, then it is perturbed to pass through that pixel's center. See figure 2 for an illustration. Notice this key aspect of snap rounding: a kink is added to an ursegment only where a vertex of the arrangement lies on the segment, or where the ursegment passes "very near" an integral point which will become a vertex of the rounded arrangement. Figure 1(right) shows snap rounding for the arrangement of figure 1(left) — it is evident that fewer kinks have been added. We call this snap rounding because all ursegments passing through a hot pixel are snapped to pass through that pixel's center.

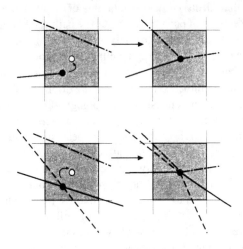

Figure 2: The rules of snap rounding.

3 A topological analysis of snap rounding

It is first of all trivial to prove that after snap rounding, an ursegment and the corresponding polysegment are quite near each other.

Lemma 1 *After snap rounding, the polysegment σ corresponding to ursegment s is contained within the Minkowski sum of s with a pixel (unit square) centered at the origin.*

Understanding the relation between the topology of the original and rounded arrangements is much more interesting. In his original manuscript, Greene defined and proved a number of "topological consistency" properties of his perturbation. His argument was phrased in a context allowing more general pixel shapes than we have here, but it was involved and required a global analysis using graph-theoretic arguments. In this paper we give a very different topological analysis of snap rounding. We show topological consistency between the original and rounded arrangements by giving an *explicit continuous deformation* that takes an arrangement of segments into snap-rounded form. During that deformation features of the arrangement may collapse (as they must in any rounding scheme), but they never invert — in the sense that no vertex of the arrangement ever crosses through one of the segments. We call the latter property the *non-penetration condition*. Our deformation is easy to visualize and the proof of non-penetration is entirely local and straightforward. In addition, because of the clear understanding of the topological transformation going on that our deformation provides, we are able to easily prove certain additional lemmata that are useful in the implementation of our incremental variant of snap rounding.

We now proceed to define a deformation $\mathcal{D}$ that starts from the original arrangement and continuously transforms it to its snap-rounded form. We initially break up each ursegment into a number of subsegments, by introducing a breakpoint, or *node*, whenever the ursegment crosses the boundary of a hot pixel. Note that if an ursegment crosses a pixel boundary separating two hot pixels, then two nodes will be placed there, with a zero-length subsegment between them. See figure 3 for an illustration.

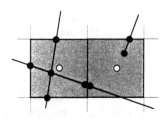

Figure 3: Segment fragmentation by nodes at hot pixel boundaries.

Our deformation $\mathcal{D}$ proceeds in two stages. In the first stage every hot pixel contracts simultaneously and at the same rate in the x direction, towards the vertical axis through its center. If at time $t = 0$ we have the original arrangement, then each hot pixel is linearly scaled down in the x-direction until, at time say $t = 1$, all hot pixels have collapsed to their vertical axis. All nodes on the hot pixel boundaries follow the pixel motion. Each ursegment thus is continuously deformed to a polysegment, the polysegment joining the current locations of its nodes for each time t. At time $t = 1$ each hot pixel has become a hot "stick," with nodes marked on it. In the second stage of the deformation, say from $t = 1$ to $t = 2$, the hot sticks vertically contract simultaneously and at the same rate towards their center. Again, the polysegments are defined by just tracking the corresponding nodes. See Figure 4 for an illustration of this process.

It is not hard to show that at time $t = 2$ our polysegments are exactly those defined by snap rounding. Note that, during the deformation $\mathcal{D}$, our nodes partition each polysegment into fragments of two kinds: *internal* — those inside a hot pixel, and *external* — those connecting nodes on the boundary of

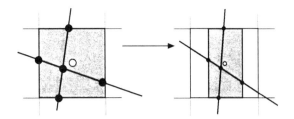

Figure 4: Halfway through the first stage of the explicit deformation.

two different hot pixels. At the end of the deformation $\mathcal{D}$, internal fragments collapse to pixel centers and only external fragments remain. No node ever crosses over a fragment during this deformation. In particular, an endpoint of a polysegment can never move over another polysegment, nor does the orientation of an elementary triangle formed by three polysegments ever invert — see figure 5.

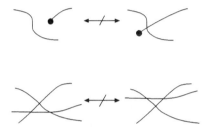

Figure 5: Some of the topological consistency conditions.

Theorem 1 *During the deformation $\mathcal{D}$, no node ever crosses over a fragment.*

Proof. It is first of all clear that there can be no trouble among all the internal fragments of a particular hot pixel: they all occur in a region of the plane that undergoes a uniform x- or y-scaling transformation. The crucial part of the proof is the invariant that no external fragment ever touches or enters a contracting hot pixel (other than those at its endpoints). To see this, we can argue as follows. Consider, for example, the x-part of the deformation (first stage). If an external fragment is to enter a hot pixel, then it must overtake during the deformation one of the corners of that pixel's boundary. Furthermore, at the point of overtaking, the external fragment and the corner must be moving in the same direction in x (i.e., both left or both right), as the hot pixel is contracting. But by the way we have defined the x-deformation, the x-velocity of the pixel corners is always maximal in magnitude among all points on a pixel boundary. And in addition, a point along a fragment is moving in x by a velocity which is a convex combination of the velocities of its endpoints, which are nodes following other hot pixel boundaries. An entirely analogous argument holds for the y-part of the deformation. Thus a fragment can never overtake a deforming hot pixel. □

As a consequence of the above, we can deduce many topological consistency properties between the original line segment arrangement and its snap rounding.

Corollary 1 *After snap rounding:*

- *No fragments intersect except at their endpoints.*

- *If ursegment r intersects ursegment s and then ursegment t, then polysegment ρ cannot intersect polysegment τ before polysegment σ.*

- *If a vertical line ℓ through pixel centers intersects ursegment s and then ursegment t, then ℓ cannot intersect polysegment τ before polysegment σ.*

4 Data Structures

Here we briefly summarize the data structures we use. The vertical cell decomposition's top-level data structures are vertices and two types of line segments, fragments and vertical attachments. Figure 6(left) illustrates these structures; a vertex is represented by a filled circle, a fragment with a solid line, and a vertical attachment by a dotted line with an arrow that points to the fragment at one end of the attachment. Note that the x-shaped configuration of fragments at the top of the figure is not two fragments intersecting at a vertex, which is impossible by corollary 1, but four fragments that share an endpoint. For convenience, we bound the VCD with a rectangle of four ursegments with integral endpoints, as is standard.

An ursegment is stored with the fragments that make up its polysegment. Snap rounding can cause two or more ursegments to share the same fragment, so each fragment maintains a list of ursegments that it represents. Extensive experimentation has shown that these lists rarely have more than five ursegments. An ursegment also has pointers to the vertices at its endpoints.

Snap rounding can produce "degenerate" configurations that the VCD must be able to represent, such as the vertical fragment in the upper right of Figure 6(left). Such degenerate configurations can make the vertical boundaries of a trapezoid arbitrarily complicated, as shown in Figure 6(right).

The VCD's data structures make it easy to move vertically. Moving vertically from a vertex, vertical fragment, or vertical attachment is straightforward because each can have only one structure above and one below. Nonvertical fragments, however, can have a sequence of structures above and below them, which we represent with two doubly linked lists called the *ceiling* (for the structures above) and *floor* (for those below) lists of the fragment. In addition to facilitating vertical motion from a nonvertical fragment, they make it easy to move along the fragment.

Figure 7 illustrates the ceiling and floor lists of some nonvertical fragments in a simple VCD. On the left is a VCD with four nonvertical fragments. We shall refer to a fragment by

193

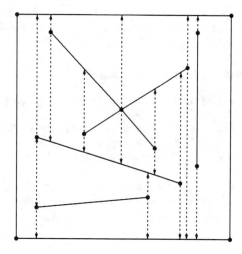 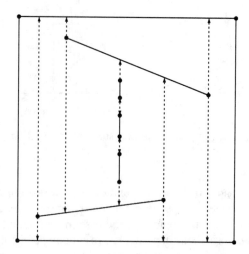

Figure 6: Left, top-level structures in the VCD; right, a degenerate vertical boundary.

the number that appears to the left of its left endpoint. Figure 7(right) shows the ceiling and floor lists of the fragments schematically. Each fragment is represented by a horizontal line segment with filled circles at its endpoints; the number of the fragment appears to the left of the line. The numbers and arrowheads above a fragment's line correspond to the fragments and vertical attachments above the fragment as we move from its left endpoint to its right. Pointers to these fragments and vertical attachments are stored on the fragment's ceiling list. The numbers and arrowheads below each fragment's correspond to the fragments and vertical attachments below the fragment, pointers to which are stored on the fragment's floor list.

The vertical attachments on a floor or ceiling list define a partition of the interval in x occupied by the fragment. The open intervals in each list, where another fragment is above or below the fragment, we represent with structures called *xspans*. Each trapezoid in the VCD is bounded below by a ceiling xspan (belonging to the fragment at the bottom of the trapezoid) and above by a floor xspan (belonging to the fragment at the top). For example, the trapezoid at the center of Figure 7(left) is bounded below by the second xspan on 2's ceiling list, and above by the first xspan on 3's floor list. The floor and ceiling xspans that bound a trapezoid point to each other, which makes it possible to move vertically across a trapezoid. To cross from the floor of a nonvertical fragment to its ceiling (or vice versa), we maintain with each fragment a third doubly linked list consisting of all vertical attachments on the floor and ceiling lists.

A simple algorithm for locating a point illustrates how these structures are used. We first find the xspan containing the x coordinate of the point by linearly searching the ceiling list of the horizontal fragment that defines the bottom of the VCD's bounding rectangle. We search upwards in y, using xspans to cross trapezoids, for example, until we find the structure that contains the point. We discuss an algorithm for locating points that exploits the hierarchy of VCDs in section 6.

5 Inserting a new ursegment

Inserting a new ursegment s requires three different searches. First, we must determine the new hot pixels created by s by locating its endpoints and detecting its intersections with other ursegments. Second, we must detect the existing hot pixels through which s passes. Third, we must detect the existing ursegments that pass through the new hot pixels (and perturb the ursegments accordingly).

The polysegment σ of s is defined by the vertices of the new and existing hot pixels found in the first two of these searches. Once the vertices of σ are known, we insert those of its fragments not already in the VCD (with the new ursegment the only member of its list of ursegments); by corollary 1 these fragments intersect no others, which makes the insertion simple. For a fragment of σ already in the VCD, we need only add the new ursegment to its list of ursegments.

Figure 8 illustrates the problem of detecting s's intersections with existing ursegments. The ursegments are the dashed lines, the fragments are the solid ones, and the vertices are filled circles. The grid of pixel boundaries are the gray lines. The new ursegment s is the near-horizontal dashed line that begins at the left of the figure. Three existing ursegments and their polysegments are shown.

Each ursegment and its polysegment define a closed but not necessarily simple polygon. We call these polygons *slivers*. The existing ursegments at the left and the center of the figure depict the most common situation. In each case, s passes entirely through a sliver, so that s intersects both the existing ursegment and one of its fragments. (In general, s may intersect any number of the fragments.) Detecting intersections with such ursegments is easy, since we need only find intersections with fragments, and test the ursegments to

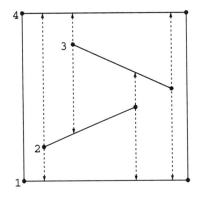
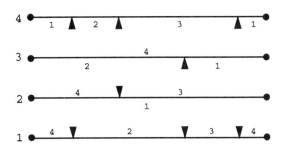

Figure 7: Left, a simple VCD; right, the fragments' ceilings and floors.

which the fragments belong.

The rightmost existing ursegment in the figure depicts a situation that can only arise at an endpoint of s. Here s intersects the existing ursegment, but because an endpoint of s lies inside the sliver, s does not intersect any of the existing ursegment's fragments. Detecting intersections with existing ursegments in this situation is not quite as simple. However, it is not hard to show that a vertex of the existing ursegment's polysegment is guaranteed to be nearby, in the sense that the vertex can be no more than three or four pointers away from the endpoint's location.

To find the ursegments that intersect s, we use the following algorithm, which we call *VCD traversal*. First, locate the left endpoint of s. Test the ursegments that belong to fragments at nearby vertices to see whether they intersect with s. Then, follow s through the VCD. When s intersects a fragment, test the fragment's ursegments to see whether they intersect with s. When the right endpoint of the new ursegment is reached, perform the same tests as at the left. The new hot pixels are those containing the endpoints of s and its intersections with existing ursegments.

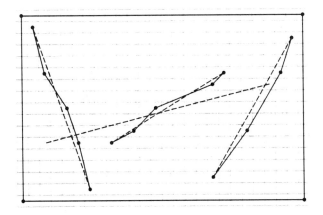

Figure 8: Detecting a new ursegment's intersections with existing ursegments.

The second of the three searches is for existing hot pixels through which s passes. Let the *sausage region* of s be the Minkowski sum of s with a pixel centered at the origin. If s passes through a hot pixel, the vertex at the pixel's center must lie in the sausage region of s. To find the hot pixels through which s passes, we visit each cell that intersects the sausage region and test whether the vertices that lie on the cell boundary are inside the sausage region. We call this search *cell sweep*.

The algorithm for cell sweep is as follows. Find the list of cells that intersect s; call this the sweep list. The idea is to sweep upwards from these cells to the top of the sausage region, then downwards to the bottom. To sweep upwards, mark each cell on the sweep list and test whether each hot pixel center on the left boundary of the cell is inside the sausage region. If so, put it on a list of hot pixels. Remove the first cell from the sweep list; call it cell i. Replace it on the sweep list with a list of the cells above cell i that have not yet been marked, and mark these cells. For cell j to be above cell i, part of cell j's bottom boundary must coincide with part of cell i's top boundary. Exclude from the sweep list cells that are completely outside the sausage region. When a cell is placed on the sweep list, test whether the hot pixel centers that lie on its left boundary are inside in the sausage region and if they are, add them to the hot-pixels list.

Keep removing the first cell on the sweep list and replacing it with the ones above until the sweep list is empty. Perform an analogous sweep downwards. Unmark all marked cells and return the list of hot pixels. These are the hot pixels through which s passes. It is clear that the cost of the cell sweep is proportional to the number of cells that intersect the sausage region of s.

The third search is for existing ursegments that pass through a new hot pixel. Figure 9 shows an example of this search at one hot pixel. The top left panel shows the VCD at the beginning of the search; the hot pixel is the shaded square near the center. This VCD consists of two ursegments (aside from those making up the bounding rectangle), each with a single-fragment polysegment, so each ursegment coincides with its fragment. Both ursegments pass through the new hot pixel. The first step is to insert a vertex at the hot pixel. This is shown in the top right panel. The new vertex does not lie on

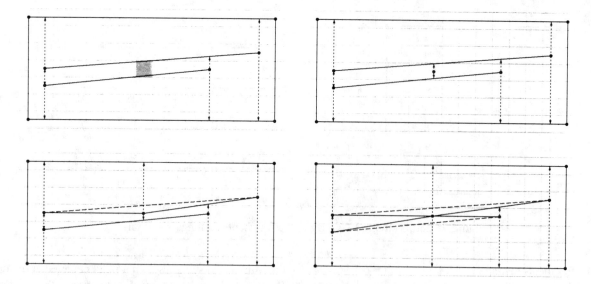

Figure 9: The search for existing ursegments that pass through a new hot pixel.

a fragment, and it has vertical attachments extending up and down to the vertically adjacent fragments. (Such isolated vertices are another degenerate configuration supported by our data structures.)

To find these ursegments, we perform a *vertical range search*. First, we search up the vertical attachment extending above the new vertex. Whenever a fragment is encountered, its ursegment is tested to see whether it passes through the hot pixel. If it does, the fragment is perturbed to pass through the new vertex. This perturbation is illustrated in the bottom left panel of Figure 9. Here an ursegment on the ursegments list of the fragment above the new vertex does pass through the hot pixel; the fragment has been split in two, so that the ursegment's polysegment now passes through the new vertex. Note that the polysegment no longer coincides with the ursegment; we now see the ursegment as a dashed line. The vertical attachment above the new vertex now points to the fragment above the one that was just perturbed. In this case, this fragment is the upper boundary of the VCD's bounding rectangle. Its ursegment does not pass through the new hot pixel, and therefore the search upwards can be terminated. This algorithm works because, by corollary 1, snap rounding ensures that the order of fragments along any vertical line through pixel centers contains no reversals of the order of the ursegments to which they belong. An analogous search is performed downwards. The bottom left panel of Figure 9 shows the VCD after the search downwards has terminated.

6 Deletions and the hierarchy

To delete an ursegment, visit the fragments of s's polysegment σ and remove s from each fragment's list of ursegments. When a list becomes empty, delete the fragment. Next, delete each vertex of σ in a hot pixel that became hot only because s ended or intersected another ursegment in the pixel; this can be decided efficiently by using the order in which the ursegments enter or leave that pixel. These operations leave the VCD in the state in which it would have been had s never been inserted.

As mentioned earlier, our algorithm produces a hierarchy of VCDs, where the lowest level (level zero) contains all the ursegments, and each successively higher level contains a subset of the ursegments present one level down. Adjacent levels are linked through their hot pixels and ursegments; each hot pixel or ursegment at level i has a *descent* pointer down to the corresponding hot pixel or ursegment at level $i-1$.

To locate a point at the bottom level of the hierarchy of VCDs, first locate it in the top (least detailed) VCD in the hierarchy, say at level i, using the point location algorithm described in section 4. Next, find a nearby hot pixel and use its descent pointer to locate the corresponding hot pixel in the VCD one level down, at level $i-1$. (This nearby hot pixel is never more than a pointer or two away, because every cell has at least two hot pixels on its boundary.) To locate the point from a vertex at level $i-1$, trace a straight line through the VCD from the hot pixel to the point. Repeat this process until the bottom level of the hierarchy has been reached.

A new ursegment is inserted into the l bottommost levels of the hierarchy, where l is computed independently for each ursegment by flipping a coin until tails is obtained. To insert the ursegment, locate an endpoint at all levels of the hierarchy as above, and then insert the ursegment at the bottommost l levels independently using the algorithm described in section 5. At each level, the descent pointers of the ursegment and of any new vertices created by the insertion are linked to the corresponding ursegments and vertices one level down.

An ursegment is deleted from the hierarchy by deleting it independently from each level.

These algorithms are quite similar to those proposed by

Mulmuley [13]. In the Mulmuley algorithm, of course, there are no hot pixels, so there is no need to decide whether a vertex is still in a hot pixel after deleting an ursegment, as we must.

7 Analysis of the algorithm

Let n be the number of ursegments we have and let A denote the combinatorial complexity (say, the number of vertices) of their ideal arrangement $\mathcal{A}$. It is well known that Mulmuley's randomized incremental algorithm builds $\mathcal{A}$ in expected time $O(n \log n + A)$ [13]. Let $\mathcal{R}$ denote the rounded arrangement of these same segments and let R denote the combinatorial complexity of $\mathcal{R}$; it is clear that $R \leq A$. The reduction in complexity from $\mathcal{A}$ to $\mathcal{R}$ is due to the fact that in the latter several features (e.g., vertices) collapse to the same feature. It is useful to think of R as follows. Let $\mathcal{H}$ denote the set of hot pixels; for a hot pixel h, let $|h|$ denote its combinatorial size, i.e. the number of ursegments intersecting it. Then it is clear that

$$R = O(\sum_{h \in \mathcal{H}} |h|).$$

(R can actually be less, as when many tiny segments appear within the same hot pixel). In our representation of the VCD, we store with each fragment a list of all the ursegments that gave rise to it. The additional storage required for these lists is easily bounded by $O(\sum_{h \in \mathcal{H}} |h|)$ as well. We note that the latter sum can be either larger or smaller than A (the sum of hot pixel sizes can exceed the true arrangement complexity if many ursegments cross through many hot pixels, but without giving rise to features of $\mathcal{A}$ there). For our analysis we also need the following quantity: define a pixel as being *warm* if it contains the endpoint of a vertical attachment in the ideal VCD of $\mathcal{A}$. We let $\mathcal{W}$ denote the set of warm pixels and $\mathcal{W}'$ the set of pixels that are either warm or are neighbors (kingwise) of warm pixels. Let C denote the sum $\sum_{w \in \mathcal{W}'} |w|$.

The most delicate issue in the analysis of our incremental snap rounding algorithm is the effect of pixel size on performance. If the pixel size is extremely small compared to separation of the vertices in $\mathcal{A}$, then we expect our algorithm to behave the same as the ideal randomized algorithm working over the reals. As the pixel size gets larger, two opposite effects come into play. On the one hand the VCD becomes coarser and its size can drop significantly — and so can the cost of traversing it. On the other hand, we lose any knowledge of the structure of $\mathcal{A}$ *within* the now large hot pixels, and as a result discovering intersections between existing ursegments and a newly added one can become expensive. In particular, when a few pixels cover all of $\mathcal{A}$, our algorithm naturally reduces to the naive quadratic algorithm that checks all pairwise ursegment intersections.

Consider an existing polysegment σ and its corresponding ursegment s. The vertices of σ define a sequence of hot pixels which both σ and s pass through in the same sequence. The following obvious lemma is critical for our analysis.

Lemma 2 *During the incremental construction, if a new ursegment t intersects the fragment of an existing polysegment σ between hot pixels h_1 and h_2, then one of the following three situations holds:*

- *t also intersects s, the ursegment corresponding to σ, between (and outside) h_1 and h_2, or*
- *t terminates in the sliver between s and σ, or*
- *t enters at least one of the hot pixels h_1 and h_2.*

This lemma allows us to estimate the cost of the VCD search for a new ursegment t. The cost of this search, over and above that of the ideal Mulmuley algorithm, is in checking for the possible but non-existing or already discovered intersection between t and other ursegments associated with fragments crossed by t. It is clear that t may cross several of the fragments defining σ, while it can cross s only once, or not at all. Case (a) of the above lemma is the favorable one — where a fragment crossing has a corresponding ursegment crossing; this is paid for by A. The cost of case (b) is $O(1)$ per endpoint, given the local search around the endpoints that we perform, as explained in section 5. Finally note that each time case (c) holds, ursegments s and t must pass though the same hot pixel (h_1 or h_2). This observations leads to the following final analysis for our algorithm. We denote by C the total number of ursegments that pass within half-a-pixel above or below the non-hot-pixel endpoints of all vertical attachments.

Theorem 2 *The complexity of our incremental rounded arrangement construction procedure is*

$$O(n \log n + A + \sum_{h \in \mathcal{H}} |h|^2) + C.$$

Proof. The key is to understand the insertion cost of a new ursegment t. The above paragraph takes care of the VCD search for the intersections between t and other existing ursegments t; these intersections, together with the endpoints of t define the new hot pixels created by t. The total cost of these searches, summed over all insertions, is captured by the bound in the theorem; over the history of the algorithm, a hot pixel h can lead to a cost of type (c) in the above lemma at most as many times as the number of pairs of ursesegments crossing it.

Any existing ursegments crossing the new hot pixels need also to be snap-rounded to these new pixel centers. The vertical range search method for this given in section 5 takes time proportional to the number of such ursegments (plus one, to be exact), as the search (down or up) stops as soon as we encounter a non-crossing ursegment. The total cost for these searches is then $O(\sum_{h \in \mathcal{H}} |h|)$.

It remains to deal with the cell sweeps. During the insertion of t, the cell sweep finds any existing hot pixels crossed by the new ursegment t. Without loss of generality, we confine our attention to the upwards sweep; the analysis for the downwards one will be symmetric. We remarked in section 5

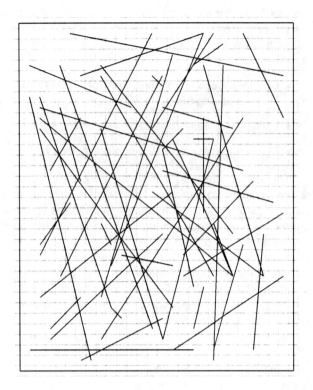 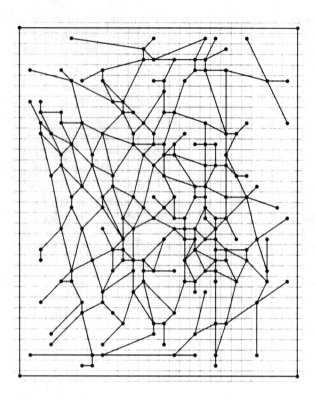

Figure 10: Left, an arrangement of 50 ursegments; right, the snap-rounded form of the arrangement.

that the cost of this cell sweep is proportional to the number of trapezoids contained in the sausage of t. Note that each trapezoid of the VCD has at least two vertices which are hot pixels. If a trapezoid is fully contained in the sausage of t, then at least one of its vertices is a hot pixel center, and t crosses that hot pixel. This leaves unaccounted for trapezoids which partially intersect the sausage and all of whose vertices in the sausage are endpoints of vertical attachments. It is easy to check that in that case each such vertex must be within a 1-neighborhood of some warm pixel, and that the ursegment t must intersect this neighborhood. Thus such vertices and their trapezoids are paid for by the term C. □

Although we have not observed this in practice, the quadratic term above (the sum of the squares of the hot pixel complexities) can be quite large. However, it is easy to give a modification to our insertion algorithm (not implemented), that would reduce this complexity. The idea is to exploit the natural ordering of the ursegments entering and exiting each hot pixel h. In the rounded arrangement, we can order the lists of ursegments we keep with each fragment out of h according to this circular ordering out of h (recall that fragments do not cross outside hot pixels, so all ursegments giving rise to a fragment between hot pixels h_1 and h_2 must be ordered consistently around the two pixels). With these circular orderings around the hot pixels we can process efficiently new ursegments crossing hot pixels. When, as in case (c) of Lemma 2, we determine that the new ursegment

t might intersect some existing ursegments inside hot pixel h, we can find all k such ursegments (k may be 0) in time $O(\log |h| + k)$. This can be done by a standard two-level 1-d range search structure, as in [10, 13]. This more elaborate approach reduces the overall insertion complexity to

$$O(n \log n + A + \sum_{h \in \mathcal{H}} |h| \log |h|) + C.$$

We note that we have been unable to obtain a more refined analysis of the C term in the above cost.

We omit a detailed analysis of the deletion procedure. We only remark that we can implement the test for whether a vertex of a polysegment σ becomes deletable (after the deletion of the corresponding ursegment s) in time proportional to the logarithm of the number of ursegments crossing the hot pixel of that vertex.

8 Implementation and Verification

We have implemented the algorithms and data structures described above in Allegro Common Lisp on a Silicon Graphics Indigo Elan. We began by constructing a naive and inefficient implementation of snap rounding. Each component of the more sophisticated implementation, which takes advantage of the algorithms described earlier, was developed and tested independently by replacing the corresponding component in the naive implementation. To verify the new "partially sophisticated" implementation, we generated large numbers of test cases and, using software to test whether two VCDs were identical, compared the output of the new implementation to that of the naive one.

For example, to implement the search for existing ursegments that pass through the new hot pixels, we used exhaustive versions of the search for new hot pixels and of the search for the existing hot pixels through which the new ursegment passes. We implemented the more efficient local search only for the existing ursegments that pass through the new hot pixels discovered by the exhaustive search.

This process quickly revealed errors in earlier versions of our algorithms. Usually the errors arose because we did not anticipate certain configurations of ursegments that have a very low probability of occurring. Only after many random trials were the errors exposed. To understand how these configurations caused our algorithm to fail, we wrote visualization tools tailored to our data structures. Using these tools, we could explore the problematic configurations in detail, and in most cases it rapidly became obvious why the algorithm failed – although not always how to fix it!

A number of factors were key to our implementation and verification of the software. The existence of a naive implementation simple enough for us to be confident that it was producing correct results provided a benchmark against which more sophisticated implementations could be tested. Our strategy of constructing the more sophisticated implementation by incrementally replacing components of the naive implementation made it much easier to identify the errors in our algorithms. The software that automated comparing one VCD to another made it possible for us to test the incremental algorithms with very large numbers of cases by comparing the resulting VCDs to those constructed by the naive algorithm. Finally, our tools to visualize and inspect local structures in the VCD helped us to better understand the algorithms, especially when they were producing incorrect results.

9 Conclusions

Trapezoidal decompositions, flat or hierarchical, are now part of the standard machinery of cuttings in Computational Geometry. Yet when we try to perturb and round these structures so as to make all their features exactly representable in finite arithmetic, many subtle issues arise. This is especially so in a dynamic context, as many of the invariants that guarantee the correctness and good performance of the infinite precision algorithms do not hold when we try to emulate these algorithms using the finite-precision structures we can actually store.

The key contribution of our paper has been to show how to marry the planar VCD rounding with dynamic operations on the arrangement of line segments it represents—while maintaining as much as possible the performance of the ideal algorithms. In our implementation we saw a true interplay between theory and practice. We have been able to employ simpler algorithms for operating on the rounded VCDs because we could go back and theoretically derive additional properties of snap rounding which then enabled us to prove the correctness of these simpler methods, as well as to analyze their performance.

We expect that this combination of theory, analysis, and implementation should apply to many other finite-precision models of geometric data structures and algorithms. We also believe that this type of algorithm will necessitate a new kind of analysis that elucidates the interaction between metric (distance, area) and combinatorial measures on geometric configurations.

Acknowledgements: We thank Dan Greene for useful discussions. Leonidas Guibas wishes to acknowledge support by NSF grants CCR-9215219 and IRI-9306544.

Our email addresses are guibas@cs.stanford.com and marimont@parc.xerox.com.

References

[1] J. L. Bentley and T. A. Ottmann, "Algorithms for reporting and counting geometric intersections," *IEEE Trans. Comput.*, v. C-28, pp. 643–647, 1979.

[2] B. Chazelle and H. Edelsbrunner, "An optimal algorithm for intersecting line segments in the plane," *J. Assoc. Comput. Mach.*, v.39, pp. 1–54, 1992.

[3] K. Clarkson and P. Shor, "Applications of random sampling in computational geometry II", *Discrete Comput. Geom.*, v.4, pp. 387-421, 1989.

[4] S. Fortune, "Stable maintenance of point-set triangulation in two dimensions," unpublished manuscript, AT&T Bell Laboratories. (An abbreviated version appeared in *Proc. 30h Ann. Symp. on Foundations of Computer Science*, pp. 494-499, 1989.)

[5] S. Fortune and V. Milenkovic, "Numerical stability of algorithms for line arrangements," *Proc. Seventh Ann. Symp. on Computational Geometry*, pp. 334-341, 1991.

[6] Daniel H. Greene, "Integer Line Segment Intersection," unpublished manuscript.

[7] Daniel H. Greene and Frances F. Yao, "Finite-Resolution Computational Geometry," *Proc. 27th Ann. Symp. on Foundations of Computer Science*, pp. 143-152, 1986.

[8] John D. Hobby, "Practical Segment Intersection with Finite Precision Output," submitted for publication.

[9] C. Hoffman, J. Hopcroft, and M. Karasick, "Towards implementing robust geometric computation," *Proc. Fourth Ann. Symp. on Computational Geometry*, pp. 106–117, 1988.

[10] J. Matoušek, "Geometric range searching," Technical Report B-93-09, Fachbereich Mathematik und Informatik, Free Univ. Berlin, 1993.

[11] V. Milenkovic, *Verifiable Implementations of Geometric Algorithms using Finite Precision Arithmetic*, Ph.D. Thesis, Carnegie-Mellon, 1988. Also Technical Report CMU-CS-88-168, Carnegie Mellon University, 1988.

[12] Ketan Mulmuley, "A fast planar partition algorithm I", *J. Symbolic Comput.*, v. 10, pp. 253-280, 1990.

[13] Ketan Mulmuley, *Computational Geometry: An Introduction Through Randomized Algorithms*, Prentice Hall (Englewood Cliffs, NJ), 1994.

[14] L. Guibas, D. Salesin, and J. Stolfi, "Epsilon Geometry: building robust algorithms from imprecise calculations," *Proc. Fifth Ann. Symp. on Computational Geometry*, pp. 208–217, 1989.

[15] K. Sugihara and M. Iri, *Geometric Algorithms in finite-precision arithmetic*, Research Memorandum RMI 88-10, University of Tokyo, September 1988.

A New Technique for Analyzing Substructures in Arrangements

Boaz Tagansky*

Abstract

We present a simple but powerful new probabilistic technique for analyzing the combinatorial complexity of various substructures in arrangements of piecewise-linear surfaces in higher dimensions. We apply the technique (a) to derive new and simpler proofs of the known bounds on the complexity of the lower envelope, of a single cell, or of a zone in arrangements of simplices in higher dimensions, and (b) to obtain improved bounds on the complexity of the vertical decomposition of a single cell in an arrangement of triangles in 3-space, and of several other substructures in such an arrangement (the entire arrangement, all nonconvex cells, and any collection of cells). The latter results also lead to improved algorithms for computing substructures in arrangements of triangles.

1 Introduction

The study of arrangements of curves or surfaces is an important area of research in computational and combinatorial geometry. This is due to the fact that many geometric problems in diverse areas can be reduced to problems involving arrangements. We assume in this paper familiarity of the reader with basic terminology and results concerning arrangements, and refer to [14, 18, 20] for more details.

In this paper we present a new technique for analyzing the combinatorial complexity of various substructures in arrangements of piecewise-linear surfaces in higher dimensions. The technique uses a simplified variant of the Clarkson-Shor probabilistic analysis technique [11], and combines it with local geometric analysis of the arrangement under consideration. The technique is a simplification of a previous technique that was used for analyzing the complexity of lower envelopes and other substructures in arrangements of more general algebraic surfaces (see [2, 19, 24]). Our technique also somewhat resembles the inductive technique used in [3, 5, 16] for analyzing arrangements of piecewise-linear surfaces, but is more general and more powerful.

We first exemplify the technique in Section 2 on the problem of bounding the maximum complexity of the lower envelope of n line segments in the plane. We obtain a bound $O(n \log n)$, which is slightly weaker than the tight $\Theta(n\alpha(n))$ bound of [21, 26]. The proof however is very short and simple. (Intuitively, the new technique is not 'sensitive' enough to yield bounds with factors like $\alpha(n)$.)

We then describe, in Section 3, the technique in a general abstract setting, and then apply this general methodology to several specific problems.

The first application, given in Section 4, is to arrangements of n $(d-1)$-simplices in $\mathbb{R}^d$. We obtain new (and simpler) proofs of the known bounds for the maximum complexity of the lower envelope, of a single cell, and of a zone in such an arrangement of simplices. For lower envelopes, the bound is $\Theta(n^{d-1}\alpha(n))$; previous proofs are given in [15, 23, 25]. For a single cell or a zone of a fixed-degree algebraic surface or an arbitrary convex surface, the bound is $O(n^{d-1} \log n)$; a previous, and considerably more involved proof is given in [5].

We then derive, in Section 6, an improved bound on the complexity of vertical decompositions in arrangements of n triangles in 3-space. The previous bound in [13] for the complexity of the vertical decomposition of a single cell in such an arrangement was $O(n^{2+\varepsilon})$, for any $\varepsilon > 0$, where the constant of proportionality depends on ε. We improve this bound to $O(n^2 \log^2 n)$ (actually, this bound applies also to the vertical decomposition of the zone of any fixed-degree algebraic surface or of any convex surface). We then extend the result to several more complex portions of the arrangement: We show that the complexity of the vertical decomposition of the entire arrangement is $O(K + n^2\alpha(n)\log n)$, where K is the complexity of the undecomposed arrangement; the previous bound of [13] was $O(K + n^{2+\varepsilon})$, for any $\varepsilon > 0$. We also obtain a bound of $O(n^{2.5})$ for the complexity of the vertical decomposition of all nonconvex cells of the arrangement, and a bound of $O(m^{1/3}n^2 + n^2 \log^2 n)$ for the complexity of the vertical decomposition of any m cells of the arrangement. We are not aware of any previous subcubic bounds for these two latter complexities. Still, we do not know whether the last two bounds are

---

*School of Mathematical Sciences, Tel Aviv University, Tel Aviv 69978, Israel.

close to optimal; the actual complexities of the (undecomposed) cells in questions are known to be smaller.

Using the randomized algorithm of [12] and applying our new bounds in its analysis, we obtain improved bounds for constructing the relevant substructures. For example, (the vertical decomposition of) a single cell in an arrangement of n triangles in 3-space can be computed in $O(n^2 \log^3 n)$ time. Also, we obtain an algorithm for constructing all nonconvex cells of the arrangement, whose running time is close to $O(n^{2.5})$. This yields a simple motion planning algorithm, of the same complexity, for an arbitrary polyhedron translating in a 3-D polyhedral environment.

This paper does not exhaust all currently known applications of our new technique. In several companion papers [8, 9] we apply the technique to obtain tight complexity bounds for the union of axis-parallel hypercubes in higher dimensions, and tight or nearly-tight bounds for various generalized Voronoi diagrams in higher dimensions. Other recent applications of our technique, still in progress, are a new proof of the $O(n^4 \log n)$ bound on the complexity of the vertical decomposition of an arrangement of n hyperplanes in $\mathbb{R}^4$ (the original proof is given in [17]), and improved bounds on the complexity of the union of convex polyhedra in 3-space, improving by a logarithmic factor earlier bounds given in [6, 7].

2 The Lower Envelope of Segments

Let $\mathcal{L}$ be a set of n nonvertical line segments in the plane, in general position. Let $\mathcal{A}(\mathcal{L})$ denote the arrangement of $\mathcal{L}$. An intersection point p between two segments of $\mathcal{L}$ is said to be a *k-level inner vertex* of $\mathcal{A}(\mathcal{L})$ if there are exactly k segments in $\mathcal{L}$ passing strictly below p. (We refer to such vertices as *inner vertices*, to distinguish them from the endpoints of the segments.) Let $C_k(\mathcal{L})$ denote the number of k-level inner vertices in $\mathcal{A}(\mathcal{L})$, and let $C_k(n)$ denote the maximum of $C_k(\mathcal{L})$ over all sets of n segments in the plane, in general position. We want to bound $C_0(n)$, the maximum number of inner vertices on the lower envelope of $\mathcal{L}$. It is well known that $C_0(n) = \Theta(n\alpha(n))$ [21, 26]; here we will only prove a weaker upper bound of $O(n \log n)$.

Our technique consists of two parts. The first part involves geometric analysis that enables us to bound the number of 0-level inner vertices in terms of the number of 1-level inner vertices, plus some excess that we can bound using a more direct analysis (which is trivial in the case of segments). It proceeds as follows.

Let p be a 0-level inner vertex of $\mathcal{A}(\mathcal{L})$. We sweep a vertical line from p to the right, and stop the sweeping as soon as we encounter one of the following two types of events (see Figure 1):

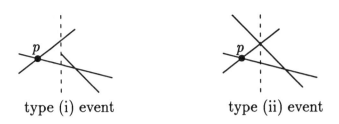

Figure 1: Sweeping from a 0-level inner vertex: Terminal events of type (i) and (ii)

(i) We reach an endpoint of one of the segments.

(ii) We reach a 1-level vertex.

An important observation is that the sweeping line can never reach a 0-level inner vertex before any of these events are encountered, so we can reach every such event only from (at most) one 0-level vertex. Let $D(\mathcal{L})$ denote the number of events of type (i); we clearly have $D(\mathcal{L}) \leq 2n$, which leads to the inequality

$$C_0(\mathcal{L}) \leq C_1(\mathcal{L}) + D(\mathcal{L}) \leq C_1(\mathcal{L}) + 2n. \qquad (1)$$

The second part of our technique applies a simple variant of the random-sampling analysis technique of Clarkson and Shor [11]. Together with the inequality (1), this gives a recurrence relation whose solution yields a bound on $C_0(n)$.

This part proceeds as follows. We take a random sample $\mathcal{R}$ of $n-1$ segments of $\mathcal{L}$, that is, we obtain $\mathcal{R}$ by removing a random segment from $\mathcal{L}$. Each 0-level inner vertex p of $\mathcal{A}(\mathcal{L})$ will appear as a 0-level vertex of $\mathcal{A}(\mathcal{R})$ if and only if the two intersecting segments containing p are in $\mathcal{R}$. This will happen with probability $(n-2)/n$. Similarly, each 1-level inner vertex p' of $\mathcal{A}(\mathcal{L})$ will appear as a 0-level vertex of $\mathcal{A}(\mathcal{R})$ if and only if the segment in $\mathcal{L}$ below p' is not chosen in $\mathcal{R}$. This will happen with probability $1/n$. No other inner vertex of $\mathcal{A}(\mathcal{L})$ can appear as a 0-level vertex of $\mathcal{A}(\mathcal{R})$. Hence

$$\mathbf{E}(C_0(\mathcal{R})) = \frac{n-2}{n} C_0(\mathcal{L}) + \frac{1}{n} C_1(\mathcal{L}), \qquad (2)$$

where $\mathbf{E}$ denotes expectation with respect to the random sample $\mathcal{R}$. Putting equations (1) and (2) together, we get:

$$\frac{n-1}{n} C_0(\mathcal{L}) = \frac{n-2}{n} C_0(\mathcal{L}) + \frac{1}{n} C_0(\mathcal{L})$$

$$\leq \frac{n-2}{n} C_0(\mathcal{L}) + \frac{1}{n}(C_1(\mathcal{L}) + 2n) \leq \mathbf{E}(C_0(\mathcal{R})) + 2.$$

Since $\mathbf{E}(C_0(\mathcal{R})) \leq C_0(n-1)$, we obtain the following recurrence:

$$\frac{n-1}{n} C_0(n) \leq C_0(n-1) + 2, \qquad (3)$$

and $C_0(2) = 1$. The solution of this recurrence is easily seen to be $C_0(n) \leq 1 + 2n \log_e(n-1)$, so we obtain:

Theorem 2.1 *The number of inner vertices on the lower envelope of n segments in the plane is at most $2n \log_e n$.*

3 The Technique in an Abstract Setting

In this section we describe our analysis technique in a more general and abstract manner, so as to facilitate its subsequent application to several different problems. It would be helpful to compare the general setting with the specific example given above.

The general setup is similar to that used by Clarkson [10], Clarkson and Shor [11], Mulmuley [22], and others. Let $\mathcal{L}$ be a set of n (geometric) objects. Let $\mathcal{A}(\mathcal{L})$ be a set of 'features', where each feature is defined in terms of d objects of $\mathcal{L}$. We refer to d as the *dimension* of the problem. In addition to the d objects defining a feature p, there may be other objects that 'conflict' with p. If no object conflicts with p, then p is said to be *exposed*, or to lie at level 0. Otherwise, we define the *level* of p to be the smallest number of objects whose removal causes p to become exposed (observe that none of the removed objects can be any of the d objects defining p). Let $\mathcal{A}_k(\mathcal{L})$ denote the subset of features in $\mathcal{A}(\mathcal{L})$ whose level is exactly k. Our goal is to bound $|\mathcal{A}_0(\mathcal{L})|$ as a function of n. (In most applications, the minimal set of objects whose removal exposes a feature p is unique, and consists of all objects that conflict (individually) with p. In some cases, though, exposing p can be done in more than one way. As an example, consider the case where $\mathcal{L}$ is again a set of segments in the plane, and where a vertex of the arrangement of $\mathcal{L}$ is exposed if it lies on the boundary of the face of the arrangement that contains the origin. In this case it may be possible to expose a vertex by removing several different subsets of segments.)

Our analysis technique consists of two parts. The first part, which is specific for the problem under consideration, uses geometric analysis to bound the number of 0-level features in terms of the number of 1-level features plus some known excess. The result of this part is an inequality of the form

$$\rho C_0(\mathcal{L}) \leq C_1(\mathcal{L}) + D(\mathcal{L}), \quad (4)$$

for some constant $\rho > 0$, where $C_k(\mathcal{L})$ is the number of k-level features in $\mathcal{A}(\mathcal{L})$, and $D(\mathcal{L})$ is the excess. We assume that $D(\mathcal{L})$ can be bounded, using separate analysis, by some function $D(n)$ of n.

The second part of our technique is based, as in the case of segments, on a simple variant of the random-sampling technique of Clarkson and Shor [11]. Together with the above inequality, this gives us a recurrence relationship for $C_0(n)$, the maximum of $C_0(\mathcal{L})$ over all collections $\mathcal{L}$ of n objects of the type that we consider here. The solution of this recurrence yields the desired bound.

As in the case of segments, we take a random sample $\mathcal{R} \subseteq \mathcal{L}$ of size $n - 1$, that is, we obtain $\mathcal{R}$ by removing a random object from $\mathcal{L}$. Each 0-level feature p of $\mathcal{A}(\mathcal{L})$ will appear as a 0-level feature of $\mathcal{A}(\mathcal{R})$ if (and only if) the d objects defining p are chosen in $\mathcal{R}$. This will happen with probability $(n-d)/n$. For each 1-level feature p' of $\mathcal{A}(\mathcal{L})$, there exists at least one conflicting object whose removal makes p' a 0-level feature. Hence the probability that p' will appear as a 0-level feature of $\mathcal{A}(\mathcal{R})$ is at least $1/n$. Hence we have:

$$\mathbf{E}(C_0(\mathcal{R})) \geq \frac{n-d}{n} C_0(\mathcal{L}) + \frac{1}{n} C_1(\mathcal{L}), \quad (5)$$

where $\mathbf{E}$ denotes expectation with respect to the random sample $\mathcal{R}$. Putting equations (4) and (5) together, we obtain:

$$\frac{n-d+\rho}{n} C_0(\mathcal{L}) = \frac{n-d}{n} C_0(\mathcal{L}) + \frac{\rho}{n} C_0(\mathcal{L})$$

$$\leq \frac{n-d}{n} C_0(\mathcal{L}) + \frac{1}{n}(C_1(\mathcal{L}) + D(\mathcal{L})) \leq \mathbf{E}(C_0(\mathcal{R})) + \frac{1}{n} D(\mathcal{L}).$$

Since $\mathbf{E}(C_0(\mathcal{R})) \leq C_0(n-1)$, we get the following recurrence:

$$\frac{n-d+\rho}{n} C_0(n) \leq C_0(n-1) + \frac{1}{n} D(n). \quad (6)$$

This recurrence may have different solutions, depending on the possible values of ρ and $D(n)$:

Proposition 3.1 *The solution $F(n)$ of the recurrence*

$$\frac{n-t}{n} F(n) \leq F(n-1) + \frac{1}{n} D(n), \quad (7)$$

for any fixed real $t \geq 0$, satisfies:

$$F(n) = O\left(n^t \cdot \left[1 + \sum_{j=\lceil t \rceil+1}^{n} \frac{D(j)}{j^{t+1}}\right]\right). \quad (8)$$

In particular,

(a) *If $D(n) = O(n^{t'})$, for any $t' < t$, then $F(n) = O(n^t)$.*

(b) *If $D(n) = O(n^t \log^c n)$, for any $c \geq 0$, then $F(n) = O(n^t \log^{c+1} n)$.*

(c) *If $D(n) = O(n^{t'})$, for any $t' > t$, then $F(n) = O(n^{t'})$.*

Proof. Omitted in this version. It is trivial for integer t (divide the recurrence by $(n-1)(n-2)\cdots(n-t)$), but is more complicated when t is not an integer. □

4 Lower Envelopes in Arrangements of Simplices

Let $\mathcal{S}$ be a set of n $(d-1)$-simplices in $\mathbb{R}^d$, and let $\mathcal{A}(\mathcal{S})$ denote the arrangement of $\mathcal{S}$. In this section we consider the problem of bounding the combinatorial complexity of the lower envelope of $\mathcal{S}$. That is, if we regard each simplex in $\mathcal{S}$ as a partial linear $(d-1)$-variate function, the lower envelope is the (graph of the) pointwise minimum of these functions. The lower envelope is a polyhedral set, and we measure its complexity by the number of its vertices (the number of all other faces is easily seen to be proportional to the number of vertices, if the simplices are in general position). Note that the pointwise minimum of simplices is generally not continuous.

We derive here a new, and simpler proof of the known tight bound on the complexity of the envelope, previously proved in [15, 23, 25]:

Theorem 4.1 *The maximum number of vertices of the lower envelope of n $(d-1)$-simplices in $\mathbb{R}^d$, for $d \geq 2$, is $\Theta(n^{d-1}\alpha(n))$.*

Proof. Let us draw a vertical hyperplane through each $(d-2)$-face of every simplex in $\mathcal{S}$, and add those hyperplanes to our arrangement. An *outer* vertex in the augmented arrangement is a vertex that lies in one of these vertical hyperplanes, whereas an *inner* vertex lies in the intersection of exactly d simplex interiors and avoids these hyperplanes.

We assume that the simplices of $\mathcal{S}$ are bounded and in *general position*. It is eeasily seen that neither assumption involves any real loss of generality (see also the previously cited proofs).

The lower bound has been shown, e.g., in [23]; it follows easily from the lower bound $\Omega(n\alpha(n))$ for arrangements of segments in the plane, as given in [26].

We prove the upper bound by induction on d. The bound holds for $d = 2$ (the case of line segments in the plane), as follows from the results of [21]. Fix $d \geq 3$, assume that the upper bound holds for all $2 \leq d' \leq d-1$, and let $\mathcal{S}$ be a collection of n $(d-1)$-simplices in $\mathbb{R}^d$.

A vertex $v \in \mathcal{A}(\mathcal{S})$ is said to be a k-*level* vertex if there are exactly k simplices of $\mathcal{S}$ passing strictly below v. Let $C_k(\mathcal{S})$ denote the number of k-level inner vertices in $\mathcal{A}(\mathcal{S})$, and let $C_k(n, d)$ denote the maximum of $C_k(\mathcal{S})$ over all possible sets $\mathcal{S}$ of n $(d-1)$-simplices in $\mathbb{R}^d$ in general position. Let $D(\mathcal{S})$ denote the number of outer vertices of the lower envelope, and let $D(n,d)$ denote the maximum of $D(\mathcal{S})$ over all sets $\mathcal{S}$ of n simplices as above. Our goal is thus to bound $C_0(n,d)$ and $D(n,d)$, which together serve as our measure of the combinatorial complexity of the lower envelope.

The outer vertices are easy to deal with. Let f be a $(d-2)$-face of a simplex s, and let h_f be the vertical hyperplane passing through f. We intersect h_f with all the simplices in $\mathcal{S}$, thereby obtaining a $(d-1)$-dimensional arrangement of $O(n)$ $(d-2)$-simplices within h_f, which we denote by $\mathcal{A}_f$. We also denote by $\mathcal{A}_f^-$ the arrangement $\mathcal{A}_f$ after removing from it the intersection $s \cap h$. Then each outer vertex of the lower envelope of $\mathcal{A}(\mathcal{S})$ within h_f can be identified with an (inner or outer) vertex of the lower envelope of either $\mathcal{A}_f$ or $\mathcal{A}_f^-$. Since there are $nd(d-1)/2$ such $(d-2)$-faces f, we get:

$$D(n,d) \leq \frac{nd(d-1)}{2} \cdot 2 \Big(C_0(O(n), d-1) + D(O(n), d-1) \Big)$$

and the induction hypothesis implies that $D(n,d) = O(n^{d-1}\alpha(n))$.

Next we bound the number of inner vertices, using our new technique. Each 0-level inner vertex v of $\mathcal{A}(\mathcal{S})$ is incident to exactly d edges of $\mathcal{A}(\mathcal{S})$, each of which emanates from v and is hidden from below by one of the simplices incident to v. We trace each of these edges from v, and stop the tracing as soon as we first encounter one of the following types of events:

(i) We reach a vertical $(d-1)$-dimensional slab h, spanned by a $(d-2)$-face of some simplex in $\mathcal{S}$.

(ii) We meet a new simplex at a point v', which is necessarily a 1-level inner vertex of $\mathcal{A}(\mathcal{S})$. In this case we say that v' and v are neighbors (in the arrangement $\mathcal{A}(\mathcal{S})$).

An important observation is that we can never reach a 0-level vertex during any of these d tracings, before one of these types of events is encountered. Hence any of these 'terminal' events can be reached at most once along any of its incident edges.

If we intersect the $nd(d-1)/2$ vertical slabs and hyperplanes that arises in case (i) with the simplices in $\mathcal{S}$, we get $nd(d-1)/2$ arrangements, each consisting of $O(n)$ $(d-2)$-simplices in $\mathbb{R}^{d-1}$. Each of the terminal events of type (i) is easily seen to be either a 1-level or a 2-level inner vertex of one of these arrangements. We will charge each such vertex by, say 2 units from each of the two possible directions in which it can be reached, for a total of 4 units. The total charge of all terminal events of types (i) and (ii) is thus bounded by

$$\frac{nd(d-1)}{2} \cdot 4 \Big(C_1(O(n), d-1) + C_2(O(n), d-1) \Big).$$

Using a standard application of the Clarkson-Shor probabilistic technique [11], we easily obtain

$$C_1(O(n), d-1) + C_2(O(n), d-1) = O(C_0(O(n), d-1)).$$

Thus, by the induction hypothesis, the total charge of all terminal events of type (i) is bounded by $O(n^{d-1}\alpha(n))$.

In case of a type (ii) terminal event, we will charge the 1-level inner vertex v' that we reach. Let $m(v')$ denote the number of 0-level neighbors of v', which is the number of times v' will be charged. We will give v' a total of 1 unit of charge, and it will pay $1/m(v')$ units to each 0-level neighbor. A 1-level inner vertex can be charged by type (ii) events up to d times, thus the charge it can pay to a 0-level neighbor will be at least $1/d$ units. This implies that any 0-level inner vertex v receives at least 1 unit of charge.

We next claim that every 0-level inner vertex v receives at least $4/3$ units of charge. By construction, this is the case when at least one of the tracings from v reaches a terminal event of type (i), so we will only consider vertices v for which all d tracings from v terminate at type (ii) events.

We will represent each inner vertex of $\mathcal{A}(\mathcal{S})$ by the (unordered) tuple of the d simplices incident to the vertex. Let v be a 0-level inner vertex represented by the tuple $(1, 2, \ldots, d)$.

For any simplex x, not incident to v, let V_x be the collection of all 1-level neighbors of v incident to x, and let S_x be the set of simplices hiding some vertex in V_x. We have the following two claims:

(a) Any 0-level neighbor of any 1-level vertex $v' \in V_x$ must be incident to all the simplices in S_x.

(b) For any $k \in \{1, \ldots, d\}$, only members of one set V_x can have 0-level neighbors not incident to simplex k.

Consider e.g., the case $k = 3$, and assume to the contrary that $v_1' \in V_x$ has a 0-level neighbor v_1 represented by $(1, 2, x, 4, \ldots, d)$, and that $v_2' \in V_y$ has a 0-level neighbor v_2 represented by $(1, 2, y, 4, \ldots, d)$, and $x \neq y$. (It is easily verified that these are the only possible vertices with these properties.) The three vertices v, v_1, v_2 lie on the same line segment e, formed by the intersection of the simplices $1, 2, 4, \ldots, d$, and none of them is hidden from below by any simplex of $\mathcal{S}$. Suppose, with no loss of generality, that v_1 is the vertex in the middle. Then the hyperplane π_x containing the simplex x must pass strictly below one of v, v_2; say, below v_2. Consider the polygonal path $p: v_1 v_1' v v_2' v_2$. By construction, p starts on simplex x, and does not meet any vertical hyperplane through any $(d-2)$-face of x. Thus p must reach v_2 while it is above the simplex x, a contradiction that establishes our claim. See Figure 2. Similar arguments apply when v or v_2 is the middle vertex.

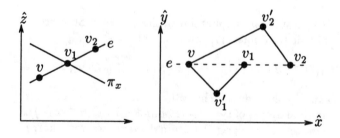

Figure 2: The polygonal path used in the proof of Claim (b); with a side view (left) and a top view (right).

Suppose that the 1-level neighbors of v are grouped into m subsets $V_{x_1}, \ldots, V_{x_m}$, for appropriate distinct simplices $x_1, \ldots, x_m$. Let a_i denote the size of V_{x_i}, and let b_i be the maximal number of 0-level neighbors, other than v, of any member of V_{x_i}, for $i = 1, \ldots, m$. We clearly have $\sum_{i=1}^{m} a_i = d$. Claim (a) is easily seen to imply that $a_i + b_i \leq d$, for each i, and Claim (b) is easily seen to imply that $\sum_{i=1}^{m} b_i \leq d$. The number of 0-level neighbors of each member of V_{x_i} is at most $b_i + 1$, so each of these members will pay to v at least $1/(b_i + 1)$ units. Summing this over all 1-level neighbors of v, we conclude that the total charge that v will receive is at least $\sum_{i=1}^{m} a_i/(b_i + 1)$. We thus want to minimize this sum, subject to the constraints that a_i and b_i are positive integers satisfying (i) $\sum_{i=1}^{m} a_i = d$, (ii) $a_i + b_i \leq d$, for each $i = 1, \ldots, m$, and (iii) $\sum_{i=1}^{m} b_i \leq d$. A simple exercise shows that the above sum is always $\geq 4/3$ when $d \geq 3$. Hence each 0-level inner vertex receives at least $4/3$ units of charge. This implies that

$$\frac{4}{3} C_0(\mathcal{S}) \leq C_1(\mathcal{S}) + O(n^{d-1}\alpha(n)).$$

Using our technique, we obtain the recurrence

$$\frac{n - d + 4/3}{n} C_0(n) \leq C_0(n-1) + O(n^{d-2}\alpha(n)),$$

whose solution is $C_0(n) = O(n^{d-1}\alpha(n))$ (see Proposition 3.1). This completes the proof of the theorem. □

5 Single Cells and Zones in Arrangements of Simplices

Let $\mathcal{S}$ be a set of n $(d-1)$-simplices in $\mathbb{R}^d$ in general position. let P be a fixed algebraic surface of dimension $\leq d - 1$ and of constant algebraic degree, or the boundary of an arbitrary convex set. We denote by $Z_P(\mathcal{S})$ the *zone* of P in $\mathcal{A}(\mathcal{S})$, namely the collection of all cells of $\mathcal{A}(\mathcal{S})$ whose intersection with P is nonempty. We want to bound the combinatorial complexity of $Z_P(\mathcal{S})$, which we can measure, up to a constant of proportionality, by the number of vertices of the boundary of $Z_P(\mathcal{S})$. If P is a singleton, $Z_P(\mathcal{S})$ is the cell of $\mathcal{A}(\mathcal{S})$ that contains P, so our analysis applies to single cells as well.

We obtain a new and simple proof of the following result; earlier proofs were given in [3, 5]:

Theorem 5.1 *The maximum number of vertices in a single cell or in a zone of a fixed-degree algebraic surface, or of the boundary of any convex set, in an arrangement of n $(d-1)$-simplices in $\mathbb{R}^d$ is $O(n^{d-1} \log n)$.*

Proof. Let $\overline{Z_P(\mathcal{S})}$ be the closure of the union of $Z_P(\mathcal{S})$. A k-face of $\mathcal{A}(\mathcal{S})$ is called *inner* if it lies in the intersection of the relative interiors of exactly $d - k$ simplices of $\mathcal{S}$ and avoids simplex boundaries. A k-face is called *outer* if it is contained in the relative boundary of at least one simplex in $\mathcal{S}$. An inner k-face f is said to be *popular* if $f \subseteq \overline{Z_P(\mathcal{S})}$ but $f \not\subseteq \partial \overline{Z_P(\mathcal{S})}$. For instance, every cell of $Z_P(\mathcal{S})$ is popular, and a popular facet is one that touches cells of $Z_P(\mathcal{S})$ on both sides. This notation is borrowed from [5].

For each simplex $\sigma \in \mathcal{S}$, let h_σ be the hyperplane containing σ. For each inner vertex v of $\mathcal{A}(\mathcal{S})$, the triple (v, R, F) is called a *0-level k-border* of $Z_P(\mathcal{S})$ if the two following conditions are satisfied: (i) R is a closed cone of $\mathbb{R}^d$ whose interior is a connected component of $\mathbb{R}^d \setminus \bigcup_{i=1}^d h_{\sigma_i}$, where $\sigma_1, \ldots, \sigma_d$ are the d simplices incident to v. Each inner vertex v induces 2^d cones R of this form. (ii) F is an intersection of $d-k$ hyperplanes from $\{h_{\sigma_1}, \ldots, h_{\sigma_d}\}$ (F is a k-flat containing v). If $k = d$ then F must be $\mathbb{R}^d$.

Let (v, R, F) be any 0-level k-border and let f be the unique k-face $f \subset F$ of $\mathcal{A}(\mathcal{S})$, having v incident to $f \cap R$. f is said to be the face of (v, R, F) in $\mathcal{A}(\mathcal{S})$. If f is a popular face we say that (v, R, F) is a *0-level popular k-border*.

For a simplex $\sigma \in \mathcal{S}$ we say that the quadruple (v', R', F', σ) is a *1-level popular k-border* if (v', R', F') is not a 0-level popular k-border of $Z_P(\mathcal{S})$, but it is a 0-level popular k-border of $Z_P(\mathcal{S} \setminus \{\sigma\})$. Let $C_i^{(k)}(P, \mathcal{S})$ be the number of i-level popular k-borders, for $i = 0, 1$ and $k = 0, 1, \ldots, d$, and let $C_i^{(k)}(P, n)$ be the maximum of $C_i^{(k)}(P, \mathcal{S})$ over all sets $\mathcal{S}$ of n $(d-1)$-simplices in $\mathbb{R}^d$ in general position. The quantity $C_0^{(d)}(P, \mathcal{S})$ counts all the inner vertices on the boundary of every cell in $Z_P(\mathcal{S})$, where a vertex is multiply counted, once for every 0-level d-popular border that it generates within $Z_P(\mathcal{S})$. Since the number of all outer vertices in $\mathcal{A}(\mathcal{S})$ is $O(n^{d-1})$, it suffices to show that $C_0^{(d)}(P, \mathcal{S}) = O(n^{d-1} \log n)$.

Let $b_0 = (v_0, R_0, F)$ be a 0-level popular k-border of $Z_P(\mathcal{S})$, and let f_0 be the (popular) k-face of b_0 in $\mathcal{A}(\mathcal{S})$. Let e_1 be an edge of f_0 incident to v_0 and within R_0, and let σ_1 be the (unique) simplex containing v_0 but not containing e_1. Let e'_1 be the other edge in $\mathcal{A}(\mathcal{S})$ incident to v_0 but not contained in σ_1 (e_1 and e'_1 are adjacent edges on the same line), and let v_1 be the other endpoint of e'_1. See Figure 3 for an illustration.

If v_1 is an inner vertex then let R_1 be the unique 'cone' induced by v_1 and containing R_0, and let f_1 be the face (whether popular or not) of (v_1, R_1, F) in $\mathcal{A}(\mathcal{S})$. Let g_1 be the $(k-1)$-face of $(v_0, R_0, F \cap h_{\sigma_1})$ in $\mathcal{A}(\mathcal{S})$. Note that if v_1 is an inner vertex then e'_1 is incident to f_1, $v_0 \in R_1$, g_1 is a $(k-1)$-face incident to f_0 and f_1, and if we remove σ_1 from $\mathcal{S}$ then the faces f_0 and f_1 become part of a larger inner k-face f'_1, which is clearly also popular. Note that it is possible that $f_0 = f_1$. We conclude that if v_1 is an inner vertex then (v_1, R_1, F) is a 0-level popular k-border of $Z_P(\mathcal{S} \setminus \{\sigma_1\})$. In any case, one of the following three types of configurations must arise:

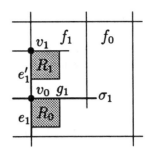

Figure 3: Moving from an inner vertex v_0 to vertex v_1 along an edge transversal to an incident simplex

(i) v_1 is an outer vertex.

(ii) f_1 is a popular face (in $Z_P(\mathcal{S})$). In this case g_1 is also a popular face and $b' = (v_0, R_0, F \cap h_{\sigma_1})$ is a 0-level popular $(k-1)$-border of $Z_P(\mathcal{S})$.

(iii) v_1 is an inner vertex and f_1 is not a popular face. Then $b_1 = (v_1, R_1, F, \sigma_1)$ is a 1-level popular k-border of $Z_P(\mathcal{S})$. We shall say that b_0 and b_1 are neighbors in $\mathcal{A}(\mathcal{S})$.

We repeat this analysis for each of the k edges of f_0 incident to v_0 and within R_0.

For type (i) configurations, we charge the outer vertex v_1 by 2 units. It is easy to see that v_1 can be charged only $O(1)$ times, so the total charge for type (i) events is $O(n^{d-1})$, where the constant of proportionality depends on d.

For type (ii) configurations, we charge the popular $(k-1)$-border b' in the configuration by 2 units. A 0-level popular $(k-1)$-border (v_0, R_0, G) can be charged only by 0-level popular k-borders of the form (v_0, R_0, F') where $G \subset F'$. Since there are $d-k+1$ such k-flats F', it follows that (v_0, R_0, G) can be charged at most $d-k+1$ times in type (ii) configurations, and the total charge for this type of events is at most $2(d-k+1)C_0^{(k-1)}(P, \mathcal{S})$.

For type (iii) configurations, we charge the 1-level popular k-border b_1; we may also charge some outer vertices of $Z_P(\mathcal{S})$ (see case (b.i) below). Let $m(b_1)$ denote the number of 0-level popular k-border neighbors of b_1, which is the number of times b_1 will be charged (in type (iii) events). Since b_1 can be charged only along the k edges of f_1 incident to v_1 and within R_1, we have $m(b_1) \leq k$. We will give b_1 a total of 1 unit of charge, and it will pay $1/m(b_1)$ units to each 0-level neighbor. Thus the charge it can pay to a 0-level neighbor will be at least $1/k$ units. This implies that any 0-level popular k-border receives at least 1 unit of charge, and we show that, for $k \geq 3$, it receives more than 1 unit. It suffices to consider popular k-borders for which all these charging configurations are of type (iii).

Let $b_0 = (v_0, R_0, F)$ be a 0-level popular k-border, with the inner vertex v_0 represented (as in section 4) by the tuple $(1, 2, \ldots, d)$. Let $b'_1 = (v'_1, R'_1, F, 1)$ and $b'_2 = (v'_2, R'_2, F, 2)$ be two distinct 1-level k-popular neighbors of b_0, with v'_1 represented by $(x, 2, \ldots, d)$ and with v'_2 represented by $(1, y, 3, \ldots, d)$, for a simplices $x, y \in \mathcal{S}$.

(a) $x = y$: In this case it is easily seen that b'_1 (and also b'_2) can have at most $k - 1$ 0-level neighbors. (A proof is given in the full version) It follows that b'_1 can pay at least $1/(k-1)$ units of charge to b_0, so b_0 gets a total of at least $\frac{1}{k-1} + \frac{k-1}{k} = 1 + \frac{1}{k(k-1)} > 1$ units of charge.

(b) $x \neq y$: Suppose that b'_1 (resp. b'_2) has a 0-level neighbor $b_1 = (v_1, R_1, F)$, (resp. $b_2 = (v_2, R_2, F)$).

(b.i) Suppose that v_1 and v_2 are both obtained by replacing the same simplex, say simplex 3, by the respective hiding simplices 1 and 2. That is, v_1 is represented by $(x, 1, 2, 4, \ldots, d)$ and v_2 is represented by $(y, 1, 2, 4, \ldots, d)$. It is easily seen, by construction, that both v_1 and v_2 must belong to R_0. Moreover, the three vertices all lie on the same line segment s, formed by the intersection of the $d-1$ simplices $1, 2, 4, \ldots, d$. Since v_0 is the apex of R_0, it follows that v_0 cannot be the middle vertex along s. Suppose, with no loss of generality, that the middle vertex is v_1, and consider the triangle τ with vertices v_0, v_2 and v'_2. See Figure 4 for an illustration. Let Δ be the 2-flat containing the intersection of the simplices $1, 4, \ldots, d$, and thus containing τ. By construction, the edges $v_0 v'_2$ and $v_2 v'_2$ of τ are not crossed by any simplex of $\mathcal{S}$. On the other hand, v_1, on the edge $v_0 v_2$, is the intersection of s with the simplex x. We conclude that f'_2, the 2-face of (v'_2, R'_2, Δ) in $\mathcal{A}(\mathcal{S})$, in not convex and thus has an outer vertex p' as one of its vertices. We will charge p' by 4 units, to be divided between b_0 and b_2. One can show that p' can be charged only via f'_2 and only by the pair b_0 and b_2, hence b_0 receives at least 2 units of charge in this case. Since the number of outer vertices is $O(n^{d-1})$, the total charge of this kind is also $O(n^{d-1})$.

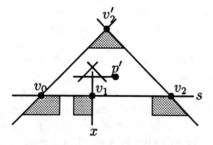

Figure 4: Charging an outer vertex in the triangle τ

(b.ii) Suppose next that the preceding subcase does not occur. Suppose that v_1 is not incident to simplex k, for some $k \in \{3, \ldots, d\}$ (that is, v_1 is obtained by replacing simplex k by simplex 1 in the tuple defining v'_1). Then v_2 must be incident to k. This implies that $m(b'_1) + m(b'_2) \leq k + 2$, so at least one of $m(b'_1)$, $m(b'_2)$ must be strictly smaller than k (for $k \geq 3$), which means

that b_0 will receive, as in case (a), at least $\rho_k = 1 + \frac{1}{k(k-1)}$ units of charge.

Hence, for $k \geq 3$

$$\rho_k C_0^{(k)}(P, \mathcal{S}) \leq C_1^{(k)}(P, \mathcal{S}) + O(n^{d-1} + C_0^{(k-1)}(P, \mathcal{S}))$$

and, for $k = 2$,

$$C_0^{(2)}(P, \mathcal{S}) \leq C_1^{(2)}(P, \mathcal{S}) + O(n^{d-1} + C_0^{(1)}(P, \mathcal{S})).$$

We now apply our probabilistic analysis technique. Even though 1-level borders are not quite the same as 0-level borders, it still follows from our analysis that the expected number of 0-level popular k-borders of $Z_P(\mathcal{R})$ is $\frac{n-d}{n} C_0^{(k)}(P, \mathcal{S}) + \frac{1}{n} C_1^{(k)}(P, \mathcal{S})$. Thus, using our technique, we obtain, for $k \geq 3$,

$$\frac{n - d + \rho_k}{n} C_0^{(k)}(P, n) \leq C_0^{(k)}(P, n-1) +$$
$$+ \frac{1}{n} O(n^{d-1} + C_0^{(k-1)}(P, n)), \qquad (9)$$

and, for $k = 2$,

$$\frac{n - d + 1}{n} C_0^{(2)}(P, n) \leq C_0^{(2)}(P, n-1) +$$
$$+ \frac{1}{n} O(n^{d-1} + C_0^{(1)}(P, n)). \qquad (10)$$

As argued in [3, 5], one can show that the number of popular edges is $O(n^{d-1})$, from which it follows that $C_0^{(1)}(P, n) = O(n^{d-1})$. Using (10), we get, by Proposition 3.1, $C_0^{(2)}(P, n) = O(n^{d-1} \log n)$. Then, for $k \geq 3$, we proceed by induction on k, and use (9). Applying again Proposition 3.1, we obtain $C_0^{(k)}(P, n) = O(n^{d-1} \log n)$, for $k = 3, \ldots, d$. □

6 Vertical Decompositions in Arrangements of Triangles in $\mathbb{R}^3$

In this section we obtain the main new result of the paper: improved bounds on the complexity of vertical decompositions in arrangements of triangles in 3-space (improving and extending previous bounds in [13]).

Let $\mathcal{T}$ be a set of n triangles in $\mathbb{R}^3$, in general position. Let $\mathcal{A}(\mathcal{T})$ denote the arrangement of $\mathcal{T}$. Let P be a fixed point set in $\mathbb{R}^3$. We denote by $Z_P(\mathcal{T})$ the zone of P in $\mathcal{A}(\mathcal{T})$, as defined in the previous section. Let e_1, e_2 be two edges of $\mathcal{A}(\mathcal{T})$ intersecting a common vertical line ℓ. The open segment $s \subset \ell$ between e_1 and e_2 is said to be a k-level visibility segment, and the triple (e_1, e_2, s) is said to be a k-level visibility configuration, if exactly k triangles of $\mathcal{T}$ intersect s, and s intersects a cell in $Z_P(\mathcal{T})$. We shall distinguish between outer visibility configurations, where at least one of the edges e_1, e_2 is a portion of the boundary of some triangle, and the complemetary *inner* configurations. Denote by $C_k(P, \mathcal{T})$ (resp. $D_k(P, \mathcal{T})$) the number of inner (resp. outer) k-level visibility configurations. Denote by $C_k(P, n)$ (resp. $D_k(P, n)$) the maximum of $C_k(P, \mathcal{T})$ (resp. $D_k(P, \mathcal{T})$) over all sets $\mathcal{T}$ of n triangles in general position.

We want to analyze the combinatorial complexity of the *vertical decomposition* of $Z_P(\mathcal{T})$. This is a decomposition of the cells of $Z_P(\mathcal{T})$ into subcells of constant complexity, obtained as follows. We take each edge e of each cell C of $Z_P(\mathcal{T})$, and extend from each point on e a vertical segment up and down into C until it meets the boundary of C again. The resulting collection of vertical faces partitions each cell of $Z_P(\mathcal{T})$ into subcells, each of which has a unique top face and a unique bottom face, each being a portion of some triangle in $\mathcal{T}$. The complexity of each of these subcells need not be constant, so we further refine each subcell, slicing it by planes parallel to the y-axis, into prism-like subcells of constant complexity. We refer the reader to [13] for more details concerning vertical decompositions in arrangements of triangles. As shown there, the complexity of the vertical decomposition of $Z_P(\mathcal{T})$ is proportional to the sum of the number of 0-level visibility configurations and the number of vertices of $Z_P(\mathcal{T})$, so we will concentrate on bounding the number $C_0(P, \mathcal{T}) + D_0(P, \mathcal{T})$ of 0-level visibility configurations.

We begin by estimating the number of outer visibility configurations. Let e be a boundary edge of a triangle in $\mathcal{T}$, and let h be the vertical strip spanned by e (it is the union of all vertical lines intersecting e). Let h^+, h^- denote, respectively, the portions of h that lie above and below e. We intersect h with all the other triangles in $\mathcal{T}$, thereby obtaining a 2-dimensional arrangement of segments within h. Let $\mathcal{A}_e^+$, $\mathcal{A}_e^-$ denote, respectively, the portions of this arrangement within h^+ and within h^-. Each 0-level outer visibility configuration involving e can be identified with a vertex of either the lower envelope of the arrangement $\mathcal{A}_e^+$ or the upper envelope of the arrangement $\mathcal{A}_e^-$. Since the complexity of any such envelope is $O(n\alpha(n))$ [21], and since there are $3n$ boundary edges of the triangles of $\mathcal{T}$, the total number of 0-level outer visibility configurations is $D_0(P, n) = O(n^2 \alpha(n))$, even if P is the entire 3-space. Similarly, each 1-level outer visibility configuration can be identified with a 1-level vertex in some arrangement $\mathcal{A}_e^+$ or $\mathcal{A}_e^-$. Using a simple variant of the Clarkson-Shor probabilistic technique [11], it is easily seen that we also have $D_1(P, n) = O(n^2 \alpha(n))$. Consider next the (more involved) case of inner visibility configurations. Let $v = (e_1, e_2, s)$ be a 0-level inner visibility configuration, where e_1 (resp. e_2) is a portion of the intersection of two triangles t_1, t_2 (resp. of t_3, t_4) of $\mathcal{T}$, and where e_1 is above e_2. The visibility segment s lies fully within a single cell, which is necessarily a cell of $Z_P(\mathcal{T})$, so, in particular, e_1 and e_2 are edges of $Z_P(\mathcal{T})$. We slide

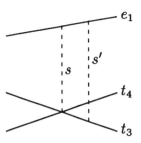

Figure 5: Sliding a vertical segment from a 0-level visibility configuration

a vertical segment s', starting from s, so that the top endpoint of s' moves along e_1 and its bottom endpoint moves along t_3, so that s' crosses t_4. There is always a unique direction of movement of s' where these conditions are satisfied; see Figure 5. We stop the sliding process as soon as we first encounter one of the following types of events:

(i) s' reaches an endpoint of e_1, which is either an outer or an inner vertex (in the notation of the preceding sections) of some cell of $Z_P(\mathcal{T})$.

(ii) s' reaches a boundary edge e_3 of t_3, at the 1-level outer visibility configuration (e_1, e_3, s').

(iii) s' passes through a boundary of a triangle between t_3 and e_1, at an appropriate 1-level or 0-level outer visibility configuration.

(iv) s' reaches an intersection edge e', between t_3 and a new triangle, at the corresponding 1-level inner visibility configuration (e_1, e', s').

We apply this sliding process four times, twice by sliding along e_1 and along one of the triangles t_3, t_4 (as described above), and twice by sliding along e_2 and along one of the triangles t_1, t_2, in a fully symmetric manner.

An important observation is that the sliding process can never reach a 0-level inner visibility configuration before any of these terminal events are encountered, so we can reach any terminal event at most once in each possible sliding direction.

In any of the above cases, we will charge the appropriate terminal configuration or vertex by one unit. Thus each 0-level inner visibility configuration receives 4 units of charge.

Each vertex v of some cell C of $Z_P(\mathcal{T})$ can be charged at most six times in events of type (i), and each (0-level or 1-level) outer visibility configuration can be charged at most twice in events of type (ii). Hence, if we denote by $M(P, \mathcal{T})$ the complexity of $Z_P(\mathcal{T})$, we can bound the total amount of charge made in cases (i), (ii), and (iii) by $O(n^2 \alpha(n) + M(P, \mathcal{T}))$.

A 1-level inner visibility configuration may be charged at type (iv) events up to 4 times. We will give each

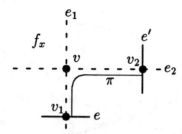

Figure 6: Constructing a path from a 1-level visibility configuration; the dashed edges are the projections of e_1, e_2 onto t_x

such configuration 2 units of charge, so, if more than two charges are made to such a configuration, we need to pass the extra charges to other features.

Let $v = (e_1, e_2, s)$ be a 1-level inner visibility configuration, with e_1 lying above e_2, whose visibility segment s stabs the triangle t_x at a face f_x of $\mathcal{A}(\mathcal{T})$. Let t_1, t_2 (resp. t_3, t_4) be the two triangles whose intersection contains e_1 (resp. e_2). If v is charged more than twice then it must be charged by at least one 0-level inner visibility configuration v_1 of the form (e_1, e, σ) and by at least one such configuration v_2 of the form (e', e_2, σ'). We will say that the first (resp. second) charge is made by sliding towards v 'from above' (resp. 'from below'). Then e must be a portion of the intersection of t_x with either t_3 or t_4, and e' must be a portion of the intersection of t_x with either t_1 or t_2. With no loss of generality, assume that $e \subseteq t_x \cap t_3$ and that $e' \subseteq t_x \cap t_1$. It is important to notice that e and e' are edges of the face f_x, and that f_x is a *popular* face of $Z_P(\mathcal{T})$.

Construct a path $\pi = \pi(v)$ within f_x that connects the points $\sigma \cap t_x$ and $\sigma' \cap t_x$ by two straight links with $s \cap t_x$ as a common point; see Figure 6. Since v_1 and v_2 both charge v, it follows by definition that no event of type (i), (ii), (iii) or (iv) can occur between v_1 and v or between v_2 and v (except at v itself). This implies that the relative interior of π does not pass directly below or above any edge of $\mathcal{A}(\mathcal{T})$, except for e_1 and e_2 Also, π does not intersect any triangle, except for lying on t_x.

The path $\pi(v)$ is not necessarily unique. (there can be up to four such paths), but we will construct for each v as above only one path $\pi(v)$, and pass it up to 2 units of charge, leaving v with only 2 units of charge. This will produce a collection of paths within f_x, each connecting two points on its boundary. The above analysis is easily seen to imply that no pair of those paths can cross each other. We repeat this construction over all popular faces f_x of $Z_P(\mathcal{T})$.

Our next goal is to bound the number of paths $\pi(v)$. The system of these paths within a fixed popular face f_x can be regarded as a plane embedding of a planar graph $G(f_x)$, whose nodes are represented by the edges of f_x and whose edges are represented by the paths $\pi(v)$. By

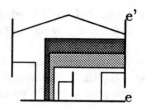

Figure 7: Constructing a planar graph from the paths $\pi(v)$; two faces of degree 2 are shaded.

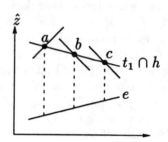

Figure 8: The vertical envelope induced by 'parallel' paths starting from the edge e

Euler's formula, the number of such paths is bounded by three times the number of nodes (edges of f_x) plus the number of graph-faces of degree 2; each such face is bounded by two paths $\pi(v)$, $\pi(v')$, connecting between the same pair of edges, e, e' of f_x, and by appropriate portions of the edges e and e'. We thus need to bound the number of these degree-2 faces; see Figure 7.

Suppose there are $z + 1$ adjacent paths that connect between the same pair of edges e, e' of f_x and create z degree-2 faces between them, where $z \geq 2$. Suppose that, as above, $e \subseteq t_x \cap t_3$ and $e' \subseteq t_x \cap t_1$. By construction, each of these paths must then start on e at a 0-level inner visibility configuration whose other edge lies on t_1 above e. In other words, consider the vertical strip h spanned by e and define in the upper portion of h the 2-dimensional arrangement $\mathcal{A}_e^+$, as above. Then the $z + 1$ paths above induce $z + 1$ inner vertices of the lower envelope of $\mathcal{A}_e^+$ that lie all on the segment $s_1 = t_1 \cap h$. However, as is easily seen, for any triple a, b, c of such vertices appearing in this order along s_1, there exists an endpoint of some segment in $\mathcal{A}_e^+$ that appears on the lower envelope between a and c. See Figure 8. It follows that there are at least $z' = \lfloor z/2 \rfloor$ such endpoints over the portion of e bounding our z degree-2 faces. Each of these endpoints induces a 0-level *outer* visibility configuration, and we can charge our z faces to these z' outer visibility configurations, concluding that the number of such faces is at most three times the number of 0-level outer visibility configurations involving edges of f_x. This analysis fails when $z = 1$, but in such a case we simply ignore the degree-2 face, replace the two adjacent paths bounding it by a single path, and note that none of the two faces adjacent to this path has degree 2.

All this implies that the number of paths drawn within a face f_x is bounded by six times the number of edges of f_x plus three times the number of 0-level outer visibility configurations involving edges of f_x. Summing these bounds over all (popular) faces of $Z_P(T)$, we conclude that the overall amount of excess charges made to 1-level inner visibility configurations (over 2 units of charge per configuration) is $O(M(P,T)+n^2\alpha(n))$. In other words, we have $4C_0(P,T) \leq 2C_1(P,T) + O(M(P,T)+n^2\alpha(n))$, or

$$2C_0(P,T) \leq C_1(P,T) + O(M(P,T) + n^2\alpha(n)). \quad (11)$$

Theorem 6.1 *Let P be an algebraic surface in $\mathbb{R}^3$ of dimension 0,1 or 2 and of constant degree, or the boundary of an arbitrary convex set. Then the vertical decomposition of the zone of P in the arrangement $\mathcal{A}(T)$ of a set T of n triangles in $\mathbb{R}^3$ in general position consists of $O(n^2 \log^2 n)$ cells.*

Proof. Since the complexity of the zone of P is $O(n^2 \log n)$ (see [3, 5] and Section 5), equation (11) becomes

$$2C_0(P,T) \leq C_1(P,T) + O(n^2 \log n).$$

We now plug this into our probabilistic analysis technique, and note that the abstract dimension of the problem is 4, because each inner visibility configuration is defined in terms of 4 triangles. Then (6) becomes

$$\frac{n-2}{n}C_0(P,n) \leq C_0(P,n-1) + \frac{1}{n}O(n^2 \log n),$$

whose solution is $C_0(P,n) = O(n^2 \log^2 n)$ (see Proposition 3.1). As noted above, this implies the asserted bound. □

If P is the entire 3-space, we can modify the above analysis as follows. Let $\mathcal{S} \subset T$ be a random sample of m triangles. Let $K(m) = \mathbf{E}_\mathcal{S}(M(\mathbb{R}^3, \mathcal{S}))$, where $\mathbf{E}_\mathcal{S}$ denotes expectation with respect to the random sample $\mathcal{S}$. Every inner vertex of $\mathcal{A}(T)$ has probability $\binom{n-3}{m-3}/\binom{n}{m}$ to appear in $\mathcal{A}(\mathcal{S})$, and the number of outer vertices in $\mathcal{A}(\mathcal{S})$ is $O(m^2)$. Thus

$$K(m) = O\left(\frac{m^3}{n^3}M(\mathbb{R}^3,T) + m^2\right).$$

Let $\overline{C_k}(m) = \mathbf{E}_\mathcal{S}(C_k(\mathbb{R}^3, \mathcal{S}))$. Using equation (11) in our new technique we get, by a simple averaging calculation, whose details are omitted here

$$\frac{m-2}{m}\overline{C_0}(m) = \overline{C_0}(m-1) + O\left(\frac{m^2}{n^3}M(\mathbb{R}^3,T) + m\alpha(m)\right).$$

Using Proposition 3.1, this recurrence solves to

$$\overline{C_0}(m) = O\left(\frac{m^3}{n^3}M(\mathbb{R}^3,T) + m^2\alpha(m)\log m\right).$$

Putting $m = n$, we thus conclude:

Theorem 6.2 *The vertical decomposition of the entire arrangement of a set of n triangles in $\mathbb{R}^3$ in general position consists of $O(K + n^2\alpha(n)\log n)$ cells, where K is the complexity of the arrangement.*

Remark: Using the algorithm given in [13] and Theorem 6.2, the vertical decomposition of the entire arrangement can be computed in $O(K \log n + n^2\alpha(n)\log^2 n)$ time. Using the randomized algorithm given in [12], the vertical decomposition of the zone of P (as defined in theorem 6.1) can be computed in $O(n^2 \log^3 n)$ expected time.

Theorem 6.3 (a) *If P is a connected curve that has L intersection points with the triangles of T then the complexity of the vertical decomposition of the zone of P is $O(n^{3/2}L^{1/2} + n^2 \log^2 n)$.*

(b) *The complexity of the vertical decomposition of all the nonconvex cells of $\mathcal{A}(T)$ is $O(n^{5/2})$.*

(c) *The complexity of the vertical decomposition of any m cells of $\mathcal{A}(T)$ is $O(m^{1/3}n^2 + n^2 \log^2 n)$.*

Remarks: (1) Assertion (b) follows from (a) by taking P to be spanning path of the union of the triangle boundaries; in this case $L = O(n^2)$.
(2) The bounds in the preceding theorem are significantly larger than the actual complexity of the relevant cells. For example, as shown in [1], the complexity of all nonconvex cells is $O(n^{7/3})$, and the complexity of any m cells is $O(m^{2/3}n + n^2 \log n)$. Nevertheless, all bounds stated in the theorem appear to be new.

Proof. We start with the proof of (a). If we apply the analysis technique obtained above, we get the recurrence

$$\frac{n-2}{n}C_0(P,T) \leq \mathbf{E}(C_0(P,\mathcal{R})) + O\left(\frac{1}{n}M(P,T) + n\alpha(n)\right),$$

where P is the given curve, and where $\mathcal{R}$ is a random sample of $n-1$ triangles of T. If we divide the recurrence by $(n-1)(n-2)$, and put $V(P,T) = \frac{C_0(P,T)}{|T|(|T|-1)}$, we obtain

$$V(P,T) \leq \mathbf{E}(V(P,\mathcal{R})) + O\left(\frac{1}{n^3}M(P,T) + \frac{\alpha(n)}{n}\right).$$

We unfold this recurrence, and replace $V(P,T)$ by the maximum value $V(P,n)$ of this quantity, over all collections of n triangles in $\mathbb{R}^3$, to obtain

$$V(P,n) \leq \sum_{j=1}^{n} \frac{M(P,j)}{j^3} + \sum_{j=1}^{n} \frac{\alpha(j)}{j}, \quad (12)$$

where $M(P,j)$ is the expected complexity of the zone of P in an arrangement of a random sample of j triangles of T. We note that the expected number of crossings of P with the triangles in such a random sample is Lj/n, so the expected number of cells in the zone of P in the sample arrangement is also Lj/n. Using the bound in [1], we can easily obtain $M(P,j) = O((Lj/n)^{2/3}j + j^2 \log j)$. We substitute this bound into (12) for $j > (L/n)^{1/2}$, and use the trivial bound $M(P,j) = O(j^3)$ for $j \leq (L/n)^{1/2}$, to comclude that the first sum in (12) is $O\left((L/n)^{1/2} + O(\log^2 n)\right)$. The second sum in (12) is only $O(\alpha(n) \log(n))$. This is easily seen to imply that $C_0(P,n) = O(L^{1/2} n^{3/2} + n^2 \log^2 n)$, thus proving (a).

The proof of (c) is very similar, but we omit it here due to a lack of space. □

Remark: The bound in Theorem 6.3(b) can be used to obtain a lazy randomized incremental algorithm, based on the technique of [12], of comparable efficiency, for constructing all nonconvex cells of $\mathcal{A}(\mathcal{T})$. It is useful for translational motion planning in 3-space: Let B be an arbitrary polyhedron translating among polyhedral obstacles. Using a standart approach (see e.g.,[7]), the problem reduces to one involving a collection $\mathcal{T}$ of n 'contact' triangles in 3-space, and two points Z_1, Z_2; we wish to determine whether Z_1 and Z_2 can be connected by a path that avoids all triangles. We first precompute all nonconvex cells of $\mathcal{A}(\mathcal{T})$, and process them for efficient point location. Then we test whether at least one of Z_1, Z_2 lies in a nonconvex cell of $\mathcal{A}(\mathcal{T})$. If so, then both must lie in the same nonconvex cell, so the point locations already determine whether a desired motion between Z_1 and Z_2 exists. Otherwise, Z_1 and Z_2 must lie in the same convex cell of $\mathcal{A}(\mathcal{T})$, which we verify by testing whether the straight segment $Z_1 Z_2$ intersects any triangle in $\mathcal{T}$. We thus get an algorithm that takes about $O(n^{5/2})$ preprocessing time and storage, where n is the number of 'contact' triangles, and can perform motion planning queries of the above kind in $O(n)$ time per query. See the full version for more details.

References

[1] P.K. Agarwal and B. Aronov Counting Facets and Incidences, *Discrete Comput. Geom.* 7 (1992), 359–369.

[2] P.K. Agarwal, O. Schwarzkopf and M. Sharir, The overlay of lower envelopes in three dimensions and its applications, Manuscript, 1994.

[3] B. Aronov, M. Pellegrini and M. Sharir, On the zone of a surface in a hyperplane arrangement, *Discrete Comput. Geom.* 9 (1993), 177–186.

[4] B. Aronov and M. Sharir, Triangles in space, or: Building (and analyzing) castles in the air, *Combinatorica* 10 (2) (1990), 137–173.

[5] B. Aronov and M. Sharir, Castles in the air revisited, *Discrete Comput. Geom.* 12 (1994), 119–150.

[6] B. Aronov and M. Sharir, The union of convex polyhedra in three dimensions, to appear in *SIAM J. Computing*.

[7] B. Aronov and M. Sharir, On translational motion planning in three dimensions, *Proc. 10th ACM Symp. on Computionl Geometry* (1994), 21–30.

[8] J.D. Boissonnat, M. Sharir, B. Tagansky and M. Yvinec, Voronoi diagrams in higher dimensions under certain polyhedral distance functions, *Proc. 11th ACM Symp. on Computationl Geometry* (1995), to appear.

[9] L.P. Chew, K. Kedem, M. Sharir, B. Tagansky and E. Welzl, Voronoi diagrams of lines in three dimensions under a polyhedral convex distance function, *Proc. 6th ACM-SIAM Symp. on Discrete Algorithms* (1995), 197–204.

[10] K. Clarkson, New applications of random sampling in computational geometry, *Discrete Comput. Geom.* 2 (1987), 195–222.

[11] K. Clarkson and P. Shor, Applications of random sampling in computational geometry II, *Discrete Comput. Geom.* 4 (1989), 387–421.

[12] M. de Berg, K. Dobrindt and O. Schwarzkopf, On lazy randomized incremental construction, *Proc. 26th Annu. ACM Sympos. Theory Comput.* (1994), 105–114.

[13] M. de Berg, L. Guibas and D. Halperin, Vertical decomposition for triangles in 3-space, *Proc. 10th ACM Symp. on Computational Geometry* (1994), 1–10.

[14] H. Edelsbrunner, *Algorithms in Combinatorial Geometry*, Springer-Verlag, Heidelberg 1987.

[15] H. Edelsbrunner, The upper envelope of piecewise linear functions: Tight bounds on the number of faces, *Discrete Comput. Geom.* 4 (1989), 337–343.

[16] H. Edelsbrunner, R. Seidel and M. Sharir, On the zone theorem for hyperplane arrangements, *SIAM J. Comput.* 22 (1993), 418–429.

[17] L. Guibas, D. Halperin, J. Matoušek and M. Sharir, On vertical decomposition of arrangements of hyperplanes in four dimensions, to appear in *Discrete Comput. Geom.*

[18] L. Guibas and M. Sharir, Combinatorics and algorithms of arrangements, in *New Trends in Discrete and Computational Geometry*, (J. Pach, Ed.), Springer-Verlag, 1993, 9–36.

[19] D. Halperin and M. Sharir, New bounds for lower envelopes in three dimensions with applications to visibility of terrains, *Discrete Comput. Geom.* 12 (1994), 313–326.

[20] D. Halperin and M. Sharir, Arrangements and their applications in robotics: Recent developments, *Proc. Workshop on Algorithmic Foundations of Robotics*, 1994.

[21] S. Hart and M. Sharir, Nonlinearity of Davenport–Schinzel sequences and of generalized path compression schemes, *Combinatorica* 6 (1986), 151–177.

[22] K. Mulmuley, *Computational Geometry: Introduction through Randomized Algorithms*, Prentice Hall, New York, 1993.

[23] J. Pach and M. Sharir, The upper envelope of piecewise linear functions and the boundary of a region enclosed by convex plates: Combinatorial analysis, *Discrete Comput. Geom.* 4 (1989), 291–309.

[24] M. Sharir, Almost tight upper bounds for lower envelopes in higher dimensions, *Discrete Comput. Geom.* 12 (1994), 327–345.

[25] M. Sharir and P. Agarwal, *Davenport-Schinzel Sequences and Their Geometric Applications*, Cambridge University Press, to appear.

[26] A. Wiernik and M. Sharir, Planar realization of nonlinear Davenport–Schinzel sequences by segments, *Discrete Comput. Geom.* 3 (1988), 15–47.

An optimal algorithm for finding segments intersections

Ivan J. Balaban

Keldysh Institute of Applied Mathematics, 125047, Miusskaya Sq. 4, Moscow, Russia.

Abstract

This paper deals with a new deterministic algorithm for finding intersecting pairs from a given set of N segments in the plane. The algorithm is asymptotically optimal and has time and space complexity $O(N \log N + K)$ and $O(N)$ respectively, where K is the number of intersecting pairs. The algorithm may be used for finding intersections not only line segments but also curve segments.

INTRODUCTION

Reporting of segments set intersections is one of the fundamental problem of computational geometry. It is known, that within the model of algebraic decision tree any algorithm solving this problem needs $\Omega(N \log N + K)$ time [1, 5, 10]. The well-known Bentley and Ottman's algorithm based on the plane sweep has $O((N + K) \log N)$ time and $O(N)$ space complexity [2]. Later, Chazelle obtained an $O(N \log^2 N / \log \log N + K)$, $O(N + K)$ algorithm [4]. Time optimal algorithm $O(N \log N + K)$ was proposed by Chazelle and Edelsbrunner with space requirement $O(N + K)$ [5]. Mulmuley using randomized approach offered an algorithm with expected running time $O(N \log N + K)$ and space requirement $O(N + K)$ [9]. Algorithm of Clarkson and Shor also based on the randomized approach has expected running time $O(N \log N + K)$ and space requirement $O(N)$ [6].

In this paper we present a deterministic, asymptotically optimal for both time $O(N \log N + K)$ and space $O(N)$ algorithm for finding intersecting segments pairs. As by product, we construct an $O(N \log^2 N + K)$, $O(N)$ algorithm that could be of practical interest due to its relative simplicity.

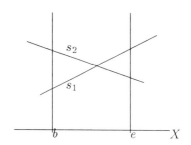

Figure 1: $s_1 <_b s_2$, $s_2 <_e s_1$.

PRELIMINARIES

Suppose we have a set S_0 consisting of N segments in the plane. The problem is to find all intersecting pairs. The set of these pairs we denote as $Int(S_0)$ and $|Int(S_0)|$ we denote as K.

To find $Int(S_0)$ we use a collection of vertical strips on the plane organized in a tree structure, connect each of the strips with some subset of S_0 and introduce procedures for finding intersections of such subsets. These procedures require appropriate ordering of the subsets. To obtain this ordering we use a method analogous to plane sweep but instead of sweeping the plane by a vertical line we use preorder traversal of the strips tree (i.e. sweeping the tree by its branch).

So, the primary objects of our algorithm are segments, ordered and unordered sets of segments, pairs of intersecting segments and vertical strips.

Let $\langle b, e \rangle$ denote the *vertical strip* $b \leq x \leq e$ and let l and r be abscissae of endpoints of segment s ($l < r$). We call the segment s

- *spanning the strip* $\langle b, e \rangle$ if $l \leq b < e \leq r$,
- *inner for the strip* $\langle b, e \rangle$ if $b < l < r < e$,
- *crossing the strip* $\langle b, e \rangle$ if $[l, r[\cap [b, e[\neq \emptyset$.

Two segments s_1 and s_2 are called *intersecting within the strip* $\langle b, e \rangle$ if their intersection point lies within this strip.

For two segments sets S and S' we denote by $Int(S, S')$ a set $\{\{s, s'\} | s \in S, s' \in S'$ and s intersect $s'\}$. Notations $Int_{b,e}(S)$ and $Int_{b,e}(S, S')$ will be used to describe subsets of $Int(S)$ and $Int(S,S')$ consisting of segments pairs intersecting within the strip $\langle b, e \rangle$. Hereafter the brackets {} are used to define unordered sets, and the brackets () are used to define ordered sets.

We shall use the following order relation among segments [2, 10]: $s_1 <_a s_2$ if segments s_1 and s_2 intersect a vertical line $x = a$ and the intersection point with s_1 lies under the intersection point with s_2 (see Fig.1).

Staircase D is the pair $(Q, \langle b, e \rangle)$, where segments set Q possesses the following properties

- $\forall s \in Q$ s spans the strip $\langle b, e \rangle$.
- $Int_{b,e}(Q) = \emptyset$.
- Q is ordered by $<_b$.

Intersections of segments of Q with $\langle b, e \rangle$ are called *stairs* of D. Intersection of the staircase D with a segments set S denoted as $Int(D, S)$ is the set $Int_{b,e}(Q, S)$.

We call the staircase D *complete relative to a set S* if each segment of S either does not span the strip $\langle b, e \rangle$ or intersects one of the stairs of D.

Lemma 1 *If a staircase D is complete relative to a set S with S consisting of segments crossing the strip $\langle b, e \rangle$ then*

$$|S| \leq Ends_{b,e}(S) + |Int(D, S)|,$$

where $Ends_{b,e}(S)$ is the number of endpoints of S within the strip $\langle b, e \rangle$.

If a point p of a segment s is placed above the ith and below the $i + 1$th stairs of D, then the number i is called *location of s on the staircase D* and denoted as $Loc(D, s)$. (see Fig.2).

Let $Loc(D, s) = i$. To find $Int(D, \{s\})$ it is necessary to reach the first stair that does not intersect s moving down from the ith stair and then to repeat this procedure moving up from the $i + 1$th stair. This procedure has time complexity $O(1 + |Int(D, \{s\})|)$.

Thus, if we knew $Loc(D, s)$ for all $s \in S$ then we could find $Int(D, S)$ in $O(|S|+|Int(D,S)|)$ time. It is easy to find $Loc(D, s)$ for any segment $s \in S$ in $O(log|Q|)$ time by binary search. If the set $S = (s_1, ..., s_n)$ is ordered by $<_x$, where $x \in [b, e]$, it is possible to find $Loc(D, s_i)$, $i = 1, ..., n$, in $O(|Q| + |S|)$ time sequentially scanning the stairs of D and the segments of S. It follows from this observation that

Proposition 1 *Given a staircase $D = (Q, \langle b, e \rangle)$ and a segments set S, the set $Int(D, S)$ can be found in $O(|S|\log|Q| + |Int(D,S)|)$ time. If S is ordered by $<_x$, $x \in [b, e]$, it is possible to find $Int(D, S)$ in $O(|S| + |Q| + |Int(D,S)|)$ time.*

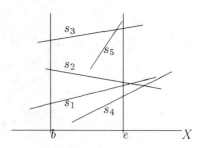

Figure 2: $D = ((s_1, s_2, s_3), \langle b, e \rangle)$, $Loc(D, s_4) = 0$, $Loc(D, s_5) = 2$ or 3, $Int(D, \{s_4, s_5\}) = \{\{s_3, s_5\}\}$

INTERSECTIONS IN A STRIP

Consider the following problem: *Let L be a set of segments spanning the strip $\langle b, e \rangle$ and ordered by $<_b$. We wish to find $Int_{b,e}(L)$ and to reorder L by $<_e$.* We denote the reordered set L by R.

Let us split the set L into subsets Q and L' so that the staircase $D = (Q, \langle b, e \rangle)$ be complete relative to L'. The intersections of segments of L is computed in two phases. The first phase is to find the intersection D with L'. The second phase is to find all intersections in L'. To find the intersections in L' we may split the set again etc.

As a result the problem of finding the intersections in L is reduced to the following two problems: how to split L efficiently and how to find intersections between D and L'.

The following procedure splits a set L ordered by $<_b$ into subsets Q and L' ordered by $<_b$ so that staircase $(Q, \langle b, e \rangle)$ is complete relative to L'.

procedure $\text{Split}_{b,e}(L, Q, L')$;
{Let $L = (s_1, ..., s_k)$, $s_i <_b s_{i+1}$}
   $L' := \emptyset$; $Q := \emptyset$;
   For $j = 1, ..., k$ **do**
     If the segment s_j doesn't intersect
     the last segment of Q within $\langle b, e \rangle$ and
     spans this strip **then**
       add s_j to the end of Q
     else
       add s_j to the end of L';
     endif;
   endfor;
end procedure.

This procedure runs in $O(|L|)$ time. Thus, given L we may find $Int_{b,e}(L)$ and R using the following recursive procedure:

procedure $\text{SearchInStrip}_{b,e}(L, R)$
   $\text{Split}(L, Q, L')$;
   If $L' = \emptyset$ **then** $R := Q$; exit; **endif**;

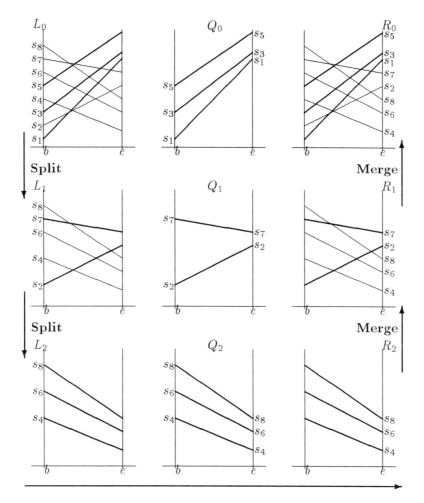

Figure 3: Example of SearchInStrip execution.

```
    Find $Int_{b,e}(Q, L')$;
    SearchInStrip$_{b,e}(L', R')$;
    $R := Merge_e(Q, R')$;
end procedure.
```

Here $Merge_x(S1, S2)$ is a procedure of merging sets $S1$ and $S2$ ordered by $<_x$. Running time of the procedure SearchInStrip equals the sum of running times of its calls. It is easy to understand that the ith procedure call runs in $O(|L_i| + |Int_{b,e}(Q_i, L'_i)|)$ time, where L_i, Q_i, L'_i are appropriate sets ($L_0 = L$, $L_{i+1} = L'_i$). Taking into account Lemma 1 we conclude that the procedure SearchInStrip$_{b,e}(L, R)$ runs in $O(|L| + |Int_{b,e}(L)|)$ time.

INTERMEDIATE ALGORITHM

Let's apply proposed approach to find $Int(S)$, where S is some set of segments. Let all segments of S be placed within a strip $\langle b, e \rangle$. Thus, at the beginning we have a pair S, $\langle b, e \rangle$. The following procedure is used.

The set S is splitted into subsets Q and S' so that the staircase $D = (Q, \langle b, e \rangle)$ be complete relative to S'. It is necessary to find the intersection D with S' and then all intersection in S'. To find the S' intersections we cut the strip $\langle b, e \rangle$ and the set S' along the vertical line $x = c$ into the strips $\langle b, c \rangle$, $\langle c, e \rangle$ and the sets S'_{ls}, S'_{rs} respectively, where c is the median of endpoints between b and e. Then, we recursively apply the procedure to the pairs S'_{ls}, $\langle b, c \rangle$ and S'_{rs}, $\langle c, e \rangle$.

The key fact is that according to Lemma 1 $|S'| \leq Ends_{b,e}(S') + Int(D, S')$, thus, the number of additional segments appearing during cutting is proportional to the number of found intersections.

Let us give a somewhat more detailed explanation of the algorithm.

Without loss of generality, we could assume that all end and intersection points have different abscissae. Within the scope of this paper the abscissae of segment's ends can be normalized by replacing each of them by its rank in their left-to-right order. We consider these abscissae as the integers in the range $[1, 2N]$. So, let p_i, i and $s(i)$ be the ith endpoint its abscissa and the segment to which it belongs, respectively.

The central point of our algorithm is a recursive pro-

213

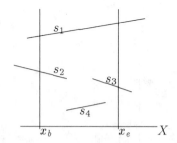

Figure 4: $S_v = \{s_1, s_2, s_3, s_4\}$, $I_v = \{s_4\}$, $L_v = (s_2, s_1)$, $R_v = (s_3, s_1)$.

cedure TreeSearch. We connect each of the procedure calls with a node in some binary tree referred to as *recursion tree*. We mark all values, sets and parameters of a call by a label of the corresponding node and identify calls and corresponding nodes. As a result we shall analyze our algorithm examining the recursion tree. Let RT be the set of all nodes of recursion tree, and V be the set of internal nodes. We shall define some values, sets and notations inside the bodies of procedures.

procedure IntersectingPairs(S_0).
  Sort the $2N$ endpoints by abscissa and
    find p_i, $s(i)$, $i = 1, ..., 2N$; $S_r := S_0$;
  TreeSearch($S_r, 1, 2N$);
end procedure.

procedure TreeSearch(S_v, b, e).
1. **If** $e - b = 1$ **then**
    $L_v :=$ sort S_v by $<_b$; SearchInStrip$_{b,e}(L_v, R_v)$; exit;
   endif;
2. Split S_v into Q_v and S'_v so that staircase
    $D_v := (Q_v, \langle b, e \rangle)$ be complete relative to S'_v;
3. Find $Int(D_v, S'_v)$;
4. $c := \lfloor (b + e)/2 \rfloor$;
5. Place segments of S'_v
    crossing the strip $\langle b, c \rangle$ into $S_{ls(v)}$ and
    the strip $\langle c, e \rangle$ into $S_{rs(v)}$;
6. TreeSearch($S_{ls(v)}, b, c$);
7. TreeSearch($S_{rs(v)}, c, e$);
end procedure.

Hereafter $ls(v)$, $rs(v)$ and $ft(v)$ denote, respectively, the left son, the right son and the father node of the node v.

Our task is to show how to perform all operations connected with node v in $O(|S_v| + |Int(D_v, S'_v)| + (e_v - b_v) \log N)$ time and to prove that $\sum_v |S_v| = O(N \log N + K)$ (obviously $\sum_v |Int(D_v, S'_v)| \leq K$).

The procedure TreeSearch is similar to the procedure SearchInStrip. The main difference is that SearchInStrip calls itself without changing the strip, and TreeSearch cuts the current strip into two parts, then, applies itself to both. Another difference is that the set S_v is not ordered as L. As a result we can't directly use

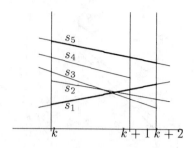

Figure 5: $D_v = ((s_1, s_5), \langle k, k + 2 \rangle)$, $L_v = (s_1, s_2, s_3, s_4, s_5)$, $L_{ls(v)} = (s_2, s_3, s_4)$, $R_{ls(v)} = (s_3, s_2, s_4)$, $L_{rs(v)} = (s_3, s_2)$, $R_{rs(v)} = (s_3, s_2)$, $R_v = (s_3, s_2, s_1, s_5)$.

the procedure Split for the efficient splitting the set S_v.

To resolve this problem we represent a set S_v as a union of three sets: a set L_v ordered by $<_b$, an unordered set I_v, and a set R_v ordered by $<_e$. We place segments of S_v intersecting the line $x = b$ into L_v, segments intersecting the line $x = e$ into R_v, segments inner for $\langle b, e \rangle$ into I_v (see Fig.4).

Now we may apply the procedure Split to the set L_v and construct Q_v in $O(|L_v|) = O(|S_v|)$ time. But we encounter a new problem. Given sets L_v, R_v and I_v, it is necessary to find corresponding sets of the sons of v.

The unordered sets $I_{ls(v)}$ and $I_{rs(v)}$ are easily constructed. The set $L_{ls(v)}$ will be found by applying the procedure Split$_{b,e}(L_v, Q_v, L_{ls(v)})$ for the third step of TreeSearch. The set $L_{rs(v)}$ can be obtained from $R_{ls(v)}$ in linear time by inserting (if p_c is the left endpoint) or deleting (if p_c is the right endpoint) the segment $s(c)$. But how to obtain $R_{ls(v)}$ from L_v, R_v and I_v without sorting?

For the leaves we execute step 1 and obtain R_v from L_v. Imagine that L_v and I_v are known and both of v sons are leaves. At first we execute the procedure Split$(L_v, Q_v, L_{ls(v)})$ and find Q_v and $L_{ls(v)}$. Now we have to find $Int(D_v, S'_v) = Int(D_v, L_{ls(v)}) \cup Int(D_v, I_v) \cup Int(D_v, R_{rs(v)})$, but we don't know $R_{rs(v)}$, and we may find $Int(D_v, L_{ls(v)}) \cup Int(D_v, I_v)$ only. Then, according to TreeSearch, we apply SearchInStrip to $L_{ls(v)}$ and receive $R_{ls(v)}$. The set $L_{rs(v)}$ is obtained from $R_{ls(v)}$ by inserting or deleting the segment $s(c)$. After that, we apply SearchInStrip to $L_{rs(v)}$ and find $R_{rs(v)}$. Now we may complete the computation $Int(D_v, S'_v)$ by computing $Int(D_v, R_{rs(v)})$ and may build R_v by merging Q_v and $R_{rs(v)}$ (See Fig.5).

The resulting procedures are listed below.

procedure IntersectingPairs(S_0).
  Sort the $2N$ endpoints by abscissa
    and find p_i, $s(i)$, $i = 1, ..., 2N$;
  $L_r := (s(1))$; $I_r := S_0 \setminus (\{s(1)\} \cup \{s(2N)\})$;
  TreeSearch($L_r, I_r, 1, 2N, R_r$);

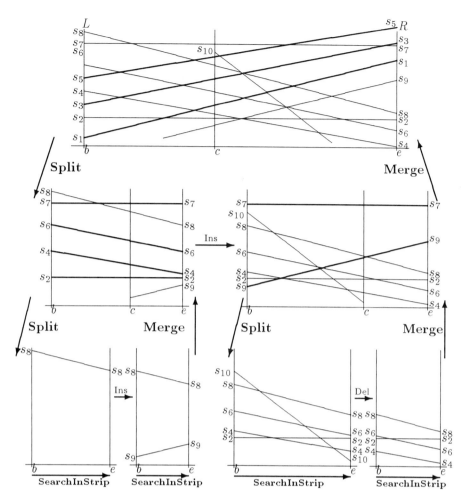

Figure 6: Example of TreeSearch execution. Ins. denote Insert, Del. denote Delete (step 7).

end procedure.

procedure TreeSearch(L_v, I_v, b, e, R_v).
1. **if** $e - b = 1$ **then**
     SearchInStrip$_{b,e}(L_v, R_v)$; exit;
   endif;
2. Split$_{b,e}(L_v, Q_v, L_{ls(v)})$; $D_v := (Q_v, \langle b, e \rangle)$;
3. Find $Int(D_v, L_{ls(v)})$;
4. $c := \lfloor (b+e)/2 \rfloor$;
5. Place segments of I_v
     inner for the strip $\langle b, c \rangle$ into $I_{ls(v)}$,
     inner for the strip $\langle c, e \rangle$ into $I_{rs(v)}$;
6. TreeSearch($L_{ls(v)}, I_{ls(v)}, b, c, R_{ls(v)}$);
7. **If** p_c is the left endpoint of $s(c)$ **then**
     $L_{rs(v)} :=$ insert $s(c)$ into $R_{ls(v)}$
   else
     $L_{rs(v)} :=$ delete $s(c)$ from $R_{ls(v)}$;
   endif;
8. TreeSearch($L_{rs(v)}, I_{rs(v)}, c, e, R_{rs(v)}$);
9. Find $Int(D_v, R_{rs(v)})$;
10. **For** $s \in I_v$ **do**
      find $Loc(D_v, s)$ using binary search;
    endfor
11. Find $Int(D_v, I_v)$ using values
      found in the previous step;
12. $R_v := Merge_e(Q_v, R_{rs(v)})$;
end procedure.

We must be careful executing the 9th step of our algorithm as the sets $R_{rs(v)}$ and $L_{ls(v)}$ could have common segments (segments spanning $\langle b, e \rangle$). We have found their intersections with D_v during the 3rd step and should not output these intersections again.

At first, we evaluate the space requirement of this algorithm. The algorithm uses the recursive procedure TreeSearch. A sequence of the procedure calls may occupy memory. The sequence can be represented by a path from the root of the recursion tree to a node. We call this node and the corresponding call *active*. The active call occupies $O(N)$ space, each of its "ancestors" retains $O(|I_v| + |Q_v|)$ memory, other structures use $O(N)$ memory. It is easy to show that for any path pt from the root of recursion tree to some of its nodes $\sum_{v \in pt}(|I_v| + |Q_v|) = O(N)$ ($|I_v| \leq e_v - b_v$, $e_v - b_v \leq$

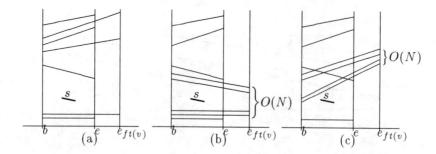

Figure 7: In the situations b,c a number of $D_f t(v)$ stairs intersect trapezoid where s are placed.

$\lceil(e_{ft(v)} - b_{ft(v)})/2\rceil$). Thus, the storage requirement of this algorithm is $O(N)$.

Lemma 2 $\forall v \in V$ $|S'_v| \leq b_v - e_v + |Int(D_v, S'_v)|$.

Proof. The statement directly results from Lemma 1 and the obvious fact that for any set $S \subset S_0$ the number of segment's ends lying in the strip $\langle b_v, e_v \rangle$ is less than $e_v - b_v$. □

Theorem 1 $\sum_{v \in V} |S'_v| \leq 2N\lceil \log N + 1 \rceil + K$.

Proof. The statement directly results from Lemma 2 and the following relation $\sum_v (e_v - b_v) \leq 2N\lceil \log N + 1 \rceil$. □

Theorem 2 $\sum_{v \in RT} |S_v| \leq N\lceil 4\log N + 5 \rceil + 2K$.

Proof. For all nodes except for the root r the following equality holds $|S_v| \leq |S'_{ft(v)}|$, hence, $\sum_{v \in RT} |S_v| \leq |S_r| + \sum_{v \in RT \setminus r} |S_{ft(v)}| \leq N + 2\sum_{v \in V} |S'_v| \leq N\lceil 4 \log N + 5 \rceil + 2K$. □

Initial sorting and initialization of the sets L_r and I_r can be performed in $O(N \log N)$ time. The running time of the procedure TreeSearch is evaluated by summing up the running time of all its calls. Each external node call adds to this sum a value $O(|L_v| + |Int_{b,e}(L_v)|)$. As for the internal node calls, let's find the time required for the step 10 and for all the other steps separately. The step 10 takes $O(|I_v| \log N)$ time, and all the others run in $O(|S_v| + |Int(D_v, S'_v)|)$ time (see Preliminaries). If we sum up these values, we come to the conclusion that our algorithm runs in time $O(N \log^2 N + K)$. Note that time complexity of the algorithm is $O(N \log N + K)$ if we don't take into account the running time of the 10th step.

Thus, to obtain time optimal algorithm we have to find $Loc(D_v, s)$ at the 10th step within $O(1)$ time bounds.

OPTIMAL ALGORITHM.

For better understanding the ideas of this section see [3, 7, 8, 11].

The main idea is to find $Loc(D_v, s)$ using information about the location of s in the staircases of sons of v. Suppose $Loc(D_v, s)$ is defined, then the value $Loc(D_{ft(v)}, s)$ is not arbitrary and should belong to some interval of $[0..|Q_{ft(v)}|]$. Let us hold lower and upper bounds of the intervals in two arrays SB_v and SE_v. If $Loc(D_v, s) = i$, then $Loc(D_{ft(v)}, s)$ belong to $[SB_v[i], ..., SE_v[i]]$ and the interval $[SB_v[i], ..., SE_v[i]]$ is a minimal one having this property. In the situation illustrated in the figure 7a the presence of these arrays may considerably simplify the search of $Loc(D_{ft(v)}, s)$, but in the situations presented in figures 7b,c the arrays are useless. Therefore it is necessary to change the method of staircase constructing to eliminate the situations given in the figures 7b,c and to make $SE_v[i] - SB_v[i]$ equal to $O(1)$ for any v and i.

Now the staircase D_v is constructed in the following way:

- Split the set L_v in two subsets Or_v and $L_{ls(v)}$ so that

  1. $\forall s \in Or_v$ s spans the strip $\langle b, e \rangle$.
  2. $Int_{b,e}(Or_v) = \emptyset$.
  3. $\forall s \in Or_v$ $|Int_{b,e}(Q_{ft(v)}, \{s\})| \leq 1$.
  4. $\forall s \in L_{ls(v)}$ $Or_v \cup \{s\}$ don't satisfy the conditions 1-3.

- Let $Q_{ft(v)} = (s_1, ..., s_k)$ and $n = \lfloor k/4 \rfloor$. Construct the set Inh_v containing those of the segments $s_4, s_8, s_{12}, ..., s_n$ which don't intersect the segments of Or_v within the strip $\langle b, e \rangle$.

- $Q_v = Or_v \cup Inh_v$. Construct staircase $D_v = (Q_v, \langle b, e \rangle)$.

Now the situation Fig.7c is impossible due to the properties of the set Or_v and the situation Fig.7b is impossible due to the properties of the set Inh_v.

Let the segments of Or_v be called the *original segments* of D_v, and their intersections with the strip $\langle b, e \rangle$ be called the *original stairs*. The segments of Inh_v are referred to as the *inherited segments*, and the corresponding stairs as the *inherited stairs*.

Theorem 3 $\forall v \; \forall i, \; SE_v[i] - SB_v[i] \leq 5$.

Proof. The demonstration is illustrated in figure 8. $[A, B]$ and $[C, D]$ are the ith and $i+1$th stairs of the staircase D_v. Obviously, $SE_v[i] - SB_v[i]$ is equal to the number of $D_{ft(v)}$ stairs intersecting the trapezoid $ABCD$. Suppose, this number is greater than five. The sides AB and CD of the trapezoid intersect not more than two stairs of $D_{ft(v)}$ (due to condition 3 during constructing Or_v). Therefore, at least four sequential stairs of $D_{ft(v)}$ don't intersect any stair of D_v that contradicts the procedure of D_v constructing. □

As a result, we may seek $Loc(D_v, s)$ sequentially scanning stairs of D_v beginning from $SB_{v'}[Loc(D_{v'}, s)]$, where v' is the son of v with $Loc(D_{v'}, s)$ already computed. In the procedure below each segment s has additional field bs for keeping the current value of $SB_{v'}[Loc(D_{v'}, s)]$. Thus, the improved procedure TreeSearch is as follows:

procedure NewTreeSearch$(L_v, I_v, Q_{ft(v)}, b, e, R_v)$.
1. **if** $e - b = 1$ **then**
    SearchInStrip$_{b,e}(L_v, R_v)$; exit;
   endif;
2. Construct the staircase D_v;
3. Construct array SB_v;
4. **For** $s \in L_{ls(v)}$ **do**
    find $Loc(D_v, s)$; $s.bs := SB_v[Loc(D_v, s)]$);
   endfor;
5. Find $Int(D_v, L_{ls(v)})$;
6. $c := \lfloor (b + e)/2 \rfloor$;
7. Place segments of I_v
    inner for the strip $\langle b, c \rangle$ into $I_{ls(v)}$,
    inner for the strip $\langle c, e \rangle$ into $I_{rs(v)}$;
8. NewTreeSearch$(L_{ls(v)}, I_{ls(v)}, Q_v, b, c, R_{ls(v)})$;
9. **If** p_c is the left endpoint of $s(c)$ **then**
    $L_{rs(v)} :=$ insert $s(c)$ into $R_{ls(v)}$
   else
    $L_{rs(v)} :=$ delete $s(c)$ from $R_{ls(v)}$;
   endif;
10. NewTreeSearch$(L_{rs(v)}, I_{rs(v)}, Q_v, c, e, R_{rs(v)})$;
11. **For** $s \in R_{rs(v)}$ **do**
    find $Loc(D_v, s)$; $s.bs := SB_v[Loc(D_v, s)]$;
    endfor;
12. Find $Int(D_v, R_{rs(v)})$;
13. **For** $s \in I_v$ **do**
    scan the stairs of D_v beginning from $s.bs$
    until we detect $Loc(D_v, s)$;
    $s.bs := SB_v[Loc(D_v, s)]$;
    endfor;
14. Find $Int(D_v, I_v)$ using values
    found in the previous step;
15. $R_v := Merge_e(Q_v, R_{rs(v)})$;
   end procedure.

When we find intersection of a staircase with some segments set, we should output intersections with original

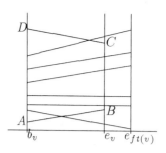

Figure 8: If more than five stairs of $D_{ft(v)}$ intersect trapezoid $ABCD$, then at least fore sequential stairs of $D_{ft(v)}$ have no intersections with stairs of D_v.

stairs only.

Obviously, the steps 4-11 of NewTreeSearch can be done in $O(|S_v| + |Int(D_v, S'_v)|)$ time (see Preliminaries). The procedure NewSplit receiving L_v and $Q_{ft(v)}$ as an input constructs Or_v and $L_{ls(v)}$ in $O(|L_v| + |Q_{ft(v)}|)$ time.

procedure NewSplit$_{b,e}(L_v, Q_{ft(v)}, Q_v, L_{ls(v)})$.
{Let $L_v = (s_1, ..., s_k)$, $Q_{ft(v)} = (s'_1, ..., s'_n)$}
  $L_{ls(v)} := \emptyset$; $Or_v := \emptyset$; $i := 1$;
  For $j = 1, ..., k$ **do**;
    While $s'_i <_b s_j$ and $i \leq n$ **do**
      $i := i + 1$;
    endwhile;
    If s_j doesn't intersect s'_{i-2}, s'_{i+1} and
    the last segment of Or_v within the strip $\langle b, e \rangle$ and
    span it **then**
      add s_j to the end of Or_v
    else
      add s_j to the end of $L_{ls(v)}$;
    endif;
  endfor;
end procedure.

The following procedure builds set Q_v and array SB_v using Or_v and $Q_{ft(v)}$ in time linear in the input.

procedure BuildQSB$_{b,e}(Or_v, Q_{ft(v)}, Q_v, SB_v)$.
{Let $Or_v = (s_1, ..., s_k)$, $Q_{ft(v)} = (s'_1, ..., s'_n)$}
  $SB_v[0] := 0$; $Q_v := \emptyset$; $i := 1$; $l := 1$;
  For $j = 1, ..., k$ **do**
    While $s'_i <_b s_j$ and $i \leq n$ **do**
      If $i \bmod 4 = 0$ and s'_i doesn't intersect s_j and
      s_{j-1} within the strip $\langle b, e \rangle$ **then**
        add s'_i to the end of Q_v;
        $l := l + 1$; $SB_v[l] := i$;
      endif;
      i:=i+1;
    endwhile;
    Add s_j to the end of Q_v; $l := l + 1$;
    If s_j intersects s'_{i-1} within $\langle b, e \rangle$ **then**
      $SB_v[l] := i - 2$

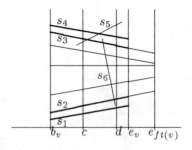

Figure 9: $Or_v = (s_1, s_4)$, $Inh_v = (s_2, s_3)$, $S = \{s_5, s_6\}$, $Int_{c,d}^{or}(D_v, S) = \{\{s_4, s_5\}\}$, $Int_{c,d}^{sn}(D_v, S) = \{\{s_3, s_5\}\}$, $Int_{c,d}^{ml}(D_v, S) = \{\{s_2, s_6\}, \{s_3, s_6\}\}$.

    else
        $SB_v[l] := i - 1$;
    endif;
   endfor;
end procedure.

It follows from this observation that

Proposition 2 *All work executed by NewTreeSearch in internal node v can be done in $O(|Q_{ft(v)}| + |S_v| + |Int(D_v, S'_v)|)$ time.*

Given a staircase $D = (Q, \langle b, e \rangle)$, a strip $\langle c, d \rangle \subset \langle b, e \rangle$ and a segment s intersecting a segment $s' \in Q$ within the strip $\langle c, d \rangle$, then the pair $\{s, s'\}$ is referred to as

- *original intersection* of the segment s with the staircase D within the strip $\langle c, d \rangle$ if s' is an original segment of D;

- *multiple inherited intersection* of the segment s with the staircase D within the strip $\langle c, d \rangle$ if s' is an inherited segment of D, and the segment s intersects at least one more inherited segment of D within the strip $\langle c, d \rangle$;

- *single inherited intersection* of the segment s with the staircase D within the strip $\langle c, d \rangle$ in other cases.

Let $Int_{c,d}^{or}(D, S)$, $Int_{c,d}^{ml}(D, S)$ and $Int_{c,d}^{sn}(D, S)$ denote the sets of the original, multiple inherited and single inherited intersections of the segments of a set S with a staircase D within a strip $\langle c, d \rangle$, respectively (see Fig.9).

Lemma 3 $|S'_v| \leq e_v - b_v + |Int_{b_v,e_v}^{or}(D_v, S'_v)| + |Int_{b_v,e_v}^{or}(D_{ft(v)}, S'_v)| + |Int_{b_v,e_v}^{ml}(D_{ft(v)}, S'_v)|$.

Proof. Let us suppose that for some node v the above stated property is not true. Then set S'_v must contain a segment s spanning the strip $\langle b_v, e_v \rangle$, having no original intersections with the staircases D_v and $D_{ft(v)}$ and intersecting at most one inherited stair of $D_{ft(v)}$ (inside the strip). This contradicts the condition 4 during D_v construction.□

Lemma 4 $|Int_{c,d}^{ml}(D_v, \{s\})| \leq \frac{1}{2}(|Int_{c,d}^{or}(D_{ft(v)}, \{s\})| + |Int_{c,d}^{ml}(D_{ft(v)}, \{s\})|)$.

Proof. Let $|Int_{c,d}^{ml}(D_v, \{s\})| = n$, $n \geq 2$ by the definition. Since Q_v could inherit each 4th segment of $Q_{ft(v)}$, $|Int_{c,d}(Q_{ft(v)}, \{s\})| \geq 4n - 3$ (see Fig.9). But $|Int_{c,d}(Q_{ft(v)}, \{s\})| = |Int_{c,d}^{or}(D_{ft(v)}, \{s\})| + |Int_{c,d}^{sn}(D_{ft(v)}, \{s\})| + |Int_{c,d}^{ml}(D_{ft(v)}, \{s\})|$ and $|Int_{c,d}^{sn}(D_{ft(v)}, \{s\})| \leq 1$. As a result, we have $|Int_{c,d}^{or}(D_{ft(v)}, \{s\})| + |Int_{c,d}^{ml}(D_{ft(v)}, \{s\})| \geq 4n - 4 \geq 2|Int_{c,d}^{ml}(D_v, \{s\})|$.□

Let $H(i)$ be the set of the recursion tree nodes with depth i. Let h_m be the height of the tree minus one and $anc(v, i)$ be the ancestor of the node v having depth i. We denote the sums $\sum_{v \in H(i)} |Int_{b_v,e_v}^{or}(D_{anc(v,j)}, S'_v)|$ and $\sum_{v \in H(i)} |Int_{b_v,e_v}^{ml}(D_{anc(v,j)}, S'_v)|$ as $M_{or}(i, j)$ and $M_{ml}(i, j)$, respectively. According to Lemma 4 $M_{ml}(i, j) \leq 1/2(M_{or}(i, j-1) + M_{ml}(i, j-1))$. Evidently, we have $M_{ml}(i, 1) = 0$.

Lemma 5 $\sum_{i=j+1}^{h_m} M_{or}(i, i-j) \leq K \; \forall j \in [0, .., h_m - 1]$.

Proof. The sum presents the intersecting pairs (possibly not all) counted in some way. To prove the statement it is sufficient to show that each pair is counted at most once. $\sum_{i=j+1}^{h_m} M_{or}(i, i-j) = \sum_{i=j+1}^{h_m} \sum_{v \in H(i)} |Int_{b_v,e_v}^{or}(D_{anc(v,i-j)}, S'_v)|$. Assume that the pair $\{s1, s2\}$ intersecting in a point p is counted twice and belongs to $Int_{b_v,e_v}^{or}(D_{anc(v,h_v-j)}, S'_v)$ and $Int_{b_w,e_w}^{or}(D_{anc(w,h_w-j)}, S'_w)$ simultaneously, where h_v and h_w are the depth of nodes v and w, respectively. Then, the point p should belong to the strips $\langle b_v, e_v \rangle$ and $\langle b_w, e_w \rangle$ simultaneously. Therefore, either the node v is an ancestor of w or the node w is an ancestor of v. W.l.o.g. assume the latter is the case, then, the node $anc(w, h_w - j)$ is an ancestor of the node $anc(v, h_v - j)$. Since $Or_{anc(v,h_v-j)} \cap Or_{anc(w,h_w-j)} = \emptyset$, the pair $\{s1, s2\}$ cannot belong to $Int_{b_v,e_v}^{or}(D_{anc(v,h_v-j)}, S'_v)$ and $Int_{b_w,e_w}^{or}(D_{anc(w,h_w-j)}, S'_w)$ simultaneously.□

Lemma 6 $\sum_{i=j+1}^{h_m} M_{ml}(i, i-j) \leq K \; \forall j \in [0, .., h_m - 1]$.

Proof. $\sum_{i=j+1}^{h_m} M_{ml}(i, i-j) \leq \frac{1}{2} \sum_{i=j+2}^{h_m} (M_{or}(i, i-(j+1)) + M_{ml}(i, i-(j+2))) \leq \frac{1}{2} \sum_{i=j+2}^{h_m} M_{or}(i, i-(j+1)) + \frac{1}{4} \sum_{i=j+3}^{h_m} (M_{or}(i, i-(j+2)) + M_{ml}(i, i-(j+2))) \leq ... \leq \sum_{n=1}^{h_m-j} \frac{1}{2^n} \sum_{i=j+n+1}^{h_m} M_{or}(i, i-(j+n)) \leq \sum_{n=1}^{h_m-j} \frac{1}{2^n} K \leq K$.□

Theorem 4 $\sum_{v \in V} |S'_v| \leq 2N \lceil \log N + 1 \rceil + 3K$.

Proof. $\sum_{v \in V} |S'_v| \leq \sum_{v \in V} (b_v - e_v + |Int_{b_v,e_v}^{or}(D_v, S'_v)| + |Int_{b_v,e_v}^{or}(D_{ft(v)}, S'_v)| + |Int_{b_v,e_v}^{ml}(D_{ft(v)}, S'_v)|) \leq 2N \lceil \log N + 1 \rceil + \sum_{i=1}^{h_m} M_{or}(i, i) + \sum_{i=2}^{h_m} M_{or}(i, i-1) + \sum_{i=2}^{h_m} M_{ml}(i, i-1) \leq 2N \lceil \log N + 1 \rceil + 3K$.□

Theorem 5 $\sum_{v \in V} |Int(D_v, S'_v)| \leq 2N\lceil \log N + 1 \rceil + 5K$.

Proof. $|Int(D_v, S'_v)| = |Int^{or}_{b_v, e_v}(D_v, S'_v)| + |Int^{sn}_{b_v, e_v}(D_v, S'_v)| + |Int^{ml}_{b_v, e_v}(D_v, S'_v)|$. According to the definition of the single inherited intersection $|Int^{sn}_{b_v, e_v}(D_v, S'_v)| \leq |S'_v|$, hence, $\sum_{v \in V} |Int(D_v, S'_v)| \leq \sum_{v \in V} (|Int^{or}_{b_v, e_v}(D_v, S'_v)| + |S'_v| + |Int^{ml}_{b_v, e_v}(D_v, S'_v)|)$. To prove the theorem it is sufficient to apply Lemmas 5, 6 and theorem 4 to the last sum. □

Theorem 6 $\sum_{v \in V} |S_v| \leq N\lceil 4\log N + 5 \rceil + 6K$.

Proof. For all nodes except for the root r $|S_v| \leq |S'_{ft(v)}|$, hence, $\sum_{v \in RT} |S_v| \leq |S_r| + \sum_{v \in RT \setminus r} |S_{ft(v)}| \leq N + 2\sum_{v \in V} |S'_v| \leq N\lceil 4 log N + 5 \rceil + 6K$. □

As $Or_v \in S_v$, hence, $\sum_{v \in V} |Or_v| \leq N\lceil 4 log N + 5 \rceil + 6K$ too.

Theorem 7 $\sum_{v \in V} |Q_v| \leq 2N\lceil 4\log N + 5 \rceil + 12K$.

Proof. Let us designate the sums $\sum_{v \in H(i)} |Or_v|$ and $\sum_{v \in H(i)} |Q_v|$ through $F(i)$ and $G(i)$, respectively. According to construction $G(1) = F(1)$. Each internal node has two sons, each son inherits not more than a quarter of the father stairs, hence, $G(i+1) \leq F(i+1) + \frac{1}{2}G(i)$. Therefore, $\sum_{i=1}^{h_m} G(i) \leq 2\sum_{i=1}^{h_m} F(i)$. □

Theorem 8 *The procedure IntersectingPairs(S_0) with new version of the procedure TreeSearch runs in time $O(N \log N + K)$.*

Proof. Results from the proposition 2 and the theorems 5, 6, 7.

Theorem 9 *The memory requirement of resulting algorithm is $O(N)$.*

Proof. Evidently the active call occupies $O(N)$ memory each of its "ancestors" retains $O(|I_v| + |Q_v|)$ memory, other structures use $O(N)$ memory. It is sufficient to prove that for any path pt from the root of recursion tree to some of its node $\sum_{v \in pt}(|I_v| + |Q_v|) = O(N)$. We have $|I_v| \leq e_v - b_v$, $\sum_{v \in pt}(e_v - b_v) \leq N$, hence $\sum_{v \in pt} |I_v| \leq N$. If $v \in pt$ and $w \in pt$ then $Or_v \cap Or_w = \emptyset$, hence $\sum_{v \in pt} |Or_v| \leq N$. According to construction $|Q_v| \leq |Or_v| + \frac{1}{4}|Q_{ft(v)}|$, hence $\sum_{v \in pt} |Q_v| \leq \frac{4}{3} \sum_{v \in pt} |Or_v| \leq \frac{4}{3}N$. □

CONCLUSIONS

In this work we have proposed two algorithms for finding all K intersecting pairs among N segments in the plane. They have $O(N \log^2 N + K)$- and $O(N \log N + K)$-time complexity and require $O(N)$ space. The algorithms can operate not only with line segments. Segments may be any connected 2D objects which have at the most one intersection with any vertical line and permit the following operations to be performed in $O(1)$ time

- Given a segment and a vertical line. Find their intersection point.

- Given a vertical strip and a pair of segments. Determine if the segments intersect within the strip.

In this case the algorithms have time and space complexity stated above, the only difference is that K denote the number of intersection point (in assumption that all intersection points are different).

References

[1] Ben-Or, M. Lower bounds for algebraic computation trees. In *Proceedings of the 15th Annual ACM Symposium Theory of Computing.* (1983), 80–86.

[2] Bentley, J.L., and Ottman, T. Algorithms for reporting and counting geometric intersections. *IEEE Trans. Comput. C-28* (1979), 643-647.

[3] Chazelle, B. Filtering search: A new approach to query-answering. *SIAM J. Comput. Vol.15* (1986), 703-725.

[4] Chazelle, B. Reporting and counting segment intersections. *J. Comput. Sys. Sci. 32* (1986), 156-182.

[5] Chazelle, B., and Edelsbrunner, H. An optimal algorithm for intersecting line segments in the plane. *Journal of the ACM, Vol. 39* (1992), 1-54.

[6] Clarkson, K.L., and Shor, P.W. Applications of random sampling in computational geometry, II. *Discrete and Computational Geometry, Vol. 4* (1989), 387–421.

[7] Edelsbrunner, H., Guibas L.J., Stolfi J. Optimal point location in monotone subdivision. *SIAM J. Comput. Vol.15(2)* (1986), 317-340.

[8] Lueker, G. A data structure for orthogonal range queries. In *Proceedings of the 19th annual IEEE Symposium on Foundations of Computer Science.*(1979), 28–34.

[9] Mulmuley, K. A fast planar partition algorithm. In *Proceedings of the 29th annual IEEE Symposium on Foundations of Computer Science.*(1988), 580–589.

[10] Preparata, F.P., and Shamos, M.I. *Computational geometry: An introduction.* Springer-Verlag, New York, 1985.

[11] Willard, D. Poligon retrieval. *SIAM J. Comput. Vol.11* (1982), 149-165.

Triangulations Intersect Nicely

Oswin Aichholzer
Franz Aurenhammer
Michael Taschwer
Institute for Theoretical Computer Science
Graz University of Technology
Klosterwiesgasse 32/2, A-8010 Graz, Austria
e-mail: {oaich,auren}@igi.tu-graz.ac.at

Günter Rote
Institut für Mathematik
Technische Universität Graz
Steyrergasse 30
A-8010 Graz, Austria
e-mail: rote@ftug.dnet.tu-graz.ac.at

Abstract

We prove that two different triangulations of the same planar point set always intersect in a systematic manner, concerning both their edges and their triangles. As a consequence, improved lower bounds on the weight of a triangulation are obtained by solving an assignment problem. The new bounds cover the previously known bounds and can be computed in polynomial time. As a by-product, an easy-to-recognize class of point sets is exhibited where the minimum-weight triangulation coincides with the greedy triangulation.

1 Introduction

The aim of this paper is to prove and discuss some surprising and rather general intersection properties of planar triangulations.

Given two triangulations of a point set, we can find a matching between their edge sets such that matched edges either cross or coincide. This theorem and a few related statements will be proved in Section 2.

The remaining part of the paper deals with applications of this result to the computation of minimum-weight triangulations. In Section 3 we identify special cases of point sets for which the minimum-weight triangulation can be computed efficiently, see Figure 2. Section 4 offers several lower bounds on the weight of a triangulation, and Section 5 describes some algorithms to compute these bounds. In Section 6 we will discuss possible applications of our results and some open questions.

2 The Matching Theorems

Let P be a set of n points in the plane. We assume that not all points lie on one line. Consider the set E of all line segments connecting two points of P and containing no other points of P. (If P is in general position then $|E| = \binom{n}{2}$.) The elements of E are called *edges*. Two distinct edges are said to cross iff they intersect in their interiors. In particular, two edges sharing only one endpoint are not considered as crossing. A *triangulation* T of P is a maximal set of non-crossing edges. It dissects the convex hull of P into triangular faces. We start by recalling a well-known fact on triangulations.

Lemma 1 *Every triangulation of P consists of $m = 3n - 3 - b$ edges, where b is the number of points in P which lie on the boundary of the convex hull of P. Every set of non-crossing edges consists of at most m edges.* □

The following is the main theorem of this paper.

Theorem 1 *Let P be a finite set of points in the plane and consider two triangulations R and B of P. There exists a perfect matching between R and B, with the property that matched edges either cross or are identical.*

Proof. For notational simplicity, let us color the edges in B as blue and the edges in R as red. We consider the intersection graph, G, of $R \cup B$. G represents each edge in R by a red node and each edge in B by a blue node. (Edges in $R \cap B$ are represented by two nodes.) Two nodes are connected by an arc if the corresponding edges either cross or are identical. Clearly, there are no arcs between nodes of the same color as edges in a single triangulation do not cross. Hence G is bipartite. Note that, by Lemma 1, the number of red nodes equals the number of blue nodes.

We prove that G contains a perfect matching by showing that G fulfills the Hall condition of the marriage theorem (see, e.g., [B]). This condition requires that, for each subset X of red nodes of G, the number of blue neighbors in G is at least $|X|$.

Let $R_1 \subseteq R$, and let B_1 collect each blue edge that either crosses or coincides with an edge in R_1. With this notation, the Hall condition reads $|B_1| \geq |R_1|$. Let B_2 be the complement of B_1 in B. We claim that $B_2 \cup R_1$ is a non-crossing set of edges. The edges of R_1 do not cross each other, since $R_1 \subseteq R$, and likewise, the edges of B_2 do not cross each other. An edge b of B_2 cannot cross an edge of R_1 because b would belong to B_1, otherwise. For the same reason, R_1 and B_2 are disjoint, which gives

$$|R_1| + |B_2| = |R_1 \cup B_2| \leq m = |B|,$$

by Lemma 1. Thus we get $|B_1| = |B| - |B_2| \geq |R_1|$, and the Hall condition is proved. $\square$

It seems natural to try to prove Theorem 1 in a direct way, without resorting to the marriage theorem. For example, any triangulation of a planar point set can be changed into the Delaunay triangulation (and thus into an arbitrary other one) by repeated application of so-called edge flips, see [Law]. An edge flip exchanges the diagonals of a convex quadrilateral in the current triangulation and thus naturally corresponds to a match of the involved edges. We did not succeed in a proof of Theorem 1 based on the flipping paradigm, however.

Corollary 1 *Let P be a finite set of points in the plane. Let R be a set of non-crossing edges between points of P and let T be a triangulation of P. Then there is a matching between the edges in R and some edges in T (an injective mapping from R to T) such that every edge of R is either matched with the identical edge in T or with an edge which crosses it.*

Proof. Since R can be extended to a triangulation the corollary follows from Theorem 1. $\square$

Our proof mainly exploits the property of triangulations expressed in Lemma 1. A version of Lemma 1 concerning triangles instead of edges is obvious.

Lemma 2 *Every triangulation of P dissects the plane into $2n - 2 - b$ interior triangular faces plus the exterior face, where b is the number of points in P which lie on the boundary of the convex hull of P. Every set of non-crossing edges dissects the plane into at most $2n - 1 - b$ connected components.*

This enables us to prove, with exactly the same technique as above, the next theorem on triangles.

Theorem 2 *Let P be a finite set of points in the plane and consider two triangulations R and B of P. There exists a perfect matching between the set of triangles of R and the set of triangles of B, with the property that matched triangles either overlap or are identical.*

Figure 1 displays a perfect triangle matching for two triangulations of a convex polygon.

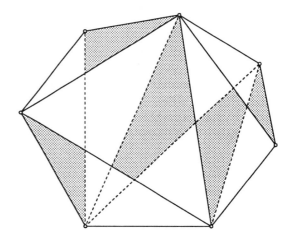

Figure 1: Perfect matching of overlapping triangles between a "solid" and a "broken" triangulation of a 7-gon. Every shaded area is the intersection between two matched triangles.

Our matching theorems have been found independently by Cheng and Xu [CX]. They generalized the result to the framework of arbitrary independence systems.

An *independence system* $\mathcal{I}$ is a non-empty collection of subsets of a (finite) ground set E which is closed under taking subsets: If $A \in \mathcal{I}$ and $B \subset A$ then $B \in \mathcal{I}$. The elements of $\mathcal{I}$ are called the *independent* sets, the remaining subsets of E are called *dependent*. A *circuit* of $\mathcal{I}$ is a minimal dependent set.

In our example of triangulations, a set of non-crossing edges (or of non-overlapping triangles) may be considered independent. The circuits of this independence system have two elements; they are the pairs of crossing edges (or of overlapping triangles, respectively).

Theorem 3 *Let $R \in \mathcal{I}$ be any independent set, and let $B \in \mathcal{I}$ be an independent set of maximum cardinality in $\mathcal{I}$. Then there is an injective mapping $g: R \to B$ such that for every edge $e \in R$ we have $g(e) = e$, or $\{g(e), e\}$ is contained in a circuit.*

The proof shows that the Hall condition is fulfilled, following the lines of the proof of Theorem 1. Theorems 1 and 2 are corollaries of Theorem 3.

Cheng and Xu [CX] also strengthened Theorem 2 by requiring in addition that matched triangles have at least one vertex in common. The proof shows the Hall condition by an argument about sums of angles.

Let us now briefly take the algorithmic standpoint and ask how expensive it is to compute the perfect matchings whose existence is predicted in Theorems 1 and 2. We restrict attention to edge matchings; the situation for triangles is similar. As before, assume we are given a red triangulation R and a blue triangulation B of a set P of n points. The first approach that comes into mind is to use a standard bipartite matching algorithm for the intersection graph defined by R and B.

In a first pass, edges showing up in both triangulations are eliminated since their matches are determined uniquely. In particular, edges bounding the convex hull of P are those. The intersection graph, G, of the remaining edges consists of $O(n)$ nodes and $k = O(n^2)$ arcs, where k is the total number of crossings between R and B. We compute a perfect matching in G by applying the bipartite matching algorithm of Hopcroft and Karp [HK]; see also [T]. A direct application would result in $O(k\sqrt{n})$ time and $O(k)$ space. We can reduce the space requirement to $O(n)$ by extracting the adjacency lists of G's nodes on-line. In an $O(n)$-time preprocessing step we locate, for each point $p \in P$, the incident red edges among incident blue triangles by merging two sorted lists of angles. To obtain the neighbors of, say, a red node we simply proceed along the red edge it represents, as long as this edge intersects blue triangles.

This result is quite satisfactory for situations where the number k of crossings is not substantially larger than the size n of the underlying set of points. If $k = \Theta(n^2)$, a $\Theta(n^2\sqrt{n})$ time algorithm is obtained. As both the input (two triangulations) and the output (the matched edge pairs) are of size $O(n)$, a solution that exploits the geometric nature of the problem in a better way is desirable.

The idea of constructing a perfect matching in a greedy manner suggests itself: After matching identical edges, we match each unmatched red edge with the first yet unmatched blue edge crossing it which we find. From the theory of bipartite graphs it is known that the greedy method always matches at least half of the nodes of the graph, provided it contains a perfect matching. Hence at least half of the matches among crossing edges will be found. However, the illusion that geometry would be strong enough to enforce a perfect matching is destroyed by an easy example.

By observing that the (at least three) convex hull edges are matched by the greedy method anyway, we get:

Lemma 3 *Any greedy matching leaves strictly less than half of the edges unmatched.*

We do not further pursue the issue of finding a perfect matching here. We rather concentrate on applications of the edge matching theorem to minimum-weight triangulations which we consider —apart from the matching theorems themselves— as the most interesting part of our paper.

3 Light Triangulations

The intersection properties of planar triangulations expressed in the foregoing section should have various applications in combinatorial and computational geometry. In the present paper, two applications of Theorem 1 (perfect matching between edges) to minimum-weight triangulations are demonstrated.

In this section, we exhibit a class of planar point sets where the minimum-weight triangulation coincides with the greedy triangulation and thus can be computed easily and in polynomial time. An efficient algorithm for recognizing such point sets is given in Section 5.

As before, let P be a set of n points in the plane, and let E be the set of all edges defined by P. The length (weight) of an edge $e = pq \in E$ is just the Euclidean distance between p and q and

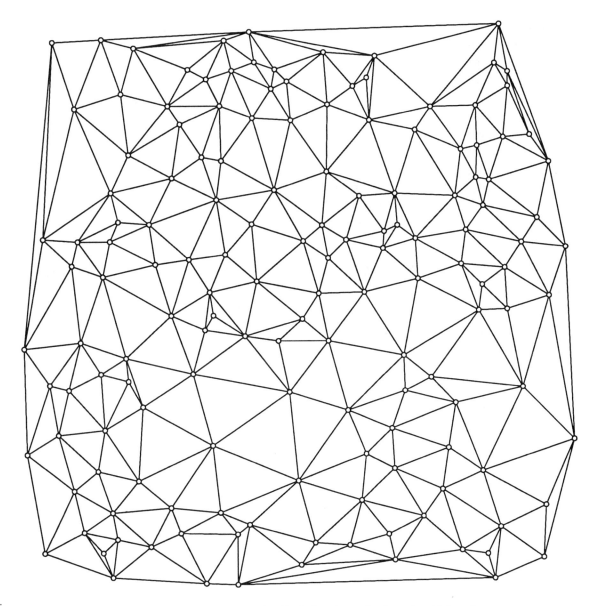

Figure 2: Minimum weight (light) triangulation for 150 points.

it will be denoted by $|e|$ or $|pq|$. To keep things simple, we will henceforth assume general position so that all $\binom{n}{2}$ distances between points of P are distinct. (Otherwise, if a coincidence of two edges with the same length happens, we may for example give preference to the edge which comes first in some given prespecified numbering.) The weight $w(R)$ of a set R of edges is the sum of the lengths of its edges.

The *minimum-weight triangulation* T^* is defined to have $w(T^*) \leq w(T)$ for all triangulations T of P. Computing a minimum-weight triangulation is an important and interesting problem, whose complexity status is unknown. It is one of the open problems listed at the end of Garey and Johnson's book about NP-completeness [GJ].

We first reconsider a type of triangulations that has already been looked at from various points of view. (We refer the reader to [Lam] for a comprehensive survey on triangulations.) A *locally short triangulation* of the point set P is one which includes, for each convex quadrilateral Q it contains, the shorter diagonal of Q. Five points are enough to see that the locally short triangulation is not unique. Any given triangulation of P can be made locally short by finitely many diagonal flips, as each flip decreases the total edge length. Unfortunately, a locally short triangulation fails to fulfill length optimization criteria. It is well known that the flipping procedure can get caught in a local optimum: It need not yield the minimum-weight triangulation. On the other hand note that a minimum-weight triangulation must of course be locally short.

Our interest here is in a subclass of locally short triangulations. Let us call an edge $e \in E$ *light* if there is no edge in E that crosses e and is shorter than e. Light edges obviously do not cross (under the above general position assumption), so the set L of light edges can form at most a triangulation of P. If L actually is a triangulation then it has to be unique and, of course, locally short. We then call L the *light triangulation* of P. See Figure 2 for an example. (In fact, it is quite easy to draw a light triangulation by 'well' distributing the points).

In the light of Theorem 1, it is easy to prove length optimality.

Theorem 4 *If a planar point set P admits a light triangulation then this is the minimum-weight triangulation for P.*

Proof. Let T be any triangulation of P. We show $w(T) \geq w(L)$, for L being the light triangulation of P. Consider a perfect matching as in Theorem 1 between the edges of T and the edges of L. For each matched pair of edges $e \in T$ and $e' \in L$, either $e = e'$ or e crosses e', in which case we know that $|e| > |e'|$ by the lightness of L. Summing over all edges gives $w(T) \geq w(L)$. □

A popular kind of locally short triangulations is the *greedy triangulation*. This structure is defined procedurally. It is obtained by iteratively inserting the shortest edge of E that does not cross previously inserted edges. While, in general, a locally short triangulation need not contain all light edges (Figure 3), the greedy triangulation does so. A light edge e never can be blocked by previously inserted edges as E does not contain any shorter edge crossing e. So, if a light triangulation exists for P, it can be found quickly by applying one of the various known greedy triangulation algorithms. More importantly, by Theorem 4 the minimum-weight triangulation of P can be computed in polynomial time in this case, in fact in $O(n^2)$ time and $O(n)$ space, see Levcopoulos and Lingas [LL] and Wang [W]. See also Goldman [G] for an easy-to-implement $O(n^2 \log n)$ time variant, and [DDMW, DRA] for fast expected-time algorithms.

Selecting all light edges from a given greedy triangulation takes polynomial time, naively $O(n^3)$. We will improve this bound to $O(n^2 \log n)$ in Section 5.

It is noteworthy that, in general, not all light edges occur in a minimum-weight triangulation T^*.

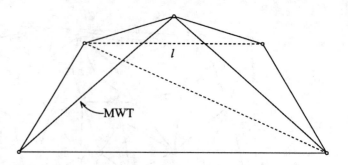

Figure 3: A light edge l not included in the minimum-weight triangulation T^*.

Figure 3 gives an example. However, recent results [K, YXY] imply that certain subsets of light edges always have to be there. Keil [K] shows that T^* always includes the so-called $\sqrt{2}$-skeleton of P. This structure consists of all edges pq such that the two circles through both p and q and with diameter $\sqrt{2}|pq|$ are empty of points in P; see Figure 4. It

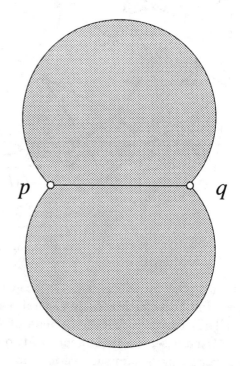

Figure 4: Keil's empty region defining the $\sqrt{2}$-skeleton.

is not difficult to verify that this condition forces $|pq|$ to be light. The $\sqrt{2}$-skeleton is a subset of the Gabriel graph (1-skeleton) which itself is known to be not part of the minimum-weight triangulation.

Yang et al. [YXY] recently proved that a condition stronger than lightness implies inclusion of pq in T^*: For any edge xy crossing pq, $|pq| \leq \min\{|px|, |py|, |qx|, |qy|\}$. By the quadrilateral inequality, xy then has to be the longest of the six edges involved. Hence pq is a light edge, and we obtain the region in Figure 5 whose emptiness implies $pq \in T^*$.

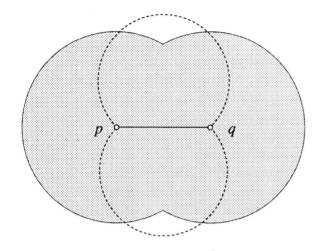

Figure 5: Empty region of Yang et al.

4 Lower Bounds for $w(T^*)$

In this section, we offer new methods for computing lower bounds for the weight of planar triangulations. Apart from the trivial lower bound (the sum of the $|T^*|$ shortest edges in E), all bounds known to us require the knowledge of a set of edges which is a subset of T^*. Though there is recent insight into this problem [K, YXY], the quality of such bounds usually suffers from the small cardinality of such subsets. Theorem 1 allows us to prove a lower bound for $w(T^*)$ in a completely different way by solving an assignment problem.

Theorem 5 *Let R be a non-crossing set of edges and let T^* be the minimum-weight triangulation. Let $X(R, E)$ denote the set of all injective mappings (matchings) $g: R \to E$ with the properties required in Theorem 1: $g(e) = e$ or $g(e)$ crosses e. Then we have the following lower bound.*

$$w(T^*) \geq \min_{g \in X(R,E)} \sum_{e \in R} |g(e)|. \quad (1)$$

Proof. Let $g^* \in X(R, E)$ be a matching $g^*: R \to T^*$ whose existence is guaranteed by Theorem 1 and Corollary 1. We have

$$w(T^*) = \sum_{e \in T^*} |e| \geq \sum_{e \in R} |g^*(e)| \geq \min_{g \in X(R,E)} \sum_{e \in R} |g(e)|.$$

□

The set $X(R, E)$ is just the set of complete matchings in a bipartite graph, and hence we can optimize over this set in polynomial time. Note that edges in the set $g(R) = \{ g(e) \mid e \in R \}$ may cross each other. Solving the assignment problem with the constraint that $g(R)$ is non-crossing yields a solution to the original minimum-weight triangulation problem. (However, in its general form the assignment problem can be shown to be NP-complete by reduction from independent set [GJ].)

Stated the other way, when R is a triangulation and $g_{opt}(R)$ happens to be non-crossing for the optimal assignment g_{opt} in (1) then $g_{opt}(R)$ must be minimum-weight.

Corollary 2 *Let R be a triangulation and let $g_{opt} \in X(R, E)$ satisfy*

$$\sum_{e \in R} |g_{opt}(e)| = \min_{g \in X(R,E)} \sum_{e \in R} |g(e)|.$$

If the set $g_{opt}(R)$ is non-crossing then it is a minimum-weight triangulation.

The optimal assignment g_{opt} has another nice property.

Theorem 6 *The optimal assignment matches no edge e to a longer edge, that is, $|g_{opt}(e)| \leq |e|$ for all $e \in R$.*

Proof. Suppose that a matching $g \in X(R, E)$ has $|g(e)| > |e|$ for some edge e. Then we can improve g by assigning e to itself, i.e., by setting $g(e) = e$ and leaving g unchanged for the remaining edges. Since no other edge in R crosses e we still obtain a matching, and hence the original matching g cannot be optimal. □

If we relax the requirements on e in Theorem 5 and permit all injective mappings from R to E we get the trivial lower bound: the sum of the $|R|$ shortest edges in E. This shows that our bound can never be worse than the trivial one.

We may weaken the bound of Theorem 5 in a different way, by neglecting the fact that the matching g must assign different edges of R to different edges of T^*. The resulting bound is very easy to compute once some auxiliary information on the edges in R has been precomputed. For an edge $e \in E$, let us define its *excess* $\varepsilon(e)$ as

$$\varepsilon(e) = \max\{0, |e| - \lambda(e)\},$$

where $\lambda(e)$ is the length of the shortest edge crossing e. If e is crossed by no edge (for example, if e is an edge of the convex hull) then we set $\lambda(e) = \infty$, hence $\varepsilon(e) = 0$. Edges which have no crossing edges are called *unavoidable* edges; they belong to every triangulation, see Xu [X]. An edge has no shorter edge which crosses it iff its excess vanishes. We have called such edges light in the preceding section.

Theorem 7 *Let R be a non-crossing set of edges and let T^* be a minimum-weight triangulation. Then*

$$w(T^*) \geq w(R) - \sum_{e \in R} \varepsilon(e).$$

Proof. Let $g^*: R \to T^*$ be a matching as in the proof of Theorem 5. Then for all $e \in R$, either $g^*(e)$ crosses e or $g^*(e) = e$. In the first case, we have

$$|e| - \varepsilon(e) \leq |g^*(e)| \qquad (2)$$

by the definition of the excess $\varepsilon(e)$. In the second case, the same inequality holds because $\varepsilon(e) \geq 0$. By summing (2) over all $e \in R$ we get

$$w(R) - \sum_{e \in R} \varepsilon(e) \leq \sum_{e \in R} |g^*(e)| \leq \sum_{e \in T^*} |e| = w(T^*).$$

□

The consequences of this theorem for the set $L \subset E$ of light edges are particularly interesting. Since L is a non-crossing set and since $\varepsilon(e) = 0$ for all edges $e \in L$, we have:

Corollary 3

$$w(T^*) \geq w(L).$$

We stress the fact that, despite $L \not\subseteq T^*$ in general, summing up the light edges provides a valid lower bound. This bound covers the bounds in [K, YXY] as the subsets of T^* considered there are built of light edges only; see Section 3. Generally, if $R' \subset R$, then the bound given by R in Theorem 7 is stronger than the bound given by R'. Thus it pays off to complete the non-crossing set R (or L) to a triangulation before applying Theorem 7. Note finally that if L happens to be a triangulation then $w(L) = w(T^*)$ and we obtain Theorem 4 of Section 3.

5 Algorithmic Issues

We now address the algorithmic issues raised in Sections 3 and 4.

5.1 Computing excesses and light edges

Theorem 4 asks for a method to decide whether a given planar n-point set P admits a light triangulation, or in other words, whether the light edges triangulate P. Also, to compute the lower bound in Corollary 3 all light edges realized by P are required. The greedy triangulation contains all light edges. So we only need to test greedy edges for lightness. More generally, the bound in Theorem 7 needs knowledge of the excess of all edges in a given non-crossing set.

We solve these problems in time $O(n^2 \log n)$ and space $O(n)$ by giving an algorithm for the following problem: Given an arbitrary triangulation T of P, compute $\lambda(e)$ for each $e \in T$, where $\lambda(e)$ is the length of the shortest edge crossing e. As a by-product, all unavoidable edges spanned by P are detected as those getting $\lambda(e)$ never assigned.

We describe an $O(n \log n)$-time routine which, for a given point $p \in P$, computes

$$\lambda_p(e) = \min_{q \in P}\{\,|pq|\mid pq \text{ crosses } e\,\}$$

for all $e \in T$. Calling the routine for all $p \in P$ and maintaining the minimum for each e then gives $\lambda(e)$.

Given the point p, we set up a semi-dynamic data structure for a point set Q and for the following type of queries: For a wedge V with apex p and angle less than π, return the closest point to p in $Q \cap V$. We initialize Q with P and allow only deletions from Q. The data structure we use is a binary search tree whose leaves store Q in cyclic order around p and whose interior nodes store minimum distances from p to all points in the subtree.

We then query the structure with wedges spanned by edges e in the triangulation T. The queries and deletions are carried out in a specific order. This guarantees that, when we query for the wedge belonging to edge e, Q contains no point that lies between e and p and includes all points that lie opposite to p with respect to e.

To this end, a topological order for the edges of T with respect to the in-front/behind relation as seen from p is necessary. Problems may arise if this relation contains cycles. (The class of regular triangulations, and thus the Delaunay triangulation, is acyclic in this sense [DFPN, E], but a general triangulation is not.) We circumvent this difficulty by cutting T with a ray emanating from p, and re-triangulating where necessary, in $O(n)$

time. For an edge e cut into e' and e'' we clearly have $\lambda_p(e) = \min\{\lambda_p(e'), \lambda_p(e'')\}$.

A topological order now is obtained in $O(n)$ time by starting with the set of triangles incident to p, and adding adjacent triangles one by one while keeping the invariant that the boundary of their union U is star-shaped as seen from p. At each point in time, and the search tree stores exactly the points of P exterior to U. As queries and deletions take $O(\log n)$ time each, the claimed $O(n \log n)$ time bound is obtained.

Theorem 8 *Let P be a set of n points in the plane.*

(i) Given a triangulation T of P, the excesses for all edges in T can be computed in $O(n^2 \log n)$ time and $O(n)$ space.

(ii) The excesses for all edges spanned by P can be computed in $O(n^3 \log n)$ time and $O(n^2)$ space.

(iii) The set of light edges spanned by P can be computed in $O(n^2 \log n)$ time and $O(n)$ space.

Proof. Statement (i) follows from the foregoing discussion. Statement (ii) follows from (i) since we can cover all edges by n triangulations: We simply connect a fixed point p to all other points and complete this to a triangulation. Repeating this for all points p, we will have covered every edge between two points of P at least once. For proving (iii) it is sufficient to note that the greedy triangulation of P can be computed in the claimed time and space bounds. □

5.2 Computing the strong optimal assignment bound

The problem whose solution is required for applying Theorem 5 is an assignment problem between two edge sets R and E, where R is a given triangulation and E is the set of all edges between points of P. (In fact, R could be any non-crossing set but we restrict our attention to a triangulation.) The straightforward application of the standard algorithm for the assignment problem leads to a complexity of $O(n^4)$ time and $O(n^3)$ space.

Note, however, that the weights are in fact not associated with the *arcs* of the bipartite graph in which we find the matching but with the *node* set E. The subsets of nodes in a graph which are matchable (for which there exists a matching which covers all of them) is a matroid, in our case of a bipartite graph a transversal matroid, see [La, Corollary 7.4.3, p. 272, and Section 5.4, p. 192]. (This is in contrast to the independence system of the arc sets which form a matching.) It follows from basic matroid theory (see [La, T]) that it is possible to solve the problem by a greedy algorithm on the set E: We start with the empty matching and process the edges of E in order of increasing weight. For each edge we try to augment the current matching to include the corresponding node. A node which is matched in this process will never become unmatched. We may stop if we have $|R|$ matching pairs.

Let us focus on the form of an augmenting path which we are looking for when processing an edge e_0. The path goes from e_0 to an adjacent node r_1 (an edge of R), from r_1 to the node e_1 to which r_1 is matched, from e_1 to an adjacent node r_2, from r_2 to the node e_2 to which r_2 is matched, and so on until it terminates in an unmatched node $r_i \in R$, $i \geq 1$. The augmentation consists of exchanging the $i - 1$ matching arcs with the i non-matching arcs on this path.

We may search for such an augmenting path starting from e_0 using any graph search method, for example breadth-first search. We make several observations:

(i) We only need to explore the list of adjacent nodes for the edges of E; when we are at a node of R we simply proceed to the node of E to which it is matched. Since the neighbors of $e \in E$ are the edges of the triangulation R which cross e, we can find these neighbors by walking through the triangulation, without any need to store adjacency lists.

(ii) If a node of E cannot be matched we can ignore it in the future. We need not remember to which nodes it is adjacent. Thus if $e \in E$ is not matched in the optimal solution, its incident arcs are explored only once during the whole algorithm.

(iii) If we have processed a node in R or E during an unsuccessful search for an augmenting path we know that no unmatched node of R can be reached from this node, and we may skip the search from this node in all subsequent searches. This situation remains unchanged as long as no augmentation occurs. Thus, between two successful augmentations, every edge of the *current graph* is visited at most once. The current graph is the graph spanned by the nodes R and the subset E' of nodes of E which are currently matched.

(iv) There are $|R| = O(n)$ successful augmentation steps. Hence, if $E_{opt} = \{\, g_{opt}(e) \mid e \in R \,\}$ denotes the set of matched nodes of E in the optimal matching g_{opt}, each of the arcs between R and E_{opt} is visited at most $O(n)$ times.

The only operation which we have not discussed is the generation of all edges in E according to length. Dickerson et al. [DDS] have shown that the m shortest edges in an n-point set P may be generated in $O((m+n)\log n)$ time and $O(m+n)$ space. By trying successively the values $m = n, 2n, 4n, 8n, \ldots$ we can make sure that we never generate more than twice as many edges as we actually need. The $O(\log n)$ overhead for each of the m edges generated is also sufficient to locate the first triangle in the triangulation R which is cut by that edge. Summarizing, we have the following theorem.

Theorem 9 *Let $E_{opt} = \{\, g_{opt}(e) \mid e \in R \,\}$ denote the set of edges to which the edges of R are assigned in the optimal matching g_{opt} of (1) in Theorem 5. Let $K = O(n^2)$ denote the number of crossings between E_{opt} and R.*

Furthermore let $E_s \subseteq E$ be the set of edges which are not longer than the longest edge of E_{opt}, and let $L \leq |E_s| \cdot |R| = O(n^3)$ denote the number of crossings between E_s and R.

Then the optimal matching g_{opt} can be computed in $O(Kn + L + |E_s|\log n) = O((K + |E_s|)n) = O(n^3)$ time and $O(n + |E_s|) = O(n^2)$ space.

Since we do not know the largest edge of E_{opt} beforehand we may take, by Theorem 6, the largest edge weight of R in the definition of E_s in order to get an a-priori upper estimation. In either case, unavoidable edges such as the boundary edges, which belong to every triangulation, can be ignored when computing the longest edge of E_{opt} or R.

The worst-case time bound of $O(n^3)$ is rather high, but the explicit parameters of the complexity indicate that for a good starting triangulation R the complexity might be quite good. (Within the worst-case time of $O(n^3)$, E_s can also be enumerated in $O(n)$ space instead of $O(n+|E_s|)$ by a more primitive method.)

Theorem 6 also implies that we may skip every arc between $e \in E$ and a node $r \in R$ which is shorter than e. However, we can make use of this only at the expense of storing the current graph.

6 Conclusion and Open Problems

We are planning to use our bounds in a branch-and-bound algorithm to compute the minimum-weight triangulation for arbitrary point sets. The bounds of Theorems 5 and 7 can be strengthened for subproblems where some specified edges are forced into the solution or excluded from the solution.

We hope that a practically efficient algorithm for computing the minimum-weight triangulation may make experiments possible which lead to a better understanding of the properties of minimum-weight triangulations, with the ultimate goal of resolving the complexity status of the problem.

Various open questions are raised by the results in the present paper. Below is a list of what seemed most interesting to us.

The point sets for which the light edges form a triangulation are interesting. What good properties do they and their associated triangulations have?

Given a planar n-point set P, can it always be extended by adding, say, $O(n)$ points so that it admits a light triangulation? Can we find such "enlightening" points in polynomial time? These questions might be interesting from the point of view of engineering applications, where only some points on the boundary rather than the complete set of points of the triangulation are fixed in advance.

Can we bound the quality of the bounds of Theorems 5 and 7 when R is, for instance, the greedy triangulation or the Delaunay triangulation? (What is the maximum ratio between the two sides of the inequality?)

Theorems 1 and 2 are easily seen to hold for triangulations of arbitrary polygonal regions, possibly with holes. The latter theorem (perfect matching between overlapping triangles) defines a mapping of a polygonal region onto itself. Do such mappings have applications to geometric searching problems? What properties do such mappings have in the limit, when the number of points approaches infinity?

A triangulation can be viewed either as a set of edges or as a set of triangles. When we take the weight of a triangle to be its perimeter, the minimum-weight triangulation problem is identical for both formulations. (In the triangle formulation, all edges are counted twice except for the boundary edges, which are counted once but are of fixed length.) Nevertheless, we get two different greedy algorithms from the two formulations. We are not aware of investigations of the "triangle-greedy" algorithm. Since there are $O(n^3)$ triangles but only $O(n^2)$ edges it seems plausible that it should beat the usual "edge-greedy" algorithm in practice. (We have examples which show that none of the two greedy algorithms beats the other for all problems.) How fast can the triangle-greedy triangulation be computed?

Acknowledgements. We thank Xu Yin-Feng for sharing his ideas in enlightening discussions. Thanks also go to Scot Drysdale, Herbert Edelsbrunner, and Gerhard Wöginger for discussions on the presented topic.

This reasearch was supported by the Spezialforschungsbereich F 003, *Optimierung und Kontrolle*. Research of Oswin Aichholzer was partially supported by the Austrian Ministry of Science and the Jubiläumsfond der Österreichischen Nationalbank.

References

[B] B. Bollobás, *Graph Theory. An Introductory Course*, Springer Verlag, 1979.

[CX] S.-W. Cheng and Y.-F. Xu, *Constrained independence system and triangulations of planar point sets*, Manuscript, Dept. of Comput. Sci., The Hong Kong University of Science & Technology, 1994; submitted to COCOON'95, First Ann. Int. Computing and Combinatorics Conf., Xi'an, China, August 24-26, 1995.

[DFPN] L. de Floriani, B. Falcidieno, C. Pienovi, and G. Nagy, *On sorting triangles in a Delaunay tessellation*, Report, Ist. Mat. Appl., Consiglio Nazionale delle Ricerche, Genova, Italy, 1988.

[DDMW] M. Dickerson, R. L. Drysdale, S. McElfresh, and E. Welzl, *Fast greedy triangulation algorithms*, Proc. 10th Ann. Symp. Computational Geometry (1994), 211-220.

[DDS] M. Dickerson, R. L. Drysdale, and J.-R. Sack, *Simple algorithms for enumerating interpoint distances and finding k nearest neighbors*, Int. J. Computational Geometry & Appl. 3 (1992), 221-239.

[DRA] R. L. Drysdale, G. Rote, O. Aichholzer, *A simple linear time greedy triangulation algorithm for uniformly distributed points*, Manuscript, 1994.

[E] H. Edelsbrunner, *An acyclicity theorem for cell complexes in d dimensions*, Combinatorica **10** (1990), 252-260.

[GJ] M. Garey and D. Johnson, *Computers and Intractability. A Guide to the Theory of NP-completeness*, Freeman, 1979.

[G] S. Goldman, *A space efficient greedy triangulation algorithm*, Inf. Process. Lett. **31** (1989), 191-196.

[HK] J. E. Hopcroft and R. Karp, *An $n^{\frac{5}{2}}$ algorithm for maximum matchings in bipartite graphs*, SIAM J. Comput. **2** (1973), 225-231.

[K] M. Keil, *Computing a subgraph of the minimum weight triangulation*, Computational Geometry: Theory and Applications **4** (1994), 13-26.

[LL] C. Levcopoulos and A. Lingas, *Fast algorithms for greedy triangulation*, BIT **32** (1992), 280-296.

[Lam] T. Lambert, *Empty-shape triangulation algorithms*, Ph. D. Thesis, Dept. of Comput. Sci., Univ. of Manitoba, 1993.

[La] E. Lawler, *Combinatorial Optimization: Networks and Matroids*, Holt, Rinehart, and Winston, New York, 1976.

[Law] C. L. Lawson, *Software for C^1 surface interpolation*, In: Mathematical Software III, J. Rice (ed.), Academic Press, New York, 1977.

[T] R. E. Tarjan, *Data Structures and Network Algorithms*, SIAM Press, Philadelphia 1987.

[W] C. A. Wang, *Efficiently updating constrained Delaunay triangulations*, BIT **33** (1993), 238-252.

[X] Y. Xu, *Minimum weight triangulation problem of a planar point set*, Ph. D. thesis, Institute of Applied Mathematics, Academia Sinica, Beijing, 1992.

[YXY] B.-T. Yang, Y.-F. Xu, and Z.-Y. You, *A chain decomposition algorithm for the proof of a property on minimum weight triangulations*, In: Algorithms and Computation, Proc. Conf., Beijing 1994, Lecture Notes in Computer Science 834, Springer-Verlag, 1994, pp. 423-427.

How to cut pseudo-parabolas into segments

Hisao Tamaki* Takeshi Tokuyama *

Abstract

Let Γ be a collection of unbounded x-monotone Jordan arcs intersecting at most twice each other, which we call pseudo-parabolas, since two axis parallel parabolas intersects at most twice. We investigate how to cut pseudo-parabolas into the minimum number of curve segments so that each pair of segments intersect at most once. We give an $\Omega(n^{4/3})$ lower bound and $O(n^{5/3})$ upper bound. We give the same bounds for an arrangement of circles. Applying the upper bound, we give an $O(n^{23/12})$ bound on the complexity of a level of pseudo-parabolas, and $O(n^{11/6})$ bound on the complexity of a combinatorially concave chain of pseudo parabolas. We also give some upperbounds on the number of transitions of the minimum weight matroid base when the weight of each element changes as a quadratic function of a single parameter.

1 Introduction

Arrangement of curves in a plane is a major research target in computational geometry. Combinatorial complexities of parts of arrangements such as a cell, many cells, k-levels, $\leq k$-levels, and x-monotone chains play key roles in designing algorithms on geometric optimization and motion planning problems [4, 8, 13, 14].

Although arrangements of lines and line segments are most popular, an arrangement of curves which satisfy the condition that each pair of

---

*IBM – Research Division, Tokyo Research Laboratory, 1623-14, Shimo-tsuruma, Yamato, Kanagawa, 242, Japan.

curves intersect at most s times for a given constant s, is also important in both theory and applications[10, 15]. When $s = 1$ and the curves are x-monotone and unbounded, such an arrangement is known as an arrangement of pseudo-lines, to which many results on an arrangement of lines generalizes. For example, the complexity of the k-level of an arrangement of n pseudo-lines is known to be $O(\sqrt{k}n)$ [17, 4].

We focus on the case $s = 2$ in this paper. A familiar example of such an arrangement is that of axis-parallel parabolas (Figure 1), in which two curves intersect at most twice. An arrangement of axis-parallel parabolas is used in dynamic computational geometry, since it shows the transition of L_2 distances among a set of linearly moving points. The complexity of the lower envelope and the k-level of an arrangement of parabolas gives the number of combinatorial changes on the nearest pair and the nearest k pairs, respectively [1]. Also, the complexity of topological change on the L_2 minimum spanning tree (and maximum spanning tree) can be formulated into a problem on an arrangement of parabolas [11].

More generally, we consider an arrangement of unbounded x-monotone Jordan curves intersecting each other at most twice. Such an arrangement is called an arrangement of 2-intersecting curves in the literature; however, for convenience's sake, we call it an arrangement of *pseudo-parabolas*.

It is often more difficult to analyze the complexity of an arrangement of curves than an arrangement of lines or pseudo-lines. For example, the only upper bound previously known on the k-level complexity of an arrangement of parabolas is $O(kn)$, which is the same as the bound for $\leq k$-levels[15].

The aim of this paper is to link the complexity of an arrangement of pseudo-parabolas to that of an arrangement of pseudo-lines. Our approach is to split pseudo-parabolas by cut points, generating an arrangement of *pseudo-segments* in which each pair of pseudo-segments intersect at most once. For example, the arrangements of Figure 1 can be made into an arrangement of pseudo-segments by giving seven cuts (Figure 2). We call the minimum number of cuts to make an arrangement Γ into arrangement of pseudo-segments the *cutting number* of Γ.

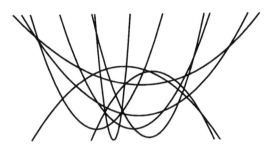

Figure 1: Arrangement of parabolas

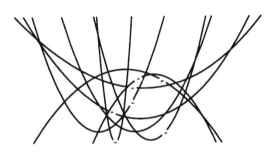

Figure 2: Cut points

Our main results are the following two theorems:

Theorem 1 *There exits an arrangement of axis parallel parabolas whose cutting number is $\Omega(n^{4/3})$.*

Theorem 2 *An arrangement of pseudo-parabolas can be cut into an arrangement of pseudo-segments with $O(n^{5/3})$ cuts.*

We also give the same bounds for the cutting number of an arrangement of circles.

The lower bound of Theorem 1 is derived from the Edelsbrunner-Welzl's lower bound example [5] on the complexity of n cells in an arrangement of lines.

The upper bound of Theorem 2 is derived from an inequality of Lovász's used in the proof of his fractional covering theorem [12], combined with extremal graph theory [2] and a probabilistic method [3, 15]. A greedy algorithm outputs cuts attaining this upper bound.

Combining Theorem 2 with the known upper bound on the level complexity of an arrangement of pseudo-lines, we derive a non-trivial $O(n^{23/12})$ upper bound on the complexity of a level of an arrangement of pseudo-parabolas. The technique used here is such that any improved upper bound for pseudo-lines will lead to an improved upper bound for pseudo-parabolas. Thus, Theorem 2 establishes an important link between the complexities of arrangements of these two types.

We also give some upperbounds on the number of transitions of the minimum weight matroid base when the weight of each element changes as a quadratic function of a single parameter t.

2 Preliminaries

Let Γ be an arrangement of pseudo-parabolas. The arrangement makes a subdivision of the plane into faces. We use the terms *cell*, *edge*, and *vertex* for 2, 1, and 0 dimensional faces, respectively. A *lens* of the arrangement is the boundary of a closed region bounded by two pseudo-parabolas; we say that these pseudo-parabolas form a lens.

The boundaries of the shaded regions in Figure 3 are lenses. We say a lens is a 1-lens if no curve cross the lens. Consequently, a 1-lens consists of exactly two edges of the arrangement. There exists no 1-lens in Figure 3.

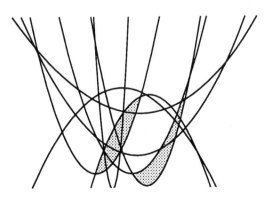

Figure 3: Lenses

We have the following lemma, which will be used

in the proof of the lower bound:

Lemma 3 *The cutting number of Γ is not less than the number of 1-lenses.*

Proof: One of the two edges in each 1-lens must be cut, and an edge is contained in at most one 1-lens. Thus, the lemma follows. □

We define a hypergraph $H(\Gamma)$, whose node set is the set of edges of the arrangement Γ. A set of nodes of $H(\Gamma)$ forms a hyperedge if and only if its corresponding set of edges of the arrangement forms a lens.

The node covering of a hypergraph H is a subset of the node set of H such that every hyperedge contains at least one node of the set.

The node covering with minimum size (number of nodes) is called the minimum covering. The size of the minimum covering is called the covering number.

The following is a key lemma for our upper bound of the cutting number:

Lemma 4 *The cutting number of Γ equals the covering number of $H(\Gamma)$.*

Proof: Given a minimum covering C of $H(\Gamma)$, we cut all edges of Γ associated with nodes in C. Then all lens are cut, that means all pair of curve segments after the cut intersect at most once. On the contrary, given a minimum cut of Γ, consider the collection of the edges cut. Then associated node set of $H(\Gamma)$ is a covering. □

3 Lower bound

We give a proof of Theorem 1 in this section. Let us recall the following well-known fact:

Theorem 5 *[5] The total complexity of m cells in an arrangement of lines is $\Omega(m^{2/3}n^{2/3})$ in the worst case if $m > n^{1/2}$.*

In particular, the total complexity of n cells is $\Omega(n^{4/3})$. We show the same lower bound for the cutting number of an arrangement of axis-parallel parabolas using a similar technique.

Proposition 6 *There exists an arrangement of axis-parallel parabolas whose cutting number is $\Omega(n^{4/3})$.*

Proof: To show the lower bound example of the complexity of n cells in an arrangement of lines, Edelsbrunner and Welzl [5] construct a highly degenerate arrangement $\mathcal{A}'$ which has a set V of n vertices with total degree $\Omega(n^{4/3})$. Also, due to that construction, we can assume that each line of the arrangement has a positive slope which is not larger than $\pi/4$.

We consider a very skinny axis-parallel parabola γ, and draw a copy $\gamma(v)$ which has its peak at v for each v in V. Now, we translate each line, so that the degeneracy of $\mathcal{A}'$ is resolved, and each line through a point v in V is translated so that it is tangent to $\gamma(v)$. Since the point of contact of a line to a given parabola is determined by the slope of the line, the translation is unique for a line with a given slope. Hence, the translation does not generate global inconsistency.

We consider the union of $\mathcal{A}'$ and the skinny parabolas. Then, the number of vertices in this arrangement at which two curves are tangent is $\Omega(n^{4/3})$. Thus, by perturbing this arrangement, we obtain an arrangement of parabolas containing $\Omega(n^{4/3})$ 1-lenses. From Lemma 3, this arrangement needs $\Omega(n^{4/3})$ cuts. □

4 Upper bound

It suffices to give an upper bound for the covering number of hypergraph $H(\Gamma)$. We recall some notations on hypergraphs. The degree $d(x)$ of a node x of a hypergraph is the number of hyperedges containing x. The maximum degree in a hypergraph H is denoted by $d(H)$.

A k-matching of H is a collection $\mathcal{M}$ of hyperedges (the same edge may occur more than once) such that each node belongs to at most k of them. A k-matching is simple if no edge occurs in it more than once. The maximum number of hyperedges in a simple k matching is denoted by $\nu_k(H)$ (this is denoted by $\tilde{\nu}_k(H)$ in Lovász [12]). Note that $\nu_{d(H)}(H)$ is the number of hyperedges in H. We remove the argument H from functions d and ν, if no confusion arises. A greedy algorithm for computing a covering is the following:

1: Find the node of maximum degree;
2: Insert the node in the covering,
and remove the node and all edges containing it from H;
3: If all edges are covered, Exit; Else GOTO 1;

Lovász [12] shows that the greedy algorithm achieves a covering of size at most $\log d(H) + 1$ times the covering number of H. The following is the key inequality in his proof. Let t be the number of covering of H obtained by a greedy algorithm. Then,

$$t \leq \frac{\nu_1}{1 \cdot 2} + \frac{\nu_2}{2 \cdot 3} + \cdots + \frac{\nu_{d-1}}{(d-1) \cdot d} + \frac{\nu_d}{d} \quad (1)$$

Consequently, the minimum covering number of H is also bounded by the righthand side of (1). Therefore, we want to estimate $\nu_k(H(\Gamma))$ for $k = 1, 2, ..d$.

Suppose we have a simple 1-matching $\mathcal{M}$ of $H(\Gamma)$ of size M. Recall that a hyperedge in $H(\Gamma)$ is a lens in Γ.

We define a bipartite graph $G(\mathcal{M})$. The vertex set is $S_1 \cup S_2$, where S_1 and S_2 are disjoint, $|S_1| = |S_2| = |\Gamma| = n$, with associated bijections $\gamma_1 : S_1 \to \Gamma$ and $\gamma_2 : S_2 \to \Gamma$.

We draw an edge between a node u of S_1 and v of S_2 if and only if the associated curves $\gamma_1(u)$ and $\gamma_2(v)$ form a lens which is associated with a hyperedge in $\mathcal{M}$, and $\gamma_1(u)$ is above $\gamma_2(v)$ within the lens. Here, a curve γ is *above* another curve μ within their lens L if a vertical downward ray from a point on $\gamma \cup L$ intersects μ.

By definition, $G(\mathcal{M})$ has $2n$ vertices. It is clear that the number of edges in $G(\mathcal{M})$ is the size of the matching $\mathcal{M}$.

First, we consider the size $\nu_1(H(\Gamma))$ of the maximum simple 1-matching.

Lemma 7 *Suppose $\mathcal{M}$ is a simple 1-matching of $H(\Gamma)$. Then, $G(\mathcal{M})$ does not contain $K_{3,4}$ as a subgraph.*

Proof: Assume that $G(\mathcal{M})$ contains a copy of $K_{3,4}$. Then, we have three curves C_1, C_2, C_3 and four curves D_1, D_2, D_3, D_4 such that each pair (C_i, D_j) makes a lens, in which C_i is above D_j, for $1 \leq i \leq 3, 1 \leq j \leq 4$. Furthermore, because $\mathcal{M}$ is a 1-matching, no edge in the arrangement constructed from these seven curves is contained in more than one such lense. Let $A(C)$ denote the arrangement of $\{C_1, C_2, C_3\}$ and $A(D)$ that of $\{D_1, D_2, D_3, D_4\}$. If an edge e is located on D_1, and e is below two curves C_1 and C_2, e must be on both of two lenses (C_1, D_1) and (C_2, D_1). This means that the two arrangements $A(C)$ and $A(D)$ intersect each other only at points that are on the upper envelope of $A(C)$. Similarly, those intersection points must be on the lower envelope of $A(D)$ at the same time. Since there must be 12 lenses, the number of intersections must be at least 24. However, the upper envelope of $A(C)$ has at most 5 edges, and the lower envelope of $A(C)$ has at most 7 edges. Because each pair of curves intersects at most twice, the number of intersecting points cannot exceed 22, which is a contradiction. □

Remark. We can also show that $G(\mathcal{M})$ does not contain $K_{3,3}$ with more careful analysis.

We use the following result in extremal graph theory, which can be found in Bollobas [2] in a more general form (page 73, Lemma 7).

Lemma 8 *Let G be a bipartite graph with n vertices on each side which does not contain $K_{s,t}$ as a subgraph. Suppose G contains $m = yn$ edges. Then, $n\binom{y}{s} \leq (t-1)\binom{n}{s}$.*

Theorem 9 *$G(\mathcal{M})$ contains $O(n^{5/3})$ edges. Hence, $\nu_1(H(\Gamma)) = O(n^{5/3})$.*

Proof: We substitute $s = 3$ and $t = 4$ in the above lemma, and obtain $y = O(n^{2/3})$. □

Next, we consider $\nu_k(H(\Gamma))$, for $k \leq \sqrt{n}$. We apply the probabilistic argument similar to the one Sharir [15] used for analyzing the complexity of $\leq k$ level of curves.

Suppose we have a simple k-matching $\mathcal{M}$, and associated set $\mathcal{L}(\mathcal{M})$ of lenses. Assume $\mathcal{M}$ has ν_k hyperedges (i.e. $\mathcal{L}(\mathcal{M})$ has ν_k lenses).

For each lens L bounded by two curves C_1 and C_2, its edge is called *extremal* if it contains one of the intersection points of C_1 and C_2. Obviously, there are at most four extremal edges associated with L.

Now, we choose a random sample Y of n/k curves from Γ. We say a lens L is a *near 1-lens* of the sample if (1) L is a lens consisting of two

233

curves of Y and (2) for each extremal edge e of L, L is the only lens which consists of two curves of Y and contains e.

Let us consider the set of near 1-lenses of sample Y, and consider the associated matching $\mathcal{M}_0$ of $H(Y)$.

Lemma 10 $\mathcal{M}_0$ *is a 1-matching of* $H(Y)$.

Proof: Suppose an edge e of the arrangement Y is contained in two near 1-lenses L and L'. Let C be the curve on which e is located. Then, L and L' contain intervals I and I' of C. Both intervals must contain e; thus, at least one endpoint of either I or I' must be contained in the other interval. W.l.o.g., we assume an endpoint of I is contained in I'. This means that an extremal edge of L is contained in L', which contradicts the definition of near 1-lens. □

Now, analyze the expected number of near 1-lenses. A lens L of $\mathcal{L}(\mathcal{M})$ becomes a near 1-lens if (1) both of its bounding curves are in the sample and (2) no other curve contributing a lens containing an extremal edge of L is in the sample. Since $\mathcal{M}$ is a k-matching, the number of such curves in (2) is at most k for each extremal edge.

Thus, the probability that L becomes a near 1-lens is at least $k^{-2}(1-1/k)^{4k}$, hence the expected number of near 1-lenses is at least $k^{-2}(1-1/k)^{4k}\nu_k$. Note that $(1-1/k)^{4k} > 1/256$.

Since the number of 1-matchings in the sample is $O((n/k)^{5/3})$, $\nu_k = O(k^2(n/k)^{5/3}) = O(n^{5/3}k^{1/3})$.

Now, let us compute the righthand side of the Lovász's inequality (1).

$$\frac{\nu_1}{1 \cdot 2} + \frac{\nu_2}{2 \cdot 3} + \cdots + \frac{\nu_{d-1}}{(d-1) \cdot d} + \frac{\nu_d}{d}$$
$$= O(n^{5/3}[\sum_{k=1}^{\sqrt{n}} \frac{k^{1/3}}{k(k+1)}] + \sum_{k=\sqrt{n}}^{n} \frac{n^2}{k(k+1)})$$
$$= O(n^{5/3}).$$

This proves Theorem 2.

5 Applications

Level complexity

Let Γ be an arrangement of n pseudo-parabolas. The *level* of an edge e of the arrangement Γ is the number of edges which intersect with the vertical half line downward from an internal point on e. Level is well-defined since the above number is independent of the choice of the internal point of e. It is well-known that the union of all edges with a given level k is a connected chain of curve and separates the plane. This chain is called the k-level of Γ. The complexity of k-level of Γ is the number of edges whose level is k.

Theorem 11 *The complexity of k-level of Γ is $O(n^{23/12})$.*

Proof: Without loss of generality, we assume that the arrangement is simple, that means no three curves intersect at a point. We consider the set P of cutting points of Γ into pseudo-line arrangements. We know that $m = |P| = O(n^{5/3})$. We subdivide the plane into $m+1$ slabs with m vertical lines through points of P. Inside a slab, the arrangement can be considered as an arrangement of pseudo-lines.

Let X_i be the number of vertices of the arrangement located in the i-th slab S_i. Then, $\sum_{i=1}^{m+1} X_i = O(n^2)$.

Suppose that exactly n_i curves contribute to k-level in S_i. The k-level of Γ inside S_i is a level of these n_i curves.

Suppose a curve γ is on k'-level at the left end of the slab S_i, and contributes to k-level in S_i. Then, γ must have at least $|k - k'|$ vertices on it in S_i. Therefore, the arrangement must have at least $n_i^2/2$ vertices in S_i. This means that $n_i = O(\sqrt{X_i})$.

The complexity of a level of n_i pseudo-lines is $O(n_i\sqrt{n_i})$, which is $O(X_i^{3/4})$.

Thus, the complexity of the k-level of Γ is $O(\sum_{i=1}^{m+1} X_i^{3/4}) = O(m(n^2/m)^{3/4})$. Since $m = O(n^{5/3})$, we obtain $O(n^{23/12})$. □

Note that this result improves the known bound of $O(kn)$ [15] when $k > n^{11/12}$. Also note that the bound would be automatically improved further, if we had a better bound either on the level complexity of an arrangement of pseudo-lines or on the cutting number of pseudo parabolas.

Complexity of combinatorial concave chains

Given an arrangement $\mathcal{F}$ of x-monotone curves, a combinatorial concave chain c is a x-monotone chain in the arrangement $\mathcal{F}$ satisfying the following concave condition: If two curves γ_1 and γ_2 intersects at a point p, and the chain c lies on γ_1 to the left of p and on γ_2 to the right of p, respectively. Then, γ_1 stabs γ_2 from below at p (curves are directed from left to right). See Figure 4

Figure 4: Combinatorial concave chain

Complexity of a combinatorial concave chain is $O(n)$ for an arrangement of n lines. But, this fails for an arrangement of pseudo parabolas.

Theorem 12 *The worst-case complexity of a combinatorial concave chain of n pseudo-parabolas is $\Omega(n^{4/3})$ and $O(n^{11/6})$.*

Proof:
The lower bound can be obtained from the degenerate arrangement of lines of Edelsbrunner and Welzl [5] which was used in Proposition 6. This arrangement has $O(n)$ lines and $\Omega(n)$ "heavy" nodes to each of which $\Omega(n^{1/4})$ lines are incident. We rotate the arrangement with an infinitesimally small positive angle. Each "heavy" vertex is replaced by a tiny concave chain by using suitable perturbation. We connect the rightmost vertex of this tiny chain with the leftmost vertex of the tiny chain of the next (from top to bottom) lattice vertex in the same column with a very steep parabola. Moreover, we connect the topmost lattice point in a column with the bottom lattice point in the next column using a very sharp parabola, so that the parabola can joint two concave chains into one concave chain. Then, this arrangement has a combinatorial concave chain of length $\Omega(n^{4/3})$.

The upper bound can be obtained similarly to Theorem 11. □

In contrast to the above $O(n^{11/6})$ upper bound, the following theorem implies that no nontrivial upper bound is likely to hold if we permit that each pair of curves intersects three times.

Theorem 13 *The worst-case complexity of a combinatorial concave chain of n curves where each pair of curves intersects at most three times is $\Omega(n^2/\log n)$.*

Proof: We can make any x-monotone chain in a line arrangement into a combinatorial concave chain of n 3-intersecting curves by using local changes. Thus, the theorem follows from the Matoušek's $\Omega(n^2/\log n)$ lower bound on the length of x-monotone chain [14]. □

Transitions of minimum matroid base and MST

Let E be a finite set and $\mathcal{B}$ a family of subsets of E. The pair $(E, \mathcal{B})$ is called a *matroid* $M(E, \mathcal{B})$, and the elements of $\mathcal{B}$ are the *bases* of $M(E, \mathcal{B})$, if the following two axioms hold [16]:

(A1) For any $B, C \subset E$ with $B \neq C$, if $B \in \mathcal{B}$ and $C \subset B$, $C \notin \mathcal{B}$.

(A2) For any $B, B' \subset \mathcal{B}$ with $B \neq B'$ and for any $e \in B - B'$, there exists $e' \in B' - B$ such that $(B - \{e\}) \cup \{e'\} \in \mathcal{B}$.

For instance, let $\mathcal{T}$ be a set of spanning trees in an undirected connected graph $G = (V, E)$; then $(E, \mathcal{T})$ forms a matroid and $\mathcal{T}$ is a set of bases [16].

The number $|B|$ of elements of a base $B \in \mathcal{B}$ is independent of the choice of B [16], and is denoted by p. Let $m = |E|$, and assume the elements of E to be indexed from 1 through m. We assume that each element i has a real-valued weight $w_i(t)$ that is a function in the parameter t. The minimum (resp. maximum) weight base is the one in which the sum of weights of elements is minimum (resp. maximum).

If the weight functions of two elements have a constant number of intersections, we have an $O(m^2)$ trivial upper bound on the number of transitions of the minimum (resp. maximum) weight base of $M(E, \mathcal{B})$. If $w_i(t)$ is linear, this was improved to $O(m \min\{\sqrt{p}, \sqrt{m-p}\})$ by [7, 9]. Recently, Eppstein constructed a matroid with linear weights for which the number of transitions of the minimum matroid base is $\Omega(m^{2/3}p^{2/3})$[6].

If the weight functions are quadratic, it is not even clear if $O(pm)$ upperbound holds. From Theorem 12, an $O(pm^{11/6})$ bound is obtained for the case where all $w_i(t)$ are quadratic. The above bound is nontrivial only if $p \leq m^{1/6}$. We have a bound shown below that works for p closer to n (we omit the proof in this version):

Theorem 14 *When all $w_i(t)$ are quadratic in t, and $p < m^{1-\delta}$ for some constant $\delta \leq 0.3$, the number of transitions is $O(m^{2-\delta/15})$*

Corollary 15 *Let G be a graph with m edges and p nodes, and each edge has a weight function which is quadratic in a parameter t. We assume that G is dense satisfying $p < m^{1-\delta}$ for some constant $\delta \leq 0.3$. Then, the number of transitions in its minimum (maximum) spanning tree is $O(m^{2-\delta/15})$.*

Corollary 16 *Let S be a set of n linearly moving points in d dimensional space, where d is a constant. Then, the number of transitions of its Euclidean maximum spanning tree is $O(n^{3.98})$.*

Proof: In the Euclidean maximum spanning tree, $m = n(n-1)/2$, and $p = n-1$. □

Note. For the number of transitions of an Euclidean minimum spanning tree, an $O(n^3 2^{\alpha(n)})$ bound is known [11], where α is the inverse Ackerman function. In order to improve that bound, we need a subquadratic bound where $p = \Omega(m)$.

6 Related topics

Arrangement of circles

We are given an arrangement of n circles. Although a pair of circles intersects at most twice, a circle is not a simple curve. However, we can cut each circle with its horizontal diameter, and divide it into an upper half-circle and lower half-circle. We can connect two vertical downward (resp. upward) rays to an upper (resp. lower) half-circle at its two endpoints, and obtain a simple curve separating the plane. It is easy to see that every pair of curves intersects at most twice.

Thus, we have a family of pseudo-parabolas. We can now apply our upper bound results in the previous section. Also, the lower bound example for the cutting number of pseudo-parabolas in Section 3 can be easily adapted to an arrangement of circles.

Thus, we have the following:

Theorem 17 *Using $O(n^{5/3})$ cuts, an arrangement of circles can be transformed to an arrangement of pseudo-segments. There exists an example for which $\Omega(n^{4/3})$ cuts are required.*

Cutting a $2t$ intersecting family

It is desired to extend the upper bound on cutting number to that for an arrangement of curves intersecting t times each other. Unfortunately, in the worst case, we need to cut at $\Omega(n^2)$ points in order to make an arrangement of 3-intersecting curves into an arrangement of pseudo-segments. The lower bound example is obtained as follows: Consider an arrangement of n lines. At each vertex, we locally replace one of the lines meeting there with a curve intersecting the other line three times to make a pair of two 1-lenses. The resulting arrangement has $\Omega(n^2)$ 1-lenses.

More generally, $\Omega(n^2)$ cuts are needed to make $(2t+1)$-intersecting family to $2t$-intersecting family in the worst case. On the other hand, for a family with even intersecting-numbers, we have the following result (we omit the proof):

Theorem 18 *Given an arrangement of curves in which every pair of curves intersects at most $2t$ times, we can cut it at $O(n^{2-1/\beta(t)})$ points to make it an arrangement of curve segments, in which every pair of segments intersects at most $2t-1$ times. Here, $\beta(t)$ is the minimum positive number y satisfying $y^2 \geq 4t\lambda_{2t}(y)$, where $\lambda_{2t}(y)$ is the length of Davenport-Schinzel sequence of degree $2t$ on y characters.*

Acknowledgement. The authors thank Naoki Katoh for pointing out an error in an earlier version of the paper.

References

[1] M. Atallah, Some Dynamic Computational Geometry Problems, *Comp. and Math. with Appls.* **11** (1985), 1171-1181.

[2] B. Bollobas, *Graph Theory*, Springer Verlag, 1979.

[3] K. Clarkson and P. Shor, Applications of Random Sampling in Computational Geometry II, *Disc. & Comput. Geom.* **4** (1989), 387-421.

[4] H. Edelsbrunner, *Algorithms in Combinatorial Geometry* Springer Verlag, 1986.

[5] H. Edelsbrunner and E. Welzl, On the maximum number of edges in many faces in an arrangement. *J. Combin. Theory Ser. A* **41** (1986), 159-166.

[6] D. Eppstein. Geometric Lower Bounds for Parametric Matroid Optimization. To Appear, *Proc. 27th STOC*, 1995.

[7] D. Gusfield. Bounds for the Parametric Spanning Tree Problem. *Proc. Humbolt Conf. on Graph Theory, Combinatorics and Computing.* Utilitas Mathematica, 1979, 173-183.

[8] L. Guibas and M. Sharir, Combinatorics and Algorithms of Arrangements, *in New Trends in Discrete and Computational Geometry*, J. Pach ed., Springer-Verlag 1991, pp.9-33.

[9] N. Katoh and T. Ibaraki, On the Total Number of Pivots Required for Certain Parametric Combinatorial Optimization Problems, Working Paper No. 71, Institute of Economic Research, Kobe Univ. of Commerce, 1983.

[10] K. Kedem and M. Sharir, An Efficient Motion Planning Algorithm for a Convex Polygonal Objects Moving in 2-dimensional Space, *Disc. & Comput. Geom.* **5** (1990), 43-75.

[11] N. Katoh, T. Tokuyama, and K. Iwano, On Minimum and Maximum Spanning Trees of Linearly Moving Points. *Proc. 33rd FOCS* (1992) 396-405

[12] L. Lovász, On the Ratio of Optimal Integral and Fractional Covers, *Discrete Mathematics* **13** (1975), 383-390.

[13] J. Matoušek, On Geometric Optimization with Few Violated Constraints *Proc. 10th ACM Comput. Geom. Symp.* (1994), 312-321.

[14] J. Matoušek, Lower bounds on the Length of Monotone Paths in Arrangements *Disc. & Comput, Geom* **6** (1991), 129-134.

[15] M. Sharir, On k-Sets in Arrangements of Curves and Surfaces, *Disc. & Comput. Geom.* **6** (1991) 593-613.

[16] D.J.A. Welsh, *Matroid Theory*, Academic Press, London, 1976.

[17] E. Welzl, More on k-Sets of Finite Sets in the Plane. *Disc. & Comput. Geom.* **1** (1986), 95-100.

New Algorithms and Empirical Findings on Minimum Weight Triangulation Heuristics

EXTENDED ABSTRACT

Matthew T. Dickerson*   Scott A. McElfresh[†]   Mark Montague*

1 Introduction

The weight of a triangulation is the sum of the lengths of all the edges in the triangulation. A *Minimum Weight Triangulation* (MWT) of a point set S is a triangulation that minimizes weight over all possible triangulations. Wang and Aggarwal [19] and Baraquet and Sharir [1] use the MWT of simple polygons to reconstruct three dimensional objects from two dimensional contours (or slices). The MWT also arises in numerical analysis in a method suggested by Yoeli [20] for numerical approximation of bivariate data. Though it has been shown how to compute the MWT in time $O(n^3)$ for the special case of n-vertex simple polygons [9], there are no known efficiently computable algorithms for the MWT of general point sets. We therefore seek efficiently computable *approximations* to the MWT, of which there are a number of heuristics.

We report on the practical efficiency and effectiveness of various of these heuristics including multiple algorithms for the greedy triangulation, other approaches to *locally* minimal triangulations, and some new algorithms that perform comparatively well. Our empirical findings come from a number of different tests. We report on triangulations of general points sets over random distributions, as well as of simple polygons taken from an application to medical image processing[1]. In addition to a comparison of run times and weights, we also analyze certain primitive operations such as the number of edges flipped and the number of pairs of points examined. One important aspect of this paper is an empirical report on triangulations that meet the property of local minimality. This general property has not previously been studied in the literature, and we show in this paper that local minimality works well as a goal in approximation algorithms.

In Section 2 we describe the platform and algorithms, and comment on issues arising in the implementation. In Section 3 we give a basic comparison of the effectiveness and efficiency of the algorithms. In Section 4 we present a more *detailed* analysis of some of the algorithms including the number of calls to various primitive operations. In the case of multiple-phase algorithms, we present some data on the division of labor between phases. In the final section of the paper, we present some conclusions and summary comments regarding this work.

We note that in this report we present only a small portion of the data collected. We implemented far more algorithms than we are able to report on here, and thus we have chosen only the most interesting and promising results. In order to save space, we have also shrunk numerous tables by eliminating many intermediate values of n. In most cases, data given for random distributions of general point sets is an average of 50 independent trials. In all cases, when algorithms are compared, they have been tested on the exact same point sets.

1.1 Notation

Throughout the paper, we use MWT for the minimum weight triangulation, with MWT(S) being the MWT of a specific point set S. Likewise, we use GT for the greedy triangulation, DT for the Delaunay triangulation, and LMT for any *locally minimal* triangulation. For any triangulation T(S), we use $\mathcal{W}$(T(S)) for the actual weight of the triangulation.

2 Platform and Algorithms

We begin with a general description of the platforms and algorithms chosen for the tests. We comment on our

---

*Middlebury College, Middlebury VT. This study was supported in part by the funds of the National Science Foundation, NSF CCR-9301714

[†]Dartmouth College, Hanover NH.

[1]One application of the MWT is to the reconstruction of three dimensional polyhedral arrangements from two dimensional polygonal slices [1] as is done, for example, in medical imaging. Thanks to Gill Barequet for providing as sample test data a large set of these simple polygons coming from cross-sectional images of a hip, jaw, retina, and lung.

reasons for these choices as well as on some drawbacks that were discovered and on other issues arising from our implementations including how the code was optimized.

2.1 Platforms and General Implementation Issues

We used two distinct platforms to test our algorithms, and we report on them both.

Smalltalk/V on a Macintosh Quadra 800 Algorithms were initially implemented in Smalltalk/V for the Macintosh. They were run on a Quadra 800, with a MC68040 CPU running at 33 MHz (including a FPU, though we ran most tests using integer point sets only), 32 Megabytes of RAM, and a 1 Megabyte VRAM.

The Smalltalk language was chosen for a number of reasons. For prototyping, we wanted to use a true object-oriented language. The initial object library of Smalltalk/V is extensive, as is its integrated debugger and development environment. The choice of Smalltalk/V proved successful in one critical area. The implementation process in the Smalltalk/V language with its excellent development environment proved very fast. Over the course of a few months, we actually implemented far more (rather than fewer) algorithms than we had originally hoped. Thus we were able to test and compare a large number of different algorithms. In this test, we used point sets with integer coordinates to avoid floating point error. Uniform random point sets in a square were generated on a 1000 × 1000 integer grid. Thus only in the distance computation to compute the final weight of the triangulation was real arithmetic required. In contrast, our test data for simple polygons was drawn from an actual set of cross-sectional slices of a heart, lung, retina, and hip and included a total of approximately 550 polygons with over 19000 vertices. Polygons ranged from 4 to 102 vertices each. The coordinates of the polygon vertices were real, however they had only 3 significant digits to the right of the decimal point, and so were converted to integer points without loss by multiplying the entire set by a scale factor of 1000.

The major drawback to the platform proved to be speed. For large numbers of tests on numerous algorithms, we were not able to progress beyond sets of 7000 points (a relatively small number in computational geometry).

C on a Decstation 5000 Because of the slow speed of the Smalltalk/V implementation, we also implemented some algorithms on a faster platform in order to test them on significantly larger data sets. We include in our report data from an implementation in C, with the programs run on DecStation 5000. The gnu compiler was used. With some pre-existing code for graphics in X windows in place, graphical debugging was not cumbersome to include in the code. The use of the C language also enabled us to utilize code already written, debugged, and tuned for certain algorithms. (For instance, there exist many versions of code for computing the Delaunay triangulation in C.) The point sets were uniformly generated random points on an integer grid in a square of dimensions ranging from 1000 × 1000 to $10^6 \times 10^6$. The standard rand function in Unix C was used to generate the coordinates of the points. Point sets ranged in size of 10 up to 50000.

2.2 General Comments on Implemented Algorithms

We now describe the algorithms implemented in this experiment, including comments on their optimization. We chose from the literature a number of different algorithms to implement. Early conjectures mentioned both the Delaunay (DT) and greedy (GT) triangulations as approaches yielding an exact MWT. Though the conjecture was wrong in both cases [14, 13], much research has focussed on these two triangulations and we include these in our report. One of the few provable properties of the GT is that they do produce triangulations which are locally minimal [8, 3] (also called local optimality). That is, there are no edges in a GT that may be flipped with the other edge in the same quadrilateral to produce a triangulation of lower weight. Thus at the local level, the GT may be said to be minimal. Surprisingly, this property of local minimality has not been studied in and of itself. An emphasis and contribution of this paper is therefore the empirical study of some algorithms producing triangulations with the property of local minimality. We look at a number of different algorithms, described below in Section 2.4, that produce LMTs. These include edge-flipping approaches, and incremental insertions.

Numerous other heuristics for approximating the MWT have also been developed. Of polynomial time algorithms known so far, those of Lingas [11] and Heath and Pemmaraju [8] have been shown empirically in at least one report [8] to produce the lowest weight triangulations of general point sets. Their approaches make use of the convex hull and a spanning tree to create a single cell, and then use a minimal *cell* triangulation algorithm generalized from a dynamic programming approach for computing the MWT of a convex polygon [6]. Unfortunately, both of these algorithms require $O(n^3)$ time which is impractical for large point sets. In [8], the authors only report on data for sets of size 50 and smaller. We also implemented the algorithm of [8], and compared results for similarly small point sets, but as we were not able to run the algorithm on sufficiently large sets we do not include it in this report.

2.3 Greedy Triangulations

We now give more detail on one of the more well known MWT heuristics, the greedy triangulation. Work on the GT has progressed along two entirely different tracks. The straightforward approach follows the "generate-and-test" paradigm of Gilbert [6], where edges are generated in order and tested one at a time. Edges are never added to the triangulation and then removed. Gilbert gave an $O(n^2 \log n)$ time and $O(n^2)$ space GT algorithm. Manacher and Zobrist [15] gave an $O(n^2)$ expected time and $O(n)$ space greedy triangulation algorithm that makes use of a probabilistic *pretest* for compatibility. More recently, Dickerson, Drysdale, McElfresh, and Welzl [3] presented a GT algorithm requiring $O(n \log n)$ time in the expected case for uniform random distributions over any convex region.

An alternate approach to the "generate-and-test" paradigm is to begin with another more easily computed triangulation and modify it until it becomes the greedy triangulation. Goldman [7] and Lingas [12] independently discovered a way to do this by computing the Delaunay triangulation constrained to contain the current set of known GT edges. The next edge to be added to the GT can be found in linear time from the constrained DT, which must then be updated to include the new edge. This gives an $O(n^2 \log n)$ time and $O(n)$ space algorithm. Recently Levcopoulos and Lingas, and independently Wang, have shown how to do the update step in $O(n \log n)$ time, leading to an $O(n^2)$ time and $O(n)$ space algorithm in the worst case [10, 17]. Lingas [12] shows that his method runs in $O(n \log^{1.5} n)$ expected case time for points chosen uniformly from the unit square. Finally, Wang has very recently claimed an algorithm using this approach with a worst-case running time of $O(n \log n)$ [18]. However, though these methods are very elegant, they are also somewhat confusing and very complicated to implement. We chose therefore to focus on the algorithm of [3].

2.3.1 Greedy Triangulation Algorithms of [DDMW]

We implemented Algorithm 2 of [3], which is nearly self-contained, requiring only a fixed-radius near neighbors search which we implemented efficiently using the bucketing approach of Bentley, Stanat, and Williams [2]. The algorithm proved simple and easy to implement. The approach has two phases. In the first phase, we generate all pairs of points with radius less than a certain value d, and then pre-test these *candidate* pairs to see if they are *plausible*. The plausibility test for pair (p, q) checks the disc of radius γ centered at the midpoint of $\overline{pq}$. The disc is split by segment pq into two half-discs, at least one of which must be empty if pq is in the GT. All plausible pairs are then sorted and tested for compatibility, and added to the triangulation if compatible. After all pairs have been tested, we move to the second phase. Those points to which no new edges can be added are removed from the list of points. All remaining pairs of points are generated, sorted, and tested.

The algorithm was optimized by determining the best value of the initial search distance d. For points in a convex region of unit area, $d = 2B/\gamma$ where $B = \sqrt{(c \ln n)/(n-2)}$ and $\gamma = 1/\sqrt{5}$. Running numerous tests for various values of n, we found good values for the constant c to be between 0.055 and 0.1 [2] A more interesting result is that for point sets of size $n \leq 25000$, we found the algorithm works no slower and in some cases significantly faster if we altogether eliminate the plausibility test, though this test was necessary for the proof of the algorithm's asymptotic running time. Elimination of the test adds work to the sorting stage, and adds to the number of pairs tested for compatibility, but the added work is compensated for by the savings.

Note that the $O(n \log n)$ time expected case analysis of the GT algorithm of [3] does not apply to non-uniform distributions. In particular, this algorithm is not expected to be as efficient for points distributed on a circle (i.e. all of the points fall on the convex hull). There are two reasons for this. One is that a much smaller number of points will be completed in the first phase, leaving considerably more work for the second phase. Second, since there are no interior points, the plausibility test will never eliminate any candidate pairs. We present some data on non uniform distributions also.

2.3.2 A Fast New Incremental Greedy Algorithm

An analysis of the actual implementation of the GT algorithm of [3] led to the development of a new GT algorithm by Dickerson and Montague [4].[3] We extend the two-phase approach to an arbitrary k-phase approach, where at each phase we eliminate the completed points, increment the search radius d, generate and sort a new set of pairs, and continue. We give a few additional details regarding this algorithm, which we call the *incremental greedy* (IGT), since—unlike other algorithms implemented in this report—the IGT does not appear in the previously published literature.

---
[2] Thanks to Mobina Hashmi, who performed many of the test runs which enabled us to determine these values.

[3] Aichholzer, et al., [5] have also proposed a fast new variant of the [3] idea. They present a GT algorithm with $O(n)$ expected case running time on uniform distributions. Some of the ideas are similar; based on the proofs in [5], we believe the this new algorithm presented here not only performs well in practice but may be *provably* linear time in the expected case.

This new IGT algorithm raised a number of interesting implementation questions, such as how to increment the search radius at each phase, and how and when to rebucket. We experimented with a number of different approaches, and found some interesting results as we fine-tuned the algorithm. There were three basic binary variables we manipulated to form $2^3 = 8$ different variants of the algorithm: 1) The search radius at the ith iteration for $i \geq 1$ could be id or $2^{i-1}d$ where d is the initial search radius as used in [3]; 2) After each iteration, triangulated points could either be removed from existing buckets, or we could rebucket entirely with the new search radius; and finally, 3) we experimented with making use of the existing cell structure so that rather than searching by radius, we search in a spiralling pattern through surrounding buckets. That is, we avoid recomputing all shorter pairs at each round, but at the ith round we look *only* at points in buckets a distance of $i - 1$ from the search point.

Experimental results showed that the fastest variant used a linearly growing search radius of id at round i, rebucketted all uncompleted points at each round using buckets of size id, and did not try to make use of the cell structure but simply searched all pairs of points closer than id, discarding those closer than $(i - 1)d$.

This new *iterative* algorithm proved faster than the original two-phase approach of [3] for point sets of size $n \leq 6000$. For example, for a uniform random set of 450 points, the algorithm of [3] required 224 seconds in Smalltalk/V, while the same algorithm without the plausibility test required only 176 seconds, and the new algorithm of [4] only 114 seconds. A further report on this new algorithm is given in [4].

2.4 Locally Minimal Triangulations

One obvious approach to global minimality (an exact MWT) is to strive for local minimality. The GT has been proven to accomplish the property of being locally minimal [3, 8]. In addition to the GT, we also experimented with three other approaches to LMTs of a general point set. Though we were not aware of these methods having been previously studied in the literature for the purposes of generating LMTs, they seemed like obvious candidates.

Edge Flipping One approach is to generate some triangulation (either a fast triangulation, or one with nice global properties like the DT) and to continue flipping any edge which is not locally minimal. We tried this approach on general point sets, convex polygons, and simple polygons. We experimented with both the Delaunay triangulation and with other faster triangulations as our starting point. We also experimented with flipping edges in sorted and random order.

Incremental Triangulations Another approach to local minimality is to iteratively add new points to the triangulation, and test all new edges for this criterion. Any edge that fails the local minimality test is flipped, and the diagonal edges of all the newly formed convex quadrilaterals are recursively tested. There are two approaches. One is to sort all of the points with respect to their x-coordinate, then add them in sorted order connecting each with all visible points on the convex hull. New internal edges may then be checked for any local property, such as minimality. Any edge that is not minimal is flipped, requiring that the new edges are recursively checked.

The other approach is a randomized incremental approach. This reduces the expected number of edges to be flipped, but adds the problem of point location. As will be shown in later sections, this proved to be the fastest family of algorithms in our test.

2.5 Comments on Simple Polygon Heuristics

We now comment on our algorithms for triangulation simple polygons. It has been shown how to compute the exact MWT in time $O(n^3)$ for the special case of n-vertex polygons [9]. However $O(n^3)$ is slow and impractical for large values of n. For example, the asymptotic bottleneck in the algorithm of Barequet and Sharir [1] for piecewise-linear interpolation between polygon slices is the substep of the $O(n^3)$ time MWT algorithm–though interestingly the exact MWT is itself only a heuristic used for "good interpolations". Unfortunately, the asymptotically more efficient GT algorithms from the literature do not work for arbitrary simple polygons.

For simple polygons, we began with an implementation of the exact MWT algorithm. We also implemented a fast simple polygon triangulation algorithm [16] requiring $O(n)$ time in the expected case (using bucket sort) and $O(n \log n)$ time in the worst case. We then made three changes to the GT algorithm of [3] to adapt it to simple polygons. The plausibility test may not be used, the compatibility test must be revised, and there is a simpler termination condition. Specifically, each edge must be tested to make sure it is interior to the polygon, and the algorithm may be halted when the triangulation is complete (has $2n - 3$ edges). We also modified the incremental insertion and edge-flipping approaches to produce LMTs of simple polygons.

3 Basic Results: an overall comparison of algorithms

We now provide an overall comparison of the run times of the various algorithms, with the weights of the tri-

n	(RI)		(RI) Flip to LMT		(SI) DT		(RI) DT		(SI) LMT		(RI) LMT	
	time	wt.	time	wt.	time	wt.	time	wt.	time	wt.	time	wt.
250	5.21	1449	101.13	571	71.32	582	67.43	582	18.28	571	13.61	573
500	12.55	2414	260.49	811	162.69	828	142.77	828	43.50	813	29.71	814
750	22.36	3281	430.43	990	258.93	1009	216.57	1009	70.24	991	45.56	993
1000	32.09	4002	608.64	1138	360.14	1160	290.68	1160	98.80	1139	61.47	1142
1250	43.91	4805	821.19	1274	464.58	1299	365.30	1299	128.68	1276	77.64	1278
1500	56.04	5325	1028.84	1389	571.06	1415	438.86	1415	160.00	1391	97.07	1394
1750	70.38	6074	1232.39	1496	678.85	1526	512.70	1526	191.60	1498	117.38	1502
2000	85.06	6585	1410.83	1597	788.12	1626	587.49	1626	223.61	1599	137.76	1602

Figure 1: Incremental Triangulations

3.1 Incremental Insertions

angulations produced. This basic information provided much of the focus for continued efforts.

We found that our fastest family of algorithms were our incremental insertion algorithms. The table in Figure 1 gives the running times in seconds (using the Smalltalk/V implementation) of a number of incremental triangulation algorithms, along with the weights (in arbitrary units) of the triangulations produced. The algorithms, ordered from left to right, are: random incremental insertion (RI) (with no edge or triangle criteria), random incremental followed by edge-flipping to make it LMT, sorted incremental (SI) DT, random incremental (RI) DT, sorted incremental (SI) LMT, and random incremental (RI) LMT. We used integer points in a uniform distribution in a square, with n ranging from 250 to 2000. Values given are averages over 50 trials, with all algorithms run on the same point sets.

We see that the randomized incremental triangulation (with no criteria on edges) is the fastest. It is also a horrible heuristic, producing triangulations with weights nearly 4 times that of the other triangulations. The next fastest triangulation is the randomized incremental LMT, which produces triangulations of relatively low weight. Suprisingly, the sorted incremental LMT produced triangulations of slightly smaller weight than the randomized version, as did the edge-flipped LMT, though the later did so at a significant cost in running time. The DT algorithm not only produced higher weight triangulations than did the LMT algorithms, but were substantially slower.

3.2 Overview of General Point Set Methods

For a number of the triangulations, we ran tests on much larger sets (using the DecStation 5000 implementation) to compare weights. Figure 2 shows a table of the weights of the triangulations produced by various algorithms on random point sets uniformly distributed

n	GT	DT	(RI) LMT	(SI) LMT
500	16170	16513	16243	16199
1000	23205	23615	23330	23293
2500	35484	36357	35743	35646
5000	49527	50687	49834	49762
7500	60584	61967	60874	60769
10000	69628	71334	70086	69957

Figure 2: Weights: General Point Sets

over a square of $10^6 \times 10^6$ integer points. We include the GT computed by Algorithm 2 of [3], the random incremental DT, the random incremental LMT, and a sorted incremental LMT.

We see that the GT, though slower than the incremental insertion methods, does produce triangulations of lower weight, and thus may be preferable in some instances. As before, both LMT algorithms are faster than the DT and produce triangulations of lower weights.

3.3 Convex Polygon using General Point Set Methods

We also ran some general point set algorithms on a different random distribution: points uniformly distributed on a circle. (Thus all points in the set S lie on the convex hull of S.) We tested the the GT algorithm of [3], the random incremental DT, and the random incremental LMT. This circle distribution is particularly interesting for two reasons. It is a distribution where the DT is not likely to be effective at minimizing weight; in fact, since all points are cocircular, the DT is not uniquely defined and the edges may be chosen randomly. Second, this is a distribution where the GT algorithm of [3] was conjectured to perform poorly since only a constant fraction of points will be "completed" after the first phase. The tests were run on the DecStation 5000 implementation, with real points and a circle of radius 5000. They are reported in Figure 3.

The tests verify what one suspects. The GT algo-

	GT		(RI) DT		(RI) LMT	
n	time	wt.	time	wt.	time	wt.
500	2815	23	3726	2	2881	1
750	3031	75	4192	2	3083	2
1000	3213	99	4565	3	3219	2
1250	3300	163	4862	8	3327	4
1500	3367	406	4422	11	3412	5
1750	3465	715	4546	14	3489	7
2500	N/A	N/A	4865	29	3655	13
5000	N/A	N/A	5741	118	3977	49
7500	N/A	N/A	6237	265	4166	115
10000	N/A	N/A	6938	1314	4298	193

Figure 3: Points on a Circle

rithm of [3] does run very slowly on this distribution. In fact, for point sets of size 2000 and larger we ceased testing of this algorithm because it become combersome. However for smaller point sets, it still produced the triangulation of smallest weight. The DT, though faster than the GT, produced triangulations whose weight was much worse than the other two. The LMT is both fast, and nearly as effective as the GT. For this distribution, the random incremental LMT is clearly the algorithm of choice!

3.4 Overview of Simple Polygon Methods

The table in Figure 4 gives the runnings times of various algorithms for a set of approximately 550 simple polygons. For reasons of brevity, we include only a small sample of the numerous algorithms tested. We have chosen those algorithms which performed best by one of the two measures, or which proved most promising as a good heuristic. These are the fast simple polygon triangulation algorithm [16] requiring linear time in the expected case, the same triangulation with edges flipped to be locally minimal, an incremental approach to the LMT, an incremental approach to the DT, and the exact MWT.

The polygons on which the algorithms were tested are taken from a file of cross-sectional slices of a hip, jaw, retina and lung used in an application to 3D surface reconstruction. Individual polygons in these files have as many as 102 vertices, with a total of over 19000 vertices in all polygons. The files contained numerous polygons for each value of n. (For example, there were 11 polygons with 9 vertices, 10 polygons with 28 vertices, 5 polygons with 59 vertices, and only one polygon with 102 vertices.) Thus again for reasons of brevity, we give a sample of the polygon data for a small set of values of n. Note that for simple polygons we were able to produce the exact minimum weight triangulation using an $O(n^3)$ algorithm. Since the sample polygons varied considerably in area, we do not present the absolute weights of the triangulations, but instead present for each triangulation T the *ratio of effectiveness* $\mathcal{W}(T)/\mathcal{W}(MWT)$.

Not surprisingly, the simple triangulation algorithm was the fastest as it enforced no criteria on the edges or triangles. However for polygons even as small as 20 vertices, it may be worse than the exact MWT by a factor of 2 in weight. Both LMT algorithms, on the other hand, performed very well at approximating the exact MWT as expected. The incremental LMT was the faster of the two and produced triangulations averaging within 10% of the optimal. At a slight cost in time, the triangulation produced by computing a fast triangulation and then edge-flipping to local minimality performed even better, averaging approximately 2% of optimal. As with general point set methods, the incremental DT was both slower than the incremental LMT and greater in weight, though it is faster than the edge-flipping approach. As was well-known, the exact MWT is very slow. The cubic running time degrades quickly, though for small values of n ($n \leq 12$) it is practical and is actually faster than the edge-flipping algorithm.

4 More Detailed Analysis: some primitive operations

4.1 Incremental Insertion

One interesting comparison is between various triangulations produced by random incremental insertion. As was shown in Section 3, the LMT produced by the random incremental approach not only proves to be a better MWT heuristic than the DT, but is also significantly faster. We were curious about the differences in speed, as the only difference in the algorithms is the criteria by which edges are flipped. The local minimality test requires a comparison of two distances, whereas the incircle test for the DT computes the determinant of a 4×4 integer matrix. We ran the algorithms for point sets ranging from 500 to 6500 points, and compared the number of edges flipped and the running times of the two different triangulations. This data is given in the table in Figure 5.

We see again that the DT is only approximately 1.3% to 1.4% greater in weight than the LMT, but for sets of under 7000 points takes roughly 4 times as long to compute. What is surprising is that the LMT flips significantly fewer edges (more than 20% fewer in these samples) than the DT. This 20% difference is not enough to account for the larger discrepency in running times—the test for local minimality is clearer a much faster test than the incircle test—but is nonetheless a surprising and interesting result. This could be interpreted as indicating that a random incremental triangulation

n	Fast Simple		Edge Flipped		Inc. LMT		Inc. DT		MWT
	time	ratio	time	ratio	time	ratio	time	ratio	time
10	2.15	1.10	3.89	1.01	1.14	1.01	1.35	1.03	2.99
19	3.82	1.98	9.92	1.01	4.39	1.02	4.79	1.04	23.22
28	5.34	1.68	15.08	1.01	7.59	1.02	8.97	1.08	74.41
47	5.10	1.79	27.11	1.03	20.61	1.11	20.35	1.23	378.45
59	6.41	2.07	39.98	1.03	26.77	1.07	28.35	1.19	718.39
69	10.49	1.67	41.16	1.01	33.01	1.05	34.01	1.14	793.40
76	21.82	2.17	63.51	1.03	38.58	1.09	42.50	1.22	1320.03
97	27.90	2.28	80.58	1.02	64.52	1.11	67.25	1.26	3211.29
102	31.90	2.19	94.05	1.02	59.05	1.10	69.28	1.25	3682.28

Figure 4: Simple Polygon Triangulation Algorithms

	total	DT		LMT	
n	edges	time	flipped	time	flipped
500	1480	144	1457	31	1321
1500	4475	438	4368	97	3864
2500	7470	743	7358	176	6422
3500	10473	1047	10338	266	9013
4500	13473	1337	13295	362	11572
5500	16467	1657	16453	466	14307
6500	19462	1975	19465	579	16920

Figure 5: Random Incremental: DT vs. LMT

	First Phase			Second Phase		
n	cand.	plau.	add.	cand.	plau.	add.
50	264	260	123	256	186	12
100	654	637	266	634	334	18
150	1128	1087	410	721	230	23
200	1499	1451	546	1831	615	34
250	2074	1992	704	1060	256	30
300	2571	2487	849	1572	394	36
350	3092	2973	993	2057	461	39
400	3560	3415	1142	2379	419	41
450	4202	3988	1281	3611	654	47

Figure 7: Phases of Greedy Algorithm

is actual closer to being locally minimal than it is to being Delaunay.

Similar results also held for larger point sets and different distributions. Figure 6 (following page) shows weights, running times, and the number of edges flipped for a random incremental DT, a random incremental LMT, and a sorted incremental LMT for points distributed on a circle (that is, on the perimeter or circumference of the circle, not on a disc or the interior of a circle). Data is from the DecStation implementation. We see that the same conclusions hold for this distribution as well. It is also interesting to note how many more edges need to be flipped for the sorted incremental as opposed to the random incremental. This accounts for the slower running time of the sorted incremental, despite not having to do any point location algorithm. Interestingly enough, the sorted incremental produces a lower weight triangulation for this distribution as well.

4.2 More Details on the Greedy Algorithm

Implementation of the greedy triangulation algorithm of [3] also yielded interesting information. We examinined the balance of work done during the two phases of the algorithms. For the optimal constant $c = 0.055$ chosen in the Smalltalk/V implementation, the table in Figure 7 shows the balance of edges examined in the two phases. We show the number of candidate edges (the total number of edges examined in that phase), the number of plausible edges (candidate edges that passed the plausibility test), and the number of edges actually added to the triangulation (plausible edges that passed the compatibility test).

This data suggests a number of interpretations. The optimal running time seems to come from balancing the number of candidate edges between the two phases, and not the number of plausible edges. Remember that the number of candidate edges is the number of *tests for plausibility* whereas the number of plausible edges is the number of edges that get sorted and tested for compatibility. Thus our data suggests that the slowest aspect of the algorithm is not the sorting or the compatibility test, but rather the plausibility test. This interpretation was verified in a subsequent experiment: altogether removing the plausibility test sped up the algorithm significantly.

Looking closer at the data in Figure 7 gives further information. Note that in the first round, the number of plausible edges is very close to the number of candidate edges. That is, the plausibility test is not removing many edges from candidacy; we are doing considerable work with little gain. In the second phase, however, the plausibility test successfully removes a much higher

	(RI) Delaunay			(RI) LMT			(SI) LMT		
n	wt	time	flipped	wt	time	flipped	wt	time	flipped
500	3726	2	1718	2881	1	353	2826	2	4387
1000	4565	3	2932	3219	2	708	3192	4	9880
1500	4422	11	4277	3412	5	1009	3362	10	15525
2000	4600	19	5650	3552	8	1319	3509	17	21534
2500	4865	29	7022	3655	13	1666	3608	25	27675
5000	5741	118	17843	3977	49	3202	3926	92	60882
7500	6237	265	25811	4166	115	4717	4115	205	95676
10000	6938	1314	40022	4298	193	6235	4251	375	132062

Figure 6: Points on a Circle: Various Incremental Algorithms

percentage of candidate edges. This suggests that the plausibility test might be effectively used in the second phase only. We tested this hypothesis on some larger point sets using the DecStation 5000 implementation. We compared the original version of the algorithm to a modified version with no plausibility test, and a version with the plausibility test in the second round only. For sets of size 5000 to 15000, the data was inconclusive. The version with the plausibility test in the second round seems slightly faster than the version with no plausibility test at all. We hope to to examine this further in our next round of testing.

4.3 Variants of the Greedy

As was shown earlier, the GT algorithm of [3] was particularly slow for points on a circle. For this distribution (and others like it), its failure to close off a large enough number of points in the first phase was one of the things that led to the development of the new IGT algorithm [4]. Also, if all points are strictly on the convex hull then all pairs are plausible and thus the plausibility test is clearly not going to gain anything while costing a considerable amount of time. This is a case where removing the plausibility test seemed especially likely to improve the running time of the GT algorithm of [3]. Using the Smalltalk/V implementation, we tested the GT algorithm both with and without the plausibility test, and the IGT algorithms on some small sets of points on a circle. For comparison, we also include the random incremental LMT. The results are given in Table 8. Conclusions are not difficult to draw, nor are they suprising. Removing the plausibility test from the GT algorithm was a significant improvement. However the IGT is much faster than even the modified GT algorithm. And for this distribution, all of the GT algorithms were significantly slower than the randomized incremental LMT. For these small point sets, the ratio of difference in weights remained approximately 3%. If you need the fastest heuristic, and can tolerate a 3% increase in weights, use the randomized LMT.

Finally, we compare the fastest variant of the new IGT algorithm of [4] and the two variants of the GT algorithm of [3] (with and without the plausibility test) with the fastest LMT algorithm (randomized incremental approach) for general uniformaly distributed point sets. This data is given in Table 9. Values are averages of 15 trials on a uniform distribution of integer points on a 1000 x 1000 square, run on the Smalltalk/V implementation.

We see again that the randomized incremental LMT is significantly faster than even the fastest GT algorithm, though it does not do as well at approximating the MWT. Of the GT algorithms, the fastest is the IGT variant of [4].

5 Summary

Surprising, the general question of the effectiveness of only the criterion of local minimality does not seem to have been studied apart from the GT. Though we did not focus on proving expected case theoretical properties of locally minimal triangulations, our experiments yielded considerable evidence to support the thesis that local minimality is a good goal to aim for. For general point sets, any LMT consistently gives better weight than the DT. And while the GT provides the best weight of all *feasible* algorithms tested, the other LMT algorithms gave weights comparably close while the random incremental LMT had much faster running times than even the best GT algorithm. The picture for simple polygon triangulations was very similar.

We also showed that the GT algorithm of [3] can be made to run significantly faster without the plausibility test, and our data supports a conjecture that the fastest implementation may have a plausibility test in the second phase only. In examining the algorithm of [3], we came up with a faster iterative (or multi-phase) GT algorithm and found that it works best if points are rebucketed every round, and if the search distance is incremented linearly rather than geometrically.

	Times				Weights	
n	LMT	GT w/ plaus.	GT no plaus.	IGT	GT and IGT	LMT
25	0.50	3.36	2.91	3.19	2160	2238
50	1.11	10.55	7.17	6.38	2761	2862
75	1.65	20.35	14.67	11.74	3115	3201
100	2.15	37.29	24.12	18.29	3412	3535
125	2.76	44.60	35.55	22.56	3641	3755
150	3.49	66.38	39.81	28.95	3776	3901
175	3.90	94.60	54.38	37.81	3945	4050
200	4.48	109.45	68.45	40.87	4121	4256
225	5.37	120.59	87.26	44.72	4213	4327
250	6.03	154.78	90.93	63.92	4314	4427
275	6.64	183.02	105.39	68.51	4424	4558
300	7.37	194.47	124.20	68.28	4532	4652

Figure 8: Circle: LMT and variants of the GT

	Times				Weights	
n	LMT	GT w/ plaus.	GT no plaus.	IGT	GT and IGT	LMT
50	2.18	12.11	9.35	8.72	23558	23671
100	4.90	31.51	24.96	20.42	35385	35531
150	7.78	54.89	42.96	32.12	44004	44230
200	10.62	76.95	60.33	44.10	50644	50976
250	13.59	99.25	78.52	57.76	57476	57829
300	16.60	128.17	100.27	74.15	62676	63086
350	19.76	158.63	126.02	86.81	67787	68260
400	22.67	186.72	149.42	100.74	72391	72907

Figure 9: General Point Sets: GT and LMT.

6 Acknowledgements

This paper benefitted from numerous discussions with Scot Drysdale.

References

[1] G. Barequet and M. Sharir, "Piecewise-linear interpolation between polygonal slices." *Proceedings of the Tenth Annual Symposium on Computational Geometry* (1994) 93–102.

[2] J. Bentley, D. Sanat and E. Williams Jr., "The complexity of finding fixed-radius near neighbors." *Information Processing Letters* 6 (1977) 209–213.

[3] M. Dickerson, R.L. Drysdale, S. McElfresh, and E. Welzl, "Fast Greedy Triangulation Algorithms." *Proceedings of the Tenth Annual Symposium on Computational Geometry* (1994) 211-220.

[4] M. Dickerson and M. Montague, "Fast new algorithms for locally minimal triangulations–and a platform for testing them." *manuscript* (1994).

[5] S. Drysdale, G. Rote, and O. Aichholzer, "A simple linear time greedy triangulation algorithm for uniformly distributed points." *manuscript* (1994).

[6] P. Gilbert, "New results in planar triangulations." MS Thesis, University of Illinois, Urbana, IL, 1979.

[7] S. Goldman, "A Space Efficient Greedy Triangulation Algorithm." *Information Processing Letters* 31 (1989) 191–196.

[8] L. Heath and S. Pemmaraju, "New results for the minimum weight triangulation problem." *Virginia Polytechnic Institute and State University, Department of Computer Science, TR 92-30* (1992). To appear in *Algorithmica*.

[9] G. Klincsek, "Minimal triangulations of polygonal domains." *Ann. Discrete Math.* 9 (1980) 121–123.

[10] C. Levcopoulos and A. Lingas, "Fast Algorithms for Greedy Triangulation." *BIT* 32 (1992) 280–296.

[11] A. Lingas, "A new heuristic for the minimum weight triangulation." *SIAM Journal of Algebraic and Discrete Methods* 8 (1987) 646–658.

[12] A. Lingas, "Greedy triangulation can be efficiently implemented in the average case." *Proceedings of Graph-Theoretic Concepts in Computer Science* (1988) 253–261.

[13] E. Lloyd, "On triangulations of a set of points in the plane." *Proceedings of the 18th FOCS* (1977) 228–240.

[14] G. Manacher and A. Zobrist, "Neither the greedy nor the Delaunay triangulation of the planar set approximates the minimal triangulation." *IPL* 9 (1979) 31–34.

[15] G. Manacher and A. Zobrist, "Probabilistic methods with heaps for fast-average-case greedy algorithms." *Advances in Computing Research* vol. 1 (1983) 261–278.

[16] J. O'Rourke "Computational Geometry in C." Cambridge University Press, 1994.

[17] C. A. Wang, "Efficiently updating the constrained Delaunay triangulations." *BIT* 33 (1993) 238–252.

[18] C. A. Wang, "An optimal algorithm for greedy triangulation of a set of points" *Proc. Sixth Canadian Conference of Computational Geometry* (1994).

[19] Y.F. Wang and J.K. Aggarwal, "Surface reconstruction and representation of 3-D scenes." *Pattern Recognition* 19 (1986) 197–207.

[20] P. Yoeli, "Compilation of data for computer-assisted relief cartography." In *Display and Analysis of Spatial Data* J.C.Davis and M.J. McCullagh, editors, John Wiley & Sons, NY (1975).

Computing the Visibility Graph via Pseudo–triangulations

Michel Pocchiola[*] Gert Vegter[†]

Abstract

We show that the k free bitangents of a collection of n pairwise disjoint convex plane sets can be computed in time $O(k + n \log n)$ and $O(n)$ working space. The algorithm uses only one advanced data structure, namely a splittable queue. We introduce (weakly) greedy pseudo–triangulations, whose combinatorial properties are crucial for our method.

1 Introduction

Consider a collection $\mathcal{O}$ of pairwise disjoint convex objects in the plane. We are interested in problems in which these objects arise as obstacles, either in connection with visibility problems where they can block the view from an other geometric object, or in motion planning, where these objects may prevent a moving object from moving along a straight line path. The *visibility graph* is a central object in such contexts. For polygonal obstacles the vertices of these polygons are the nodes of the visibility graph, and two nodes are connected by an arc if the corresponding vertices can see each other. [9] describes the first nontrivial algorithm for computing the visibility graph of a polygonal scene with a total of n vertices in $O(n^2)$ time. [4] presents an optimal $O(n \log n + k)$ algorithm, where k is the number of arcs of the visibility graph. A practically feasible $O(k \log n)$ algorithm is contained in [6].

In this paper we present an optimal—with respect to both time and working space—algorithm that computes the *tangent visibility graph* of $\mathcal{O}$. Recall that a *bitangent* is a closed line segment whose supporting line is tangent to two obstacles at its endpoints; it is called *free* if it lies in *free space* (i.e., the complement of the union of the relative interiors of the obstacles). The endpoints of these bitangents split the boundaries of the obstacles into a sequence of arcs; these arcs and the free bitangents are the edges of the tangent visibility graph, see Figure 1.

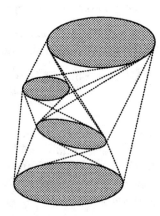

Figure 1: The tangent visibility graph.

In [7] we described an optimal method for computing the so-called visibility complex of the collection $\mathcal{O}$. Just as the algorithm of Ghosh and Mount, see [4], it is based on complicated data structures (e.g. the split–find structure of Gabow and Tarjan, see [3]). Therefore it is not suitable for a practical implementation.

We give two practical, yet efficient methods to compute the tangent vibility graph of a collection of n disjoint convex sets in the plane in output sensitive time. The first algorithm is very simple, and uses

---

[*]e-mail: pocchiol@dmi.ens.fr
[†]e-mail: gert@cs.rug.nl

$O(k \log n)$ time, where k is the number of arcs of the tangent visibility graph. Throughout the paper we assume that the complexity of the objects is $O(1)$, that is, the common bitangents of any pair of objects can be computed in constant time. With each unit vector in the plane we associate a subdivision of free space, which we call the *greedy pseudo–triangulation* associated with this vector. The algorithm maintains the greedy pseudo–triangulation as the unit vector rotates over an angle of π. The basic operation that updates the pseudo–triangulation is a *flip* of a free bitangent with smallest slope greater than the slope of the rotating unit vector. Relaxing the order in which bitangents are flipped we obtain an optimal algorithm, using $O(k + n \log n)$ time and $O(n)$ working space. (To the best of our knowledge, even for the case of line segments, this is the first optimal algorithm that uses linear working space.)

If the obstacles are points, our second method—translated into dual space—is an alternative for the topological sweep algorithm for arrangements of lines, of Edelsbrunner and Guibas, see [2]. Our pseudotriangulations replace their (upper and lower) horizon trees.

It turns out that, in general, our second method can be interpreted as a topological sweep of the visibility complex, introduced in [7]. This point is briefly discussed in the last section.

2 Greedy pseudo–triangulations

Definition and basic properties

Let $\mathcal{O} = \{O_1, O_2, \ldots, O_n\}$ be a family of n pairwise disjoints convex sets (obstacles for short). A *pseudo-triangulation* of a set of obstacles is the subdivision of the plane induced by a maximal (with respect to inclusion) family of pairwise noncrossing free bitangents. It is clear that a pseudo–triangulation always exists and that the bitangents of the boundary of the convex hull of the obstacles are edges of any pseudo-triangulation. A pseudo-triangulation of a collection of four obstacles is depicted in Figure 2. The subdivision owes its name to the special shape of its regions. A *pseudotriangle* is a simply connected subset T of the plane, such that (i) the boundary ∂T consists of three convex chains, that share a tangent at their common endpoint, and (ii) T is contained in the triangle formed by the three endpoints of these convex chains. These three endpoints will be called the *cusps* of T. (In this paper a chain is an alternating sequence of free bitangents and arcs, such that successive elements share a common endpoint, at which the bitangent is tangent to the arc.) Without proof we mention the following result (it is easy to prove using Euler's relation for planar graphs, see the full version).

Lemma 1 *The bounded free faces of any pseudo-triangulation are pseudotriangles. Furthermore the number of pseudotriangles (of a pseudo-triangulation of a collection of n obstacles) is $2n - 2$ and the number of bitangents is $3n - 3$.*

Consider a unit vector u in the plane. The u–slope of a directed line segment b is defined as the *positive* (counterclockwise) angle over which we have to rotate u in order to obtain a vector parallel to b. The *greedy* pseudo–triangulation, associated with a unit vector u, is the pseudo–triangulation induced by the family $B(u) = \{b_1, b_2, \ldots, b_{3n-3}\}$ of bitangents, recursively defined as follows.
1. b_1 is the bitangent with smallest u–slope in the set of free bitangents.
2. b_{i+1} is the bitangent with smallest u–slope in the set of free bitangents disjoint from $b_1, b_2, \ldots, b_i$.
Figure 2 depicts a greedy pseudo-triangulation. The greedy pseudo-triangulation associated with u, will be denoted by $\mathcal{T}(u)$.

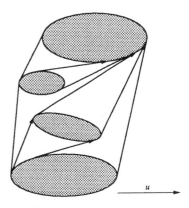

Figure 2: The greedy pseudo-triangulation $\mathcal{T}(u)$ (with respect to u).

If t is a free bitangent, then either $t \in B(u)$, or t intersects at least one bitangent of $B(u)$ whose u–slope is less than the u–slope of t. This property holds for *all* bitangents in $B(u)$, intersecting t:

Proposition 2 *The u–slope of a free bitangent t, $t \notin B(u)$, is larger than the u–slope of every bitangent in the sequence $B(u)$, intersecting t.*

Proof. Suppose the result does not hold. Let t be a free bitangent of minimal u–slope, intersecting a bitangent in $B(u)$ of larger u–slope. As we have just observed, there also is a bitangent in $B(u)$, intersecting t, of smaller u–slope than t. In particular, there are $b, b' \in B(u)$ intersecting t, such that (i) the u–slope of t is greater than the u–slope of b, but less than the u–slope of b', and (ii) there is no bitangent in $B(u)$ intersecting t between its points of intersection with b and b'. In other words: b and b' are in the boundary of the same pseudo–triangle in $\mathcal{T}(u)$. Let us denote this pseudotriangle by T.

Consider the point $q \in \partial T$ whose tangent line is parallel to t. If t intersects b before (after) b', the point q lies to the left (right) of t, see Figure 3. Let p be the tail (head) of b.

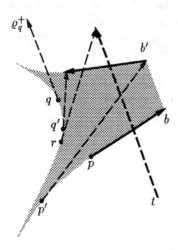

Figure 3: Proof of the basic property, in case t intersects b before b'.

For a point x in the boundary of T let ϱ_x^+ (ϱ_x^-) be the directed free line segment starting (terminating) at x, and extending in forward (backward) direction along the tangent line of ∂T at x, until it hits some obstacle. This object is called the visibility of x along the ray.

As x moves from p to q along ∂T, let p' be the first and q' be the last point on ∂T for which the corresponding ray intersects the bitangent b'. Note that p and p', as well as q and q', may coincide. Furthermore ϱ_x^+ (ϱ_x^-) intersects b' for all points $x \in \partial T$ between p' and q'. As x travels along ∂T from p' to q', the u–slope of ϱ_x^+ is increasing, so in particular it is less than the u–slope of t. We shall argue that, as a point x moves along ∂T from p' to q', the object visible from x along ϱ_x^+ (ϱ_x^-) changes. Suppose we know this is true, then there is a point $x \in \partial T$ between p' and r, such that ϱ_x^+ contains a point of tangency y. Therefore xy is a free bitangent intersecting b', whose u–slope is larger than the u–slope of b'. Since t is the free bitangent with minimal u–slope satisfying this property, the u–slope of xy is smaller than the u–slope of t. But we just observed that the u–slope of ϱ_x^+, and hence the u–slope of xy, is smaller than the u–slope of t. This is a contradiction.

So it remains to prove that the visibility along ϱ_x^+ is not constant. We only do so in the case where t intersects b before b'. The other case is treated similarly. Assume that we see the same object, O' say, along $\varrho_{p'}^+$ and $\varrho_{q'}^+$ (otherwise we're done). Let I be the open line segment connecting the endpoints of these rays, then $I \subset O'$, so in particular I and t are disjoint.

Let $r \in \partial T$ be the point where the tangent line l_r of ∂T through r contains the endpoint of t. Obviously r lies between p' and q on ∂T. It even lies between p' and q'. Indeed, if this were not the case then $\varrho_{q'}^+$ would end at the head of b', which would therefore be a point of I. Then the line supporting $\varrho_{q'}^+$ would intersect I before t. On the other hand, the line supporting $\varrho_{p'}^+$ obviously intersects t before I. Since the line supporting ϱ_x^+ intersects both t and I, for all $x \in \partial T$ between p' and q', the segments t and I would not be disjoint. This contradiction proves that r lies between p' and q'.

Let O'' be the object containing the endpoint of t. Since I and t are disjoint, the line l_r intersects O'' before O', so the object visible from r along ϱ_r^+ is different from O'. □

Lemma 3 *The greedy pseudo–triangulation of a collection of n disjoint convex obstacles in the plane with respect to some unit vector can be computed in $O(n \log n)$ time.*

Proof. Omitted from this version. The construction is based on a standard rotational sweep à la Bentley-Ottmann, from direction 0 to direction π, during which we maintain the visibility map associated to the current direction. The $O(n)$ events correspond to the detection of free bitangents of the greedy pseudo–triangulation. □

3 The greedy flip algorithm

The idea of the first version of the algorithm is very simple: just maintain $\mathcal{T}(u)$ as u rotates over an angle of π, starting from the horizontal direction u_0. It is obvious that $\mathcal{T}(u)$ remains constant as long as it is not parallel to any of the free bitangents of $\mathcal{O}$. It turns out that we can obtain all greedy pseudo–triangulations of the collection $\mathcal{O}$ by *flipping* the bitangent of minimal slope with respect to the current unit vector. To make this idea more precise, consider two pseudotriangles T_1 and T_2 that share a bitangent,

b say. We obtain a new pseudo-triangulation by *flipping* b, i.e. by replacing b by the common bitangent of T_1 and T_2. (To see that this common bitangent is unique, observe that two distinct tangent lines of ∂T_i cross inside T_i.) E.g. in Figure 4, flipping b_1 amounts to replacing it by b^* (here T_1 and T_2 are the shaded regions incident upon $b = b_1$). Flipping a bitangent in the boundary of the convex hull boils down to reverting its direction.

Lemma 4 *Let b be the bitangent of $T(u)$ of minimal u–slope. Let u' be a unit vector obtained by infinitesimally rotating u beyond b. Then $T(u')$ is obtained from $T(u)$ by flipping the bitangent b.*

Proof. Let $B(u) = \{b_1, \cdots, b_{3n-3}\}$ and $B(u') = \{b'_1, \cdots, b'_{3n-3}\}$. Furthermore let b^* be the free bitangent obtained by flipping b_1. First assume that b_1 is an internal bitangent, i.e. it is not in the boundary of the convex hull of the collection of objects. Then there is an index i, with $1 \leq i < 3n - 3$, such that $b'_j = b_{j+1}$, for $1 \leq j < i$, and $b'_i \neq b_{i+1}$. We shall successively prove:
(i) b'_i intersects b_1.
(ii) u–slope$(b'_i) \leq u$–slope(b^*).
(iii) $b'_i = b^*$.
(iv) $b'_j = b_j$, for $i < j \leq 3n - 3$.
This will obviously prove the lemma. To prove (i), assume that b'_i and b_1 are disjoint. Since $T(u')$ is a greedy pseudo-triangulation, that b'_i is the bitangent with smallest u'–slope in the set of free bitangents disjoint from $b_2, \ldots, b_{i-1}$, and hence also in the set of free bitangents disjoint from $b_1, \ldots, b_{i-1}$. But then $b'_i = b_i$, since $T(u)$ is a greedy pseudo-triangulation. This contradiction proves (i).

Since b^* is disjoint from all bitangents in $\{b_2, \ldots, b_i\} = \{b'_1, \ldots, b'_{i-1}\}$, and b'_i is the free bitangent of smallest u'–slope among the free bitangents that are disjoint from $b'_1, \ldots, b'_{i-1}$, we see that u–slope$(b'_i) \leq u$–slope(b^*), which proves (ii).

To prove (iii), assume that $b^* \neq b'_i$. Since b'_i intersects b_1, b'_i must intersect the boundary of the quadrangle Q obtained by merging the pseudotriangles of $T(u)$, incident upon b_1, see figure 4.

Note that the bitangents in ∂Q, whose u–slope is larger than the u–slope of b^*, lie either between the heads or between the tails of b_1 and b^*. The crucial observation is that b'_i intersects only bitangents in the boundary of ∂Q whose u–slope is *less than* the u–slope of b'_i, see proposition 2. In particular b'_i is disjoint from this part of ∂Q (note that it also can't be tangent to this part of ∂Q, since its u–slope does not exceed the u–slope of b^*). But then b'_i intersects b^* from right to left (note that, in view of (i), it intersects b_1 from right to left). This is a contradiction with (ii),

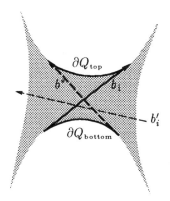

Figure 4: The pseudo-quadrangle Q, and its diagonals b_1 and b^*. Bitangent b^* is obtained by flipping b_1.

so $b^* = b'_i$.

Finally (iv) is an immediate consequence of (i), (ii) and (iii). The case in which b_1 is in the boundary of the convex hull is obvious, since flipping b_1 here amounts to reverting its direction. □

Lemma 4 suggests a simple algorithm: conceptually we rotate a unit vector u, starting at position u_0, over an angle π. We maintain the set of bitangents in the current pseudo-triangulation in a priority queue, where the weight of a bitangent in the queue is its u–slope. As long as the queue is non-empty, extract the minimal weight bitangent, flip it, and insert the new bitangent into the queue if its u_0–slope is less than π (so it has not yet been detected). We shall call this method the *greedy flip algorithm*.

In this way the total time for the operations on the queue is $O(k \log n)$, since every free bitangent is deleted from the queue exactly once. The total cost of all flips is $O(k)$ (amortized). This will become clear in the next section, where we prove a more general result. Summarizing we have:

Theorem 5 *The greedy flip algorithm computes the tangent visibility graph of a collection of n disjoint convex objects in the plane in $O(k \log n)$ time, where k is the number of free bitangents.*

4 The weakly greedy flip algorithm

In this section we improve the time complexity of the algorithm by relaxing the constraint that bitangents are flipped in order of increasing slope. To achieve this goal we enlarge the class of pseudo-triangulations, and replace the linear order, induced by the slope of the bitangents, by a *partial order* on the set of bitangents in the pseudo-triangulation, such that the set of bitangents that are candidates for flipping can be maintained in constant time per flip. Furthermore, the class of pseudo-triangulations should be invariant under flipping of candidate edges. The crucial feature of this class is the property proved in proposition 2, which we now require to hold *by definition*. We then prove invariance under flipping, and describe an efficient implementation of the flip-operation, whose amortized cost is finally analyzed.

Weakly greedy pseudo–triangulations

First we need some terminology. Let B be a set of free bitangents. For a subset A in the plane the set of elements of B intersecting A is denoted by B_A. So if B is the set of all bitangents of a pseudo-triangulation $\mathcal{T}$, and T is a pseudotriangle of $\mathcal{T}$, the set B_T consists of all bitangents in ∂T. In this case the pseudotriangle of $\mathcal{T}$ incident upon $b \in B$ and —locally— to the left (right) of b is denoted by $ltriang(b)$ ($rtriang(b)$).

Consider a pseudotriangle T, and fix some (directed) bitangent $b_T \in B_T$. The direction of the tangent line in a point of ∂T is uniquely determined by the requirement that its b_T-slope is less than π. This b_T-slope is also called the *slope* of this point. The *base-point* of T, denoted by p_T, is the tail of b_T, if $T = rtriang(b_T)$, or the head of b_T, if $T = ltriang(b_T)$. (It is the unique point on ∂T at which the slope is not well-defined; by definition, we set this slope equal to 0). The positive (negative) orientation of ∂T corresponds to increasing (decreasing) slope. A subsegment of ∂T with positive (negative) orientation will be called a *walk* (*reverse walk*) along ∂T. In particular, the walk starting at the base-point of T defines a linear order on the set of bitangents B_T, called the *slope order* (with respect to b_T). The successive cusps we pass during a walk starting at the base-point of T, are denoted by x_T, y_T and z_T. The *forward* and *backward* view of point p in ∂T are the points of intersection of ∂T with the tangent line at p, lying ahead and behind p, respectively. The point whose forward (backward) view is p_T, if $T = rtriang(b_T)$ ($T = ltriang(b_T)$) is denoted by q_T.

Definition 6 *A pseudo–triangulation $\mathcal{T}$ is called* **weakly greedy** *if there is a partial order $\prec$ on its set B of bitangents such that*
(i) for every pseudotriangle T in $\mathcal{T}$ the restriction of $\prec$ to B_T is a linear order $\prec_T$, that corresponds to the slope order with respect to the minimal element of B_T.
(ii) Every free bitangent t can be given a unique direction that is compatible with the slopes of both its endpoints, such that all bitangents in B_t intersect t from left to right. This unique direction will be called the **canonical direction** of t (with respect to $\prec$).

Obviously proposition 2 tells us that every greedy pseudo-triangulation is weakly greedy. If T is a pseudotriangle of a weakly greedy pseudo-triangulation $\mathcal{T}$ with partial order $\prec$, we denote the minimal element of B_T with respect to $\prec$ by b_T. We say that T is a weakly greedy pseudotriangle if it is a pseudotriangle in a weakly greedy pseudo-triangulation. For later use we isolate a simple, but crucial feature of weakly greedy pseudotriangles.

Lemma 7 *Let T be a weakly greedy pseudotriangle.*
1. If $z_T \neq p_T$, then the part of ∂T between z_T and p_T is an arc.
2. If y_T lies between x_T and q_T, then the part of ∂T between y_T and q_T is an arc (i.e. it contains no bitangents).

Proof. We shall prove that no bitangent $t \in B_T$ has forward an backward views of smaller slope. This will prove 1, since all points on the segments $z_T p_T$ have both forward and backward view of smaller slope. A similar argument proves 2.

To prove the claim, suppose that both the backward and forward view, p_0 and p_1 say, of t have smaller slopes than t. We only consider the case in which p_0 has smaller slope than p_1, see Figure 5. Then $T = ltriang(t)$, and the part of ∂T between p_0 and p_1 lies completely to the left of the line supporting t.

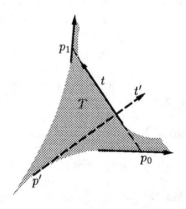

Figure 5: Backward and forward views p_0 and p_1 of t can't both have smaller slope than t.

Observe that the object containing $tail(t)$ is different from the one containing $head(t)$. Arguing as in the proof of proposition 2, we can show that there is a free bitangent t', intersecting t, whose tail p' is a point on ∂T between p_0 and p_1. But t intersects t'

from right to left, in contradiction with the weakly greedyness of the pseudo-triangulation. This proves the lemma. □

Flipping $\prec$-minimal bitangents

If we work in the class of weakly greedy pseudo-triangulations we can, in general, flip more bitangents than just the one with minimal u_0-slope, without disturbing the weakly greedyness. (From now on u_0 will be a fixed, say horizontal, direction.) More precisely, we shall prove that any $\prec$-minimal bitangent can be flipped.

The partial order $\prec^*$

To introduce the partial order on the new pseudo-triangulation, consider a $\prec$-minimal bitangent b, with $R = rtriang(b)$ and $L = ltriang(b)$. Let b^* be the bitangent obtained by flipping b, and let $\mathcal{T}^*$ be the pseudo-triangulation after the flip. The right and left pseudotriangles of b^* are denoted by R' and L', respectively. The partial order $\prec^*$ on the set of bitangents of $\mathcal{T}^*$ is the transitive closure of the relation, defined by the collection of linear orders $\prec_T^*$, for $T \in \mathcal{T}^*$. If $T \neq R', L'$, then T is a pseudotriangle in both $\mathcal{T}$ and $\mathcal{T}^*$, and we take $\prec_T^*$ equal to $\prec_T$. So it remains to define $\prec_T^*$ for $T = R', L'$.

First consider the pseudotriangle R'. Let b_R' be the $\prec$-successor of b in B_R. The $\prec^*$-minimal element of $B_{R'}$ is one of the bitangents b_R' and b^*, viz the one with minimal b-slope. So $b^* = \min B_{R'}$, if $p^* = tail(b^*)$ lies between b and b_R', and $b_R' = \min B_{R'}$, otherwise. Hence there are three basic cases, that will return throughout this section, see Figure 6.

Case 1 b and b_R' are not separated by a cusp of R.
Then $R' = rtriang(b_R')$, and p^* doesn't lie on the arc between b and b_R'. Therefore $\min B_{R'} = b_R'$.
Case 2 b and b_R' are separated by a cusp of R, but p^* doesn't lie on the arc between b and b_R'.
Then $R' = ltriang(b_R')$ and $\min B_{R'} = b_R'$. (Note: in this case $x_R = head(b_R')$, as in Figure 6, or $x_R = head(b)$.)
Case 3 b and b_R' are separated by a cusp of R, but p^* lies on the arc between b and b_R'.
Then $R' = rtriang(b^*)$ and $\min B_{R'} = b^*$.

The restriction of $\prec^*$, restricted to $B_{L'}$, is defined similarly.

To make sure that the flipping terminates, we only flip bitangents whose u_0-slope is less than π. This condition, as it turns out, guarantees that the partial order, restricted to B_T, for $T \in \mathcal{T}$, is compatible with the linear order according to increasing u_0-slope.

Lemma 8 *Let $(\mathcal{T}, \prec)$ be a weakly greedy pseudo-triangulation. Let b be a $\prec$-minimal bitangent of $\mathcal{T}$, whose u_0-slope is less than π. Then the pseudo-triangulation $(\mathcal{T}^*, \prec^*)$, obtained by flipping b, is again weakly greedy. Furthermore, the canonical direction of all free bitangents t, $t \neq b$, doesn't change due to the flip, whereas the canonical direction of b is reversed.*

Proof. The proof is built from ingredients of the proof of lemma 4. We refer to figure 4 for an illustration of the proof, with the understanding that b is identified with b_1.

Since the u_0-slope of b is less than π, the relation $\prec^*$ is compatible with the order according to increasing u_0-slope, so its transitive closure is a partial order. Therefore it remains to prove that for all free bitangents $t \neq b, b^*$, the canonical direction of t with respect to $\prec^*$ is well-defined, and that the 'left-to-right' property holds. To this end it is sufficient to prove that for all such bitangents $t \neq b, b^*$, having exactly one endpoint on $\partial L \cup \partial R$, the slope at that endpoint doesn't change due to flipping b. Consider a bitangent t, having exactly one endpoint on ∂R. Assume, by contradiction, that the slope at this endpoint changes upon flipping b. Now all points of ∂R, whose slope is reversed after flipping b, lie on the arc c between b and its successor b_R' in B_R. Therefore t is tangent to c. In particular the slope of t is less

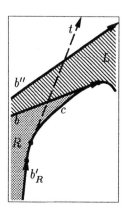

Figure 7: All free bitangents, $\neq b$, keep the same canonical direction.

than the slope of b_R', and hence less than the slope of all bitangents in $B_R \setminus \{b\}$. By definition 6.(ii), t is therefore disjoint from all bitangents in $B_R \setminus \{b\}$, so t intersects ∂R in a point of b. Since, again by definition, b intersects t from left to right, we conclude that $tail(t) \in c$, see Figure 7. Since all points of ∂R, whose slope is reversed, lie between the basepoint of R and the basepoint of $R' = rtriang(b^*)$, even $slope(t) < slope(b^*)$.

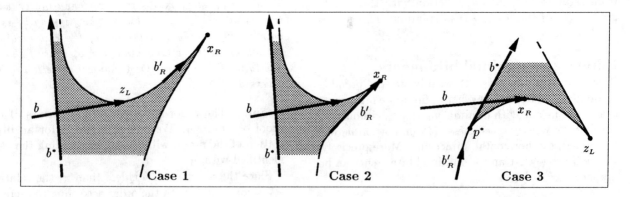

Figure 6: The partial order $\prec^*$, restricted to the bitangents of $\partial R'$. (Note that in case 2 either $x_R = head(b'_R)$, as in the Figure, or $x_R = head(b)$.)

Now all bitangents t', satisfying (i) $t' \cap b \neq \emptyset$, (ii) $tail(t') \in \partial R$, and (iii) $slope(t) < slope(t')$, have their head to the left of the line supporting t, see Figure 7. Therefore $head(b^*)$ lies to the left of this line. On the other hand t is different from both b and b^*, so it intersects ∂L in a bitangent b'', different from b. Since b'' also intersects t from left to right, all points of ∂L to the left of t have slope between $slope(b)$ and $slope(b'')$. Therefore $slope(b^*) < slope(b'')$, and hence $slope(b^*) < slope(t)$. This contradiction proves that the slope at $tail(t)$ is not reversed. Similarly one can prove that the slope at $head(t)$ is not affected by the flipping of b.

We finally have to prove that either b^* and t are disjoint, or b^* intersects t from left to right. So assume $b^* \cap t \neq \emptyset$. As in the proof of lemma 4, let Q be the quadrangle obtained by merging the pseudo-triangles L and R, see figure 4. Let ∂Q_{top} ($\partial Q_{\text{bottom}}$) be the part of ∂Q between the heads (tails) of b and b^*. Since $\mathcal{T}$ is weakly greedy, t is disjoint from $\partial Q_{\text{top}} \cup \partial Q_{\text{bottom}}$, since otherwise it would intersect the bitangents in this subset of ∂Q from left to right. Since t intersects b from right to left, it therefore also intersects b^* from right to left. The preceeding argument also shows that the 'left-to-right' property holds. □

The pseudotriangles R' and L'
We now consider the pseudotriangle R' in more detail, in particular its cusps $x_{R'}$, $y_{R'}$ and $z_{R'}$. (The story for L' is completely similar.) To this end we consider each of the cases 1–3 introduced above see also Figure 6.

Case 1 $R' = rtriang(b'_R)$.
In this situation b and b'_R are not separated by a cusp, so $x_{R'} = x_R$. Furthermore, if p^* lies between x_R and y_R, then the second cusp $y_{R'}$ is equal to p^*, otherwise it is equal to y_R, see Figure 8a. Similarly the third cusp $z_{R'}$ is equal to y_L, if q^* lies between x_L and y_L, otherwise it is equal to q^*, see Figure 8b.

Case 2 $R' = ltriang(b'_R)$ and $b'_R = \min B_{R'}$.
In this case the basepoint of R' is $head(b'_R)$, which lies between x_R and y_R. Therefore the first cusp $x_{R'}$ is is equal to p^*, if p^* lies between x_R and y_R, otherwise it is equal to y_R, see Figure 8a. Similarly the second cusp $z_{R'}$ is equal to y_L, if q^* lies between x_L and y_L, otherwise it is equal to q^*, see Figure 8b. Finally the third cusp $z_{R'}$ is equal to z_L, if $head(b) = x_R$, otherwise it is equal to x_R, see Figure 8c.

Case 3 $R' = rtriang(b^*)$ and $b^* = \min B_{R'}$.
In this case $head(b) = x_R$, and the tail p^* of b^* lies on the arc of ∂R separating b and b'_R. Therefore the basepoint of R' is p^*, which is also equal to the third cusp $z_{R'}$, see the left part of Figure 8a. Since in this case x_R is a cusp of R, the second cusp is equal to z_L, see the left part of Figure 8c. Finally the first cusp is equal to y_L or q^*, depending on whether q^* lies between y_L and z_L or between x_L and y_L, see figure 8b. Figure 9 summarizes the previous discussion.

The algorithm

Every pseudotriangle in a weakly greedy pseudo-triangulation has a unique minimal bitangent. If a bitangent is minimal for both its left and right pseudo-triangles, lemma 8 guarantees that it can be flipped. Such bitangents are called candidates.

Definition 9 *A bitangent, belonging to a weakly greedy pseudo-triangulation $(\mathcal{T}, \prec)$, is called a **candidate** if it is a minimal element with respect to $\prec$, and its u_0-slope is less than π.*

Lemma 8 suggests a very simple algorithm. It maintains the set of candidates in a *set $\mathcal{C}$*:

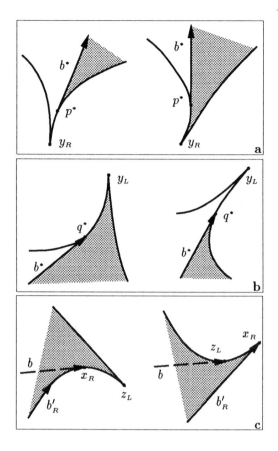

Figure 8: The cusps of R'.

1. compute the greedy pseudo-triangulation with respect to the horizontal direction u_0;
2. put all candidate edges in a *set* $\mathcal{C}$
3. while $\mathcal{C} \neq \emptyset$ do
4.   select a bitangent b from $\mathcal{C}$;
5.   flip b;
6.   update $\mathcal{C}$;

The major improvement is that we abandoned the priority queue in favor of any simple data structure for the representation of sets, that allows us to insert and delete an element in $O(1)$ time. Of course we still have to prove that the algorithm is correct, and that the total time needed for flipping (viz step 5)

	$x_{R'}$	$y_{R'}$	$z_{R'}$
Case 1	x_R	y_R or p^*	y_L or q^*
Case 2	y_R or p^*	y_L or q^*	z_L or x_R
Case 3	y_L or q^*	z_L	p^*

Figure 9: The cusps of R'.

and updating the set of candidates (viz step 6) is $O(k)$. We shall say that the algorithm *detects* a free bitangent at the moment it is flipped (in step 5). The correctness of the algorithm follows from:

Lemma 10 *The weakly greedy algorithm detects every free bitangent (or, equivalently, every edge of the tangent visibility graph).*

Proof. Let $(\mathcal{T}, \prec)$ and $(\mathcal{T}^*, \prec^*)$ be the initial and final pseudo-triangulations, respectively. For every free bitangent t the canonical direction of t with respect to $\prec$ lies between u_0 and $-u_0$. whereas its canonical direction with respect to $\prec^*$ lies between $-u_0$ and u_0. Therefore lemma 8 implies that t has been flipped. □

The splittable queue Awake[T]

Conceptually the flipping can be done by walking—in positive direction, starting at the basepoint—along the boundaries of the triangles L (left) and R (right), incident upon the flipped bitangent b, with one leg in every triangle, such that at any moment the tangent lines at the points underneath our left and right legs are parallel. We keep walking until these tangent lines coincide. At that point we have found b^*. This is too expensive, since some bitangents may be passed during many walks involved in the flip operations. To cut the budget, we shall need an auxiliary data structure, that enables us to start the walk at a more favorable point.

Observe that the tail p^* of b^* lies between the first cusp x_R and the point q_R, whose tangent contains the base-point $tail(b)$ of R. Similarly q^* lies between x_L and q_L. For a pseudotriangle T, a point in ∂T is called *awake* if it lies between x_T and q_T. Note that the points of ∂R that are awake have forward view of smaller slope, whereas the points awake in L have backward view of smaller slope, see Figure 10. Lemma 7 tells us that the set of points that are awake is a sequence of arcs and bitangents on a convex chain, possibly followed by a single arc between y_T and q_T (in case q_T does not lie between x_T and y_T).

If b and its successor b'_R in B_R are not separated by the cusp x_R (so case 1 occurs), the point p^* lies even between q'_R and q_R, where q'_R is the point whose tangent contains $tail(b'_R)$, see Figure 10.

So the walk along ∂R starts at q'_R in case 1, and in x_R, otherwise. Similarly the walk along ∂L starts in q'_L or in x_L. Now x_T can be determined in $O(1)$ time, but how do we determine q'_T efficiently, for $T = L, R$? To this end we consider the segment $x_T q_T$ of points in ∂T that are awake as an alternating sequence of bitangents and arcs, or atoms for short, where the

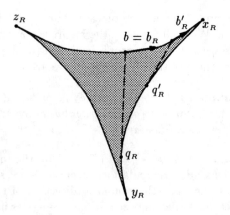

Figure 10: The set of points that are *awake* in R is the segment $x_R q_R$. When the algorithm flips $b = b_R$, the walk on ∂R starts in q'_R (case 1), or in the cusp x_R (cases 2 and 3).

atoms are in slope order. This sequence will be represented by a *splittable queue*, denoted by `Awake[T]`, a data structure for ordered lists that allows for the following operations:

(i) *enqueue* an atom, either at the head or at the tail of the list;
(ii) *dequeue* the head or the tail of the list;
(iii) *split* the sequence at an atom x; this split is preceeded by a *search* for the atom x.

A few comments on the split operation are in order. We assume that the initial search for the atom x is guided by a real-valued function, f say, defined for atoms in the sequence, that is monotonic with respect to the order of the atoms in the sequence. Now a *split* amounts to determining the atom x for which $f(x) = 0$, and successively splitting the sequence (destructively) into the subsequences of atoms with negative f-values and those with positive f-values. More specifically, to find the point q'_T (in case 1) we do a split operation in `Awake[T]`, where the search for q'_T is guided by the position of $tail(b'_T)$ with respect to the tangent lines at the endpoints of an atom.

Lemma 11 *There is a data structure, implementing a splittable queue, such that an enqueue or dequeue operation takes $O(1)$ amortized time, and a split operation at an atom x on a queue of n atoms takes $O(\log \min(d, n-d))$ amortized time, where d is the rank of x in the sequence represented by the queue.*

Moreover, a sequence of m enqueue, dequeue and split operations on a collection of n initially empty splittable queues is performed in $O(m)$ time.

Splittable queues can be implemented using red-black trees with parent-pointers, where atoms are stored in the leaves. Splittable queues are in fact a special case of finger trees, implemented as level-linked red-black trees (see [5], and also [1] for similar ideas). In our case we don't need level links, however, since the search implicit in the split operation can be implemented as a dovetailing search up the ridges of the tree, starting from the minimal and maximal leaf of the tree. For more details and a sketch of the proof we refer to the full version of this paper.

We now describe in some detail (i) how to compute b^*, using `Awake[R]` and `Awake[L]`, and (ii) how to compute the queues `Awake[R']` and `Awake[L']`. We shall argue that doing all flips and maintaining the collection of queues `Awake[T]`, $T \in \mathcal{T}$, cost $O(k)$ enqueue, dequeue and split operations. Hence the total cost of (i) and (ii) is $O(k)$, see Lemma 11.

Construction of b^*.
If b and its successor b'_R in B_R are not separated by the cusp x_R of R, then the walk along ∂R starts in q'_R. In this case we *split* `Awake[R]` at q'_R into `AwakeMin[R]` and `AwakeMax[R]`, where the atoms in the former queue have smaller slope than the atoms in the latter queue. Otherwise, viz if b and b'_R are separated by the cusp x_R, the walk along ∂R starts in x_R, and we set `AwakeMin[R]` $\leftarrow \emptyset$ and `AwakeMax[R]` $\leftarrow$ `Awake[R]`. Here $\emptyset$ denotes the empty queue. In either case p^* lies on an arc, represented by an atom in the queue `AwakeMax[R]`. We similarly initialize the splittable queues `AwakeMin[L]` and `AwakeMax[L]`.

The simultaneous walk along ∂R and ∂L can be implemented by *dequeueing* atoms from `AwakeMax[R]` and `AwakeMax[L]`, until the atoms (arcs) are found that contain p^* and q^*, respectively. Obviously, this sequence of synchronous dequeue-operations takes time proportional to the number of dequeued atoms. So we construct b^* at the cost at most one split on `Awake[R]` and at most one split on `Awake[L]`, followed by a number of successive dequeue operations.

We finally adjust the first atoms in the queues `AwakeMax[R]` and `Awake[L]`, (viz the atoms containing p^* and q^*, respectively), by replacing their endpoints of smaller slope with p^* and q^*, respectively. After this final operation the splittable queues `AwakeMax[R]` and `AwakeMax[L]` represent the segments $p^* q_R$ of ∂R and $q^* q_L$ of ∂L, respectively. We shall use these queues in the construction of the queues `Awake[R']` and `Awake[L']`.

Construction of `Awake[R']` and `Awake[L']`
We only describe the construction of `Awake[R']` if for both R' and L' case 1 occurs, viz when $R' = rtriang(b'_R)$ and $L' = ltriang(b'_L)$. In this situation $head(b) = z_L = q_L$, see Figure 6a, and also $tail(b) = z_R = q_R$. Furthermore, the basepoint of R'

is $tail(b'_R)$, so we have $q_{R'} = q'_R$. Hence, by definition, all points that are awake in R' lie between $x_R \, (= x_{R'})$ and q'_R, so we set `Awake[R']` ← `AwakeMin[R]`. Similarly, set `Awake[L']` ← `AwakeMin[L]`.

If case 2 or 3 occurs for R or L, the computation of `Awake[R']` and `Awake[L']` is even simpler: It requires only a number of dequeue and *at most two* enqueue operations on the splittable queues `AwakeMax[R]` or `AwakeMax[L]`. (The full version contains further details.)

As for the amortized time complexity, observe that the initial collection of splittable queues—one for each pseudotriangle in the greedy pseudotriangulation we start out with—can be computed in $O(n \log n)$ time (for instance simply by enqueueing the bitangents and arcs, that are awake in the boundary of each pseudotriangle). This amounts to $O(n)$ enqueue–operations. As we have just indicated, doing all flips and maintaining the collection of queues `Awake[T]`, $T \in \mathcal{T}$, cost $O(k)$ further enqueue, dequeue and split operations. Note that at any time the storage needed for all these queues is $O(n)$, see Lemma 1. Together with Lemma 11 this observation implies our main result.

Theorem 12 *The weakly greedy flip algorithm is optimal: it computes the tangent visibility graph of a collection of n disjoint convex objects in the plane in $O(n \log n + k)$ time and $O(n)$ working storage, where k is the number of free bitangents.*

5 Topological sweep of the visibility complex

It is worth noting that, translated in "dual space" our algorithm implements an efficient "topological sweep" of the visibility complex, introduced in [7]. We explain this briefly. Recall that the underlying space $|X|$ of the visibility complex X is the quotient space of the space of free rays $\mathcal{F} \times \mathcal{R}$ under the equivalence relation $(p, u) \sim (q, u)$ iff the line segment $[p, q]$ lies in free space $\mathcal{F}$ and the slope of the line (pq) is equal to u modulo π. The topology of $|X|$ induces a natural structure of a 2–dimensional regular cell complex on $|X|$ (see [7]). In particular there is an onto–mapping $b \mapsto |b|$ from the set X_0 of vertices of X to the set of free bitangents. (The preimage of the bitangent $b = [p, q]$ and direction u is the set of rays $(p, u + k\pi)$, $k \in \mathcal{Z}$.) Let x be a face (= vertex, edge or facet) of X. We define $\sup x$ ($\inf x$) to be the ray [1] with maximal (minimal) slope in the closure of x. We turn X into a poset $(X, \prec)$ by taking the transitive closure of the relation $\inf x \prec x \prec \sup x$.

---
[1] By a slight abuse of terminology a point in $|X|$ is still called a ray.

(See e.g. [8], chapter 3, for terminology on posets.) A cut of X is a maximal antichain of $(X \setminus X_0, \prec)$. A cut depends only on its subset of edges, and there is exactly one edge per oriented obstacle. We extend $\prec$ to the set $\mathcal{C}$ of cuts by setting $\Phi \prec \Phi'$ iff $E_i \prec E'_i$, for all i, where E_i is the edge in the cut Φ associated with the oriented obstacle i. One can check that if Φ' covers Φ then $\Phi \setminus \Phi'$ and $\Phi' \setminus \Phi$ are composed of 2 edges and 1 facet incident to the same vertex b, i.e. $\sup \Phi \setminus \Phi' = \inf \Phi' \setminus \Phi = b$; we will say that Φ' covers Φ via b. Now the intuitive notion of a topological sweep of X is formally defined as a topological sorting of $(X, \prec)$ or, equivalently, as a maximal chain of the poset $(\mathcal{C}, \prec)$. The following theorem asserts that the flip algorithm realizes a topological sweep of the visibility complex.

Theorem 13 *Let Φ be a cut of X. Then $\sup \Phi$ is a weakly greedy pseudo–triangulation. Furthermore a vertex b is minimal in $\sup \Phi$ iff Φ is covered by some Φ' via b. In that case $\sup \Phi'$ is obtained from $\sup \Phi$ by flipping b.*

References

[1] B. Chazelle and H. Edelsbrunner. An optimal algorithm for intersecting line segments in the plane. *J. ACM*, 39:1–54, 1992.

[2] H. Edelsbrunner and L. J. Guibas. Topologically sweeping an arrangement. *J. Comput. Syst. Sci.*, 38:165–194, 1989. Corrigendum in 42 (1991), 249–251.

[3] H. N. Gabow and R. E. Tarjan. A linear-time algorithm for a special case of disjoint set union. *J. Comp. Sys. Sci.*, 30:209–221, 1985.

[4] S. K. Ghosh and D. M. Mount. An output-sensitive algorithm for computing visibility graphs. *SIAM J. Comput.*, 20:888–910, 1991.

[5] K. Hoffmann, K. Mehlhorn, P. Rosenstiehl, and R. E. Tarjan. Sorting Jordan sequences in linear time using level-linked search trees. *Inform. Control*, 68:170–184, 1986.

[6] M. H. Overmars and E. Welzl. New methods for computing visibility graphs. In *Proc. 4th Annu. ACM Sympos. Comput. Geom.*, pages 164–171, 1988.

[7] M. Pocchiola and G. Vegter. The visibility complex. In *Proc. 9th Annu. ACM Sympos. Comput. Geom.*, pages 328–337, 1993.

[8] Richard P. Stanley. *Enumerative Combinatorics*, volume 1. Wadsworth & Brooks/Cole, 1986.

[9] E. Welzl. Constructing the visibility graph for n line segments in $O(n^2)$ time. *Inform. Process. Lett.*, 20:167–171, 1985.

Searching for the Kernel of a Polygon — A Competitive Strategy

Christian Icking* Rolf Klein*

Abstract

We present a competitive strategy for walking into the kernel of an initially unknown star-shaped polygon. From an arbitrary start point, s, within the polygon, our strategy finds a path to the closest kernel point, k, whose length does not exceed 5.48 times the distance from s to k. This is complemented by a general lower bound of $\sqrt{2}$. Our analysis relies on a new result about an interesting class of curves which are *self-approaching* in the following sense. For any three consecutive points a, b, c on the curve the point b is closer to c than a to c.

Keywords. Competitive strategy, simple polygon, kernel, curves with increasing chords, self-approaching curves.

1 Introduction

Suppose that a mobile robot equipped with a 360 degree vision system "wakes up" in an unknown environment. Its task is to go to some location from which the whole environment is visible. This should be achieved as efficiently as possible.

In attacking this problem we are using the following model. The robot is a point, its environment is a simple polygon, P. Clearly, we have to assume that P is star-shaped, or no location with the required property would exist; see Section 2 for definitions. At each point p, the robot is provided with the visibility polygon vis(p). The cost of the robot's motion is measured by the arc length of its path; we ignore the cost of planning the path, which will in fact turn out to be negligible.

If the robot knew a map of the polygon and its start position beforehand, it could employ one of the classic algorithms for computing the kernel, see [12, 1]. Then it could determine the kernel point, k, closest to its start position, s, and go straight from s to k; note that by

---

*FernUniversität Hagen, Praktische Informatik VI, Elberfelder Str. 95, 58084 Hagen, Germany, Christian.Icking@FernUni-Hagen.de, Rolf.Klein@FernUni-Hagen.de

definition each point of the kernel can see any other point of P, so that the line segment from k to s cannot be blocked. This would result in a perfect solution at cost $d(k, s)$, where d denotes the Euclidean distance.

But since the polygon P is not known to the robot a priori, its actions are bound to contain elements of try and error which cause extra cost. For example, in the situation depicted in Figure 1 a detour cannot be avoided; see Lemma 15.

In this paper we present a strategy that guarantees a path length bounded by 5.48 times $d(k, s)$. In general, a strategy for a problem class PC is called *competitive* with competitive factor C if each instance P of PC can be solved at a cost not exceeding C times the cost of a perfect solution of P with full information in advance.

Competitive strategies for autonomous robots have recently received considerable interest in computational geometry. For example, the problem of finding a goal in an unknown environment has been studied in [9, 11, 2, 7, 6, 13]. In [3, 4] the task of drawing a map is addressed. How to decide on the robot's current location if a map is available has been investigated in [5, 10]. A special case of the present problem, where P contains only one reflex vertex, has been optimally solved in [8].

Competitive strategies offer many advantages. They are superior to heuristic approaches because they come with a proven performance guarantee. Though, they can often be made as simple as heuristic rules, as the present paper demonstrates. Sometimes competitive strategies exist even for problems whose optimal solution would be NP-hard; see [4, 5]. However, analyzing a competitive strategy is often difficult such that the proven bound is considerably bigger than the real competitive factor.

This paper is organized as follows. In Section 2 we briefly recall some definitions and state a simple fact on visibility polygons. Then, in Section 3, we describe our strategy CAB (= Continuous Angular Bisector) for finding the closest kernel point. CAB is very easy to implement. We show that each path generated by CAB consists of $O(n)$ many pieces of ellipses or hyperbolae (including circular arcs and line segments as special cases), where n denotes the number of vertices of P. See Figure 3 for an example of a complete path generated by CAB. If the polygon P is not star-shaped, the robot notices this and ends its motion. Otherwise it will arrive at the closest kernel point, k. The robot's path from s to k has the following crucial property. As the

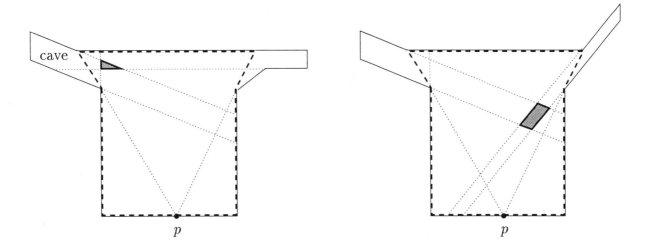

Figure 1: Identical visibility polygons but completely different kernels.

robot proceeds, it will always get closer to all the future points on its path. We call such curves *self-approaching*; see Figure 6 and Figure 7 for further examples.

In Section 4 we prove that for self-approaching curves there is an upper bound of 5.48 for their *maximum detour*, i.e. the ratio of their arc length divided by the distance between their endpoints. Therefore, we have the same value as an upper bound for the competitive factor of strategy CAB. Curves that are self-approaching in *both* directions have been called *curves with increasing chords*. Recently, Rote [14] has solved an old open problem by proving a tight bound for this subclass of self-approaching curves. Unfortunately, it seems that his technique cannot be applied to our non-symmetric case. Our proof for the upper bound is based on the observation that the arc length of a self-approaching curve does not exceed the perimeter of its convex hull. This result is complemented by an example of a self-approaching curve whose ratio equals approximately 4.21.

Finally, in Section 5, we prove that there is a lower bound of $\sqrt{2}$ for the competitive factor of *any* possible strategy for finding the kernel. For our particular strategy CAB, there is a situation where $\pi + 1 \approx 4.14$ is achieved (a case of only one reflex vertex). For this particular situation we have in [8] presented an optimal competitive strategy with factor 1.21218 which may be useful for further improvement of CAB.

2 Visibility in a polygon

In this section we briefly recall some elementary facts on visibility in simple polygons. As before, let P denote a simple planar polygon, interior area plus boundary.

Definition 1 Let p, q be two points in P. Then q is *visible* from p if the line segment $\overline{pq}$ is contained in P. The set of all points of P that are visible from p is called the *visibility polygon* of p, vis(p).

Definition 2 The *kernel* of P, ker(P), is the set of all points in P from which each point of P is visible. If ker(P) is not empty, then P is called *star-shaped*.

Clearly, visibility is a symmetric relation. Only for convex polygons does $P = \ker(P)$ hold. Otherwise the polygon has at least one *reflex vertex*, i.e. one whose internal angle is greater than $180°$, and the kernel is a proper subset of P which can be obtained in the following way. Each edge e of P defines two halfplanes, an inner one which locally contains points of the interior of P, and an outer one. The kernel ker(P) is known to be the intersection of all inner halfplanes. In particular, the kernel is convex. In Figure 1, the kernels are represented by shaded areas.

Each visibility polygon vis(p) is star-shaped and contains p in its kernel. Its boundary consists of (segments of) original edges of P, and of edges that do not belong to P. These *spurious* edges are segments of lines emanating from p that touch a reflex vertex of P before they hit the boundary of P. A spurious edge separates the part of P that is visible from p from a part which is not, a so-called *cave*; see Figure 1, where the visibility polygons are depicted by dashed lines.

In the two pictures of Figure 1 the visibility polygons are identical, whereas the kernels are quite different. Thus, from point p the robot has only little information about the kernel of P. But from any point in P, one can at least identify the kernel of the actual visibility polygon! The following fact shows how these sets are related.

Lemma 3 Let p, q be points in P such that vis(q) $\subseteq$ vis(p). Then we have the following inclusions.

$$\ker(P) \subseteq \ker(\mathrm{vis}(p)) \subseteq \ker(\mathrm{vis}(q))$$

Proof. The kernel of vis(p) is the intersection of all inner halfplanes defined by

(1) all edges of P that are (at least partially) visible from p, and

(2) the spurious edges of vis(p).

The edges of (1) are a subset of all edges of P. Moreover, the inner halfplane of any spurious edge e properly contains the intersection of the inner halfplanes of the two edges of P that are adjacent to the reflex vertex supporting e. Therefore, ker(vis(p)) is defined by less severe or fewer restrictions than ker(P), proving the first inclusion.

By assumption, point q can see no more edges of P than p does. If e_q is a spurious edge of vis(q), two cases arise. If its cave is not entirely visible from p this gives rise to a spurious edge e_p in vis(p). Then the inner halfplane of e_q contains the intersection of the inner halfplanes of e_p and f, where f denotes the edge visible from q that is adjacent to the supporting vertex of e_q; see Figure 2. If the whole cave is visible from p then there is one edge, denoted by g in Figure 2, that yields, together with f, an even more severe restriction than e_p does. This proves the second inclusion. □

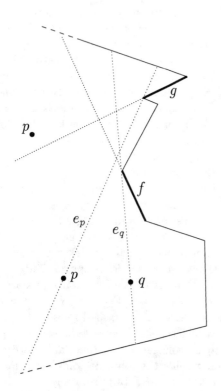

Figure 2: Two possible locations of p discussed in the proof of Lemma 3.

3 A strategy for the kernel problem

Throughout this section we assume that P is a star-shaped simple polygon.

At the start point, s, the robot checks if vis(s) contains spurious edges. If not, it can conclude that s lies in the kernel of P, and stop. Otherwise, the robot should proceed from s on such a path that it gains insight into all the caves, and never lets out of sight points of P it has already seen, so that the visibility polygon grows monotonically.

The robot can only gain insight into a cave if it walks into the inner halfplane defined by the cave's spurious edge, causing the edge to rotate about its supporting reflex vertex *into* the cave. The robot can only keep an eye on what it has so far seen if, in addition, it never leaves the inner halfplane of a visible edge of P; in particular, it must stay within P. This leads to the following definition. We assume that the intersection of an empty set of halfplanes equals the full plane.

Definition 4 Let p be a point in P.
(i) Let $G(p)$ denote the *gaining wedge* that results from intersecting the inner halfplanes of the spurious edges of vis(p). The reflex vertices of P associated with the two halflines bounding $G(p)$ are called the *maximal constraint vertices* at p.
(ii) Let $K(p)$ denote the *keeping wedge* that results from intersecting the inner halfplanes of all edges of P that contain p, or are visible from and colinear with p.

A point p belongs to the kernel of P if and only if $G(p)$ equals the full plane. For almost all points of P the keeping wedge $K(p)$ is equal to the full plane.

In Figure 3, for each point p of the segment between p_2 and p_3 of curve M, there is only one constraint vertex, namely v_5, because this is the only visible reflex vertex that causes a cave. In this case, $G(p)$ has an angle of 180°. Otherwise, there are at least two constraint vertices two of which are maximal, e. g. v_1 and v_5 between p_1 and p_2. At p_2, the keeping wedge, $K(p_2)$ is the inner halfplane of the former spurious edge through v_1.

Lemma 5 *Within a sufficiently small neighborhood of a point p in P the set $G(p) \cap K(p)$ equals ker(vis(p)) and is nonempty. If the robot moves into this set then vis(p) grows, unless p belongs already to ker(P).*

Proof. The equality is immediate. The set cannot be empty because it contains ker(P), due to Lemma 3. □

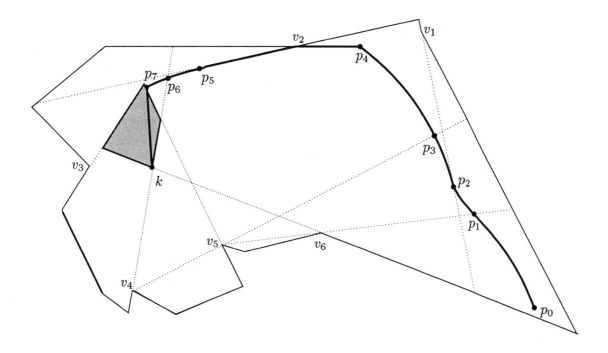

Figure 3: A path generated by strategy CAB.

Now we can formulate the robot's strategy CAB.

Strategy CAB
$p := s$;
WHILE vis$(p) \neq P$ DO
    compute gaining wedge $G(p)$;
    $m := $ angular bisector of $G(p)$;
    IF m leaves the keeping wedge $K(p)$
        THEN walk in direction
            of the projection of m
            along the boundary of $K(p)$
        ELSE walk along m
    END-IF
END-WHILE;
go straight to the point $k \in \ker(P)$ closest to s
END

This strategy is meant to be continuous, so the "steps" should be infinitesimal. It could easily be adapted to polygons that are not star-shaped: As soon as $G(p) \cap K(p)$ becomes empty, the robot knows that $\ker(P)$ is empty, and stops moving.

Figure 4 illustrates a case where the THEN-branch applies. Here the following lemma holds.

Lemma 6 *If the angular bisector m of $G(p)$ leaves the wedge $K(P)$ then it forms an angle $\mu < 90°$ with exactly one of the two bounding halflines of $K(p)$. Therefore, the projection onto $K(p)$ is well-defined. Moreover, $G(p) \cap K(p)$ includes an angle $\beta \leq 90°$.*

Proof. Using the notations of Figure 4, we have $\mu = \gamma/2 - \beta \leq 90° - \beta \leq 90°$. Equality could hold for *both* halflines bounding $K(p)$ only if $G(p)$ and $K(p)$ are both 180° wedges with disjoint interiors. Since $\ker(P)$ is contained in their intersection, this means that p can already look into the cave(s) responsible for $G(p)$. This is a contradiction to Definition 4 because an edge is spurious only if it separates a non-empty part of P from vis(p). Finally, we have $\beta \leq 90°$ because μ is bigger than 0, by assumption. □

Figure 3 shows an example where the robot walks along boundaries of keeping wedges. Between p_3 and p_4, the maximal constraint vertices are v_4 and v_5. At point p_4, it hits the extension of the upper edge incident to reflex vertex v_2. The robot follows this extension to v_2. Here, $K(p)$ is defined by both edge extensions. Now the robot follows the extension of the lower edge incident to v_2 until it arrives at p_5. Still, v_4 and v_5 are the maximal constraint vertices. From p_5 on, the robot resumes following the angular bisector which no longer crosses the edge extension.

As soon as the robot arrives in the kernel of P at p_7, it knows the whole polygon. According to CAB, it then determines which kernel point k would have been closest to its start point, s, and walks to k within $\ker(P)$. In practice, one might skip the last step because the robot's job can be considered done as soon as it reaches *some* point of the kernel. Since this does not lead to a smaller competitive factor, we can as well account for the last step, too.

Now we study the path generated by CAB.

Lemma 7 *The path M generated by strategy CAB consists of segments of hyperbolae and ellipses.*

Proof. If there are two maximal constraint vertices, the two cases depicted in Figure 5 may arise. In (i), the

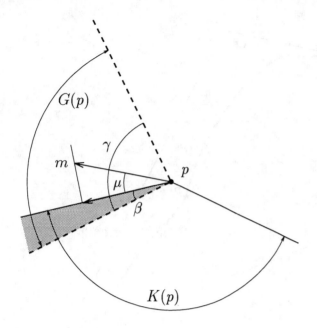

Figure 4: If the angular bisector m of the gaining wedge, $G(p)$, leaves the keeping wedge, $K(p)$, the robot walks along the boundary of $K(p)$.

boundary of $G(p)$ consists of halflines originating from p that pass through their reflex vertices v_1 and v_2. If the robots makes a small step of angle α with respect to $\overline{pv_1}$ then its distance to v_1 changes at the rate $\cos\alpha$. Since the angular bisector forms an angle α with $\overline{pv_2}$, too, the distance to v_2 changes at the same rate. Thus, $d(p, v_1) - d(p, v_2)$ remains constant. The locus of these points is a hyperbola. If the difference equals 0 one obtains a line, as a special case (line segments are also generated while the robot follows the boundary of a keeping wedge).

In (ii), the gaining wedge is formed by one halfline originating from p that runs through its reflex vertex, v_1, and another one, whose prolongation beyond p runs through v_2. Let w_2 denote the image of v_2 when reflected at p. By the previous discussion, the distances to v_1 and w_2 change at the same rate, as the robot follows their angular bisector a small step. But the distance to w_2 decreases at the same rate as $d(p, v_2)$ increases, and vice versa. Therefore, $d(p, v_1) + d(p, v_2)$ remains constant. This defines an ellipse. A circle arises as a special case if there is only one constraint vertex. □

Lemma 8 *The number of segments the path M generated by CAB consists of is $O(n)$, where n denotes the number of vertices of P.*

Proof. While the ELSE-branch of CAB is executed, a segment of M can only terminate if one of the maximal constraint vertices (mcv, for short) changes. This happens whenever a new cv becomes visible whose associated inner halfplane is going to intersect the old gaining wedge, like vertex v_5 at p_1 in Figure 3, or when one of the former mcvs ceases to exist because its cave becomes fully visible, like vertex v_1 at point p_2. For each reflex vertex of P, each of these events can happen only once. Otherwise no mcv can change, because visible constraint vertices cannot take turns in being maximal.

Additional (line) segments of M result whenever the robot hits an extension of an edge of P, in the THEN-branch of CAB. But the robot will not hit the same extension again unless one of the mcvs has changed, due to the convexity of the angular bisector curves proven in Lemma 7.

This shows that the total number of segments in M is in $O(n)$. □

Now we prove that the arc length of M does not exceed $d(s, k)$, times a constant factor. Analyzing the segments of M explicitly would seem quite a difficult task. Fortunately, we can proceed in a different way, thanks to the following global property of M.

Theorem 9 *Let M be the oriented path created by strategy CAB from the start, s, to the closest kernel point, k. Then for each point p on M, the remaining part of M lies fully in front of the normal of M at p.*

Proof. First, we concentrate on the part M_0 of M from s to the first kernel point, f.

We show that the normal of M_0 at p supports $\ker(\mathrm{vis}(p))$ but does not intersect its interior. For points p that lie on an angular bisector segment of M_0, the boundary of the gaining wedge $G(p)$ locally consists of two adjacent edges of the convex set $\ker(\mathrm{vis}(p))$, or of a single edge containing p, due to Lemma 5. Thus, the angular bisector enters $\ker(\mathrm{vis}(p))$ at an interior angle of $\leq 90°$ on either side. For any linear segment of M_0 that lies on an edge extension we know from Lemma 6 that the edges of $\ker(\mathrm{vis}(p))$ meeting at p include an angle $\leq 90°$, and that M follows one of them; see Figure 4.

The rest of the proof is quite simple. At each point p, the robot walks into $\ker(\mathrm{vis}(p))$. Due to Lemma 5, $\mathrm{vis}(p)$ grows monotonically. Lemma 3 implies that $\ker(\mathrm{vis}(p))$ shrinks monotonically. Therefore, the future part of M_0 from p on is contained in $\ker(\mathrm{vis}(p))$, which lies in front of the normal.

Now we consider the last straight line segment of M, leading from f to k. It is contained in $\ker(P) \subset \ker(\mathrm{vis}(p))$, for each point p on M_0, so that the normal through p cannot cross it. On the other hand, no normal through a point p of $\overline{fk}$ crosses the remaining part of this line segment. This proves the theorem for the full path M. □

From Lemma 12 (ii) in Section 4 it will follow that M is self-approaching and by Theorem 14 (see the remark after the proof) we will have an upper bound of 5.48 for the maximum detour of an *arbitrary* curve that enjoys this property. This yields the following result.

Theorem 10
Strategy CAB is competitive with factor 5.48.

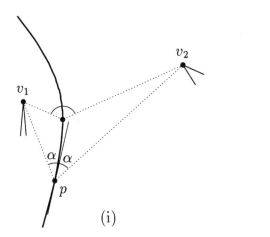 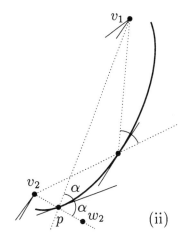

Figure 5: Following the angular bisector generates (i) hyperbolae or (ii) ellipses.

Remark. Strategy CAB can easily be extended to determine if the polygon is or is not star-shaped, but it will be not competitive for this task.

4 Self-Approaching Curves

The curves considered in this section are assumed to be rectifiable, i.e. one can determine their finite lengths, but they are not necessarily smooth. For a curve C and a point $a \in C$, any line passing through point a and not crossing, at this point, to the other side of C is called a *tangent* to C at a, and any line through point a perpendicular to a tangent to C at a is called a *normal* to C at a.

As before, let $d(a, b)$ denote the Euclidean distance between two points a and b. For two points $a \leq b$ on a directed curve C, $C^{\geq a}$ denotes the part of C from a to the end, $C[a, b]$ means the part of C between a and b, and length($C[a, b]$) means its arc length. Let CH(C) denote the convex hull of curve C and per(C) the length of the perimeter of CH(C).

Definition 11 An oriented planar curve C is called *self-approaching* if the inequality

$$d(a, c) \geq d(b, c)$$

is fulfilled for any three consecutive points a, b, c on the curve. The curve has *increasing chords* if

$$d(a, e) \geq d(b, c)$$

holds for any four consecutive points a, b, c, e on the curve. The quantity

$$\sup_{\substack{a,b \in C \\ a \neq b}} \frac{\text{length}(C[a,b])}{d(a,b)}$$

is called the *maximum detour* of a curve.

The *minimum growth rate* as used in [14] is the reciprocal of the *maximum detour*. The following lemma shows a relation between self-approaching curves and those with increasing chords.

Lemma 12
(i) The class of curves with increasing chords is equal to the class of curves which are self-approaching in both directions.
(ii) A curve C is self-approaching iff any normal to C at any point a does not cross $C^{>a}$.
(iii) A curve C has increasing chords iff any normal to a curve C at any point a does not cross C except at that point.

In [14], a tight upper bound of $\frac{2}{3}\pi \approx 2.09$ for the maximum detour of any curve with increasing chords was shown. For the strictly bigger class of self-approaching curves, one can expect a bigger upper bound, if at all. Unfortunately, it seems that the technique of [14] for an optimal bound cannot be applied to this case. Nevertheless, a bound of 5.52 will be shown in Theorem 14, which can be further improved to 5.48.

Example. The logarithmic spiral, directed to the center, is an interesting example for such a curve. It is self-approaching if the (always constant) angle, α, between the tangent and the radius fulfills

$$\alpha \leq \arctan\left(\frac{3\pi}{2\,W(\frac{3}{2}\pi)}\right) \approx 74.66°$$

in which W denotes Lambert's W function defined by the functional equation $W(x)\,e^{W(x)} = x$. Figure 6 shows the limiting case where the normal at any point is tangent to the rest of the curve. This special curve is in a sense the narrowest self-approaching logarithmic spiral and has the additional property that it is its own involute. One can show that its maximum detour equals $1/\cos\alpha_{\max} \approx 3.78$, but despite its "optimized" form there are other self-approaching curves with a bigger maximum detour.

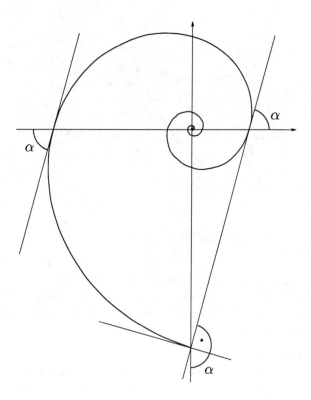

Figure 6: The narrowest self-approaching logarithmic spiral

Such a curve is shown in Figure 7. From the start point, a, to the end, e, it consists of a circular arc of radius 2 and angle β, a half circle of radius 1, and a line segment of length 1. The length of the curve equals $2\beta + \pi + 1$ while $d(a,e) = \sqrt{5 - 4\cos\beta}$. The quotient takes on its maximum value of ≈ 4.38 for $\beta \approx 13.98°$.

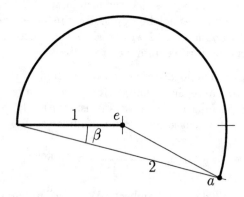

Figure 7: This self-approaching curve has a maximum detour of 4.38.

With the help of a neat trick, this can still be improved. The cycloid is known to be a curve which has another cycloid as its involute. The curve in Figure 7 remains self-approaching, if we replace the line segment of length 1 with a sequence of pieces of cycloids whose height can be arbitrarily small, see Figure 8. Then this part of the curve has length 2 (instead of 1), and the maximum detour is

$$\max_{\beta \in [0..\frac{\pi}{2}]} \frac{2\beta + \pi + 2}{\sqrt{5 - 4\cos\beta}} \approx 5.33$$

Conjecture. We think that the proof of Theorem 14 can be refined to this bound, in other words we conjecture this to be a tight upper bound for the maximum detour of all self-approaching curves.

Figure 8: Pieces of cycloids replace the line segment.

Now we prove the central result of this section.

Theorem 13 *The arc length of a self-approaching curve C is less or equal to the perimeter of its convex hull.*

Proof. The arc length of a rectifiable curve C is, by definition, the supremum of the lengths of all polygonal chains with vertices on C in the same order. Therefore, an upper bound for the length of all such chains is also an upper bound for the length of C.

We take an arbitrary polygonal chain Q whose vertices lie on C in the same order. By induction on the number of vertices of Q, we will prove that Q is shorter than the perimeter, $\mathrm{per}(Q)$, of its convex hull, $\mathrm{CH}(Q)$, which is not greater than $\mathrm{per}(C)$. Note that the vertices of $\mathrm{CH}(Q)$ are also vertices of Q and are therefore points on C.

The assertion is true for Q being a line segment, so let us assume that Q has at least three vertices, the first two are called a and b. The induction hypothesis is that $\mathrm{length}(Q^{\geq b}) < \mathrm{per}(Q^{\geq b})$.

We distinguish two cases depending on whether b lies on the boundary of $\mathrm{CH}(Q)$ or not.

Case 1. The point b is on the boundary of $\mathrm{CH}(Q)$. We have a situation as depicted in Figure 9. Some vertices of $\mathrm{CH}(Q^{\geq b})$, in this example c and e, are not vertices of $\mathrm{CH}(Q)$.

For these vertices, we construct auxiliary points as follows. The point c' lies at distance $d(c, b)$ from c on the line $l(e, c)$, the point e' at distance $d(e, c')$ from e on $l(f, e)$, and finally f' lies at distance $d(f, e')$ from f on $l(f, a)$. In other words, the sequence of circular arcs from b via c' and e' to f' is the involute of the chain $b c e f$.

Let c'', e'' be the points of C on the lines $l(e, c)$ and $l(f, e)$, respectively. Since $C[a, b]$ approaches c we have

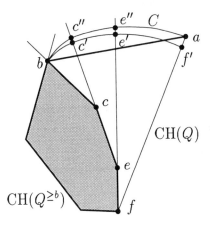

Figure 9: $d(f,a) \geq d(f,e) + d(e,c) + d(c,b)$.

$d(c, c'') \geq d(c, b) = d(c, c')$. It also approaches e, therefore

$$\begin{aligned} d(e, e'') &\geq d(e, c'') \\ &= d(e, c) + d(c, c'') \\ &\geq d(e, c) + d(c, c') \\ &= d(e, c) + d(c, b) \end{aligned}$$

Another application of this argument shows that $d(f, a) \geq d(f, e) + d(e, c) + d(c, b)$.

The bigger convex hull $\mathrm{CH}(Q)$ exceeds in perimeter $\mathrm{CH}(Q^{\geq b})$ by $d(a,b) + d(f,a) - (d(f,e) + d(e,c) + d(c,b))$ which is bigger than $d(a,b)$, the difference between length(Q) and length$(Q^{\geq b})$.

Notice that this argument generalizes to any number of vertices of $\mathrm{CH}(Q^{\geq b})$, like c and e.

Case 2. The point b is not on the boundary of $\mathrm{CH}(Q)$. Then a must lie in the wedge formed by the prolongations of the adjacent edges of $\mathrm{CH}(Q^{\geq b})$ at b, see Figure 10. The neighbouring vertices of b in $\mathrm{CH}(Q^{\geq b})$ are called c and e. W.l.o.g. we can assume that c appears before e on C. Then $d(b,e) \geq d(c,e)$ holds, and we have $\varphi \leq 90°$ since $\overline{ce}$ is not the longest edge of the triangle bce.

Now consider a point p moving along $\overline{ba}$. We want to show that at each step per$(Q^{\geq b} \cup \{p\})$ grows by at least the same amount as the step length. For an infinitesimal step of length x of p the distance $d(e,p)$ increases by $x\cos\alpha$ whereas $d(c,p)$ increases by $x\cos(\varphi-\alpha)$. Therefore, the value of perimeter minus length changes by

$$x\cos\alpha + x\cos(\varphi - \alpha) - x = x(\cos\alpha + \cos(\varphi - \alpha) - 1)$$

This factor is nonegative, since $0 \leq \alpha \leq \varphi \leq \frac{1}{2}\pi$. This argument holds for each small step along $\overline{ba}$ and also for the case that more vertices of $\mathrm{CH}(Q^{\geq b})$ than only b do not reappear as vertices in $\mathrm{CH}(Q)$ which concludes the proof. □

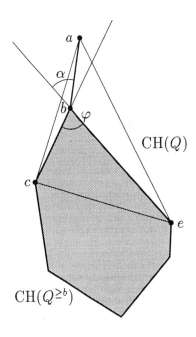

Figure 10: $d(a,c) + d(a,e) - d(b,c) - d(b,e) \geq d(a,b)$

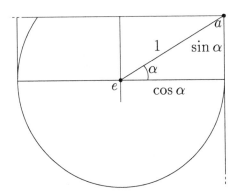

Figure 11: Estimating the maximum detour of a self-approaching curve.

Theorem 14 *The maximum detour of any self-approaching curve C is less than $\sqrt{4 + (2+\pi)^2} \approx 5.52$.*

Proof. Let a, e denote the first resp. last point of C. With an argument just like the one we have seen at the beginning of Case 2 in the proof of Theorem 13 it can be shown that the whole of C lies inside a wedge at a of $90°$. We use a coordinate system with axes parallel to this wedge and scale to $d(a,e) = 1$, see Figure 11. From Theorem 13 it follows that an upper bound for per(C) also bounds its maximum detour.

We may assume that $0 \leq \alpha \leq \frac{1}{4}\pi$, the other case is symmetric. When C leaves the bounding box of a and e, its distance to e is at most $\max(\cos\alpha, \sin\alpha) = \cos\alpha$. The self-approaching property guarantees that C cannot leave the circle with radius $\cos\alpha$ around e. Therefore, an upper bound for per(C) is $2\sin\alpha + (2+$

$\pi)\cos\alpha$. This function takes on a maximum value of $\sqrt{4 + (2+\pi)^2} \approx 5.52$, as some analysis shows. □

Remark. In the analysis above, per(C) was shown to be less than $2\sin\alpha + (2+\pi)\cos\alpha$, i.e. the length, $\pi \cos\alpha$, of the half circle plus the length, $2\sin\alpha + 2\cos\alpha$, of three sides of the rectangle above. But if we consider the exact intersection of the wedge at a and the interior of the circle, its perimeter is a little shorter because of the shortcut in the upper left corner. This new function generates slightly less pleasant formulae, but can be shown to give an improved upper bound of 5.48.

5 A lower bound

A lower bound for any strategy for the kernel problem can be derived in the following way.

Lemma 15 *No strategy for the kernel problem can guarantee a maximum detour of less than $\sqrt{2}$.*

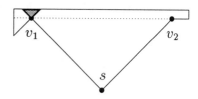

Figure 12: Any strategy can be forced to make a detour of $\sqrt{2}$.

Proof. The polygon P shown in Figure 12 is a rectangular, isosceles triangle with two small caves at the upper corners. We allow only two cases, the kernel may be near the left corner, as shown in the picture, or near the right corner, i.e. if we flip left and right corners in the picture.

The robot starts at the lower corner, s, and before reaching the line $\overline{v_1 v_2}$ no strategy can determine which of the two cases occurs. If $\overline{v_1 v_2}$ is reached to the left of its midpoint, then the kernel can be at the right corner and vice versa. But then a detour of $\sqrt{2}$ is unavoidable. □

References

[1] R. Cole and M. T. Goodrich. Optimal parallel algorithms for polygon and point-set problems. *Algorithmica*, 7:3–23, 1992.

[2] A. Datta and C. Icking. Competitive searching in a generalized street. In *Proc. 10th Annu. ACM Sympos. Comput. Geom.*, pages 175–182, 1994.

[3] X. Deng, T. Kameda, and C. Papadimitriou. How to learn an unknown environment. In *Proc. 32nd Annu. IEEE Sympos. Found. Comput. Sci.*, pages 298–303, 1991.

[4] X. Deng, T. Kameda, and C. H. Papadimitriou. How to learn an unknown environment I: the rectilinear case. Technical Report CS-93-04, Department of Computer Science, York University, Canada, 1993.

[5] G. Dudek, K. Romanik, and S. Whitesides. Localizing a robot with minimum travel. In *Proc. 6th ACM-SIAM Sympos. Discrete Algorithms*, 1995.

[6] K. Ghosh and S. Saluja. Optimal on-line algorithms for walking with minimum number of turns in unknown streets. Technical Report TCS-94-2, Tata Institute of Fundamental Research, Bombay, 1994.

[7] C. Icking. *Motion and visibility in simple polygons*. PhD thesis, Department of Computer Science, FernUniversität Hagen, Germany, 1994.

[8] C. Icking, R. Klein, and L. Ma. How to look around a corner. In *Proc. 5th Canad. Conf. Comput. Geom.*, pages 443–448, Waterloo, Canada, 1993.

[9] R. Klein. Walking an unknown street with bounded detour. *Comput. Geom. Theory Appl.*, 1:325–351, 1992.

[10] J. M. Kleinberg. The localization problem for mobile robots. In *Proc. 35th Annu. IEEE Sympos. Found. Comput. Sci.*, 1994.

[11] J. M. Kleinberg. On-line search in a simple polygon. In *Proc. 5th ACM-SIAM Sympos. Discrete Algorithms*, pages 8–15, 1994.

[12] D. T. Lee and F. P. Preparata. An optimal algorithm for finding the kernel of a polygon. *J. ACM*, 26:415–421, 1979.

[13] A. López-Ortiz and S. Schuierer. Going home through an unknown neighbourhood. Technical report, Department of Computer Science, University of Waterloo, Canada, 1994.

[14] G. Rote. Curves with increasing chords. *Mathematical Proceedings of the Cambridge Philosophical Society*, 115(1):1–12, 1994.

Stabbing Triangulations by Lines in 3D*

Pankaj K. Agarwal[†]   Boris Aronov[‡]   Subhash Suri[§]

Abstract

Let S be a set of (possibly degenerate) triangles in $\Re^3$ whose interiors are disjoint. A triangulation of $\Re^3$ with respect to S, denoted by $T(S)$, is a simplicial complex in which each face of $T(S)$ is either disjoint from S or contained in a higher dimensional face of S. The line stabbing number of $T(S)$ is the maximum number of tetrahedra of $T(S)$ intersected by a segment that does not intersect any triangle of S. We investigate the line stabbing number of triangulations in several cases—when S is a set of points, when the triangles of S form the boundary of a convex or a nonconvex polyhedron, or when the triangles of S form the boundaries of k disjoint convex polyhedra. We prove almost tight worst-case upper and lower bounds on line stabbing numbers for these cases. We also estimate the number of tetrahedra necessary to guarantee low stabbing number.

---

*The work by the first author was supported by National Science Foundation Grant Grant CCR-93-01259, an NYI award, and by matching funds from Xerox Corp. The work by the second author was supported by National Science Foundation Grant CCR-92-11541.

[†]Department of Computer Science, Box 90129, Duke University, Durham, NC 27708-0129.

[‡]Computer Science Department, Polytechnic University, Six MetroTech Center, Brooklyn, NY 11201.

[§]Department of Computer Science, Washington University, Campus Box 1045, One Brookings Drive, St. Louis, MO 63130.

Permission to copy without fee all or part of this material is granted provided that the copies are not made or distributed for direct commercial advantage, the ACM copyright notice and the title of the publication and its date appear, and notice is given that copying is by permission of the Association of Computing Machinery.To copy otherwise, or to republish, requires a fee and/or specific permission.
11th Computational Geometry, Vancouver, B.C. Canada
© 1995 ACM 0-89791-724-3/95/0006...$3.50

1 Introduction

1.1 Problem Statement

A *triangulation* of $\Re^3$ is a simplicial complex that covers the entire 3-space.[1] Let S be a simplicial complex consisting of n triangles, segments, and/or vertices in $\Re^3$. A *constrained triangulation* of $\Re^3$ with respect to S is a triangulation $T(S)$ of $\Re^3$ in which the interior of any k-face ($k = 0, 1, 2, 3$) of $T(S)$ is either disjoint from S or contained in the interior of a k'-face of S, with $k' \geq k$. For the sake of brevity, we will refer to $T(S)$ as a *triangulation with respect to S*. A vertex of $T(S)$ is called a *Steiner point* if it is not a vertex of S.

For a line segment γ and a triangulation T with respect to S, let $\sigma(S, T, \gamma)$ denote the number of (closed) simplices of T intersected by γ. Define

$$\sigma(S, T) = \max \sigma(S, T, \gamma),$$

where maximum is taken over all segments γ that do not intersect S; $\sigma(S, T)$ is called the *line stabbing number* of $T(S)$. Finally, define $\sigma(S) = \min \sigma(S, T)$, where minimum is taken over all triangulations with respect to S; $\sigma(S)$ is called the *minimum stabbing number* of S. In this paper we prove upper and lower bounds on the minimum stabbing numbers of triangulations with respect to S in several cases, including when S is a set of points in $\Re^3$, when the triangles of S form the boundary of a convex or a nonconvex polyhedron, and when the triangles of S form the boundaries of k disjoint convex polyhedra. By an $f(n)$ lower bound, we mean an explicit example of a set of n objects whose *any* triangulation has stabbing number at least $f(n)$. A $g(n)$ upper bound is established by presenting an algorithm that gives a triangulation of $\Re^3$ with respect to S whose stabbing number is at most $g(n)$. For the lower bounds on stabbing numbers we do not make any assumptions on the size of the triangulation, but for the upper bounds

---

[1]For the purposes of this paper, a *simplicial complex* is a collection of (possibly unbounded) tetrahedra, triangles, edges, and vertices, in which any two objects are either disjoint or meet along a common face (vertex, edge, or triangle).

we prove that the size of the triangulation produced by the algorithm is not very large.

1.2 Motivation and Previous Results

The line stabbing problem has relevance to several application areas including computer graphics and motion simulation. In computer graphics, for instance, realistic image-rendering methods use the ray tracing technique to compute light intensity at various parts of scene. Given a three-dimensional scene, modeled by polyhedral objects, a triangulation having a low line stabbing number could act as a simple, yet efficient, data structure for ray tracing—we just walk through the triangulation complex, visiting only the tetrahedra that are intersected by the directed line that represents the query ray. Indeed, the best data structures for the ray tracing problem in a two-dimensional scene utilize exactly this methodology: Hershberger and Suri [13] give a triangulation-based method with the best theoretical performance, while Mitchell, Mount, and Suri [15] propose a quad-tree based method with a more practical bend.

Other potential applications of triangulations with low stabbing number are in computer simulations of fluid dynamics or complex motion. Simulating a motion in a complicated and obstacle-filled environment requires visibility-checks to detect collisions. These checks can be made using a ray shooting method: find the first intersection point between a ray and the set of obstacles. Although the motion of the flying object (e.g., a camera) in general may be non-linear, one can approximate it using piecewise linear curves. In simulating fluid dynamics, the finite element method is one of the most popular method: it subdivides the domain into quad-cells; flow values are "measured" at cell vertices, and interpolated at all other points. In some applications, it is desirable to "integrate" the flow along a line. This involves computation over all those cells that are intersected by the line, and a triangulation (or quad cell partition) that minimizes the number of intersected cells is of obvious interest.

Because of the wide-spread use of triangulations in mesh generation, surface reconstruction, robotics, and computer graphics, the topic of triangulation has attracted a lot of attention within computational geometry. Most of the work to date, however, has concentrated on computing either an arbitrary triangulation or a triangulation that optimizes certain parameters related to shapes of triangles, e.g., angles, edge lengths, heights, volumes. See the survey paper by Bern and Eppstein [3] (and the references therein) for a summary of known results on triangulations. We are not aware of any paper that studies the line stabbing number of 3-dimensional triangulations.

Chazelle and Welzl [9] studied the hyperplane stabbing number of spanning trees of point sets. Specifically, given a set S of n points in $\Re^d$ and a spanning tree T of S, the *hyperplane stabbing number* of T is the maximum number of edges of T crossed by a hyperplane. The hyperplane stabbing number of S is the minimum hyperplane stabbing number over all spanning trees of S. They proved that the hyperplane stabbing number of n points in $\Re^d$ is $\Theta(n^{1-\frac{1}{d}})$ in the worst case; see also [18]. An analogous result for matchings is also discussed in [9], and a related problem is considered in [1]. Modifying the argument of Chazelle and Welzl, we can prove a lower bound of $\Omega(\sqrt{n})$ on the line stabbing number of a triangulation of the plane with respect to a set of n points. Hershberger and Suri [13] showed that a simple polygon can be triangulated into linear number of triangles, so that the stabbing number of the triangulation is $O(\log n)$. Combining their result with that of Chazelle and Welzl, we can compute a linear-size triangulation of the plane with respect to a set S of n disjoint segments, whose stabbing number is $O(\sqrt{n}\log n)$. If S forms the edges of k disjoint convex polygons, the stabbing number can be improved to $O(\sqrt{k}\log n)$. A result of Dobkin and Kirkpatrick [10, 11] implies that the interior of a convex polyhedron with n vertices can be triangulated with $O(n)$ tetrahedra, whose line stabbing number is $O(\log n)$. See also [6] for some related results.

1.3 Summary of Results

Throughout the paper, we assume that the vertices of S are in general position, namely, no four points are coplanar. The nondegeneracy assumption is important for our problem, since the lower bound arguments become trivial if degenerate point sets were allowed. (In fact, the points used in our main lower bound construction in Section 3.2 do not even lie on any fixed-degree algebraic surface.) We triangulate the entire affine 3-space—in order to allow unbounded tetrahedra, we think of each tetrahedron as the intersection of at most four half-spaces.

We first discuss some bounds on stabbing numbers for the case when S forms the boundary of a convex polytope; these bounds follow easily from hierarchical representations of convex polyhedra (Section 2).

Secondly, if S is a set of n points in $\Re^3$ in general position, we show that the worst-case line stabbing number for a non-Steiner triangulation is $\Theta(n)$. On the other hand, if we allow Steiner points, the lower bound on the stabbing number is $\Omega(\sqrt{n})$ even if the points are in general position. Notice that if points are in degenerate position—all of them are coplanar—the $\Omega(\sqrt{n})$ lower bound follows immediately from the known two-dimensional results. We describe an algorithm for computing a triangulation with respect to S of $O(n)$ size and $O(\sqrt{n}\log n)$ stabbing number.

Next, we consider the case when S forms the boundary of a nonconvex polyhedron. It is known that there are nonconvex polyhedra that cannot be triangulated without using Steiner points [16], and that it is NP-hard to determine whether a polyhedron can be triangulated without adding Steiner points [17]. Therefore we consider only Steiner triangulations in this case. We prove an $\Omega(n)$ lower bound on the stabbing number, and show how to construct a triangulation of $O(n^2)$ size and $O(n \log n)$ stabbing number. A result of Chazelle [5] shows that the size of the triangulation cannot be improved in the worst case. If S forms a boundary of a polyhedron with few reflex edges, the stabbing bound for triangulations of its *interior* can be improved, but we omit this improvement from the abstract.

Finally, we consider the case when S forms the faces of k disjoint convex polyhedra. The lower bound result for the general polytope can be modified to obtain an $\Omega(k + \log n)$ lower bound on the stabbing number. We describe an algorithm that computes a triangulation of size $O(nk \log n)$ and $O(k \log^2 n)$ stabbing number. If $k < \sqrt{n}$, the size can be improved: we can compute a triangulation of $O(n \log n + k^3 \log k)$ size and $O(k \log^2 n)$ stabbing number, or a triangulation of $O(n + k^3 \log k)$ size and $O(k \log^3 n)$ stabbing number.

2 Preliminaries

In this section, we discuss some conventions used in our triangulations and review a method for triangulating a convex polytope or its exterior using polyhedral hierarchies to achieve polylogarithmic line stabbing number.

Inner hierarchy Given a convex polytope P of n vertices, the *inner hierarchy* of P is a sequence of convex polytopes $P = P_1, P_2, \ldots, P_k$ where P_{i+1} is obtained by removing a constant fraction of constant-degree vertices of P_i and taking the convex hull of the remaining vertices. In going from P_i to P_{i+1}, we remove an independent set of polyhedral caps of P_i, each of constant size; for a vertex q of P_i, the *cap* of q with respect to P_i is the closure of $P_i \setminus CH(V(P_i) \setminus \{q\})$, where $V(P_i)$ is the vertex set of P_i, and CH denotes the convex hull. We can triangulate each of these caps with $O(1)$ tetrahedra without using any Steiner points. The union of all these triangulations is a non-Steiner triangulation $T(P)$ of P.

Lemma 2.1 (DK85) *The line stabbing number of $T(P)$ is $O(\log n)$.*

Proof: Each polytope P_i, $i = 1, 2, \ldots, k$, in the inner hierarchy of P is a convex polytope, so a line ℓ intersects it in at most two points. Since the caps at each level i belong to independent vertices, the line ℓ intersects at most two caps between P_i and P_{i+1}. Each cap has only a constant number of tetrahedra. Since there are $O(\log n)$ levels in the hierarchy, the total number of tetrahedra intersected by ℓ is $O(\log n)$. $\square$

We thus obtain the following result.

Lemma 2.2 *The interior of a convex polytope P of n vertices can be triangulated into $O(n)$ tetrahedra, without using Steiner points, so that the line stabbing number of the triangulation is $O(\log n)$. Moreover, the stabbing number of any triangulation (with or without Steiner points) of P is $\Omega(\log n)$.*

Proof: The upper bound follows from Lemma 2.1. To establish the lower bound, we fix a point o in the interior of P and consider the following ray-shooting problem: determine the face hit by a query ray ρ emanating from o. Given a triangulation $T(P)$ of P with the stabbing number $f(n)$, we can solve our ray-shooting query in time $O(f(n))$, by simply tracing ρ through $T(P)$. However, there is an obvious information-theoretic lower bound of $\Omega(\log n)$ for our ray-shooting problem, since there are $\Omega(n)$ distinct possible solutions to the query, one for each face of the polytope. This proves that $f(n)$ is $\Omega(\log n)$. $\square$

Outer hierarchy The *outer hierarchy* of P is a sequence of convex polytopes $P = P_1, P_2, \ldots, P_k$, where P_k is a tetrahedron, $P_{i+1} \supseteq P_i$, and P_{i+1} is obtained from P_i by removing a constant fraction of faces, each of constant size, and taking the intersection of the appropriate halfplanes defined by the remaining faces. These removed faces, along with the faces of P_k, will be referred to as the *walls* of the outer hierarchy. Each face of P contributes exactly one wall to the outer hierarchy. The outer hierarchy of P can be used to triangulate the exterior of P, as follows. The space between P_i and P_{i+1} can be filled with $O(|P_i| - |P_{i+1}|)$ tetrahedra, where $|P|$ denotes the number of edges, vertices, and faces of P. We will refer to the resulting decomposition as the *unrefined hierarchical decomposition* of the exterior of P. The families of tetrahedra in this decomposition, however, use Steiner points as their vertices, and do not necessarily form a simplicial complex.

If we consider a wall w, constructed during the growth from P_i to P_{i+1}, and take any tetrahedron t of the unrefined hierarchical decomposition, then either $w \cap t$ is null, or it is a face of t—this is guaranteed by the construction so far. (A two-dimensional analog of the unrefined decomposition is schematically depicted in Figure 1.) In order to convert this convex decomposition into a simplicial complex, we treat each wall as a zero-volume convex polytope and triangulate its interior using Lemma 2.2. In particular, let w be a wall of the outer hierarchy that was removed to obtain P_{i+1} from P_i. We regard w as a two-sided convex polygon—w_u (resp. w_l) denoting the side that lies outside (resp. inside) P_i. Some of the edges of the tetrahedra introduced

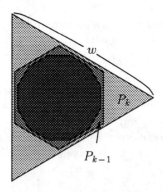

Figure 1: Outer hierarchy for a 2D polygon.

to fill $P_{i+1} \setminus P_i$ may lie in w_u. If both endpoints of any of these edges lie on the boundary of w_u, we split e into two edges by adding a Steiner point p in the relative interior of e. The tetrahedra that have e as an edge are also split at p. Let M_u be the resulting planar subdivision of w_u. Similarly, we define the planar subdivision M_l induced by the edges of tetrahedra touching w_l at one of its faces. We convert w into a thin convex polytope P_w, with the boundary of w as its silhouette and M_u, M_l as the projections of the two halves of P_w onto w. Finally, we triangulate the interior of P_w using Lemma 2.2.

It is easily checked that the size of the triangulation is linear. To bound the stabbing number of the triangulation, observe that any line ℓ meets $O(\log n)$ levels of the outer hierarchy and thus $O(\log n)$ cells of the unrefined decomposition. By Lemma 2.2, ℓ intersects $O(\log n)$ tetrahedra of thin polytopes constructed at the boundaries of each of these cells. Hence, the line stabbing number of the overall triangulation is $O(\log^2 n)$. We omit the details, and summarize the result in the following lemma.

Lemma 2.3 *The exterior of a convex polytope with n vertices can be triangulated into $O(n)$ tetrahedra, using Steiner points, so that the line stabbing number of the triangulation is $O(\log^2 n)$. Any triangulation of the exterior has stabbing number $\Omega(\log n)$.*

Nested convex polytopes Given two convex polytopes P and Q, with $P \subseteq Q$ and $|P|+|Q| = n$, the region between P and Q can be triangulated into $O(n \log n)$ simplices using Lemmas 2.2 and 2.3 (see also [4]): Compute a triangulation $T(Q)$ of the interior of Q using inner hierarchy, compute the unrefined hierarchical decomposition $T(P)$ of the exterior of P, overlay the two decompositions, and clip the overlay within $Q \setminus P$. Each cell of the resulting decomposition, Δ, has constant size. Decompose each cell of Δ into $O(1)$ tetrahedra, and finally, convert the resulting convex decomposition into a triangulation, as above. It is easily seen that any line intersects $O(\log n)$ cells of Δ, so the line stabbing number of the resulting triangulation is $O(\log^2 n)$. Each vertex of Δ is a vertex of $T(P)$, a vertex of $T(Q)$, or an intersection point of an edge of $T(P)$ (resp. $T(Q)$) with a face of $T(Q)$ (resp. $T(P)$), so Δ has $O(n \log n)$ vertices. Since the refinement of Δ increases its size by a constant factor, the claim follows.

Chazelle and Shouraboura [8] show that, by combining the inner hierarchy of Q and the outer hierarchy of P more carefully, instead of just overlaying the two, a triangulation of $Q \setminus P$ of $O(n)$ size can be computed. The stabbing number, however, increases to $O(\log^3 n)$, because their algorithms consists of $O(\log n)$ steps and each step constructs a convex decomposition similar to the outer hierarchy. Omitting some details, which can be found in [8], we summarize the results of this section in the following lemma.

Lemma 2.4 *Given two convex polytopes P, Q, with $P \subseteq Q$ and $|P| + |Q| = n$, the region between P and Q can be triangulated using $O(n \log n)$ tetrahedra so that the line stabbing number of the triangulation is $O(\log^2 n)$. The number of tetrahedra can be reduced to $O(n)$ at the expense of increasing the stabbing number to $O(\log^3 n)$.*

3 Triangulations for Point Sets

In this section we assume that S is a set of n points in $\Re^3$ in general position. We prove sharp bounds on the minimum line stabbing number of S for both Steiner and non-Steiner triangulations. The most interesting result of this section is the $\Omega(\sqrt{n})$ lower bound on the line stabbing number of any Steiner triangulation with respect to a set of n points in $\Re^3$.

3.1 Non-Steiner Triangulations

We first consider non-Steiner triangulations of S. For the sake of simplicity, we triangulate the convex hull of S; the exterior of the convex hull of S can easily be triangulated using $O(n)$ unbounded tetrahedra.

It is easily seen that the convex hull of any set of n points S in $\Re^3$ in general position can be triangulated using $O(n)$ tetrahedra [12]. A simple algorithm works as follows. Compute the convex hull of S. Triangulate the boundary of the convex hull. Pick the bottommost point b of S, and join b to each triangle on the convex hull boundary that is not already incident to b. This introduces $O(n)$ tetrahedra that together partition the space inside the convex hull. Now, consider the remaining points of S one at a time: each such point lies in the interior of an existing tetrahedron; decompose that tetrahedron into four tetrahedra by joining the new points to the four facets of the containing tetrahedron.

We now turn to the lower bound. Let $p_1, \ldots, p_{n-2}$ be a set of $n-2$ points lying on the curve $y = 0$, $z = x^2 + 1$. We perturb these points slightly so that the set is no longer degenerate, but the plane determined by any triple still has the points $q_1 = (0, -1, 0)$ and $q_2 = (0, +1, 0)$ on opposite sides. See Figure 2. We prove that any non-Steiner triangulation with respect to the point set $S = \{p_1, \ldots, p_{n-2}, q_1, q_2\}$ has line stabbing number $\Omega(n)$.

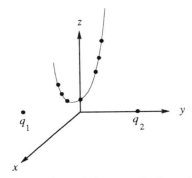

Figure 2: The lower bound for non-Steiner triangulation.

Let T be such a triangulation of S. Consider any point p_i, for $1 < i < n - 2$, and let p_i^- denote the point obtained by translating p_i downward (in the negative z-direction) by a suitably small distance. The point p_i^- lies in the convex hull of S and therefore in some tetrahedron of T. We claim that p_i^- lies in a tetrahedron $\Delta_i \in T$ whose four vertices are p_i, p_j, q_1, q_2, for some $j \neq i$.

To prove the claim, observe that any tetrahedron touching p_i must have p_i as a vertex, as T is a simplicial complex. By construction, p_i^- lies below any tetrahedron defined by four vertices from the set $S \setminus \{q_1, q_2\}$. Finally, p_i^- also lies outside any tetrahedron defined by $\{p_j, p_k, p_\ell, q_1\}$ or $\{p_j, p_k, p_\ell, q_2\}$. Thus, the tetrahedron containing p_i^- is defined by p_i, q_1, q_2 and a fourth vertex p_j, for $j \neq i$. Repeating this argument for all i, where $1 < i < n - 2$, we conclude that at least $(n-4)/2$ tetrahedra of T are determined by quadruples of the form $\{q_1, q_2, p_i, p_j\}$. All of these tetrahedra can be stabbed by a line $\ell : y = 0, z = \varepsilon$, for a suitably small $\varepsilon > 0$. Thus, the stabbing number of T is $\Omega(n)$, establishing the following theorem.

Theorem 3.1 *There exists a set of n points in $\Re^3$ whose minimum stabbing number is $\Omega(n)$. Moreover, for every set of n points in $\Re^3$ in general position, there exists a triangulation of linear size whose line stabbing number is $O(n)$.*

Remark. If S is not in general position, then there exist a set of n points such that any triangulation of S has $\Omega(n^2)$ simplices: Place $\lfloor n/2 \rfloor$ points on the x-axis and $\lceil n/2 \rceil$ points on the line $\{x = 1, z = 1\}$.

3.2 Steiner Triangulations

In this subsection, we show that by allowing Steiner points the upper bound on line stabbing number can be reduced to $O(\sqrt{n} \log n)$. We also prove a lower bound of $\Omega(\sqrt{n})$. As mentioned in the introduction, the line stabbing number for two-dimensional point sets is $\Omega(\sqrt{n})$ in the worst case [9]. Then, why doesn't it imply a similar lower bound in three dimensions? This is where the non-degeneracy assumption of no four points being coplanar is crucial. It seems difficult to make lower-bound arguments for a two-dimensional construction *perturbed* to remove degeneracies, since we have no control over the positions of Steiner points. Actually, since a line has co-dimension two in 3-space and one in the plane, one might think that the line stabbing number should *decrease* as one increases the dimension of the ambient space. The standard lower bound example in two dimensions for $\Omega(\sqrt{n})$ line stabbing number is a $\sqrt{n} \times \sqrt{n}$ grid. A 3-dimensional grid, however, yields a lower bound of only $\Omega(n^{1/3})$ on the line stabbing number.

The upper bound Let $S \subseteq \Re^3$ be a set of n points. Without loss of generality, assume that no two points have the same x and y coordinates; otherwise first perform a rotation of the space. Orthogonally project the points onto the plane $\pi : z = 0$. Let $\bar{S}$ denote the projection of S. Construct a Steiner triangulation $\bar{T}$ of π with respect to $\bar{S}$ with line stabbing number $O(\sqrt{n} \log n)$ [9, 13]. Lift this triangulation to 3-space: each point of $\bar{S}$ maps to its corresponding point in S, and we lift each Steiner point so that it lies in the convex hull of S. This results in a triangulated, piecewise linear, xy-monotone surface Σ in 3-space, with $O(n)$ triangles. We add two more Steiner points at infinity, at $z = \infty$ and $z = -\infty$, and connect each triangle of Σ to these points. This is a triangulation T of $\Re^3$. In this triangulation, all tetrahedra are unbounded, and each tetrahedron $\tau \in T$ is uniquely associated with a triangle $t \in \bar{T}$: the triangle t is the projection of τ.

The line stabbing number of T is $O(\sqrt{n} \log n)$, which can be proved as follows. Consider a line ℓ, and let $H(\ell)$ denote the vertical plane passing through ℓ. $H(\ell)$ intersects precisely those tetrahedra of T whose associated triangles in $\bar{T}$ are intersected by the line $H(\ell) \cap \pi$. Since the line stabbing number of $\bar{T}$ is $O(\sqrt{n} \log n)$, the number of tetrahedra intersected by ℓ is also $O(\sqrt{n} \log n)$.

The lower bound We exhibit a non-degenerate set of n points $S \subseteq \Re^3$ every triangulation of which, with or without Steiner points, has line stabbing number $\Omega(\sqrt{n})$.

Consider the following set of n points:

$$\bar{S} = \{(i, j, ij) \mid i, j = 1, \ldots, k\},$$

where we assume for simplicity that $n = k^2$ for an integer k. These points all lie on the hyperbolic paraboloid $\sigma : z = xy$. See Figure 3 for an illustration. We perturb each point $p \in \bar{S}$ arbitrarily inside a box of side length ε centered at p, for a sufficiently small $\varepsilon > 0$. This perturbation produces the desired nondegenerate set of points S.

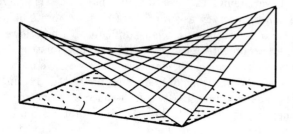

Figure 3: The surface $\sigma : z = xy$ and the lattice lines.

In our proof, we make use of the pre-perturbation set $\bar{S}$, whose points are arranged in a $k \times k$ "grid" on σ. The lines of the grid have the form (1) $x = i, z = iy$ and (2) $y = i, z = ix$, for $i = 1, 2, \ldots, k$. Let $\mathcal{L}$ denote this set of "lattice" lines. $\bar{S}$ is exactly the set of pairwise intersections of lines of $\mathcal{L}$.

Let T denote an arbitrary Steiner triangulation of the point set S. Consider the point $\bar{p} \in \bar{S}$ with coordinates (i, j, ij). The point $\bar{p}$ has four *primary* neighbors (north, east, south, west) and four *secondary* neighbors (north-east, south-east, south-west, north-west). The *north neighbor* of $\bar{p}$ is denoted $\bar{p}_n$ and has coordinates $(i, j+1, i(j+1))$, while the *north-east neighbor* is $\bar{p}_{ne} = (i+1, j+1, (i+1)(j+1))$. The remaining neighbors are defined similarly. The following lemma states a key property of our construction.

Lemma 3.2 Let $\Delta \in T$ be a tetrahedron, and let $\bar{p} = (i, j, ij)$ be some point of $\bar{S}$ contained in Δ. Then, at least one of the four secondary neighbors of $\bar{p}$ lies outside Δ.

Proof: We prove the lemma by contradiction. Suppose that all four secondary neighbors $\bar{p}_{ne}, \bar{p}_{nw}, \bar{p}_{sw}$, and $\bar{p}_{se}$ lie in Δ. Then, by the convexity of Δ, their convex hull C is also contained in Δ. Let B be the box of side length ε centered at $\bar{p}$. As $\bar{p}$ lies in the interior of C, for an appropriately small ε, we have $B \subset C \subseteq \Delta$. However, this contradicts our assumption that Δ is a tetrahedron in a triangulation of S, because B contains in its interior a point of S. This completes the proof of the lemma. $\square$

We will show that one of the lattice lines (a line of $\mathcal{L}$) stabs $\Omega(\sqrt{n})$ tetrahedra of T. Our lower bound proof rests on the following lemma, which says that the "volume" of each tetrahedron in terms of the number of points of $\bar{S}$ contained in it is proportional to the number of lattice lines intersecting it. We introduce some notation to facilitate the proof. Given a tetrahedron $\Delta \in T$, let $\mathcal{L}(\Delta)$ denote the set of lines in $\mathcal{L}$ that intersect Δ. Define $\bar{S}(\Delta) = \bar{S} \cap \Delta$, and $\text{vol}(\Delta) = |\bar{S}(\Delta)|$. Since every point of $\bar{S}$ lies in some tetrahedron of T, we have

$$\sum_{\Delta \in T} \text{vol}(\Delta) \geq k^2. \qquad (1)$$

Finally, let $m(T)$ denote the stabbing number of T. Then,

$$m(T) \geq \frac{1}{|\mathcal{L}|} \sum_{\Delta \in T} |\mathcal{L}(\Delta)| = \frac{1}{2k} \sum_{\Delta \in T} |\mathcal{L}(\Delta)|. \qquad (2)$$

Lemma 3.3 For any tetrahedron $\Delta \in T$,

$$\text{vol}(\Delta) \leq 4|\mathcal{L}(\Delta)|.$$

Proof: By the convexity of Δ, a lattice line ℓ intersects Δ in a line segment. In particular, the set $\bar{S} \cap \ell$, if nonempty, consists of a block of consecutive lattice points. We call $\bar{p} \in \bar{S}(\Delta)$ a *boundary point* if at least one of its primary neighbors (north, east, south, or west) is outside Δ. The number of boundary points is at most $2|\mathcal{L}(\Delta)|$: every line of $\mathcal{L}(\Delta)$ contains at most two of them, one where it enters Δ and one where it leaves.

We call the non-boundary points of $\bar{S}(\Delta)$ its *kernel points*. We claim that no lattice line has three or more kernel points in Δ, which implies that the number of kernel points is also at most $2|\mathcal{L}(\Delta)|$.

We first argue that kernel points form blocks of consecutive lattice points along every lattice line. Indeed, without loss of generality, suppose that there are two kernel points $\bar{p} = (i, j, ij)$ and $\bar{q} = (i, j', ij')$ in Δ, where $j' > j+1$. In particular, the points $(i \pm 1, j, ij)$ and $(i \pm 1, j', ij')$ lie in Δ. Then, by convexity, the points $(i', j'', i'j'')$ for $i-1 \leq i' \leq i+1, j \leq j'' \leq j'$, also lie in Δ; see Figure 4(a). Thus the lattice points between $\bar{p}$ and $\bar{q}$ are also kernel points, as claimed.

Consider three consecutive kernel points, starting with $\bar{p}$. We now have the situation that all four secondary neighbors of $(i, j+1, i(j+1))$, namely, $(i-1, j, (i-1)j), (i+1, j, (i+1)j), (i-1, j+2, (i-1)(j+2))$, and $(i+1, j+2, (i+1)(j+2))$ are contained in Δ (see Figure 4(b)), which contradicts Lemma 3.2. This completes the proof of the lemma. $\square$

By combining the inequality in the preceding lemma with eqs. (1) and (2), we obtain:

$$m(T) \geq \frac{1}{2k} \sum_{\Delta \in T} |\mathcal{L}(\Delta)| \geq \frac{1}{8k} \sum_{\Delta \in T} \text{vol}(\Delta) \geq k/8.$$

Since $k = \sqrt{n}$, the stabbing number of T is $\Omega(\sqrt{n})$. Since T was assumed to be an arbitrary triangulation of S, we have proved the following theorem.

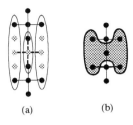

Figure 4: Illustration to the proof of Lemma 3.3.

Theorem 3.4 *There exists a nondegenerate set of n points in $\Re^3$, so that the line stabbing number of any triangulation with respect to S is $\Omega(\sqrt{n})$. Moreover, for any set S of n points in $\Re^3$ there is a Steiner triangulation with respect to S whose line stabbing number is $O(\sqrt{n}\log n)$.*

4 Triangulations for General Polyhedra

As mentioned in the introduction, a general polyhedron cannot always be triangulated without using Steiner points. A well-known example is the Schönhardt polytope [16] (see Figure 5). Therefore, we will discuss only Steiner triangulations for general polyhedra.

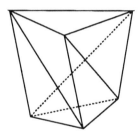

Figure 5: A polyhedron that cannot be triangulated without Steiner points.

The upper bound Every polyhedron with n facets or, more generally, every set of n non-intersecting triangles in $\Re^3$ can be Steiner-triangulated so that the line stabbing number is $O(n \log n)$. Briefly, we construct a decomposition of space by drawing vertical planes through all edges, which partitions the space into convex, polyhedral cells; this step is borrowed from the triangulation algorithm of Chazelle and Palios [7]. The analogous construction for an arbitrary collection of possibly intersecting triangles is described by Aronov and Sharir [2]. Each cell is then triangulated using the inner polyhedral hierarchy (cf. Section 2). It can be shown that a line intersects $O(n)$ cells of the convex decomposition, and thus has stabbing number $O(n \log n)$.

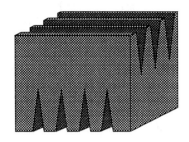

Figure 6: A non-convex polyhedron with $\Omega(n)$ line stabbing number.

The lower bound Our lower bound construction uses a version of the polyhedron S introduced by Chazelle in [5]. See Figure 6. This polyhedron is obtained from the rectangular box of dimensions $[0, n+1] \times [0, n+1] \times [0, n^2+1]$ by removing two families of n narrow notches. Notches of one family start at the bottom of the cube: the top edges of these notches lie along the lines $\{x = i, z = iy - \varepsilon/2\}$ for $i = 1, \ldots, n$, for an appropriate choice of $\varepsilon > 0$. The second family starts from the top face and is orthogonal to the first family. The bottom edges of the top family of notches lie along lines $\{y = i, z = ix + \varepsilon/2\}$, for $i = 1, \ldots, n$. It is shown in [5] that any convex decomposition of this polyhedron requires $\Omega(n^2)$ polyhedra. We show that any triangulation of this polyhedron has line stabbing number $\Omega(n)$. (Again, the degeneracies can be removed by perturbing the vertices of the polyhedron slightly, without decreasing the stabbing number.)

Our proof is similar to the one used in the previous section for the case of point sets. We consider the set of points $\bar{S} = \{(i, j, ij) \mid i, j = 1, \ldots, n\}$ lying on the surface $\sigma: z = xy$. The remainder of the proof is essentially unchanged, except that in the polyhedron case, the grid of points $\bar{S}$ has size $n \times n$, rather than $\sqrt{n} \times \sqrt{n}$ as in the case of points. Thus, the line stabbing number of any triangulation with respect to S is $\Omega(n)$.

Theorem 4.1 *There exists a set S of n disjoint triangles in $\Re^3$ forming the boundary of a polyhedron in $\Re^3$ such that any triangulation with respect to S has line stabbing number $\Omega(n)$, and such that the size of any triangulation of S is $\Omega(n^2)$. Moreover, one can always triangulate the space with respect to a given set of n disjoint triangles into $O(n^2)$ simplices so that the line stabbing number of the triangulation is $O(n \log n)$.*

Remark. If S forms the boundary of a polyhedron with r reflex edges, then there exists a triangulation of the *interior* of the polyhedron of $O(nr)$ size and $O(r \log n)$ stabbing number. On the other hand if the triangles of S intersect, there exists a $T(S)$ of $O(n^3)$ size and $O(n \log n)$ stabbing number.

5 Triangulations for Sets of Convex Polyhedra

The triangulations described so far represent two ends of the stabbing number spectrum: $O(\log^2 n)$ at one end for the boundary of one convex polytope, and $O(n \log n)$ at the other end for a set of arbitrary disjoint triangles. In this section we show that it is possible to interpolate the stabbing number bound between these two extremes. Specifically, we show that, if the triangles of S form the boundaries of k disjoint convex polyhedra, we can construct a triangulation $T(S)$ of $O(nk \log n)$ size and $O(k \log^2 n)$ stabbing number. If $k < \sqrt{n}$, using the previous bound and Lemma 2.4, we can obtain a slightly better bound on the size of the triangulation: we can construct a triangulation of $O(n + k^3 \log k)$ size and $O(k \log^3 n)$ stabbing number, or a triangulation of $O((n + k^3) \log n)$ size and $O(k \log^2 n)$ stabbing number, depending on which bound of Lemma 2.4 we use. These bounds are close to optimal, because a lower bound of $\Omega(n + k^2)$ on size and $\Omega(k + \log n)$ on the stabbing number can be obtained by modifying the construction described in the previous section.

The upper bound Let $\mathcal{P} = \{P_1, P_2, \ldots, P_k\}$ be the family of k disjoint convex polyhedra, whose faces constitute the input set S. Let the *silhouette* of P_i with respect to the z-axis be the set of points of vertical tangency. Our triangulation algorithm proceeds in two steps: in the first step, we triangulate the closed common exterior of the polyhedral set $\mathcal{P}$; in the second step, we triangulate the interiors of P_i's in such a way that the inner and outer triangulations agree along their common boundaries.

We begin by decomposing the common exterior $U = cl\left(\bigcup ext P_i\right)$ into simply connected regions. More precisely, we add "walls" that are the union of maximal vertical (i.e., parallel to z-axis) segments passing through the points of silhouettes and contained in U. In addition, we draw a plane parallel to yz-plane through each x-extreme vertex of a polyhedron and thus form at most $2k$ more "walls" consisting of all points of U lying in such planes. It is easily checked that the additional complexity introduced by this construction is $O(nk)$ and that it partitions U into "cylinders," each of which is bounded by one polyhedron on the top, one on the bottom, and by vertical walls elsewhere. (A cylinder may be unbounded if the top or bottom polyhedron is missing; we will pay no special attention to these unbounded cylinders in our discussion since they do not cause any conceptual difficulties.) We first compute a tetrahedral decomposition of each cylinder, and then refine the tetrahedra to convert the resulting decomposition of U into a triangulation of U.

We cut each cylinder C by a plane π separating the top polyhedron from the bottom one. The cross section $K = \pi \cap C$ is a simple polygon; the top and bottom halves of the cylinder are denoted C^+ and C^-. In order to triangulate the top half C^+, we take a linear-size triangulation $T(K)$ of K that guarantees $O(\log n)$ stabbing number [13] and erect an infinite vertical prism on each triangle. We also take the *unrefined* outer hierarchical decomposition of the exterior of the ceiling polytope of C^+. It has stabbing number $O(\log n)$. We overlay the two partitions and clip it within C^+, obtaining a convex decomposition of T, each cell of which has constant complexity. Since the stabbing number of each partition is $O(\log n)$, arguing as in Section 2, we can show that the size of the overlay is $O(|C^+| \log n)$ and that its stabbing number is $O(\log n)$. The overlay is further triangulated so that the tetrahedra are consistent on two sides of every vertical face separating adjacent prisms, but they may not match on two sides of a wall of the outer hierarchy. This problem will be addressed by a later step of our construction. We compute a similar triangulation $T(C^-)$ of the bottom half C^- by overlaying the prisms erected on $T(K)$ and the unrefined hierarchical decomposition. Repeating these steps over all cylinders, we obtain a tetrahedral decomposition Ξ of U. We next convert Ξ into a triangulation of U.

First, we patch up Ξ along the vertical boundaries of cylinders. We can assume that at most one face of each tetrahedron of Ξ is supported by a vertical boundary. (If two or more faces of a tetrahedron touch the vertical boundaries, we can split it into four tetrahedra by adding a Steiner point in its interior so that the desired property holds.) Let w be a maximal connected portion of a vertical boundary of a cylinder contained in a plane. We regard w as a two-sided wall. The tetrahedra incident on each side of w induce a planar map. We overlay these two planar maps and retriangulate each face of the overlay. This gives a triangulation Π_w of w. For every tetrahedron Δ one of whose faces, say, f, is contained in w, we split Δ into smaller tetrahedra, each of which has a triangle of $\Pi_w \cap f$ as the base and v as the apex, where v is the vertex of Δ not contained in w. The number of new tetrahedra added in this step is proportional to the number of vertices in Π_w. Each vertex of Π_w is either a vertex of a tetrahedron incident to w or the intersection point of a pair of edges of the tetrahedra incident on the opposite sides of w. Let n_w be the number of tetrahedra incident to w. Then there are at most $3n_w$ vertices of the first type and $O(n_w \log n)$ vertices of the second type because the line stabbing number of Π_w is $O(\log n)$. Repeating this procedure for all vertical boundaries of cylinders erected in the first step, we ensure that the faces of tetrahedra match along these boundaries. Since $\sum_w n_w = O(nk)$, the total size of the refined decomposition, Ξ', is $O(nk \log n)$. Moreover, the stabbing number of Ξ' is $O(k \log^2 n)$ — any line ℓ intersects at most $4k+1$ cylinders, $O(\log n)$ tetrahedra of Ξ within each cylinder, and the above step may subdivide such a tetrahedron further, increasing stabbing number by a factor of $O(\log n)$.

Next, we patch up Ξ' within each cylinder. Fix a cylinder C. Let w be a wall in the unrefined hierarchical decomposition of the floor or ceiling polytope of C, clipped within C. The vertical prisms induce a subdivision of w into convex faces. Let f be such a face. We patch up Ξ' around f, treating it as a thin convex polytope, as in Section 2, to ensure that the faces of the tetrahedra touching f from opposite sides match. This step increases the size of the decomposition by a constant factor and adds $O(\log n)$ to the number of tetrahedra meeting any line that passes through (or near) f. As no line passes through (or nearly parallel to) more than $O(\log n)$ walls in the unrefined outer hierarchy of the floor or ceiling of each of at most $4k+1$ cylinders it meets, the increase in the stabbing number is at most $O(k \log^2 n)$. Finally, we patch up Ξ' along the cross section K: Let τ be a triangle of $T(K)$. Since no edge of Ξ' crosses the boundary of τ, we convert τ into a thin polytope and triangulate it, as above. After performing these two steps, we obtain a $O(nk \log n)$-size triangulation $T(U)$ of U whose line stabbing number of $O(k \log^2 n)$.

Finally, we triangulate the interior of each polytope using Lemma 2.2. If there is a tetrahedron, one of whose faces contains a Steiner point of $T(U)$ in its relative interior, we refine Δ in the same way as we refined a tetrahedron incident to a boundary of a cylinder. This step adds no more than $O(nk \log n)$ simplices. The stabbing number of a segment lying inside a polytope is easily seen to be $O(k \log^2 n)$. Putting everything together, we thus obtain a triangulation of $\Re^3$ with respect to $\mathcal{P}$ whose size and stabbing number are $O(nk \log n)$ and $O(k \log^2 n)$, respectively.

If $k \leq \sqrt{n}$, we can compute a triangulation with fewer simplices, as follows. We separate every pair of polyhedra by a plane and use the resulting planes to build a family of k disjoint polyhedra with a total of k^2 facets, each containing one the input polyhedra. We then triangulate the region outside the larger polyhedra using the above method, giving $O(k^3 \log k)$ size and $O(k \log^2 k)$ stabbing number, and triangulate the regions between each larger polyhedron and the smaller one contained within it, using methods described in Section 2. Finally, we patch the triangulations together in the same fashion as in the previous algorithm. We thus obtain a triangulation of $O(n+k^3 \log k)$ size and $O(k \log^3 n)$ stabbing number, or of $O((n+k^3) \log n)$ size and $O(k \log^2 n)$ stabbing number.

The lower bound It can be shown that $\Omega(\log n)$ is a lower bound on line stabbing for the triangulation of the exterior of a single convex polytope (with or without Steiner points)—the proof is similar to the one for the triangulation of the interior of a polytope. Next, we mimic the construction of Section 4 to produce an arrangement of k polytopes that resembles our lower-bound polyhedron there. In particular, we take our convex polytopes to be triangular notches that are cut out of the box in Section 4. Essentially the same proof shows that any triangulation of the complement of the notches has line stabbing number $\Omega(k)$. Combining the two lower bounds, we get a lower bound of $\Omega(k + \log n)$ for the line stabbing number for a Steiner triangulation of k convex polytopes. We summarize in the following theorem. To summarize:

Theorem 5.1 *If S forms the boundary of k convex polyhedra, then there exists a triangulation $T(S)$ of $O(nk \log n)$ size and $O(k \log^2 n)$ stabbing number, of $O((n+k^3) \log n)$ size and $O(k \log^2 n)$ stabbing number, or of $O(n+k^3 \log k)$ size and $O(k \log^3 n)$ stabbing number. Moreover, there exists a set of n triangles forming the boundary of k disjoint convex polyhedra such that any triangulation $T(S)$ has size $\Omega(n + k^2)$ and $\Omega(k + \log n)$ stabbing number.*

Remark. If the polyhedra in $\mathcal{P}$ are intersecting, we can compute a triangulation with respect to S whose size is $O(nk^2 \log^2 n)$ and whose line stabbing number is $O(k \log^2 n \log k)$. This is close to optimal, as the complexity of the overlay of k polyhedra can be as high as $\Theta(nk^2)$ and the lower bound of $\Omega(k + \log n)$ on the stabbing number applies here.

6 Conclusion

We have obtained nearly sharp bounds on the line stabbing number of triangulations in several cases. Although the hyperplane stabbing problem has been studied previously, the line stabbing number has not received much attention. In several applications, however, the line stabbing appears quite naturally, such as in computer graphics. In the paper, we assumed the domain of triangulation to be the entire 3-space, but this is not a severe restriction—most of our results continue to hold even if we restrict the triangulation domain to a relevant subset of the space such as the convex hull of S or a enclosing simplex surrounding S.

Several open problems are suggested by our work; we are currently investigating some of them: (i) minimize the number of Steiner points, (ii) obtain tight bounds in all the cases considered here, (iii) obtain sharp bounds when S forms a terrain, (iv) define *average line stabbing number* and obtain bounds on the average line stabbing number, (v) extend the results to yield lower bounds on ray-shooting in a more general model.

References

[1] B. Aronov and J. Matoušek. On stabbing triangles by lines in 3-space. *Commentationes Mathematicae Universitatis Carolinae,* to appear.

[2] B. Aronov and M. Sharir, Triangles in space, or building (and analyzing) castles in the air, *Combinatorica* 10(2) (1990), 137–173.

[3] M. Bern and D. Eppstein, Mesh generation and optimal triangulation, in: *Computing in Euclidean Geometry*, (D. Du and F. Hwang, eds.), 1992, World Scientific, Singapore.

[4] M. Bern. Compatible tetrahedralizations. In *Proceedings of the 9th ACM Symposium on Computational Geometry*, pages 281–288, 1993.

[5] B. Chazelle. Convex partitions of polyhedra: A lower bound and worst-case optimal algorithm. *SIAM J. of Computing*, 13:488–508, 1984.

[6] B. Chazelle, H. Edelsbrunner, and L. Guibas. The complexity of cutting complexes. *Discr. Comput. Geom.*, 4:139–181, 1989.

[7] B. Chazelle and L. Palios. Triangulating a nonconvex polytope. *Discr. Comput. Geom.*, 5:505–526, 1990.

[8] B. Chazelle and N. Shouraboura. Bounds on the size of tetrahedralizations. In *Proceedings of the 10th ACM Symposium on Computational Geometry*, pages 231–239, 1994.

[9] B. Chazelle and E. Welzl. Quasi-optimal range searching in spaces with finite VC-dimension. *Discr. Comput. Geom.*, 4:467–489, 1989.

[10] D. P. Dobkin and D. G. Kirkpatrick. Fast detection of polyhedral intersections. In *Proceedings of 9th Intl. Conf. Automata, Languages, and Programming, Lecture Notes in Computer Science* vol. 140, Springer-Verlag, 1982, pages 154–165.

[11] D. Dobkin and D. Kirkpatrick. A linear algorithm for determining the separation of convex polyhedra. *J. of Algorithms*, 6:381–392, 1985.

[12] H. Edelsbrunner. Triangulations. Course Notes CS 497. University of Illinois, Urbana-Champaign, 1991.

[13] J. Hershberger and S. Suri. A pedestrian approach to ray-shooting: shoot a ray, take a walk. In *Proceedings of the Fourth Annual ACM-SIAM Symposium on Discrete Algorithms*, pages 54–63, 1993. To appear in *Journal of Algorithms*, 1995.

[14] J. Matoušek. Efficient partition trees. *Discrete Comput. Geom.*, 8:315–334, 1992.

[15] J. Mitchell, D. Mount and S. Suri. Query-sensitive ray shooting, *Proc. 10th ACM Symp. Computational Geometry* 1994, 359–368. To appear in *International Journal of Computational Geometry and Applications*.

[16] J. O'Rourke. *Art Gallery Theorems and Algorithms.* Oxford University Press, 1987.

[17] J. Ruppert and R. Seidel. On the difficulty of tetrahedralizing 3-dimensional non-convex polyhedra. In *Proceedings of the 5th ACM Symposium on Computational Geometry*, pages 380–392, 1989.

[18] E. Welzl. On spanning trees with low crossing numbers. Tech. Rept. B92-02, Free University, Berlin, 1992.

A Complete and Practical Algorithm for Geometric Theorem Proving
(Extended Abstract)

Ashutosh Rege [*]

Computer Science Division
University of California
Berkeley, CA 94720

Abstract

This paper describes a complete and practical algorithm for the problem of geometric theorem proving. The algorithm works over algebraically closed fields as well as over the reals and takes care of degenerate cases. Our work is motivated by several recent improvements in algorithms for sign determination and symbolic-numeric computation. Based on these, we provide an algorithm for solving triangular systems efficiently using straight-line program arithmetic. The report concludes with a description of an implementation and provides preliminary benchmarks from the same.

1 Introduction

In this paper, we describe a practical algorithm (and its implementation) for automatic geometric theorem proving and related problems. Earlier methods for geometric theorem proving were based on a synthetic geometric reasoning approach using natural deduction, forward and backward chaining and the like [7]. Following Wu's seminal work ([17], [18]), a significant amount of the recent work has focussed on an algebraic approach involving determining the feasibility of systems of polynomial equations [3], [10], [9]. Our work is based on this latter methodology and is motivated by recent improvements in algorithms for symbolic-numeric computation with polynomials [13]. The papers by Wu and others showed how a subclass of geometric theorems can be proved by expressing the hypotheses and conclusions as polynomial equations. A geometric theorem can then be proved by essentially determining the algebraic set corresponding to the hypotheses and checking whether the algebraic sets of the conclusions contain it. Different methods have been proposed for doing this - Wu's method involves reduction of the hypotheses system of polynomials to a triangular form (i.e. one in which each successive polynomial has at most one extra variable) and then using successive pseudo-division to determine the feasibility of the conclusion. Other work has focussed on using Gröbner bases to determine whether a system of polynomials (obtained from the original polynomials) has a common zero [10], [9].

Wu's original method is incomplete in the sense that the feasibility of the conclusions is determined over an algebraically closed field whereas geometric theorems are often assertions about real numbers. It is not possible to refute a conjecture over the reals using this approach and in general, confirmation of a theorem can not be obtained automatically from a proof over an algebraically closed field.

Another important consideration in geometric theorem proving is the notion of *genericity*. It is necessary to rule out *degenerate* cases such as non-distinct points or zero-radius circles. Wu's method and others generate a set of subsidiary polynomials representing such cases so that the conclusion polynomial vanishes at the common zeros of the hypotheses which are not zeros of the subsidiary polynomials.

From the above discussion, it seems important that a geometric theorem prover should satisfy the following criteria

1. It should work over the reals as well as the complex field.

2. It should handle degenerate cases.

3. It should be practicable.

In this paper we discuss a method for geometric theorem proving which satisfies the first two criteria and based on an initial implementation, seems to be rather efficient in practice.

---

[*] E-mail: rege@cs.Berkeley.edu. Fax: +1-510-642-5775. Supported by a David and Lucile Packard Foundation Fellowship and by NSF P.Y.I. Grant FD93-20588

The basic problem we consider is the following : *given a system of hypotheses polynomial equations in triangular form, determine whether the set of their common zeros over the reals is contained in the set of real zeros of the conclusion polynomial.* (One can use any standard algorithm for triangulating if the hypotheses are not triangular to begin with, see e.g. [4]). The observation here is that, in order to prove a theorem or refute it, one is only interested in the *sign* (+, - or 0) of the conclusion polynomial at the real zeros of the hypotheses polynomials. We provide an efficient method to encode the roots of a triangular system using the Sylvester resultant. To determine the signs, we make use of Sturm sequences which enable us to determine the number of real roots of the hypotheses where the conclusion is non-zero. Thus our algorithm can determine the truth or falsehood of an assertion over the reals.

The problem of genericity is handled by instantiating the independent variables (or parameters) of the given theorem with random values. Probabilistically, over $\mathbb{C}$ this takes care of degenerate cases without affecting the validity of the conclusion. Over $\mathbb{R}$ it is harder to get a necessary condition for the theorem to be true - the algorithm for checking the necessary condition is given in section 7.

Computation in the prover is done over the field of reals (or rationals) extended by variables. The numbers in this extension field are then rational functions in these variables. It is, however, very expensive to compute with explicit rational functions. We circumvent this difficulty by using *straight-line programs* or SLP's to represent all intermediate calculations. SLP's are extremely useful for various types of computations with polynomials over field extensions. In the algorithms we need for theorem-proving, one is primarily interested in the signs of the coefficients of polynomials over the extension field. These can be determined efficiently when the computation is done over the field of SLP's.

In order to make our prover more efficient, we use "mixed" arithmetic for algebraic computations : the idea is to represent a number as a structure with two values, the first is the integer value of the number modulo a prime p and the second is a floating-point approximation to that number. We use the finite field part for equality tests and the floating point part for relative ordering of two numbers and sign-determination.

Geometric theorem proving falls in the more general class of problems defined by the first-order theory of the reals. The work described in the subsequent sections is part of ongoing research at Berkeley aimed at developing an efficient practicable toolkit for solving more general problems. We plan on applying the tools and techniques described in this paper to a broader scope of problems including solving systems of polynomial equations, point-location in varieties, existential theory of the reals etc. (See [13] for an overview.) Further, one can use these techniques to extend our prover to work with polynomial inequalities and also to solve more general problems in constraint-based reasoning.

The paper is organized as follows : Section 2 gives the formulation of the problem of theorem proving we shall use. In section 3 we give an outline of the prover implemented by Chou [4], since we shall use that as a benchmark to compare our prover with. Section 4 gives an overview of our theorem prover. Section 5 sketches the algorithm for solving triangular systems. Sections 6 and 7 discuss theorem proving and refutation over $\mathbb{R}$ and $\mathbb{C}$. We conclude with Section 8 which describes the implementation and give benchmarks on some problems.

2 Geometric Theorem Proving - Formulation of the Problem

This section provides the formulation we shall use of the questions of what constitutes a geometric theorem and what it means to prove it. Several formulations of these two questions have been proposed [7], [8],[10]. See the book by Chou [4] for a discussion of the various formulations and the limitations of each. We will use the formulation used by Chou [4] which seems to be the one that is the most satisfactory :

By a *geometric statement* (S) we shall mean a collection of polynomial equations

$$h_1(y_1, \ldots, y_m) = \ldots = h_n(y_1, \ldots, y_m) = 0$$

called the *hypotheses* and a *conclusion* $g(y_1, \ldots, y_m) = 0$ where the h_i and g are in $K[y_1, \ldots, y_m]$. Here K is some field under consideration, usually $\mathbb{Q}$.

Let $V = V(h_1, \ldots, h_n)$ denote the set of common zeroes in E^m, where E is an extension of K. Thus an instance of the theorem, i.e. a configuration which satisfies the given hypotheses, corresponds to an element of V and vice versa. Note, however, that V contains the so-called degenerate cases which happen to satisfy the hypotheses. In order to distinguish the degenerate cases from the nondegenerate ones, we use the variables $u_1, \ldots, u_d \equiv \mathbf{u}$, called the *parameters*, among the y_i to denote the nonzero coordinates of those points which can be chosen arbitrarily and we use the variables $x_1, \ldots, x_n \equiv \mathbf{x}$, to denote the rest of the y_i. These latter are called *dependent variables*.

If the extension field E is *algebraically closed* a basic fact of algebraic geometry is that every algebraic set has a unique decomposition as a union of irreducible algebraic sets such that no one contains another. Thus the set V can be expressed as a union,

$$V = V_1 \cup \ldots \cup V_s \cup V_1^* \cup \ldots \cup V_t^*$$

where the V_i are all those (*nondegenerate*) components on which the u_j are algebraically independent and the $V_k{}^*$ are those (*degenerate*) components on which they are dependent. We call the V_i components which are *general for u*.

If E is *not algebraically closed*, the decomposition is, in general, not an irreducible one. However, the u_j are algebraically dependent on the algebraic sets $V_k{}^*$ and the latter still correspond to the degenerate cases. We will still call the V_i components *general for u*. (Note, however, that in this case any V_i can be reducible or empty.)

We say that a geometric statement (S) is *generally true* if g vanishes on all nondegenerate components of V. Else it is *not generally true*. A geometric statement is *generally false* if g vanishes on none of the nondegenerate components of V.

3 The Ritt-Wu-Chou Prover

In this section, we provide a brief description of the prover implemented by Chou [4] based on the work of Ritt and Wu. Among all the approaches that have been used and implemented, Chou's prover seems to be most complete and efficient and we will use this prover as the basis of comparison for our prover. This section will therefore try to give a basic knowledge of the algorithms and techniques used by Chou's prover so we can better understand its strengths and limitations.

In 1978, Wu introduced an elegant method for automated theorem proving for algebraic geometry [17]. The basis of his work was a triangulation procedure which was implicit in the work of Ritt [15], [16] and which was given in detail by Wu. Wu's approach was implemented with various modifications and improvements by Chou in his prover [3], [4].

The basic algorithm for the prover implemented by Chou is the following (assuming the notation introduced in the previous section) :

- **Triangulation** The hypotheses are converted into a triangular form
$$f_1(\mathbf{u}, x_1), f_2(\mathbf{u}, x_1, x_2) \cdots, f_n(\mathbf{u}, x_1, \ldots, x_n)$$

- **Successive pseudo-division** Letting $g = R_n$, the remainder R_0 is computed by taking successive pseudo-remainders : at the ith step $(i = n \ldots 0)$ the remainder computed is $prem(R_{i+1}, f_{i+1}, x_{i+1}) = R_i$. Depending on whether $R_0 = 0$ or not, and depending on other conditions (discussed below) the prover can conclude whether the theorem is generally true or not.

We will now discuss the various issues that come up in the process of triangulation and successive pseudo-division and how Chou's prover deals with them.

3.1 Triangulation

A triangular system can be obtained rather easily from the original system of hypotheses polynomials by using pseudo-remainders to eliminate variables. A simple algorithm for triangulation is the following (see e.g. [4], [5]) :

Assume we are given a system of hypotheses polynomials as before,
$$h_1(\mathbf{u}, \mathbf{x}) = \cdots = h_n(\mathbf{u}, \mathbf{x}) = 0$$

In order to transform it into a triangular form, we first eliminate x_n from $n - 1$ polynomials to obtain
$$p_1(\mathbf{u}, x_1, \ldots, x_{n-1}) = \cdots = p_{n-1}(\mathbf{u}, x_1, \ldots, x_{n-1})$$
$$= p_n(\mathbf{u}, x_1, \ldots, x_n) = 0$$

We now apply the same procedure to eliminate x_{n-1} from $n - 2$ polynomials and so on. The algorithm eliminates x_n by taking pseudo-remainders as required (see [5]).

At the end of this procedure we obtain a system of the form
$$f_1(\mathbf{u}, x_1) = f_2(\mathbf{u}, x_1, x_2) = \cdots = f_n(\mathbf{u}, x_1, \ldots, x_n) = 0$$

3.2 Successive pseudo-division

Once a triangular form $f_1 \ldots f_n$ is obtained, the next step is to do successive pseudo-division as explained earlier. This results in a final pseudo-remainder R_0 which we denote by $prem(g, f_1, \ldots, f_n)$. It is easy to show the following remainder formula :

$$I_1{}^{s_1} \cdots I_n{}^{s_n} g = Q_1 f_1 + \cdots + Q_n f_n + R_0$$

where the I_j are the leading coefficients or *initials* of the f_jr; $s_1, \ldots, s_n$ are non-negative integers and $Q_1, \ldots, Q_n$ are polynomials.

If the final remainder $R_0 = 0$ then we can say that $g = 0$ if $I_1 \neq 0, \ldots, I_n \neq 0$. Thus we get *subsidiary conditions*, namely $I_1 \neq 0, \ldots, I_n \neq 0$ which are included in the hypotheses of the original geometric statement (S). These subsidiary conditions are usually connected with nondegeneracy.

3.3 Problems with the above algorithms

The main problem with the simple triangulation procedure given above is that the variety of the triangular system obtained at the conclusion of the procedure could be strictly larger than the one we started out with due to the process of taking pseudo-remainders. If it turns out that the conclusion polynomial g vanishes on this new larger variety, then the validity of the theorem holds on the smaller variety *a fortiori*. It could easily be the case, however, that g does not vanish on the larger variety. This problem is resolved in the Ritt-Wu-Chou algorithm by testing if the triangular system obtained above is *irreducible* that is, whether each f_i is irreducible in the polynomial ring $K(u_1, \ldots, u_n)[x_1, \ldots, x_i]/(f_1, \ldots, f_{i-1})$. If the system is irreducible and $prem(g, f_1, \ldots, f_n) = 0$, the following can be shown [4]

1. The geometric statement (S) is generally true in all fields

2. For all fields, $(h_1 = 0, \ldots, h_n = 0, I_1 \neq 0, \ldots, I_n \neq 0) \Rightarrow g = 0$ where the I_k are the initials of the f_k.

So given an irreducible system, $prem(g, f_1, \ldots, f_n) = 0$ gives us a sufficiency condition for the validity of the theorem in question. For the converse, we first define

$$P = \{ g \mid g \in K[u, x] \text{ and } prem(g, f_1, \ldots, f_n) = 0\}$$

Given an irreducible system, $prem(g, f_1, \ldots, f_n) = 0$ is a necessary condition for (S) to be generally true if one of the following is satisfied

1. $V(P)$ is of degree d.

2. E is an algebraically closed field.

3. The system $f_1, \ldots, f_n$ has a generic point in E.

The first question that arises is testing for irreducibility. If it turns out that $deg(f_i, x_i) = 1$ for all i, then the triangular system is trivially irreducible and real generic points trivially exist. In this case, $prem(g, f_1, \ldots, f_n) = 0$ gives a necessary and sufficient condition to determine whether the theorem is generally true or not. Several of the simpler theorems in the collection solved by Chou's prover fall in this category ([4], p. 56). If the degree of a polynomial is not 1 in its primary variable, Chou's prover uses factoring over extension fields to check for irreducibility. This approach, however, has been implemented only for *quadratic* extension fields. If the degree of some f_i in x_i is greater than 2, Chou's prover can not handle this case (unless $i = 1$ in which case only factoring over $\mathbb{Q}$ is required). Further, even in the case of quadratic polynomials, the prover works in general only over *algebraically closed fields*. For the prover to work over the *reals*, the existence of a real generic point needs to be shown as required by the sufficiency condition given above. As far as we know, a mechanical approach to this has not been implemented in Chou's prover ([4], p. 56).

Next, we come to the question of what happens when the triangular system is reducible. If this happens, in general Chou's prover uses Ritt's decomposition algorithm to obtain all irreducible components of the original variety. This also involves factoring over field extensions and is computationally expensive for high degrees.

Again, this has been implemented only for quadratic cases.

Finally, we address the question of successive pseudo-division. As shown in the earlier section, we have the following remainder formula

$$I_1^{s_1} \cdots I_n^{s_n} g = Q_1 f_1 + \cdots + Q_n f_n + R_0$$

If the remainder $R_0 \neq 0$ then we can conclude that the theorem is not generally true since the right side would be non-zero at any instance of the theorem implying $g \neq 0$ at that instance. If, however, the remainder is zero, g could still be non-zero unless $I_j \neq 0$ for all j. These then become the additional subsidiary hypotheses under which the theorem is true. It could be true however that $prem(I_j, f_1, \ldots, f_{j-1}) = 0$ in which the hypotheses would become inconsistent. To avoid this situation, in general, one has to use the complete method in Chou's prover which involves triangulation using the Ritt decomposition algorithm.

To summarize, Chou's prover, though efficient and applicable in the case of a large class of geometric theorems, has the following limitations :

- Applicable in general to systems of degree quadratic or less.

- Does not work in all cases over $\mathbb{Q}$.

- Uses significant amount of computational resources for some problems of degree 2 ([4], p. 69).

4 Overview of the Theorem Prover

We will give a schematic overview of our theorem prover in this section. The notation used in this and subsequent sections is the same as introduced earlier. The prover takes as input a set of hypotheses polynomials $h_1, \ldots, h_n$ and a conjecture or conclusion polynomial g in $\mathbb{Q}[u_1, \ldots, u_m, x_1, \ldots, x_n]$. Here, as before, the x_i's represent the dependent variables and the u_j's, the independent variables. In what follows, we will abuse notation and write the hypotheses as polynomials over the x_i's only. It returns "true" or "false" according to whether the theorem is generally true or not over the reals. The prover can also be used to work over $\mathbb{C}$. The modules of the prover can be summarized as follows :

- **Random instantiation of parameters** The parameters u_i are set to random values to get the hypotheses as polynomials in $x_1, \ldots, x_n$ only. (See the implications of this step for the reals in Section 7).

- **Triangulation** The hypotheses are first triangulated using the triangulation algorithm described earlier to obtain the polynomials

$$f_1(x_1), f_2(x_1, x_2), \ldots, f_n(x_1, \ldots, x_n)$$

- **Root representation for triangular systems**
  This module takes as input the triangular system

  $$f_1(x_1) = f_2(x_1, x_2) = \ldots = f_n(x_1, \ldots, x_n) = 0$$

  and returns a *symbolic* representation of the roots of the system : the output is a univariate polynomial $p(s)$ and rational functions $r_1(s), \ldots, r_n(s)$. If the roots of $p(s) = 0$ are $\alpha^{(i)} \in \mathbb{C}$, for $i = 1, \ldots, N$, then

  $$\xi^{(i)} = r(\alpha^{(i)})$$

  One can think of $r(s)$ as a parametric curve in $\mathbb{C}^n$ which passes through all the roots $\xi^{(i)}$, $i = 1, \ldots, n$. The values of s at which it passes through a solution of $f_1 = f2 = \ldots = f_n$ are precisely the roots of $p(s) = 0$. The fact that we have a symbolic representation of the roots allows us to determine correctly the feasibility of the conjecture polynomial given the hypotheses polynomials.

- **Substitution** We put $x_i = r_i(s)$, $1 \leq i \leq n$, in the conjecture polynomial g. This reduces g to a univariate polynomial $g(s)$.

- **Sign Determination** This module determines the sign, i.e. $+$, $-$ or 0, of the instantiated conjecture $g(s)$ at the real roots of $p(s) = 0$ using Sturm sequences.

- **Theorem proving or refutation :** The conjecture should vanish at all of the common real roots of the hypotheses polynomials. Thus the sign determination module should return 0 for the sign of $g(s)$ at every root. If not, there exists a root where the conjecture does not hold and is therefore false OR the triangulation process introduced extra roots. Section 6 discusses what to do in that case.

5 An Algorithm for Solving Triangular Systems

This section describes the basic algorithm for solving triangular systems used in our geometric theorem prover. Section 5.1 sketches the algorithm from a theoretical viewpoint, section 5.2 goes into more details of the algorithm especially the use of straight-line programs. We conclude with section 5.3 which gives a complexity analysis of the algorithm. Complete details on the algorithm including implementation issues can be found in the full version of the paper [11]. It should be noted that our algorithm for triangular systems is probabilistic - this arises from the fact that we use straight-line programs for representing the rational functions which make up the coefficients of the polynomials generated in the algorithm. However, most algorithms for polynomials require us to verify that the leading coefficient is not identically zero. This is done by giving random values to the variables in the corresponding SLP and evaluating the SLP at those values.

5.1 Theoretical Algorithm

The input to the algorithm is a system of triangular polynomial equations in $\mathbb{Q}(x_1, \ldots, x_n)$,

$$f_1(x_1) = f_2(x_1, x_2) = \cdots = f_n(x_1, \ldots, x_n) = 0$$

Let $\xi^{(i)} \in \mathbb{C}^n$ denote the solutions to the given system.

As described earlier, the algorithm returns a *symbolic* representation of the roots of the system i.e. the output is a univariate polynomial $p(s)$ and rational functions $r_1(s), \ldots, r_n(s)$. If the roots of $p(s) = 0$ are $\alpha^{(i)} \in \mathbb{C}$, for $i = 1, \ldots, n$, then

$$\xi^{(i)} = r(\alpha^{(i)})$$

Our algorithm is based on an approach for computing (p, r) for more general systems due to Renegar [14]. We use his basic method but take advantage of certain properties of triangular systems.

The algorithm proceeds iteratively; at the ith step it adds the polynomial f_i and eliminates the extra variable x_i introduced by the polynomial to get a rational univariate representation $r_1(s_i), \ldots, r_n(s_i)$ of the roots of the first i polynomials. More precisely, at the start of the ith step, the algorithm has computed

$$r_1(s_{i-1}), \ldots, r_{i-1}(s_{i-1}) \text{ and } p_{i-1}(s_{i-1})$$

The algorithm now introduces $f_i(x_1, \ldots, x_i)$. It sets $x_1 = r_1, \ldots, x_{i-1} = r_{i-1}$ in f_i. For the new variable, x_i, the algorithm sets,

$$x_i = \frac{s_i - l(\mathbf{x})}{l_i} = \frac{s_i - l_0 - l_1 x_1 - \cdots - l_{i-1} x_{i-1}}{l_i}$$

where $x_1, \ldots, x_{i-1}$ are instantiated to $r_1, \ldots, r_{i-1}$ respectively. Thus we obtain from f_i a rational function g_i in s_{i-1} and s_i. We denote by g_{i_n} the numerator polynomial of g_i and by g_{i_d} the denominator.

The algorithm computes the (univariate) Sylvester resultants $R(g_{i_n}, p_{i-1})$ and $R(g_{i_d}, p_{i-1})$ with respect to the variable s_{i-1}. We have :

Lemma 5.1 $R(g_{i_d}, p_{i-1})$ *divides* $R(g_{i_n}, p_{i-1})$ *exactly.*

Proof Given below.

Thus we can set

$$R = \frac{R(g_{i_n}, p_{i-1})}{R(g_{i_d}, p_{i-1})}$$

If the polynomial $l(\mathbf{x}) = l_0 + l_1 x_1 + \cdots l_{i-1} x_{i-1}$ is specialized to a random linear polynomial L, then with probability one, the values $l(\xi^{(j)})$ will be distinct for distinct $\xi^{(j)}$. The roots of the resultant R will also be distinct. The algorithm sets $p_i(s_i) = R$.

We obtain the functions r_j by differentiating R. We define r_j as

$$r_j(s_i) = \left. \frac{\left(\frac{dR}{dl_j}\right)}{\left(\frac{dR}{dl_0}\right)} \right|_{l=L} \quad j = 1, \ldots, i$$

It can be verified that the r_j's have the correct values at $s_i = \alpha^{(k)}$.

At termination the algorithm eliminates all the variables $x_1, \ldots, x_n$ and returns rational functions $r_j(s)$ in the variable $s = s_n$ and the polynomial $p(s) = p_n(s_n)$. (Note : for the first step, the algorithm only needs to set $x_1 = (s_1 - l_0)/l_1$ in f_1 to get $R = p_1(s_1)$, i.e. $p_0 \equiv 0$.)

Sketch of proof (Lemma 5.1) The proof is based on the following well-known property of resultants - the resultant of $n+1$ polynomials f_i, $(i = 1, \ldots, n+1)$, with the first n polynomials having common roots $\xi^{(j)}$, $j = 1, \ldots, N$, is given by $\prod_{j=1}^{N} f_{n+1}(\xi^{(j)})$. In particular if $f_{n+1} = s - l_0 - l_1 x_1 - \cdots - l_n x_n = s - l(\mathbf{x})$, then the resultant of the f_i's is $R(s, l) = \prod_{j=1}^{N}(s - l(\xi^{(j)}))$. Further, let $r_{i_n}(s, l) = (dR/dl_i)$ and $r_d(s, l) = (dR/dl_0)$ and let $\xi^{(j,k)}$ denote the kth coordinate of the jth root. Then we have

$$r_{i_n}(s,l) = \frac{dR}{dl_i} = \sum_{j=1}^{N}(s - l(\xi^{(1)})) \cdots$$

$$(s - l(\xi^{(j-1)}))(-\xi^{(j,i)})(s - l(\xi^{(j+1)})) \cdots (s - l(\xi^{(N)}))$$

It follows from the above property that the resultant of R and $r_{i_n}(s,l)$ w.r.t. s is (by evaluating $r_{i_n}(s,l)$ at the roots of R)

$$\prod_{j=1}^{N}(l(\xi^{(j)}) - l(\xi^{(1)})) \cdots (l(\xi^{(j)}) - l(\xi^{(j-1)}))(-\xi^{(j,i)})$$

$$(l(\xi^{(j)}) - l(\xi^{(j+1)})) \cdots (l(\xi^{(j)}) - l(\xi^{(N)}))$$

Similarly the resultant of R and $r_d(s,l)$ can be shown to be

$$\prod_{j=1}^{N}(l(\xi^{(j)}) - l(\xi^{(1)})) \cdots (l(\xi^{(j)}) - l(\xi^{(j-1)}))(-1)$$

$$(l(\xi^{(j)}) - l(\xi^{(j+1)})) \cdots (l(\xi^{(j)}) - l(\xi^{(N)}))$$

Thus we have

$$Resultant(R, r_d) \mid Resultant(R, r_{i_n}) \quad (1)$$

To use these results in our algorithm, we note that the two step process, at the ith step, of instantiation of f_i with the rational functions and computing the resultants $R(g_{i_n}, p_{i-1})$ and $R(g_{i_d}, p_{i-1})$ corresponds exactly to computing the resultant of $f_1, \ldots, f_i$ and $s - l_0 - l1x1 - \cdots - l_i x_i$, since the input system of equations was triangular. Finally, to complete the proof, we apply the same approach as we did above for showing (1) to $R(g_{i_n}, p_{i-1})$ and $R(g_{i_d}, p_{i-1})$. □

5.2 Algorithm Details

We now give a more complete description of the algorithm based on the theoretical version given in section 5.1. The algorithm performs all its computations representing the polynomials involved as *univariate* polynomials in the *primary variable* with coefficients which are rational functions in the remaining variables. These coefficients are represented by *straight-line programs* (see 8.1 for a summary of SLP properties; [12] discusses issues with SLP computations in greater detail).

As mentioned in 5.1, the first step of the algorithm corresponds to simply computing the resultant of f_1 and $s - l_0 - l_1 x_1$ by setting $x_1 = (s - l_0)/l_1$ in f_1 and then taking derivatives of this resultant w.r.t l_0 and l_1 to get the rational function representation for x_1. The ith step of the algorithm proceeds as follows :

1. Instantiation At the start of the ith step the algorithm the algorithm has computed

$$r_1(s_{i-1}), \ldots, r_{i-1}(s_{i-1}) \text{ and } p_{i-1}(s_{i-1})$$

These are represented as quotients of polynomials in s_{i-1}, the primary variable for the ith iteration (the coefficients of these polynomials are simply numbers as will be seen at the last step of the loop). The algorithm sets

$$x_i = \frac{s_i - l_0 - l_1 x_1 - \cdots - l_{i-1} x_{i-1}}{l_i}$$

with $x_1, \ldots, x_{i-1}$ instantiated to the rational functions r_j. Thus, all of these rational functions including the one for x_i are represented as quotients of univariate polynomials in s_{i-1} whose coefficients are rational functions in s_i and the l variables and represented by SLP's.

2. Substitution The rational functions are substituted in the input polynomial f_i. We therefore get f_i as a rational function which is a quotient of polynomials in s_{i-1}.

3. Resultant computation. The resultants of the numerator and the denominator of the instantiated f_i with the previous resultant p_{i-1} are now computed w.r.t s_{i-1}. These two computations are identical to univariate computations since the cost of doing SLP arithmetic is constant time per operation. The output of this step will be the two resultants. However, since the resultant is a polynomial in the coefficients of the two input polynomials, the resultants so obtained will be both SLP's in s_i and the l_k's.

4. Division We divide the two resultants in SLP form by simply creating a divide node with the two SLP's as children. Call this resultant R.

5. Differentiation To compute the rational functions r_j required for the next step, we need to compute the derivatives of R w.r.t. the l_k's. We do this by differentiating the SLP corresponding to R with respect to each variable as described in section 8.1. We now have a collection of SLP's in s_i and the l_k's.

6. Random instantiation We set each l_k to a random value, to get the random hyperplane corresponding to the L. Our collection of SLP's is now defined only over s_i.

7. Interpolation For the start of the next iteration, we need the resultant $R = p_i$ and the numerators and denominator of the rational functions r_j as *polynomials* in s_i. To do this we interpolate the SLP's representing them into polynomials over s_i. This can be done with straight-forward univariate polynomial interpolation. The degree of the polynomial an SLP represents is obtained by a single depth-first traversal of the SLP. The value of the polynomial at $s_i = q \in \mathbb{Q}$ is obtained by simply evaluating the SLP with s_i set to q, an operation also requiring a depth-first traversal of the SLP. At the end of this process we have the p_i and the numerators and denominator of the r_j's as polynomials in s_i as required by the first step.

5.3 Complexity Analysis

We now analyze the time and space requirements of the algorithm in 5.2. We will look at the final loop of the algorithm (when x_n is eliminated) since that step dominates the time and space complexity. Note that the space complexity corresponds to the number of operations we will perform over coefficients of univariate polynomials, since instead of actually performing these operations we store them in SLP's.

Let d_i be the total degree of f_i. Then it is straightforward to show that the **degree of the final resultant** $p_n(s_n)$ will be $D = d_1 \cdots d_n$. As explained in 5.2, Steps 1 and 2 in the algorithm (instantiation and substitution) have the same complexity as substitution with univariate polynomials. The polynomials we substitute with (i.e. the numerators and denominator of the r_j's) have degree $d_1 \cdots d_{n-1}$ in s_{n-1}. The time and space complexity of these steps is therefore $\mathcal{O}((d_1 \cdots d_n)^2)$. The resultant computation now takes as input two polynomials with degree $\leq D$ and the time and space required is therefore $\mathcal{O}((d_1 \cdots d_n)^2)$. The next step, differentiation takes as input the resultant SLP obtained in the previous step which has size $\mathcal{O}((d_1 \cdots d_n)^2)$. The worst case while differentiating arises when we encounter a division node which gives rise to 5 SLP nodes and we have to take derivatives w.r.t $n+1$ variables. Thus the time required is $\mathcal{O}(5(n+1)(d_1 \cdots d_n)^2)$ and space required is $\mathcal{O}(5(d_1 \cdots d_n)^2)$ (there are $n+1$ derivatives we need to take but we can by perform the differentiation and interpolation w.r.t. a given l_k one after the other and then re-use the space for the next l_k).

Finally we interpolate these SLP's. Each SLP has to be evaluated a number of times equal to the degree of the polynomial (in s_n) it represents. This degree being D, it follows that the time spent in evaluating an SLP's is $\mathcal{O}((d_1 \cdots d_n)^3)$ and since there are $n+1$ SLP's, the total time for interpolation is $\mathcal{O}(n(d_1 \cdots d_n)^3)$ The space required for univariate interpolation is linear in the degree of the input using any standard interpolation algorithm ([19]) and we can re-use the space from the previous interpolation as explained before. Hence the space required for interpolation is $\mathcal{O}((d_1 \cdots d_n))$. We have n such iterations of the algorithm, however the space required in one iteration can be re-used in the other since once the SLP's have been interpolated we no longer need the space to represent them. From this analysis it follows that :

Theorem 5.2 *The (probabilistic) time complexity of the above algorithm for determining the root representation of a triangular system is $\mathcal{O}(n^2(d_1 \ldots d_n)^3)$ where d_i is the total degree of f_i. i.e. $\mathcal{O}(n^2 d^{\mathcal{O}(n)})$ where d is an upper bound on each d_i. The space required is $\mathcal{O}((d_1 \ldots d_n)^2)$ or $\mathcal{O}(d^{\mathcal{O}(n)})$*

In contrast the worst-case complexity of computing characteristic sets as required by the original algorithm of Ritt with subsequent modifications by Wu is non-elementary. However, it was shown in [6] that the degree of the polynomials in a characteristic set for a zero-dimensional ideal $I \subseteq \mathbb{Q}[x_1, \ldots, x_n]$ is $\mathcal{O}(d^{\mathcal{O}(n)})$ and the complexity of computing these characteristic sets is $\mathcal{O}(n^{3.376} d^{\mathcal{O}(n^3)})$. (The complexity of the algorithm used by Chou's prover, as far as we know, is not known). Thus, at even relatively small degrees, $d \geq 4$, the computation required by the modified Wu's method can be rather expensive. On the other hand, our algorithm is probabilistic as mentioned in the introduction to section 5.

6 Sign Determination and Theorem Proving

Once we have a symbolic representation (p, r) of the roots of the triangular system of hypotheses, it is relatively straightforward to determine whether the theorem is true or not. The first step is to reduce the conjecture polynomial c to a univariate one by substituting $x_i = r_i(s), 1 \leq i \leq n$. This reduces c to a univariate polynomial $g(s)$. We now need to compute the sign of $g(s)$ at the real roots of $p(s)$. We can use Sturm sequences to do that as follows :

If $f(s)$ and $g(s)$ are polynomials, let the Sturm sequence of f and g be denoted $r_0, r_1, \ldots, r_k$, where $r_0(s) = f(s)$, $r_1(s) = g(s)$, and the intermediate remainders are computed via $r_{i+1} = q_i r_i - r_{i-1}$. where q_i is the pseudo-quotient of the polynomial division of r_{i-1} by r_i. For a real value v, let $SA(f,g,v)$ denote the number of changes in sign in the sequence $r_0(v), r_1(v), \ldots, r_k(v)$ and $SC(f,g)$ denote $SA(f,g,+\infty) - SA(f,g,-\infty)$. The classical Sturm theorem states that $SC(f, f')$ equals the number of real roots of $f(s)$.

We now compute $r(s) = p(s)/(gcd(p(s), g(s))$. Assuming we didn't introduce any extra roots in the triangulation procedure, we have the following conditions

- For theorem proving over the $\mathbb{C}$, the theorem is true iff r(s) is identically 1.

- For theorem proving over the reals, if r has no real roots then the theorem is true. However, if r does have real roots we need to check if those roots correspond to degenerate conditions, i.e. where the u_j's are algebraically dependent. and this is discussed in section 7. For the sufficiency condition, we compute $SC(r, r')$. Assuming r is square-free, if r has no real roots, $SC(r, r')$ will be 0. Section 7 discusses how to get a necessary condition on r.

We could have the problem discussed earlier of having introduced extra roots in the triangulation procedure. To check for this eventuality, we instantiate the original hypotheses polynomials h_j with the rational functions $r_i(s)$ as we did for g. We can now do sign-determination for the system $h_1(s), \ldots, h_n(s)$ at the roots of $r(s) = 0$ using the algorithm of [2]. This procedure yields for each root of $r(s)$ an ordered sign sequence in $\{+, -, 0\}^{n+1}$ where the ith element in the sign sequence corresponds to the sign of the ith polynomial in $h_1(s), \ldots, h_n(s)$ at that root of $r(s)$. Thus we can find out if the any of the roots of $r(s)$ are actual roots of the original system. If none are then we are done and the theorem is true. Else, over $\mathbb{C}$ the theorem is false. Over $\mathbb{R}$ we need to use the necessary condition as discussed in section 7.

7 Genericity and Completeness

We now turn our attention to the implications of the fact that we are actually determining the feasibility of conjectures based on hypotheses defined in $K[u_1, \ldots, u_d, x_1, \ldots, x_n]$ rather than just $K[x_1, \ldots, x_n]$. In the latter case, the algorithms given above work in a straightforward manner – one uses simple triangulation followed by the algorithm for triangular systems and then that

for sign determination. In the former case, things are not quite as nice. In the algorithm given in section 4, we eliminate the u_i's in step 1 by setting them to random values. This approach works when computing over $\mathbb{C}$ since with probability 1 we will not run into a degenerate situation. However, when the field is not algebraically closed, the situation becomes complicated. Consider, for example, the polynomial equation $x^2 + u^2 - 1 = 0$. For values of u such that $-1 \leq u \leq 1$, the equation has real solutions for x. However for other values of u there are no real solutions. If u were a parameter and the equation was an input hypothesis polynomial, random instantiation for u could lead to a wrong conclusion about the validity of the conjecture over the reals.

In order to disprove conjectures over the reals, we need to ensure the existence of a *real generic point*. This can be done as follows : we run the same algorithms as before (i.e. triangulation and triangular system solution) but instead of random instantiation for the u_j's we do our computation over $K(u_1, \ldots, u_d)$. The coefficients of the various polynomials will be rational functions in $\mathbf{u}$ and are represented by straight-line-programs. At the end of this process we obtain $r(s)$ as in section 6 except that r is in $K[\mathbf{u}, s]$ instead of $K[s]$. Now we need to find those components of the variety of r on which the u_j's are algebraically independent - if there are any, then we know that a real generic point exists and the theorem is generically false.

To get this necessary condition we do the following : First, we compute the resultant of $r(s, \mathbf{u})$ and $dr(s, \mathbf{u})/ds$ with respect to s. Computing the resultant involves determining the Sturm sequence of the two polynomials. The resultant is obtained as a polynomial in $\mathbf{u}$, say $R(\mathbf{u})$. Geometrically, this operation corresponds to projecting the variety $V \subseteq \mathbb{R}^{d+1}$ of $r(s, \mathbf{u})$ along the s direction to get the variety $W \subseteq \mathbb{R}^d$ of $R(\mathbf{u})$. Consider now the connected components of the complement of W in $\mathbb{R}^d$. Note that these components are necessarily open sets in $\mathbb{R}^d$. Any point in one of these connected components corresponds to a selection of the parameters $\mathbf{u}$. For each such selection, we get a certain number of solutions in s for $r(s, \mathbf{u})$. The observation is that the number of solutions remains constant within every connected component of W^c, the complement of W. (Geometrically, this corresponds to looking at the fiber above a point in that connected component). Therefore every connected component can be represented by a "witness point" which describes the behavior of the entire component. Further the fiber for each such witness point tells us what the behavior of the variety V is over that component. If the variety has a real zero in the fiber of that witness point then we know that all values of $\mathbf{u}$ in that connected component give rise to real zeroes. This means that a generic real solution exists for r and the necessary condition would imply that the theorem was false. If no connected component has a real zero in the fiber of its witness point then all the real solutions of r correspond to degenerate conditions on $\mathbf{u}$.

Now all we need to do is find the witness points and check if they correspond to real zeroes : The complement of W can be represented by the polynomial condition : $R' = (R + \epsilon)(R - \epsilon)$ where ϵ represents an infinitesimal variable. We can now compute the witness points in each connected component of W^c as follows : We choose a linear map $\pi : \mathbb{R}^d_{|W^c} \to \mathbb{R}$ and find all the critical points of this map. Let $u'_2, \ldots, u'_d$ be a coordinate basis for the kernel of π which is a $d-1$ dimensional linear space. We now solve the following system in the limit as $\epsilon \to 0$:

$$R' = 0 \qquad \frac{\partial R}{\partial u'_2} = \cdots = \frac{\partial R}{\partial u'_d} = 0 \qquad (2)$$

The algorithm for solving this system is described in [1] and it involves computing the u-resultant of the system, arranging it in powers of ϵ and retaining the lowest degree coefficient. The result is (as in our triangular system algorithm) a polynomial H(t) and rational functions $(h_1(t), \ldots, h_d(t))$ describing the witness points in the connected components as required. Given this description, we need to figure out the behavior of the variety V at these witness points. To do that, we use the Sturm sequence which was computed to get the resultant $R(\mathbf{u})$. This is a sequence of polynomials $p_1(s), \ldots, p_k(s)$ with coefficients which are polynomials in $\mathbf{u}$. As per the Sturm theorem the number of real roots of r is given by $SC(r, r')$, that is, we look at the leading coefficients and determine the sign variations. In our case, we are interested in the number of real roots at a given witness point as obtained above - so we instantiate the leading coefficients $l_1(\mathbf{u}), \ldots, l_k(\mathbf{u})$ of the polynomials $p_1(s), \ldots, p_k(s)$ with $u_i = h_i(t)$ to get polynomials $l_1(t), \ldots, l_k(t)$. The signs of these polynomials at the roots of the primitive element polynomial $H(t)$ are exactly the signs of these polynomials at the witness points. We now apply the sign determination algorithm of [2] to determine the signs of the leading coefficients of the p_i's at the witness points. Each sign sequence gives us the number of roots in the fiber corresponding to that witness point. As mentioned earlier, those connected components which have no real roots in the fiber for their witness points correspond to degenerate values of $\mathbf{u}$ for which the hypotheses are not valid. If however a component does have real roots, then we have an open set of $\mathbf{u}$ values on which the u_j's are independent and and hence the theorem is false. It can be shown that the worst case complexity of the above procedure is $\mathcal{O}(d^{\mathcal{O}(n^2)})$. This algorithm has *not* been implemented in our prover yet - more general algorithms for determining whether a semi-algebraic set is not empty will be implemented in our toolkit in the near future [13].

8 Implementation

In this section, we briefly describe a first implementation of the above algorithms. Most of the modules are part of a larger project involving the development of a toolkit for solving problems in non-linear algebra. (This toolkit will soon be available in the public domain). The basic modules perform polynomial arithmetic over different fields such as finite fields, rationals, infinitesimal extensions, "mixed" arithmetic and most importantly straight-line programs or SLP's.

8.1 Straight-line Programs

All our computations are done over the field of arithmetic straight-line programs. An SLP can be represented by a directed acyclic graph, with each node representing an operation such as addition, and a value. *All the polynomials are thus represented as polynomials over the field of SLP's, i.e. the coefficients are straight-line programs and not numbers.* As mentioned in the introduction, this allows us to represent coefficients which are rational functions in the parameters $u_1 \ldots u_m$. In other words the extension field $\mathbb{Q}(u_1, \ldots, u_m)$ has a 1-1 correspondence with SLP's defined over $u_1, \ldots, u_m$.

Arithmetic with SLP's is easy - for example to add two SLP's one creates a new node with the operation "+" and two edges directed from the new node to the two operand nodes. The value of an SLP node (over the base field such as $\mathbb{Q}$), can be obtained by simply adding the values of its children nodes. Thus evaluation involves a depth-first search of the SLP graph and can be done in time linear in the size of the SLP.

Computing derivatives as required by the algorithm in Section 5 is particularly straightforward, as is determination of the signs of the coefficients of various polynomials (required by the Sturm sequence algorithm). To compute the derivative SLP of an SLP node one simply creates a new SLP using the usual rules of differentiation. This has roughly double the size of the original SLP. For sign determination, one uses successive differentiation and evaluation of SLP's. The details of how this works can be found in the complete version of the paper.

8.2 Some benchmarks

The table below gives some benchmarks on some standard problems. A more complete set can be found in [11]. All the benchmarks were run on a Sun Sparc 10 machine. The code is written in C. These benchmarks were obtained from a first implementation and we expect to reduce the running time by using various optimizations but the initial benchmarks seem to be rather promising. (The times given for Chou's prover are from

[4] and were obtained on a Symbolics 3600. The geometric statements for these theorems are also from [4].)

Geometric Theorem	Total deg	No. of vars	Chou's Prover	This Prover
Pappus	1	7	1.52s	0.11
Pappus Dual	1	7	1.45s	0.12
AMS-1	1	13	4.05s	0.33s
Pascal Conic-1	1	15	14.43s	0.56s
Simson	2	11	-	0.47s
Ptolemy	8	6	-	2.9
Steiner	8	17	-	33.0

9 Code Availability and Acknowledgements

The code for the prover and the toolkit described in [13] as well as papers on the same can be obtained by anonymous ftp from *robotics.eecs.berkeley.edu*. See the file pub/rege/README for details.

I would like to thank my advisor John Canny for directing me to this problem and for some very useful and insightful discussions regarding this and other topics. I would also like to thank Phil Liao for providing me with the theorems in the form required by the code and also both him and Richard Fateman for useful discussions.

10 References

[1] J.F. Canny. Computing roadmaps of general semi-algebraic sets. In *AAECC-91*, 1991. New Orleans.

[2] J.F. Canny. An improved sign determination algorithm. In *AAECC-91*, 1991. New Orleans.

[3] S.-C. Chou. Proving elementary geometry theorems using Wu's algorithm. In W.W. Bledsoe and D.W. Loveland, editors, *Theorem Proving : After 25 Years*. American Mathematical Society, 1984.

[4] S.-C. Chou. *Mechanical Geometry Theorem Proving*. Mathematics and its applications. D. Reidel, Holland, 1988.

[5] J. Little D. Cox and D. O'Shea. *Ideals, Varieties, and Algorithms*. Undergraduate Texts in Mathematics. Springer-Verlag, New York, 1992.

[6] G. Gallo and B. Mishra. Efficient algorithms and bounds for Wu-Ritt characteristic sets. In T. Mora and C. Traverso, editors, *Effective Methods in Algebraic Geometry*, volume 94 of *Progress in Mathematics*, pages 119–142. Birkhäuser, 1991.

[7] H. Gelernter. Realization of a geometry theorem proving machine. In E.A. Feigenbaum and J.E. Feldman, editors, *Computers and Thought*. McGraw-Hill, New York, 1963.

[8] D. Kapur. Geometry theorem proving using Hilbert's Nullstellensatz. In *Symposium on Symbolic and Algebraic Computation*, pages 202–208, 1986. Waterloo, Ontario.

[9] D. Kapur. A refutational approach to geometry theorem proving. In D. Kapur and J.L. Mundy, editors, *Geometric Reasoning*. MIT Press, 1988.

[10] B. Kutzler and S. Stifter. On the application of Buchberger's algorithm to automated geometry theorem proving. *J. Symbolic Comput.*, 2:409–420, 1986.

[11] A. Rege. A complete and practical algorithm for geometric theorem proving. Manuscript for journal submission and Tech Report (UC Berkeley), 1995.

[12] A. Rege and J. Canny. Straight-line programs for real field extensions and perturbations. To be submitted to special issue of *International Journal of Computational Geometry and Applications*, 1995.

[13] A. Rege and J. Canny. A toolkit for algebra and geometry. Submitted to the *International Symposium on Symbolic and Algebraic Computation*, Montreal, 1995.

[14] J. Renegar. On the computational complexity and geometry of the first-order theory of the reals, parts I, II and III. Technical Report 852,855,856, Cornell University, Operations Research Dept., 1989.

[15] J.F. Ritt. *Differential Equations from the Algebraic Standpoint*. Number 14 in AMS Colloquium publications. American Mathematical Society, New York, 1932.

[16] J.F. Ritt. *Differential Algebra*. Number 33 in AMS Colloquium publications. American Mathematical Society, New York, 1950.

[17] W. Wu. On the decision problem and mechanization of theorem proving in elementary geometry. *Sci. Sinica*, 21:150–172, 1978.

[18] W. Wu. Some recent advances in mechanical theorem proving of geometries. In W.W. Bledsoe and D.W. Loveland, editors, *Theorem Proving : After 25 Years*. American Mathematical Society, 1984.

[19] R. Zippel. *Effective Polynomial Computation*. Kluwer Academic publishers, Massachusetts, 1993.

Monte Carlo Approximation of Form Factors with Error Bounded a Priori

M. Pellegrini*

Abstract

The exchange of radiant energy (e.g. visible light, infra-red radiation) in simple macroscopic physical models is sometimes approximated by the solution of a system of linear equations (energy transport equations). A variable in such a system represents the total energy emitted by a discrete surface element. The coefficients of these equations depend on the *form factors* between pairs of surface elements. A form factor is the fraction of energy leaving a surface element which directly reaches an other surface element. Form factors depend only on the geometry of the physical model. Determining good approximations of form factors is the most time consuming step in these methods, when the geometry of the model is complex due to occlusions.

In this paper we introduce a new characterization of form factors based on concepts from integral geometry. Using this characterization, we develop a new and asymptotically efficient Monte Carlo method for the simultaneous approximation of all form factors in an *occluded* polyhedral environment. This is the first algorithm for which an asymptotic time bound *and* a bound on the absolute approximation error has been proved. This algorithm is one order of magnitude faster than methods based on the hemi-sphere paradigm, for typical scenes.

Let A be a set of convex non-intersecting polygons in R^3 with a total of n edges and vertices, covering the facets of the input polyhedra. Let ϵ be the error parameter and δ be the confidence parameter. We compute an approximation of each *non-zero* form factor such that with probability at least $1 - \delta$ the absolute approximation error is less than ϵ. The expected running time of the algorithm is $O((\epsilon^{-2} \log \delta^{-1})(n \log^2 n + K \log n))$, where K is the expected number of *regular intersections* for a random projection of A. The number of regular intersections can range from 0 to quadratic in n, but for typical applications it is much smaller than quadratic. The expectation is with respect to the random choices of the algorithm and the result holds for any input.

1 Introduction

Energy Transport problems in Graphics and Heat Transfer. The determination of radiation interchange between surfaces is a central problem in heat transfer, illumination engineering an applied optics [SH92]. Complex radiation interchange problems are simplified using a variety of assumptions and approximations. One of the simplest models in heat transfer is that in which (i) each surface is "black" (i.e. there is no reflected radiation and all emitted energy is diffuse) and (ii) each surface emits diffuse radiation uniformly in each direction and from each point of the surface (see [SH92] for an extended treatment). The balance of energy in the model is described by a system of linear equations (energy transport equations). For a surface i in a set of n surfaces, let E_i be the emissivity (i.e. the energy produced by the surface i), ρ_i the diffuse

---

*Department of Computer Science, King's College, London. WC2R 2LS England. e-mail: marco@dcs.kcl.ac.uk.

Permission to copy without fee all or part of this material is granted provided that the copies are not made or distributed for direct commercial advantage, the ACM copyright notice and the title of the publication and its date appear, and notice is given that copying is by permission of the Association of Computing Machinery.To copy otherwise, or to republish, requires a fee and/or specific permission.
11th Computational Geometry, Vancouver, B.C. Canada
© 1995 ACM 0-89791-724-3/95/0006...$3.50

reflectance (i.e. the proportion of received energy that is re-emitted) and B_i the total energy emitted by surface i. Denoting with F_{ij} the form factor between surface i and surface j, the energy transport equation for surface i is:

$$B_i = E_i + \rho_i \sum_j F_{ij} B_j,$$

where the summation is extended to all surfaces different from i. All the transport equations form a linear system in the variables $B_1, .., B_n$. The values of ρ_i and E_i are known for a given model. The form factor (also called *configuration factor* or *view factor*) is defined as *the fraction of the diffuse energy leaving one surface that directly reaches another surface*. Under the conditions of the model, form factors depend only on the geometry of the surfaces and are independent from all the other parameters of the model. The determination of the form factor is an essential preliminary step to the solution of the system.

In Graphics, the goal of synthesizing realistic images has been pursued by computing the balance light reaching each surface. Such approach is known as *radiosity computation* and has been studied extensively (e.g. [CG85, WRC88, WEH89, SP89, HSA91] [Mit91]). When we consider only the effect of diffuse light we can use assumptions similar to those used in the study of heat transfer and we obtain the same model described above (in Graphics known as Lambertian model). Again, the determination of form factors is a central computational problem.

Exact analytical computation of form factors is difficult except for restricted special cases, none of which considers occlusions. Thus, several approximation schemes have been devised to cope with realistic environments [CG85, HSA91]. One of the most popular lines of attack is the definition of a recursive divide and conquer refinement algorithm in which some local error condition determines the end of the recursion. Not surprisingly, the issue of bounding or estimating the quality of such approximations has been brought forward recently [LSG94].

Within the discipline of design and analysis of algorithms, approximation methods are compared according to two main measures: one is the computational resources (time, storage) used, the second is the quality of the approximation, i.e. a bound on the error of the approximation. The two measure are not independent in general. Until now the form factors problem has been unyielding to this type of analysis.

Current approximation algorithms. There are currently several methods available for computing approximations of form factors. Historically, the first method used was the so called *Monte Carlo ray tracing* method [EJLA91] which is based on performing a large number of ray-shooting operations. This method is simple but it is considered inefficient even for polyhedral scenes of moderate complexity and several other schemes have been proposed. One of the most popular one is the hemicube algorithm [CG85]. This algorithm is an elaboration of the hemi-sphere method of Nusselt [SH92]. Hanrahan et at. [HSA91] have recently proposed methods for computing form factors based on defining each form factor as a matrix of sub-form-factors and on an efficient hierarchical block decomposition of such matrices. A more detailed description of these algorithms is given in the full version of this paper. References to analytic solutions for special cases or to computer programs for general cases can be found in the book of Siegel and Howell [SH92].

New results in this paper. Some currently known algorithm are hard to analyze and therefore the correlation between running time and approximation error has been investigated by means of experiments, rather than algorithmic analysis. An other approach has been to produce error analysis *a posteriori*, that is an error analysis based on values computed by the algorithm itself. Last but not least, many error analysis published in the literature at my knowledge use additional assumptions, justifiable on pragmatic grounds, but not usually made in computational geometry. In this paper we present the following results:

(1) We give a integral geometric interpretation to the form factors and we show that the analytic and geometric formulations previously used are specializations of this paradigm.

(2) We derive the first Monte Carlo algorithm to compute approximations of form factors for which we have *simultaneously* a worst case deterministic asymptotic upper bound on the running time *and* an exact *a priori* upper bound, which holds with high probability, on the absolute approximation error of each form factor.

(3) The algorithm has better asymptotic running time, for typical scenes, than other methods for which a comparable analysis of the running time is possible (in particular, Nusselt's hemisphere

method).

More precisely the algorithmic result in this paper is the following. Let A be a set of convex non-intersecting polygons in R^3 with a total of n edges and vertices, covering the facets of the input polyhedra. We fix an error bound ϵ and a confidence δ. Let $K_r(A, v)$ be a certain subset of the intersection points among the orthogonal projections of the edges of A in direction v, which Mulmuley calls *regular intersections*. We denote with K the expected size of $K_r(A, v)$ over a choice of v uniform at random in the set of all directions.

Theorem 1 *We can compute in expected time $O((\epsilon^{-2} \log \delta^{-1})(n \log^2 n + K \log n))$ an approximation of each non-zero form-factor such that with probability $1 - \delta$ the absolute error on each form factor is less than ϵ.*

The number of regular intersections can range from 0 to quadratic in n, but for typical applications it is much smaller than quadratic. A typical input is a set of polygons in which polygons are rather fat or, to put it negatively, not long and thin. This is a typical input because in these conditions the assumptions of the Lambertian model are more likely to be realistic.

What is a form factor? The first task we set out to accomplish is to give a satisfactory *geometric* definition of a form factor. The original definition, as the fraction of diffuse energy leaving a surface that reaches directly another surface, belongs more to physics than to geometry. A further elaboration [SH92] equates the form factor to a certain double area integral in which the integrand function contains trigonometric functions (cosines), and thus has more an analytic flavour than a geometric one. Nusselt gives a geometric interpretation of form factors in terms of central projections on the surface of a sphere, which holds only when one of the two surfaces is infinitesimally small (a so called differential area).

In this paper we show that form factors have a natural interpretations in the classical theory of integral geometry [San76]. After we completed this research we learned that the connection between form factors and integral geometry was noted by Sbert [Sbe93], however his algorithm and error estimates follow a quite different approach from the one used in this paper.

We will show that form factors can be interpreted as the ratio between the "volume" of certain sets of lines in 3-space. We will show that the characterization of form factors previously used in literature are equivalent to the new one for specific choices of line coordinates.

A new general method. The definition of form factor via integral geometric theory is abstract in the sense that it is not immediately computable and also in the sense that it is independent of any coordinatization of lines in 3-space. When we decide for a specific coordinatization of lines we obtain an equivalent formulation that is, in principle, computable.

The next step in our research hinges on the choice of a coordinatization of the lines so that the resulting concrete formulation for the form factor contains quantities that can be computed easily and efficiently using computation geometry techniques [PS85]. In our case we use only orthogonal projections and the area of planar convex polygons, for this reason we will refer to our method as the *orthogonal projection method*. This formulation involves still an integral, which is simpler than those previously used.

An efficient algorithm. We use Monte Carlo integration to approximate the value of the integral defining the form factor as derived by the previous analysis. The integration domain is the unit sphere of directions. This domain has a finite measure and this is an important property that we use to derive an exact bound on the variance of the integrand function and consequently a bound on the absolute error of the Monte Carlo integration.

Finally, we use the vertical cylindrical decomposition of Mulmuley [Mul91] to compute efficiently the terms of the summations in the Monte Carlo approximation of the integral. Such data structures are built via plane sweep and dynamic maintenance of polygonal planar maps.

Comparisons with other algorithms. The algorithms developed for computing form factors come under many different families, some are randomized (Monte Carlo), some deterministic. A large variety of approximations in various aspects of the algorithm are used. Usually, neither the time bounds nor the error bounds are derived by analysis. For this reason we will not attempt to describe these methods as originally described in literature.

Still, it is instructive to compare our method of orthogonal projections with the other known methods. A comparison is shown in the full version of the paper which is based on the following criteria:

(i) We consider the total cost of computing all form factors, that is, a form factor for each pair of surfaces, in presence of occlusions, for an input set of disjoint convex polygons in 3-space with n vertices and edges.

(ii) We classify algorithms into families according to the formulation of the form factor as an integral that is used as guideline of the computation.

(iii) We consider in each such family the Monte Carlo method in which the domain of integration is sampled uniformly at random and the integrand function is computed *exactly* in the real-RAM model.

(iv) For the exact computations of the integrand function we will refer currently known method with known worst case or expected time bounds.

(v) We are not be able to bound the absolute error, therefore we consider the less demanding requirement that the convergence rate of the error, as derived from simple analysis of the Monte Carlo method in [DR75] is at least $O(\epsilon)$ on each form factor. Here we rule out the usual assumption done in the hemi-cube algorithm that the centre of a surface is sufficient to characterize form factors for the whole surface.

Based on the above assumptions we find that our method is more efficient than comparable algorithms based on the hemi-sphere approach. It turns out that the hemisphere approach ranges in time complexity under the above assumptions from $O(\epsilon^{-2}n^2)$ to $O(\epsilon^{-2}n^3)$ depending on the input, while our method ranges roughly from $O(\epsilon^{-2}n)$ to $O(\epsilon^{-2}n^2)$.

The hierarchical decomposition method of Hanrahan et al. [HSA91] falls somewhat outside our Monte Carlo framework and the comparison of the running time of this algorithm with the running time of the method of orthogonal projections is more problematic.

Organization of the paper. In Section 2 we give the two characterization of form factors more frequently used in literature. In Section 3 we introduce an "abstract" characterization via integral geometry. In Section 4 we derive a new "concrete" integral definition of form factor. In Section 5 we derive the error bound on the Monte Carlo evaluation of such integral. In Section 6 we give the algorithm to compute all form factors simultaneously.

2 Currently used concrete characterizations

The algorithms known in literature to compute form factors are mainly based on two formulations of form factors. In the first formulation (which is sometimes taken as the definition tout court) the form factor is a weighted double area integral [SH92]. Given a set of disjoint surfaces[1] $\mathcal{S}$ in 3-space, the form factor between two surfaces S_i and S_j in $\mathcal{S}$ is:

$$F_{ij} = \frac{1}{A_i} \int_{p \in S_i} \int_{q \in S_j} f(p,q) dp dq, \quad (1)$$

where:

$$f(p,q) = \frac{\cos \theta_i(p,q) \cos \theta_j(p,q)}{\pi |pq|^2} V(p,q,\mathcal{S});$$

A_i is the area of the surface S_i; $\theta_i(p,q)$ is the angle between the normal vector to S_i at the point p and the line through p and q; $\theta_j(p,q)$ is the angle between the normal vector to S_j at the point q and the line through p and q; $|pq|$ is the distance between point p and point q; and $V(p,q,\mathcal{S})$ is a predicate that has value 1 if the open segment pq does not meet any other surface, and is 0 otherwise.

The second characterization due to Nusselt [SH92] defines the form factor between a differential surface of area dA_i and a finite surface A_j geometrically as follows. Consider the set of points of A_j visible from a point $p \in dA_i$ and the central projection from p of such set onto the surface of a sphere centered at p. Orthogonally project such set of points on the sphere onto the plane tangent to dA_i at p. The area of the set of points obtained from the second projection is the value of the (differential) form factor between dA_i and A_j. An analytic, closed form of Nusselt's characterization for polygonal objects in 3-space can be found in [Arv94].

3 An abstract characterization

3.1 Unoccluded case

We give a new abstract characterization of form factors starting with the case in which we have only two polygons S_1 and S_2 and no occlusion between

---
[1] In this paper we consider mainly polygonal surfaces, but most observations hold for a larger class of surfaces.

the them. Let us go back to the original intuitive meaning of form factor as *the fraction of the diffuse energy leaving one surface that directly reaches another surface.*

In radiation models with non-participating mediums energy is transferred along linear trajectories, i.e. along lines. Thus we can measure the total energy $\mathcal{I}_1$ leaving one surface S_1 by summing the energy carried by each single line. Since there are infinite lines meeting a given surface S_1 the measure is more precisely defined by an integral:

$$\mathcal{I}_1 = \int_{L \cap S_1 \neq \emptyset} I(L) dL, \quad (2)$$

where $I(L)$ is the density of the energy on the line L. Under the assumptions of the Lambertian model $I(L)$ is a constant $\bar{I}$. Thus we are left with a purely geometric integral to compute

$$\int_{L \cap S_1 \neq \emptyset} dL = m(\{L : L \cap S_1 \neq \emptyset\}). \quad (3)$$

We introduce the notation $m(\mathcal{L})$ to denote such integral evaluated over the domain $\mathcal{L}$. It is convenient to look at formula 3 using the Theory of Integral Geometry as exposed in the book by Santalo [San76]. One of the concerns of Integral Geometry is measuring sets of geometric object (i.e. associating a positive real number to a set) in such a way that the measure is invariant under rigid transformations of the space. In our case we want the measure of the set of lines meeting the surface S to be invariant under rigid transformation of the Euclidean 3-dimensional space. The differential form dL denotes a differential for which this property of invariance holds. For any given parameterization of the lines in 3-space there is a corresponding formulation of dL that is unique up to constant multiplicative factors [San76]. Thus formula 3 is nothing but the (invariant) measure of the set of lines meeting S_1.

The amount of energy $\mathcal{I}_{12}$ leaving a surface S_1 and reaching a surface S_2, if there are no occlusions, is given by the following integral over the lines meeting S_1 and S_2:

$$\mathcal{I}_{12} = \int_{L \cap S_1 \neq \emptyset \wedge L \cap S_2 \neq \emptyset} I(L) dL \quad (4)$$

Since again $I(L)$ is constant and equal to $\bar{I}$ we are interested in the measure of the set of lines meeting both surfaces given by the integral:

$$m(\{L : L \cap S_1 \neq \emptyset \wedge L \cap S_2 \neq \emptyset\}) \quad (5)$$

The form factor F_{12} is given by the ratio of the two intensities: 2 and 4:

$$F_{12} = \frac{\mathcal{I}_{12}}{\mathcal{I}_1} = \frac{\int_{L \cap S_1 \neq \emptyset \wedge L \cap S_2 \neq \emptyset} dL}{\int_{L \cap S_1 \neq \emptyset} dL}. \quad (6)$$

We will from now on take 6 as our integral geometric definition of form factor in absence of occlusions.

From a classical result of Integral geometry adapted to polygons [San76, page 246] we have that

$$m(\{L \in \mathcal{L} | L \cap S_1 \neq \emptyset\}) = \pi A_1$$

where A_1 is the area of S_1. Since we can easily compute the area of a convex polygon, we concentrate our attention on the numerator of the fraction in 6. Formula 6 is valid for any coordinatization of the lines. In order to obtain a more concrete formula (as opposed to the somewhat abstract formulation in 6) we need to decide what coordinates we use for parameterizing the line L. As a consequence we will have a unique expression for dL as a differential form in the chosen coordinates.

It is easy to prove that, when we use as coordinates of a line L the two intersection points of L with the surfaces S_1 and S_2, the formulation 6 becomes formulation 1 [San76, page 230]. It is now clear thus formula 1 is linked to a particular choice of line coordinates. The hemisphere characterization of Nusselt can be derived by formula 6 when we use as coordinates of a line the intercept of L on a fixed plane (the plane tangent to dA_1) and the direction of L [San76, page 211]. In Section 4 we use a different parameterization of lines in 3-space and the correspondent invariant differential form to derive a new concrete formulation of 6.

3.2 Occluded case

Now we consider the general case in which we want to define the form factor F_{12} between S_1 and S_2 in presence of other objects $S_3, ..., S_k$. First of all to simplify the situation we notice that we need only consider the objects clipped within the convex hull of S_1 and S_2 since (portions of) objects outside $CH(S_1 \cup S_2)$ cannot possibly occlude a line meeting S_1 and S_2.

Therefore we want to compute the measure of the set of lines *meeting* S_1 and S_2, and *missing* all the clipped polygons $S_3^c, ... S_k^c$. Let $\mathcal{L}_{12} = \{L | (L \cap S_1 \neq \emptyset) \wedge (L \cap S_2 \neq \emptyset) \wedge \bigwedge_{i=3}^{k} (L \cap S_i^c = \emptyset)\}$ be such set of lines. The new abstract formula for the form factor is:

$$F_{12} = \frac{m(\mathcal{L}_{12})}{m(\{L|L \cap S_1 \neq \emptyset\})} \qquad (7)$$

4 A new concrete characterization of form factors

4.1 Unoccluded case

Let us consider two disjoint convex planar polygons in 3-space, namely S_1 and S_2. From the abstract definition in 6 we just need to evaluate the integral:

$$m(\{L|L \cap S_1 \neq \emptyset \wedge L \cap S_2 \neq \emptyset\})$$

From Santalo [San76] we have that one of the possible parameterizations of the line L is by giving the plane $P(L)$ through the origin and orthogonal to L, and the intersection point $L \cap P(L)$. Using this coordinatization the form of the invariant density for lines in 3-space is $dL = d\sigma \wedge du$, where $d\sigma$ is the invariant density of points on a plane through the origin and orthogonal to L, and du is the invariant density of unoriented planes through the origin (equivalent to the points on the unit 2-dimensional sphere U centered at the origin where opposite points are identified).

We fix a point $u \in U$ and we integrate over the density $d\sigma$ of the plane through the origin and orthogonal to the vector $\vec{Ou}$, which we call $P(u)$. A line L orthogonal to $P(u)$ meets both S_1 and S_2 if and only if the intersection point $L \cap P(u)$ is in the intersection of the projection of S_1 onto $P(u)$ and the projection of S_2 onto $P(u)$. But the integral of the density of points in a given planar set is just the area of that set. So, denoting with $A_{12}(u)$ the area of the intersections of projections of S_1 and S_2 onto $P(u)$ we have that:

$$\int_{\{L|L\cap S_1 \neq \emptyset \wedge L\cap S_2 \neq \emptyset\}} dL = \int_{u \in U} A_{12}(u) du \qquad (8)$$

Here we consider the case of two objects only and therefore there is no issue of occlusions. We can compute approximately the integral in formula 8 by using Monte Carlo integration [BGS+66, DR75]. For a fixed direction u we can compute $A_{12}(u)$ in time $O(n)$, where n is the number of vertices of the polygons S_1 and S_2, for example by using a sweeping line approach. We choose uniformly at random on the unit sphere U a set of N points $u_1, ..., u_N$, then the Monte Carlo approximation we obtain is:

$$m(\{L|L \cap S_1 \neq \emptyset \wedge L \cap S_2 \neq \emptyset\}) \approx \frac{2\pi}{N} \sum_{i=1}^{N} A_{12}(u_i)$$

We can find an approximation to the form factor of two unoccluded polygons in 3-space in time $O(Nn)$. An upper bound on the error developed in section 5 for the occluded case holds also in the unoccluded case. Here we show an high level pseudocode to compute the approximate form factor between two polygons in 3-space without occlusions.

```
proc ff(A,B:3D-polygon; eps,de:real):real
begin
  N = floor(1/(eps * eps * de));
  ff = 0;
  for i = 1 to N do
    u = random-direction;
    A1 = projection of A along u;
    B2 = projection of B along u;
    ff = ff+(area(intersection(A1,B2));
  end for;
  return (2 * ff)/(N * area(A));
end
```

4.2 Importance sampling

We can improve on the basic idea of algorithm ff by filtering out efficiently the directions for which the contribution to the summation is zero. We sketch here the method. We assume, without loss of generality, that the plane spanning S_1 does not meet S_2 and viceversa[2]. Consider the convex hull $C = CH(S1 \cup S_2)$. Assuming general position, a facet f of C, except S_1 and S_2, is incident to an edge e of S_1 and to a vertex v of S_2, or viceversa. Moreover the plane spanning f leaves S_1 and S_2 on the same side. We find, using a sort of binary search, the plane incident to e that instead leaves S_1 and S_2 on *opposite* sides (we call *positive* the side containing S_1). We find such planes for every facet f and we translate them to the origin. We take the intersection of all the translated positive half-spaces. We obtain a double cone with apex at the origin, which we call, for lack of a better name, the *anti-convex-hull* of S_1 and S_2. It is easy to see that, for a projection direction u that falls outside the anti-convex-hull, the projections of the surfaces cannot meet. Instead, if the direction u falls within the anti-convex-

---
[2]This assumption can be eliminated easily by splitting each surface in at most two parts.

hull, then the projections have a non-empty intersection. The convex hull and the anti-convex-hull can be built in time $O(n \log n)$. After preprocessing we can decide whether a sampled direction is inside the anti-convex-hull in time $O(\log n)$, since the problem is equivalent to locating a point in a planar convex polygon of $O(n)$ sides.

4.3 Occluded case

We compute the measure of the set of lines in formula 7 by integrating the motion invariant unit element dL discussed in the previous sub-section. We indicate with S_i^c the surfaces clipped within $CH(S_1 \cup S_2)$. We indicate with the superscript $''$ to a set, the set obtained by orthogonal projection in direction u. We denote with A'_{12} the area of the set $(S_1'' \cap S_2'')/\bigcup_{j=3}^{k}(S_j^c)''$. The numerator of the fraction 7 is thus equal to:

$$m(\mathcal{L}_{12}) = \int_{u \in U} A'_{12}(u) du. \qquad (9)$$

In Section 6 we will see how to compute the terms $A'_{ij}(u)$ simultaneously for all pairs ij for a fixed direction u. In the next section we estimate an upper bound to the error one incurs in a Monte Carlo approximation of integral 9.

5 Error estimate

5.1 Bounding the variance

In on order to estimate the error of our method we use an elementary analysis of Monte Carlo methods as described in [BGS+66] and [DR75]. This type of analysis relies on a bound on the variance of the integrand function over the domain of integration. This bound is hard to obtain for most of the integrand functions used in previous methods for form factor computation. Luckily, in our case a bound on the variance is obtained very easily. We recall that the measure of the set U is 2π. We introduce a function $\zeta(u)$:

$$\zeta(u) = \frac{2A'_{12}(u)}{A_1}.$$

We have that the expected value E of $\zeta(u)$ over all directions U is the value of the form factor F_{12}:

$$\frac{1}{2\pi} \int_{u \in U} \zeta(u) du = \frac{1}{2\pi} \int_{u \in U} \frac{2A'_{12}(u)}{A_1} du = F_{12}.$$

Now we estimate the variance D of the function ζ over the set of directions U:

$$D(\zeta(u)) = E(\zeta(u)^2) - [E(\zeta(u))]^2 =$$
$$[\frac{1}{2\pi} \int_{u \in U} \frac{4(A'_{12}(u))^2}{A_1^2} du] - F_{12}^2 =$$
$$\frac{2}{2\pi} \int_{u \in U} (\frac{2A'_{12}(u)}{A_1})(\frac{A'_{12}(u)}{A_1}) du - F_{12}^2 \leq$$
$$2F_{12} - F_{12}^2 \leq 1.$$

where we exploit the fact that $A'_{12}(u) \leq A_{12}(u)$ is always smaller than or equal to A_1.

5.2 A bound obtained using Chebyshev's inequality

For any random variable Y and $t > 0$, Chebyshev's inequality [Fel68, page 232] is:

$$Prob(|Y| \geq t) \leq E(Y^2)/t^2,$$

where $E()$ indicate the expected value. Let $S_N = \sum_{i=1}^{N} \zeta(u_i)$, $\mu = E(\zeta)$ and let σ^2 be the variance of $\zeta(u_i)$. We apply Chebyshev's inequality, obtaining:

$$Prob(|(S_N - N\mu)/\sqrt{N}\sigma| \geq t) \leq 1/t^2.$$

Rearranging and dividing by N, we obtain:

$$Prob(|(S_N/N - \mu)/| \geq t\sigma/\sqrt{N}) \leq 1/t^2.$$

Equivalently, if we call the absolute error ϵ, we have $\epsilon = t\sigma/\sqrt{N}$ and $t = \epsilon\sqrt{N}/\sigma$. The probability of exceeding the error bound is $\delta = 1/t^2 = \sigma^2/\epsilon^2 N$. The probability of being within the error bound is conversely $1 - \delta$.

In the previous section we proved $\sigma^2 \leq 1$, thus we can derive the number of samples attaining the desired precision as a function of ϵ and δ. We have: $N = \lceil 1/\epsilon^2\delta \rceil$. We have proved the following lemma:

Lemma 1 *The Monte Carlo approximation of formula 9 results in an approximation of every form factor within an absolute error ϵ, with probability at least $1 - \delta$ when we choose $N = \lceil 1/\epsilon^2\delta \rceil$.*

5.3 Reducing the number of samples

From the previous section we have that the number of samples needed for an (ϵ, δ) approximation algorithms is $N = \lceil 1/\epsilon^2\delta \rceil$. It is possible to reduce the number of samples to $N = O((1/\epsilon)^2 \log(1/\delta))$ by using a trick reported in [KKK+90], originally used in [JVV86].

For completeness we describe this method. We use the algorithm described in Section 6 with a constant value for the confidence parameter, say $\delta = 1/4$. We run the algorithm K times, and we take the *median* value as the outcome of the algorithm for F_{ij}. We obtain an (ϵ, δ)-approximation by choosing $K = c \log(1/\delta)$ for an appropriate constant c. We give a sketch of the proof in the rest of this subsection.

We call a value X_i good if it is with a distance ϵ from the expected value $E(X)$, and bad otherwise. In our experiment a sample will be bad with probability 1/4 and good with probability 3/4. If the median is bad, then there are at least $\lceil K/2 \rceil$ bad samples. Therefore the probability of a bad median is less than the probability of having at least $\lceil K/2 \rceil$ bad samples. To bound this second probability we use Chernoff bound in a form reported in [Spe87]. Let $Y_i = 1$ if X_i is bad and $Y_i = 0$ if X_i is good. We have that:

$$Prob(\sum_{i=1,K} (Y_i - E(Y)) > a) < e^{-2a^2/K}.$$

Using $E(Y) = 1/4$ and $a = K/4$ we obtain:

$$Prob(\sum_{i=1,K} Y_i > K/2) < e^{-K/8}.$$

Thus, to reach a confidence level δ we just need to repeat the algorithm $K = O(log(1/\delta))$ times.

6 Measuring vertical visibility

The arguments in the previous sections have reduced the problem of computing form factors to the problem of computing areas of intersections of planar polygons on an Euclidean plane. We compute such areas using an algorithm for vertical space partition as algorithmic skeleton. A good asymptotic performance is obtained using the cylindrical partition of Mulmuley [Mul91]. Let $\mathcal{S}$ be a set of interior disjoint convex polygons in 3-space and let $\mathcal{E}(\mathcal{S})$ be the set of edges of polygons in $\mathcal{S}$. We choose u as the vertical direction. Let q be an intersection among the orthogonal vertical projections of $e_1 \in \mathcal{E}(\mathcal{S})$ and $e_2 \in \mathcal{E}(\mathcal{S})$. Such a point is called a *regular* intersection point if the vertical line through

q does not intersect any polygon of $\mathcal{S}$ between e_1 and e_2. If we call k_r the number of regular intersections and k the number of intersections, we have that $0 \leq k_r \leq k \leq n^2$. In typical situations though k_r will be much smaller than n^2. The vertical decomposition $H(\mathcal{S})$ for the set $\mathcal{S}$ is a decomposition of $R^3/\mathcal{S}$ into maximal vertical prisms[3] such that: (i) each prism does not intersect $\mathcal{S}$, (ii) the two bases of the prism are subsets of two polygons in $\mathcal{S}$.

Clearly, for every pair of polygons S_i and S_j, the value of A'_{ij} is obtained by summing the area of the projection of cylinders with bases in S_i and S_j. The cylindrical partition $H(\mathcal{S})$ has size $O(n + k_r)$ and is built in time $O(n \log^2 n + k_r \log n)$ where k_r is the number of *regular intersection points* [Mul91].

6.1 Overall algorithm (simple version)

We summarize here the whole algorithm to compute form factors (simpler version). C_{ij} is a counter for the pair of surfaces with indices $i < j$, which is initialized to 0. We organize such counters using an array indexed from 1 to n, holding pointers to dynamic binary search trees. The counters are at the leaves of the trees. Clearly we need to store only counters C_{ij} which, during the construction, are found to have a value different from 0. If a counter is not in the data structure its value is by default 0. In this way we can exploit the sparsity of the visibility structure and we avoid using quadratic storage if we do not need to.

In the external loop we select randomly and uniformly for $N = \lceil 1/\epsilon^2 \delta \rceil$ times a plane through the origin (i.e. a vertical direction). We compute the vertical cylindrical decomposition of Mulmuley for that choice of the vertical direction. For each prism whose bases are supported by S_i and S_j, we add the area of the projection of the prism to the counter C_{ij}.

When we want to compute the approximation of F_{ij} we access C_{ij} and we divide $2C_{ij}$ by NA_i. To approximate F_{ji} we divide by NA_j. We can find easily all non-zero approximate form factors by visiting the data structure. The size of the data structure is proportional to the number of non-zero approximations to form-factors.

---

[3] A prism here is a solid obtained by intersecting an infinite prism with two half-spaces.

7 Observations

(1) By choosing as value of the confidence parameter $n^2\delta$ we obtain that, with probability $1-\delta$, all the non-zero form factors have a simultaneous good approximation. The running time is increased only by a factor $O(\log n)$.

(2) Our algorithm is particularly suitable in scenes with large emitting surfaces (e.g. a window). Sometimes scenes have small sources of large radiant emission. In this case our algorithm might estimate to 0 the form factor between this small source and small distant visible objects. On the other hand the fact that the source is small causes no trouble in computing form factors between the source and large close visible surfaces.

(3) It is possible to customize the algorithm so to compute only the form factors involving a specific polygon S_i by restricting the computation to the vertical cylinder based on S_i. The running time in this case would depend on the average number of regular intersections within the projection of S_i. This is desirable if we want to place more stringent error conditions on a form factor involving a bright surface S_i.

(4) Unlike methods based on formula 1 our method does not have any divergence problem due to surfaces too close to each other.

(5) Our algorithm does not spend time in computing *zero* form-factors, thus it is particularly suited for scenes where the form factor matrix is sparse. Moreover, if a form factor is zero, its approximation is also zero.

(6) Implementing the algorithm for the case of two surfaces without occlusions should be easy. For the general problem, the difficult part is the computation of the vertical cylindrical decomposition. The 3-dimensional sweeping algorithm of Mulmuley, which gives us the best bound, is challenging to implement since it relies on the dynamic maintenance of planar maps [PT90]. If we are willing to relax the time bound we can instead project all the surfaces onto a plane and use as algorithmic skeleton the line sweep algorithm of Bentley and Ottman [BO79].

8 Acknowledgments

I wish to thank Nina Amenta and the Geometry Centre at the University of Minnesota, where this research was initiated.

References

[Arv94] J. Arvo. The irradiance jacobian for partially occluded polyhedral sources. *Computer Graphics*, pages 343–350, 1994. Proceedings of SIGGRAPH 94.

[BGS+66] N.P. Buslenko, D.I. Golenko, I.M. Sobol', V. G. Sragovič, and Ju. A. Šreider. *The Monte Carlo Method*. Pergamon Press, Oxford, 1966. Yu. A. Shrider (Ed.). Also *The method of statistical trials (The Monte Carlo method)*, Fizmatgiz, Moscow, 1962.

[BO79] J. L. Bentley and T. A. Ottmann. Algorithms for reporting and counting geometric intersections. *IEEE Trans. Comput.*, C-28:643–647, 1979.

[CG85] M.F. Cohen and D.P. Greenberg. The hemicube: a radiosity solution for complex environemnts. *Computer Graphics*, 19(3):31–40, 1985. Proceedings of SIGGRAPH '85.

[DR75] P.J. Davis and P. Rabinivitz. *Methods of numerical integration*. Academic Press, New York, 1975.

[EJLA91] A.F. Emery, O. Johansson, M. Lobo, and A. Abrous. A comparative study of methods for computing the diffuse radiation viewfactors for complex structures. *Journal of Heath Transfer*, 113:413–421, 1991.

[Fel68] W. Feller. *An introduction to probability theory and its applications*. John Wiley & Son, New York, 1968. Third edition.

[HSA91] P. Hanrahan, D. Salzman, and L. Aupperle. A rapid hierarchical radiosity algorithm. *Computer Graphics*, 25(4):197–206, 1991. Proceedings of SIGGRAPH '91.

[JVV86] M. Jerrum, L. Valiant, and V. Vazirani. Random generation of combinatorial structures from a uniform distribution. *Theoretical Computer Science*, 43:169–188, 1986.

[KKK+90] N. Karmarkar, R. Karp, R. Karp, L. Lovasz, and M. Luby. A monte-carlo algorithm for estimating the permanent. Technical Report TR-90-063, Interantional Computer Science Institute, Berkeley, 1990.

[LSG94] D. Lishinski, B. Smith, and D.P. Greenberg. Bounds and error estimates for radiosity. *Computer Graphics*, pages 67–74, 1994. Proceedings of SIGGRAPH 94.

[Mit91] D.P. Mitchell. Spectrally optimal sampling for distribution ray tracing. *Computer Graphics*, 25(4):157–162, 1991. Proceedings of SIGGRAPH '91.

[Mul91] K. Mulmuley. Hidden surface removal with respect to a moving point. In *Proc. 23rd Annu. ACM Sympos. Theory Comput.*, pages 512–522, 1991.

[PS85] F.P. Preparata and M.I. Shamos. *Computational Geometry: an Introduction*. Springer Verlag, 1985.

[PT90] F. P. Preparata and R. Tamassia. Dynamic planar point location with optimal query time. *Theoret. Comput. Sci.*, 74:95–114, 1990.

[San76] L. A. Santaló. *Integral Geometry and Geometric Probability*. Addison-Wesley, Reading, MA, 1976.

[Sbe93] M. Sbert. An integral geometry based method for fast form-factor computation. In R. J. Hubbold and R. Juan, editors, *Eurographics 93*, pages 409–420, 1993. Also in Computer Graphics Forum, vol 12, num 3, 1993.

[SH92] R. Siegel and J.R. Howell. *Thermal Radiation Heat Transfer*. Hemisphere Pu. Co., Washington, 1992. Third edition.

[SP89] F. Sillion and C. Puech. A general two-pass method integrating specular and diffuse reflection. *Computer Graphics*, 23(3):335–344, 1989. Proceedings of SIGGRAPH '89.

[Spe87] J. Spencer. *Ten lectures on the probabilistic method*, volume 52 of *CBMS-NSF Regional Conference series in Applied Mathematics*. SIAM, 1987.

[WEH89] J.R. Wallace, K. A. Elmquist, and E.A. Haines. A ray tracing algorithm for progressive radiosity. *Computer Graphics*, 23(3):3115–324, 1989. Proceedings of SIGGRAPH '89.

[WRC88] G.J. Ward, F.M. Rubinstein, and R.D. Clear. A ray tracing solution for diffuse interreflection. *Computer Graphics*, 22(4):85–92, 1988. Proceedings of SIGGRAPH '88.

Strategies for Polyhedral Surface Decomposition: An Experimental Study[*]

BERNARD CHAZELLE[†]　　DAVID P. DOBKIN[‡]　　NADIA SHOURABOURA[§]　　AYELLET TAL[¶]

1 Introduction

Convex shapes are easiest to represent, manipulate, and render: Even though they form the building blocks of bottom-up solid modelers, it is more often the case that the convex structure of a geometric shape is lost in its representation. We are then presented, not with the solid-modeling problem of putting together primitive convex objects, but with the reverse problem of *extracting* convexity out of a complex shape.

The classical example is that of cutting up a 3-polyhedron into convex pieces. This is often a useful, sometimes a required, preprocessing step in graphics, manufacturing, and mesh generation. The problem has been exhaustively researched in the last few years [1]—[17]. Despite its practical motivation, however, little of that research has gone beyond the theoretical stage. One possible explanation is that even the most naive solutions are programming challenges. We observe, however, that in practice one often need not partition the polyhedron itself but only its *boundary*. In other words, it often suffices to decompose a polyhedral surface into a small number of convex patches. By abuse of terminology, we call a surface *convex* if it lies entirely on the boundary of its convex hull. We mention some applications briefly.

In rendering, the coherence provided by convex patches can be exploited to speed up radiosity calculations. For example, from the closed-form expressions recently found for the form factors between two polygons [18], it is possible to derive faster iterative methods for handling multiple pairs of facets that are known to lie on one or a few convex patches. Similar speed-ups can be obtained for shading, clipping, hit detection, etc. Snyder et al. [19] point out the importance and difficulty of exploiting polyhedral coherence in collision detection: Because polyhedral hierarchies can be defined on arbitrary convex patches (solid polyhedra are not needed), intersection primitives can be greatly speeded up when convex surface decompositions are available [13].

Another application can be found in multimedia animations. Bitmap animations, such as *flics* and cell animations, are the simplest to play back, but they are tremendously wasteful of storage. They also make it hard to build in interactive user-response features. An alternative is to store background bitmaps and represent moving objects as wire frames [3]. Besides the savings in storage, the main advantage of this approach is to allow for real-time user input. Unfortunately this has an adverse effect on speed, not to mention the added development time associated with more complicated rendering and playback algorithms. We believe that convex surface decomposition techniques can alleviate this problem. Work is underway to test this assumption.

Besides simplicity, decomposing the boundary of a polyhedron into convex patches has other advantages. For example, while a polyhedral solid decomposition can suffer a quadratic blow-up [8], boundary decompositions are always linear in size. Better than that, the number of convex patches can be kept within a constant factor of the number of reflex angles [11]. We have gathered empirical evidence suggesting that even highly complex surfaces typically consist of only a handful of patches. For example, a standard drinking glass, regardless of its description size, might involve no more than a dozen convex patches. Intuitively, one should expect only surfaces of little coherence, such as crumpled, fractal-like sheets or heavily twisted surfaces, to give rise to many patches.

Contributions of this paper: We present a comparative study of simple heuristics for convex surface decomposition. The research reported here is mostly of an experimental nature. The standout exception is the obligatory first step: motivating the search

---

[*]Work by Bernard Chazelle, David Dobkin and Ayellet Tal has been supported in part by NSF Grant CCR-93-01254 and The Geometry Center, University of Minnesota, an STC funded by NSF, DOE, and Minnesota Technology, Inc. Work by Nadia Shouraboura has been supported in part by NSF Grant PHY-90-21984.

[†]Department of Computer Science, Princeton University

[‡]Department of Computer Science, Princeton University

[§]Program in Applied and Comput. Math., Princeton University

[¶]Department of Computer Science, Princeton University

for heuristics by proving that the problem is NP-complete. Because of space limitations we give only an outline in §2 and provide the details in the full paper. In §3 we describe the three classes of heuristics investigated: *space partitioning*, *space sweep*, and *flooding*. Within each class we examine several sub-heuristics and compare their relative effectiveness. Experimental results are reported and conclusions are drawn in §4.

Finding meaningful test data was an important component of this work. In the present case, randomly generated data is all but worthless. So, we collected several hundred real-life polyhedral models from manufacturing companies and industrial AutoCAD users. This catalog confirms our working hypothesis that most objects admit of small-size boundary decompositions. We implemented various decomposition schemes from each of the three classes of heuristics; each implementation was tested and benchmarked against a representative sample of objects from our library. From this experimentation it appears that flooding heuristics are the most efficient as well as the easiest to implement. We propose a scheme, called *flood-and-retract*, which seems the method of choice among the heuristics investigated.

Unsurprisingly, space partitioning techniques fared the worst. The main motivation for including them in our investigation was that many users are equipped with space partitioning software, so it was of practical relevance to assess their effectiveness. Space-sweep heuristics were also natural candidates. Their asymptotic performance is guaranteed to be linear [11], but even the simplest ones are quite difficult to implement.

An innovative aspect of this work has been the use of animations to guide our search for good heuristics. Runs and benchmarks produce numbers that tell us how good or how bad a given heuristic is. But they do not help us to design better heuristics. Visualizing the decompositions does. Physicians swear by (and live off) the adage that lab results cannot supplant clinical examination. Likewise, we found that nothing was more useful than the ability to look at a decomposition in three dimensions as though we held it in our hands. Through this means, weaknesses of the heuristics came to light and ways to overcome them suggested themselves. In addition to our in-house geometric animation system GASP, we also used the visualization package GEOMVIEW developed at the Geometry Center. After it became apparent that flooding heuristics were the most competitive, we animated several of them and visually analyzed their most obvious flaws. These animations enabled us to converge rapidly towards the most promising heuristic, ie, flood-and-retract.[1]

---

[1] To illustrate this process of iterative improvements we have produced a video: this supplement plays the same role as traditional figures but with motion and color added.

2 The Complexity of Convex Surface Decomposition

Let S be a polyhedral surface with n vertices, and let $S_1, \ldots, S_k$ be disjoint convex patches whose union gives S. We show that minimizing the number k is NP-complete. In our terminology, a polyhedral surface is a compact piecewise-linear 2-manifold with boundary. We make no assumption on its orientability or its Euler characteristic. Obviously, convex patches are always orientable, but again their Euler characteristics are let unrestricted; in other words, the patches may be multi-connected. It is natural, however, to require that convex patches be connected.

Note that this is only one of many possible variants of the problem: for example, we could place bounds on the Euler characteristics of the patches; we could require that facets not be split; we could seek a cover and not a partition; we could relax the connectivity requirement; in the case of orientable surfaces, we could distinguish between convexity and concavity, etc.

2.1 Membership in NP

It is easy to see that the problem is a special case of clique partitioning. To begin with, observe that the facets of any patch can always be assumed to originate from the three-dimensional arrangement formed by the planes defined by all triplets of vertices of S. Indeed, for any decomposition that does not satisfy this requirement, we can always extend any facet (if necessary) until its bounding edges lie in one of those planes. This gives us up to $O(n^9)$ candidate facets from which a minimum decomposition can be guessed.

To test if a set of facets forms a convex patch, we compute its convex hull and verify that all the facets lie on the boundary. An equivalent certificate can be formed by expressing the fact that for each pair of facets, f, g, all the vertices of f lie on the same side of the plane supporting g, and vice-versa. Let G be the graph whose nodes represent the facets, and whose edges connect pairs of nodes whose corresponding facets satisfy the certificate in question. Let H be the graph with the same node-set, but whose edges connect nodes with adjacent associated facets. A minimum decomposition is simply a min-size partition of the node-set of G into cliques whose induced subgraphs in H are connected. Checking whether a given decomposition size k is achievable can thus be done nondeterministically in polynomial time.

2.2 NP-Hardness

Let $x_1, \ldots, x_n$ be n Boolean variables; an instance of SAT consists of n clauses $C_1, \ldots, C_n$, each of them a disjunction of literals x_i or $\bar{x}_i$. Here is a brief overview of our geometric model: Each variable x_k is associated with a closed polygonal curve L_k zigzagging in planes parallel to xy. There are exactly two optimal ways of cutting L_k into convex pieces: each of them corresponds to a different truth assignment of x_k. A clause is modeled by a vertical line segment that connects the curves corresponding to its literals. The contact between the vertical segment and L_k takes place at a local y-maximum or y-minimum depending on whether x_k is negated or not in the clause in question. The remainder of the proof involves transforming the curves into thin strips and then arguing that the formula is satisfiable if and only if the minimum decomposition is below a certain size.

3 Three Classes of Heuristics

All the surfaces in our library are orientable. The reason is that they usually originate from the boundary of some polyhedron. This suggested distinguishing between convex and concave patches; the latter being convex patches (in the old sense) all of whose edges exhibit reflex angles. We shall specify below which sub-heuristics make this distinction.

Space Partitioning. The strategy is to use binary space partitioning [14] to split up the surface into convex patches. Recall that the method builds a tree by recursively dividing space by a (well-chosen) cutting plane. Each node v of the tree is associated with a convex polyhedron P_v. The idea is to explore the two children of v if and only if the portion of the surface within P_v is not convex. The advantage of this method is that space partitioning code is widely available, so implementing it is by and large effortless. One drawback is that facets get split up in the middle: this produces Steiner points, which cause roundoff errors and can be undesirable. Furthermore the efficiency of the heuristic is highly sensitive to the input surface. An obvious improvement we use is to cut along reflex edges only. In the worst case the number of patches is quadratic. In practice it appears that such a blowup is unlikely, especially in view of our edge selection rule.

Space Sweep. We implemented a simplified version of the method in [11]. In particular, we modified the algorithm to prevent the creation of Steiner points. Although the linearity of the output size is no longer (theoretically) guaranteed, the simplification seems to have no adverse effect; on the other hand, it makes coding much easier. The idea is to sweep space with a plane: at any given time, the cross-section of the surface consists of simple polygonal curves which are decomposed into convex pieces. The heuristic attempts to maintain each curve as long as possible while moving the plane, thus producing convex patches in the process. Each time a convexity violation is found, we relent from including the violating facet and start up a new convex curve (and hence, a new patch).

Flooding. Let H be the dual graph of the surface, where nodes represent facets and arcs join nodes associated with adjacent facets. The class of *flooding* heuristics refers to the incremental strategy of starting from some node and traversing the graph H, collecting facets along the way as long as they form a convex patch. We distinguish between two sub-heuristics: *Greedy flooding* involves collecting facets until no adjacent facets can be found that does not violate the convexity of the current patch. A new patch must then be started, at which point the traversal can resume. *Controlled flooding* includes other stopping rules besides convexity violation. Our animations have revealed cases where flooding in a greedy fashion is arbitrarily bad: off the optimal by a factor proportional to the input size. On the other hand, producing smaller patches by controlled flooding can sometimes lead to near-optimal decompositions.

Given a convex patch, consider adding a new facet to it. The new patch may fail to be convex for one of two reasons:

1. **Local failure:** The edge at which the facet is attached to the patch exhibits non-convexity.

2. **Global failure:** The new patch is locally convex everywhere, but some facet fails to be on the boundary the new convex hull.

Global failures are rare in practice. Geometrically, they are associated with twisted shapes, as in a spiral or a drill. Note that global failures are the only reason the surface decomposition problem is hard. Without global failures, *any* greedy flooding heuristic produces an optimal decomposition (in the version of the problem where concave patches are ruled out). Furthermore, covers and partitions are indistinguishable. In other words, without global failures, trying to cover the surface into convex patches instead of partitioning it cannot produce fewer pieces. In the presence of global failures, however, just the opposite is true. This suggests producing controlled-flooding partitions in two steps. First, flood the surface by covers: this means that when restarting a new patch, allow the traversal of old facets as well as new ones. Then, in a second pass, transform the covers into partition. We pursue this approach below.

Flood-and-Retract. The first phase, flooding the surface, produces convex patches which might overlap. The purpose of the second phase is to remove the overlapping by "retracting" each patch. The main pitfall to avoid is breaking up the patch into several connected components. In Figure 1, for example, two patches cover the surface and can be easily modified into a 2-patch partition. However, a naive approach might end up producing up to $\Omega(n)$ convex patches. Plate 3 gives a three-dimensional picture of this phenomenon. To avoid this trap, we adopt the following strategy: for each patch in turn, retract it as much as possible as long as it remains connected. The retraction takes place along the portions of the boundary of the patch that lie inside other patches. The data structure is quite simple: we keep a queue of facets to be retracted. We iterate on the following process: remove the top of the queue and remove the facet from the patch, unless it disconnects it. After the removal, check which adjacent facets should be inserted into the queue, and insert them in arbitrary order. Once each patch has been retracted in this manner, transform the resulting cover into a partition in arbitrary manner.

4 Experimental Results

The tables below present the results of our experiments. We implemented 8 heuristics and ran them on the objects of the library. For the sake of concreteness, we limit our discussion to a sample of 12 representative objects. See Plates 1,2. Next to each object, we indicate the number of vertices V, the number of *convex* edges and *reflex* edges. Some edges are incident to coplanar facets: their count is *par*. Such edges originate in one of two ways: either the dihedral angle was too close to π to decide on convexity, or the edge was introduced to triangulate a facet, in which case the angle was exactly π. The heuristics chosen were:

- **BSP:** We adapted existing Binary Space Partitioning code by pruning the tree as explained earlier. One difficulty was to handle the facets that lay within cutting planes. For this reason we found it easiest to allow disconnected convex patches. No distinction is made between convex and concave patches.

- **Space-Sweep:** Both convex and concave patches are admissible in *sweep_cc*. On the other hand only convex patches are allowed in *sweep_c*.

- **Flooding:** We consider greedy flooding produced by breadth-first search with convex-concave patches (*bfs_cc*) and convex patches only (*bfs_c*). We also present results on depth-first search with convex-concave patches (*dfs_cc*) and depth-first search with convex patches only (*dfs_c*). Plates 1,2 show the *bfs_cc* decompositions of the 12 sample objects.

	name	V	convex	reflex	par	bsp
1.	epcot	195	384	192	0	344
2.	mushroom	227	365	142	165	567
3.	glass	81	124	34	49	98
4.	pawn	155	216	96	144	303
5.	bishop	251	352	118	274	671
6.	rook	131	152	80	152	381
7.	spring	910	1626	875	220	2144
8.	flashlight	388	549	145	415	532
9.	eyeball	966	1683	24	934	642
10.	torso	321	619	234	162	547
11.	book	88	118	34	88	131
12.	hand	309	525	234	162	931

Table 1.

	name	V	sweep_cc	sweep_c
1.	epcot	195	129	129
2.	mushroom	227	66	64
3.	glass	81	21	21
4.	pawn	155	41	40
5.	bishop	251	39	33
6.	rook	131	34	31
7.	spring	910	389	420
8.	flashlight	388	62	78
9.	eyeball	966	45	27
10.	torso	321	99	89
11.	book	88	48	57
12.	hand	309	146	91

Table 2

	name	V	bfs_cc	dfs_cc	bfs_c	dfs_c
1.	epcot	195	129	129	129	129
2.	mushroom	227	56	48	56	49
3.	glass	81	12	12	13	12
4.	pawn	155	28	28	31	29
5.	bishop	251	20	19	20	19
6.	rook	131	18	18	18	18
7.	spring	910	277	247	256	234
8.	flashlight	388	18	18	35	35
9.	eyeball	966	4	4	4	4
10.	torso	321	69	67	70	69
11.	book	88	13	13	12	12
12.	hand	309	89	86	93	93

Table 3

There is a remarkable consistency in the way flooding outperforms its competitors. Binary space partitioning fares so poorly it probably is worse than doing nothing. Space sweep, on the other hand, provides flooding a fairly close race. On *epcot* it produces the optimal decomposition, as does any kind of flooding. But this is an exception. On average space sweep loses to flooding by at least 50% and sometimes much more. On *book* it loses by a factor of 4. In general, the differences between convex-only and convex-concave are surprisingly small. Note that because of the nondeterministic nature of these heuristics, convex-concave does not always outperform convex-only.

Flooding performs uniformly well. The differences between depth-first search and breadth-first search are small enough to be negligible, although DFS appears to have a slight edge. Note that visually it might seem that an inordinately high number of patches are produced: but recall that from a computational standpoint, coplanarity is an elusive property. We chose to make the program err on the side of robustness. This means that the patches produced are guaranteed to be convex, but it could be that near-flat convex edges are classified as reflex because of roundoff errors. Flood-and-retract is not included in the tables because the results essentially do not differ from breadth-first-search, except on a few objects, eg, *spring*. The reason is that the retract part of the heuristic "kicks in" only when global failure is detected, something which happens rarely.

Plate 3 shows an object, a bracelet, where greedy flooding (and all the previous heuristics for that matter) performs very poorly. Flood-and-retract avoids the obvious pitfalls and provides a near-optimal decomposition. The upper-left figure displays the input surface. BFS and DFS decompositions are shown right below. Note that both produce an additional (unnecessary) patch for half the fingers of the bracelet. The right-hand side figures show flood-and-retract in action. First, the algorithm floods the bracelet with a BFS run (dark patch). Then it repeats the same operation starting outside that patch. Since the dark patch is ignored, we end up with two overlapping patches. Finally the retraction moves back one of the patches to produce the final two-patch partition of the bracelet.

It would be a mistake to think this is an isolated case. The strategy of covering first and then retracting is meant to deal with a general shortcoming of greedy flooding, ie, *fragmentation*. This phenomenon is similar to what happens when partitioning a tree with fixed-size subtrees. If we start at the root and proceed top-down we run into the risk of ending up with a linear number of fragments near the leaves, each requiring a distinct tree. Of course, in that case, we know better not to start at the root but instead to proceed bottom-up. A similar phenomenon occurs with flooding. The difference, however, is that unlike tree partitioning flooding has no natural starting place. The idea of covering first and retracting later is meant to overcome that intrinsic difficulty by trying to make the starting place immaterial.

5 Conclusions

This paper has investigated the practical issues behind convex surface decomposition, a problem shown to be NP-complete. We have not addressed issues of implementation, robustness, and computation time. Robustness is (as always) a thorny problem. Computation time, however, is not a serious bottleneck, because all our heuristics run either in $O(n \log n)$ time or in time linear in the input/output size.

Our conclusion is that the flood-and-retract heuristic should be the method of choice in practice. Of course, as we observed, in many cases the covering produced in the first phase of the algorithm will actually be a partition (in which case flood-and-retract reduces to greedy flooding). In such cases, the decomposition is optimal or near-optimal and there is no need to pursue the heuristic further, ie, DFS does the job. It remains to be seen if more complex variants can produce significantly better decompositions.

Acknowledgments: We thank Herb Voelcker and Jovan Zagajac of Cornell University for kindly making their BSP code available to us, and generally bringing us the benefit of an engineering perspective on the issues discussed in this paper.

References

[1] Aronov, B., Sharir, M. *Triangles in space or building (and analyzing) castles in the air*, Combinatorica, 10 (1990), 137–173.

[2] Aronov, B., Sharir, M. *Castles in the air revisited*, Proc. 8th Ann. ACM Symp. Comput. Geom. (1992), 146–156. To appear in Disc. Comput. Geom.

[3] *AutoCAD Reference Manual*, Autodesk, Inc. Publication 100625 (1992).

[4] Bajaj, C.L., Dey, T.K. *Convex decompositions of polyhedra and robustness*, SIAM J. Comput., 21 (1992), 339–364.

[5] Bern, M. *Compatible tetrahedralizations*, Proc. 9th Ann. ACM Symp. Comput. Geom. (1993), 281–288.

[6] Bern, M., Eppstein, D. *Mesh generation and optimal triangulation*, in: Computing in Euclidean Geometry, 1, World Scientific, ed. D. Z. Du and F. K. Hwang (1992), 23–90.

[7] Bern, M., Eppstein, D., Gilbert, J. *Provably good mesh generation*, Proc. 31st Ann. IEEE Symp. Foundat. Comput. Sci. (1990), 231–241.

[8] Chazelle, B. *Convex partitions of polyhedra: a lower bound and worst-case optimal algorithm*, SIAM J. Comput., 13 (1984), 488–507.

[9] Chazelle, B., Palios, L. *Triangulating a nonconvex polytope*, Disc. Comput. Geom., 5 (1990), 505–526.

[10] Chazelle, B., Palios, L. *Decomposition algorithms in geometry*, in Algebraic Geometry and its Applications, C. Bajaj, Ed., Chap.27, Springer-Verlag, 1994, pp. 419–447.

[11] Chazelle, B., Palios, L. *Decomposing the boundary of a nonconvex polytope*, Proc. 3rd Scandinavian Workshop on Algorithm Theory (1992), 364–375.

[12] Chazelle, B., Shouraboura, N. *Bounds on the size of tetrahedralizations*, Proc. 10th Ann. ACM Symp. Comput. Geom. (1994), 231–239.

[13] Dobkin, D.P., Kirkpatrick, D.G. *Fast detection of polyhedral intersection*, Theoret. Comput. Sci., 27 (1983), 241–253.

[14] Fuchs, H., Kedem, Z.M., Naylor, B. *On visible surface generation by a priori tree structures*, Proc. SIGGRAPH '80, Comput. Graph. 14 (1980), 124–133.

[15] O'Rourke, J. *Art Gallery Theorems and Algorithms*, Oxford Univ. Press, New York, NY (1987).

[16] Mitchell, S., Vavasis, S. *Quality mesh generation in three dimensions*, Proc. 8th Ann. ACM Symp. Comput. Geom. (1992), 212–221.

[17] Ruppert, J., Seidel, R. *On the difficulty of triangulating three-dimensional non-convex polyhedra*, Disc. Comput. Geom., 7 (1992), 227–253.

[18] Schröder, P., Hanrahan, P. *On the form factor between two polygons*, Proc. SIGGRAPH 93, ACM Press (1993), 163–164.

[19] Snyder, J.M., Woodbury, A.R., Fleischer, K., Currin, B., Barr, A.H. *Interval methods for multipoint collisions between time-dependent curved surfaces*, 321–334.

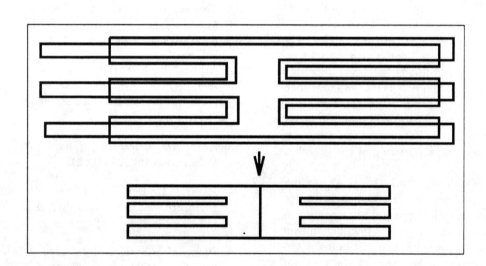

Figure 1

Plate 1

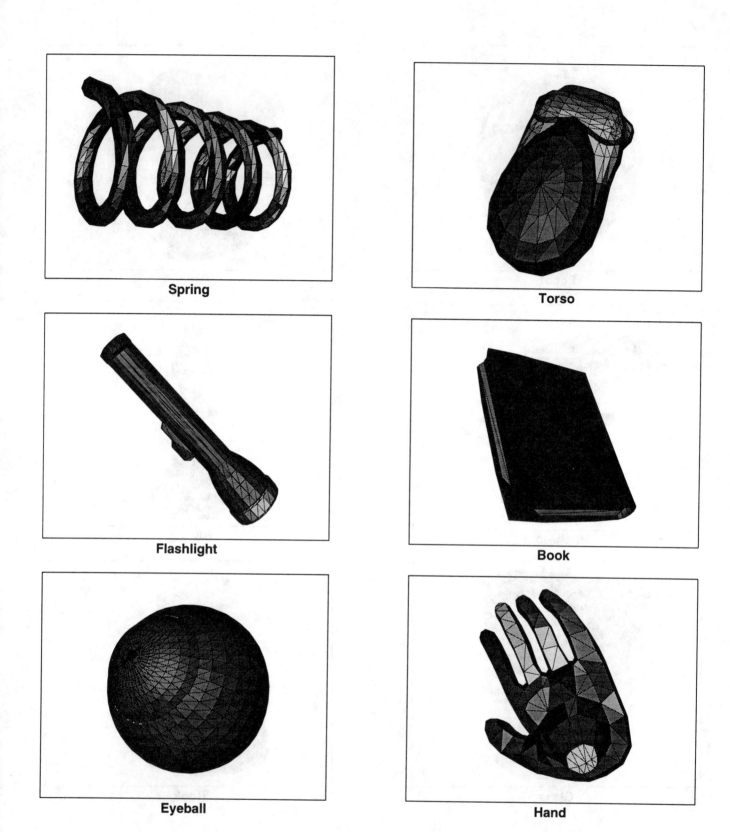

Plate 2

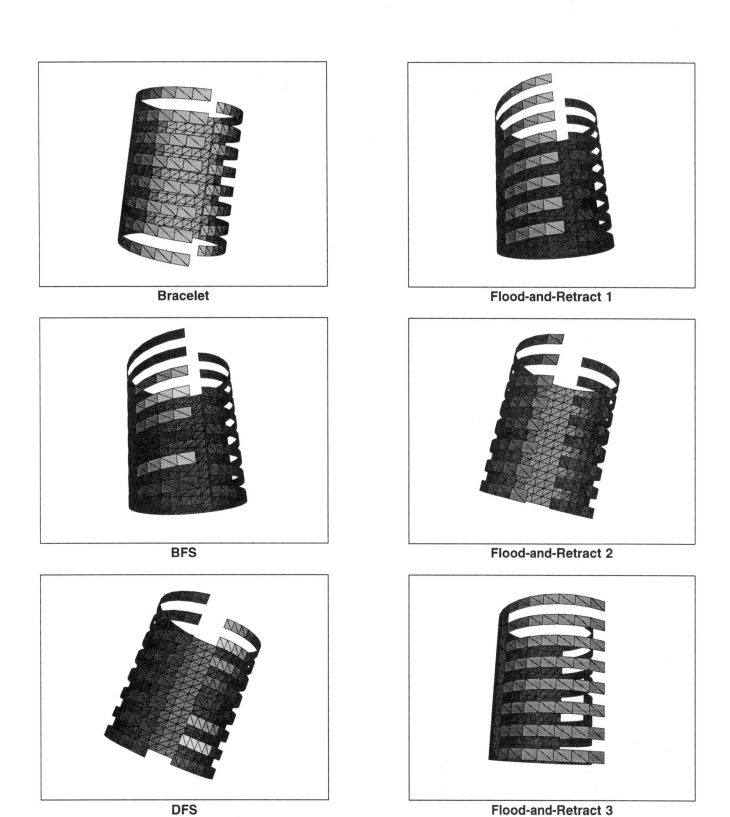

Plate 3

An Experimental Comparison of Three Graph Drawing Algorithms*

(Extended Abstract)

Giuseppe Di Battista[†]
dibattista@iasi.rm.cnr.it

Ashim Garg[‡]
ag@cs.brown.edu

Giuseppe Liotta[§]
liotta@dis.uniroma1.it

Roberto Tamassia[‡]
rt@cs.brown.edu

Emanuele Tassinari[§]
tassinar@dis.uniroma1.it

Francesco Vargiu[§]
vargiu@dis.uniroma1.it

[†] D. I. F. A.
Univ. della Basilicata
85100 Potenza
Italy

[‡] Dept. of Computer Science
Brown University
Providence, RI 02912–1910
USA

[§] Dip. Informatica e Sistemistica
Univ. di Roma "La Sapienza"
00198 Roma
Italy

Abstract

In this paper we present an extensive experimental study comparing three general-purpose graph drawing algorithms. The three algorithms take as input general graphs (with no restrictions whatsoever on the connectivity, planarity, etc.) and construct orthogonal grid drawings, which are widely used in software and database visualization applications. The test data (available by anonymous ftp) are 11,582 graphs, ranging from 10 to 100 vertices, which have been generated from a core set of 112 graphs used in "real-life" software engineering and database applications. The experiments provide a detailed quantitative evaluation of the performance of the three algorithms, and show that they exhibit trade-offs between "aesthetic" properties (e.g., crossings, bends, edge length) and running time. The observed practical behavior of the algorithms is consistent with their theoretical properties.

---

*Research supported in part by the US National Science Foundation, by the US Army Research Office, by the US Office of Naval Research and the Advanced Research Projects Agency, by the NATO Scientific Affairs Division, by the "Progetto Finalizzato Sistemi Informatici e Calcolo Parallelo (Sottoprogetto 6, Infokit)" and Grant 94.23.CT07 of the Italian National Research Council (CNR), and by the ESPRIT II Basic Research Actions Program of the European Community (project ALgorithms and Complexity).

Permission to copy without fee all or part of this material is granted provided that the copies are not made or distributed for direct commercial advantage, the ACM copyright notice and the title of the publication and its date appear, and notice is given that copying is by permission of the Association of Computing Machinery. To copy otherwise, or to republish, requires a fee and/or specific permission.
11th Computational Geometry, Vancouver, B.C. Canada
© 1995 ACM 0-89791-724-3/95/0006...$3.50

1 Introduction

Graph drawing algorithms construct geometric representations of abstract graphs and networks. Because of the direct applications of graph drawing to advanced graphic user interfaces and visualization systems, and thanks to the many theoretical challenges posed by the interplay of graph theory and geometry, an extensive literature on the subject [12, 46] has grown in the last decade.

Various graphic standards have been proposed for the representation of graphs in the plane. Usually, vertices are represented by points or simple geometric figures (e.g., rectangles, circles), and each edge (u,v) is represented by a simple open Jordan curve joining the points associated with the vertices u and v. A drawing is *planar* if no two edges cross. A graph is planar if it admits a planar drawing. An *orthogonal* drawing maps each edge into a chain of horizontal and vertical segments (see Figures 2–3). A *grid* drawing is embedded in a rectilinear grid such that the vertices and bends of the edges have integer coordinates. Orthogonal drawings are widely used for graph visualization in many applications, including database systems (Entity-Relationship diagrams), software engineering (Data-Flow diagrams), and circuit design (circuit schematics).

1.1 Previous Experimental Work in graph Drawing

Many graph drawing algorithms have been implemented and used in practical applications. Most papers in this area show sample outputs, and some also provide limited experimental results on small (with fewer than 100 graphs) test suites (see, e.g., [9, 18, 20, 24, 25, 28] and the experimental papers in [46]). However, in order to evaluate the practical performance of a graph drawing algorithm in visualization applications, it is essential to perform extensive experimentations with input graphs derived from the application domain.

The performance of four planar straight-line drawing algorithms [6, 7, 10, 38, 48] is compared in [23]. These algorithms have been implemented and tested on 10,000 randomly generated maximal planar graphs. The standard deviations in angle size, edge length, and face area are used to compare the quality of the planar straight-line drawings produced. Since the experiments are limited to randomly generated maximal planar graphs, this work gives only partial insight on the performance of the algorithms on general planar graphs.

Himsolt [21] presents a comparative study of twelve graph drawings algorithms, including [6, 10, 18, 26, 41, 42, 50, 52]. The algorithms selected are based on various approaches (e.g., force-directed, layering, and planarization) and use a variety of graphic standards (e.g., orthogonal, straight-line, polyline). Only three algorithms draw general graphs, while the others are specialized for trees, planar graphs, Petri nets, and graph grammars. The experiments are conducted with the graph drawing system GraphEd [22]. Many examples of drawings constructed by the algorithms are shown, and various objective and subjective evaluations on the aesthetic quality of the drawings produced are given. However, statistics are provided only on the edge length, and details on the experimental setting (e.g., provenance, composition, and size of the test suite) are not provided. The charts on the edge length have marked oscillations, suggesting that the number of test graphs is not very large. This work provides an excellent overview and comparison of the main features of some popular drawing algorithms. However, it does not give detailed statistical results on their performance.

1.2 Our Results

In this paper we present an extensive experimental study comparing three general-purpose graph drawing algorithms. The three algorithms take as input general graphs (with no restrictions whatsoever on the connectivity, planarity, etc.) and construct orthogonal grid drawings. The test data (available by anonymous ftp) are 11,582 graphs, ranging from 10 to 100 vertices, which have been generated from a core set of 112 graphs used in "real-life" software engineering and database applications. The experiments provide a detailed quantitative evaluation of the performance of the three algorithms, and show that they exhibit trade-offs between "aesthetic" properties (e.g., crossings, bends, edge length) and running time. The observed practical behavior of the algorithms is consistent with their theoretical properties.

The contributions of this work can be summarized as follows:

- We have generated a large test suite of graphs derived from "real-life" visualization applications. We believe that it will be useful to other researchers interested in experimental graph drawing.

- We have implemented three algorithms with solid theoretical foundations that construct orthogonal grid drawings of arbitrary input graphs.

- We have presented the first extensive experimental study of general-purpose graph drawing algorithms, with test graphs derived from software engineering and database applications.

1.3 Organization of the Paper

The rest of this paper is organized as follows: In Section 2, we overview the drawing system used for our study. The three drawing algorithms analyzed are described in Section 3. Details on the experimental setting are give in Section 4. In Section 5, we summarize our experimental results in nine charts and perform a comparative analysis of the performance of the algorithms. Finally, open problems are addressed in Section 6.

2 Diagram Server

Our experimental study was conducted using *Diagram Server* [13], a network server for client-applications that use diagrams (drawings of graphs). Diagram Server offers to its clients an extensive set of facilities to represent and manage diagrams through a multiwindowing environment. One of the most important facilities is a library of sophisticated automatic graph drawing algorithms. Diagram Server can be fully customized according to different application contexts and graphic environments. The main features of Diagram Server are listed below.

Independence from the client-applications: Because of the client-server interaction between Diagram Server and each one of its clients, both the development of the clients and the customization of Diagram Server do not require any knowledge of the internal structure of Diagram Server. The clients can specify the visual appearance of symbols and connections, the graphic standard, the aesthetic requirements that diagrams should satisfy, and the whole interface configuration.

Automatic graph drawing: Diagram Server has an automatic graph drawing facility [4] that is based on a large modular library of graph drawing algorithms and on a tool that, given the requirements of an application, selects the suitable algorithms for such requirements. Both the set of algorithms and the set of requirements can be easily extended or customized, and the result is an automatic graph drawing facility with high flexibility and extensibility, that combines the efficiency of special purpose algorithms with the variety of the possible application contexts.

A graph drawing algorithm is fully specified in the automatic graph drawing facility by an *algorithmic path*, which describes the sequence of steps and intermediate representations (e.g., planar embedding, orthogonal shape, visibility representation) produced by the algorithm.

Multiple representations: A graph can be represented by several diagrams, and different portions of a diagram can be displayed on different windows. The consistency between different representations of the same object is automatically maintained by Diagram Server.

Multi-user environment: Diagram Server supports diagram visualization activities that are shared among networked client-applications. Each client takes user input and submits requests to the server. In this way, the user gains access to capabilities not usually available locally on its desktop computer, such as an effective usage of a huge library of automatic graph drawing algorithms and the possibility of handling large diagrams.

Message-passing protocol: Diagram Server communicates to the client each action that the final user performs on a diagram through the user interface. All the exchanged data are enclosed into messages. This implies that no common memory is shared by the applications and that the system behavior is independent from the operating system and the hardware environment.

Diagram Server has been implemented and tested on a wide variety of client-applications including information systems design, project management and reverse software engineering. Also, Diagram Server has been successfully used to test new graph drawing algorithms and to evaluate their performance with respect to several aesthetic criteria.

3 The Drawing Algorithms Under Evaluation

The three drawing algorithms considered in this paper, denoted Bend-Stretch, Column, and GIOTTO, take as input general graphs (with no restrictions whatsoever on connectivity, planarity, etc.) and construct orthogonal drawings.

Algorithms Bend-Stretch and GIOTTO are based on a general approach where the drawing is incrementally specified in three phases (see Fig. 1): The first phase, *planarization*, determines the topology of the drawing. The second phase, *orthogonalization*, computes an orthogonal shape for the drawing. The third phase, *compaction*, produces the final drawing. This approach allows homogeneous treatment of a wide range of diagrammatic representations, aesthetics and constraints (see, e.g., [28, 43, 47]) and has been successfully used in industrial tools.

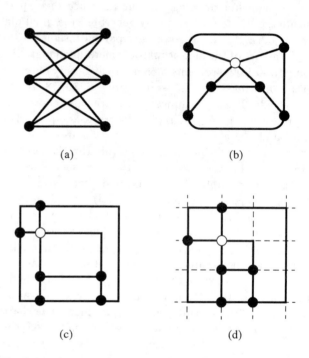

Figure 1: A general strategy for orthogonal grid drawings. (a) Input graph. (b) Planarization. (c) Orthogonalization. (d) Compaction.

The main difference between the two algorithms is in the orthogonalization phase: Algorithm GIOTTO uses a network-flow method that guarantees the minimum number of bends but has quadratic time-complexity [42]. Algorithm Bend-Stretch adopts the "bend-stretching" heuristic [45] that only guarantees a constant number of bends on each edge but runs in linear time. More specifically, algorithms GIOTTO and Bend-Stretch are described by the following two algorithmic paths of the automatic graph drawing facility of Diagram Server:

```
GIOTTO:
{ multigraph
  { MakeConnected connected
    { MakePlanar connectedplanar
      { Make4planar fourplanar
        { Step3giotto orthogonal
          { Step4giotto manhattan }}}}}}
```

Bend-Stretch:
```
{ multigraph
  { MakeConnected connected
    { MakeBiconnected biconnected
      { MakePlanar biconnectedplanar
        { Step1TaTo89 visibilityrepresentation
          { Step2TaTo89 orthogonal
            { Step4TaTo89 orthogonal
              {Step4Giotto manhattan}}}}}}}}
```

In the above notation, upper-case identifiers denote algorithmic components, and lower-case identifiers denote intermediate representations (such as graphs and drawings), as shown in Table 1.

Algorithm Column, is an extension of the orthogonal drawing algorithm by Biedl and Kant [5] to graphs of arbitrary vertex degree. More specifically, algorithm Column is described by the following algorithmic path of the automatic graph drawing facility of Diagram Server:

Column:
```
{ multigraph
  { MakeConnected connected
    { MakeBiconnected biconnected
      { BiedlKant manhattan }}}}
```

Note that the connectivity and biconnectivity augmentation steps were introduced in order to use algorithms designed for connected and bicconneceted graphs, respectively. The augmentation edges are not displayed in the final drawing.

Examples of "typical" drawings generated by Bend-Stretch, Column, and GIOTTO are shown in Figures 2–3. The drawings are all orthogonal grid drawings and have been differently scaled to fit in the page.

Let N be the number of vertices of the input graph, and C be the number of crossings in the drawing constructed. The worst-case asymptotic time complexity is $O(N+C)$ for algorithms Bend-Stretch and Column, and $O((N+C)^2 \log(N+C))$ for algorithm GIOTTO. Note that it is NP-hard to compute the minimum number of crossings, so that all the three algorithms heuristically attempt at reducing the number of crossings. However, algorithm GIOTTO guarantees to construct a planar drawing if the input graph is planar.

In addition to the three algorithms we have implemented and compared (Bend-Stretch, Column, and GIOTTO), several other orthogonal drawing algorithms (mostly for planar graphs) have been reported in the literature. Because of the techniques used, we expect the behavior of the algorithms in [14, 16, 17, 11, 27, 33, 34, 35, 32, 40] to be similar to that of one of Bend-Stretch, Column, or GIOTTO.

It would be interesting to perform further experiments on the practical performance of the recent algorithm by Papakostas and Tollis [37], and of separator-based methods [15, 31, 49] that were originally developed for VLSI layout.

multigraph	general multigraphs accepted as input
MakeConnected	connectivity testing and augmentation
connected	connected graph
MakeBiconnected	biconnectivity testing and augmentation
biconnected	biconnected graph
MakePlanar	planarization phase [3, 36]
connectedplanar	crossing are now replaced by dummy vertices
Make4planar	expansion of vertices with degree greater than 4 into rectangular symbols
fourplanar	viewing the rectangular symbols as cycles of dummy vertices, the graph has now maximum degree 4
Step3giotto	orthogonalization phase [42]
step1TaTo89	construction of a visibility representation [39, 44, 51])
step2TaTo89	fast orthogonalization [45]
step4TaTo89	bend-stretching transformations, which remove bends by local layout modifications [45]
orthogonal	orthogonal representation, describing shape of the drawing in term of its angles
Step4giotto	compaction phase [3]
BiedlKant	orthogonal drawing algorithm [5]
manhattan	the final output is an orthogonal grid drawing

Table 1: Algorithmic components and intermediate representations used by algorithms Bend-Stretch, Column, and GIOTTO in the automatic graph drawing facility of Diagram Server.

4 Experimental Setting

4.1 Quality Measures Analyzed

The following quality measures of a drawing of a graph have been considered:

Area: area of the smallest rectangle with horizontal and vertical sides covering the drawing;

Cross: total number of crossings;

TotalBends: total number of bends;

TotalEdgeLen: total edge length;

MaxEdgeBends: maximum number of bends on any edge;

MaxEdgeLen: maximum length of any edge;

UnifBends: standard deviation of the number of bends on the edges;

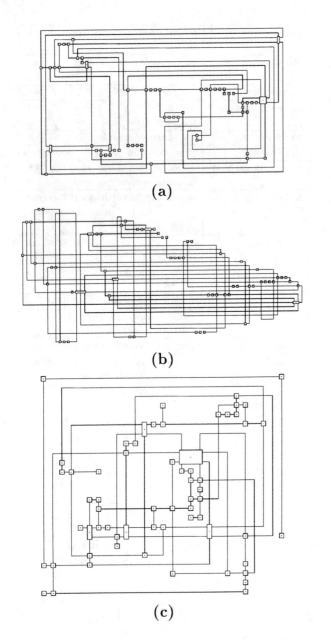

Figure 2: Drawings of the same 63-vertex graph produced by algorithms (a) Bend-Stretch, (b) Column, and (c) GIOTTO.

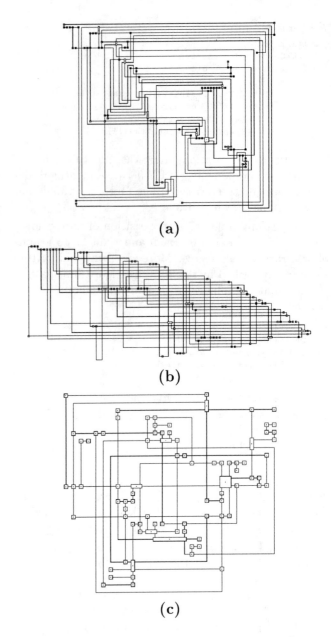

Figure 3: Drawings of the same 85-vertex graph produced by algorithms (a) Bend-Stretch, (b) Column, and (c) GIOTTO.

UnifLen: standard deviation of the edge length;

ScreenRatio: deviation from the optimal aspect ratio, computed as the difference between the width/height ratio of the the best of the two possible orientations (portrait and landscape) of the drawing and the standard 4/3 ratio of a computer screen.

It is widely accepted (see, e.g., [12]) that small values of the above measures are related to to perceived aesthetic appeal and visual effectiveness of the drawing.

4.2 Generation of the Test Graphs

Since we are interested in evaluating the performance of graph drawing algorithms in practical applications, we have disregarded approaches completely based on random graphs.

Our test graph generation strategy is as follows. First, we have focused on the important application area of database and software visualization, where Entity-Relationship diagrams and Data-Flow diagrams are usually displayed with orthogonal drawings.

Second, we have collected 112 "real life" graphs with number of vertices between 10 and 100, from now on called *core graphs*, from the following sources:

- 54% of the graphs have been obtained from major Italian software companies (especially from *Database Informatica*) and large government organization (including the Italian Internal Revenue Service and the Italian National Advisory Council for Computer Applications in the Government (*Autorità per l'Informatica nella Pubblica Amministrazione*)).

- 33% of the graphs were taken from well-known reference books in software engineering [19] and database design [1], and from journal articles on software visualization in the recent issues of *Information Systems* and the *IEEE Transactions on Software Engineering*.

- 13% of the graphs were extracted from theses in software and database visualization written by students at the University of Rome "La Sapienza".

Third, we have generated the 11,582 test graphs as variations of the core graphs. This step is the most critical, since we needed to devise a method for generating graphs "similar" to the core graphs.

Our approach is based on the following scheme. We defined several primitive operations for updating graphs, which correspond to the typical operations performed by designers of Entity-Relationship and Data-Flow Diagrams, and attributed a certain probability to each of them. More specifically, the updating primitives we have used are the following: *InsertEdge*, which inserts a new edge between two existing vertices; *DeleteEdge*, which deletes an existing edge; *InsertVertex*, which splits an existing edge into two edges by inserting a new vertex; *DeleteVertex*, which deletes a vertex and all its incident edges; and *MakeVertex*, which creates a new vertex and connects it to a subset of vertices.

The test graphs were then generated in several iterations starting from the core graphs by applying random sequences of operations with a "genetic" mechanism. Namely, at each iteration a new set of test graphs was obtained by applying a random sequence of operations to the current test set. Each new graph was then evaluated for "suitability", and those found not suitable were discarded. The probability of each primitive operation was varied at the end of each iteration.

The evaluation of the suitability of the generated graphs was conducted using both objective and subjective analyses. The objective analysis consisted of determining whether the new graph had similar structural properties with respect to the core graph it was derived from. We have taken into account parameters like the average ratio between number of vertices and number of edges and the average number of biconnected components. The subjective analysis consisted in a visual inspection of the new graph and an assessment by expert users of Entity-Relationship and Data-Flow diagrams of its similarity to a "real-life" diagram. For obvious reasons, the subjective analysis has been done on a randomly selected subset of the graphs.

Figure 4 depicts the distribution of the test graphs with respect to the number of vertices, and the average number of edges of the test graphs with N vertices, for $N = 10, \cdots, 100$. Note that at least 50 graph for each vertex cardinality between 10 and 100 have been generated. The 11,582 test graphs are available by anonymous ftp from infokit.dis.uniroma1.it:public.

Sparsity and "near-planarity" are typical properties of graphs used in software engineering and database applications [2]. As expected, the test graphs turn out to be sparse (the average vertex degree is about 2.7, see Fig. 4.b) and with low crossing number (the average crossing number is no more than about 0.7 times the number of vertices, see Fig. 6.b). We did not include graphs with more than 100 vertices because they are rarely displayed in full in the above applications (clustering methods are typically used to hierarchically display large graphs).

5 Analysis of the Experimental Results

The Bend-Stretch, Column, and GIOTTO algorithms have been executed on each of the 11,582 test graphs. Figures 6–8 show the average values of the nine quality measures Area, Cross, ScreenRatio, TotalBends, MaxEdgeBends, UnifBends, TotalEdgeLen, MaxEdgeLen, UnifLen achieved by the three algorithms on test graphs with N vertices, for $N = 10, \cdots, 100$. Figure 5 shows the line styles used in Figures 6–8 for the three algorithms.

We make the following observations on the experimental results, which show a clear trade-off between running time and aesthetic quality of the drawings:

Area: GIOTTO outperforms both Bend-Stretch and Column, which behave about the same. For $N > 70$, the difference becomes significant for the user (see, e.g., Fig. 3).

Cross: Bend-Stretch and GIOTTO behave more or less the same, especially for $10 < N < 60$, and $95 < N < 100$. This happens even though Bend-Stretch has a MakeBiconnected step before the MakePlanar step. The behavior of Column is considerably worse.

ScreenRatio: The behavior of GIOTTO is very good in the whole interval. The behavior of Bend-Stretch is about the same of GIOTTO between 40 and 100. The behavior of Column is unsatisfactory.

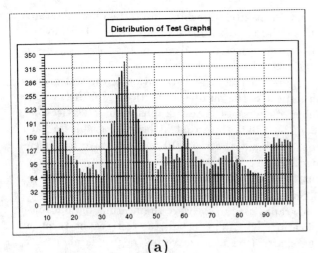

(a)

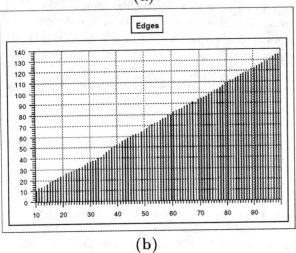

(b)

Figure 4: (a) Distribution of the test graphs with respect to the number of vertices. (b) Average number of edges of the test graphs versus number of vertices.

TotalBends: The experimental results sharply fit the theoretical results. Namely, GIOTTO has the minimum number of bends; Bend-Stretch and Column have a number of bends that, also for the constants, is essentially the one predicted by the theoretical analysis [45, 5]. Observe that the performance of Bend-Stretch on TotalBends deeply affects also the performance of such algorithm on Area.

MaxEdgeBends: Column has the best behavior. Also, since the number of bends introduced by GIOTTO on each edge is unbounded, and since Bend-Stretch guarantees for planar graphs at most two bends on each edge, we could expect a better behavior of Bend-Stretch with respect to GIOTTO. However, since the edges that are involved in crossings are split by the MakePlanar algorithm into several pieces, the result of Bend-Stretch becomes largely unsatisfactory.

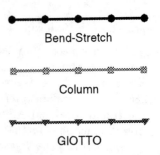

Figure 5: Line-styles in Figures 6-8 for the three algorithms Bend-Stretch, Column and GIOTTO respectively.

UnifBends: GIOTTO has the best behavior here and outperforms both Bend-Stretch and Column. It is interesting to observe that Bend-Stretch is better than Column for $N < 35$, and vice versa Column is better than Bend-Stretch for $N > 35$. Also, note that Column has a perfectly constant behavior.

TotalEdgeLen, MaxEdgeLen, and UnifLen: For all these parameters, GIOTTO is better than the other algorithms. Column and Bend-Stretch behave about the same for $N < 45$, while for $N > 45$ Column behaves clearly better.

Time: The experiments have been performed on a Sun Sparc-10/50 workstation. The implementation of the algorithms uses methods that are efficient in practice but may not be asymptotically optimal. All the three algorithms have shown good running time in practice. However, Column is definitely the fastest: it takes no more than 0.5 seconds of CPU time for any test. Bend-Stretch comes second, with an average of 0.5 seconds for $N = 30$, and about 1 second for $90 \leq N \leq 100$. Finally, GIOTTO is the most time critical: an average of 0.9 seconds for $N = 30$, and 3 seconds for $90 \leq N \leq 100$.

6 Open Problems

The experiments performed are an interesting source of both theoretical and practical open problems:

- It would be interesting to compare the three above algorithms with a very recent algorithm for orthogonal drawings by Papakostas and Tollis [37]. The theoretical analysis shows that the drawings produced by such algorithm have at most $0.8N^2$ area, that is about the practical behavior shown by Bend-Stretch.

- It would be interesting to perform further experiments on the practical performance of separator-based methods [15, 31, 49] that were originally developed for VLSI layout.

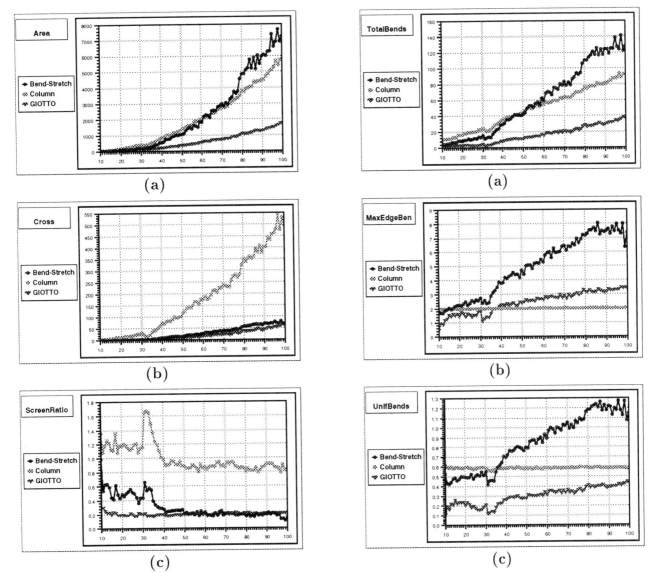

Figure 6: (a) Average area versus number of vertices. (b) Average number of crossings versus number of vertices. (c) Average deviation from the screen ratio versus number of vertices.

Figure 7: (a) Average total number of bends versus number of vertices. (b) Average maximum number of bends on any edge versus number of vertices. (c) Average standard deviation of the number of bends on the edges versus number of vertices.

- The behavior of Bend-Stretch could be improved by using, instead of the classical algorithms of [39, 44, 51], the algorithm by Kant [29] for constructing compact visibility representations.

- The performance of GIOTTO and Bend-Stretch is affected by the number of crossings introduced by the planarization phase. Can a more sophisticated heuristics (for example based on the work of Juenger and Mutzel [24] on the computation of the maximum planar subgraph) dramatically improve the behavior of such algorithms?

- The performance of algorithms Bend-Stretch and Column is affected by the biconnectivity augmentation step (MakeBiconnected). How much will it improve if we use a more sophisticated biconnectivity augmentation technique that preserves planarity (e.g., [30])?

- The computational bottleneck of the GIOTTO algorithm is the bend minimization step (Step3giotto), which has quadratic time complexity [42]. It would be interesting to improve on the time complexity of bend minimization.

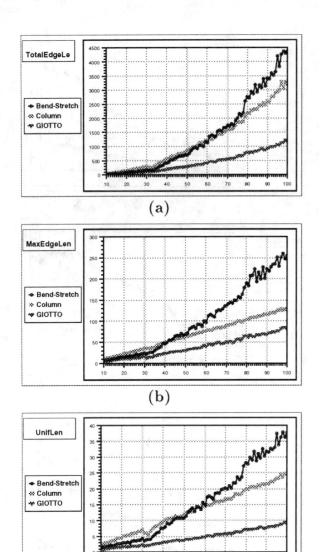

Figure 8: (a) Average total edge length versus number of vertices. (b) Average maximum edge length versus number of vertices. (c) Average standard deviation of the edge length versus number of vertices.

- It would be interesting to devise practical algorithms for orthogonal drawings in the 3-dimensional space. For recent results, see [8].

- Extensive experiments on algorithms for constructing other types of drawings (e.g., straight-line, polyline, upward) should be conducted.

References

[1] C. Batini, S. Ceri, and S. B. Navathe. *Conceptual Database Design, an Entity-Relationship Approach*. Benjamin Cummings, 1992.

[2] C. Batini, L. Furlani, and E. Nardelli. What is a good diagram? A pragmatic approach. In *Proc. 4th Internat. Conf. on the Entity Relationship Approach*, 1985.

[3] C. Batini, E. Nardelli, and R. Tamassia. A layout algorithm for data-flow diagrams. *IEEE Trans. Softw. Eng.*, SE-12(4):538–546, 1986.

[4] M. Beccaria, P. Bertolazzi, G. Di Battista, and G. Liotta. A tailorable and extensible automatic layout facility. In *Proc. IEEE Workshop on Visual Languages (VL'91)*, pages 68–73, 1991.

[5] T. Biedl and G. Kant. A better heuristic for orthogonal graph drawings. In *Proc. 2nd Annu. European Sympos. Algorithms (ESA '94)*, volume 855 of *Lecture Notes in Computer Science*, pages 24–35. Springer-Verlag, 1994.

[6] N. Chiba, K. Onoguchi, and T. Nishizeki. Drawing planar graphs nicely. *Acta Inform.*, 22:187–201, 1985.

[7] M. Chrobak and T. H. Payne. A linear time algorithm for drawing a planar graph on a grid. Technical Report UCR-CS-90-2, Dept. of Math. and Comput. Sci., Univ. California Riverside, 1990.

[8] R. F. Cohen, P. Eades, T. Lin, and F. Ruskey. Three-dimensional graph drawing. In R. Tamassia and I. G. Tollis, editors, *Graph Drawing (Proc. GD '94)*, volume 894 of *Lecture Notes in Computer Science*, pages 1–11. Springer-Verlag, 1995.

[9] R. Davidson and D. Harel. Drawing graphs nicely using simulated annealing. Technical report, Department of Applied Mathematics and Computer Science, The Weizmann Institute of Science, Rehovot, 1989.

[10] H. de Fraysseix, J. Pach, and R. Pollack. Small sets supporting Fary embeddings of planar graphs. In *Proc. 20th Annu. ACM Sympos. Theory Comput.*, pages 426–433, 1988.

[11] H. de Fraysseix and P. Rosenstiehl. Structures combinatoires pour des traces automatiques de reseaux. In *Proc. 3rd European Conf. on CAD/CAM and Computer Graphics (Paris)*, pages 332–337. Hermes, 1984.

[12] G. Di Battista, P. Eades, R. Tamassia, and I. G. Tollis. Algorithms for drawing graphs: an annotated bibliography. *Comput. Geom. Theory Appl.*, 4:235–282, 1994.

[13] G. Di Battista, A. Giammarco, G. Santucci, and R. Tamassia. The architecture of diagram server. In *Proc. IEEE Workshop on Visual Languages (VL'90)*, pages 60–65, 1990.

[14] G. Di Battista, G. Liotta, and F. Vargiu. Spirality of orthogonal representations and optimal drawings of series-parallel graphs and 3-planar graphs. In *Proc. Workshop Algorithms Data Struct.*, volume 709 of *Lecture Notes in Computer Science*, pages 151–162. Springer-Verlag, 1993.

[15] D. Dolev, F. T. Leighton, and H. Trickey. Planar embedding of planar graphs. In F. P. Preparata, editor, *Advances in Computing Research*, volume 2, pages 147–161. JAI Press, Greenwich, Conn., 1985.

[16] S. Even and G. Granot. Rectilinear planar drawings with few bends in each edge. Technical Report 797, Computer Science Dept., Technion", 1994.

[17] S. Even and G. Granot. Grid layouts of block diagrams — bounding the number of bends in each connection. In R. Tamassia and I. G. Tollis, editors, *Graph Drawing*

[17] (Proc. GD '94), volume 894 of *Lecture Notes in Computer Science*, pages 64–75. Springer-Verlag, 1995.

[18] T. Fruchterman and E. Reingold. Graph drawing by force-directed placement. *Softw. - Pract. Exp.*, 21(11):1129–1164, 1991.

[19] G. Gane and T. Sarson. *Structured Systems Analysis*. Prentice Hall, 1979.

[20] E. R. Gansner, S. C. North, and K. P. Vo. DAG – A program that draws directed graphs. *Softw. - Pract. Exp.*, 18(11):1047–1062, 1988.

[21] M. Himsolt. Comparing and evaluating layout algorithms within graphed. Manuscript, Fakultat fur Mathematik und Informatik, Univ. Passau, 1994.

[22] M. Himsolt. GraphEd: a graphical platform for the implementation of graph algorithms. In R. Tamassia and I. G. Tollis, editors, *Graph Drawing (Proc. GD '94)*, volume 894 of *Lecture Notes in Computer Science*, pages 182–193. Springer-Verlag, 1995.

[23] S. Jones, P. Eades, A. Moran, N. Ward, G. Delott, and R. Tamassia. A note on planar graph drawing algorithms. Technical Report 216, Department of Computer Science, University of Queensland, 1991.

[24] M. Juenger and P. Mutzel. Maximum planar subgraphs and nice embeddings: Practical layout tools. *Algorithmica*, to appear.

[25] T. Kamada. *Visualizing Abstract Objects and Relations*. World Scientific Series in Computer Science, 1989.

[26] T. Kamada and S. Kawai. An algorithm for drawing general undirected graphs. *Inform. Process. Lett.*, 31:7–15, 1989.

[27] G. Kant. Drawing planar graphs using the lmc-ordering. In *Proc. 33th Annu. IEEE Sympos. Found. Comput. Sci.*, pages 101–110, 1992.

[28] G. Kant. *Algorithms for Drawing Planar Graphs*. PhD thesis, Dept. Comput. Sci., Univ. Utrecht, Utrecht, Netherlands, 1993.

[29] G. Kant. A more compact visibility representation. In *Proc. 19th Internat. Workshop Graph-Theoret. Concepts Comput. Sci. (WG'93)*, 1993.

[30] G. Kant and H. L. Bodlaender. Planar graph augmentation problems. In *Proc. 2nd Workshop Algorithms Data Struct.*, volume 519 of *Lecture Notes in Computer Science*, pages 286–298. Springer-Verlag, 1991.

[31] C. E. Leiserson. Area-efficient graph layouts (for VLSI). In *Proc. 21st Annu. IEEE Sympos. Found. Comput. Sci.*, pages 270–281, 1980.

[32] Y. Liu, P. Marchioro, and R. Petreschi. A single bend embedding algorithm for cubic graphs. Manuscript, 1994.

[33] Y. Liu, P. Marchioro, R. Petreschi, and B. Simeone. Theoretical results on at most 1-bend embeddability of graphs. Technical report, Dipartimento di Statistica, Univ. di Roma "La Sapienza", 1990.

[34] Y. Liu, A. Morgana, and B. Simeone. General theoretical results on rectilinear embeddability of graphs. *Acta Math. Appl. Sinica*, 7:187–192, 1991.

[35] Y. Liu, A. Morgana, and B. Simeone. A linear algorithm for 3-bend embeddings of planar graphs in the grid. Manuscript, 1993.

[36] E. Nardelli and M. Talamo. A fast algorithm for planarization of sparse diagrams. Technical Report R.105, IASI-CNR, Rome, 1984.

[37] A. Papakostas and I. G. Tollis. Improved algorithms and bounds for orthogonal drawings. In R. Tamassia and I. G. Tollis, editors, *Graph Drawing (Proc. GD '94)*, volume 894 of *Lecture Notes in Computer Science*, pages 40–51. Springer-Verlag, 1995.

[38] R. Read. New methods for drawing a planar graph given the cyclic order of the edges at each vertex. *Congressus Numerantium*, 56:31–44, 1987.

[39] P. Rosenstiehl and R. E. Tarjan. Rectilinear planar layouts and bipolar orientations of planar graphs. *Discrete Comput. Geom.*, 1(4):343–353, 1986.

[40] J. A. Storer. On minimal node-cost planar embeddings. *Networks*, 14:181–212, 1984.

[41] K. Sugiyama, S. Tagawa, and M. Toda. Methods for visual understanding of hierarchical systems. *IEEE Trans. Syst. Man Cybern.*, SMC-11(2):109–125, 1981.

[42] R. Tamassia. On embedding a graph in the grid with the minimum number of bends. *SIAM J. Comput.*, 16(3):421–444, 1987.

[43] R. Tamassia, G. Di Battista, and C. Batini. Automatic graph drawing and readability of diagrams. *IEEE Trans. Syst. Man Cybern.*, SMC-18(1):61–79, 1988.

[44] R. Tamassia and I. G. Tollis. A unified approach to visibility representations of planar graphs. *Discrete Comput. Geom.*, 1(4):321–341, 1986.

[45] R. Tamassia and I. G. Tollis. Planar grid embedding in linear time. *IEEE Trans. on Circuits and Systems*, CAS-36(9):1230–1234, 1989.

[46] R. Tamassia and I. G. Tollis, editors. *Graph Drawing (Proc. GD '94)*, volume 894 of *Lecture Notes in Computer Science*. Springer-Verlag, 1995.

[47] H. Trickey. Drag: A graph drawing system. In *Proc. Internat. Conf. on Electronic Publishing*, pages 171–182. Cambridge University Press, 1988.

[48] W. T. Tutte. How to draw a graph. *Proceedings London Mathematical Society*, 3(13):743–768, 1963.

[49] L. Valiant. Universality considerations in VLSI circuits. *IEEE Trans. Comput.*, C-30(2):135–140, 1981.

[50] J. Q. Walker II. A node-positioning algorithm for general trees. *Softw. - Pract. Exp.*, 20(7):685–705, 1990.

[51] S. K. Wismath. Characterizing bar line-of-sight graphs. In *Proc. 1st Annu. ACM Sympos. Comput. Geom.*, pages 147–152, 1985.

[52] D. Woods. *Drawing Planar Graphs*. PhD thesis, Department of Computer Science, Stanford University, 1982. Technical Report STAN-CS-82-943.

Visibility with Reflection

Boris Aronov[1]
Alan R. Davis[2]
Tamal K. Dey[3]
Sudebkumar P. Pal[3]
D. Chithra Prasad[3]

Abstract

We extend the concept of the polygon visible from a source point S in a simple polygon by considering visibility with two types of reflection, *specular* and *diffuse*. In specular reflection a light ray reflects from an edge of the polygon according to Snell's law: the angle of incidence equals the angle of reflection. In diffuse reflection a light ray reflects from an edge of the polygon in all inward directions. Several geometric and combinatorial properties of visibility polygons under these two types of reflection are revealed, when at most one reflection is permitted. We show that the visibility polygon $Vs(S)$ under specular reflection may be non-simple, while the visibility polygon $Vd(S)$ under diffuse reflection is always simple. We present a $\Theta(n^2)$ worst case bound on the combinatorial complexity of both $Vs(S)$ and $Vd(S)$ and describe simple $O(n^2 \log^2 n)$ time algorithms for constructing the sets.

1 Introduction

Visibility problems have a long history, with one of the first problems, the "art gallery problem," posed in 1973 by Klee [KW91], who asked how many stationary guards are needed to guarantee that the union of their visibility polygons cover the entire polygonal room. Linear time algorithms for constructing point visibility polygons were presented by ElGindy and Avis [EA81], by Lee [Lee83], and by Guibas et al. [GHL+87].

Many aspects of this problem have been considered, including the use of different types of guards, with different powers of vision. Horn and Valentine introduced the concept of k-link visibility, defining a point D to be k-link-visible to a point S if there exists a k-link polygonal path between them that remains inside the polygon [HV49]. Link visibility has also more recently been investigated, among others, by Suri [Sur87, Sur90] and by Ke [Ke89]. O'Rourke's book on art gallery theorems [O'R87] is a comprehensive foundation, while Shermer's survey [She92] provides a more recent bibliography on art gallery problems.

Billiard paths and illumination of regions bounded by mirrors are two more variations of visibility problems. A good description and summary of some of the open problems can be found in the book by Klee and Wagon [KW91]. Much of the work on billiard paths has been done in the field of ergodic theory, and the reader can find a full introduction in the paper by Gutkin [Gut86].

Visibility has been studied in computer graphics in relation to ray tracing and hidden surface elimination algorithms. One of the early discussions of the problem can be found in [FKN80]. Arvo and Kirk developed a five-dimensional ray classification scheme for ray-tracing acceleration [AK87]. Reif, Tygar, and Yoshida have investigated the difficulty of ray tracing in optical systems [RTY94].

This paper investigates the geometric and combinatorial structures of visibility polygons under assumption of a single reflection. We model two different types of reflection, specular and diffuse. In specular reflection we assume that energy is reflected along a specific direction and that it abides by the standard law of reflection (Snell's law). This type of reflection occurs from a perfectly mirror-like surface. In diffuse reflection, energy is reflected in all directions. If the reflecting surface is rough, this type of reflection is more common.

We show that the (point) visibility polygon under specular reflection $Vs(S)$ may be non-simple while the

---

[1] Computer Science Department, Polytechnic University, Brooklyn, NY 11201, USA. Work on this paper by Boris Aronov has been supported by NSF Grant CCR-92-11541.
[2] Div. of Computer Science, Math. and Science, St. Johns University, Jamaica, NY 11439, USA.
[3] Dept. of Computer Science and Engineering, Indian Institute of Technology, Kharagpur, Kharagpur 721302, India. Work on this paper by S.P. Pal is partially supported by a research grant from the Jawaharlal Nehru Center, Bangalore, India.

(point) visibility polygon under diffuse reflection $Vd(S)$ is always simple. We prove a tight $\Theta(n^2)$ bound on the worst-case complexity of both types of visibility polygons and describe simple $O(n^2 \log^2 n)$ algorithms for computing them.

One possible application of visibility path problems is in the design of wireless local area networks.

The remainder of this paper is divided into six sections. Section 2 contains basic definitions. Section 3 describes and defines the important features of $Vs(S)$. Section 4 analyzes the complexity of $Vs(S)$. Section 5 describes and defines the important features of $Vd(S)$ and gives a bound on its complexity. Section 6 presents an algorithm for constructing $Vs(S)$ and $Vd(S)$. Section 7 contains conclusions and discussion.

2 Definitions

Let $P = (v_0, v_1, ..., v_{n-1})$ be a simple polygon with no three collinear vertices. Let $int(X)$ and $bd(X)$ denote the relative interior and boundary of a region X, respectively. Two points in P are said to be *visible* if the interior of the line segment joining them lies in $int(P)$. A point p is said to be *weakly visible* from a line segment $st \subset P$, if there is a point z in the interior of st such that p and z are visible.

In *specular* reflection a light ray is reflected according to Snell's law: the angle of incidence equals the angle of reflection. Light rays that hit a vertex of P will terminate without reflection. A point D is said to be *indirectly visible* from the source point S under *specular* reflection, if there exists a point p lying in the interior of an edge of P, visible from S and D, and such that Sp and pD make the same angle with the normal to $bd(P)$ at p. We define a *visibility path* to be any such polyline SpD between two points, with each segment being a *link* in the path.

In *diffuse* reflection a light ray incident on a point is assumed to be reflected equally in all inward directions. A point D is said to be *indirectly visible* from a point S under *diffuse* reflection, if there exists a point p lying in the interior of an edge of P and visible from S and D. In Figure 1, D_1 is indirectly visible from S under

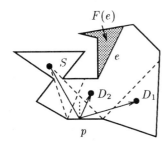

Figure 1: Indirect visibilities

diffuse reflection, D_2 is indirectly visible from S under specular reflection. However, D_1 is not visible from S under specular reflection.

We define the *extended specular visibility region*, $Vs(S)$, to be the set of all points D in P directly or indirectly visible to S under specular reflection. The *extended diffuse visibility region*, $Vd(S)$, is the set of all points D in P directly or indirectly visible to S under diffuse reflection.

3 Specular Visibility Regions

In this section, all reflections are assumed to be specular, unless stated otherwise.

Let e_i be an edge of P weakly visible from S. Let m_i be the portion of e_i actually visible from S. We will refer to m_i as a *mirror*. Define the *mirror visibility polygon* of m_i, $V_i(S)$, as the set of points of P that can see S after reflecting off of m_i. Let the image of S under reflection about e_i be S^i. Let Δ^i be the triangle formed by S^i and m_i. Then $V_i(S) = P \cap V(S^i, P \cup \Delta^i)$. (More precisely, $P \cup \Delta^i$ denotes here the "Riemann surface" obtained by gluing P to Δ^i along e_i, which handles the possibility that, considered as a subset of the plane, Δ^i overlaps P.) This is the dark-shaded region in Figure 2. If $a \leq n$ is the number of edges weakly visible from S,

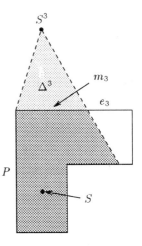

Figure 2: Mirror visibility polygon

we let $m_0, m_1, \ldots, m_{a-1}$ denote the set of mirrors.

By definition, $Vs(S) = V(S) \cup \bigcup_{i=0}^{a-1} V_i(S)$. In Figure 3, $V(S) \subset Vs(S)$ is shaded darker, $Vs(S) - V(S)$ is shaded lighter. We can see in the figure that $Vs(S)$ may extend $V(S)$ without completely covering P. We define these non-indirectly-visible regions to be blind spots. More precisely, a *blind spot* is a connected component of $P - Vs(S)$. It is these blind spots which must be described in order to have a complete description of $Vs(S)$.

As a finite union of polygons, $Vs(S)$ is a polygonal region contained in P. $Vs(S)$ is connected since every point in it is connected to S by a path consisting entirely

317

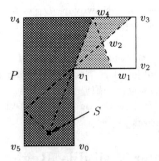

Figure 3: $Vs(S)$

of points visible or indirectly visible to S. Any segment whose interior lies in the interior of P will be called an *interior segment*, such as w_1w_2 in Figure 3. A vertex generated by the intersection of interior segments will be called an *interior vertex* of $Vs(S)$, such as w_2. The boundary of $Vs(S)$ may consist of edges of P, such as v_0v_1; subedges of P, such as v_1w_1; and interior segments of P, such as w_1w_2.

The interior segments that bound $Vs(S)$ come from the interior segments bounding $V(S)$ and the $V_i(S)$. We have assumed that visibility terminates at a vertex, so we define a *(specular) vpath limit* to be any 2-link path from S to $bd(P)$ that obeys the reflection property but may pass through a reflex vertex or an endpoint of a mirror. The portion of a vpath limit beyond the occluding vertex is not visible from S, and the region adjacent to one side of that portion of the vpath limit is in shadow. These subsegments will be called *shadow edges*, and may become part of the boundary of $Vs(S)$. However, some portions of shadow edges or entire shadow edges are made visible from other directions, such as v_1w_2 in Figure 3, and therefore do not become boundary edges of $Vs(S)$.

We can label vpath limits according to whether the shadow is inside (**in**) or outside (**out**) of the angle of reflection, see Figure 4 and Figure 5. In addition, a

Figure 4: Specular vpath limit types (I)

vpath limit can touch a reflex vertex **before** or **after** reflection, or it can touch the endpoint (**end**) of the edge it reflects off of. In the figures, small triangles

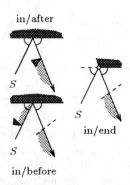

Figure 5: Specular vpath limit types (II)

indicate reflex vertices of P. Circles emphasize that the shadow is generated because the reflection occurs at the end of an edge of P. Shadow edges are defined locally, according to the effects along a single path. They may be lit from other directions, as indicated by the dashed lines. We give without proof the following fundamental facts.

FACT 3.1 *In a simple polygon, the portion of a segment visible from a point is either empty or forms a connected subsegment.*

FACT 3.2 *There can be no part of $bd(P)$ inside any region bounded by a curve completely contained in P.*

3.1 Blind Spots

Portions of $bd(V(S))$ that do not lie on $bd(P)$ are called *windows*. In the terminology of the previous section, they are the shadow edges of before-type vpath limits that begin at the reflex vertex and end at the point of reflection. They cut $V(S)$ out of P. For example, v_1w_4 in Figure 3 is a window.

By placing small mirrors oriented appropriately along edges of our polygon we can produce thin mirror visibility polygons that intersect to create blind spots in $Vs(S)$ that are disjoint from $bd(P)$. We define these blind spots to be *interior blind spots*. This shows that $Vs(S)$ is not necessarily simply connected. The interior blind spot, labeled 1 in Figure 6, was created by encircling it with thin mirror visibility polygons in an otherwise non-visible region.

We define a *boundary blind spot* to be a connected region of $P - Vs(S)$ which has part of $bd(P)$ as its boundary. Some of the vertices of a boundary blind spot lie on $bd(P)$ and some are interior vertices. A boundary blind spot with a single interior vertex is called a *simple boundary blind spot* (labeled 4 in the figure). As seen in Figure 6 some boundary blind spots, with no interior vertices are generated at the far ends of mirror visibility polygons (labeled 2). But we can also have boundary blind spots created outside of mirror visibility polygons by shadow edges and windows (labeled 3).

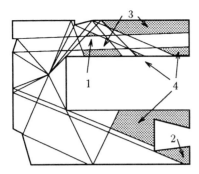

Figure 6: Blind Spot Types

We define a direction along a vpath according to increasing distance from S along the path. If two directed line segments cross at x, we call the quadrant formed by the two edges leaving x the *top quadrant* and the vertex x a *top vertex* of the region containing the top quadrant near x. This is q_4 in Figure 7(a). The quadrant formed by the two edges entering x is the *bottom quadrant* (q_2 in the figure) and the vertex x a *bottom vertex* of the corresponding region. We define *left* (q_1) and *right* (q_3) quadrants and vertices similarly. See Figure 7.

LEMMA 3.3 *If x is a point not visible to S on the intersection of two vpath limits, then the source S must be in either the closed left or the closed right quadrant.*

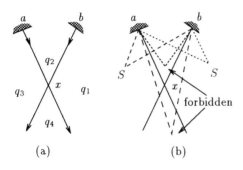

Figure 7: Position of Source 1

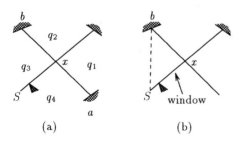

Figure 8: Position of Source 2

PROOF. Consider two segments of vpath limits that intersect at x determining the four quadrants labeled $q_1, ..., q_4$. If they are both second links of vpath limits we can without loss of generality indicate the first reflections, a and b, of the vpath limits as in Figure 7(a). If we tried to place the source S of the vpath limits in q_2, then x would be visible to S since S and x would be opposite vertices of the non-convex quadrilateral $Saxb$ whose boundary is completely contained in P. The case placing S in q_4 is similar. Therefore, the only valid placements for S are in q_3 or q_1 as shown in Figure 7(b), as desired.

If one of the intersecting segments is the first link of a vpath limit, i.e., a window, then there must be a reflex vertex of P on that link between x and S, as in Figure 8(a). The reflecting edge for the other vpath limit can not lie at a since that would enclose the reflex vertex inside $\triangle xaS$ bounded by subsegments of vpaths, contradicting Fact 3.2. Therefore, up to symmetry, the only valid configuration is as shown in Figure 8(b), as claimed. Finally, two windows cannot meet, so x cannot lie on two windows. □

We will now give a more formal description of an interior blind spot.

LEMMA 3.4 *An interior blind spot R is a closed convex polygon with each vertex on two vpath limits, and containing exactly one top and one bottom vertex.*

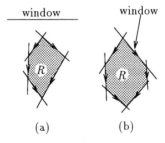

Figure 9: Interior blind spots

PROOF. As $Vs(S)$ is polygonal, $P - Vs(S)$ is polygonal and so is R. If A is on $bd(R)$, then A is on a vpath limit. Because R is in $P - V(S)$, A must occur after the reflex vertex of the vpath limit so A is not directly or indirectly visible since it is blocked from S by a vertex. Then the boundary of R is formed by shadow edges of vpath limits and R contains its boundary. If A is a vertex of R then it is on two vpath limits which intersect at A. Since only one side of a vpath limit is in shadow, only one of the four sectors formed by the two segments can be in shadow, as it will not be lit by either beam of light. Therefore A is a convex vertex of R, so R is convex.

Since R lies outside $V(S)$, there is a window of $V(S)$ between S and R and all vpath limits appearing on $bd(R)$ must intersect that window. The perpendicular distance from the window to any point on a vpath limit

319

increases as we traverse the vpath limit away from S and thus away from the window. Therefore the direction of the edges of R is consistent with increasing perpendicular distance from the line containing the window. As R is convex, it thus must have a unique minimum (top vertex) and a unique maximum (bottom vertex) while the remaining vertices must be side vertices. Figure 9(a) illustrates the case where R does not touch the window, while Figure 9(b) shows the case when the window overlaps the boundary of R. □

LEMMA 3.5 *A boundary blind spot has no bottom vertices.*

PROOF. Consider a bottom vertex x. By definition, it cannot lie on a window. Let y and z be the points where the (second links of) the two vpath limits containing x intersect the window which separates x from $V(S)$. By definition of a bottom vertex and of a window, the points immediately outside $\triangle xyz$ lie in $Vs(S)$, so the blind spot containing x must be fully contained in $\triangle xyz$ and must thus be an interior blind spot. □

LEMMA 3.6 *The following figures indicate the only possible locations for the reflex vertices and edge endpoints needed to generate the indicated blind spot vertices. Blind spot vertices lying on windows are not shown.*

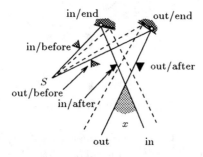

Figure 10: Top vertex types

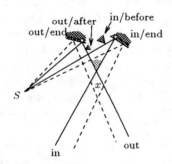

Figure 11: Bottom vertex types

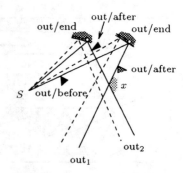

Figure 12: Left vertex types

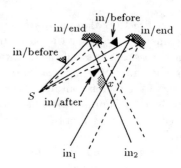

Figure 13: Right vertex types

PROOF. In figures 10 through 13, solid lines show the vpath limits involved in the blind spot vertices. The dashed lines indicate the extent of the "light beams" (i.e., mirror visibility regions) whose interaction produces the vertices in question. Reflex vertices of P indicated by shaded wedges are not affected by the rest of the mirror visibility polygon. Those reflex vertices indicated by black triangles must be placed outside the "light beam" boundaries indicated by the dashed lines.

Consider Figure 12, which deals with a left vertex of a blind spot. According to Lemma 3.3, S may only lie in the left or right quadrant with respect to x; we place S to the left; the other case is a mirror image of Figure 13. According to the labeling scheme of Figures 4 and 5, both vpath limits must be labeled **out**. While the vpath limit labeled out_1 permits all three placements of a reflex vertex allowed by as shown in Figure 12, the other does not. The **out/before** option is not possible along the second path because such placement of the reflex vertex would contradict Fact 3.2. The reflex vertices indicated by small triangles must meet the additional requirement that they be placed outside the other mirror visibility polygon as indicated by dashed lines. The placement of the reflex vertices in the other figures is determined in a similar manner. Note that in contrast to the other three cases, all possible combinations of placements from are allowed for top vertices; see Figure 10. □

LEMMA 3.7 Two mirror visibility polygons can not intersect to generate two bottom vertices of two blind spots.

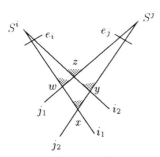

Figure 14: One bottom vertex per pair of mirrors

PROOF. If two mirror visibility polygons intersect to generate vertices of two blind spots, then there must be either three or four vpath limits involved. The four vpath limits are labeled i_1, i_2, j_1, j_2, as in Figure 14. The four possible intersections are labeled x, y, z, and w.

If the two vertices are produced by three vpath limits, without loss of generality assume that the pair of blind spot vertices w, x occurs on the single vpath limit i_1 from edge e_i, and that, for this case, i_2 need not exist. The pair of vpath limits from edge e_j intersected with i_1 form a "region" wxS^j whose boundary is contained in P. More precisely, $\triangle wxS^j$ can be folded across e_j to form a union of a triangle and a quadrilateral whose interior may not contain a reflex vertex of P by Fact 3.2. This vertex x can not be a bottom vertex of a blind spot, as there is no way to produce a shadow along the appropriate side of j_2 at x. Therefore two bottom vertices cannot lie along a single vpath limit.

Suppose now that the two bottom vertices do not share a vpath limit. There are only two such pairs in the figure, w and y, or x and z. In either case we now have two "regions" whose boundaries are contained in P, zyS^j and wzS^i. As before, these regions can not have a blind spot in their interiors so y and w can not be bottom vertices of blind spots. Therefore we can not have a pair of bottom vertices generated by two mirrors.

Note that, if $bd(P)$ intersects the vpath limits such that i_1 and j_2 do not intersect, then the above analysis still rules out the possibility that both w and y are bottom vertices. □

LEMMA 3.8 Two intersecting mirror visibility polygons can not generate two right (left) blind spot vertices.

PROOF. Omitted, similar to Lemma 3.7. □

4 Complexity of $Vs(S)$

Recall the use of thin mirror visibility polygons in Figure 6 to create a blind spot. By placing $\Theta(n)$ small mirrors appropriately along the boundary of P we can produce a $Vs(S)$ with $\Theta(n^2)$ components; see Figure 15. This implies an $\Omega(n^2)$ lower bound on the worst case complexity of $Vs(S)$ and on the number of blind spots.

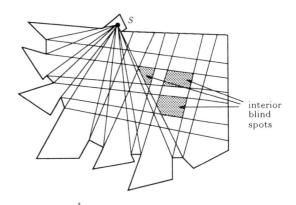

Figure 15: Lower bound construction for $Vs(S)$

We bound the combinatorial complexity of $Vs(S)$ by counting the vertices generated by its three parts: $V(S)$, the interior blind spots, and the boundary blind spots. Let P have n edges and r reflex vertices. Let $V(S)$ have a edges that are edges or subedges of P (these are exactly the mirrors) and $b \leq a$ windows. $V(S)$ has $a + b$ edges so it has $a + b$ vertices each of which is a boundary vertex. A boundary vertex is one of the n original vertices of P or is generated by the endpoints of the vpath limits from each of the $V_i(S)$. A vpath limit must pass through one of the r reflex vertices of P or the two endpoints of a mirror. Therefore we can have at most $a(r + 2) + a + b + n$ boundary vertices. Interior vertices can only be generated by intersections of vpath limits occurring in $P - V(S)$ and must be of type top, bottom, left or right.

Lemma 3.7 implies that any bottom vertex can be uniquely charged to the pair of mirrors generating it, and that no other bottom vertex can be charged to the same two mirrors. Therefore we can label a bottom vertex as (i, j) and see that the number of such vertices is $O(a^2)$. Lemma 3.8 similarly implies that the number of left (right) vertices is also $O(a^2)$. Every top vertex must be incident to a left or a right vertex unless it is the top vertex of a simple boundary blind spot. So the number of all interior vertices, not counting top vertices of simple boundary blind spots, is $O(a^2)$. A simple boundary blind spot has a top vertex and exactly two boundary vertices, and there are at most $a(r + 2)$ such boundary vertices. Therefore there can be at most $O(ar)$ top vertices of simple boundary blind spots. Combining these estimates, we obtain an $O(ar + n)$ upper bound on the complexity of $Vs(S)$.

THEOREM 4.1 The combinatorial complexity of $Vs(S)$ is $O(ar+n) = O(an)$, so it is $\Theta(n^2)$ in the worst case.

5 Diffuse Visibility Regions

We now consider the extended diffuse visibility region $Vd(S)$.

The notion of a mirror remains the same. The mirror visibility polygon $V_i(S)$ of mirror m_i is just the weak visibility polygon of m_i, i.e., the set of points that can see at least one point in the interior of m_i. $Vd(S) = V(S) \cup \bigcup_{i=0}^{a-1} V_i(S)$. The boundary of $Vd(S)$ consists of portions of windows and portions of the boundaries of the $V_i(S)$. Similar to the case of specular reflection, $bd(V_i(S))$ comes from limiting cases of visibility paths which we call *(diffuse) vpath limits*. The types of segments (besides windows) that arise in the vpath limits are illustrated in Figure 16 (see, for example, [GHL+87]).

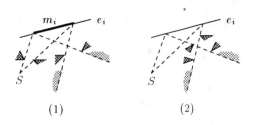

Figure 16: Diffuse vpath limit types

The second link of a diffuse vpath limit either (1) emanates from an endpoint of m_i and passes through a single reflex vertex, or (2) emanates from an interior point of m_i and passes through two reflex vertices. In the former case the line containing the second link has the mirror and the reflex vertex on the same side. In the latter case, it separates the two reflex vertices. See the figure. The shadow edge of the diffuse vpath limit starts after the *last* reflex vertex on the path and this vertex determines on which side the shadow lies. We define this vertex to be the *key reflex vertex* for the diffuse vpath limit.

For spectral reflections, distinct vpaths reflecting from a common mirror must diverge. However, for diffuse reflections distinct vpaths reflecting from a given edge may intersect. Analysis of blind spot vertices using diffuse vpath limits will therefore need to consider cases involving intersecting diffuse vpath limits from a given mirror. It appears that these are the only new cases that require a change of the arguments we have used for specular reflection.

LEMMA 5.1 *If two distinct diffuse vpath limits reflecting from a single mirror intersect, they may not meet their respective key reflex vertices until after that intersection.*

Figure 17: Intersecting diffuse vpath limits

PROOF. Let two diffuse vpath limits from S through points y and z of mirror m_i intersect at x as in Figure 17.

Without loss of generality consider the diffuse vpath limit through y.

If y is an endpoint of m_i, the corresponding key reflex vertex must occur on the same side of the vpath limit as m_i. If it occurred before x it would be inside $\triangle yxz$ contradicting Fact 3.2.

Assume y is interior to m_i and we attempt to place the key reflex vertex before x. Placing it inside $\triangle yxz$ contradicts Fact 3.2, so we must try outside. If the key reflex vertex is placed on the outside of $\triangle yxz$, the corresponding first reflex vertex must be placed on the inside, again contradicting Fact 3.2.

Therefore in all cases the key reflex vertices generating the diffuse vpath limits must occur after the intersection at x. □

Note that no interior vertex of a blind spot can be formed by two vpath limits reflected off the same edge, as $V_i(S)$, which is the region of weak visibility from m_i, is cut out of P by disjoint chords (see, e.g., [GHL+87]). A single vpath limit intersected with a window can not generate a blind spot vertex by Fact 3.2; the details are contained in the full paper.

LEMMA 5.2 $Vd(S)$ *has no interior blind spots.*

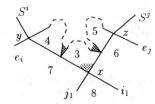

Figure 18: No interior blind spots

PROOF. We prove that there are no bottom vertices. If x is a bottom vertex of a blind spot it must lie at the intersection of two vpath limits i_1 and j_1. As already observed, these vpath limits must come from two different edges of P. Let these vpath limits intersect e_i and e_j at y and z respectively. As x is a bottom vertex of a blind spot, it must be shaded locally as in Figure 18. Then there must exist two key reflex vertices as indicated in the figure.

Now consider the placement of the source S. The boundary of P can not cross a vpath limit, or cross itself, so the only possible configuration for it is shown as the dashed line in the figure. S can not be placed in regions 3, 4, or 7 since then z would not be directly visible to S. S can not be placed in regions 3, 5, or 6 since then y would not be directly visible to S. S can not be placed in region 8 since then x would be directly visible from S and could not be a blind spot vertex.

Since there is no valid placement for S, there can be no bottom vertex of a blind spot. Therefore there can be no blind spots in $Vd(S)$, as following the argument of Lemma 3.4, a diffuse interior blind spot would have to be convex and thus would have to contain a bottom vertex. □

We now show how to count the other types of blind spot vertices in a manner similar to that used for specular reflections. The possibility of intersection of vpath limits reflected from the same edge increases the number of cases we need to consider, however Lemma 5.1 will reduce this number.

LEMMA 5.3 *Interaction of two mirror visibility polygons can not generate two left blind spot vertices.*

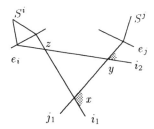

Figure 19: Two left vertices cannot share a vpath limit

PROOF. If the two vpath limits generated through a single mirror do not intersect each other, the proof is essentially the same as in the previous development for spectral reflection. This is because we need only consider the key reflex vertex on the vpath in order to get a contradiction—we never directly used Snell's law in any of the proofs. We now consider the new cases involving intersecting vpath limits.

A single vpath limit can not contain two left blind spot vertices. For a contradiction, assume it does as in Figure 19. The key reflex vertex for diffuse vpath limit i_1 can not occur before the intersection of i_1 and i_2 at z by Lemma 5.1. Since it must be on the same side as the shadow along i_1 at x it would have to be placed after x and inside $\triangle xyz$ contradicting Fact 3.2. Therefore two left blind spot vertices cannot lie along the single vpath limit j_1.

Therefore each left blind spot vertex must be on a separate vpath limit. If there are two vpath limits from e_j, j_1 and j_2, one of them must generate the left blind spot on i_1. This creates $\triangle xyz$ in the figure and does not permit x to be a right blind spot vertex.

Therefore, regardless of how the vpath limits are generated from edge e_j we can not create two right blind spot vertices from the pair of vpath limits out of e_i. □

Therefore a left blind spot vertex can be associated with a unique pair of mirrors, and no pair is charged more than once, so there are at most a quadratic number of left blind spot vertices. Similarly for right vertices.

LEMMA 5.4 *A boundary blind spot can not have more than one top vertex.*

PROOF. Let R be a boundary blind spot. First of all, as $Vd(S)$ is connected and R is a component of $P - Vd(S)$, it must be cut out of P by a single (unclosed) polygonal line γ. Following the same argument as that of Lemma 3.4, we conclude that γ is convex in the sense that it makes only clockwise turns if one traverses it keeping R to the right. For γ to contain two top vertices, it would have to contain a bottom vertex between them, contradicting Lemma 5.2. □

THEOREM 5.5 *The complexity of $Vd(S)$ is $\Theta(n^2)$, in the worst case.*

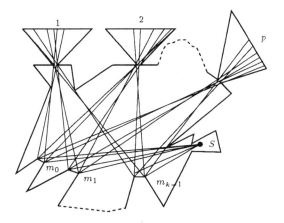

Figure 20: Lower bound construction for $Vd(S)$

PROOF. In Figure 20, each of the $p = \Theta(n)$ pockets generates $k = \Theta(n)$ boundary blind spots, for a total of $\Theta(n^2)$ blind spots, implying an $\Omega(n^2)$ lower bound on worst-case complexity of $Vd(S)$.

$Vd(S)$ is a polygon whose interior vertices with respect to P can only be left, right, or top vertices of boundary blind spots. By Lemma 5.3, we can only have $O(a^2)$ left vertices. Similarly for right vertices. A simple argument shows that the number of boundary vertices is $O(an)$. Since any boundary blind spot can have at most one top vertex by Lemma 5.4, there are at most

$O(an)$ top vertices. Therefore the complexity of $Vd(S)$ is $O(an + a^2 + n) = O(an) = O(n^2)$. □

6 Algorithms

In order to compute the extended visibility polygon from point S in P:

1. Compute $V(S)$.
2. Compute $V_i(S)$, for $i = 0, \ldots, a-1$.
3. Compute $P_i = V(S) \cup V_i(S)$, for $i = 0, \ldots, a-1$.
4. Merge P_i's in pairs, then merge the resulting sets in pairs, etc., obtaining the desired set in $\lceil \log a \rceil$ rounds.

Each of the $a+1$ sets in steps (1) and (2) can be computed in linear time by a standard visibility polygon algorithm, or weak visibility for diffuse mirror visibility polygons [EA81, Lee83, GHL+87]. The pairwise merges in steps (3) and (4) are accomplished by a sweepline algorithm [BO79], at the cost of $O(\log n)$ per vertex of the polygons being merged and per vertex of their union. Fact 3.1 easily implies that each P_i has linear complexity, so computing all P_i's takes $O(an \log n)$ time.

Let $T(k)$ be the amount of time it takes to produce the union of k P_i's. It is easy to verify that, with no modification, the proofs of Theorems 4.1 and 5.5 imply that the union of $k \geq 1$ P_i's has complexity $O(kn)$. Thus $T(k)$ has to satisfy

$$T(1) = O(n), \text{ and}$$
$$T(k) = 2T(k/2) + O(kn \log n).$$

Recalling that there are a mirrors altogether, we conclude that the entire algorithm runs in time

$$T(a) = O(an \log a \log n).$$

Thus we conclude

THEOREM 6.1 Both $Vs(S)$ and $Vd(S)$ can be constructed in $O(an \log a \log n)$ time, which is $O(n^2 \log^2 n)$ in the worst case. Here a is the number of edges of P visible from S.

7 Conclusions

We have defined the concepts of an extended specular visibility region $Vs(S)$ and an extended diffuse visibility region $Vd(S)$ based on the type of reflection that occurs along 2-link visibility paths. We have shown that $Vs(S)$ may be non-simple with interior blind spots, while the $Vd(S)$ is always simple. We have presented a $\Theta(n^2)$ worst case bound on the combinatorial complexity of both $Vs(S)$ and $Vd(S)$ and described simple $O(n^2 \log^2 n)$ time algorithms for constructing both sets.

The art gallery problem with reflections is to find a set of a minimum number of point light sources positioned at vertices of a polygon that can illuminate the entire polygon. Lee and Lin [LL86] have shown that this problem is NP-hard for direct visibility. In a companion paper [DPP95] Dey, Pal and Prasad show that this problem remains NP-hard for visibility with reflection as well.

A variant of the art gallery problem of finding the location (if any) of a single source which can illuminate the entire polygon arises in practical applications. Let kernel K denote the set of points (if any), each of which can illuminate the entire polygon after one level of reflection. Observe that visibility with one level of reflection is a special case of link-2 visibility [LPS+88, Ke89]. The kernel problem has been solved in [LPS+88, Ke89] for general link-k visibility. These algorithms use the crucial property of the link kernel that it is the intersection of link-k neighborhoods of vertices of the given polygon. Unfortunately this is no longer true for visibility with reflections. Let $N = \bigcap_v Vd(v)$, with the intersection taken over all vertices v of polygon P. Clearly $K \subseteq N$. However, we can construct examples showing that $K \neq N$; an analogous statement holds for spectral reflection. This poses a real difficulty in computing kernels under reflections. The only progress we could make in this regard is to characterize the kernel under diffuse reflection as follows.

Let $S(e)$ denote the set of all points that are visible directly or indirectly by one reflection from every point of an edge e. We call $S(e)$ the strong visibility region of e. We can establish that the kernel under diffuse reflection coincides with the intersection of strong visibility regions of all edges. Details of these results on art gallery problem and its relatives are reported in [DPP95].

In another direction of research we consider multiple levels of reflection. The results obtained in this direction is reported in a companion paper under preparation [ADD+95].

Acknowledgment

The authors wish to thank Thomas Shermer and Joseph O'Rourke for a helpful discussion.

References

[ADD+95] B. Aronov, A. R. Davis, T. K. Dey, S. P. Pal, and D. C. Prasad. Visibility with multiple reflection. In preparation, 1995.

[AK87] J. Arvo and D. B. Kirk. Fast ray tracing by ray classification. In Maureen C. Stone, editor, Computer Graphics (SIGGRAPH'87 Proceedings), volume 21, pages 55–64, July 1987.

[BO79] J. L. Bentley and T. A. Ottmann. Algorithms for reporting and counting geometric intersections. *IEEE Trans. Comput.*, C-28:643–647, 1979.

[DPP95] T. K. Dey, S. P. Pal, and D. C. Prasad. Art gallery problems for visibility with reflection. In preparation, 1995.

[EA81] H. ElGindy and D. Avis. A linear algorithm for computing the visibility polygon from a point. *J. Algorithms*, 2:186–197, 1981.

[FKN80] H. Fuchs, Z. M. Kedem, and B. Naylor. On visible surface generation by a priori tree structures. *Comput. Graph.*, 14:124–133, 1980.

[GHL+87] L. J. Guibas, J. Hershberger, D. Leven, M. Sharir, and R. E. Tarjan. Linear-time algorithms for visibility and shortest path problems inside triangulated simple polygons. *Algorithmica*, 2:209–233, 1987.

[Gut86] E. Gutkin. Billiards in polygons. *Physica D*, 19:311–333, 1986.

[HV49] A. Horn and F. Valentine. Some properties of *l*-sets in the plane. *Duke Mathematics Journal*, 16:131–140, 1949.

[Ke89] Y. Ke. An efficient algorithm for link-distance problems. In *Proc. 5th Annu. ACM Sympos. Comput. Geom.*, pages 69–78, 1989.

[KW91] V. Klee and S. Wagon. *Old and New Unsolved Problems in Plane Geometry and Number Theory*. Mathematical Association of America, 1991.

[Lee83] D. T. Lee. Visibility of a simple polygon. *Comput. Vision Graph. Image Process.*, 22:207–221, 1983.

[LL86] D. Lee and A. Lin. Computational complexity of art gallery problems. *IEEE Trans. Inform. Theory*, 32:276–282, 1986.

[LPS+88] W. Lenhart, R. Pollack, J.-R. Sack, R. Seidel, M. Sharir, S. Suri, G. T. Toussaint, S. Whitesides, and C. K. Yap. Computing the link center of a simple polygon. *Discrete Comput. Geom.*, 3:281–293, 1988.

[O'R87] J. O'Rourke. *Art Gallery Theorems and Algorithms*. Oxford University Press, New York, NY, 1987.

[RTY94] J. Reif, J. Tygar, and A. Yoshida. Computability and complexity of ray tracing. *Discr. & Comput. Geom.*, 11:265–288, 1994.

[She92] T. C. Shermer. Recent results in art galleries. *Proc. IEEE*, 80(9):1384–1399, September 1992.

[Sur87] S. Suri. *Minimum link paths in polygons and related problems*. Ph.D. thesis, Dept. Comput. Sci., Johns Hopkins Univ., Baltimore, MD, 1987.

[Sur90] S. Suri. On some link distance problems in a simple polygon. *IEEE Trans. Robot. Autom.*, 6:108–113, 1990.

Efficient Randomized Algorithms for Some Geometric Optimization Problems*

Pankaj K. Agarwal[†]   Micha Sharir[‡]

Abstract

In this paper we first prove the following combinatorial bound, concerning the complexity of the vertical decomposition of the minimization diagram of trivariate functions: Let $\mathcal{F}$ be a collection of n totally or partially defined algebraic trivariate functions of constant maximum degree, with the additional property that, for a given pair of functions $f, f' \in \mathcal{F}$, the surface $f(x,y,z) = f'(x,y,z)$ is xy-monotone (actually, we need a somewhat weaker property—see below). We show that the vertical decomposition of the minimization diagram of $\mathcal{F}$ consists of $O(n^{3+\varepsilon})$ cells (each of constant complexity), for any $\varepsilon > 0$. In the second part of the paper we present a general technique that yields faster randomized algorithms for solving a number of geometric optimization problems, including (i) computing the width of a point set in 3-space, (ii) computing the minimum-width annulus enclosing a set of n points in the plane, and (iii) computing the 'biggest stick' inside a simple polygon in the plane. Using the above result on vertical decompositions, we show that the expected running time of all three algorithms is $O(n^{3/2+\varepsilon})$, for any $\varepsilon > 0$.

---

*Work on this paper by the first author has been supported by NSF Grant CCR-93-01259, an NYI award, and matching funds from Xerox Corporation. Work on this paper by the second author has been supported by NSF Grants CCR-91-22103 and CCR-93-11127, by a Max-Planck Research Award, and by grants from the U.S.-Israeli Binational Science Foundation, the Israel Science Fund administered by the Israeli Academy of Sciences, and the G.I.F., the German-Israeli Foundation for Scientific Research and Development.

[†] Department of Computer Science, Box 90129, Duke University, Durham, NC 27708-0129

[‡] School of Mathematical Sciences, Tel Aviv University, Tel Aviv 69978, Israel, and Courant Institute of Mathematical Sciences, New York University, New York, NY 10012, USA

Permission to copy without fee all or part of this material is granted provided that the copies are not made or distributed for direct commercial advantage, the ACM copyright notice and the title of the publication and its date appear, and notice is given that copying is by permission of the Association of Computing Machinery.To copy otherwise, or to republish, requires a fee and/or specific permission.
11th Computational Geometry, Vancouver, B.C. Canada
© 1995 ACM 0-89791-724-3/95/0006...$3.50

1 Introduction

In this paper we present a general technique that yields faster randomized algorithms for the following problems:

1. Computing the width of a set of points in $\mathbb{R}^3$.

2. Computing an annulus of minimum width that contains a given set of points in the plane.

3. Computing a longest segment that can be placed inside a simple polygon in the plane.

In order to achieve a fast implementation of our technique, we use the following combinatorial result, which is derived in the first part of the paper. Let $\mathcal{F}$ be a collection of n totally or partially defined algebraic trivariate functions of constant maximum degree, with the following additional *xy-monotonicity* property: For any pair $f, f' \in \mathcal{F}$, the xy-plane can be decomposed into a constant number of faces, each of constant description complexity, such that, for every face c, the surface $f(x,y,z) = f'(x,y,z)$ is the graph of a continuous bivariate function (of x and y) over the interior of c. The *lower envelope* $E_\mathcal{F}$ of $\mathcal{F}$ is the pointwise minimum

$$E_\mathcal{F}(x,y,z) = \min_{f \in \mathcal{F}} f(x,y,z),$$

and the *minimization diagram* $M_\mathcal{F}$ is the projection of the graph of $E_\mathcal{F}$ onto $\mathbb{R}^3$. That is, $M_\mathcal{F}$ is a decomposition of $\mathbb{R}^3$ into relatively-open connected cells of dimension 0, 1, 2, and 3, so that, over each cell, $E_\mathcal{F}$ is attained by a fixed subset of functions of $\mathcal{F}$ (and/or of function boundaries). It is known [22] that $M_\mathcal{F}$ has $O(n^{3+\varepsilon})$ faces.

We prove that the *vertical decomposition* of $M_\mathcal{F}$ also consists of $O(n^{3+\varepsilon})$ cells, each of constant description complexity. See below and [7, 9, 13, 23] for the definition of vertical decompositions. Briefly, this is the only known general-purpose technique for decomposing cells of arrangements of low-degree algebraic surfaces in higher dimensions into a reasonably small number of subcells of constant complexity. Such a decomposition is a prerequisite to many randomized incremental or divide-and-conquer algorithms involving arrangements of this kind. Unfortunately, the known upper bounds on the number of resulting subcells are much higher than the actual complexity of the cells being decomposed, and this affects adversely (the upper bounds one can prove on) the complexity of the relevant algorithms. Hence any result, like the one we prove here, which establishes nearly-tight bounds for the size of vertical decompositions, is significant, as indeed will be demonstrated below.

Our bound on the vertical decomposition immediately leads to a data structure, of size $O(n^{3+\varepsilon})$, for efficient point location queries in the region below $E_{\mathcal{F}}$: For a point $\mathbf{x} = (a_1, a_2, a_3, a_4)$ in $\mathbb{R}^4$, we can determine in $O(\log n)$ time whether $a_4 < E_{\mathcal{F}}(a_1, a_2, a_3)$. The technique for constructing this data structure crucially relies on the existence of a vertical decomposition of this region with near-cubic complexity.

We next observe that each of the three optimization problems mentioned above can be reduced to the problem of computing a closest (or farthest) pair between two sets of objects in $\mathbb{R}^d$, under some appropriate (pseudo-)distance function. That is, we define two sets of objects A, B and a function $\delta : A \times B \to \mathbb{R}^+ \cup \{0\}$, and reduce the original optimization problem to that of computing $\delta^* = \min_{a \in A, b \in B} \delta(a,b)$. Actually, we need to solve several instances of the closest-pair problem, but we show that the overall running time is still bounded by (a polylogarithmic factor times) the time complexity of the algorithm for computing δ^*. We use a randomized divide-and-conquer approach to compute δ^*, which is inspired by the Clarkson–Shor algorithm [10] for computing the diameter of a set of n points in 3-space. The merge and the divide steps of our algorithm require a data structure for point location in the minimization diagram of a set of trivariate functions that satisfy the aforementioned properties. Surprisingly, the xy-monotonicity property, which might be regarded as a somewhat restrictive condition, is also satisfied for each of the three optimization problems under consideration. Using our bounds on the complexity of vertical decompositions, we show that the expected running time of our algorithms, for all three problems, is $O(n^{3/2+\varepsilon})$, for any $\varepsilon > 0$. The previously best known algorithms for these problems are due to Agarwal et al. [1], and are based on Megiddo's *parametric search* technique (see also [3, 6]). The expected running time of the algorithms in [1] is $O(n^{17/11+\varepsilon})$.

We consider the main contributions of this paper to be the general algorithmic technique itself and the bound on the size of the vertical decomposition, both of which might be useful in other problems. Even though Megiddo's parametric search technique is a very powerful paradigm, it typically leads to quite complicated algorithms. Recently, there have been several attempts [5, 11, 18, 19, 20] to present simpler and more direct algorithms for some of the problems that have traditionally been solved using parametric searching. Our technique can be viewed as another step in this direction.

The paper is organized as follows. We first establish in Section 2 our bound on the complexity of the vertical decomposition. We then present in Section 3 the general algorithmic technique for computing a closest pair, and exemplify it by applying it to the width problem. We briefly discuss the minimum width annulus and the biggest stick problems in Sections 4 and 5, respectively.

2 Complexity of the Vertical Decomposition

Let $\mathcal{F}$ be a collection of n totally or partially-defined algebraic trivariate functions of constant maximum degree b that satisfy the following properties:

(F1) If a function $f \in \mathcal{F}$ is partially defined, then we require that the set of points where f is undefined have measure 0, in the following strong sense: There is a constant number of algebraic arcs of constant maximum degree in the xy-plane (these arcs depend on f), so that, for each point (x, y) not lying on any of these arcs, f is defined at all points (x, y, z), for any $z \in \mathbb{R}$.

(F2) For any pair of functions $f, f' \in \mathcal{F}$, the surface $f(x, y, z) = f'(x, y, z)$ is xy-monotone, that is, every z-vertical line (not passing through any curve where f or f' is undefined) crosses this surface in exactly one point. Actually, we somewhat relax this assumption, requiring only that, for each such surface σ, the xy-plane can be decomposed into a constant number of regions, each of constant description complexity (i.e., described by a constant number of polynomial equalities and inequalities of constant maximum degree), so that, for each of these regions c, the surface σ is the graph of a continuous bivariate function (of x and y) over the interior of c.

These assumptions are rather restrictive, but, as we will show below, and rather surprisingly, they hold for several of the current main applications of lower envelopes in 4-space, as listed in the introduction and studied recently in [1, 3, 6]. We also assume that the functions in $\mathcal{F}$ are in *general position*, as defined, e.g., in [22]; it is easy to show, using a variant of the argument given in [22], that this assumption does not involve any loss of generality, and that our results also hold for collections not in general position. Under this assumption, for $j = 0, \ldots, 3$, the envelope $E_{\mathcal{F}}$ is attained by at most (or, if the functions in $\mathcal{F}$ are totally defined, exactly) $4 - j$ functions of $\mathcal{F}$ over any j-dimensional cell of $M_{\mathcal{F}}$. We will use the terms vertex, edge, face, and cell to denote, respectively, 0-dimensional, 1-dimensional, 2-dimensional, and 3-dimensional cells of $M_{\mathcal{F}}$. Each vertex, edge, face, or cell c of $M_{\mathcal{F}}$ will be labeled by the corresponding set of functions of $\mathcal{F}$ attaining $E_{\mathcal{F}}$ over c. In particular, the term f-cell will refer to a (3-dimensional) cell of $M_{\mathcal{F}}$ over which $E_{\mathcal{F}}$ is attained by the function f.

The *vertical decomposition* of $M_{\mathcal{F}}$ is defined in the following standard manner. In the first decomposition stage, we erect, for each edge e of $M_{\mathcal{F}}$, a z-vertical wall from e, which is the union of all maximal z-vertical relatively-open segments passing through points of e and not meeting any other vertex, edge or face of $M_{\mathcal{F}}$. The collection of these walls partitions the cells of $M_{\mathcal{F}}$ into subcells, so that each subcell c is bounded from above and from below (in the z-direction) by (portions of) a fixed pair of faces of $M_{\mathcal{F}}$; c may also extend to infinity in either direction. In the second decomposition step, we take each of these subcells c and project it onto the xy-plane. We construct the 2-dimensional vertical decomposition of the projection c^*, by erecting a maximal y-vertical segment, contained in the closure of c^*, from each vertex of c^* and each (locally) x-extremal point on ∂c^*. The collection of these segments partitions c^* into 'pseudo-trapezoidal' subcells. Each of these subcells τ induces a subcell of c, obtained by intersecting c with the vertical cylinder $\tau \times \mathbb{R}$ over τ. The resulting collection of subcells constitutes the vertical decomposition of $M_{\mathcal{F}}$, which we denote by $M_{\mathcal{F}}^*$. Each of these cells has constant description complexity, in the sense that it is defined by a constant number of polynomial equalities and inequalities of constant maximum degree (depending on the maximum degree b of the functions of $\mathcal{F}$). See [7, 9, 23] for more details concerning vertical decompositions.

Theorem 2.1 *If $\mathcal{F}$ is a collection of trivariate functions satisfying the assumptions made above, then the number of subcells of $M_{\mathcal{F}}^*$ is $O(n^{3+\varepsilon})$, for any $\varepsilon > 0$, where the constant of proportionality depends on ε and on the maximum degree b.*

Proof: Let $\mathcal{F}$ be a collection of n trivariate functions satisfying the above assumptions. It is easily seen that, in general, the second vertical decomposition step does not increase the complexity of the decomposition by more than a constant factor, so it suffices to bound the increase in the complexity of $M_{\mathcal{F}}$ caused by the first vertical decomposition step. In other words, we want to count the number of pairs (e, e') of edges of $M_{\mathcal{F}}$, both bounding the same cell c, such that there exists a z-vertical segment connecting a point on e to a point on e' and fully contained in c. (We actually want to count the number of these vertical segments, but, by assumption, this number is larger than the number of pairs (e, e') by only a constant factor, depending on the maximum degree b.) We say that such a pair (e, e') of edges are *vertically visible*. Suppose c is an f_0-cell, for some $f_0 \in \mathcal{F}$ (note that assumption (F1) implies that there is no 3-dimensional cell of $M_{\mathcal{F}}$ over which $E_{\mathcal{F}}$ is undefined). Then e must be either a portion of an intersection curve of the form $f_0 = f_1 = f_2$, for some pair of functions $f_1, f_2 \in \mathcal{F}$, or a portion of the boundary of an xy-monotone piece of a surface $f_0 = f$, for some $f \in \mathcal{F}$. Similarly, e' must also be a portion of an intersection curve or of a boundary curve of the above forms.

We estimate the number of vertically-visible pairs of edges for which the vertical segment connecting the edges crosses an f_0-cell, separately for each fixed $f_0 \in \mathcal{F}$. Recall that, by assumption, each surface $f_0 = f$ can be decomposed into a constant number of xy-monotone pieces, so that the xy-projections of these pieces are pairwise disjoint. Consider the collection $\Sigma(f_0)$ consisting of all these xy-monotone portions of surfaces of the form $f_0 = f$, for $f \in \mathcal{F}$. We regard each such portion as a partially defined function of x and y. It follows from assumption (F1) that, for each surface $\sigma \in \Sigma(f_0)$ contained in the graph of $f_0 = f$, for some $f \in \mathcal{F}$, either all points lying vertically above σ (in the z-direction) satisfy $f_0 > f$ or all such points satisfy $f_0 < f$ (this property may fail at points lying on the boundary $\partial \sigma$ of σ, but this limit behavior does not affect our analysis). Let $\Sigma^+(f_0)$ (resp. $\Sigma^-(f_0)$) denote the subset of surfaces $\sigma \in \Sigma(f_0)$ for which the corresponding function f satisfies $f_0 > f$ (resp. $f_0 < f$) for all points lying vertically above σ (note that a function f may contribute surfaces to both collections $\Sigma^+(f_0)$, $\Sigma^-(f_0)$, over pairwise-disjoint portions of the xy-plane). It is then clear that the union of all f_0-cells is the same as the region enclosed between the upper envelope of $\Sigma^-(f_0)$ and the lower envelope of $\Sigma^+(f_0)$. It then follows from Theorem 3.2 of [2], concerning the complexity of the region enclosed between two envelopes in 3-space, that the complexity of the vertical decomposition of all f_0-cells is $O(n^{2+\varepsilon})$, for any $\varepsilon > 0$. Repeating this argument over all functions $f_0 \in \mathcal{F}$, we obtain the bound asserted in the theorem. □

Let $C_\mathcal{F}$ be the cell in the arrangement of $\mathcal{F}$ lying below $E_\mathcal{F}$. The vertical decomposition of $C_\mathcal{F}$, denoted as $C_\mathcal{F}^*$, can be obtained by lifting each cell $\tau \in M_\mathcal{F}^*$ to the cell

$$\hat{\tau} = \{(\mathbf{x}, z) \mid \mathbf{x} \in \tau, -\infty \leq z \leq E_\mathcal{F}(\mathbf{x})\}.$$

Since each cell of $M_\mathcal{F}^*$ contributes exactly one cell to $C_\mathcal{F}^*$, the latter also has $O(n^{3+\varepsilon})$ cells. Similarly, it follows that the vertical decomposition of the cell lying above the graphs of all functions in $\mathcal{F}$ also has $O(n^{3+\varepsilon})$ cells.

Remark: An obvious open problem is to extend this bound to vertical decompositions of minimization diagrams of more general trivariate functions. Using recent analysis techniques, as those in [2, 22], we can obtain an $O(n^{4+\varepsilon})$ bound for the general case of partially-defined trivariate low-degree algebraic functions, but we conjecture that the correct bound is near-cubic.

3 Width in 3-Space

The *width* of a set S of n points in $\mathbb{R}^3$ is the smallest distance between a pair of parallel planes such that the closed slab between the planes contains S. Although the width of a set of n points in the plane can be computed in $O(n \log n)$ time [17], the problem becomes considerably harder in three dimensions. Houle and Toussaint [17] gave a simple $O(n^2)$-time algorithm for computing the width in $\mathbb{R}^3$, and raised the open problem of obtaining a subquadratic solution. Recently, Chazelle et al. [6] presented an $O(n^{8/5+\varepsilon})$-time algorithm, for any $\varepsilon > 0$, which was subsequently improved by Agarwal et al. [1] to $O(n^{17/11+\varepsilon})$. As observed in [6], and further exploited in [1], the problem of computing the width in 3-space can be reduced to the following *bichromatic closest line-pair* problem: Given a set L of m 'red' lines and another set L' of n 'blue' lines in $\mathbb{R}^3$, such that all red lines lie above all blue lines,[1] compute the closest pair of lines in $L \times L'$, where the distance between a pair of lines $\ell, \ell' \in \mathbb{R}^3$ is

$$d(\ell, \ell') = \min_{p \in \ell, q \in \ell'} d(p, q).$$

---

[1] For any pair of nonparallel and nonvertical lines $\ell \in L$, $\ell' \in L'$, we say that ℓ lies above ℓ' if the vertical line passing through the intersection point of the xy-projections of ℓ and ℓ' intersects ℓ above ℓ'. It is interesting to note that the requirement that all red lines lie above all blue lines crucially affects the analysis of the complexity of the resulting algorithm.

Let $d(L, L') = \min_{\ell \in L, \ell' \in L'} d(\ell, \ell')$ denote the distance between a closest pair in $L \times L'$.

Before presenting the algorithm, we need to describe some geometric transforms, which will be crucial for our algorithm. We can map each line $\ell \in L$, not parallel to the yz-plane, to a point $\psi(\ell) = (a_1, a_2, a_3, a_4)$ in $\mathbb{R}^4$, where $y = a_1 x + a_3$ is the equation of the xy-projection of ℓ, and $z = a_2 x + a_4$ is the equation of the xz-projection of ℓ. For any fixed real parameter $\delta \geq 0$, we can also map a line $\ell' \in L'$ to a surface $\gamma(\ell')$, which is the locus of all points $\psi(\ell)$ such that $d(\ell, \ell') = \delta$ and ℓ lies above ℓ'. We refer to the coordinates of this parametric space as $\xi_1, \xi_2, \xi_3, \xi_4$. Observe that any line parallel to the ξ_4-axis intersects $\gamma = \gamma(\ell')$ in at most one point. If the corresponding lines ℓ in $\mathbb{R}^3$ lie in a vertical plane parallel to ℓ' and not containing ℓ', then the intersection point may not exist. It follows that γ can be partitioned into a constant number of surface patches, each of constant description complexity, such that, for each patch $\tilde{\gamma}$, all points of $\mathbb{R}^4$ lying vertically above $\tilde{\gamma}$ represent lines ℓ in $\mathbb{R}^3$ that lie above ℓ' and $d(\ell, \ell') > \delta$, and all points lying below $\tilde{\gamma}$ represent lines ℓ that either lie below ℓ', pass through ℓ', or lie above ℓ' and $d(\ell, \ell') < \delta$. In other words, $\gamma(\ell')$ is the graph of a partially defined function $\xi_4 = f_{\ell'}(\xi_1, \xi_2, \xi_3)$. For a point $\psi(\ell) = (a_1, a_2, a_3, a_4)$, such that ℓ lies above ℓ', if $f_{\ell'}(a_1, a_2, a_3)$ is defined, then $a_4 > f_{\ell'}(a_1, a_2, a_3)$ if and only if $d(\ell, \ell') > \delta$, and $a_4 < f_{\ell'}(a_1, a_2, a_3)$ if and only if $d(\ell, \ell') < \delta$. Let $\mathcal{F}$ be the collection $\{f_{\ell'} \mid \ell' \in L'\}$, and let $U_\mathcal{F}$ denote the upper envelope of $\mathcal{F}$. For a line $\ell \in L$ with $\psi(\ell) = (a_1, a_2, a_3, a_4)$, we have $a_4 \geq U_\mathcal{F}(a_1, a_2, a_3)$ if and only if $d(\{\ell\}, L') \geq \delta$. It is easily checked that the functions $f_{\ell'}$ are all partially-defined, algebraic functions of constant maximum degree.

Lemma 3.1 *(a) For any line $\ell' \in L'$ and for any fixed ξ_1, ξ_2, so that ξ_1 is not equal to the ξ_1-coordinate of ℓ', the function $f_{\ell'}(\xi_1, \xi_2, \xi_3)$ is defined for all ξ_3. (b) For any pair of non-parallel lines $\ell_1', \ell_2' \in L'$ and for any fixed ξ_1, ξ_2, the equation*

$$f_{\ell_1'}(\xi_1, \xi_2, \xi_3) = f_{\ell_2'}(\xi_1, \xi_2, \xi_3) \qquad (1)$$

has a unique solution ξ_3, except when (ξ_1, ξ_2) lies on a certain critical line $\lambda(\ell_1', \ell_2')$ that depends on ℓ_1' and ℓ_2', or when ξ_1 is equal to the ξ_1-coordinate of ℓ_1' or of ℓ_2'.

Proof: Part (a) is trivial: If ξ_1, ξ_2 are fixed, then the spatial orientation of the corresponding line ℓ' is fixed. If the xy-projections of ℓ and ℓ' do not have the same orientation, then $F_{\ell'}(\xi_1, \xi_2, \xi_3)$ is defined for all ξ_3. (If these projections have the same orientation, then $F_{\ell'}(\xi_1, \xi_2, \xi_3)$ is defined only when ξ_2 is equal to the ξ_2-coordinate of ℓ and when ξ_3 is such that ℓ and ℓ' lie in the same vertical plane.)

As to part (b), let ξ_3 be a solution of (1), and let $\xi_4 = F_{\ell_1'}(\xi_1, \xi_2, \xi_3) = F_{\ell_2'}(\xi_1, \xi_2, \xi_3)$. The line ℓ^*, parametrized by $(\xi_1, \xi_2, \xi_3, \xi_4)$, thus lies in the vertical plane $\pi(\xi_3)$: $y = \xi_1 x + \xi_3$, and, as is easily checked, its slope in that plane, with respect to the coordinate frame (u, z), where u is the axis orthogonal to the z-axis, is equal to $\xi_2/\sqrt{1+\xi_1^2}$. Moreover, by definition, ℓ^* is a common upper tangent line to the two cylinders C_1, C_2 of radius δ, whose symmetry axes are the lines ℓ_1', ℓ_2', respectively. Let $K_i = K_i(\xi_3) = C_i \cap \pi(\xi_3)$, for $i = 1, 2$. The sets K_1 and K_2 are two ellipses, and the line ℓ^* must be a common upper tangent to K_1 and K_2 in the plane $\pi(\xi_3)$ (this holds provided that ξ_1 is not equal to the ξ_1-coordinate of ℓ_1' or of ℓ_2'). As ξ_3 varies, the plane $\pi(\xi_3)$ translates parallel to itself, and the two ellipses $K_i(\xi_3)$ also translate within that plane, so that the positions of their centers are given by two linear functions of ξ_3. Moreover, for $i = 1, 2$, let $w_i = w_i(\xi_3)$ denote the point on $K_i(\xi_3)$ so that the line tangent to K_i at w_i has slope $\xi_2/\sqrt{1+\xi_1^2}$ and lies above K_i. It follows that, as ξ_3 varies, $w_i(\xi_3)$ moves within the plane $\pi(\xi_3)$ as a linear function of ξ_3, for $i = 1, 2$. Thus, ξ_3 solves (1) if and only if the line connecting $w_1(\xi_3)$ and $w_2(\xi_3)$ has slope $\xi_2/\sqrt{1+\xi_1^2}$ in $\pi(\xi_3)$. This equation is linear in ξ_3, as easily follows from the above arguments, and so has either one solution, no solutions, or infinitely many solutions.

To analyze when this equation has no solution, or has infinitely many solutions, we represent the above geometric reasoning in an algebraic form. It is easily verified that the existence of a unique solution to (1) is not affected if we translate ℓ_1' and ℓ_2' by any amounts (such a translation only changes the constant term in the resulting linear equation), so we may assume, with no loss of generality, that both lines pass through the origin. Let (a_1, b_1, c_1), (a_2, b_2, c_2) be two unit vectors lying, respectively, on the lines ℓ_1', ℓ_2'. The intersection $s_1(\xi_3)$ of ℓ_1' with $\pi(\xi_3)$ is a point $(a_1 t, b_1 t, c_1 t)$ that satisfies the equation $b_1 t = \xi_1 a_1 t + \xi_3$, so we have $t = \xi_3/(b_1 - a_1\xi_1)$, which implies that $s_1(\xi_3)$ is the point

$$\left(\frac{a_1 \xi_3}{b_1 - a_1 \xi_1}, \frac{b_1 \xi_3}{b_1 - a_1 \xi_1}, \frac{c_1 \xi_3}{b_1 - a_1 \xi_1} \right),$$

and, similarly, the intersection $s_2(\xi_3)$ of ℓ_2' with $\pi(\xi_3)$ is the point

$$\left(\frac{a_2 \xi_3}{b_2 - a_2 \xi_1}, \frac{b_2 \xi_3}{b_2 - a_2 \xi_1}, \frac{c_2 \xi_3}{b_2 - a_2 \xi_1} \right).$$

(As above, these points are well-defined only when ξ_1 is not equal to the ξ_1-coordinate of ℓ_1' or of ℓ_2'.) For $i = 1, 2$, the point $w_i(\xi_3)$ is a translated copy of $s_i(\xi_3)$ by a fixed vector, independent of ξ_3. The coefficient of ξ_3 in the equation (1) is thus easily seen to be (proportional to)

$$\frac{c_2}{b_2 - a_2 \xi_1} - \frac{c_1}{b_1 - a_1 \xi_1} - \xi_2 \left(\frac{a_2}{b_2 - a_2 \xi_1} - \frac{a_1}{b_1 - a_1 \xi_1} \right).$$

Hence, the equation (1) does not have a unique solution only when this expression is 0. That is,

$$\frac{c_2 - a_2 \xi_2}{b_2 - a_2 \xi_1} = \frac{c_1 - a_1 \xi_2}{b_1 - a_1 \xi_1},$$

which is easily seen to be a linear equation in ξ_1 and ξ_2 (it does not vanish identically, unless ℓ_1' and ℓ_2' are parallel). This completes the proof of the lemma. $\square$

Lemma 3.1 implies that the collection $\mathcal{F}$ satisfies the assumptions (F1), (F2) of Theorem 2.1. Let $C_\mathcal{F}$ denote the cell in the arrangement of $\mathcal{F}$ that lies above the upper envelope of $\mathcal{F}$. In view of Lemma 3.1, Theorem 2.1, the above discussion, and standard point-location techniques, such as those in [7, 8], we obtain

Corollary 3.2 *The vertical decomposition $C_\mathcal{F}^*$ of $C_\mathcal{F}$ consists of $O(n^{3+\varepsilon})$ cells, for any $\varepsilon > 0$. Moreover, $C_\mathcal{F}$ can be preprocessed in time $O(n^{3+\varepsilon})$ into a data structure of size $O(n^{3+\varepsilon})$, for any $\varepsilon > 0$, so that, for any query point p, we can determine in $O(\log n)$ time whether $p \in C_\mathcal{F}$.*

We are now in position to describe the algorithm for computing $d(L, L')$. We will first present an outline of the algorithm, and then describe each of the nontrivial steps in some detail.

ALGORITHM: CLOSEST-PAIR

1. Let n_0 be a sufficiently large constant, whose value will be fixed later. If $n \leq n_0$, then we compute $d(\ell, \ell')$ for every pair $(\ell, \ell') \in L \times L'$, in $O(m)$ time, and return the minimum distance.

2. Assume that $n > n_0$. Randomly choose a line $\ell_0 \in L$ and compute $\delta_0 = d(\{\ell_0\}, L')$, in $O(n)$ time.

3. Set $r = \lceil m^{3/8}/n^{1/8} \rceil$. We partition L into $k+1$ subsets $L_0, L_1, \ldots, L_k$, with the following properties:

   (i) $k = O(r^{3+\varepsilon})$, for any $\varepsilon > 0$;
   (ii) if $r = 1$, then $k = 1$, $L_0 = \emptyset$, $L_1 = L$;
   (iii) for each $1 \leq i \leq k$,

   $$|\{\ell' \in L' \mid d(\{\ell'\}, L_i) < \delta_0 \}| \leq \frac{n}{r};$$

   (iv) $L_0 \subseteq \{\ell \in L \mid d(\{\ell\}, L') < \delta_0\}$; L_0 may be empty (as is the case when $r = 1$).

4. For each $1 \leq i \leq k$, we compute a set L'_i of size at most n/r such that

$$L'_i \supseteq \{\ell' \in L' \mid d(\{\ell'\}, L_i) < \delta_0\};$$

if $r = 1$ and $k = 1$, we put $L'_1 = L'$. Set $m_i = |L_i|$ and $n_i = |L'_i|$.

5. For each $1 \leq i \leq k$, we do the following: If $n_i = 0$, we set $d(L_i, L'_i) = +\infty$. Otherwise, we compute $\delta^*_i = d(L_i, L'_i)$ directly, using a different algorithm (detailed below). We then compute $\delta_1 = \min_i \delta^*_i$.

6. If $L_0 \neq \emptyset$, we compute $\delta_2 = d(L_0, L')$ recursively.

7. Return $\min\{\delta_0, \delta_1, \delta_2\}$ as $d(L, L')$.

Next, we explain Steps 3–5 in detail, and analyze their expected running time; the other steps are trivial and need no further explanation. We will then conclude the analysis by proving the correctness of the algorithm.

Steps 3–4: We compute L_i, L'_i, for $1 \leq i \leq k$, using a divide-and-conquer approach. We construct a tree T, each of whose nodes v is associated with a subset $L_v \subseteq L$ and another subset $L'_v \subseteq L'$. The root of the tree is associated with L and L' themselves. The subsets associated with the leaves of T will correspond to the sets L_i and L'_i.

If $r = 1$ then T consists of a single node; we set $k = 1$, $L_1 = L$, $L'_1 = L'$, and $L_0 = \emptyset$. Next, assume that $r > 1$. Let s be some sufficiently large constant. We choose a random subset $X \subseteq L'$ of size $c_1 s \log s$, where c_1 is an appropriate constant independent of s, and compute C^*_X, the vertical decomposition of the cell C_X lying above the graphs of all the functions $\{f_{\ell'} \mid \ell' \in X\}$ (defined in terms of the parameter δ_0 computed in Step 2). By Corollary 3.2, C^*_X has $O((s \log s)^{3+\varepsilon})$ cells. For each cell $\tau \in C^*_X$, we compute the set $L'_\tau \subseteq L'$ of lines ℓ' such that $\gamma(\ell')$ intersects τ. By standard ε-net theory [16], we have, with high probability, $|L'_\tau| \leq n/s$ for every $\tau \in C^*_X$. If $|L'_\tau| > n/s$ for some $\tau \in C^*_X$, we choose another random subset and repeat the above steps. Otherwise, for each $\tau \in C^*_X$ we compute the subset $L_\tau \subseteq L$ of lines ℓ such that $\psi(\ell) \in \tau$. Set $m_\tau = |L_\tau|$ and $n_\tau = |L'_\tau|$. If $L_\tau \neq \emptyset$, we create a child v_τ of the root corresponding to τ. We associate L_τ, L'_τ with v_τ. If $|L'_\tau| \leq n/r$, then v_τ is a leaf. Otherwise, v_τ is an internal node of T, and we expand T further at v_τ by applying the same procedure recursively to L_τ, L'_τ.

By construction, the depth of T is at most $\lceil \log_s r \rceil$. Since each node has at most $O((s \log s)^{3+\varepsilon})$ children, the total number of leaves in T is $k \leq c_2 r^{3+\varepsilon'}$, for any $\varepsilon' > \varepsilon$ and for some constant c_2 independent of s and r (but depending on $\varepsilon, \varepsilon'$). We set L_i, L'_i to be the subsets associated with the ith leaf of T, for $i = 1, \ldots, k$. Finally, we set $L_0 = L - \bigcup_{i=1}^k L_i$. Note that a line ℓ is placed in L_0 only when its image $\psi(\ell)$ lies below the upper envelope of some collection $\{f_{\ell'} \mid \ell' \in X\}$, for some $X \subseteq L'$. Hence, by definition, all lines $\ell \in L_0$ satisfy $d(\{\ell\}, L') < \delta_0$. In particular, $\ell_0 \notin L_0$, so $|L_0| < m$, a property that we will use below when proving the correctness of the algorithm. This also shows that L_0 satisfies property (iv) of Step 3.

The sets L_i, for $1 \leq i \leq k$, are pairwise disjoint, and $|L'_i| \leq n/r$, for all $1 \leq i \leq k$. It thus remains to show that

$$L'_i \supseteq \{\ell' \in L' \mid d(\{\ell'\}, L_i) < \delta_0\}.$$

In fact, the following stronger claim is true, and follows easily by construction.

Lemma 3.3 *For any node v_τ in T,*

$$L'_\tau \supseteq L''_\tau = \{\ell' \in L' \mid d(\{\ell'\}, L_\tau) < \delta_0\}.$$

Proof: We prove this by induction on the depth of v_τ in T. The claim obviously holds for the root of T. Suppose it holds for the parent v_ζ of a node v_τ. Since $L_\tau \subseteq L_\zeta$, obviously $L''_\tau \subseteq L''_\zeta$. Let C^*_X be the set of cells that we constructed at v_ζ. Then $\tau \in C^*_X$ and, for every $\ell \in L_\tau$, we have $\psi(\ell) \in \tau$. Let $\ell' \in L''_\tau$ and let ℓ be a line in L_τ satisfying $d(\ell, \ell') < \delta_0$. Then, by definition, the point $\psi(\ell)$ lies below the surface $\gamma(\ell')$. Since τ is unbounded in the $+\xi_4$-direction, it follows that $\gamma(\ell')$ intersects τ. Moreover, by the induction hypothesis, $\ell' \in L'_\zeta$, which implies that $\ell \in L'_\tau$, and thus the claim is true for τ as well. $\square$

Hence, the sets L_i, L'_i, for $1 \leq i \leq k$, satisfy the desired properties of Steps 3 and 4.

Next, we analyze the expected time spent in computing these subsets. Let $f(a, b)$ denote the maximum expected time spent by the recursive algorithm for Steps 3–4, where expectation is with respect to the choices of random samples by the algorithm, and where the maximum is taken over all sets L, L' of lines, as above, of respective sizes a, b. At each level of recursion, X is chosen, with high probability, only once, and we spend $O((s \log s)^{3+\varepsilon}(a+b))$ time to compute all the sets L_τ, L'_τ, so we obtain the following recurrence:

$$f(a, b) \leq \sum_{\tau \in C^*_X} f(a_\tau, b_\tau) + c(s \log s)^{3+\varepsilon}(a + b),$$

where $|C^*_X| \leq c'(s \log s)^{3+\varepsilon}$, $\sum_{\tau \in C^*_X} a_\tau \leq a$, $b_\tau \leq b/s$, and c, c' are constants (depending on ε). The recursion stops when $b \leq n/r$, so $f(a, b) = O(1)$ for $b \leq n/r$.

The solution of the above recurrence is

$$f(a,b) \leq A\left(a \log b + \frac{b^{3+\varepsilon'}}{n^2}r^2\right),$$

for any $\varepsilon' > \varepsilon$; here $A = A(\varepsilon')$ is a sufficiently large constant depending on the value of ε.

We prove this by induction on b. The inequality obviously holds for $b \leq n/r$. For larger values of b, we obtain, by the induction hypothesis,

$$\begin{aligned}f(a,b) &\leq \sum_{\tau \in C_X^*} A\left(a_\tau \log b_\tau + \frac{b_\tau^{3+\varepsilon'}}{n^2}r^2\right) + \\ &\quad c(s \log s)^{3+\varepsilon}(a+b) \\ &\leq A \sum_{\tau \in C_X^*}\left(a_\tau \log \frac{b}{s} + \left(\frac{b}{s}\right)^{3+\varepsilon'}\cdot\frac{r^2}{n^2}\right) + \\ &\quad c(s \log s)^{3+\varepsilon}(a+b) \\ &\leq Aa \log b + a\left[c(s \log s)^{3+\varepsilon} - A \log s\right] + \\ &\quad Ab^{3+\varepsilon'}\frac{r^2}{n^2}\left[c's^{\varepsilon-\varepsilon'}\log^{3+\varepsilon}s + \right. \\ &\quad\quad \left.\frac{c}{A}\frac{n^2/r^2}{b^2}(s\log s)^{3+\varepsilon}\right] \\ &\leq A\left(a \log b + b^{3+\varepsilon'}\frac{r^2}{n^2}\right),\end{aligned}$$

because $b > n/r$, $\varepsilon' > \varepsilon$, and A is chosen sufficiently large. Since $r = \lceil m^{3/8}/n^{1/8}\rceil$, and initially $a = m$ and $b = n$, we obtain

$$f(m,n) = O(m^{3/4+\varepsilon'}n^{3/4+\varepsilon'} + m^{1+\varepsilon'} + n^{1+\varepsilon'}).$$

Step 5: For each $1 \leq i \leq k$, We compute $d(L_i, L_i')$ using a somewhat simpler version of the randomized algorithm described by Agarwal et al. [1]. We give a brief sketch of this variant.

(i) Let n_0 be some sufficiently large constant. If $n_i \leq n_0$, we compute $d(\ell, \ell')$ for all pairs $\ell \in L_i$, $\ell' \in L_i'$, in $O(m_i)$ time, and return the minimum distance.

(ii) Assume $n_i > n_0$. Choose a random subset $A \subseteq L_i'$ of size $\lceil n_i/2 \rceil$; each subset of size $\lceil n_i/2 \rceil$ is chosen with equal probability.

(ii) Recursively compute $\delta' = d(L_i, A)$.

(iii) Compute the set

$$B = \{\ell' \in L_i' - A \mid d(L_i, \{\ell'\}) < \delta'\}.$$

(iv) Compute $d(\ell, \ell')$ for all pairs $\ell \in L_i$, $\ell' \in B$, and return the minimum distance (or output δ' if B is empty).

The correctness of the algorithm is obvious (see also [1]), so we now analyze its expected running time. For a line $\ell \in L_i$, let

$$B^{(\ell)} = \{\ell' \in L_i' - A \mid d(\ell, \ell') < \delta'\}.$$

Using a standard probabilistic argument, it can be shown that the expected size of $B^{(\ell)}$ is $O(1)$. Since $B = \bigcup_{\ell \in L_i} B^{(\ell)}$, the expected size of B is $O(m_i)$, and the expected running time of Step 5(iv) is $O(m_i^2)$.

By reversing the direction of the z-axis, setting $\delta = \delta'$, and using Corollary 3.2, L_i can be preprocessed into a data structure of size $O(m_i^{3+\varepsilon})$, so that, for each $\ell' \in L_i' - A$, we can determine in $O(\log m_i)$ time whether $\ell' \in B$. The time spent in Step 5(iii) is thus $O(m_i^{3+\varepsilon} + n_i \log m_i)$, which subsumes the expected cost of Step 5(iv). The running time of both steps can be improved, by a standard batching technique, to $O(m_i n_i^{2/3+\varepsilon} + n_i^{1+\varepsilon})$; see [1]. Let $\varphi(m_i, n_i)$ denote the maximum expected running time for computing $d(L_i, L_i')$ by this algorithm, where the maximum is taken over all sets L_i, L_i' of sizes m_i, n_i, respectively. Then we obtain the following recurrence

$$\varphi(m_i, n_i) \leq \varphi(m_i, \lceil n_i/2 \rceil) + O(m_i n_i^{2/3+\varepsilon} + n_i^{1+\varepsilon}),$$

whose solution is easily seen to be

$$\varphi(m_i, n_i) = O(m_i n_i^{2/3+\varepsilon} + n_i^{1+\varepsilon}).$$

By the choice of the parameter r, the expected time spent in Step 5 is thus

$$\begin{aligned}\sum_{i=1}^{k}\varphi(m_i,n_i) &= \sum_{i=1}^{k} O(m_i n_i^{2/3+\varepsilon} + n_i^{1+\varepsilon}) \\ &= O\left(\left(\frac{n}{r}\right)^{2/3+\varepsilon}\sum_{i=1}^{k}m_i + \left(\frac{n}{r}\right)^{1+\varepsilon}\cdot r^{3+\varepsilon'}\right) \\ &= O(m^{3/4+\varepsilon'}n^{3/4+\varepsilon'} + m^{1+\varepsilon'} + n^{1+\varepsilon'}),\end{aligned}$$

The total expected time spent by the algorithm, excluding the time spent in the recursive call, is thus $O(m^{3/4+\varepsilon'}n^{3/4+\varepsilon'} + m^{1+\varepsilon'} + n^{1+\varepsilon})$.

Recall that all lines $\ell \in L_0$ satisfy $d(\{\ell\}, L') < \delta_0$. Recall also that ℓ_0 was chosen randomly in Step 2. If we sort the lines $\ell \in L$ in the nondecreasing order of their distances $d(\{\ell\}, L')$, then the probability that ℓ is the ith item in this list is $1/m$, and in this case we must have $|L_0| < i$. Let $T(m,n)$ denote the maximum expected time for the algorithm to compute $d(L, L')$, where the maximum is taken over all sets L, L' of sizes m and n, respectively. The arguments just given imply that

$$T(m,n) \leq \begin{cases} c_1 m & \text{for } n \leq n_0, \\ c_2 n^{1+\varepsilon} & \text{for } n > n_0, m < n^{1/3}, \\ \frac{1}{m}\sum_{i=0}^{m-1} T(i,n) + \\ A(m^{3/4+\varepsilon'} n^{3/4+\varepsilon'} + m^{1+\varepsilon'} + n^{1+\varepsilon'}) \\ & \text{for } n > n_0, m > n^{1/3}. \end{cases}$$

The solution of the above recurrence is

$$T(m,n) \leq B(m^{3/4+\varepsilon'} n^{3/4+\varepsilon'} + m^{1+\varepsilon'} \log n + n^{1+\varepsilon'}),$$

for any $\varepsilon' > \varepsilon$ and for some constant $B = B(\varepsilon')$.

To complete the analysis, we finally show:

Lemma 3.4 *Algorithm* CLOSEST-PAIR *computes the distance* $\delta^* = d(L, L')$ *correctly.*

Proof: We prove the lemma by double induction on m and n. If $n < n_0$ then the correctness is trivial (see Step 1). If $m \leq n^{1/3}$ (i.e., $r = 1$), then the algorithm is also correct, by the analysis of Step 5 given above. Suppose that $m > n^{1/3}$ (so $r > 1$), and that the algorithm computes $\delta^* = d(L, L')$ correctly for all sets of lines L and L' such that $|L| < m$ or $|L| = m$ and $|L'| < n$. Since the algorithm returns the distance between a line of L and a line of L', it always returns a number at least as large as δ^*.

If $\delta^* = \delta_0$, there is nothing to prove, because Step 7 returns the minimum of δ_0, δ_1, and δ_2. Suppose $\delta^* < \delta_0$. Let $\ell \in L, \ell' \in L'$ be a pair of lines with $d(\ell, \ell') = \delta^*$. If $\ell \in L_0$, then, $d(L, L') = d(L_0, L') = \delta_2$. Since $|L_0| < m$, as argued above, the algorithm computes $d(L_0, L')$ correctly, by the induction hypothesis, so $d(L, L')$ is also correctly computed.

If $\ell \in L_i$ for some $1 \leq i \leq k$, then, by construction, $\ell' \in L_i'$, and therefore $d(L, L') = d(L_i, L_i') = \delta_1$. Since $|L_i| \leq m$ and $|L_i'| \leq n/r < n$, the claim follows again by the induction hypothesis. This completes the proof of the lemma. □

Hence, we can conclude

Theorem 3.5 *Given a set L of m red lines and a set L' of n blue lines in $\mathbb{R}^3$, such that all red lines lie above all blue lines, $d(L, L')$ can be computed by a randomized algorithm in $O(m^{3/4+\varepsilon} n^{3/4+\varepsilon} + m^{1+\varepsilon} + n^{1+\varepsilon})$ expected time, for any $\varepsilon > 0$.*

Combining this result with the observations in [6], which we omit here, and which show how the width itself can be computed using Algorithm CLOSEST-PAIR as a subroutine, we obtain:

Corollary 3.6 *The width of a set of n points in $\mathbb{R}^3$ can be computed by a randomized algorithm whose expected running time is $O(n^{3/2+\varepsilon})$, for any $\varepsilon > 0$.*

4 Minimum Width Annulus

Given a set $S = \{p_1, \ldots, p_n\}$ of n points in the plane, we want to compute an annulus of the smallest width that contains S. That is, we want to compute two concentric circles, centered at a point ξ, of radii r_1, r_2, respectively, such that $r_2 - r_1$ is minimized, subject to the constraints $r_1 \leq d(\xi, p_i) \leq r_2$, for each $1 \leq i \leq n$. For a given point $\mathbf{x}$, let $\omega(\mathbf{x})$ denote the smallest width of an annulus containing S and centered at $\mathbf{x}$. Let p_c, p_f be the nearest and the farthest neighbors of $\mathbf{x}$ in S, respectively. Then

$$\omega(\mathbf{x}) = d(\mathbf{x}, p_f) - d(\mathbf{x}, p_c).$$

Let $Vor_c(S)$ (resp. $Vor_f(S)$) denote the closest (resp. farthest) point Voronoi diagram of S. Ebara et al. [14] observed that the center of a minimum width annulus containing S is a vertex of $Vor_c(S)$, a vertex of $Vor_f(S)$, or an intersection point of an edge of $Vor_c(S)$ and an edge of $Vor_f(S)$ (see also [24]). Based on this observation, they gave a rather simple $O(n^2)$-time algorithm for computing a minimum width problem, which was improved by Agarwal et al. [3] to $O(n^{8/5+\varepsilon})$, and subsequently by Agarwal et al. [1] to $O(n^{17/11+\varepsilon})$.

We first make the following observation, which was missed in the earlier treatments of the problem cited above, and which transforms the problem into a width-like problem in $\mathbb{R}^3$. We lift the points of S to the paraboloid $z = x^2 + y^2$ by the standard lifting map $(x, y) \mapsto (x, y, x^2 + y^2)$, then a circle C of radius r with center (a, b) is mapped to the plane $C^* : z = 2ax + 2by + (r^2 - a^2 - b^2)$. A point $p \in \mathbb{R}^2$ lies inside (resp. on, outside) C if and only if its lifted image p^* lies below (resp. on, above) the plane C^*. Thus an annulus with center (a, b) and radii $r_1 < r_2$ and containing S is mapped to a pair of parallel planes

$$z = 2ax + 2by + (r_1^2 - a^2 - b^2) \text{ and}$$
$$z = 2ax + 2by + (r_2^2 - a^2 - b^2),$$

such that the image S^* of S is fully contained in the slab bounded between these two planes. In other words, the minimum-width annulus problem reduces to the problem of finding a pair of parallel planes

$$z = \kappa_1 x + \kappa_2 y + \kappa_3', \quad z = \kappa_1 x + \kappa_2 y + \kappa_3$$

that supports S^*, such that

$$\sqrt{\kappa_3' + (\kappa_1^2 + \kappa_2^2)/4} - \sqrt{\kappa_3 + (\kappa_1^2 + \kappa_2^2)/4} \quad (1)$$

is minimized.

This width-like problem can be solved by applying the algorithm of Section 3 almost verbatim, except for the definition of the distance function $d(\ell, \ell')$ between pairs of lines in $\mathbb{R}^3$. In fact, the algorithm itself is essentially independent of the definition of $d(.,.)$, which enters only into the analysis.

An explicit expression for $d(\ell, \ell')$ can be obtained as follows. Let the equations defining ℓ, ℓ' be, as in Section 3,

$$\ell: \quad y = a_1 x + a_3, \qquad z = a_2 x + a_4,$$
$$\ell': \quad y = b_1 x + b_3, \qquad z = b_2 x + b_4.$$

The direction of the common normal to both lines is

$$\mathbf{n} = (a_1 b_2 - a_2 b_1,\ a_2 - b_2,\ -(a_1 - b_1)),$$

and the planes orthogonal to $\mathbf{n}$ and containing ℓ, ℓ', respectively, are

$$(\mathbf{x} - (0, a_3, a_4)) \cdot \mathbf{n} = 0, \text{ and } (\mathbf{x} - (0, b_3, b_4)) \cdot \mathbf{n} = 0,$$

or

$$z = \frac{a_1 b_2 - a_2 b_1}{a_1 - b_1} x + \frac{a_2 - b_2}{a_1 - b_1} y - \frac{a_3(a_2 - b_2) - a_4(a_1 - b_1)}{a_1 - b_1},$$
$$z = \frac{a_1 b_2 - a_2 b_1}{a_1 - b_1} x + \frac{a_2 - b_2}{a_1 - b_1} y - \frac{b_3(a_2 - b_2) - b_4(a_1 - b_1)}{a_1 - b_1}.$$

Plugging this into (1), we thus can write

$$d(\ell, \ell') = \sqrt{\mu a_3 + a_4 + \nu} - \sqrt{\mu b_3 + b_4 + \nu},$$

where μ, ν depend only on a_1, a_2, b_1, b_2.

We can now obtain the appropriate variant of Lemma 3.1. First, we note that if ℓ' is fixed then $d(\ell, \ell')$ is defined whenever the ξ_1-coordinate of ℓ is different than that of ℓ'. In this case $d(\ell, \ell')$ monotonically increases with a_4, which implies that the appropriate variant of the functions $f_{\ell'}$ is well-defined (whenever the ξ_1-coordinates of ℓ and of ℓ' are different). To establish the second part of the lemma, for a fixed parameter δ, let ℓ'_1, ℓ'_2 be two given lines, let ξ_1 and ξ_2 be fixed (with ξ_1 different from the ξ_1-coordinates of ℓ'_1 and of ℓ'_2), and consider the equation

$$f_{\ell'_1}(\xi_1, \xi_2, \xi_3) = f_{\ell'_2}(\xi_1, \xi_2, \xi_3). \qquad (2)$$

We seek a line ℓ that lies above ℓ'_1 and ℓ'_2 and satisfies $d(\ell, \ell'_1) = d(\ell, \ell'_2) = \delta$. This can be written as

$$\sqrt{\mu_1 \xi_3 + \xi_4 + \nu_1} = \sqrt{\mu_1 b_3^{(1)} + b_4^{(1)} + \nu_1} + \delta$$
$$\sqrt{\mu_2 \xi_3 + \xi_4 + \nu_2} = \sqrt{\mu_2 b_3^{(2)} + b_4^{(2)} + \nu_2} + \delta,$$

where $b^{(1)}, b^{(2)}$ are the 4-tuples defining ℓ'_1, ℓ'_2, respectively, and $\mu_1, \nu_1, \mu_2, \nu_2$ are independent of ξ_3, ξ_4. We thus get a linear system of equations in ξ_3, ξ_4, which can be easily reduced to a linear equation in ξ_3 of the form $(\mu_1 - \mu_2)\xi_3 = \alpha$. This shows that (2) has a unique solution, unless $\mu_1 = \mu_2$, or

$$\frac{\xi_2 - b_2^{(1)}}{\xi_1 - b_1^{(1)}} = \frac{\xi_2 - b_2^{(2)}}{\xi_1 - b_1^{(2)}},$$

which is the equation of a line in the $\xi_1 \xi_2$-plane.

We have thus shown that conditions (F1) and (F2) of Section 2 are satisfied in this case too, so the analysis of the preceding section applies here as well, and we can conclude:

Theorem 4.1 *Given a set S of n points in the plane, an annulus of smallest width that contains S can be computed by a randomized algorithm in $O(n^{3/2+\varepsilon})$ expected time, for any $\varepsilon > 0$.*

5 Biggest Stick in a Polygon

The *biggest stick* problem is to find the longest segment that can be placed inside a simple n-gon P in the plane. Chazelle and Sharir gave an $O(n^{1.99})$-time algorithm, which was improved by Agarwal et al. [1] to $O(n^{17/11+\varepsilon})$; see also [3] for an intermediate bound. If the endpoints of the segment are restricted to be at vertices of P, the problem becomes considerably easier, and can be solved in $O(n \log^3 n)$ time [4].

Following the same idea as in [3], we use a divide-and-conquer approach. Partition P into two simple polygons P_1, P_2 by a chord c, so that each of P_1, P_2 has at most $2n/3$ vertices. Let l_c be the line passing through c. We recursively compute the longest segment that can be placed within P_1 and within P_2. The merge step requires computing the longest segment having one endpoint in P_1 and the other in P_2. An easy perturbation argument shows that such a longest segment has to touch two vertices of P, say, v_1, v_2. The difficult case is when $v_1 \in P_1$ and $v_2 \in P_2$. Agarwal et al. [3] showed that finding such a segment can be reduced to the following problem: We have a set L_1 of a lines, where each line $\ell_i \in L_1$ is dual to some vertex v_i of P_1 and has an edge $e_i \in P_1$ associated with it. Similarly, we have another set L_2 of b lines, where each line $\ell_j \in L_2$ is dual to some vertex v_j of P_2 and has an edge e_j of P_2 associated with it. L_1 and L_2 satisfy the following property: For any pair $\ell_i \in L_1, \ell_j \in L_2$, the segment $g_{ij} = a_i a_j$ lies inside P, where a_i (resp. a_j) is the intersection point of the line passing through v_i, v_j with e_i (resp. e_j); see Figure 1. The goal is to compute the longest segment g_{ij}, over all pairs $\ell_i \in L, \ell_j \in L_2$. If we define the distance function $d(\ell_i, \ell_j)$ as the length of the segment g_{ij}, then the goal is to compute a farthest pair in $L \times L'$.

We can show that L_1, L_2 can be parametrized by 4 real parameters (two for the associated vertex of P and two for the associated edge), so that the setup of Section 3 arises here as well; in particular, the resulting collection of trivariate functions satisfies conditions (F1) and (F2). This allows us to apply the

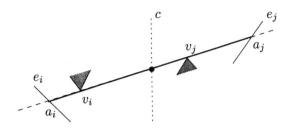

Figure 1: Illustration to the distance function $d(\ell_i, \ell_j)$

algorithm described in Section 3. Omitting all further details here, we obtain:

Theorem 5.1 *Given a simple n-gon P, the longest segment that can be placed inside P can be computed by a randomized algorithm in $O(n^{3/2+\varepsilon})$ expected time, for any $\varepsilon > 0$.*

References

[1] P. Agarwal, B. Aronov and M. Sharir, Computing lower envelopes in four dimensions with applications, *Proc. 10th Annual Symp. on Computational Geometry*, 1994, 348–358.

[2] P. Agarwal, O. Schwarzkopf and M. Sharir, The overlay of lower envelopes in three dimensions and its applications, *this proceedings*.

[3] P. Agarwal, M. Sharir, and S. Toledo, New applications of parametric searching in computational geometry, *J. Algorithms* 17 (1994), 292–318.

[4] A. Aggarwal and S. Suri, The biggest diagonal in a simple polygon, *Inf. Proc. Letters* 35 (1990), 13–18.

[5] H. Brönnimann and B. Chazelle, Optimal slope selection via cuttings, *Proc. 6th Canadian Conf. on Computational Geometry*, 1994, 99–103.

[6] B. Chazelle, H. Edelsbrunner, L. Guibas, and M. Sharir, Diameter, width, closest line pair, and parametric searching, *Discrete Comput. Geom.* 10 (1993), 183–196.

[7] B. Chazelle, H. Edelsbrunner, L. Guibas, and M. Sharir, A singly exponential stratification scheme for real semi-algebraic varieties and its applications, *Proc. 16th Int. Colloq. on Automata, Languages and Programming*, (1989), 179–192. Lecture Notes in Computer Sciences, vol. 371, Springer-Verlag, Berlin. (Also in *Theoretical Computer Science* 84 (1991), 77–105.)

[8] B. Chazelle and M. Sharir, An algorithm for generalized point location and its applications, *J. Symbolic Computation* 10 (1990), 281–309.

[9] K. Clarkson, H. Edelsbrunner, L. Guibas, M. Sharir and E. Welzl, Combinatorial complexity bounds for arrangements of curves and spheres, *Discrete Comput. Geom.* 5 (1990), 99–160.

[10] K. Clarkson and P. Shor, Applications of random sampling in computational geometry II, *Discrete Comput. Geom.* 4 (1989), 387–421.

[11] M. Dillencourt, D. Mount, and N. Netanyahu, A randomized algorithm for slope selection, *Int. J. Comput. Geom. and Appls* 2 (1992), 1–27.

[12] M. de Berg, K. Dobrindt, and O. Schwarzkopf, On lazy randomized incremental construction, *Proc. 26th Annual ACM Symp. on Theory of Computing*, 1994, 105–114.

[13] M. de Berg, D. Halperin, and L. Guibas, Vertical decomposition for triangles in 3-space, *Proc. 10th Annual Symp. Computational Geometry*, 1994, 1–10.

[14] H. Ebara, N. Fukuyama, H. Nakano, and Y. Nakanishi, Roundness algorithms using the Voronoi diagrams, *First Canadian Conf. on Computational Geometry*, 1989.

[15] D. Halperin and M. Sharir, New bounds for lower envelopes in 3 dimensions, with applications to visibility in terrains, *Discrete Comput. Geom.* 12 (1994), 313–326.

[16] D. Haussler and E. Welzl, ϵ-nets and simplex range queries, *Discrete Comput. Geom.* 2 (1987), 127–151.

[17] M. Houle and G. Toussaint, Computing the width of a set, *IEEE Transactions on Pattern Anal. and Mach. Intell.* 5 (1988), 761–765.

[18] M. Katz and M. Sharir, Optimal slope selection via expanders, *Inf. Proc. Letters* 47 (1993), 115–122.

[19] M. Katz and M. Sharir, An expander-based approach to geometric optimization, *Proc. 9th Annual Symp. on Computational Geometry*, 1993, 198–207.

[20] J. Matoušek, Randomized optimal algorithm for slope selection, *Info. Proc. Letters* 39 (1991), 183–187.

[21] N. Megiddo, Applying parallel computation algorithms in the design of serial algorithms, *J. ACM* 30 (1983), 852–865.

[22] M. Sharir, Almost tight upper bounds for lower envelopes in higher dimensions, *Discrete Comput. Geom.* 12 (1994), 327–345.

[23] M. Sharir and P. Agarwal, *Davenport-Schinzel Sequences and Their Geometric Applications*, Cambridge University Press, Cambridge-New York-Melbourne, 1995.

[24] M. Smid and R. Janardan, On the width and roundness of a set of points in the plane, manuscript, 1995.

Accounting for Boundary Effects in Nearest Neighbor Searching

Sunil Arya[*]   David M. Mount[†]   Onuttom Narayan[‡]

Abstract

Given n data points in d-dimensional space, nearest neighbor searching involves determining the nearest of these data points to a given query point. Most average-case analyses of nearest neighbor searching algorithms are made under the simplifying assumption that d is fixed and that n is so large relative to d that *boundary effects* can be ignored. This means that for any query point the statistical distribution of the data points surrounding it is independent of the location of the query point. However, in many applications of nearest neighbor searching (such as data compression by vector quantization) this assumption is not met, since the number of data points n grows roughly as 2^d. Largely for this reason, the actual performances of many nearest neighbor algorithms tend to be much better than their theoretical analyses would suggest. We present evidence of why this is the case. We provide an accurate analysis of the number of cells visited in nearest neighbor searching by the bucketing and k-d tree algorithms. We assume 2^d points uniformly distributed in dimension d and the L_∞ metric. Our analysis is tight in the limit as d approaches infinity. Empirical evidence is presented showing that the analysis applies even in low dimensions.

1 Introduction

Finding nearest neighbors is a fundamental problem in computational geometry with applications to many areas such as pattern recognition, data compression, and statistics. The *nearest neighbor* problem is: given a set of n points P in d-dimensional space, and given a query point $q \in R^d$, find the point of P that minimizes the distance to q.

The problem of preprocessing a set of n points P so that nearest neighbor queries can be answered efficiently has been extensively studied. Nearest neighbor searching can be performed quite efficiently in relatively low dimensions. However, as the dimension d increases, in the worst case, either the space or time complexities increase dramatically. There are many important applications of this problem in moderate dimensions (e.g. in the range from 10 to 20), where best worst-case algorithms are of little practical value. For dimensions in this range, the most practical approaches for nearest neighbor searching are based on simple spatial subdivisions, having poor worst-case performance, but relatively good average-case performance. (Recently it has been suggested that another way to achieve good performance in practice

---

[*]Max-Planck-Institut für Informatik, Im Stadtwald, D-66123 Saarbrücken, Germany. The author was supported by the ESPRIT Basic Research Actions Program, under contract No. 7141 (project ALCOM II).

[†]Department of Computer Science and Institute for Advanced Computer Studies, University of Maryland, College Park, Maryland. The support of the National Science Foundation under grant CCR–9310705 is gratefully acknowledged.

[‡]Physics Department, Harvard University, Cambridge, MA 02138 and AT&T Bell Laboratories, 600 Mountain Avenue, Murray Hill, NJ 07974. Supported in part by the Society of Fellows at Harvard University.

is to compute approximate rather than exact nearest neighbors [1, 3, 6].)

A very simple algorithm that works well for uniformly distributed data is the *bucketing algorithm* (sometimes called *Elias's algorithm* [13]). Rivest [11] presented an analysis of the performance of this algorithm for points uniformly distributed on the vertices of a d-dimensional hypercube, and later Cleary analyzed the case of general uniformly distributed point sets [7]. A practical and more flexible approach for nearest neighbor searching in high dimensions is based on the k-d tree, as introduced by Bentley [5] and applied to the nearest neighbor problem by Friedman, Bentley and Finkel [8].

The analyses of these algorithms presented in the literature [7, 8] show that their running times contain a constant factor, which grows on the order of 2^d, assuming the L_∞ metric. The analysis in [7] assumes that the points are uniformly distributed while the analysis in [8] assumes that the points are randomly chosen from some smooth underlying distribution. The dimension d is regarded as a fixed constant, and the analysis is asymptotic as n tends to infinity. This greatly simplifies the analysis because for any query point (assuming it is chosen from the same distribution as the data points), the statistical distribution of the data points surrounding it can be assumed to be essentially independent of the location of the query point.

However, there are many important applications where the number of data points n and dimension d are related. One such application is *vector quantization*, a technique used in the compression of speech and images [9]. Samples taken from a signal are blocked into vectors of length d (typically after applying some smoothing transforms). Based on a training set of vectors, a set of codevectors is first precomputed. The technique then encodes each new vector by the index of its nearest neighbor among the codevectors. The rate r of a vector quantizer is the number of bits used to encode a sample, and it is related to n, the number of codevectors, by $n = 2^{rd}$. For the common case of $r = 1$, it follows that $n = 2^d$.

For applications in which d and n are related, the theoretical analyses may significantly overestimate the running time of the algorithm. In fact, our interest in this problem stemmed from our observations of the empirical running times for the k-d tree and related algorithms (even for more general point distributions). These algorithms regularly run faster than their predicted performance in higher dimensions. Intuitively, the reason is that when a query point lies close to the periphery of the point set, a significant amount of the data structure that would otherwise need to be searched for the nearest neighbor may be pruned away. Because exponential constant factors in dimension are one of the main obstacles to extending nearest neighbor searching to much higher dimensions (where many more important applications reside), it is of important practical interest to accurately understand the nature of these factors.

In this paper we provide a theoretical explanation of the phenomenon of these *boundary effects* in nearest neighbor searching. We analyze the bucketing algorithm for the uniform distribution, taking into account the effects of the boundary. Our results also apply to the k-d tree algorithm [7]. Because of the complexity of the analysis, we assume that points are uniformly distributed in a d-dimensional unit hypercube, and that distances are measured using the L_∞ metric. Our main result is that given 2^d points in d dimensions, as d tends infinity, the expected number of cells visited by the bucketing algorithm grows as $(0.90\ldots)(1.56594\ldots)^d$. Empirical evidence indicates that this is remarkably close, even for small d. This is significantly smaller than the growth rate of 2^d predicted by previous analyses which ignore boundary effects. (For example, a difference of roughly a factor 7.8 for dimension 8, and a factor of 55 in dimension 16.)

The remainder of the paper is organized as follows. In the next section we present some background on the k-d tree and bucketing algorithms. We present two different analyses on the number of cells visited by the bucketing algorithm. The model is described in Section 3 and analyzed in Section 4 and 5. Both the analyses are done for the limiting case of large dimensions. Section 4 provides a simple analysis which yields a fairly good upper bound. This analysis, however, relies on a technical assumption which we call the *monotonicity conjecture*. In Section 5 we present a sophisticated analysis which yields a tight bound. Finally, in Section 6 we provide the results of our empirical analysis.

2 Preliminaries

One of the the most practical and simple approaches for nearest neighbor searching in high dimensions is based on the k-d tree. Bentley [5] introduced the k-d tree as a generalization of the binary search tree in high dimensions. Each internal node of the k-d tree is associated with a hyperrectangle and a hyperplane orthogonal to one of the coordinate axis, which splits the hyperrectangle into two parts. These two parts are then associated with the two child nodes. The process of partitioning space continues until the number of data points in the hyperrectangle falls below some given threshold. Given a suitable splitting rule, the k-d tree induces a partitioning of space into cells whose sizes adapt to the local density of the data points; the partitioning is finer where the density is higher.

Friedman, Bentley and Finkel [8] gave an algorithm to find the nearest neighbor using optimized k-d trees. The internal nodes of the *optimized* k-d tree split the set of data points lying in the corresponding hyperrectangle into two equal parts, along the dimension in which the data points have maximum spread. The algorithm works by first descending the tree to find the data points lying in the cell that contains the query point. Then it examines surrounding cells if they intersect the ball B centered at the query point and having radius equal to the distance between the query point and the closest data point visited so far. Efficient implementations of the k-d tree algorithm have been give by Sproull [12] and Arya and Mount [2].

Friedman et al. [8] showed that their algorithm takes $O(\log n)$ expected time, under certain simplifying assumptions on the distribution of data and query points. They also showed that the expected number of points examined attains its minimum value when each cell contains one point, and is bounded by

$$N \leq \{[G(d)]^{1/d} + 1\}^d. \qquad (1)$$

Here d is the dimension and $G(d)$ is the ratio of the volume of a d-dimensional hypercube to the volume of the largest enclosed ball in the metric used for distance measurement. For the L_∞ metric, $G(d)$ is 1, and the expected number of points examined is bounded by 2^d.

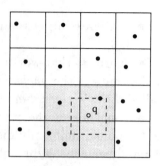

Figure 1: Bucketing Algorithm.

A much simpler algorithm that works well for uniformly distributed data is the *bucketing algorithm*, (sometimes called *Elias's algorithm* [13]). Space is divided into identical cells and for each cell, the data points inside it are stored in a list (see Fig. 1). The cells are examined in order of increasing distance from the query point and for each cell the distance is computed between the data points inside it and the query point. The search terminates when the distance from the query point to the cell exceeds the distance to the closest point visited. The algorithm was analyzed for data points uniformly distributed on the vertices of the d-dimensional hypercube by Rivest [11], and later for general uniformly distributed point sets by Cleary [7]. Cleary's analysis showed that the number of points examined was independent of n, and that the fewest number of points were examined when there was one data point per cell. His analysis also applies to the k-d tree algorithm and furnishes more accurate bounds than those obtained by Friedman et al [8].

3 Model

Consider a set of m^d data points distributed uniformly in a d-dimensional hypercube H with edge length m. We will assume that m is an integer greater than 1. The hypercube H is split into m^d cells of equal size, which gives a density of one data point per cell. Each cell stores a pointer to the data points inside it (if any). Fig. 1 shows the subdivision in two dimensions for $m = 4$. Consider that the query point q is also chosen from the uniform distribution. The bucketing algorithm visits the cells in increasing order of distance from the query point, examines the point(s) associated with each cell, and updates the closest point seen so far. The algorithm terminates when the next cell to be visited is farther than the closest point visited until then. The complexity of the algorithm is measured by the number of cells visited averaged over all possible locations of the query point.

4 Crude Analysis

In this section, we compute an upper bound on the expected number of cells visited by the algorithm. Denote an L_∞-ball of radius r centered at the query point by $B_d(r)$. This is a hypercube of side length $2r$ centered at the query point. It is easy to show that the expected volume of intersection of the nearest neighbor ball (the ball centered at the query point on whose surface lies the nearest neighbor) with hypercube H approaches unity in the large d limit. This is true no matter where the query point is located. This suggests the following model for the nearest neighbor ball, which we employ to simplify the analysis. For any query point, define r_n to be value of r such that

$$Vol(B_d(r_n) \cap H) = 1 \qquad (2)$$

We will compute an upper bound on the expected number of cells overlapped by $B_d(r_n)$, and regard this as an approximation to the true upper bound.

Let the random variable $I_i(r)$, for $1 \leq i \leq d$, denote the length of the side of $B_d(r) \cap H$ along the i-th dimension. Note that for any given r, the $I_i(r)$'s are independent, identically distributed random variables. Define the critical radius r_c to be the value of r such that $E[\log(I_i(r_c))] = 0$, where log denotes the natural logarithm function. Such a value of r must exist in the range $0.5 \leq r \leq 1$, since $I_i(r) \geq 1$ for $r \geq 1$ and $I_i(r) \leq 1$ for $r \leq 0.5$. The following lemma establishes that r_n almost always lies close to r_c, for high dimension.

Lemma 1 *For any constant $\epsilon > 0$,*

$$\lim_{d \to \infty} P[r_c - \epsilon < r_n < r_c + \epsilon] = 1. \qquad (3)$$

Proof: We only establish that $\lim_{d\to\infty} P[r_n < r_c + \epsilon] = 1$. The proof of the other part is similar and is omitted.

It is easy to see that $E[\log(I_i(r))]$ is a monotonically increasing function of r. From this it follows that, $E[\log(I_i(r_c + \epsilon))] > 0$. Using the law of large numbers, we then get

$$\lim_{d \to \infty} P\left[\frac{1}{d}\sum_{i=1}^{d} \log(I_i(r_c + \epsilon)) > 0\right] = 1 \qquad (4)$$

Rewriting this expression and taking antilogs, we get

$$\lim_{d \to \infty} P\left[\prod_{i=1}^{d} I_i(r_c + \epsilon) > 1\right] = 1 \qquad (5)$$

Since $Vol(B_d(r) \cap H)$ is given by $\prod_{i=1}^{d} I_i(r)$, we can write this as

$$\lim_{d \to \infty} P[Vol(B_d(r_c + \epsilon) \cap H) > 1] = 1 \qquad (6)$$

Since $r_n < r_c + \epsilon$ if and only if $Vol(B_d(r_c+\epsilon) \cap H) > 1$, we get the desired result. □

The following lemma gives the expected value of $Nb_d(r)$, which is defined as the number of cells overlapped by the L_∞-ball $B_d(r)$.

Lemma 2 *For $0.5 \leq r \leq 1$,*

$$E[Nb_d(r)] = \left[1 + 2r\left(1 - \frac{1}{m}\right)\right]^d \qquad (7)$$

Proof: Let the random variable $Z_i(r)$, for $1 \leq i \leq d$, denote the number of unit intervals overlapped along the i-th dimension, by the corresponding side of $B_d(r)$. Then $Nb_d(r)$ is given by $\prod_{i=1}^{d} Z_i(r)$. Since the $Z_i(r)$'s are independent, identically distributed random variables, we have

$$E[Nb_d(r)] = \prod_{i=1}^{d} E[Z_i(r)]. \qquad (8)$$

It is easy to see that for $0.5 \leq r \leq 1$, $Z_i(r)$ takes on the values 1, 2, and 3 with the following probabilities.

$$\begin{array}{cc} Z_i(r) & \text{Probability} \\ 1 & \frac{2-2r}{m} \\ 2 & \frac{(m-2)(2-2r)+2r}{m} \\ 3 & \frac{(m-2)(2r-1)}{m} \end{array} \qquad (9)$$

From this, we get the following expression for the expected value of $Z_i(r)$,

$$E[Z_i(r)] = \left[1 + 2r\left(1 - \frac{1}{m}\right)\right]. \qquad (10)$$

Substituting into Eq. (8), we get the desired result.
□

We now come to the main theorem of this section, which establishes an upper bound on the expected number of cells overlapped by the nearest neighbor ball $B_d(r_n)$. (Since Lemma 1 says that in high dimensions r_n is almost always close to r_c, one may be tempted to conclude that $E[Nb_d(r_n)]$ approaches $E[Nb_d(r_c)]$. However, we will see in Section 5 that this is in fact not true.)

Our proof of the theorem relies on the following conjecture.

Conjecture 1 (Monotonicity conjecture) *For all d, $E[Nb_d(r_n) \mid r_n = r]$ is a monotonically decreasing function of r.*

The intuition is that r_n is usually smaller when the query point is away from the boundaries of the hypercube H; this case leads to a large number of cells being intersected by the nearest neighbor ball $B_d(r_n)$. While we we have confirmed this observation in our experiments, we do not have a rigorous proof for it. We present some evidence in favor of the conjecture in Fig. 2, which shows the plot obtained experimentally of $[E[Nb_d(r_n)|r_n = r]]^{1/d}$ versus r for $d = 8$ (here $m = 2$). To compute this plot, we partitioned the radii in the range 0.5 to 0.75 into a set of intervals. We generated 10^7 points from the uniform distribution, and for each point, grew a ball having unit volume of intersection with hypercube H. We determined the interval in which the radius of the ball lies and the number of cells intersected by it. For each interval, we computed the average number of cells intersected over all balls whose radius lies in that interval, and plotted these against the interval midpoint. (The plot terminates at $r = 0.75$ since the probability of finding points where the radius of the nearest neighbor ball is higher than this is very low and our sampling technique does not find enough of them.)

Theorem 1 *Assuming Conjecture 1, for any constant $\epsilon > 0$, there exists d_0 such that for all $d > d_0$,*

$$E[Nb_d(r_n)] \leq \left[1 + 2(r_c + \epsilon)\left(1 - \frac{1}{m}\right)\right]^d. \quad (11)$$

Proof: We compute $E[Nb_d(r_n)]$ over two possible cases, $r_n < r_c + \epsilon'$ and $r_n \geq r_c + \epsilon'$ (later we specify

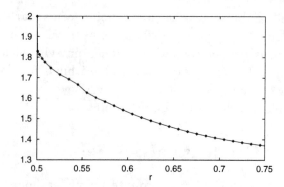

Figure 2: Monotonicity conjecture (for $d = 8$): $[E[Nb_d(r_n) \mid r_n = r]]^{1/d}$ vs. r.

the choice for ϵ').

$$E[Nb_d(r_n)] =$$
$$P[r_n < r_c + \epsilon']E[Nb_d(r_n) \mid r_n < r_c + \epsilon'] +$$
$$P[r_n \geq r_c + \epsilon']E[Nb_d(r_n) \mid r_n \geq r_c + \epsilon']. \quad (12)$$

Conjecture 1 implies that

$$E[Nb_d(r_n) \mid r_n \geq r_c + \epsilon'] \leq E[Nb_d(r_n) \mid r_n < r_c + \epsilon']$$
$$(13)$$

Substituting into Eq. (12) we get an upper bound for $E[Nb_d(r_n)]$,

$$E[Nb_d(r_n)] \leq E[Nb_d(r_n) \mid r_n < r_c + \epsilon']. \quad (14)$$

This can be written as

$$E[Nb_d(r_n)] \leq E[Nb_d(r_c + \epsilon') \mid r_n < r_c + \epsilon']$$
$$\leq \frac{E[Nb_d(r_c + \epsilon')]}{P[r_n < r_c + \epsilon']}. \quad (15)$$

From Lemma 2, substituting for $E[Nb_d(r_c + \epsilon')]$ we get,

$$E[Nb_d(r_n)] \leq \frac{\left[1 + 2(r_c + \epsilon')\left(1 - \frac{1}{m}\right)\right]^d}{P[r_n < r_c + \epsilon']}. \quad (16)$$

From Lemma 1, it follows that for sufficiently large d, $P[r_n < r_c + \epsilon']$ is arbitrarily close to 1. Choosing $\epsilon' < \epsilon$ in the above inequality, we have the desired result.

□

To compute the critical radius r_c, we first write the cumulative density for $I_i(r)$. It is easy to see that this is given by

$$P[I_i(r) \leq x] = \begin{cases} 0 & \text{if } x < r \\ \frac{2(x-r)}{m} & \text{if } r \leq x < 2r \\ 1 & \text{if } x = 2r \end{cases} \quad (17)$$

m	$[E[Nb_d(r_n)]]^{1/d}$
2	1.601
3	1.748
4	1.815
5	1.854
100	1.993

Table 1: Expected number of cells visited by the bucketing algorithm.

From this, after some calculation, we get the following expression for $E[\log(I_i(r))]$,

$$E[\log(I_i(r))] = \frac{2r}{m}(\log(2) - 1) + \log(2r). \quad (18)$$

For any m, we can equate this expression to 0, and solve it numerically to obtain the critical radius r_c. We can then use Theorem 1 to compute an upper bound on $E[Nb_d(r_n)]$, the expected number of cells visited. The results are shown in Table 1 for several different values of m. For $m = 2$, the table shows an upper bound of $(1.601\ldots + \epsilon)^d$, for any small $\epsilon > 0$. Since the average density is one point per cell, this bound also applies to the number of points visited by the algorithm. The fact that the number of points visited is much fewer than the asymptotic bound of 2^d confirms the importance of boundary effects when the number of points is on the order of 2^d.

5 Refined Analysis

In this section we achieve a tight bound for the expected number of cells visited by the bucketing algorithm. For simplicity we do this analysis for the case of $m = 2$. This analysis does not rely on the monotonicity conjecture.

Let $\mathbf{x} = (x_1, \ldots, x_d)$ denote a point in the hypercube H. Consider an L_∞ ball centered at $\mathbf{x}$. Let V denote the logarithm of its volume of intersection with hypercube H, and let $Z(\mathbf{x}; V)$ be the number of cells that this ball intersects. The quantity of interest N is then $Z(\mathbf{x}; 0)$, averaged over $\mathbf{x}$ lying anywhere inside hypercube H. Thus

$$N = \int dV \delta(V) \int_0^1 dx_1 \ldots \int_0^1 dx_d Z(\mathbf{x}; V) \quad (19)$$

where the δ-function enforces the condition that we are only interested in balls which have unit intersection volume with hypercube H. As in the previous section, such a ball is called a nearest neighbor ball. (Recall that the Dirac δ-function has the properties (a) $\delta(x) = 0$ if $x \neq 0$ and (b) $\int_{-\infty}^{\infty} g(x)\delta(x)dx = g(0)$ for all integrable $g : \mathbb{R} \to \mathbb{R}$.) Here the range of each x_i integral is taken to be 0 to 1, since the 0 to 2 interval is symmetric around 1, and in this form we do not need any normalization factors.

Eq. (19) can be expressed in terms of the radius r of the ball as $N =$

$$\int_{\frac{1}{2}}^1 dr \frac{\partial V(\mathbf{x}; r)}{\partial r} \delta(V(r)) \int_0^1 dx_1 \ldots \int_0^1 dx_d Z(\mathbf{x}; r). \quad (20)$$

The integral over r ranges from $\frac{1}{2}$ to 1, since these are the minimum and maximum values of r needed for any $\mathbf{x}$ in order to achieve a ball with unit intersection volume.

The quantity V, which is the logarithm of the volume of intersection, is the sum of the logarithms of the intercepts on all the d sides, which are given by

$$I_i(x_i; r) = \begin{cases} \log(x_i + r) & \text{for } x_i \leq r \\ \log(2r) & \text{for } x_i \geq r \end{cases} \quad (21)$$

Therefore the derivative, $\partial V/\partial r$ is the sum of d terms, each of which lies between $1/r$ and $1/2r$, i.e. between $\frac{1}{2}$ and 2. Therefore, to within a constant factor, $N \sim$

$$d \int_{\frac{1}{2}}^1 dr \delta \left(\sum_{1 \leq i \leq d} I_i(x_i; r) \right) \int_0^1 dx_1 \ldots \int_0^1 dx_d Z(\mathbf{x}; r). \quad (22)$$

For any specific choice of r and $\mathbf{x}$, the number Z of cells intersected is the product of d factors $Z_i(x_i; r)$, which are given by

$$Z_i(x_i; r) = \begin{cases} 1 & \text{for } x_i \leq 1 - r \\ 2 & \text{for } x_i > 1 - r \end{cases} \quad (23)$$

Thus Eq. (22) can be written as

$$N \sim d \int_{\frac{1}{2}}^1 dr \int_{-\infty}^{\infty} \frac{d\omega}{2\pi} \left\{ \int_0^1 dx_i H_i(x_i; r; w) \right\}^d \quad (24)$$

where $H_i(x_i; r; w) = Z_i(x_i; r) \exp[-j\omega I_i(x_i; r)]$. Here we have used the representation of the δ-function, $2\pi\delta(y) = \int d\omega \exp[-j\omega y]$. For fixed ω

and r, the integrals over all the x_i's then decouple from each other.

The term in the {}-brackets is equal to

$$F(\omega; r) = \int_{x_i=r}^{1} dx_i\, 2 \exp[-j\omega \log(2r)] + \\ \int_{x_i=1-r}^{x_i=r} dx_i\, 2 \exp[-j\omega \log(x_i+r)] + \\ \int_{x_i=0}^{x_i=1-r} dx_i\, \exp[-j\omega \log(x_i+r)] \quad (25)$$

which can be evaluated to be

$$F(\omega; r) = \\ \left\{2(1-r) + \frac{4r}{1-j\omega}\right\} \exp[-j\omega \log(2r)] - \\ \frac{1}{1-j\omega} - \frac{r}{1-j\omega} \exp[-j\omega \log(r)]. \quad (26)$$

To solve for N from Eq. (24), we first evaluate the ω integral. For large d, this can be done by the method of steepest descent [4, pages 280–302], [10, pages 287–291]. In this method, we extend the domain of ω to the complex plane, and seek for a point $\omega(r)$ where $\partial_\omega F(\omega; r) = 0$. The contour of integration for ω, which is originally along the real axis, is deformed so that it passes through the point $\omega(r)$, and such that the phase of $F(\omega; r)$ is constant along it. For any r, $F(\omega; r)$ is the Fourier transform (extended to complex ω) of $\tilde{Z}(I; r) = \int_0^1 dx_i Z_i(x_i; r)\delta(I - I_i(x_i; r))$ with respect to I. $\tilde{Z}(I; r)$ is the mean (averaged over x_i) number of cells intersected along the i-th direction by a ball of radius r, subject to the constraint that the intercept on the i-th coordinate should be I. Since $\tilde{Z}(I; r)$ is always positive, its Fourier transform F with respect to I is real and positive for imaginary ω. Also, for any line parallel to the real axis, the magnitude of F has a maximum on the imaginary axis, where all the contributions to F are in phase. Thus, if we write $\omega = jw$, where w is real, $\omega(r)$ is the point where the *real* function $F(w; r)$ has a minimum. This can be found numerically. The integral over ω is now performed over the stationary phase contour passing through $\omega(r)$; if we parameterize this curve by s, the integral over ω is of the form $[F(\omega(r); r)]^d \int ds \exp[-d\{a(r)s^2 + b(r)s^3 + \ldots\}]$, which in the large d limit is equal to $[F(\omega(r); r)]^d / \sqrt{da(r)}$. Thus, to within numerical factors, Eq. (22) can be expressed as

$$N \sim \sqrt{d} \int_{\frac{1}{2}}^{1} dr [F(\omega(r); r)]^d \quad (27)$$

where $F(\omega(r); r)$ can be evaluated numerically for any r by searching for $\omega(r)$ along the imaginary axis, and finding the value of F there [1]. Fig. 3 shows a plot of this quantity made using Mathematica.

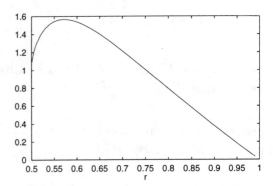

Figure 3: $F(\omega(r); r)$ vs. r.

The integral in Eq. (27), having a d in the exponent, can again be done by the method of steepest descent. For large d, the integral is dominated by the point where $F(\omega(r); r)$ attains a maximum as a function of r over the domain $\frac{1}{2}$ to 1. This can be evaluated numerically. We thus obtain, to within a constant factor,

$$N \sim [F(\omega(r_M); r_M)]^d = (1.56594\ldots)^d \quad (28)$$

where r_M is the point where $F(\omega(r); r)$ has a maximum, and is found numerically to be $r_M = 0.5715 \pm 0.0005$. The multiplicative factor in Eq. (28) is a constant to within $O(1/d)$ corrections. It is fairly straightforward to extend the proof above to calculate the constant as being approximately 1.02.

So far our analysis has been an approximation, based on the assumption presented at the start of Section 4, which states that the nearest neighbor ball has unit intersection volume with hypercube

---

[1] Note that the integral over ω that we have evaluated by the method of steepest descent is *not well defined* when r is *exactly* equal to $\frac{1}{2}$. This is because for all points **x** satisfying $0.5 \leq x_i \leq 1.5$, which is a finite fraction of the volume of the hypercube H, the radius of the nearest neighbor ball that is centered around the point is $\frac{1}{2}$. Thus the probability density that the radius of the nearest neighbor ball is r has a δ-function at $r = \frac{1}{2}$. However, since the weight of this δ-function is $\frac{1}{2}^d$, and the number of cells intersected for any point in this central region is 2^d, the contribution this singularity makes to N is $(\frac{1}{2})^d \times 2^d = 1$, which can be neglected compared to the eventual answer, obtained in Eq. (28).

H. We show that Eq. (28) is valid, to within constant factors, even without this assumption. We only require that the probability density that the nearest neighbor ball has intersection volume λ be *independent* of the location of **x**. This weaker assumption is satisfied in the limit for large dimension, for points uniformly distributed in the unit hypercube. In this case, it is easily verified that the probability that the nearest neighbor ball has intersection volume λ, has the Poisson density, $\exp[-\lambda]$. The expression in Eq. (28) for N can then be replaced by one of the form,

$$N = \int d\lambda \exp[-\lambda] N(\lambda) \qquad (29)$$

where $N(\lambda)$ is the expected number of cells when the intersection volume is λ (the quantity evaluated in Eq. (28) is $N(\lambda = 1)$). This results in an extra factor of $\exp[j\omega(\log\lambda)/d]$ in $F(\omega;r)$, which gives rise to $O((\log\lambda)/d)$ corrections to $F(\omega(r);r)$. Thus $N(\lambda)$, which is dominated by the region around r_M, is given by $N(1) \times \lambda^c$, where c is some constant. Integrating over λ with an $\exp[-\lambda]$ weight factor, this gives the same result as in Eq. (28), but with a different multiplicative factor. The multiplicative factor in this case is approximately 0.90. We summarize our main result:

Theorem 2 *Given a set of 2^d points uniformly distributed in a d-dimensional hypercube, and a query point also from the uniform distribution, the expected number of cells visited by the bucketing algorithm, N_d, satisfies*

$$\lim_{d \to \infty} \frac{N_d}{(0.90\ldots) \cdot (1.56594\ldots)^d} = 1.$$

Remark: The procedure that we have carried out above can be carried out (more easily) to evaluate the probability density $\rho(r)$ that the radius of the nearest neighbor ball is r; the only differences are those resulting from the fact that the factor of Z_i in Eq. (24) is missing. Since the expression in Eq. (27) can be written as $N = \int dr \rho(r) \overline{N}(r)$, and the corresponding expression one would get in evaluating the probability density is $1 = \int dr \rho(r)$, one can obtain the $\overline{N}(r)$, which is the expected number of cells intersected by the nearest neighbor ball for points where the nearest neighbor ball has radius r. Using Mathematica as in Fig. 3, the d-th root of

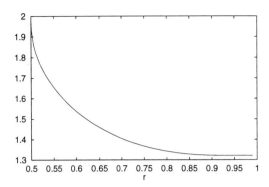

Figure 4: Evidence for the Monotonicity conjecture: $\lim_{d \to \infty} [E[N b_d(r_n) \mid r_n = r]]^{1/d}$ vs. r.

this quantity was computed for various values of r and is plotted in Fig. 4. If one believes that the plot in Fig. 4 is a faithful representation of the function, then one may regard this as a sort of "numerical proof" of the monotonicity conjecture in the large d limit for $m = 2$.

6 Experimental Results

In this section, we present experimental results which show that even for low dimensions (≤ 16), our analysis yields good bounds. Fig. 5 shows the average number of cells examined by the bucketing algorithm and the optimized k-d tree algorithm, when the dimension d ranges from 1 to 16, and the number of data points is 2^d. In each case the number of cells examined is averaged over 10 different experiments, where each experiment is performed with a different set of data points and averaged over 2,500 query points. Both data and query points are chosen from the uniform distribution. The figure shows that Eq. (28) using the multiplicative factor of 0.90 (as derived from Eq. (29)) provides an excellent fit to the number of points visited by the two algorithms. The fit is slightly worse for the k-d tree algorithm than for the bucketing algorithm. This could be because the k-d tree partition is not perfectly regular. It could also be because the k-d tree algorithm does not necessarily visit the cells in increasing order of distance, rather in an order determined by the structure of the tree. These factors could lead to a small overhead.

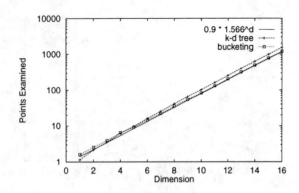

Figure 5: Average number of points examined by the bucketing and k-d tree algorithms; the total number of data points in dimension d is 2^d.

7 Conclusions

Previous analysis have neglected the effects of the boundary in order to simplify the analysis. However, there is plenty of evidence of the importance of boundary effects in realistic instances in high dimensions. We have presented analysis that takes these effects into account, thus providing a significantly more accurate analysis.

The main limitation of our analysis is that it applies only to the L_∞ metric and the uniform distribution. It would be nice to extend these results to other distance norms and other distributions. Another interesting open problem is to prove the monotonicity conjecture.

References

[1] S. Arya and D. M. Mount. Approximate nearest neighbor queries in fixed dimensions. In *Proceedings of the 4th ACM-SIAM Symposium on Discrete Algorithms*, pages 271–280, 1993.

[2] S. Arya and D. M. Mount. Algorithms for fast vector quantization. In J. A. Storer and M. Cohn, editors, *Proc. of DCC '93: Data Compression Conference*, pages 381–390. IEEE Press, 1993.

[3] S. Arya, D. M. Mount, N. S. Netanyahu, R. Silverman and A. Wu. An optimal algorithm for approximate nearest neighbor searching. In *Proceedings of the 5th ACM-SIAM Symposium on Discrete Algorithms*, pages 573–582, 1994.

[4] C. M. Bender and S. A. Orszag. *Advanced mathematical methods for scientists and engineers*. International series in pure and applied mathematics. McGraw-Hill, 1978.

[5] J. L. Bentley. Multidimensional binary search trees used for associative searching. *Communications of the ACM*, 18(9):509–517, September 1975.

[6] K. L. Clarkson. An algorithm for approximate closest-point queries. In *Proceedings of the 10th Annual ACM Symposium on Computational Geometry*, pages 160–164, 1994.

[7] J. G. Cleary. Analysis of an algorithm for finding nearest neighbors in Euclidean space. *ACM Transactions on Mathematical Software*, 5(2):183–192, June 1979.

[8] J. H. Friedman, J. L. Bentley, and R.A. Finkel. An algorithm for finding best matches in logarithmic expected time. *ACM Transactions on Mathematical Software*, 3(3):209–226, September 1977.

[9] A. Gersho and R. M. Gray. *Vector Quantization and Signal Compression*. Kluwer Academic, 1991.

[10] G. Goertzel and N. Tralli. *Some mathematical methods of physics*. McGraw-Hill, 1960.

[11] R. L. Rivest. On the optimality of Elias's algorithm for performing best-match searches. In *Information Processing*, pages 678–681. North Holland Publishing Company, 1974.

[12] R. L. Sproull. Refinements to nearest-neighbor searching in k-dimensional trees. *Algorithmica*, 6, 1991.

[13] T. Welch. Bounds on the information retrieval efficiency of static file structures. Technical Report 88, MIT, June 1971.

A Parallel Algorithm For Linear Programming in Fixed Dimension

Martin Dyer[*]
School of Computer Studies
University of Leeds, Leeds, UK.

email: dyer@scs.leeds.ac.uk

Abstract

We describe a parallel algorithm for linear programming in fixed dimension which runs in deterministic time $O(\log n (\log \log n)^{d-1})$ on an EREW PRAM. This improves a previous bound of of $O((\log n)^2)$ for this problem.

1 Introduction

Deng [9] gave a parallel algorithm for linear programming in two dimensions which runs in $O(\log n)$ time on a CREW PRAM, using $O(n/\log n)$ processors. The method does not generalise to higher dimensions, but he asked the question as to whether an algorithm with similar running time can be found. In this paper, we go part of the way to answering this question in the affirmative by describing a parallel algorithm for linear programming in fixed dimension which runs in time $O(\log n (\log \log n)^{d-1})$ on an EREW PRAM. The previous best (deterministic) time bound for this problem in general fixed dimension is, as far as we know, an $O((\log n)^2)$ algorithm obtained by derandomizing Clarkson's algorithm [6] by the method of Chazelle and Matoušek [5], but using the more explicit and simplified treatment of Goodrich [11]. We note in passing that Alon and Megiddo [3] gave a randomized parallel implementation of Clarkson's algorithm which runs in constant time on a CREW PRAM with high probability in fixed dimension.

Ajtai and Megiddo [2] have given an algorithm which runs in $O((\log \log n)^d)$ time in fixed dimension on a particular model of parallel computation which is discussed in Section 2 below. Their algorithm is a clever parallel implementation of the idea of Megiddo [13]. The algorithm described here is also a parallel implementation of Megiddo's algorithm which, while being more straightforward, appears to possess most of the power of that of Ajtai and Megiddo.

2 Background

As mentioned above, the algorithm developed here is essentially a parallel implementation of the idea of Megiddo [13]. His method is based on rapidly eliminating a proportion of the constraints using few "hyperplane queries". Megiddo's hyperplane query problem has the following form. We have some unknown point

---

[*]Supported in part by SERC Grant GR/H 77811

x^* (which we choose to be the linear program optimum), and a set of n hyperplanes,

$$H_i = \{x \in \mathbb{R} : \alpha.x_i = \beta_i\} \quad (i = 1, \ldots, n)$$

(which we choose to be the linear constraints). We have to determine

$$\text{sign}(\alpha_i.x^* - b_i),$$

for each $i = 1, \ldots, n$, where sign: $\mathbb{R} \to \{-1, 0, +1\}$. (Clearly such information is adequate to solve a linear program.) There is an oracle which, given an arbitrary hyperplane $H = \{x : \alpha.x = \beta\}$, will return $\text{sign}(a.x - b)$. Megiddo's result is that in fixed dimension we can determine the signs of a constant proportion of the hyperplanes using only a (carefully selected) constant number of queries to the oracle. In linear programming a hyperplane query is a linear program in smaller dimension. The method is recursive on dimension. For simplicity let us ignore possible degeneracies. Then we may assume without loss that $a_{i1} = 1$ for $i = 1, \ldots, n$. The dimensional recursion results from pairing hyperplanes so that each pair has one member with $a_{i2} > 0$ and one with $a_{i2} < 0$. In order to ensure that about $\frac{1}{2}n$ pairs can be formed, a rotation on the first two coordinates is performed after a median calculation.

There is a straightforward parallel implementation of Megiddo's algorithm [13] which runs in $O((\log n)^d)$ time on a EREW PRAM. This algorithm is rather inefficient in terms of processor utilisation, since at the later stages, when there are few constraints remaining, most processors are idle. In order to improve matters, it is necessary to remedy this defect, Ajtai and Megiddo [2] gave an algorithm based on the use of an expander graph to select more non-disjoint pairs to utilise the idle processors and obtain more rapid elimination. Their algorithm runs in $O((\log \log n)^d)$ time in a non-standard model of parallel computation. Their model allows $O(\log \log n)$ time median selection from n numbers using n processors. This is known to be possible [1] in Valiant's comparison model [14], but is also known to require $\Omega(\log n / \log \log n)$ even on a CRCW PRAM [4]. The Ajtai-Megiddo model also uses a non-uniform scheme for $O(\log \log n)$ time processor allocation, again based on the use of expander graphs.

Here we work within the standard PRAM model, and show that rapid elimination can be achieved without the use of expander graphs. The method is based on using groupings larger than pairs in order to achieve full processor utilisation.

3 The Search Problem

We will consider the hyperplane search problem. Suppose we have n hyperplanes and $m \geq n$ processors. Let us say a hyperplane is *located* if $\text{sign}(a_i.x^* - b_i)$ has been determined. We wish to locate a proportion of the given hyperplanes in a constant number of *rounds*. Each repetition of the basic recursion step in the location process will be called a round. Each round contains some non-trivial computations, but we will ignore this initially. However, we will ensure that there are sufficient processors available for all the computations associated with any given round.

Let us again assume, for ease of exposition, that degeneracies do not occur, and so $a_{i1} = 0$ for all $i = 1, \ldots, n$. (For a more careful treatment of the degeneracy issue in the context of this search problem, see [10].) The method is then as follows.

Choose a "suitable" *group* size $r = \max\{2, \lfloor (m/n)^{1/d} \rfloor\}$, and let $s = n/r$. For simplicity let us assume s is an integer. For $k = 1, \ldots, r$, determine the ksth largest value of a_{i2} ($i = 1, \ldots, n$), σ_k say. Let $\sigma_0 = -\infty$. These values induce a partition of the set of hyperplanes H_i into r sets, according as $\sigma_{k-1} < a_{i2} \leq \sigma_k$ ($k = 1, \ldots, r$). Now form s groups of size r so that each group has exactly one H_i from each of these r sets. (This is analogous to the pairing in Megiddo's sequential algorithm.) In each group, consider each pair of hyperplanes H_i, H_j say, where $a_{i2} \leq a_{j2}$. There will exist σ_k, σ_ℓ such that $\sigma_k \in [a_{i2}, a_{j2}]$ and $\sigma_\ell \notin [a_{i2}, a_{j2}]$ ($1 \leq k, \ell \leq r$). Make a change of variables so that

$$x'_1 = x_1 + \sigma_k x_2, \quad x'_2 = x_1 + \sigma_\ell x_2.$$

If we now eliminate each of x'_1, x'_2 in turn, we produce hyperplanes H_{ij}, H_{ji} such that if we can locate both H_{ij} and H_{ji}, we can locate at least one of H_i, H_j. (See [13].) But locating the H_{ij} is essentially a $(d-1)$-dimensional problem, since all such hyperplanes have $a'_{i2} = 0$. Note that there are only r different variables x'_1 or x'_2 in this process. Thus there are r *collections* of $(d-1)$-dimensional hyperplanes. We recursively locate a proportion of the hyperplanes in each of these collections.

We will repeat the location process at the $(d-1)$-dimensional level *eight* times, each time removing from consideration those hyperplanes already located. This is simply to boost the proportion sufficiently, and the number eight has no significance other than it is large enough for this purpose. This idea of increasing the proportion was used in [7, 10], in relation to Megiddo's algorithm. There is a difference here, however. We cannot wait for the location on each collection before we start on the next as is done in the sequential algorithm, since this uses the accumulated information to guide the query process. This is no longer feasible because the number of collections, r, is not constant if $m \gg n$. Thus we have to operate on all collections in parallel.

At the bottom level, we have zero-dimensional hyperplanes, i.e. points. We can now locate with respect to a proportion $(r-1)/r$ of these points in one round. We do this by selecting the $\lfloor jn/r \rfloor$th largest points for $j = 1, 2, \ldots, r-1$. The resulting queries will in fact require solving up to three linear programs in $(d-1)$ dimensions. Note that solving a one-dimensional linear program reduces to a simple maximum or minimum computation.

4 Analysis of the search algorithm

Let the number of rounds for location in dimension d be T_d, and let p_d denote the proportion of d-dimensional hyperplanes which are located. Each group of size r gives rise to $r(r-1)$ hyperplanes in $(d-1)$ dimensions. These are partitioned into r collections, according to the selection computations. Location at the $(d-1)$-dimensional level implies that all $r(r-1)$ hyperplanes from the same group will have been located for a fraction of at least $1 - r(r-1)(1 - p_{d-1})^8$ of the groups, since there are eight repetitions of the $(d-1)$-dimensional process.

Now, we will have located at least $(r-1)$ d-dimensional hyperplanes in each of these groups. (If this were not the case, there would be some pair of hyperplanes for which we had not located either, a contradiction.) Hence we have the recurrence

$$p_d \geq \frac{(r-1)}{r}(1 - r(r-1)(1 - p_{d-1})^8), \quad (d \geq 2).$$

We also have, by assumption, $p_1 = (r-1)/r$ and $T_1 = 1$. It is easy to check, by induction, that for $d \geq 2$,

$$p_d \geq \begin{cases} \frac{1}{4}, & r = 2; \\ \frac{r-2}{r}, & r \geq 3. \end{cases}$$

Also, clearly, $T_d = 8T_{d-1}$ for $d \geq 2$, and since $T_1 = 1$ this gives $T_d = 8^{d-1}$ rounds. Thus, in fixed dimension, only a constant number of rounds are required for the search procedure we have described.

We have only to check that there are sufficient processors for this search procedure. Now the total number of $(d-1)$-dimensional hyperplanes generated simultaneously is $n/r \times r(r-$

$1) = n(r-1)$. Thus the total number of hyperplanes simultaneously generated in one dimension is at most $n(r-1)^{d-1} \leq m$. We therefore have enough processors to allocate one to each hyperplane simultaneously in each dimension. The total number of queries generated at the bottom of the recursion is $r^{d-1}(r-1)$, since r collections are generated at each intermediate level, and $(r-1)$ queries are generated at the bottom level. We need n processors for each of these queries, since they are linear programs with n constraints. So we require that $n(r-1)r^{d-1} \leq m$, which is clearly true by choice of r.

We have merely ensured that we have sufficient processors, but we may note that the total storage requirement is only a constant factor larger (in fixed dimension).

5 The linear programming algorithm

To solve linear programs, we simply iterate the above search procedure, starting with the n hyperplanes assigned one to each of n processors.

Let n_i be the number of d-dimensional hyperplanes remaining at the start of iteration i. Thus $n_1 = n$, and the number of processors m is equal to n throughout. Thus we choose the group size r_i at iteration i so that $r_i = \lfloor (n/n_i)^{1/d} \rfloor$. Then we will have

$$n_{i+1} \leq n_i(1-p_d) \leq \begin{cases} \frac{3}{4}n_i & r = 2; \\ n_i \times 2/r_i & r \geq 3 \end{cases}.$$

It follows that

$$n_{i+1} \leq \min\{\tfrac{3}{4}n_i, 3n_i^{(d+1)/d}/n^{1/d}\}.$$

Letting $\xi_i = n_i/n$, we have

$$\xi_{i+1} \leq \min\{\tfrac{3}{4}\xi_i, 3\xi_i^{(d+1)/d}\}, \text{ with } \xi_1 = 1,$$

from which it follows easily that $\xi_i \leq 1/n$ (i.e. $n_i \leq 1$) after $O(\log \log n)$ iterations. However, each iteration involves solving (in parallel) $(d-1)$-dimensional linear programs.

Thus the total number of (nested) iterations is $O((\log \log n)^{d-1})$. The bottom (one dimensional) level requires a maximum or minimum, which is a selection, which we have disregarded so far.

Thus only $O((\log \log n)^{d-1})$ rounds are required in total (in fixed dimension). So far we have ignored the work per round. We must now account for this. Each round requires a selection operation. This can be done in $O(\log n)$ time on an EREW PRAM by, for example, sorting on the value of a_{i2} using Cole's algorithm [8]. The group formation can then also be accomplished easily. Constructing the new collections by pair formation requires broadcasting a hyperplane known by one processor to at most r others. This can be achieved in $O(\log r) = O(\log n)$ time. Finally, deleting located hyperplanes, and re-organising the unlocated hyperplanes for the next iteration can be achieved by prefix sum computation in $O(\log n)$ time. Thus the "hidden" work per round can be carried out in $O(\log n)$ time. The total time bound of $O(\log n (\log \log n)^{d-1})$ for the entire algorithm now follows.

6 Remarks

While we have described the algorithm for an EREW PRAM, we may note that the algorithm could be run in $O((\log \log n)^d)$ time on the parallel computation model of Ajtai and Megiddo [2]. (Their processor allocation technique would also be required, however.)

Observe that there is a $O(\log n)$ time lower bound on finding the maximum of n numbers on an EREW PRAM [12, p. 889]. Since one-dimensional linear programming amounts to computing a maximum, it follows that there is a $O(\log n)$ time lower bound for fixed-dimensional linear programming. Hence the $O((\log n)^{1+o(1)})$ upper bound for running time on an EREW PRAM achieved here is not too far from optimal.

References

[1] M. Ajtai, J. Komlós, W. Steiger and E. Szemerédi, Deterministic selection in $O(\log \log n)$ time, *Proc. 18th Ann. Symp. on Theory of Computing* (1986), 188-195.

[2] M. Ajtai and N. Megiddo, A deterministic poly($\log \log n$)-time n-processor algorithm for linear programming in fixed dimension, *Proc. 24th Ann. Symp. on Theory of Computing* (1992), 327-338.

[3] N. Alon and N. Megiddo, Parallel linear programming almost surely in constant time, *Proc. 31st IEEE Symp. on Foundations of Computer Science* (1990), 574-582.

[4] P. Beame and J. Hastad, Optimal bounds for decision problems on the CRCW PRAM, *Proc. 19th Ann. Symp. on Theory of Computing* (1987), 83-93.

[5] B. Chazelle and J. Matoušek, On linear-time deterministic algorithms for optimization problems in fixed dimension, *Proc. 4th ACM-SIAM Symp. on Discrete Algorithms* (1993), 281-290.

[6] K. Clarkson, A Las Vegas algorithm for linear programming when the dimension is small, *Proc. 29st IEEE Symp. on Foundations of Computer Science* (1988), 452-456.

[7] K. Clarkson, Linear programming in $O(n \times 3^{d^2})$ time, *Information Processing Letters* **22** (1986), 21-27.

[8] R. Cole, Parallel merge sort, *SIAM J. on Computing* **17** (1988), 770-785.

[9] X. Deng, An optimal parallel algorithm for linear programming in the plane, *Information Processing Letters* **35** (1990), 213-217.

[10] M. Dyer, On a multidimensional search problem and its application to the Euclidean one-centre problem, *SIAM J. on Computing* **15** (1986), 725-738.

[11] M. Goodrich, Geometric partitioning made easier, even in parallel, *Proc. 9th ACM Symp. on Computational Geometry* (1993), 73-82.

[12] R. Karp and V. Ramachandran, Parallel algorithms for shared-memory machines, in *Handbook of Theoretical Computer Science Vol. A: Algorithms and Complexity*, Ed. J. van Leeuwen, Elsevier, Amsterdam, 1990, pp. 869-941.

[13] N. Megiddo, Linear programming in linear time when the dimension is fixed, *J. of the ACM* **31** (1984), 114-127.

[14] L. Valiant, Parallelism in comparison problems, *SIAM J. on Computing* **4** (1975), 348-355.

Precision-Sensitive Euclidean Shortest Path in 3-Space *

(Extended Abstract)

Joonsoo Choi [†] Jürgen Sellen [‡] Chee-Keng Yap [§]

Abstract

This paper introduces the concept of *precision-sensitive algorithms*, in analogy to the well-known output-sensitive algorithms. We exploit this idea in studying the complexity of the 3-dimensional *Euclidean shortest path* problem. Specifically, we analyze an incremental approximation approach based on ideas in [CSY], and show that this approach yields an asymptotic improvement of running time. By using an optimization technique to improve paths on fixed edge sequences, we modify this algorithm to guarantee a relative error of $O(2^{-r})$ in a time polynomial in r and $1/\delta$, where δ denotes the relative difference in path length between the shortest and the second shortest path.

Our result is the best possible in some sense: if we have a *strongly precision-sensitive* algorithm then we can show that *USAT* (unambiguous SAT) is in polynomial time, which is widely conjectured to be unlikely.

Finally, we discuss the practicability of this approach. Experimental results are provided.

---

*Work on this paper by Sellen has been supported by a postdoctoral fellowship from the DAAD, Germany. Choi and Yap are supported by NSF grant #CCR-9402464.

[†] Courant Institute of Mathematical Sciences, New York University, 251 Mercer Street, New York, NY 10012.

[‡] Fachbereich Informatik, Universität des Saarlandes, Postfach 151150, 66041 Saarbrücken.

[§] Courant Institute of Mathematical Sciences, New York University, 251 Mercer Street, New York, NY 10012.

1 Introduction

The complexity of geometric algorithms generally falls under one of two distinct computational frameworks. In the *algebraic framework*, the (time) complexity of an algorithm is measured by the number of algebraic operations on real-valued variables, assuming exact computations. In simple cases, the input size has one parameter n corresponding to the number of input values. In the *bit framework*, (time) complexity is measured by the number of bitwise boolean operations, assuming input values are encoded as binary strings. The input size parameter n above is supplemented by an additional parameter L which is an upper bound on the bit-size of any input value. See [CSY].

Currently, practically every computational geometry algorithm is based on the algebraic model. For instance, we usually say that the planar convex hull problem can be solved in optimal $O(n \log n)$ time. This presumes the algebraic framework. What about the bit framework? One can easily deduce that the bit complexity is $O(n \log n \mu(L))$ where $\mu(L)$ is the bit complexity of multiplying two L-bit integers. However, it is not clear that this is optimal. Thus the possibility for faster planar convex hull algorithms seems wide open in the bit model. Of course, the situation with other problems in computational geometry is similar.

This paper is interested in bit complexity, and may be seen as a follow-up on [CSY]. Besides its inherent interest, there are other reasons for believing that the bit model will become more important for computational geometry in the future. As the field now begins to address implementation issues in earnest, it must focus on low-level operations (what was previously dismissed as "constant time operations"). In low-level operations, it is the bit size of numbers that is the main determinant of complexity. Second, there are reasons to think that "exact computation" (see [Ya]) will be an important paradigm for future implementations of geometric algorithms. [The emphasis is on "implementations" since exact computation is already the de facto standard in theoretical algorithms.] In exact computation, complexity crucially depends on the bit-sizes of input numbers.

The main conceptual contribution in this paper is the idea of *precision-sensitive* algorithms. Today, the concept of *output-sensitive algorithms* has become an important pillar of computational geometry. But output-sensitivity is basically a concept in the algebraic framework. We suggest that precision-sensitivity is the analogous concept in the bit framework. As in output-sensitive algorithms, we may define some implicit parameter $\delta = \delta(I)$ for any input instance I. Instead of measuring the output size, δ now measures the "precision-sensitivity" of I. We seek to design algorithms that can take advantage of this parameter δ.

The introduction of precision-sensitivity paves the way for studying problems that were previously considered hopeless or "solved". Notice that the same situation arises with the introduction of output-sensitivity. To take one example, the hidden surface elimination which is trivially $\Theta(n^2)$ in the usual complexity model (ergo "uninteresting") becomes very interesting when we consider output-sensitive algorithms. See [Berg, Bern] for some interesting results that exploit output-sensitivity in this problem.

1.1 Precision-Sensitive Approach to Euclidean Shortest Path

This paper focuses on the 3-dimensional *Euclidean shortest path* (3ESP) problem: given a collection of polyhedral obstacles in $\mathbf{R}^3$, and source and target points $s, t \in \mathbf{R}^3$, construct an obstacle-avoiding path

$$p_{min} = (s, x_1, \ldots, x_k, t), \quad (1)$$

$k \geq 0$, from s to t with minimal Euclidean length. This problem is ideal for introducing precision-sensitivity because conventional approaches are doomed to failure due to its NP-hardness, a result of Canny and Reif [CR]. It is also useless to introduce output-sensitivity here because the output-size is $O(n)$.

On the other hand, something interesting is going on in the bit model: the algebraic numbers that describe the lengths of the shortest paths can have exponential degrees (see subsection 2.2). This means that to compare the lengths of two *combinatorially distinct* shortest paths may require exponentially many bits. "Combinatorially distinct" means that the respective paths pass through different sequences of edges, and each is shortest for its edge sequence. In this paper, we use the relative difference between the length d_1 of a shortest path and the length d_2 of the combinatorially distinct next shortest path as our measure of "precision-sensitivity"

$$\delta = \delta(I) := (d_2 - d_1)/d_1. \quad (2)$$

It should be noted that δ may be 0. This is a crucial step towards a practical 3ESP algorithm, but it is not enough.

First we clarify some further aspects of 3ESP. The exponential behavior of 3ESP has two sources: not only is the bit complexity apparently exponential, the number of combinatorially distinct shortest paths can also be exponential. In fact, Canny and Reif's NP-hardness construction exploits the latter property of 3ESP. We can separate the combinatorial aspects from the algebraic aspects as follows. Define the *combinatorial 3ESP problem* which, with input as in 3ESP, asks for a *shortest edge sequence*

$$S_{min} = (e_1, \ldots, e_k), \quad (3)$$

such that $x_i \in e_i$ for $i = 1, \ldots, k$, where the x_i are the breakpoints of some shortest path p_{min} given by (1). Once S_{min} is obtained, there are effective numerical methods to zoom into the actual breakpoints $x_1, \ldots, x_k$, as we shall see. Thus the "purely" numerical part of ESP is delegated to a subsequent phase of computation.

How hard is the combinatorial 3ESP problem? Define the implicit parameter $s(I)$ of an input I to 3ESP to be $s = s(I) = |\log(|d_1 - d_2|)|$. We say an algorithm for the combinatorial 3ESP problem is *strongly precision-sensitive* if it is polynomial-time in the parameters n, L, s. By a careful analysis of the Canny-Reif proof, we show:

Theorem 1 *If there exists a strongly precision-sensitive algorithm for the combinatorial 3ESP problem then USAT can be solved in polynomial-time.*

Here USAT is the *unambiguous satisfiability problem*, commonly believed not to be in polynomial-time [Pa2, VV]. Note that the parameter $s(I)$ is an absolute measure while our sensitivity parameter $\delta(I)$ is a relative one. But this difference is not crucial. What is more important is the fact that $s(I)$ is roughly logarithmic in $\delta(I)$. In some sense, this theorem justifies our choice of $\delta(I)$.

1.2 Towards a Practical Algorithm

In hopes of developing a "practical algorithm", Papadimitriou [Pa1] introduces the *approximate 3ESP problem*. The input is as in 3ESP plus a new input parameter $\varepsilon > 0$. The problem is to compute an ε-*approximate shortest path*, i.e., one whose length is at most $(1 + \varepsilon)$ times the length of the shortest path. The bit-complexity of this approach is resolved in [CSY], yielding an algorithm with time

$$T(n, M, W) = O((n^3 M \log M + (nM)^2) \cdot \mu(W)), \quad (4)$$

where $M = O(nL/\varepsilon)$ and $W = O(\log(n/\varepsilon) + L)$. Despite initial hopes, this result is still impractical, even for small examples, because the stated complexity is, roughly speaking, achieved for every input instance. Our goal is to remedy this by introducing precision-sensitivity.

Recall that Papadimitriou's approach is to subdivide each obstacle edge into *segments* in a clever way and, by treating these segments as nodes in a weighted graph, to reduce the problem to finding the shortest path in a graph.

In order to introduce precision-sensitivity, we exploit the alternative scheme introduced in [CSY] for subdividing edges into segments. The subdivision is parameterized by a choice of $\epsilon > 0$. Our scheme has the property that the $\epsilon/2$-subdivision is a refinement of the ϵ-subdivision. The idea is to discard all segments that are provably not used by the shortest path; what remains are called *essential* segments. While it is obvious that such an implementation can drastically decrease running time in practice, we now show that – depending on the parameter δ – this improvement is also asymptotical. Assuming *non-degeneracy* (see section 2.1) of S_{min} in (3), we prove the following theorem:

Theorem 2 *There is an incremental algorithm to compute an ε-approximate shortest path in time that is polynomial in $1/\delta$ and $1/\varepsilon$. Omitting logarithmic factors, the dependency on $1/\varepsilon$ is only linear rather than quadratic.*

In case the shortest path sequence S_{min} is unique (i.e., $\delta > 0$), we can use techniques from mathematical optimization as soon as we have reached a refinement in which only S_{min} is left. The convergence depends on the spectral bounds

$$\mu, \quad \rho$$

corresponding to the minimum and maximum (respectively) eigenvalue of the Hessian H of the path length function $l(\lambda_1, \ldots, \lambda_k)$, where $\lambda_1, \ldots, \lambda_k \in \mathbf{R}$ parameterize the points $x_1, \ldots, x_k$ on S_{min}.

Theorem 3 *If $\delta > 0$, then the length of the shortest path can be approximated to relative error ε in time polynomial in $1/\delta, \log(1/\varepsilon), n, L$ and the spectral bounds μ, ρ.*

This theorem, and the remark in theorem 2 about a linear dependency on $1/\varepsilon$ are actually of practical significance. In section 5, we shall provide some experimental results, addressing the practicability of the incremental technique.

2 Preliminaries

Throughout the paper, we assume that the input is given by a source point s, a target point t, and a set of pairwise disjoint polyhedral obstacles, with a total of less than n edges. For each obstacle edge e, denote its endpoints by $s(e), t(e)$ and write $e = \overline{s(e)t(e)}$. Let $[e]$ denote the infinite line through e. We assume that s, t as well as endpoints of edges are specified by L-bit rational numbers. For any point $q \in \mathbf{R}^3$, $\|q\|$ denotes its Euclidean norm.

2.1 Basic Properties

We assume the notation in the introduction. In particular, p_{min} is a *global shortest path* from s to t in the free space FS defined by the obstacles. A breakpoint x_i of p_{min} in (1) that lies on the line between its neighboring vertices, $x_i \in \overline{x_{i-1}x_{i+1}}$, is called *redundant*. W.l.o.g. we may assume that p_{min} contains no redundant vertices.

First we fix an *edge sequence* $S = (s, e_1, \ldots, e_k, t)$. The sequence S is *degenerate* if $s \in [e_1], t \in [e_k]$, or $[e_j] \cap [e_{j+1}] \neq \emptyset$ for some $j \in \{1, \ldots, k-1\}$. We parameterize points $x_i \in [e_i]$ by a scalar λ_i according to

$$x_i = s(e_i) + \lambda_i u(e_i), \text{ with } u(e_i) = \frac{t(e_i) - s(e_i)}{\|t(e_i) - s(e_i)\|}.$$

Let $x_0 = s$ and $x_{k+1} = t$. Then the polygonal path $p = (s, x_1, \ldots, x_k, t)$ over S has length

$$l_S(\lambda_1, \ldots, \lambda_k) = \sum_{i=0}^{k} \|x_{i+1} - x_i\|.$$

We also write $|p|$ for $l_S(\lambda_1, \ldots, \lambda_k)$. Let $p_{min}(S)$ be defined to be the path p over S that minimizes the function $l_S(\lambda_1, \ldots, \lambda_k)$, *without consideration of the obstacles*.

A necessary condition for $l_S : \mathbf{R}^k \to \mathbf{R}$ to take its global minimum at $\lambda = (\lambda_1, \ldots, \lambda_k)$ is that all partial derivatives vanish at λ. This condition can be interpreted as *Snell's law*, and, as the next lemma will reveal, is also a sufficient condition to specify shortest paths:

Lemma 1 *The function $l_S : \mathbf{R}^k \to \mathbf{R}$ is convex. If the shortest path over the lines $[e_i]$ has no redundant vertices, then l_S has a unique minimizer $\zeta \in \mathbf{R}^k$.*

Proof: (1) Let $l = l_S = \sum_{i=0}^{k} l_i$, where

$$l_i(\lambda_1, \ldots, \lambda_k) := \|x_{i+1} - x_i\|.$$

We may interpret l_i as a function in 2 variables λ_i and λ_{i+1} (resp., in one variable λ_1 or λ_k).

To show that l is convex, it suffices to show that each of the l_i is convex (the sum of convex functions is convex). The convexity of l_i is a special case of a general result in convex analysis: for any norm $\|.\| : \mathbf{R}^k \to \mathbf{R}$ and any linear function $f : \mathbf{R}^m \to \mathbf{R}^k$, the function $\|f\| : \mathbf{R}^m \to \mathbf{R}$ is convex (see e.g. [Ro]).

(2) The convexity of l guarantees that every local minimum of l is a global minimum, say, d_S, and that the set of points $\lambda \in \mathbf{R}^k$ satisfying $l(\lambda) = d_S$ (the set of minimizers) is convex.

Assume that there are two distinct minimizers $\zeta_1, \zeta_2 \in \mathbf{R}^k$. Then every $\mu(t) = (\mu_1(t), \ldots, \mu_k(t)) := \zeta_1 + t(\zeta_2 - \zeta_1)$, $t \in [0,1]$, is a minimizer, and hence $l(\mu(t)) \equiv const$. But
$$l(\mu(t)) = \sum_{i=0}^{k} l_i(\mu(t)),$$
where
$$l_i(\mu(t)) = \sqrt{A_i(t - B_i)^2 + C_i}$$
with $A_i \geq 0$ and $C_i \geq 0$.

This can be constant only if each of the functions $l_i(\mu(t))$ is linear in t, i.e., $A_i = 0$ or $C_i = 0$ (this directly follows from $\partial^2 l(\mu(t))/\partial t^2 \equiv 0$).

Now let j be the first index for which $\mu_i(t)$ is not constant, i.e., $\mu_i(t) \equiv const$ $\forall$ $i = 1, \ldots, j-1$ and $\mu_j(t) \not\equiv const$ (j may be equal to 1).

The fact that $l_j(\mu(t))$ is linear then implies that $x_{j-1} \in [e_j]$: the point $x_j(t) := s(e_j) + \mu_j(t)u(e_j)$ is moving on the line $[e_j]$ while keeping distance $l_j(\mu(t))$ to the fixed point $x_{j-1} = s(e_{j-1}) + \mu_{j-1}(t)u(e_{j-1})$.

Thus x_{j-1} and $x_j(t)$ lie on the same line $[e_j]$ for all $t \in [0,1]$. From Snell's law it follows that also $x_{j+1}(t)$ must lie on this line, showing that the vertex $x_j(t)$ is redundant. □

Note that, in contrast to this proof, the known proof for the uniqueness of the shortest path [SS] (see also [Ya, appendix]) uses geometrical arguments.

Lemma 2 *If S is non-degenerate, then l_S is strictly convex, and the Hessian of l_S is positive-definite.*

Proof: The Hessian H of $l = l_S$ is a tridiagonal $k \times k$-matrix
$$H = \begin{pmatrix} a_1 & b_1 & & & \\ b_1 & a_2 & b_2 & & \\ & b_2 & \ddots & & \\ & & & & b_{k-1} \\ & & & b_{k-1} & a_k \end{pmatrix}.$$

Let H_i be the Hessian of the function $l_i = \|x_{i+1} - x_i\|$, interpreted as function over λ_i and λ_{i+1} if $1 \leq i \leq k-1$, and over λ_1 (resp., λ_k) if $i = 0$ (resp., $i = k$):
$$H_0 = (a_1^-), \quad H_i = \begin{pmatrix} a_i^+ & b_i \\ b_i & a_{i+1}^- \end{pmatrix}, \quad H_k = (a_k^+),$$
with
$$a_i^+ = \frac{\partial^2 l_i}{\partial \lambda_i^2}, \quad a_{i+1}^- = \frac{\partial^2 l_i}{\partial \lambda_{i+1}^2}, \quad b_i = \frac{\partial^2 l_i}{\partial \lambda_i \partial \lambda_{i+1}}.$$

Then H and the H_i are related by $a_i = a_i^- + a_i^+$. Abusing the notation, we may write $H = H_0 + \ldots + H_k$.

The strict convexity of l follows from the positive-definiteness of H. To prove that H is positive-definite, it remains to show that the determinant of H is not zero. Let
$$H = \begin{pmatrix} a_1 & b_1 & \\ b_1 & & \\ & & H' \end{pmatrix}, \quad H^+ = \begin{pmatrix} a_1^+ & b_1 & \\ b_1 & & \\ & & H' \end{pmatrix}.$$

Then
$$det(H) = det(H^+) + a_1^- det(H').$$

As all H_i are positive-semi-definite, also $H^+ = H_1 + \ldots + H_k$ is positive-semi-definite, and thus $det(H^+) \geq 0$. This implies
$$det(H) \geq a_1^- det(H').$$
Continuing recursively, we finally get
$$det(H) \geq a_1^- \cdot \ldots \cdot a_k^-.$$

Abbreviating
$$v_i = \frac{x_{i+1} - x_i}{\|x_{i+1} - x_i\|},$$
we have
$$a_i^- = \frac{1 - \langle v_{i-1}, u(e_i) \rangle^2}{\|x_i - x_{i-1}\|}.$$

This is strictly positive if S is non-degenerate, and hence also $det(H) > 0$. □

2.2 Bit Complexity

The goal of this subsection is to provide some background on the algebraic complexity of the problem. The techniques are basically standard, so we briefly state the results.

For a fixed edge sequence $S = (s, e_1, \ldots, e_k, t)$ and "flags" $f_1, \ldots, f_k$, which specify whether the shortest path $p_{min}(S)$ passes through the endpoints of e_i, we construct a Boolean formula
$$B_S(f_1, \ldots, f_k)$$
with free variables $\lambda_1, \ldots, \lambda_k$, which specifies that the path defined by $\lambda_1, \ldots, \lambda_k$ has no redundant vertices, and – according to the flag f_i – obeys Snell's law at e_i or passes through an endpoint. The path $p_{min}(S)$ is specified by one of these formulas. Using the fast quantifier elimination technique in [Re], and root separation results, we obtain:

Lemma 3 *Every coordinate λ_i of a parameter tuple λ satisfying $B_S(f_1, \ldots, f_k)$ has a defining polynomial h_{λ_i} of degree $n^{O(n)}$ with integer coefficients of bit-size $Ln^{O(n)}$. The polynomial h_{λ_i} together with an isolating interval for λ_i can be computed in time polynomial in L and $n^{O(n)}$.*

Lemma 4 *Let λ and λ' be solutions of $B_S(f_1, \ldots, f_k)$ and $B_{S'}(f'_1, \ldots, f'_{k'})$, respectively, and let p and p' be the corresponding paths.*
Then $|p| = |p'|$ or $|(|p| - |p'|)| = 2^{-Ln^{O(n)}}$.

In order to actually compute the global shortest path p_{min}, we have to filter out those paths $p_{min}(S)$ that would collide with obstacles. Having calculated the parameter λ satisfying $B_S(f_1, \ldots, f_k)$, this amounts to answering the queries '$\overline{x_i x_{i+1}} \in FS$?', for $i = 0, \ldots, k$. But these queries can be expressed as Tarski sentences in a fixed number of variables, and be decided in time polynomial in L and $n^{O(n)}$. We finally obtain:

Theorem 4 *It is possible to compute algebraic representations of all combinatorially distinct shortest paths in time polynomial in L and $n^{O(n)}$.*

3 Combinatorial 3ESP is as hard as USAT

Recall that the exponential complexity of the 3ESP problem has a combinatorial and an algebraic source. We give evidence that 3ESP remains intractable even after eliminating the algebraic source of complexity.

We briefly review the Canny-Reif construction ([CR], section 2.5): Given a 3SAT-formula f in conjunctive form with m clauses and n variables $b_1, \ldots, b_n$, it is possible to construct an environment $E(f)$ such that the following holds for a fixed "reference length" $l = 2^{3n}$, and $\Delta = 2^{-nm-3n-4}$: To each instantiation of $(b_1, \ldots, b_n)$, there corresponds an edge sequence $S = S(b_1, \ldots, b_n)$ such that the shortest path p over S lies in free space and satisfies

$$|p| \in \begin{cases} [l, l + \Delta] & \text{if } f(b_1, \ldots, b_n) = 1, \\ [l + 2\Delta, \infty) & \text{if } f(b_1, \ldots, b_n) = 0. \end{cases}$$

The number of edges of $E(f)$ as well as the maximal bit-size of coordinates is polynomial in n and m. Deciding the satisfiability of f is reduced to deciding if the shortest path in $E(f)$ has length $\leq l + \Delta$.

A careful analysis shows the following property of $E(f)$: if the formula f is unique satisfiable, i.e., by exactly one instantiation of $(b_1, \ldots, b_n)$, then the shortest path in $E(f)$ is unique and the gap in length between this path and any path that passes over a different edge sequence is single-exponential (i.e., $> c^{-nm}$ for some $c > 1$).

We sketch the idea: The basic construction elements in [CR] are parallel, 2-dimensional plates with (for ease of description) 1-dimensional slots. The construction is based on a scene with 2^n shortest paths, with length $l' \leq l + \Delta$. In the final step, obstacles are introduced which stretch all paths that correspond to non-satisfying instances by at least Δ. It remains to verify that there are no further suboptimal paths with length close to l'. The use of parallel plates ensures that these paths would have additional legs between slots. The spacing between plates and between the break points of the shortest paths in the slots gives a lower bound on the additional length, and is again roughly Δ. Finally, the gap Δ is single-exponential.

Now assume that we have a strongly precision-sensitive algorithm as defined in subsection 1.1. Consider the satisfiability problem restricted to 3SAT formulas that are satisfiable by at most one variable instance, known as the *unambiguous satisfiability problem USAT*. Assume we are given such a formula f. By constructing $E(f)$ and running our algorithm, we would be able to decide the satisfiability of f in polynomial time. This proves theorem 1.

4 Approximation

For simplicity, we shall describe algorithms in this section in the algebraic framework. It is important to note that the hardness result of section 3 is not valid in this model. However, as in [CSY], the technique extends to the bit framework. In particular, it suffices to compute intermediate numbers to precision $W = O(\log(n/\varepsilon) + L)$.

We review the approximation scheme in [CSY]: the algorithm mainly consists of three steps. In the first step, the edges are subdivided into segments such that the following holds with $\varepsilon' = \varepsilon/Cn$, C a suitable constant:

Lemma 5 ([CSY])
(1) Each segment σ of the subdivision satisfies $|\sigma| \leq \varepsilon' \text{dist}(s, \sigma)$.
(2) Each edge is divided into $O(L/\varepsilon')$ segments.
(3) The subdivision for $\varepsilon'/2$ is a refinement of the subdivision for ε'.

In the second step of the algorithm, the *visibility graph* $G_0 = (V_0, E_0)$ of the segments is constructed. The nodes of the graph comprise the subdivision segments including s and t. The edges comprise pairs (σ, σ') of segments that can "see each other", meaning that there exists $x \in \sigma$ and $x' \in \sigma'$ such that $\overline{xx'} \in FS$. In the third step, the visibility graph G_0 is weighted by assigning to each edge (σ, σ') the Euclidean distance between the midpoints of σ and σ'. Finally, the shortest path Σ in G_0 is computed by running Dijkstra's shortest path algorithm. This path is a *segment sequence* $\Sigma = (s, \sigma_1, \ldots, \sigma_k, t)$. Its "weight" $|\Sigma|$ satisfies $|\Sigma| \leq (1 + \varepsilon/C')|p_{min}|$ and $|p_{min}| \leq (1 + \varepsilon/C')|\Sigma|$, with C' depending on C.

4.1 An Incremental Algorithm

The above algorithm uses a fixed subdivision. In the following, we shall exploit property (3) in lemma 5 by successively halving the error bound ε, and by refining only those segments which the global shortest path potentially could use.

Let $\varepsilon_i = 2^{-i}$ and $\varepsilon'_i = \varepsilon_i/Cn^2$, with $C = 32$ (note that we divide by Cn^2 instead of Cn). Let $G_i = (V_i, E_i)$ be the weighted visibility graph for any set of segments V_i fulfilling the basic inequality (1) of lemma 5, and let l_i denote the length of the shortest path from s to t in G_i. Then the following lemma holds:

Lemma 6 *If $\Sigma = (s, \sigma_1, \ldots, \sigma_k, t)$ is a path in G_i with $|\Sigma| > (1 + \varepsilon_i/4n)l_i$, then any path p over Σ satisfies $|p| > |p_{min}|$.*

We define the *essential subgraph* $G_i^{ess} = (V_i^{ess}, E_i^{ess})$ of G_i to be the subgraph which is spanned by all (s,t)-paths Σ in G_i with $|\Sigma| \leq (1 + \varepsilon_i/4n)l_i$.

Corollary 1 *If p_{min} leads over a segment sequence $\Sigma = (s, \sigma_1, \ldots, \sigma_k, t)$ in G_i, then Σ is in G_i^{ess}.*

To approximate a shortest path p_{min} by successive refinement, we need thus only to consider the segments in V_i^{ess} in the next step.

We can compute G_i^{ess} as follows: run Dijkstra's single source shortest path algorithm on G_i twice, starting at s and starting at t, and assign to each $\sigma \in V_i$ the distances $d_s(\sigma)$ (resp., $d_t(\sigma)$) to s (resp., t) in G_i. This implies $l_i = d_s(t)$. Let the weight of edge (σ, σ') in G_i be given by $\omega(\sigma, \sigma')$. Then the following holds:

Lemma 7

$$E_i^{ess} = \{ (\sigma, \sigma') \in E_i \;;\; d_s(\sigma) + d_t(\sigma') + \omega(\sigma, \sigma') \leq (1 + \varepsilon_i/4n)l_i \}.$$

In practice, G_i^{ess} should be significantly smaller than G_i. In fact, it approaches the 1-dimensional skeleton formed by all global shortest paths as $\varepsilon_i \to 0$.

Lemma 8 *All paths $\Sigma = (s, \sigma_1, \ldots, \sigma_k, t)$ in G_i^{ess} with $k < n$ satisfy $|\Sigma| \leq (1 + \varepsilon_i)|p_{min}|$.*

Proof: It is sufficient to show that $|\Sigma| \leq (1 + \varepsilon_i/4)l_i$. The claim follows from this by $l_i \leq (1 + \varepsilon_i/4)|p_{min}|$. We first prove

(*) All edges $e = (\sigma, \sigma') \in E_i^{ess}$ satisfy

$$d_s(\sigma) + \omega(e) \leq d_s(\sigma') + (\varepsilon_i/4n)l_i.$$

Assume that $\omega(e) > d_s(\sigma') - d_s(\sigma) + (\varepsilon_i/4n)l_i$. Let w be a path from s to σ with length $|w| = d_s(\sigma)$, and w' a shortest path from σ' to t. Then $d_s(\sigma') +$ $|w'| \geq d_s(t) = l_i$, and thus $|w'| \geq l_i - d_s(\sigma')$. With the assumption, we get

$$|w| + \omega(e) + |w'|$$
$$> d_s(\sigma) + d_s(\sigma') - d_s(\sigma) + (\varepsilon_i/4n)l_i + l_i - d_s(\sigma')$$
$$= (1 + \varepsilon_i/4n)l_i.$$

But there can be no shorter path from s to t that uses edge e, and thus e can not lie in G_i^{ess}. Contradiction.

Now consider the path $\Sigma = (s, \sigma_1, \ldots, \sigma_k, t)$. We prove inductively that the subpath $\Sigma_j = (s, \sigma_1, \ldots, \sigma_j)$ satisfies

$$|\Sigma_j| \leq d_s(\sigma_j) + j(\varepsilon_i/4n)l_i.$$

For $j = 1$, the claim follows directly from property (*):

$$|\Sigma_1| = \omega(s, \sigma_1) \leq d_s(\sigma_1) + (\varepsilon_i/4n)l_i.$$

For the induction step from j to $j + 1$, (*) amounts to $\omega(\sigma_j, \sigma_{j+1}) \leq d_s(\sigma_{j+1}) - d_s(\sigma_j) + (\varepsilon_i/4n)l_i$. The claim follows from the induction hypothesis:

$$|\Sigma_{j+1}| = |\Sigma_j| + \omega(\sigma_j, \sigma_{j+1})$$
$$\leq d_s(\sigma_{j+1}) + (j+1)(\varepsilon_i/4n)l_i.$$

The case $j = k + 1$ (in which $\sigma_{k+1} = t$) proves the lemma. □

We are now ready to formulate our *incremental algorithm*, to get a relative error of $\varepsilon = 2^{-r}$:

(1) $i := 0; \varepsilon'_0 := 1/Cn^2$;
(2) Compute the initial subdivision V_0;
(3) Repeat
(4) $\quad \varepsilon'_i := 2^{-i}(1/Cn^2)$;
(5) $\quad$ Construct the visibility graph $G_i = (V_i, E_i)$;
(6) $\quad$ Compute $G_i^{ess} = (V_i^{ess}, E_i^{ess})$;
(7) $\quad$ Compute V_{i+1} by refining V_i^{ess};
(8) $\quad i := i + 1$;
(9) Until $i - 1 = r$.

4.2 Spectral Analysis

Our first goal in this subsection is to characterize the behavior of the incremental algorithm for a fixed edge sequence S.

Let $l = l_S$, and let again H be the Hessian of l with eigenvalues μ and ρ. Let ζ be the parameter tuple specifying $p_{min}(S)$, and z_j the break point of $p_{min}(S)$ on e_j, specified by ζ_j. Our goal is to show that a path p over S whose length differs only slightly from $|p_{min}(S)|$ must also have a parameter λ which is close to ζ. Taylor's theorem shows that for any $\lambda \in \mathbf{R}^k$, there exists $\tau \in \mathbf{R}^k$ such that

$$l(\lambda) - l(\zeta) = \langle \nabla l(\zeta), \lambda - \zeta \rangle + \frac{1}{2}\langle \lambda - \zeta, H(\tau)(\lambda - \zeta)\rangle.$$

The first term is never negative ($\langle \nabla l(\zeta), \lambda - \zeta \rangle$ is the directional derivative of l at ζ in the direction from ζ to λ). The second can be bounded by the "spectrum" of H:

$$\mu \|\lambda - \zeta\|^2 \leq \langle \lambda - \zeta, H(\tau)(\lambda - \zeta) \rangle \leq \rho \|\lambda - \zeta\|^2.$$

This implies

$$l(\lambda) - l(\zeta) \geq \frac{\mu}{2} \|\lambda - \zeta\|^2.$$

Thus, the parameter λ of any path p over S with $||p| - |p_{min}(S)|| \leq \tilde{\varepsilon}$ satisfies $\|\lambda - \zeta\|^2 \leq \tilde{\varepsilon}(2/\mu)$.

Lemma 9 *Let $\Sigma = (s, \sigma_1, \ldots, \sigma_k, t) \in G_i^{ess}$ be a segment sequence with $\sigma_j \subseteq e_j$, let x_j denote the vertices of $p_{min}(\Sigma)$, and z_j the vertices of $p_{min}(S)$. Then*

$$\|x_j - z_j\| \leq (4\varepsilon_i \frac{|p_{min}|}{\mu})^{\frac{1}{2}}.$$

Proof: We know that $|p_{min}(\Sigma)| \leq |\Sigma| \leq (1 + \varepsilon_i)|p_{min}| \leq (1 + \varepsilon_i)|p_{min}(S)|$. Setting $\tilde{\varepsilon} = \varepsilon_i |p_{min}(S)|$, we get

$$\|x_j - z_j\| = |\lambda_j - \zeta_j| \leq \|\lambda - \zeta\|$$
$$\leq (2\varepsilon_i \frac{|p_{min}(S)|}{\mu})^{\frac{1}{2}}.$$

The claim follows with $|p_{min}(S)| \leq (1 + \varepsilon_i)|p_{min}|$. □

Now assume we run our incremental algorithm for the fixed edge sequence S, and are in step i. Then on each edge e_j we only refine segments with distance $const \cdot \sqrt{\varepsilon_i}$ from z_j, with a constant depending on S but not on i. By construction, each segment σ on e_j has length $|\sigma| \geq a\varepsilon_i/Cn^2$ (with a the distance from e_j to the source). Thus we refine at most

$$C(S) \cdot n^2 \varepsilon_i^{-\frac{1}{2}}$$

segments on each e_j in the i-th step, with

$$C(S) = O((\frac{|p_{min}|}{a^2 \mu})^{\frac{1}{2}}).$$

Let $\varepsilon = 2^{-r}$ be the desired relative error. Summing over $i = 1, \ldots, r$ we produce a total of $O(r\sqrt{1/\varepsilon_r})$ segments. This is a significant improvement to the original scheme, which would produce $O(1/\varepsilon_r)$ segments.

In the above discussion, we considered a fixed edge sequence. The described effect will occur as soon as the shortest path is "resolved". Together with lemma 8, the definition of δ in (2) implies that all segment sequences Σ in G_i^{ess} must lie on S_{min} if $\varepsilon_i < \delta$. We get:

Lemma 10 *The essential subgraph G_i^{ess} contains less than*

$$M_i = O(n^2(L/\delta + C(S_{min}) \cdot 2^{\frac{i}{2}}))$$

segments per edge.

This should be compared to the number $M = O(nL2^r)$ of segments per edge which are produced by the algorithm in [CSY] to achieve the error bound 2^{-r}. The visibility relation between segments can be computed separately for each of the $O(r)$ refinement steps by a sweep algorithm, as described in [CSY]. The cost of this algorithm dominates the computation of G_i^{ess}. Thus, the running time of the i-th step is $T(n, M_i, W)$, with T as in equation (4). The running time of the total algorithm can be bounded by

$$O(r \cdot T(n, M_r, W)).$$

Thus, we have proven theorem 2.

4.3 Path Optimization

With the incremental approach above, we have a tool to determine S_{min} in (3) in time polynomial in $1/\delta$. As soon as there is only one possible edge sequence (or only a few combinatorially distinct sequences) left, it is however more efficient to use an optimization technique to approximate the actual shortest path. We propose to use a steepest descend method (see e.g. [Go, section C-5]):

Let the spectrum of H be bounded below by $\mu > 0$ and above by ρ (choose μ as the smallest, and ρ as the biggest eigenvalue of H). We can derive explicit values for these bounds (especially for ρ) as described in subsection 4.4.

Let $I = [\frac{1}{2\rho}, \frac{3}{2\rho}]$. Define the sequence

$$\lambda^{i+1} = \lambda^i - \kappa \nabla l(\lambda^i)$$

with $\kappa \in I$ and λ^0 the known approximation. Then this sequence converges to the unique minimizer ζ of l at the rate of a geometric progression with ratio $q = 1 - \mu/\rho$:

$$\|\lambda^{i+1} - \zeta\| \leq q\|\lambda^i - \zeta\| \leq q^{i+1}\|\lambda^0 - \zeta\|.$$

With

$$\|l(\lambda^i) - l(\zeta)\| \leq \frac{\rho}{2}\|\lambda^i - \zeta\|^2$$

and

$$\|\lambda^0 - \zeta\|^2 \leq \frac{2}{\mu}(l(\lambda^0) - l(\zeta)) \leq \frac{2\delta}{\mu} l(\zeta),$$

we get

$$\|l(\lambda^i) - l(\zeta)\| \leq \frac{\rho\delta}{\mu} q^{2i} |p_{min}|.$$

To achieve $\|l(\lambda^i) - l(\zeta)\| < 2^{-r}|p_{min}|$, it is sufficient to choose $i > N$ with

$$N = \Theta(\frac{\rho}{\mu}(r + |\log \delta| + |\log \frac{\rho}{\mu}|)).$$

Again, it is important to note that this method – because of the freedom of choice for κ – easily extends to the bit framework. The running time of the whole algorithm can be resolved as

$$O(\log(1/\delta) \cdot T(n, M_\delta, W) + Nn\mu(W)),$$

where $M_\delta = O(n^2 L/\delta)$. This finally proves theorem 3.

4.4 Spectral Bounds

In this subsection, we discuss two different methods to get bounds on the spectrum of the Hessian H of the path length function $l = l_{S_{min}}$. We shall use the notations of section 2, and assume that S_{min} is non-degenerate.

By the theory of Gerschgorin circles, the eigenvalues of H are bounded above by $\rho = max_i\{ a_i + |b_i| + |b_{i-1}| \}$, with a_i and b_i as in the proof of lemma 2.

With the help of ρ, we can directly give a bound on μ. As the determinant of H is equal to the product of all k eigenvalues of H, we get

$$\mu \geq \frac{det(H)}{\rho^{k-1}},$$

where the determinant of H satisfies the inequality

$$det(H) \geq \prod_{i=1}^{k} a_i^-.$$

Naturally, μ will be close to 0 if S_{min} is "nearly degenerate", i.e., if one of the α_i^- is small. A more crucial deficiency of the above bound is that it is not only exponential in L, but also in k, the number of intermediate vertices of p_{min}.

A bound on μ not depending on k can be obtained by a method based on the theorem of Courant-Fischer (see [Wi], pp. 101-102):

Let A, B and C be positive-semi-definite symmetric matrices with $C = A + B$, and α (resp., β) the smallest eigenvalue of A (resp., B). Then each eigenvalue γ of C satisfies $\gamma \geq \alpha + \beta$. Assuming k to be even (the case of odd k can be similarly treated), we split $H = A + B$ according to

$$A = \sum_{i=0}^{k/2} H_{2i}, \quad B = \sum_{i=1}^{k/2} H_{2i-1}.$$

The matrices A and B are block matrices, and the eigenvalues are the eigenvalues of the H_i. It follows that

$$\mu \geq min\{ \mu_i \; ; \; i = 0, \ldots, k \},$$

where μ_i is the smallest eigenvalue of H_i. This bound has the nice property that it depends only on pairs of edges of S_{min}.

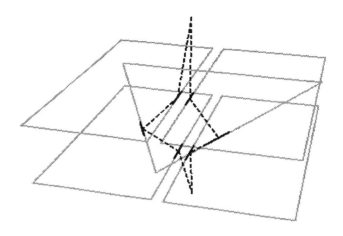

Horizontal Obstacles 1

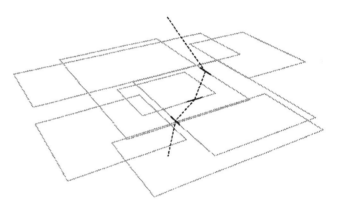

Horizontal Obstacles 2

5 Experimental Results

The preceding algorithms for the approximate 3ESP problem have a high polynomial dependency on the number of edges n or the desired error bound ε. But these theoretical bounds need not reflect the "average behavior" or practical situations. This suggests some empirical studies.

To verify the practicability of our incremental approach, we implemented a simplified version of the proposed algorithm. The simplification is based on the observation that for certain special cases, the visibility relation between segments can be replaced by the visibility of segment midpoints:

Let the obstacles be 2-dimensional facets arranged in h parallel planes separating the start and target point. Let $\varepsilon' = \varepsilon/h$, and consider the subdivision defined in section 4. Then the following holds:

Steps	0	2	4	6	8	10
Error	1.6771	0.4193	0.1048	0.0262	0.0066	0.0016
Length	2.4852	2.3746	2.3290	2.3252	2.3250	2.3250
Segments	19	76	224	186	366	728

Table 1

Steps	0	2	4	6	8	10
Error	2.2500	0.5625	0.1406	0.0352	0.0088	0.0022
Length	2.4116	2.3263	2.3085	2.3070	2.3069	2.3069
Segments	28	112	290	334	368	430

Table 2

Lemma 11 *There exists a free polygonal path p from s to t, which connects segment midpoints and satisfies $|p| \leq (1+\varepsilon)|p_{min}|$.*

Proof: The shortest path p_{min} from s to t is strictly monotone in the direction of the normal vector of the planes containing the obstacle facets. Now pick an arbitrary vertex v of p_{min} and move this vertex on the incident edge in either direction while keeping the other path vertices fixed. We continue this deformation until the (deformed) path hits another obstacle facet or until v hits a segment midpoint. In the first case, we consider the intersection point as a further path vertex, and in the latter case we continue the process by picking another path vertex until all vertices coincide with segment midpoints. It is easy to see that this process terminates after at most h deformation steps, introducing an absolute error of at most $h\varepsilon'|p_{min}|$. □

Further, we used a uniform subdivision for each edge, by starting with the edges as segments and successively halving the segments.

Tables 1 and 2 show the result of the incremental algorithm for the situation in the figures Horizontal Obstacles 1 and 2. The figures visualize the situation after 10 iteration steps: the essential segments are the solid black parts of the edges, and the shortest paths from start to goal determined so far are drawn dashed. In the first example there are two shortest paths, which are resolved after the 8th iteration step. In the second example, the unique shortest path is resolved after the 10th step.

The tables show the relative error in path length, the length of the shortest path in the current visibility graph, and the number of the essential segments. The run-time of the algorithm is mostly quadratic in this number, and was – for these examples – in the range of seconds on a state-of-the-art workstation.

The following behavior has been typical for the examples we tried: until the error bound ε_i is small enough to discard, the number of essential segments is doubled per step. Then comes a phase where the segment number does not change significantly. Once the shortest paths are resolved, the number of essential segments is doubled every 2 iteration steps, as predicted by the theoretical results. We mention a possible problem with our algorithm: the step at which discarding of segments starts depends on the input complexity (here the parameter h).

6 Final Remarks

(1) We have developed the first precision-sensitive algorithms for 3ESP. Beyond its intrinsic interest, it demonstrates a critical exploitation of precision-sensitivity. We conjecture that other previously intractable problems may likewise yield to this approach.

(2) If the sensitivity parameter δ is zero, we can modify our approach to take advantage of the "next sensitivity" parameter, namely the gap between the second and the third shortest path, etc. A general treatment of this may be interesting.

(3) We note that attention has to be paid to degenerate situations in this problem. E.g., the requirement for S_{min} to be non-degenerate in theorem 2 and 3 is quite severe. But it seems unavoidable to take this into account because degeneracy seems to be one cause of intrinsic complexity in 3ESP. Note that there has been some recent literature on degeneracy in geometric problems.

(4) The merits of the incremental 3ESP seem evident: in our examples, there would have been no chance to detect the shortest path by the exhaustive approach or by previously published methods (at least not on our machines). Concluding, our algorithm is a useful tool when a researcher needs to determine the real shortest path in a particular small environment.

References

[Berg] M. de Berg, "Efficient algorithms for ray shooting and hidden surface removal", *PhD Dissertation*, Department of Computer Science, Utrecht University, the Netherlands, 1992.

[Bern] M. Bern, "Hidden surface removal for rectangles", *J. Computer and System Sciences*, Vol. 40, 1990, pp. 49-59.

[CR] J.F. Canny, J. Reif, "New Lower Bound Techniques for Robot Motion Planning Problems", *Proc. 28th IEEE Symposium on Foundations of Computer Science*, Los Angeles 1987, pp. 49-60.

[CSY] J. Choi, J. Sellen, C.-K. Yap, "Approximate Euclidean Shortest Path in 3-Space", *Proc. 10th ACM Symposium on Computational Geometry*, 1994, pp. 41-48.

[Go] A.A. Goldstein, *Constructive Real Analysis*, Harper & Row Publishers, New York, 1967.

[Re] J. Renegar, "On the Computational Complexity and Geometry of the First-Order Theory of the Reals", Part III, School of Operations Research Report, Cornell University, 1989.

[Pa1] C.H. Papadimitriou, "An Algorithm for Shortest-Path Motion in Three Dimensions", *Information Processing Letters*, Vol. 20, 1985, pp. 259-263.

[Pa2] C.H. Papadimitriou, *Computational Complexity*, Addison-Wesley, Reading, Massachusetts, 1994.

[Ro] R.T. Rockafellar, *Convex Analysis*, Princeton University Press, 1970.

[SS] M. Sharir, A. Schorr, "On Shortest Paths in Polyhedral Spaces", *SIAM Journal on Computing*, Vol. 15 No. 1, 1986, pp. 193-215.

[VV] L.G. Valiant, V.V. Vazirani, "NP is as Easy as Detecting Unique Solutions", *Theoretical Computer Science*, 47, 1986, pp. 85-93.

[Wi] J.H. Wilkinson, *The Algebraic Eigenvalue Problem*, Clarendon Press Oxford, 1965.

[Ya] C.-K. Yap, "Towards Exact Geometric Computation", *Fifth Canadian Conference on Computational Geometry*, 1993, pp. 405-419 (to appear, *Computational Geometry: Theory and Applications*)

Approximation Algorithms for Geometric Tour and Network Design Problems

(Extended Abstract)

Cristian S. Mata *   Joseph S. B. Mitchell †

1 Introduction

In this paper, we provide a simple method to obtain provably good approximation algorithms for a variety of NP-hard geometric optimization problems having to do with computing shortest constrained tours and networks:

Red-Blue Separation Problem (RBSP): Consider the problem of finding a minimum-perimeter Jordan curve (necessarily, a simple polygon) that separates a set of "red" points, R, from a set of "blue" points, B. This problem is seen to be NP-hard, using a reduction from the Euclidean traveling salesman problem [3, 12]. (Replace each city in the TSP instance by a pair of points, one red and one blue, very close together.) While Euclidean TSP can be approximated to within a factor of 1.5 times optimal (using Christofides' heuristic [9]), the seemingly related RBSP has defied previous attempts to devise a provably good approximation algorithm. We provide an $O(\log m)$ approximation bound algorithm for RBSP, where $m < n$ is the minimum number of sides

---

*Department of Computer Science, SUNY Stony Brook, NY 11794-4400, email: cristian@cs.sunysb.edu; Supported by NSF grants ECSE-8857642 and CCR-9204585, and by a grant from Hughes Aircraft.

†Department of Applied Mathematics and Statistics, State University of New York, Stony Brook, NY 11794-3600, **email:** jsbm@ams.sunysb.edu; Partially supported by NSF grants ECSE-8857642 and CCR-9204585, and by grants from Boeing Computer Services and Hughes Aircraft.

Permission to copy without fee all or part of this material is granted provided that the copies are not made or distributed for direct commercial advantage, the ACM copyright notice and the title of the publication and its date appear, and notice is given that copying is by permission of the Association of Computing Machinery.To copy otherwise, or to republish, requires a fee and/or specific permission.
11th Computational Geometry, Vancouver, B.C. Canada
© 1995 ACM 0-89791-724-3/95/0006...$3.50

of a minimum-perimeter rectilinear polygonal separator for the $n = |R| + |B|$ input points.

A similar approximation bound also applies to the case in which R and B are sets of nonoverlapping polygons, and to the "multi-color" case in which there are $K > 2$ sets of points or polygons, and we must find a planar subdivision of minimum total length that has objects of only one color in each face.

TSP with Neighborhoods: Let S be a collection of k possibly overlapping simple polygons ("neighborhoods") in the plane. The "Geometric Covering Salesman Problem", or *TSP with Neighborhoods* (TSPN) problem asks for a shortest tour that visits (intersects) each of the neighborhoods. The special case in which the neighborhoods are singleton points is just the usual Euclidean TSP problem; thus, TSPN is also NP-hard. One can think of the TSP with Neighborhoods problem as an instance of the "One-of-a-Set TSP" (also known as the "Multiple Choice TSP", the "Covering Salesman Problem", and the "Group TSP through sets of nodes"), where the "sets" are *connected* regions in the plane. This problem is widely studied for its importance in several applications, particularly in communication network design ([18]) and VLSI routing ([26]).

The only known approximation results for TSPN are those of [2], who, using judiciously placed "representative" points in each neighborhood, give constant-factor approximation algorithms for the special case in which the neighborhoods all have diameter segments that are parallel to a common direction, and the ratio of the longest to the shortest diameter is bounded by a constant.

We provide an $O(\log k)$ approximation algorithm for the case in which the neighborhoods are arbitrary polygonal regions.

Watchman Route Problem: Let $\mathcal{P}$ be a polygonal room, possibly with "holes" (obstacles), having n vertices. The problem of computing a shortest tour in order that a mobile guard can "see" all of $\mathcal{P}$ is known as the *Watchman Route Problem* (WRP). If $\mathcal{P}$ is a simple polygon (having no holes), then WRP can be solved exactly, in time $O(n^4)$ [7, 23]. However, WRP is known to be NP-hard if $\mathcal{P}$ has holes (a simple reduction from Euclidean TSP; see [8]), even if $\mathcal{P}$ is rectilinear. But, as with RBSP, no approximation algorithms have previously been found for this problem. We give an $O(\log m)$ approximation algorithm for the WRP when the polygon $\mathcal{P}$ is rectilinear, where $m < n$ is the minimum number of edges in a shortest rectilinear watchman route.

k-MST Problem; Prize-Collecting Traveling Salesman Given a set S of n points in the plane, and an integer $k \leq n$, the k-MST problem asks for a spanning tree, possibly with Steiner points, of minimum length that joins some subset of k of the n points.

A related problem is the *Prize-Collecting Traveling Salesman Problem* (PCTSP): Consider a salesperson that must sell a given quota, k, of widgets before returning home. Each city (point in the plane) has an integer demand associated with it, indicating how many widgets can be sold there. The goal in the "Prize Collecting Salesman" problem is to find a minimum-length tour (polygon) that visits a set of cities whose demands sum to at least k.

Until recently, no approximation algorithms were known for the k-MST and PCTSP problems. Awerbuch et al. [4] have given an $O(\log^2 k)$ approximation algorithm for both problems, on graphs with non-negative edge weights.

As a simple consequence of our method, we give an $O(\log k)$ approximation algorithm for the geometric instances of both problems, even when the sites are given as polygonal regions. Our bounds match in approximation factor those obtained very recently for point sites by Garg and Hochbaum [15]. In the time since the original submission of this paper, [6] have obtained a *constant* factor approximation bound for point sites, employing an algorithm very similar to our own, together with a new geometric lemma relating the lengths of minimum-spanning trees and "division trees" (trees induced by a binary space partition).

Overview of the Method

Our method is based on a standard approach to obtaining approximation algorithms — (1) prove that any solution to the original problem can be transformed into a solution to a simpler problem (at little change in objective function value); (2) solve the simpler problem to optimality; and (3) show how to transform a solution to the simpler problem into a solution to the original problem (again, with little change in objective function value). Specifically, our steps involve: (1) transforming an optimal tour problem into an optimal subdivision problem (which, by a geometric lemma, is shown to be close in length to the optimal tour length); (2) solving a canonical instance of the optimal subdivision problem by means of a dynamic programming (DP) algorithm; and (3) showing that the DP solution can be transformed into a solution of the original problem, with a small factor increase in total length. This method is not new; e.g., it has been used by Gonzalez et al. [16] in optimal partitioning problems, and by Agarwal and Suri [1] and Mitchell [22] in other geometric separation and approximation problems.

2 Geometric Preliminaries

We restrict our attention to problems in two dimensions. We will work with planar polygonal subdivisions and triangulations, always within some bounded polygonal region, $\mathcal{R}$. The *length* of a planar subdivision, triangulation, or polygon is defined to be the sum of the Euclidean lengths of its edges. We say that a subdivision $\mathcal{S}'$ is a *refinement* of subdivision $\mathcal{S}$ if each edge of $\mathcal{S}$ is a subset of some edge of $\mathcal{S}'$. The *bounding box*, $bb(X)$, for a set X is defined to be the smallest axis-aligned rectangle that contains the set. The *grid* induced by a finite set of points is the graph consisting of all the vertices and bounded edges of the arrangement of all vertical and horizontal lines through the points.

A polygonal subdivision $\mathcal{S}$ of polygon $\mathcal{R}$ is said to be "guillotine" if it is a binary space partitioning (BSP) of $\mathcal{R}$; i.e., either there exist no edges of $\mathcal{S}$ interior to $\mathcal{R}$, or there exists a (straight) edge of $\mathcal{S}$ such that this edge is a chord of $\mathcal{R}$, dividing $\mathcal{R}$ into two regions, $\mathcal{R}_1$ and $\mathcal{R}_2$, such that $\mathcal{S}$ restricted to $\mathcal{R}_i$ is a guillotine subdivision of $\mathcal{R}_i$. (See [11] and standard references on BSP trees [24, 25].) If all faces of the subdivision are rectangles, we obtain a guillotine *rectangular* subdivision (GRS). Here, we will concern ourselves only with guillotine rectangular subdivisions.

The following geometric lemmas will be needed in our proofs and are of independent interest. The first lemma is due to Clarkson [10]; we provide a simple proof based on a recent result of [17]. The second lemma is due to Levcopoulos and Lingas [21]; again, for completeness, we provide a simple proof.

Lemma 1 *(Clarkson [10]) Given a simple polygon P, having n vertices, oone can compute, in $O(n)$ time, a triangulation of P, using $O(n)$ Steiner points (non-vertices of P), whose length is $O(\log n)$ times the length of P.*

Proof. In [17], it is shown that, for any simple polygon P, one can compute in $O(n)$ time a triangulation $\mathcal{T}$ such that (1) the vertices of $\mathcal{T}$ consist of the n vertices of P, plus a set of $O(n)$ additional Steiner points inside P; and (2) any line segment within P intersects $O(\log n)$ triangles of $\mathcal{T}$.

We claim that the low "stabbing number" triangulation $\mathcal{T}$ has the desired property for our lemma. To see this, take any edge (diagonal), e, of $\mathcal{T}$. If e is an edge of P, then we "charge" e's length to the perimeter of P. Otherwise, we charge e's length off to the perimeter of P, as follows.

If the slope of the line through e is greater than 1, then we project e to the right, onto the boundary of P. We say that a point $p \in \partial P$ on the boundary of P is in the *rightward shadow* of e if there exists a point $q \in e$, with q left of p, such that the segment qp is horizontal and lies within P.

The length of the rightward shadow of e is *at least* $|e|/\sqrt{2}$; thus, we charge the length of e to the rightward shadow of e. Since $\mathcal{T}$ has $O(\log n)$ stabbing number, any horizontal segment within P can cross at most $O(\log n)$ diagonals of $\mathcal{T}$. Thus, any point on the boundary of P can be in the rightward shadow of at most $O(\log n)$ diagonals, and can therefore be charged at most $O(\log n)$ times. It follows that the sum of the lengths of all diagonals e having slope greater than 1 is $O(\log n)$ times the length of P.

A similar argument applies to those diagonals with slope less than 1, whose length we charge to their *upward shadows*. □

Remark. Note that it is necessary to allow Steiner points in the above lemma. In fact, a "boomerang" polygon that consists of a triangle, minus a convex pocket, has a unique triangulation, whose length is $\Omega(n)$ times its perimeter. Also, note that the result is essentially tight (see Eppstein [13]).

Lemma 2 *(Levcopoulos-Lingas [21]) Given a simple rectilinear polygon P, having n vertices, one can compute in time $O(n)$ a decomposition of P into $O(n)$ axis-aligned rectangles, whose length is $O(\log n)$ times the length of P.*

Proof. (Sketch) In [5], it is shown that, using the decomposition into histograms given by [19, 20], for any simple rectilinear polygon P, one can compute in $O(n)$ time a decomposition into $O(n)$ rectangles, $\mathcal{R}$, such that any horizontal or vertical line segment within P intersects $O(\log n)$ rectangles of $\mathcal{R}$. We claim that the low "stabbing number" decomposition $\mathcal{R}$ has the desired property for our lemma.

The proof of this claim is similar to the proof of Lemma 1: Simply "charge" (project) each internal edge of the decomposition off to the boundary. No boundary point is charged more than $O(\log n)$ times. □

Remark. In fact, the bound in the above lemma is tight in the worst case. A staircase polygon can be used to show that, in general, a minimum-length rectangular subdivision of P may have length $\Omega(\log n)$ times the length of P.

Lemma 3 *Let $\mathcal{S}$ be a rectangular subdivision of a (rectilinear) polygon $\mathcal{R}$, with $\mathcal{S}$ having a total of n vertices. Then, there exists a guillotine rectangular subdivision, $\mathcal{S}_G$, that is a refinement of $\mathcal{S}$ such that the length of $\mathcal{S}_G$ is at most $O(1)$ times the length of $\mathcal{S}$.*

Proof. (Sketch) We adapt the proof of [11], who show that an optimal GRS is at most twice the length of an optimal rectangular subdivision of a given set of points. (Here, in the optimization problem, the goal is to include each point of the given set on (at least) one of the edges of the subdivision.) A careful examination of their proof reveals that it is not necessary to make any assumptions about the "optimality" of the subdivisions: One can show that an arbitrary rectangular subdivision can be transformed into a GRS, which is a refinement of the original and whose total length is at most twice that of the original. □

3 Red-Blue Separation

We begin with the problem of finding a minimum-perimeter Jordan curve (necessarily, a simple polygon) that separates a set of "red" points, R, from a set of "blue" points, B; we call such a separator a *separating simple polygon*. We let $S = R \cup B$, and $n = |S| = |R| + |B|$. We say that a subdivision is *color-conforming* if each of its open 2-faces contains points of only one color (from R or from B, but not from both).

If, instead of minimizing the Euclidean length, we want to minimize the *combinatorial* size of a separating simple polygon (i.e., the number of vertices or "links"), then an $O(\log k^*)$ approximation bound has recently been established ([1, 22]), where k^* is the size of an optimal separator; the problem is known to be NP-hard in this case [14]. Here, we give an approximation result for the problem of minimizing the Euclidean length of the separator, a problem shown to be NP-hard in [3, 12].

Consider a minimum-length separating simple polygon, P^*, with length ℓ^*. Note that P^* either contains all red points inside (or on the boundary), or all the blue points. Thus, we will solve two problems and pick the shorter length: find a minimum-length polygon that encloses red, while excluding blue; and find a minimum-length polygon that encloses blue, while excluding red.

Without loss of generality, assume that P^* surrounds the blue points and excludes the red points. Let $\mathcal{B}$ be the bounding box for B; note that P^* lies within $\mathcal{B}$ and that $\mathcal{B}$ is the bounding box of P^*.

Lemma 4 *There exists a rectilinear red-blue separating polygon, contained in $\mathcal{B}$, having $O(n)$ vertices, lying on the grid induced by S, and with length at most $\sqrt{2} \cdot \ell^*$. In particular, a minimum-length rectilinear separating polygon must have length at most $\sqrt{2} \cdot \ell^*$.*

Proof. We transform each edge uv of P^* into a "staircase" path π lying on the grid induced by S. The endpoints of the edge, u and v, must lie on the grid, since the vertices of P^* must be vertices in S. We define the rectilinear path π as the *right rectilinear envelope* of uv, a path from u to v in the grid such that it is to the right of uv and such that no vertex of the grid is inside the polygon defined by uv and π. See Figure 1.

It is not hard to show the following claim: The polygon P^*_π obtained from P^* by replacing each edge uv by the corresponding right envelope π is a (possibly degenerate) simple polygon. (proof omitted here)   □

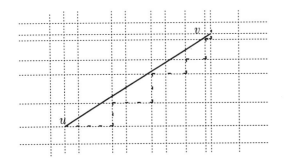

Figure 1: Right rectilinear envelope for edge uv

Among all minimum-length rectilinear separating polygons, let P^*_R be one with a minimum number, m, of vertices; clearly, $m \leq n$.

Consider the planar subdivision consisting of P^*_R and its bounding box, $\mathcal{B}$. Each face of this subdivision is a simple rectilinear polygon, with at most m vertices, containing either red points or blue points on its interior, but not both. Now, by Lemma 2, we know that this subdivision can be refined into a color-conforming rectangular subdivision, $\mathcal{S}$, of $\mathcal{B}$, whose length is $O(\log m)$ times the length of P^*_R (and thus of P^*). Then, by Lemma 3, we know that there exists a color-conforming GRS, $\mathcal{S}_G$, of $\mathcal{B}$, a refinement of $\mathcal{S}$, whose length is $O(1)$ times the length of $\mathcal{S}$, and hence is $O(\ell^* \log m)$. Below, we give a dynamic programming algorithm to compute a minimum-length color-conforming GRS of a given bounding box. Applying this to $\mathcal{B}$, we obtain a subdivision, $\mathcal{S}^*_G$, whose length is at most that of $\mathcal{S}_G$, and hence at most $O(\ell^* \log m)$.

Lemma 5 *For any color-conforming GRS of $\mathcal{B}$, there exists a separating simple polygon whose boundary is a subset of the edges of the subdivision and whose length is $O(1)$ times the length of the subdivision.*

Proof. (Sketch) Consider the binary tree associated with the BSP that defines the color-conforming GRS. Each non-leaf node in this tree has an associated guillotine cut (line segment), ξ. We consider ξ to be "attached", at one of its endpoints, to one edge of the associated bounding box. (Such an edge is a guillotine cut associated with an ancestor of the node, assuming it is not the root). The other endpoint of ξ is considered to be "touching", but not attached to, the opposite side of the bounding box; we consider there to be a tiny gap at this endpoint. In this way, the guillotine cuts of the GRS form a (degenerate, rectilinear) tree within $\mathcal{B}$. We begin with a clockwise tour that is a doubling of this tree. If an (oriented) segment uv along this tour is associated with a leaf cut $\xi = uv$ that has a (leaf) rectangular facet on its left, and this facet contains only blue points, then we replace uv with the 3-link path that detours (clockwise) around the boundary of the facet. The resulting tour encloses all blue points, excludes all red points, and has length at most twice that of the GRS.   □

A Dynamic Programming Algorithm We now give a dynamic programming algorithm to compute $\mathcal{S}^*_G$. Let $x_1 < x_2 < \cdots < x_n$ (resp., $y_1 < y_2 < \cdots < y_n$) denote the sorted x (resp., y) coordinates of the $n = |S| = |R| + |B|$ given points.

Let $V(i, j, k, l)$ denote the minimum length of a color-conforming GRS of the rectangle, $R(i, j, k, l)$, defined by x_i, x_j ($x_j > x_i$), y_k, and y_l ($y_l > y_k$). Here, in $V(i, j, k, l)$, we do not count the perimeter of $R(i, j, k, l)$ in the length of the subdivision; we only count the lengths of interior edges. Thus, we get

$V(i, j, k, l) = 0$, if all points of S interior to $R(i, j, k, l)$ are of the same color; otherwise, $V(i, j, k, l) =$

$$\min\{(y_l - y_k) + \min_{i<i'<j}[V(i, i', k, l) + V(i', j, k, l)],$$
$$(x_j - x_i) + \min_{k<k'<l}[V(i, j, k, k') + V(i, j, k', l)]\}.$$

In other words, the length of the segments making up a minimum-length partition, excluding the bounding box of the set of points, is the length of the horizontal or vertical partitioning cut plus the sum of the lengths of optimal solutions to the two subproblems defined by the cut, optimized over all possible cuts.

The running time of this algorithm is easily seen to be $O(n^5)$, since there are $O(n^4)$ subproblems (rectangles), and we optimize over $O(n)$ choices of cut (and easily check whether a rectangle has all points of the same color).

Theorem 6 *Given a set of n points in the plane, each either "red" or "blue", a red-blue separating simple polygon whose length is $O(\log m)$ times the minimum possible length can be computed in polynomial ($O(n^5)$) time, where $m < n$ is the minimum number of vertices in a minimum-length rectilinear separating polygon.*

The high running time (and space complexity) of the algorithm can be brought down to $O(n^2)$, at a cost of increasing the approximation bound to $O(\log^3 n)$. This follows from a simple observation of [22]: After transforming an optimal tour to a rectilinear polygon and then obtaining a rectangular subdivision of the bounding box, we further refine the subdivision, by decomposing each rectangle into $O(\log^2 n)$ *canonical* rectangles. A *canonical rectangle* is the direct product of two *canonical intervals*; a canonical x-interval is one whose x-coordinates are x_{i2^j} and $x_{(i+1)2^j}$, for some integers i, j ($0 \leq i < \frac{n}{2^j}$, $j \geq 0$). (The standard segment tree data structure decomposes an interval into its $O(\log n)$ canonical subintervals.) There are only $O(n)$ canonical x-intervals, and only $O(n^2)$ canonical rectangles. We can rewrite the dynamic programming algorithm to optimize only over GRS's into canonical rectangles. For a canonical rectangle, we need consider only *two* possible cuts — a bisecting horizontal or vertical cut. Further, it is easy to tabulate in advance, in time $O(n^2)$, for all $O(n^2)$ canonical rectangles, which of them contain points of a single color. Thus, the dynamic programming algorithm will require only $O(n^2)$ time and space.

Theorem 7 *Given a set of n points in the plane, each either "red" or "blue", a red-blue separating simple polygon whose length is $O(\log^3 n)$ times the minimum possible length can be computed in $O(n^2)$ time.*

Now we observe that the approach given so far for red-blue point separation can be applied more generally to the problem of building a minimum-length subdivision that separates into color classes a given set of points, each having an assigned color. Specifically, we want to compute a simple polygon P, and a decomposition of P into simple polygonal faces, such that each face in the subdivision (including the face at infinity) has points of at most one color in it, and the total length of the decomposition is as small as possible. (There may be more than one face having "red" points in it, but in any one face there can be points of only one color.) Exactly the same asymptotic approximation bounds and time bounds apply to this problem, as given in the previous two theorems.

Finally, we observe that the same method can be applied to the generalization in which, instead of a set of colored points, we have a set of colored rectilinear polygons, and we desire a minimum-length subdivision that separates them into color classes.

4 TSP with Neighborhoods

Let $S = \{R_1, \ldots, R_k\}$ be a collection of possibly overlapping simple polygons (*neighborhoods*), having a total of n vertices. We say that a tour or a subdivision *visits* S if its edges intersect every neighborhood R_i. The *TSP with Neighborhoods* (TSPN) problem asks for a minimum-length tour that visits S. We consider lengths to be measured in the Euclidean metric, but the results apply to other standard metrics as well. Here, we provide an $O(\log k)$ approximation algorithm that runs in polynomial time.

Consider a minimum-length tour, P^*, with length ℓ^*, and let $\mathcal{B}^*$ denote its bounding box. Our first lemma is a simple consequence of the local optimality of P^*:

Lemma 8 P^* *is a simple (not self-intersecting) polygon, having at most k vertices.*

Proof. Let p be the leftmost point of P^*; if there is more than one leftmost point, let p be the one of maximum y-coordinate. It is not hard to see from the local optimality of P^* that p must lie in some region, R_j. Consider a traversal of the tour, starting from p. Let p_i be the first point of R_i visited along this traversal. (Then, $p = p_j$.) It is known (and easy to show) that a Euclidean TSP tour on the k points, $p_1, \ldots, p_k$, must be a simple polygon, with vertices among the points p_i. But P^* must itself be a TSP tour on the k points p_i — if it were longer than such a TSP tour, then it could be shortened by replacing it with the TSP tour (which visits all regions R_i). Thus, P^* is a simple polygon with at most k vertices. □

Now, our goal is to convert P^* into a GRS whose edges intersect all regions R_i and whose length is comparable to P^*. We will then perturb this subdivision so that its edges lie on the grid induced by the vertices of S. A dynamic programming algorithm will yield a shortest such GRS, which will be converted into an approximately optimal TSP.

Let uv be one edge of P^*; without loss of generality, assume that uv has negative slope. We would like to replace each edge uv with a rectilinear path. If we replace uv with the 2-link ("L-shaped") path, π_L, that goes downwards out of u, then rightwards to v, then this new path may fail to stab all the neighborhoods visited by the straight segment uv. Instead, we will replace uv with a rectilinear path, π, defined as follows.

Starting at some vertex, p, of P^*, traverse the tour P^*, and let $S_{uv} \subseteq S$ be the set of regions that edge uv intersects, but have not been intersected by an edge of P^* previously during the traversal. Let T be the (right) triangle bounded by uv and π_L. Consider a region $R_i \in S_{uv}$. If R_i has a vertex in T, pick one of them, and call it r_i;

otherwise, R_i must cross π_L. We now define the rectilinear path π to be that portion between u and v of the boundary of the *upper rectilinear hull* of the points $\{u, v\}$, together with the points r_i determined by the regions $R_i \in S_{uv}$. Thus, π is a "staircase" in T, separating segment uv from the points r_i (if any) and from π_L. See Figure 2. (Note that, if there are no points r_i associated with uv, then π will equal π_L.) Since the regions R_i are connected, and they have some point on uv and some point below π, we know that π must intersect each of the regions $R_i \in S_{uv}$.

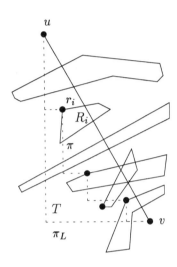

Figure 2: Construction of path π.

So, for each edge uv of P^*, we replace uv with its staircase path π. The result is a rectilinear closed walk, W, having $O(k)$ vertices, whose length is at most $\sqrt{2} \cdot \ell^*$. Note, however, that W is not necessarily simple and not necessarily lying on the grid induced by S (since u and v need not be on the grid). Let P_R^* denote a shortest *rectilinear* tour that visits all neighborhoods R_i; i.e., P_R^* is a shortest tour among all those rectilinear tours that visit all neighborhoods. Clearly, P_R^* is not longer than W. Further, it is easy to argue that P_R^* is a *simple* rectilinear polygon (since uncrossing a tour only improves its length), having $O(k)$ vertices (by a proof similar to that of Lemma 8. We select P_R^* to have a minimum number, m ($= O(k)$), of edges among all shortest rectilinear tours visiting all neighborhoods.

Using Lemmas 2 and 3, we can decompose each bounded face of the arrangement of P_R^* and $\mathcal{B} = bb(P_R^*)$ into rectangles, and then refine the resulting rectangular subdivision into a GRS, $\mathcal{S}$, of $\mathcal{B}$ whose length is $O(\ell^* \log m)$. The edges of the subdivision $\mathcal{S}$ intersect all regions $R_i \in S$, but they do not necessarily lie on the grid induced by S. The following lemma shows, however, that $\mathcal{S}$ can be transformed into a subdivision that lies on the grid, without lengthening it.

Lemma 9 $\mathcal{S}$ can be transformed to lie on the grid induced by S, while still having its edges intersect all regions $R_i \in S$, and without increasing its length.

Proof. (Sketch) First, we argue that we can transform $\mathcal{S}$ so that its bounding box lies on the grid. The details are tedious and omitted here, but the main idea is to slide the top edge of the bounding box downwards, hoping to align it with a grid line. Before this happens, though, the corner of the bounding box may "pull away" from a region, and we cannot allow this. So, if a corner c becomes incident on an edge e of some region, we then start sliding c along e, adjusting the bounding box and edges of $\mathcal{S}$ accordingly. Again, a corner may "pull away", and we may end up with all four corners constrained to slide along given edges. The crucial property is, though, that the total length of the modified subdivision is a linear function of the sliding parameter, so we can always do the sliding without lengthening $\mathcal{S}$.

Next, we must transform $\mathcal{S}$ so that all internal edges also lie on the grid. We do this perturbation one (guillotine) cut at a time. Consider a guillotine cut, ξ; assume it is vertical. Consider what happens as we slide it left/right, allowing the (horizontal) edges incident to it to lengthen/shorten. Again, one direction of sliding will only improve matters, by shortening the length of $\mathcal{S}$ (i.e., slide in the direction that has more horizontal edges incident on it). (Again, we may have to reassess which direction we are sliding when a rectangle collapses to zero area, but this presents no real problem.) Further, all regions R_i will continue to be intersected, until a vertex event, when ξ hits a vertex of S. We then recurse for the guillotine cuts on either side of ξ, eventually obtaining a GRS with all edges on the grid. □

The final result of the above discussion is that we can transform an optimal tour, P^*, into a GRS, $\mathcal{S}_G$, that lies on the grid induced by S, with total length $O(\ell^* \log m)$.

Now, by a straightforward variation of the dynamic programming algorithm given earlier (for the RBSP), we can, for each of the $O(n^4)$ possible bounding boxes $\mathcal{B}$ defined by vertices of S, compute a minimum-length GRS of $\mathcal{B}$ whose edges intersect all regions of S (if one exists — this will depend on the choice of $\mathcal{B}$). By optimizing over the choice of $\mathcal{B}$, we obtain a minimum-length GRS, $\mathcal{S}_G^*$, whose length is at most that of $\mathcal{S}_G$, and hence at most $O(\ell^* \log m)$. Finally, we need:

Lemma 10 *For any GRS of $\mathcal{B}$ whose edges intersect all regions of S, there exists a simple polygonal tour P (1) whose boundary is a subset of the edges of the subdivision, (2) that visits S, and (3) whose length is $O(1)$ times the length of the subdivision.*

Theorem 11 *Let $S = \{R_1, \ldots, R_k\}$ be a collection of possibly overlapping simple polygons, having a total of n*

vertices. *One can compute, in polynomial time, a tour P that visits S, whose length is $O(\log m)$ times the minimum length of a tour that visits S, where $m = O(k)$ is the minimum number of links in a shortest rectilinear tour visiting S.*

5 Watchman Routes

Let $\mathcal{P}$ be a (closed, connected) polygonal domain, possibly with "holes" (obstacles), having n vertices (S) and (consequently) n edges. We say that a tour (or a subdivision) is *watchman* if it "sees" all of $\mathcal{P}$; i.e., every point of $\mathcal{P}$ is seen by some point on the tour (or by some point on an edge of the subdivision). (Point $p \in \mathcal{P}$ *sees* point $q \in \mathcal{P}$ if $\mathcal{P}$ contains the segment pq.) Here, we assume that $\mathcal{P}$ is rectilinear and that no two edges are collinear. (A slight perturbation can guarantee this.) We give an $O(\log n)$ approximation algorithm for computing an optimal (shortest) watchman route. (Note that, in general, there is not a unique optimal route.)

Let P^* be an optimal watchman route, let its (Euclidean) length be ℓ^*, and let $\mathcal{B}^*$ be its bounding box. Our first lemma is a simple consequence of the local optimality of P^*: (proof omitted here)

Lemma 12 *Let $\mathcal{P}$ be an arbitrary polygonal domain (with holes) having n vertices, r of which are reflex. Then, an optimal watchman route for $\mathcal{P}$ is a simple (not self-intersecting) polygon, having at most r vertices.*

Now, our goal is, as in the TSP with neighborhoods problem, to convert P^* into a GRS, which is watchman and of comparable length to P^*, and then to perturb this subdivision so that its edges lie on the grid induced by the vertices S.

We replace each edge uv of P^* with the boundary of the maximal rectilinearly convex polygon, P_{uv}, that contains uv but is contained within uv's bounding box (i.e., $uv \subset P_{uv} \subset bb(uv)$). Such a polygon P_{uv} is unique, and is bounded by two staircase paths from u to v. Since P_{uv} surrounds uv and contains no obstacles, any point of $\mathcal{P}$ that sees uv also must see the boundary of P_{uv}. Further, the length of P_{uv} is at most $2\sqrt{2}|uv|$.

Now consider the arrangement, $\mathcal{A}$, formed by the union of all polygons P_{uv}, and the bounding box, $\mathcal{B}^*$. The total length of the edges in this arrangement is at most $2\sqrt{2}\ell^*$. Also, there exists a (rectilinear) tour, W, whose length is $O(1)$ times the length of $\mathcal{A}$, such that W includes all edges of $\mathcal{A}$, so that W is also watchman (but not simple). (Simply go around the boundary of each P_{uv} one-and-one-half times — go clockwise from u to v to u to v.) Among all minimum-length rectilinear watchman tours, let P^*_R be one with a minimum number, m, of vertices. It is not hard to see that $m \leq n$. Clearly, P^*_R is shorter than W. Further, it is easy to argue that P^*_R is a simple rectilinear polygon, and that, because of the nondegeneracy assumption on $\mathcal{P}$, P^*_R will have no two edges collinear.

Using Lemmas 2 and 3, we can decompose each bounded face in the arrangement of P^*_R and $bb(P^*_R)$ into rectangles, and then refine the resulting rectangular subdivision of $bb(P^*_R)$ into a GRS, $\mathcal{S}$, whose length is $O(\ell^* \log m)$. The subdivision $\mathcal{S}$ is watchman, since adding edges can only increase what is seen. Further, there exists a *connected* subset of *feasible* edges of $\mathcal{S}$ that suffice to see all of $\mathcal{P}$; for example, the edges of P^*_R form such a set. (A line segment is *feasible* if it lies within $\mathcal{P}$ — i.e., it does not cross any obstacles.) But $\mathcal{S}$ does not necessarily lie on the grid induced by S (since the segment endpoints, u, v, may not be grid vertices). We will show, however, that $\mathcal{S}$ can be transformed into a watchman GRS, $\mathcal{S}_G$, that lies on the grid, without lengthening it, and while maintaining the property that a connected subset of feasible edges illuminates all of $\mathcal{P}$.

A horizontal edge of $\mathcal{P}$ (outer boundary or obstacle edge) that has the interior of $\mathcal{P}$ locally above (resp., below) it will be called a *bottom* (resp., *top*) edge. Let y_{max} denote the maximum y-coordinate of any bottom edge of $\mathcal{P}$. Similarly, let y_{min} denote the minimum y-coordinate of any *top* edge of $\mathcal{P}$. Define x_{max} and x_{min} analogously.

Lemma 13 *The bounding box, $bb(P^*_R)$, of a shortest rectilinear watchman route is determined by the four coordinates $y_{max}, y_{min}, x_{max}$, and x_{min}.*

Proof. Consider a bottom edge e of $\mathcal{P}$. A watchman route must see all points on e; in order to see a point in the middle of e, the route must have some point at the same or greater y-coordinate as e. Thus, the top side of $bb(P^*_R)$ must be at or above any such e. Continuing this argument for the bottom, left, and right sides of $bb(P^*_R)$, we get that $bb(P^*_R)$ must contain the rectangle, R_0, determined by $y_{max}, y_{min}, x_{max}$, and x_{min}. (The rationale is the same as for "essential cuts", as introduced by [8].)

Now we must argue that there is no reason for P^*_R to go outside of the box R_0. Assume to the contrary that the top edge, e', of P^*_R is a horizontal edge with y-coordinate greater than y_{max}. Consider what happens as we slide e' downwards. Since e' lies above all bottom edges of $\mathcal{P}$, e' will not initially be in contact with an obstacle, so the sliding process is initially feasible. If moving e' downwards a small amount ϵ causes P^*_R to stop seeing some point, $p \in \mathcal{P}$, then it must be that p lies on a bottom edge of $\mathcal{P}$ that is collinear with e' and has y-coordinate greater than y_{max}, contradicting the definition of y_{max}. Thus, we must be able to slide e' downwards some amount $\epsilon > 0$, while maintaining feasibility of P^*_R and while maintaining the fact that P^*_R sees all of $\mathcal{P}$. But, as we slide e', P^*_R will decrease in length, contradicting its optimality. □

Corollary 14 *The bounding box, $bb(P_R^*)$, of a shortest rectilinear watchman route contains all holes (obstacles).*

Lemma 15 *S can be transformed into a watchman GRS, S_G, that lies on the grid induced by S, has length at most that of S, and has a connected subset of feasible edges illuminating $\mathcal{P}$.*

Proof. From Lemma 13, we know that $bb(P_R^*)$ must itself lie on the grid induced by S. Now, consider a guillotine cut, ξ, of $bb(P_R^*)$. Consider the process of sliding ξ parallel to itself, and adjusting the incident edges of S, making them shorter or longer, as appropriate. We can choose to slide ξ in a direction, say leftwards, that yields no increase in overall length of S. We can continue sliding ξ until (1) it hits an obstacle edge (in which case ξ now lies on the grid); or until (2) some edge of S incident on ξ shrinks to length zero (in which case we collapse and remove the zero-area rectangle and reassess which direction we should slide ξ). In the end, this process must terminate with ξ becoming collinear with an edge of $\mathcal{P}$, and we then recurse on each side of the cut. Also, during the process, the property that a connected subset of feasible edges sees all of $\mathcal{P}$ is preserved. □

Below, we give a dynamic programming algorithm to compute a minimum-length GRS of a given bounding box, with the property that all of $\mathcal{P}$ that is *inside* of the given box is illuminated by a *connected* subset of the *feasible* edges of the subdivision that are connected to a specified set of four feasible edges ("illuminators") on the boundary of the box, each of which has its endpoints at vertices of the grid induced by S. We can use this algorithm to get our approximation result, as follows.

We start by computing $bb(P_R^*)$, which, by Lemma 13, is given by simply computing y_{max}, y_{min}, x_{max}, and x_{min}. Next, we apply the dynamic programming algorithm for each of the $O(n^8)$ choices of four (feasible) illuminator segments on the boundary of $bb(P_R^*)$ — call them σ_ℓ, σ_r, σ_b, and σ_t, on the left, right, bottom, and top, respectively — and we optimize over these choices. (The four edges of P_R^* that lie on the boundary of $bb(P_R^*)$ are a witness to the fact that four illuminators suffice on the boundary of $bb(P_R^*)$; by our nondegeneracy assumption, there is *exactly* one edge of P_R^* per edge of $bb(P_R^*)$.) The result is that we obtain an optimal GRS, S_G^*, whose length is at most that of S_G, and hence at most $O(\ell^* \log m)$.

We now must argue that we can convert S_G^* into a (feasible) watchman tour that has about the same total length. We know, by the constraint imposed in the dynamic programming algorithm, that $\Sigma = \Sigma_\ell \cup \Sigma_r \cup \Sigma_b \cup \Sigma_t$ illuminates all of $\mathcal{P} \cap bb(P_R^*)$, where Σ_α is the connected component containing σ_α in the set of feasible edges of S_G^* ($\alpha \in \{\ell, r, b, t\}$).

Let H denote a shortest rectilinear tour that surrounds all obstacles and goes through σ_ℓ, σ_r, σ_b, and σ_t. (H is a rectilinear "relative convex hull", and can easily be computed.) Necessarily, H will lie within $bb(P_R^*)$, since we know that $bb(P_R^*)$ contains all holes of $\mathcal{P}$ (Corollary 14). We omit here the simple proofs of the following:

Lemma 16 *H sees all of $\mathcal{P} \setminus bb(P_R^*)$.*

Lemma 17 *The length of H is at most ℓ^*.*

Now, H must intersect Σ, since both sets touch all four sides of $bb(P_R^*)$. Thus, $F = H \cup \Sigma$ is connected, sees all of $\mathcal{P}$, and has length $O(\ell^* \log m)$. We can easily use F to construct a feasible tour that sees all of $\mathcal{P}$. One way to do this is to convert F into a tree (by making small cuts as necessary), and then to define a tour that traverses the boundary of the tree, at most doubling its length.

Theorem 18 *Given a rectilinear polygonal room with n vertices, a watchman tour whose length is $O(\log m)$ times the minimum possible length can be computed in polynomial (in n) time, where $m \leq n$ is the minimum number of edges in a rectilinear optimal watchman route.*

A Dynamic Programming Algorithm Let $x_1 < x_2 < \cdots < x_n$ (resp., $y_1 < y_2 < \cdots < y_n$) denote the sorted x (resp., y) coordinates of the n vertices, S, of $\mathcal{P}$. We now give a dynamic programming algorithm to solve the following problem:

Input: Rectangle $R(i, j, k, l)$, defined by x_i, x_j ($x_j > x_i$), y_k, and y_l ($y_l > y_k$); segments σ_ℓ ("left"), and σ_r ("right"), σ_b ("bottom"), σ_t ("top"), which are sub-segments of the top (rep., bottom, left, right) edge of $R(i, j, k, l)$, each of which has endpoints on the grid induced by S.

Objective: Compute a minimum-length GRS, S_G^*, of $R(i, j, k, l)$ such that $\Sigma_\ell \cup \Sigma_r \cup \Sigma_b \cup \Sigma_t$ illuminates all of $\mathcal{P} \cap R(i, j, k, l)$, where Σ_α is the connected component containing σ_α in the set of feasible edges of S^* ($\alpha \in \{\ell, r, b, t\}$).

We think of the segments σ_α ($\alpha \in \{\ell, r, b, t\}$) as the "illuminators"; any feasible segment that can be connected to an illuminator by a path of feasible segments also serves as an illuminator. Our goal is to illuminate all of $\mathcal{P}$ that lies inside of $R(i, j, k, l)$, by making certain that any point of $\mathcal{P} \cap R(i, j, k, l)$ can see at least one illuminator.

Let $V(i, j, k, l, \sigma_\ell, \sigma_r, \sigma_b, \sigma_t)$ be the length of an optimal GRS. Here, in $V(i, j, k, l, \sigma_\ell, \sigma_r, \sigma_b, \sigma_t)$, we do not count the perimeter of $R(i, j, k, l)$ in the length of the subdivision; we only count the lengths of interior edges. Thus, we can write the following recursion:

$V(i, j, k, l, \sigma_\ell, \sigma_r, \sigma_b, \sigma_t) = 0$, if $\mathcal{P} \cap R(i, j, k, l)$ is illuminated by $\{\sigma_\ell, \sigma_r, \sigma_b, \sigma_t\}$; otherwise,

$$V(i,j,k,l,\sigma_\ell,\sigma_r,\sigma_b,\sigma_t) =$$
$$\min\{ \quad (y_l - y_k) + \min_{i<i'<j}[\ V(i,i',k,l,\sigma_\ell,\sigma_t^\perp,\sigma_t',\sigma_b') +$$
$$V(i',j,k,l,\sigma_t^\perp,\sigma_r,\sigma_t'',\sigma_b'')],$$
$$(y_l - y_k) + \min_{i<i'<j}[\ V(i,i',k,l,\sigma_\ell,\sigma_b^\perp,\sigma_t',\sigma_b') +$$
$$V(i',j,k,l,\sigma_b^\perp,\sigma_r,\sigma_t'',\sigma_b'')],$$
$$(x_j - x_i) + \min_{k<k'<l}[\ V(i,j,k,k',\sigma_\ell',\sigma_r',\sigma_b,\sigma_\ell^\perp) +$$
$$V(i,j,k',l,\sigma_\ell'',\sigma_r'',\sigma_\ell^\perp,\sigma_t)],$$
$$(x_j - x_i) + \min_{k<k'<l}[\ V(i,j,k,k',\sigma_\ell',\sigma_r',\sigma_b,\sigma_r^\perp) +$$
$$V(i,j,k',l,\sigma_\ell'',\sigma_r'',\sigma_r^\perp,\sigma_t)]\}.$$

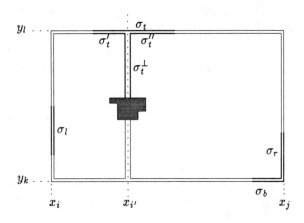

Figure 3: Notation in the recursion.

Here, $\sigma_t' \subset \sigma_t$ (resp., $\sigma_t'' \subset \sigma_t$) is that portion of σ_t that lies to the left (resp., right) of the vertical cut (line) through $x_{i'}$. Also, $\sigma_t^\perp$ is the maximal feasible (grid) segment on the vertical cut through $x_{i'}$, with its top endpoint on σ_t. See Figure 3. (We select the longest feasible segment to serve as illuminator, since having a longer illuminator can only help us, and it does not cost anything additional, since the objective function minimizes the total length of the subdivision.) Similar definitions apply to each of the other terms of the forms σ_α', σ_α'', and $\sigma_\alpha^\perp$, for $\alpha \in \{\ell, r, b, t\}$.

The running time of this algorithm is easily seen to be polynomial ($O(n^{13})$), since there are $O(n^{12})$ subproblems (rectangles, plus choice of illuminators σ_α), and we optimize over $O(n)$ choices of cut. (Determining if all of $\mathcal{P} \cap R(i,j,k,l)$ is illuminated by a given set of illuminator segments is easily tabulated, for all $O(n^{12})$ choices, within the $O(n^{13})$ time bound.)

In fact, we obtain a slightly more reasonable time bound of $O(n^9)$ by noting that the illuminators $\sigma_\alpha^\perp$ that are constructed along cuts during the algorithm will always have one endpoint at the corner of the bounding box of the subproblem and will extend to a maximal length (i.e., until an obstacle or the opposite side of the bounding box is reached). This implies that a subproblem is fully determined by its bounding box, plus a constant number of bits of extra information to tell us on which sides and attached to which corners we have illuminator segments. The exception is that portions of the original bounding box and its four illuminators may also appear on the boundary of a subproblem. In our application, the initial call to the recursion will involve a fixed rectangle, $bb(P_R^*)$, and $O(n^8)$ possible choices for illuminators σ_α. This leads to "only" $O(n^8)$ subproblems to evaluate, since a subproblem is fully determined by its bounding box, some constant number of bits, plus the appropriate portion of the (original) illuminator segments that appear on sides of the bounding box that are common to the initial box ($bb(P_R^*)$). The net result is that $O(n^9)$ time suffices for our problem.

6 k-MST

Given a set S of n points in the plane, and an integer $k \leq n$, the k-MST problem asks for a spanning tree, possibly with Steiner points, of minimum (Euclidean) length that connects some subset of k of the n points. Here, we give an $O(\log k)$ approximation algorithm for this problem, which implies a similar result for the *Prize-Collecting Traveling Salesman Problem (PCTSP)*.

Consider a minimum-length k-MST, T^*, with length ℓ^*. It is easy to check that T^* must be *simple* (non-self-crossing). Furthermore, T^* can easily be perturbed into a rectilinear tree, T_R, on the grid induced by S, without affecting its (Euclidean) length by much (factor $\leq \sqrt{2}$).

Consider the planar subdivision consisting of T_R and its bounding box, $\mathcal{B}$. Each face of this subdivision is a simple rectilinear polygon, with at most k vertices. Now, by Lemma 2, we know that this subdivision can be refined into rectangular subdivision, $\mathcal{S}$, of $\mathcal{B}$, whose length is $O(\log k)$ times the length of T_R (and thus of T^*). Then, by Lemma 3, we know that there exists a GRS, $\mathcal{S}_G$, of $\mathcal{B}$, a refinement of $\mathcal{S}$, whose length is $O(1)$ times the length of $\mathcal{S}$, and hence is $O(\ell^* \log k)$.

As in the problems already discussed, we can use dynamic programming, optimized over choice of bounding box, to obtain a minimum-length GRS (of some bounding box), $\mathcal{S}_G^*$, that visits k points of S. (In particular, we define our "value" function $V(i,j,k,l;m)$ to be the minimum length of a GRS that visits m of the points within $R(i,j,k,l)$, and then optimize over all cuts and over all partitions of the integer m.) The length of $\mathcal{S}_G^*$ is at most that of $\mathcal{S}_G$, and hence at most $O(\ell^* \log k)$. Further, any spanning tree for the k points of S on the edges of $\mathcal{S}_G^*$ is a feasible solution to the original problem, and trivially has length at most that of $\mathcal{S}_G^*$.

In fact, by a straightforward application of the ideas used to approximate the TSP with Neighborhoods problem, we can allow the set S to consist of a set of n polygons, k of which must be visited by a tree, and obtain the following result for the k-MST *with neighborhoods* problem:

Theorem 19 *Given a set S of n polygons in the plane, having a total of N vertices, and an integer $k \leq n$, one can compute, in polynomial (in N) time, a tree that visits k of the n polygons whose length is within factor $O(\log k)$ of optimal.*

Corollary 20 *The Euclidean Prize-Collecting Traveling Salesman Problem with polygonal neighborhoods can be approximated, in polynomial time, with an approximation factor of $O(\log k)$.*

7 Conclusion

There are many other problems to which our method may be applied, and we are currently pursuing some of these. The watchman route problem with arbitrary (nonrectilinear) holes can probably be solved to within a polylogarithmic factor of optimal using our methods. Further, it seems that we can obtain constant factor approximation algorithms for many of the problems we have addressed, e.g., based on the new results of [6]. The full paper will report the results of these pursuits. We are also currently investigating higher dimensional versions of these problems.

Acknowledgements

We thank Estie Arkin and Rafi Hassin for helpful discussions and introducing us to many of the problems discussed here. We thank Andrzej Lingas for discussions and pointing out some useful references.

References

[1] P. K. Agarwal and S. Suri. Surface approximation and geometric partitions. In *Proc. Fifth Annual ACM-SIAM Sympos. on Discrete Algorithms*, 1994, pages 34–43.

[2] E. M. Arkin and R. Hassin. Approximation algorithms for the geometric covering salesman problem. *Discrete Applied Math*, 55:197–218, 1994.

[3] E. M. Arkin, S. Khuller, and J. S. B. Mitchell. Geometric knapsack problems. *Algorithmica*, 10:399–427, 1993.

[4] B. Awerbuch, Y. Azar, A. Blum, and S. Vempala. Improved approximation guarantees for minimum-weight k-trees and prize-collecting salesmen. To appear, STOC 1995.

[5] M. de Berg and M. van Kreveld. Rectilinear decompositions with low stabbing number. Tech. Report RUU-CS-93-25, Dept. Comput. Sci., Utrecht Univ., 1993.

[6] A. Blum, P. Chalasani, and S. Vempala. A constant-factor approximation for the k-MST problem in the plane. To appear, STOC 1995.

[7] S. Carlsson, H. Jonsson, and B. J. Nilsson. Finding the shortest watchman route in a simple polygon. In *Proc. 4th Annu. Internat. Sympos. Algorithms Comput. (ISAAC 93)*, LNCS Vol. 762, pp. 58–67, Springer-Verlag, 1993.

[8] W. Chin and S. Ntafos. Optimum watchman routes. *Inform. Process. Lett.*, 28:39–44, 1988.

[9] N. Christofides. Worst-case analysis of a new heuristic for the traveling salesman problem. In J. F. Traub, editor, *Sympos. on New Directions and Recent Results in Algorithms and Complexity*, New York, NY, 1976. Academic Press.

[10] K. L. Clarkson. Approximation algorithms for planar traveling salesman tours and minimum-length triangulations. In *Proc. Second Annual ACM-SIAM Sympos. on Discrete Algorithms*, 1991, pages 17–23.

[11] D.-Z. Du, L.-Q. Pan, and M.-T. Shing. Minimum edge length guillotine rectangular partition. Report 02418-86, Math. Sci. Res. Inst., Univ. California, Berkeley, CA, 1986.

[12] P. Eades and D. Rappaport. The complexity of computing minimum separating polygons. *Pattern Recognition Letters*, 14:715–718, 1993.

[13] D. Eppstein. Approximating the minimum weight triangulation. In *Proc. Third Annual ACM-SIAM Sympos. Discrete Algorithms*, 1992, pages 48–57.

[14] S. Fekete. On the complexity of min-link red-blue separation. Manuscript, 1992.

[15] N. Garg and D. S. Hochbaum. An $O(\log k)$ approximation algorithm for the k minimum spanning tree problem in the plane. In *Proc. 26th Annu. ACM Sympos. Theory Comput. (STOC 94)*, pages 432–438, 1994.

[16] T. Gonzalez, M. Razzazi, and S.-Q. Zheng. An efficient divide-and-conquer approximation algorithm for hyperrectangular partitions. In *Proc. 2nd Canad. Conf. Comput. Geom.*, pages 214–217, 1990.

[17] J. Hershberger and S. Suri. A pedestrian approach to ray shooting: Shoot a ray, take a walk. In *Proc. 4th ACM-SIAM Sympos. Discrete Algorithms*, pages 54–63, 1993.

[18] E. Ihler, G. Reich, and P. Widmayer. On shortest networks for classes of points in the plane. In *Computational Geometry — Methods, Algorithms and Applications: Proc. Internat. Workshop Comput. Geom. CG '91*, LNCS Vol. 553, pp. 103–111. Springer-Verlag, 1991.

[19] C. Levcopoulos. Fast heuristics for minimum length rectangular partitions of polygons. In *Proc. 2nd Annu. ACM Sympos. Comput. Geom.*, pages 100–108, 1986.

[20] C. Levcopoulos. Heuristics for minimum decompositions of polygons. PhD diss. No. 155, Linköping University 1987.

[21] C. Levcopoulos and A. Lingas. Bounds on the length of convex partitions of polygons. In *Proc. 4th Conf. Found. Softw. Tech. Theoret. Comput. Sci.*, LNCS Vol. 181, pp. 279–295. Springer-Verlag, 1984.

[22] J.S.B. Mitchell. Approximation algorithms for geometric separation problems. Technical report, AMS Dept., SUNY Stony Brook, NY, July 1993.

[23] B.J. Nilsson. Guarding art galleries — Methods for mobile guards. Ph.D. thesis, Lund University, 1995.

[24] M. S. Paterson and F. F. Yao. Efficient binary space partitions for hidden-surface removal and solid modeling. *Discrete Comput. Geom.*, 5:485–503, 1990.

[25] M. S. Paterson and F. F. Yao. Optimal binary space partitions for orthogonal objects. *J. Algorithms*, 13:99–113, 1992.

[26] G. Reich and P. Widmayer. Beyond Steiner's problem: A VLSI oriented generalization. In *Proc. 15th Internat. Workshop Graph-Theoret. Concepts Comput. Sci.*, LNCS Vol. 411, pp. 196–210. Springer-Verlag, 1989.

Shortest Path Queries among Weighted Obstacles in the Rectilinear Plane

Danny Z. Chen[*]   Kevin S. Klenk[†]   Hung-Yi T. Tu[‡]

Abstract

We study the problems of processing single-source and all-pairs shortest path queries among weighted polygonal obstacles in the rectilinear plane. For the single-source case, we construct a data structure in $O(n \log^{3/2} n)$ time and $O(n \log n)$ space, where n is the number of obstacle vertices; this data structure enables us to report the length of a shortest path between the source and an arbitrary query point in $O(\log n)$ time, and an actual shortest path in $O(\log n + k)$ time, where k is the number of line segments on the output path. For the all-pairs case, we construct a data structure in $O(n^2 \log^2 n)$ time and space; this data structure enables us to report the length of a shortest path between two arbitrary query points in $O(\log^2 n)$ time, and an actual shortest path in $O(\log^2 n + k)$ time. Our work improves and generalizes the previously best known results on computing rectilinear shortest paths among weighted polygonal obstacles. We also apply our techniques to solve other rectilinear shortest path query problems.

1 Introduction

The problems of computing shortest paths among geometric obstacles are among the most fundamental problems in computational geometry. Given a *rectilinear* plane with disjoint weighted polygonal obstacles of n vertices altogether, we study the problems of answering shortest path queries in such an environment. The plane is rectilinear because all geometric objects (e.g., lines, obstacles, paths) in it are *rectilinear*, i.e., each line or line segment of such an object is parallel to either the

[*]Department of Computer Science and Engineering, University of Notre Dame, Notre Dame, IN 46556, USA. E-mail: chen@cse.nd.edu.

[†]Department of Computer Science and Engineering, University of Notre Dame, Notre Dame, IN 46556, USA. E-mail: Kevin.S.Klenk.1@nd.edu.

[‡]Department of Computer Science and Information Management, Providence University, Shalu, Taichung Hsien, 43309, Taiwan, R. O. C. E-mail: tu@info-server.1.edu.tw.

x- or y-axis, and distances are measured using the L_1 metric. The interior of each obstacle is disjoint from that of another obstacle. Every obstacle is *weighted*, i.e., it is associated with a nonnegative *weight* factor, so that a path in the plane, if it intersects the interior of an obstacle, is charged extra cost based on the weight of that obstacle in addition to the cost of the length of the path. A *shortest path* connecting two points in the plane is a path with the minimum total cost. A shortest path query specifies two points, s and t, in the plane and requests a rectilinear shortest path (or its weighted length) connecting s and t. If the point s is always fixed and t is arbitrary, then the query is called a *single-source* query and the fixed point s is called the *source point*. If both s and t are arbitrary points, then the query is called an *all-pairs* query. Our objective is to create data structures describing possible shortest paths in such an environment, such that shortest path queries can be answered efficiently by using the data structures. Note that the shortest paths we consider are in fact a generalization of the shortest paths (that completely avoid the interior of obstacles) in classical geometric path planning problems, because our shortest paths become obstacle-avoiding if we let the weight of each obstacle be $+\infty$.

Unlike shortest path problems in graph theory (in which both the single-source and all-pairs problems are well-studied), shortest path query problems in computational geometry, especially the all-pairs query problems, are still not well-understood. Shortest path queries in an obstacle-scattered plane cannot be handled immediately by well-known shortest path algorithms in graph theory because shortest path queries can specify arbitrary points in the plane which has uncountably infinitely many points. Furthermore, it is not clear in many situations how to obtain good solutions to the all-pairs query problems even when efficient algorithms for the corresponding single-source query problems are available. The all-pairs Euclidean shortest path problem is such an example, for which only efficient approximation results [5] are known so far.

A number of algorithms have been discovered for computing rectilinear shortest paths [2, 3, 7–11, 19, 21, 23, 25, 26, 33, 34] and shortest paths among weighted obstacles [13, 23, 28, 34]. When the plane has only a single obstacle (i.e., a simple polygon), shortest path query problems are much easier and in fact are quite well-studied (e.g., see [6, 14–17, 22]). However, shortest path query problems become substantially harder in environments with multiple obstacles. There were only a few algorithms [9, 10, 18, 22, 25–27, 30] for *single-source obstacle-avoiding* path queries in multiple obstacle environments. Results on *all-pairs* queries in such environments were even rarer; the only such algorithms we know previously are for the special case of *rectilinear* shortest paths among *rectangular* obstacles by Atallah and Chen [2, 3] and also by ElGindy and Mitra [11], and the case of rectilinear shortest paths among *rectilinear non-penetrating* obstacles [19]. Especially, Iwai, Suzuki, and Nishizeki [19] considered rectilinear shortest paths among rectilinear non-penetrating obstacles and presented an algorithm for building a data structure in $O(n^2 \log^3 n)$ time and $O(n^2 \log^2 n)$ space, which can be used to answer each all-pair length query in $O(\log^2 n)$ time. For shortest paths among weighted obstacles, very few algorithms were previously known [13, 23, 28, 34], and none of these results can handle arbitrary shortest path queries. In particular, Mitchell and Papadimitriou [28] solved the problem of determining shortest paths through a weighted planar polygonal subdivision; their algorithm takes a time of $O(n^8)$ times another factor based on the precision of the problem instance. Lee, Yang, and Chen [23] studied the problem of computing a rectilinear shortest path between two points among disjoint weighted rectilinear polygonal obstacles in the plane, and presented two algorithms for this problem; their first algorithm runs in $O(n \log^2 n)$ time and $O(n \log n)$ space and the second runs in $O(n \log^{3/2} n)$ time and space.

In this paper, we present new techniques for processing single-source and all-pairs rectilinear shortest path queries among disjoint weighted rectilinear polygonal obstacles. Using these techniques, we obtain several efficient algorithms for answering rectilinear path queries. Our results represent significant progress towards understanding shortest path queries (especially, all-pairs shortest path queries) in multiple obstacle environments. Our techniques are based on new geometric observations and the generalized visibility graphs of Lee, Yang, and Chen [23]. We introduce the idea of "gateways" which are used to control effectively the connection between query points and the generalized visibility graph and thus lead to efficient algorithms and data structures for shortest path queries. We also present methods for computing shortest path trees in the generalized visibility graphs that are more efficient in both the time and space complexities than direct applications of the best known graphic shortest path algorithm by Fredman and Tarjan [12]. Our techniques and ideas could be useful in solving other shortest path query problems.

Our main results are as follows:

- For single-source queries among weighted rectilinear obstacles in the rectilinear plane, we construct a data structure in $O(n \log^{3/2} n)$ time and $O(n \log n)$ space. This data structure enables us to determine the length of the shortest path for an arbitrary destination point in $O(\log n)$ time. An actual shortest path can be reported in $O(\log n + k)$ time, where k is the number of line segments on the output path. An immediate consequence of this result is that the previously best known algorithms of Lee, Yang, and Chen [23] for determining a rectilinear shortest path between two points among weighted obstacles can be improved by a factor of $O(\log^{1/2} n)$ in *either* the time or space complexity.

- For all-pairs queries among weighted rectilinear obstacles in the rectilinear plane, we construct a data structure in $O(n^2 \log^2 n)$ time and space. This data structure allows us to perform a length query for any pair of points in the plane in $O(\log^2 n)$ time. An actual shortest path can be obtained in $O(\log^2 n + k)$ time.

- For all-pairs rectilinear shortest path queries among disjoint arbitrary *non-penetrating* polygonal obstacles (i.e., the case studied in [7, 8, 25, 26]), we build a data structure in $O(n^2 \log^2 n)$ time and space. This data structure allows us to perform a length query for any pair of points in the plane in $O(\log n)$ time. An actual shortest path can be obtained in $O(\log n + k)$ time. In addition to the ideas used for answering all-pairs queries among weighted obstacles, the querying process for this problem makes use of the fractional cascading data structure [4] and monotone matrix searching [1]. The details are left to the full paper.

Note that compared with the result of Iwai, Suzuki, and Nishizeki [19], our algorithms work for more general situations (*weighted* rectilinear obstacles and *arbitrary* non-penetrating polygonal obstacles) and still take less time (a factor of $\log n$) than [19] in constructing data structures and answering queries.

We will only discuss the algorithms and data structures for reporting *lengths* of shortest paths because most our solutions can be easily modified to report actual shortest paths in the claimed complexity bounds. Proofs are left to the full paper.

Unless otherwise specified, all geometric objects (e.g., lines, paths, polygons) are implicitly assumed to be rectilinear; i.e., each of their segments is parallel to one of the coordinate axes.

2 Preliminaries

For two points s and t in the plane, let $\overline{st}$ denote the line segment whose end points are s and t. A path P_{st} from point s to point t in the plane is a sequence of segments $(\overline{p_0 p_1}, \overline{p_1 p_2}, \ldots, \overline{p_{m-1} p_m})$, with $p_0 = s$ and $p_m = t$. An obstacle R_i (a polygon possibly with holes) is specified by a sequence of edges on each of its outer and inner boundaries. We assume that the edge sequence on the outer boundary is given in counterclockwise order and the inner boundaries in clockwise order.

Let the x- (resp., y-) coordinate of a point s be denoted by s_x (resp., s_y). For points s and t in the plane, the (unweighted) L_1 distance $dist(s,t)$ between s and t is $dist(s,t) = |s_x - t_x| + |s_y - t_y|$. For a path P_{st} between points s and t, let $d(P_{st})$ denote its (unweighted) length, i.e., the sum of the L_1 lengths of the segments on P_{st}. If P_{st} intersects the interior of some weighted obstacles, then we define the *weighted length* of P_{st} as follows: Partition P_{st} into subpaths $A_1, B_1, A_2, B_2, \ldots, A_k, B_k$, where A_i is a subpath without intersecting the interior of any obstacle and B_i is a subpath completely within the interior of some obstacle R_i, and subpaths A_1 and B_k may be of zero length; then the weighted length of P_{st}, denoted by $d_w(P_{st})$, is $d_w(P_{st}) = \sum_{i=1}^{k}(d(A_i)+d(B_i)) + \sum_{i=1}^{k}(d(B_i) \times W(R_i))$, where $W(R_i)$ is the weight function.

For each vertex u of an obstacle R_i such that the interior angle of R_i at u is $3\pi/2$, we define the *internal projection points* of u as follows: Let $\overline{uv}$ be the horizontal (resp., vertical) edge of R_i containing u; shoot a horizontal (resp., vertical) ray from u along the direction opposite to that of v, until the ray hits the first boundary point $p_h(u)$ (resp., $p_v(u)$) of R_i after intersecting the interior of R_i. We call $p_h(u)$ (resp., $p_v(u)$) the *horizontal* (resp., *vertical*) *internal projection point* of u. Internal projection points can be used to control paths that penetrate obstacles.

The notion of visibility is generalized to environments with weighted obstacles [23]. Two points s and t are said to be *visible* to each other if $\overline{st}$ is horizontal or vertical and $\overline{st}$ either does not intersect the interior of any obstacle or is completely contained in one single obstacle. A point p is said to be *visible* from a horizontal or vertical line L iff p is visible from some point p' on L. A key component of both our single-source and all-pairs data structures is the generalized visibility graph $G = (V, E)$ defined in [23] which captures the necessary information about shortest paths among the n obstacle vertices. We denote the cost of each edge $e \in E$ by $w(e)$.

The vertex set V of the visibility graph G can be partitioned into three subsets: (i) the set V_O of obstacle vertices, (ii) the set V_I of all the internal projection points of the vertices in V_O, and (iii) the set V_S of *Steiner points*. Lee, Yang, and Chen [23] used the fol-

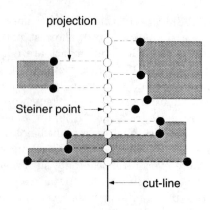

Figure 1: Creation of Steiner points on a cut-line.

lowing recursive procedure to generate Steiner points for the visibility graph:

1. Draw a vertical (resp., horizontal) line L, which we call a *cut-line*, at the median of the x- (resp., y-) coordinates of all the vertices in $V_O \cup V_I$;

2. project all the vertices in $V_O \cup V_I$ that are visible from L onto the cut-line L, as shown in Figure 1 (the projection points of $V_O \cup V_I$ on L are the Steiner points of V_S on L);

3. use the cut-line L to partition $V_O \cup V_I$ into two subsets S_1 and S_2, one on each side of L;

4. perform this procedure recursively on the vertex sets S_1 and S_2 respectively, until the size of each vertex set becomes 1.

Because the recursive procedure above has $O(\log n)$ recursion levels, it clearly generates $O(n \log n)$ Steiner points in V_S. Since $|V_O \cup V_I| = O(n)$, there are totally $O(n \log n)$ vertices in $V = V_O \cup V_I \cup V_S$. We associate with each cut-line L a *level number* $LN(L)$, which is the number of the recursion level at which L is used in the above procedure (with the root level being level 1).

The edge set E of G consists of two subsets of edges: (1) the set E_V of line segments between every Steiner point in V_S and its corresponding vertex in $V_O \cup V_I$, and (2) the set E_L of line segments between consecutive Steiner points on every cut-line. The weighted lengths of the edges in E_V can be easily determined and stored during the generation of the Steiner points. The weighted lengths of all the edges in E_L are obtained by using Widmayer's algorithm [23, 32], in $O(n \log n)$ time.

It has been proved in [23] that the generalized visibility graph G so created captures the necessary information about shortest paths among points in the plane that are represented as vertices in V. The first algorithm of Lee, Yang, and Chen [23] runs Fredman and Tarjan's shortest path algorithm [12] on G to find a shortest path between two points s and t in the plane (with both s and t being included in V), in $O(n\log^2 n)$ time and $O(n\log n)$ space.

Lee, Yang, and Chen [23] improved the time complexity of their first algorithm by using Clarkson, Kapoor, and Vaidya's idea [7] to reduce the number of vertices (precisely, Steiner points) in the graph G by increasing the number of edges; let the set of additional edges to the graph be E_S. We denote the graph so resulted by $G' = (V', E')$. G' was obtained from G in [23] as follows:

1. For every vertical (resp., horizontal) cut-line L, partition the plane into horizontal (resp., vertical) strips such that each strip contains $O(\log^{1/2} n)$ Steiner points on L;

2. within each strip, for every pair of vertices u and v in $V_O \cup V_I$ such that u and v are both visible from L and are on the opposite sides of L, add the edge (u,v) to E_S, and let the cost of (u,v) be the sum of the costs of the edges $(u,s(u))$, $(s(u),s(v))$, and $(s(v),v)$ in G, where $s(u)$ and $s(v)$ are Steiner points on L corresponding to u and v, respectively;

3. in each strip, retain in V the Steiner points that have the highest or lowest y- (resp., x-) coordinate among the Steiner points on L, and remove from G the rest Steiner points of L in that strip together with the edges adjacent to these removed Steiner points. (The retained Steiner points play the role of maintaining communication between consecutive strips.)

The graph $G' = (V', E')$ created by this procedure has $|V'| = O(n\log^{1/2} n)$ and $|E'| = O(n\log^{3/2} n)$. The second algorithm in [23] runs Fredman and Tarjan's shortest path algorithm on G', in $O(n\log^{3/2} n)$ time and space. Hence, there is a time-space trade-off between the two algorithms in [23].

Our solutions for shortest path queries make use of the graphs G and G' in the following ways: The generalized visibility graph G will be a key component of our shortest path data structures (Section 3), and the graph G' will be used for the *construction* of our shortest path data structures (Section 4). We will show later (in Section 4) that by using an *implicit* representation scheme for G', the time-space trade-off which occurs in the algorithms [23] can be eliminated, and thus the best time and space complexities of both these algorithms can be achieved simultaneously.

3 Shortest Path Queries among Weighted Rectilinear Obstacles

We discuss in this section the algorithms for answering single-source and all-pairs queries among weighted rectilinear obstacles, while deferring the complete description of the desired shortest path data structures and their construction until the next section. This will allow us to show various useful geometric structures of these shortest path query problems before presenting the actual algorithms. Note that the graph we refer to in this section is the generalized visibility graph G.

3.1 Useful Observations

In order to simplify our discussion of queries we need to first introduce some notation and useful lemmas. For each query point z, we project z vertically (resp., horizontally) onto the first point of the obstacle edge (if it exists) that is immediate above (resp., below, to the left of, to the right of) z. Let $Q(z)$ be the set of the (at most four) projection points of z on obstacle boundaries so resulted.

A path is *monotone* with respect to the x-axis (resp., y-axis) iff its intersection with every vertical (resp., horizontal) line is either empty or a contiguous portion of that line. A path is called a *staircase* if it is monotone with respect to both the x-axis and the y-axis. Note that a staircase from a point p to a point q in an obstacle-free region is a shortest path between p and q since its length equals $dist(p,q)$. We have the following lemmas.

Lemma 1 *For two points q and r, suppose there is a shortest path P_{qr} between q and r whose interior does not intersect any obstacle boundary, then P_{qr} is a staircase. Furthermore, among weighted rectilinear obstacles, either there is a shortest path between q and r that passes through an obstacle vertex, or there is a shortest path between q and r that consists of at most two line segments.*

Lemma 2 *Given two points q and r with q not on any obstacle boundary, if the interior of a shortest q-to-r path P_{qr} intersects the boundary of an obstacle, then there is a shortest q-to-r path that either goes through a point $p \in Q(q)$ (via the segment $\overline{qp}$) or goes through an obstacle vertex.*

Lemma 3 *Suppose a point q is on an obstacle boundary. Then, for any point r, there is a shortest q-to-r path P_{qr} that either consists of at most two line segments or goes through an obstacle vertex.*

The significance of the above lemmas to our shortest path queries is as follows. Given two query points s and

t, there are three possibilities: (1) some shortest path P_{st} goes through an obstacle vertex and hence through a vertex of G (by possibly going through some point in $Q(s) \cup Q(t)$), (2) P_{st} goes through some point in $Q(s) \cup Q(t)$ but not through any vertex of G, and (3) P_{st} consists of at most two line segments. In case 2, the path P_{qr}, for some $q \in Q(s)$ and $r \in Q(t)$, is one with at most two line segments.

Note that in the single-source case, it is always possible to find a shortest path from any point in the plane to the source point such that the path goes through some vertices of the graph G (e.g., the source vertex), and thus G can be used to effectively control single-source shortest paths for arbitrary query points. However, in the all-pairs case, this need not be the situation, because shortest paths between certain points in the plane need not pass any vertex in G. Therefore, our all-pairs query algorithms must account for both the path going through a vertex in G and the path not going through any vertex of G, and choose the best answer out of the two candidates. In the next subsection, we discuss the query algorithms for the situation in which a shortest path goes through a vertex of G (i.e., case 1). The situation in which no shortest path goes through a vertex of G will be handled in the full paper (cases 2 and 3).

3.2 Main Ideas and Difficulties

Clearly, we can simply focus on the all-pairs queries for shortest paths going through vertices of G in this subsection. Given two query points s and t, let $p \in Q(s) \cup \{s\}$ and $q \in Q(t) \cup \{t\}$. Based on Lemmas 2 and 3, we want to compute the length of a shortest p-to-q path through a vertex of G, for every pair of p and q.

Our main idea for computing a shortest p-to-q path that intersects vertices of G is to identify a subset $V_g(z)$ of vertices in G for each $z \in \{p, q\}$, such that a shortest p-to-q path must go through at least one pair (v_q, v_p) of vertices, with $v_q \in V_g(q)$ and $v_p \in V_g(p)$. We call the vertices of G in $V_g(q)$ the *gateways* of q. If we knew $V_g(q)$ and $V_g(p)$ and the weighted path between q (resp., p) and each vertex in $V_g(q)$ (resp., $V_g(p)$), then we can easily find a shortest p-to-q path by utilizing a data structure that maintains shortest path information for every pair of vertices in the graph G. In this section, we simply assume that our all-pairs data structure can report the length of a shortest path from one vertex to any other vertex of G in $O(1)$ time.

In order for this idea to materialize, we must overcome two difficulties: (1) identifying $V_g(q)$, and (2) determining the lengths of the weighted paths between q and the vertices in $V_g(q)$. Note that it is possible for a path between q and a vertex in $V_g(q)$ to penetrate a number of obstacles. For the efficiency of our query algorithms, it is highly desirable that $|V_g(q)|$ be small.

Indeed, we are able to show that $|V_g(q)| = O(\log n)$ and that $V_g(q)$ can be computed in $O(\log n)$ time. Furthermore, we show that the weighted paths between q and the vertices in $V_g(q)$ and their lengths can also be computed in $O(\log n)$ time.

Our idea for computing the set $V_g(q)$ of gateways for a point q is that of "inserting" q into the graph G. We treat q as if it were one of the obstacle vertices and compute the edges in G adjacent to q that would have resulted if the procedure in [23] for constructing G were applied to q. Note that such an "insertion" of a point into G does not essentially change the shortest path information contained in G (i.e., q is not a new obstacle); furthermore, the resulting "graph" (after the "insertions") contains shortest path information among the inserted points and the vertices of G, if a shortest path between the inserted points does intersect G. Also, note that our "insertion" process does not actually modify the graph G because the points are never truly inserted into G.

3.3 Characterization of Gateways

To "insert" a point q into the visibility graph G, it is necessary to project the point q onto the relevant cut-lines based on the graph construction procedure [23]. A fixed set of cut-lines are used in the construction of G. These cut-lines subdivide the plane recursively and each cut-line is associated with a particular recursion level. In consequence, all points in each region of the resulted planar subdivision can be projected onto the same subset of cut-lines. If, according to the graph construction procedure [23], q would have been projected onto a cut-line L, then we say L is a *projection cut-line* of q. Note that if a cut-line L is a projection cut-line of q, then q is visible from L. But, not every cut-line visible from q is a projection cut-line of q. It is sufficient for us to discuss geometric observations with vertical cut-lines only (those with horizontal cut-lines are symmetric). In the rest of this section, we assume all the cut-lines are vertical unless otherwise specified.

Lemma 4 *Suppose that the projection cut-lines of a point q are sorted in increasing order by their x-coordinates. Then the projection cut-lines of q that are to the right (resp., left) of q are simultaneously in the decreasing (resp., increasing) order of their level numbers.*

The set $V_g(q)$ of gateways for a point q is determined as follows: For each projection cut-line L of q, if z is the vertex of G on L that is immediately above (resp., below) the projection point of q on L, then $z \in V_g(q)$. Observe that if q were "inserted" into G, then the only edges adjacent to q in the graph would be those connecting q with its projection points. Because each projection point of q is adjacent to at most two neighboring vertices of G on its cut-line, $V_g(q)$ controls every path in G from q to any other vertex of G. Since at each recursion level, q can be projected onto at most one cut-line, there are at most $O(\log n)$ projection cut-lines for q. Along each projection cut-line of q, there can be at most two neighboring vertices of G. Therefore, we have the following two lemmas.

Lemma 5 *For any point q in the plane, $|V_g(q)| = O(\log n)$.*

Lemma 6 *For a point q and a vertex v of G, there is a shortest q-to-v path in the plane that goes through $V_g(q)$.*

For every vertex $z \in V_g(q)$, we define an "edge" (q, z) in the graph as follows: Let z be on a projection cut-line L of q; then the edge (q, z) consists of the segments $\overline{qp(q)}$ and $\overline{p(q)z}$, where $p(q)$ is the projection point of q on L (Figure 2).

By the definition of gateways, $V_g(q)$ can be obtained efficiently (e.g., in $O(\log^2 n)$ time by doing a binary search on the vertices of G along each projection cut-line of q, or in $O(\log n)$ time by using a fractional cascading data structure [4]). However, the real difficulty is that each edge connecting q with a vertex in $V_g(q)$ can penetrate many obstacles, and such methods fail to efficiently compute the weighted lengths of the edges between q and $V_g(q)$. By exploiting a number of geometric observations, we show below how to compute $V_g(q)$ and the weights of all these edges in $O(\log n)$ time.

A *gateway region* $\mathcal{R}(q)$ (Figure 2) of a point q is the area defined by an interconnection of vertices in $V_g(q)$. WLOG, we only show how the gateways in the first quadrant of q are connected together: (1) Let a vertical "pseudo" cut-line pass through q. (2) For each vertex v of $V_g(q)$ in the first quadrant of q, project v horizontally to the projection cut-line L of q that is immediately to the left of v (it can be shown that such a projection is always possible among weighted obstacles). Note that L can be the "pseudo" cut-line passing q. (3) Let the projection point of v on L be $p(v)$ and the vertex of $V_g(q)$ that is on L and in the first quadrant of q be u (if u exists). (4) Connect u and v by line segments $\overline{up(v)}$ and $\overline{p(v)v}$. (5) If v is the rightmost gateway of q, then connect v with the point (v_x, q_y) by a segment. (6) If L is the pseudo-cut line passing through q (and hence u does not exist on L), then connect v and $p(v)$ by a segment. The area in the first quadrant of q enclosed together by the polygonal chain so defined, the x-axis, and y-axis (with q as the origin) is part of $\mathcal{R}(q)$. $\mathcal{R}(q)$ is the union of the areas so defined in the four quadrants of q.

A region $\mathcal{R}$ in the plane is said to be *rectilinearly convex* if $\mathcal{R}$ is a connected region and for any horizontal or vertical line L, $L \cap \mathcal{R}$ is either a single segment or

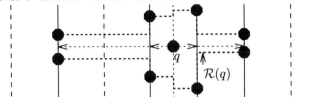

Figure 2: The gateway region $\mathcal{R}(q)$ associated with a point q.

empty. The next two lemmas characterize some crucial structures of $\mathcal{R}(q)$.

Lemma 7 *The gateway region $\mathcal{R}(q)$ is rectilinearly convex; furthermore, for any point z in $\mathcal{R}(q)$, the points (z_x, q_y) and (q_x, z_y) are both in $\mathcal{R}(q)$.*

Lemma 8 *The gateway region $\mathcal{R}(q)$ contains no vertices of $V_O \cup V_I$ in its interior.*

To further discuss the properties of gateways of a point q, we distinguish three cases: (a) q is an obstacle vertex, (b) q is not an obstacle vertex but is on a vertical obstacle edge, and (c) q is not on any vertical obstacle edge. (a) can be readily handled. Hence we only focus on (b) and (c) in the following two subsections. We first discuss (c), and then show how to reduce (b) to (c).

3.3.1 Point q Not on Any Vertical Obstacle Edge

Throughout this subsection, we assume that the point q is not on any vertical obstacle edge.

The lemma below follows from Lemma 8 and further characterizes $\mathcal{R}(q)$.

Lemma 9 *There are three possible relationships of obstacles to the gateway region $\mathcal{R}(q)$:*

1. *The interior of $\mathcal{R}(q)$ does not intersect any obstacle.*

2. *$\mathcal{R}(q)$ is completely contained within an obstacle.*

3. *$\mathcal{R}(q)$ has horizontal obstacle strips running through it (see Figure 3).*

The following lemmas provide the basis for efficient computation of $V_g(q)$ and their weighted edges to q.

Lemma 10 *Let a and b be two gateways of a point q that are both above (resp., below) the horizontal line H_q that passes q. Furthermore, let the cut-line L_a containing a have a higher level number than the cut-line L_b containing b, with L_a not being the leftmost or rightmost projection cut-line of q. Then the set of horizontal*

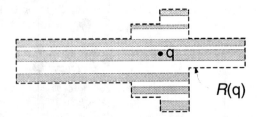

Figure 3: Horizontal obstacle strips inside the gateway region $\mathcal{R}(q)$.

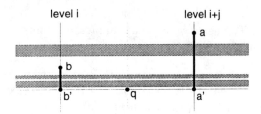

Figure 4: Illustration of Lemma 10.

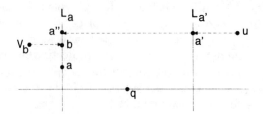

Figure 5: Illustration of Lemma 15.

obstacle strips penetrated by the segment $\overline{bb'}$ is a subset of the horizontal obstacle strips penetrated by the segment $\overline{aa'}$, where $a' = (a_x, q_y)$ and $b' = (b_x, q_y)$ (see Figure 4).

Lemma 11 Let a and b be two gateways of a point q that are both above (resp., below) the horizontal line H_q. Also, let the cut-line L_a containing a have a higher level number than the cut-line L_b containing b. Then $a_y < b_y$ (resp., $a_y > b_y$) implies that L_a is either the leftmost or rightmost projection cut-line of q and that L_a contains a vertical obstacle edge e such that e crosses the line H_q.

Corollary 1 Let a and b be two gateways of a point q that are both above (resp., below) the horizontal line H_q. Let L_a (resp., L_b) be the projection cut-line of q containing a (resp., b). Suppose that none of L_a and L_b is the leftmost or rightmost projection cut-line of q. Then L_a having a higher level number than L_b implies that $a_y \geq b_y$.

The ordering of the gateways of q by their y-coordinates is a key to our computation of $V_g(q)$ and its weighted edges to q. The next lemma shows that the gateways are "almost" sorted by their y-coordinates.

Lemma 12 Let $A(q)$ be the sequence of gateways of a point q above (resp., below) the horizontal line H_q, and let $A(q)$ be in increasing order of the level numbers of the projection cut-lines of q containing the points in $A(q)$. Then either $A(q)$ or $A(q) - \{a\}$ is in nondecreasing (resp., nonincreasing) order of the y-coordinates of the points in $A(q)$, where a is the gateway with the bigger (resp., smaller) y-coordinate among the two gateways of q that are on the leftmost or rightmost projection cut-lines of q.

Lemma 13 Let a be a gateway of a point q on the leftmost or rightmost projection cut-line of q, and a be on a vertical obstacle edge e that crosses the horizontal line H_q. Then either q is in the same obstacle that contains e or q is not contained in any obstacle.

Hence, it is easy to compute the weight of the "inserted" edge (q, a), where a is defined in Lemma 13. The following two lemmas lay the foundation for our iterative procedure for finding $V_g(q)$ and the weights of the edges between q and the vertices in $V_g(q)$.

Lemma 14 Let a be a gateway of q on the cut-line L_a, such that the level number $LN(L_a)$ of L_a is the smallest among all the projection cut-lines of q that do not contain a vertical obstacle edge e which crosses the horizontal line H_q. Let b be the point (a_x, q_y). Then one of the following two situations holds:

1. The interior of the segment $\overline{ab}$ does not intersect the interior of any obstacle.

2. The segment $\overline{ab}$ is contained in the same obstacle that contains q.

Lemma 15 Let a be a gateway of a point q on the cut-line L_a, and let a' be the gateway of q on the cut-line $L_{a'}$ such that none of L_a and $L_{a'}$ contains a vertical obstacle edge crossing H_q and that the level number $LN(L_{a'})$ of $L_{a'}$ is the smallest among the projection cut-lines of q whose level numbers are $> LN(L_a)$ (if such cut-lines exist), with both a and a' being above (resp., below) H_q. Then the following are true (Figure 5):

1. There is a Steiner point a'' on L_a such that $a'_y = a''_y$. Furthermore, there is a vertex $u \in V_O \cup V_I$ such that u is projected on both a'' and a'.

2. For any Steiner point b on L_a such that $a''_y > b_y > a_y$ (resp., $a''_y < b_y < a_y$), let $v_b \in V_O \cup V_I$ be the vertex that is projected on b. Then v_b and a' are on the opposite sides of L_a.

3. Let $p_1 = (a''_x, q_y)$ and $p_2 = (a'_x, q_y)$ be points on H_q. Then the segments $\overline{a''p_1}$ and $\overline{a'p_2}$ have the same weighted length.

Let $v_{a'} \in V_O \cup V_I$ be projected onto a'. Then $v_{a'}$ cannot be to the left of L_a; otherwise, since a' is to the right of L_a and $LN(L_a) < LN(L_{a'})$, $v_{a'}$ would have not been able to be projected onto $L_{a'}$. So $v_{a'}$ must be to the right of L_a. Then by Lemma 9 and the above claim, $v_{a'}$ must be projected onto a'' on L_a. Hence, (1) is true (with $u = v_{a'}$).

For (2), if v_b is to the right of L_a (as a'), then v_b would have been projected onto L'_a because the interior of $\mathcal{R} \cup \mathcal{R}(q)$ contains no obstacle vertex and does not intersect any vertical obstacle edge, thus contradicting that $a' \in V_g(q)$. Hence, v_b can only be to the left of L_a.

For (3), because any horizontal obstacle strip intersected by the segment $\overline{a'p_2}$ cannot stop within $\mathcal{R}$ and must run through $\mathcal{R}$, the segment $\overline{a''p_1}$ also intersects such a horizontal obstacle strip. By the same argument, if $\overline{a''p_1}$ intersects any horizontal obstacle strip, such a horizontal strip must also be intersected by $\overline{a'p_2}$. Hence, the weighted lengths of both $\overline{a''p_1}$ and $\overline{a'p_2}$ are the same. □

3.3.2 Point q on a Vertical Obstacle Edge

If the point q is on a vertical obstacle edge e, then we reduce this case to the case in which q is not on any vertical obstacle edge, as follows: Treat q as if q_x is perturbed to the left (resp., right) of e by some extremely small value $\epsilon > 0$, and then use the observations and procedures for the case with q not on any vertical obstacle edge.

In the rest of this section, we assume WLOG that q is not on any vertical obstacle edge.

3.4 Computing Gateways and Weighted Edges to Gateways

First, we need to find all projection cut-lines of a point q. These cut-lines can be identified easily by using a data structure based on trapezoidal decomposition and planar point location [29], in $O(\log n)$ time. By Lemma 4, the x-coordinates of these cut-lines can be sorted by bucket sort, in $O(\log n)$ time. Let the sequence $L_{i_1}, L_{i_2}, \ldots, L_{i_g}$ of the projection cut-lines of q be in increasing order of their level numbers, and let a_{i_j} be the gateway of q on L_{i_j} that is above the horizontal line H_q, with $g = O(\log n)$.

Next, we compute the gateways a_{i_j} and the weights of the edges (q, a_{i_j}). Let p_{i_j} be the projection point of q on the projection cut-line L_{i_j} of q that contains a_{i_j}. Since the edge (q, a_{i_j}) consists of two segments $\overline{qp_{i_j}}$ and $\overline{p_{i_j}a_{i_j}}$ and since the interior of $\overline{qp_{i_j}}$ is either contained completely in an obstacle or outside any obstacle, it is sufficient to show how to compute each a_{i_j} and the weighted length $d_w(\overline{p_{i_j}a_{i_j}})$. Note that $\overline{p_{i_j}a_{i_j}}$ may penetrate many horizontal obstacle strips in the gateway region $\mathcal{R}(q)$ of q.

We need to first find the gateways that are either on the projection cut-lines of q containing a vertical obstacle edge that crosses the line H_q or on the projection cut-line of q which has the smallest level number among the projection cut-lines that do not contain a vertical obstacle edge crossing H_q. There are at most three such gateways and they can be obtained by binary search operations on the vertices of G along the relevant projection cut-lines of q in $O(\log n)$ time. The weights of the edges from q to these gateways can be easily obtained by Lemmas 13 and 14. Next, based on Lemma 15, we show that from a_{i_j} and $d_w(\overline{p_{i_j}a_{i_j}})$, $a_{i_{j+1}}$ and $d_w(\overline{p_{i_{j+1}}a_{i_{j+1}}})$ can both be obtained in $O(1)$ time. WLOG, we assume that the sequence $A(q)$ is in sorted order based on the y-coordinates of the gateways. Our data structure for computing gateways and the weights of their edges to q contains the following items for every cut-line L.

1. For any two vertices u and v of G on L, $d_w(\overline{uv})$ can be reported in $O(1)$ time. This is done by first using Widmayer's algorithm [23, 32] to compute the weighted length between every two consecutive vertices of G on L, and then performing a prefix sum operation, along the vertices of G on L, to compute the weighted length from every vertex on L to the highest vertex on L. It is then easy to find $d_w(\overline{uv})$ from the prefix sums of u and v in $O(1)$ time.

2. Each vertex u of G on L keeps track of the vertices z in $V_O \cup V_I$ such that z is projected on u (based on the graph construction procedure [23]). Note that u can be the projections of vertices of $V_O \cup V_I$ on each side of L, and u keeps track of all such vertices.

3. For each vertex $z \in V_O \cup V_I$, maintain an array $Proj_z$ of size $O(\log n)$. If z is projected onto a cut-line of level i by the graph construction procedure [23], then $Proj_z(i)$ contains a pointer to its projection point on that line; otherwise, $Proj_z(i)$ is undefined.

4. Each vertex u of G on L keeps track of the lowest vertex v of G on L such that $v_y \geq u_y$ and v is a projection of a vertex of $V_O \cup V_I$ that is to the left (resp., right) of L (v can be u).

Note that the graph construction procedure [23] can be easily modified to compute the information in items 1–4 above in $O(n \log n)$ time and space. Now based on Lemma 15, it is an easy matter to compute $a_{i_{j+1}}$ and $d_w(\overline{p_{i_{j+1}}a_{i_{j+1}}})$ from a_{i_j} and $d_w(\overline{p_{i_j}a_{i_j}})$, in $O(1)$ time. WLOG, assume L_{i_j} is to the left of $L_{i_{j+1}}$. We perform the following steps: (1) Find the lowest vertex a'' of G on L_{i_j} such that a'' is on or above a_{i_j} and that a'' is a projection point of a vertex $z \in V_O \cup V_I$ with z being to the right of L_{i_j}; (2) let $a_{i_{j+1}} = Proj_z(LN(L_{i_{j+1}}))$ and $d_w(\overline{p_{i_{j+1}}a_{i_{j+1}}}) = d_w(\overline{p_{i_j}a_{i_j}}) + d_w(\overline{a_{i_j}a''})$.

3.5 Answering Length Queries

From the gateways, a single-source query for the length of a shortest path from a query point to the source vertex s can be easily answered in $O(\log n)$ time. An all-pairs length query can be answered in $O(\log^2 n)$ time.

4 Constructing Shortest Path Data Structures

In this section, we focus on the construction of the data structures that allow us to answer both the single-source and all-pairs queries. Both our single-source and all-pairs data structures make use of the graph G which has $O(n \log n)$ vertices and edges. Note that if Fredman and Tarjan's shortest path algorithm were directly applied to G, then computing the shortest path tree in G rooted at the source vertex (for the single-source case) would take $O(n \log^2 n)$ time and $O(n \log n)$ space, and computing $O(n \log n)$ shortest path trees in G (for the all-pairs case) would take $O(n^2 \log^3 n)$ time and $O(n^2 \log^2 n)$ space. By using an implicit graph representation scheme and divide-and-conquer paradigms, we show how the single-source data structure can be constructed in $O(n \log^{3/2} n)$ time and $O(n \log n)$ space, and how the all-pairs data structure can be constructed in $O(n^2 \log^2 n)$ time and space. Note that the graph G' (defined in Section 2) plays a crucial role in computing shortest path trees in G.

Because other components of our shortest path data structures have been discussed in the previous section, we only need to show in this section how to compute shortest path trees in G.

4.1 Distance Tables

Recall that one major difference between the graph G' and the visibility graph G is the addition of a set E_S of edges to G', with $|E_S| = O(n \log^{3/2} n)$. The edges in E_S were represented *explicitly* using $O(n \log^{3/2} n)$ space in [23]. We use an *implicit* scheme to store E_S in $O(n \log n)$ space, which still enables us to retrieve information about G' as quickly (within a constant factor) as using the explicit representation. We call such a scheme *distance tables*.

The construction of G' is based on the partition of every cut-line L into strips, with each strip containing $O(\log^{1/2} n)$ Steiner points on L. Let S be the set of Steiner points on L within a strip of L, and let $PV_l(S)$ (resp., $PV_r(S)$) denote the vertices of $V_O \cup V_I$ within the strip that are to the left (resp., right) of L and are visible from L. We associate a distance table with the strip of L that contains S. In the distance table, $PV_l(S)$ and $PV_r(S)$ are each stored in an array of size $O(\log^{1/2} n)$. Note that for every vertex $u \in PV_l(S)$ and $v \in PV_r(S)$, there is an edge (u,v) in G' that consists of segments $\overline{up(u)}$, $\overline{p(u)p(v)}$, and $\overline{p(v)v}$, where $p(z)$ denotes the projection of a vertex $z \in PV_l(S) \cup PV_r(S)$ on L. We represent the edge (u,v) implicitly. For a vertex u in $PV_l(S)$, clearly there is an edge in G' between u and every vertex in $PV_r(S)$. So the edges between $PV_l(S)$ and $PV_r(S)$ can be implicitly represented by $PV_l(S)$ and $PV_r(S)$. The weights of the edges between $PV_l(S)$ and $PV_r(S)$ can also be represented implicitly, as follows: For every $z \in PV_l(S) \cup PV_r(S)$, store $p(z)$ and the weighted length of $\overline{zp(z)}$ in the distance table; for any $p(u)$ and $p(v)$ in S, use the prefix sums of weighted lengths discussed in Subsection 3.4 to compute the weighted length of $\overline{p(u)p(v)}$ in $O(1)$ time. Hence, the weight of every edge (u,v) of G', with $u \in PV_l(S)$ and $v \in PV_r(S)$, can be computed in $O(1)$ time via the distance table. The distance table for each strip uses $O(\log^{1/2} n)$ space. The distance tables for all the $O(n \log^{1/2} n)$ strips use altogether $O(n \log n)$ space and can be built trivially in $O(n \log n)$ time. Note that running Fredman and Tarjan's shortest path algorithm on G' with the edge set E_S being represented by distance tables is similar to the explicit representation of G' [23], and it takes $O(n \log^{3/2} n)$ time and $O(n \log n)$ space.

4.2 Construction of Data Structures

The shortest path tree in the graph G rooted at a source vertex s is computed by making use of the graph G'. The procedure for computing this shortest path tree in G is based on a divide-and-conquer strategy and takes $O(n \log^{3/2} n)$ time and $O(n \log n)$ space. The details are left to the full paper.

For each of the $O(n \log n)$ vertices in G, we need to compute a shortest path tree in G rooted at that vertex. The $O(n \log n)$ shortest path trees in G together take $O(n^2 \log^2 n)$ space. Applying the algorithm for computing a single shortest path tree to each vertex of G would take $O(n^2 \log^2 n \log \log n)$ time. We achieve an $O(n^2 \log^2 n)$ time and space algorithm for computing the $O(n \log n)$ shortest path trees in G. The procedure is based on a different divide-and-conquer strategy, and the details are left to the full paper.

References

[1] A. Aggarwal, M. M. Klawe, S. Moran, P. Shor, and R. Wilber. Geometric applications of a matrix searching algorithm. *Algorithmica*, 2:209–233, 1987.

[2] M. J. Atallah and D. Z. Chen. Parallel rectilinear shortest paths with rectangular obstacles. In *Computational Geometry: Theory and Applications*, volume 1, pages 79–113. Elsevier, 1991.

[3] M. J. Atallah and D. Z. Chen. On parallel rectilinear obstacle-avoiding paths. In *Computational Geometry: Theory and Applications*, volume 3, pages 307–313. Elsevier, 1993.

[4] B. Chazelle and L. J. Guibas. Fractional cascading: I. A data structuring technique. *Algorithmica*, 1(2):133–162, 1986.

[5] D. Z. Chen. On the all-pairs euclidean short path problem. In *Proc. 6th ACM-SIAM Symp. on Discrete Algorithms*, pages 292–301, 1995.

[6] Y.-J. Chiang, F. P. Preparata, and R. Tamassia. A unified approach to dynamic point location, ray shooting, and shortest paths in planar maps. In *Proc. 4th ACM-SIAM Symp. on Discrete Algorithms*, pages 44–53, 1993.

[7] K. L. Clarkson, S. Kapoor, and P. M. Vaidya. Rectilinear shortest paths through polygonal obstacles in $O(n \log^{3/2} n)$ time. Submitted for publication.

[8] K. L. Clarkson, S. Kapoor, and P. M. Vaidya. Rectilinear shortest paths through polygonal obstacles in $O(n(\log n)^2)$ time. In *Proc. 3rd Symp. on Computational Geometry*, pages 251–257, 1987.

[9] M. de Berg, M. van Kreveld, and B. J. Nilsson. Shortest path queries in rectangular worlds of higher dimension. In *Proc. 7th Symp. on Computational Geometry*, pages 51–59, 1991.

[10] P. J. de Rezende, D. T. Lee, and Y. F. Wu. Rectilinear shortest paths in the presence of retangular barriers. *Discrete & Computational Geometry*, 4:41–53, 1989.

[11] H. ElGindy and P. Mitra. Orthogonal shortest route queries among axes parallel rectangular obstacles. *International Journal of Computational Geometry and Applications*, 4(1):3–24, 1994.

[12] M. L. Fredman and R. E. Tarjan. Fibonacci heaps and their uses in improved network optimization algorithms. *JACM*, 34(3):596–615, July 1987.

[13] L. Gewali, A. Meng, J. S. B. Mitchell, and S. Ntafos. Path planning in 0/1/∞ weighted regions with applications. In *Proc. 4th Symp. on Computational Geometry*, pages 266–278, 1988.

[14] M. T. Goodrich and R. Tamassia. Dynamic ray shooting and shortest paths via balanced geodesic triangulations. In *Proc. of the 9th Annual Symp. on Computational Geometry*, pages 318–327. 1993.

[15] L. J. Guibas and J. Hershberger. Optimal shortest path queries in a simple polygon. In *Proc. 3rd Symp. on Computational Geometry*, pages 50–63. 1987.

[16] L. J. Guibas, J. Hershberger, D. Leven, M. Sharir, and R. E. Tarjan. Linear time algorithms for visibility and shortest path problems inside triangulated simple polygons. *Algorithmica*, 2:209–233, 1987.

[17] J. Hershberger. A new data structure for shortest path queries in a simple polygon. *IPL*, 38:231–235, June 1991.

[18] J. Hershberger and S. Suri. Efficient computation of Euclidean shortest paths in the plane. In *Proc. 34th Annual Symp. on Foundations of Computer Science*, pages 508–517, 1993.

[19] M. Iwai, H. Suzuki, and T. Nishizeki. Shortest path algorithm in the plane with rectilinear polygonal obstacles (in Japanese). In *Proc. of SIGAL Workshop*, Ryukoku University, Japan, July 1994.

[20] K. S. Klenk. Rectilinear shortest path queries among weighted obstacles. Master's thesis, University of Notre Dame, November 1994.

[21] R. C. Larson and V. O. Li. Finding minimum rectilinear distance paths in the presence of barriers. *Networks*, 11:285–304, 1981.

[22] D. T. Lee and F. P. Preparata. Euclidean shortest paths in the presence of rectilinear barriers. *Networks*, 11:285–304, 1984.

[23] D. T. Lee, C. D. Yang, and T. H. Chen. Shortest rectilinear paths among weighted obstacles. *International Journal of Computational Geometry and Applications*, 1(2):109–124, 1991.

[24] J. S. B. Mitchell. A new algorithm for shortest paths among obstacles in the plane. Technical Report 832, Cornell University, School of OR/IE, Oct 1988.

[25] J. S. B. Mitchell. An optimal algorithm for shortest rectilinear paths among obstacles. In *First Canadian Conference on Computational Geometry*, 1989.

[26] J. S. B. Mitchell. L_1 shortest paths among polygonal obstacles in the plane. *Algorithmica*, 8:55–88, 1992.

[27] J. S. B. Mitchell. Shortest paths among obstacles in the plane. In *Proc. 9th Symp. on Computational Geometry*, pages 308–317, 1993.

[28] J. S. B. Mitchell and C. Papadimitriou. The weighted region problem: Finding shortest paths through a weighted planar subdivision. *JACM*, 38:18–73, 1991.

[29] F. P. Preparata and M. I. Shamos. *Computational Geometry: An Introduction*. Texts and Monographs in Computer Science. Springer-Verlag, Berlin, 1985.

[30] J. A. Storer and J. H. Reif. Shortest paths in the plane with polygonal obstacles. *JACM*, 41(5):982–1012, 1994.

[31] E. Welzl. Constructing the visibility graph for n line segments in $O(n^2)$ time. *IPL*, 20:167–171, 1985.

[32] P. Widmayer. Network design issues in VLSI. Manuscript, 1989. In D. T. Lee, C. D. Yang, and T. H. Chen. Shortest rectilinear paths among weighted obstacles. *International Journal of Computational Geometry and Applications*, 1(2):109–124, 1991.

[33] Y. F. Wu, P. Widmayer, M. D. F. Schlag, and C. K. Wong. Rectilinear shortest paths and minimum spanning trees in the presence of rectlinear obstacles. *IEEE Transactions on Computers*, pages 321–331, 1987.

[34] C. D. Yang, T. H. Chen, and D. T. Lee. Shortest rectilinear paths among weighted rectangles. *Journal of Information Processing*, 13(4):456–462, 1990.

Rectilinear Geodesics in 3-Space
(Extended Abstract)

Joonsoo Choi   Chee-Keng Yap*

Courant Institute of Mathematical Sciences

New York University

251, Mercer Street

New York, NY 10012

Abstract

Let $\mathcal{B}$ be any finite set of pairwise-disjoint, axes-parallel boxes in Euclidean 3-space. Our main theorem is that for any two points s, t not in the interior of $\cup \mathcal{B}$, there exists a shortest rectilinear $\mathcal{B}$-avoiding path from s to t that is monotone along at least one of the axes. The key concept in the proof is an appropriate notion of *pyramids*.

Exploiting this result algorithmically, we obtain: a L_1 shortest distance from a query point to a fixed source point can be computed in $O(\log n)$ time after $O(n^2 \log n)$ time preprocessing, where n is the number of boxes. This improves on known results of Clarkson et al. [CKV87] and also de Berg et al. [dBvKNO92] when their results are specialized to disjoint boxes.

1 Introduction

The *geometric shortest path* problem can be formulated as follows: given a collection $\mathcal{B}$ of polyhedral obstacles in $\mathbb{R}^d$, and source and target points $s, t \in \mathbb{R}^d$, find a shortest obstacle-avoiding path between s and t. This is a basic problem in applications such as motion planning in robotics.

We use the term *geodesic* for "geometric shortest path", and to distinguish it from shortest paths in graphs. A path that avoids the obstacles $\mathcal{B}$ is called a $\mathcal{B}$-path. In particular, we have $\mathcal{B}$-geodesics. Throughout this paper, the combinatorial complexity of $\mathcal{B}$ (*i.e.*, the number of vertices, edges, faces, etc) is assumed to be n. We also assume that s, t do not lie in the interior of any obstacle.

---

*The authors are supported by NSF grant #CCR-9402464.

There are two basic approaches to geodesic problems: one is to compute a *visibility graph* (VG) and thus reduce the problem to the graph shortest path problem for which Dijkstra's algorithm is applicable. The other is to compute a *shortest path map* (SPM) which is a subdivision of $\mathbb{R}^d$ so that all points t in a region of the subdivision have the same combinatorial path structure from a fixed source s. An SPM is useful for the *shortest path query problem*: given an arbitrary query point t, to compute a geodesic or the geodesic length from s to t.

The notion of "shortest" depends on the metric chosen. Our results concern the L_1 (equivalently *rectilinear*) *metric* and where the obstacles $\mathcal{B}$ is a collection of pairwise disjoint axes-parallel boxes in $\mathbb{R}^d$. Henceforth, unless otherwise stated, the metric is L_1 and paths are rectilinear. The most direct ancestry of our work is a paper of de Rezende, Lee and Wu [dRLW89] who considered this problem for $d = 2$. They constructed an SPM which consists of $O(n)$ rectangular regions. The preprocessing step takes $O(n \log n)$ time. The geodesic length from a query point to s is computed in $O(\log n)$ time and a geodesic can be constructed in additional $O(k)$ time, where k is the number of edges in the path. They showed that $\Omega(n \log n)$ is a lower bound on the time complexity of their problem.

The algorithm in de Rezende et al. is based on the fact that *every geodesic between two points is monotone in one of the coordinate axes*. For short, refer to this as *monotonicity property* in the plane. We extend this to 3-dimensions.

Theorem 1 (Monotonicity) *For all $s, t, \mathcal{B}$ in $\mathbb{R}^3$, there exists a $\mathcal{B}$-geodesic from s to t that is monotone along at least one of the three axes directions.*

The proof of this result require several steps, including a proper notion of "pyramids". We believe the monotonicity property holds in every dimension. The following example shows that the monotonicity property is the best possible in some sense.

EXAMPLE 1: For each $d \geq 2$, there exists a set $\mathcal{B}$ of $d - 1$ boxes such every $\mathcal{B}$-geodesic from s to t is non-monotonic along $d-1$ axes. The reader can generalize the example for $d = 3$ shown in figure 1.

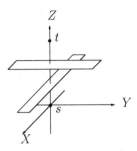

Figure 1: There is no geodesic from s to t which is monotone along X- or Y-axis.

This example can be generalized to force the geodesics to make any number of non-monotonic turns along the $d-1$ axes directions. Moreover, the boxes in $\mathcal{B}$ may not be replaced by convex polyhedrons or rectilinear convex polyhedrons.

Using the monotonicity property, we show:

Theorem 2 *Given a set of n disjoint boxes and a point s in $\mathbb{R}^3$, we can compute the geodesic length between a query point and s in $O(\log n)$ time and construct a geodesic in additional $O(k)$ time, where k is the number of edges in the path. The preprocessing takes $O(n^2 \log n)$ time.*

This result should be compared to two other results in the literature. Clarkson, Kapoor and Vaidya [CKV87] studied the case where $\mathcal{B}$ are axes-parallel but not necessarily disjoint. Their algorithm computes a geodesic between two given points in $O(n^2 \log^3 n)$ time. Berg, Kreveld, Nilsson and Overmars [dBvKNO92] studies the problem in $\mathbb{R}^d$ where obstacles are n possibly intersecting axes-parallel boxes. Their algorithm constructs a data structure in $O(n^d \log n)$ time so that each query takes $O(\log^{d-1} n)$ time. Their metric is a composite of L_1 distance and link distance. It can be specialized to the L_1 metric. They also gave an algorithm computing a geodesic between any two given points in $O(n^d \log n)$ time.

1.1 Related Results

The algorithms of de Rezende et al. has been parallelized. Exploiting the monotonicity property, ElGindy and Mitra [EM94] presented an algorithm that computes the geodesic length between two arbitrary points in $O(\log n)$ time with $O(n^2)$ preprocessing time or in $O(\sqrt{n})$ time with $O(n\sqrt{n})$ preprocessing time.

Clarkson, Kapoor and Vaidya [CKV87] considered L_1 geodesics in the plane, but with disjoint simple polygonal obstacles. Their algorithm basically constructs a visibility graph. By introducing Steiner points, they reduced the number of edges in visibility graph to $O(n \log n)$. By applying Dijkstra's algorithm, a geodesic can be computed in $O(n \log^2 n)$ time. Mitchell [Mit92] obtained a similar result by constructing an SPM with respect to s. The SPM can be constructed in $O(n \log^2 n)$ time and a geodesic can be computed in $O(\log n)$ time.

The most basic geodesic problem is to find a Euclidean (L_2) geodesic in the plane. An algorithm constructing VG in this case requires $\Omega(n^2)$ time, since VG may have $\Omega(n^2)$ edges. This bound is achieved in the algorithms of Welzl [Wel85] and Asano et al. [AAG+86]. Ghosh and Mount [GM91] gave an output-sensitive algorithm which takes $O(k + n \log n)$ time, where k is the number of edges in VG. Hershberger and Suri [HS93] obtained an $O(n \log^2 n)$ algorithm which was subsequently improved to the optimal bound of $O(n \log n)$.

2 Preliminary

The *coordinate axes* in $\mathbb{R}^d$ are labeled by $X_1, \ldots, X_d$. The $2d$ *coordinate directions* are denoted by $\pm X_1, \ldots, \pm X_d$. If ξ is an equation or inequality involving $X_1, \ldots, X_d$, then $[\xi]$ denotes the set of points in $\mathbb{R}^d$ that satisfies ξ. Thus, if ξ is $X = 0$ then $[X = 0]$ is the hyperplane normal to the X-axis and containing the origin of the coordinate system. If ϕ is a coordinate direction, then the ϕ-*semispace* is closed half-space $[\phi \geq 0]$. Thus $\mathbb{R}^d$ is the union of the ϕ-semispace and the $(-\phi)$-semispace.

Most of the time, we consider $d = 3$ in which case the coordinate axes are labeled X, Y, Z. The coordinates of a point $p \in \mathbb{R}^3$ are denoted $X(p), Y(p)$ and $Z(p)$. If u, v are two points, we say u *dominates* v, written $v \prec u$, if each coordinate of u is at least as large as the corresponding coordinate of v.

A closed interval $I \subseteq \mathbb{R}$ is denoted $[a : b]$ where $a \leq b$. A d-*box* B is a product of closed intervals $I_1 \times \cdots \times I_d \subseteq \mathbb{R}^d$. We usually call 2-boxes *rectangles*. Each of the $2d$ faces of a d-box has a unique inward normal direction that is a coordinate direction $\phi \in \{\pm X_1, \ldots, \pm X_d\}$. This face is then called a ϕ-*face*.

We consider only rectilinear paths in $\mathbb{R}^d$, simply calling them "paths". Such a path is specified by a sequence of points $\mu = (p_0, p_1, \ldots, p_k)$ where p_i and p_{i+1} differ in exactly one coordinate. A *segment* $[p_i, p_{i+1}]$ of the path when viewed as directed from p_i to p_{i+1} is denoted $\overrightarrow{p_i p_{i+1}}$. If ϕ is a coordinate direction, we say a path is ϕ-*monotone* if the $|\phi|$-coordinate of points along the directed path is monotone non-decreasing. A *monotone path* is one that is ϕ-monotone for any three orthogonal coordinate directions ϕ.

We write $\cup \mathcal{B}$ for the union of the boxes in $\mathcal{B}$. The closure of the set $\mathbb{R}^d \setminus (\cup \mathcal{B})$ is called the *free space* $\text{FP}(\mathcal{B})$ of $\mathcal{B}$. Points in $\text{FP}(\mathcal{B})$ are said to be $\mathcal{B}$-*free* or simply *free* if $\mathcal{B}$ is understood. For simplicity, we assume the boxes in $\mathcal{B}$ are non-degenerate. In this case, an $\mathcal{B}$-*path* can be simply defined to be one path that lies in $FP(\mathcal{B})$.

Without loss of generality, we assume that s is the origin of the coordinate axes, and for simplicity assume that no two faces of boxes in $\mathcal{B}$ are coplanar.

With the standard topology of $\mathbb{R}^d$, we speak of the closure $\text{Closure}(S)$, interior $\text{Int}(S)$, relative interior $\text{RelInt}(S)$ and boundary ∂S of a set $S \subseteq \mathbb{R}^d$ (See [Ede87, page 441]). For any coordinate X_i and any set $S \subseteq \mathbb{R}^d$, the X_i-*projection* of S is just the set $S' \subseteq \mathbb{R}^{d-1}$ obtained pointwise by omitting the X_i-coordinate of points in S.

3 Pyramids

The key to this paper is the pyramid structure. We will try to motivate the definitions. Then we outline the structure of our proof of theorem 1.

For each coordinate direction ϕ, we will define below the ϕ-*pyramid* of $\mathcal{B}$ (relative to s). We denote this by $\Delta_\phi = \Delta_\phi(\mathcal{B}) = \Delta_\phi(\mathcal{B}; s)$ where $\mathcal{B}, s$ are omitted when they are understood. Intuitively, the ϕ-pyramid is the region of $\mathbb{R}^3$ that cannot be reached by a monotone path but can be reached by a ϕ-monotone path.

2-D Pyramids It is best to first recall the 2-dimensional pyramid used by de Rezende et al. [dRLW89]. First, we call a path an ϕ-*path* if it is a monotone path $(p_0, p_1, \ldots, p_k)$ (possibly $p_k = \infty$) where $p_0 = s$, the source, and $\overrightarrow{p_0 p_1}$ is a maximal segment directed in the ϕ-direction. This segment is maximal in the sense that p_1 either touches some obstacle or goes to ∞ (then $k = 1$). If it touches some box B, then $\overrightarrow{p_1 p_2}$ has a direction orthogonal to ϕ and p_2 is the corner of B. The next segment $\overrightarrow{p_2 p_3}$ is again directed as ϕ and is maximal. This pattern is repeated alternately. It is easily seen that $[p_{2i-1}, p_{2i}]$ (for all $i \geq 1$) must consistently be directed along some coordinate direction ϕ'. If $\phi = +X$ then ϕ' is either $-Y$ or $+Y$. We may call the $(+X)$-path an *East-North staircase* if $\phi' = +Y$, and an *East-South staircase* if $\phi' = -Y$. The $(+X)$-*pyramid* is basically the rectilinear convex region bounded by the East-North and the East-South staircases. Similarly, we can define other 3 pyramids using appropriate staircases. See figure 2 for the pyramids (shaded region) of some set of boxes.

Thus de Rezende et al. defined pyramids by an explicit construction (the staircases). Generalizing this to higher dimensions is hardly obvious (in higher dimensions, some new phenomenon arises). What we seek is an *intrinsic* mathematical definition. That is, we should ask for properties that characterize ϕ-pyramids.

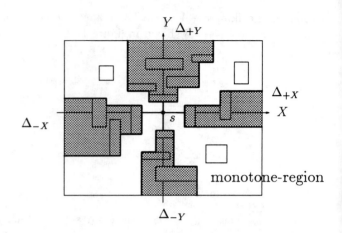

Figure 2: The four pyramids of $\mathcal{B}$ in the plane.

Definition of a pyramid Proceeding now to $\mathbb{R}^d$, for a coordinate direction ϕ, we let a ϕ-*path* μ denotes a monotone $\mathcal{B}$-path $(p_0, p_1, \ldots, p_k)$ from $s = p_0$ such that

1. if $\overrightarrow{p_{i-1} p_i}$ has direction ϕ then $i = k$, or p_i is in the relative interior of a ϕ-face of a box, or p_{i-1} is in the relative interior of a $(-\phi)$-face of a box;

2. if $\overrightarrow{p_{i-1} p_i}$ has direction orthogonal to ϕ then $[p_{i-1}, p_i]$ lies on a ϕ- or $(-\phi)$-face of a box.

This definition generalizes the staircase. A path $\pi = (p_0, \ldots, p_k)$ *belongs* to ϕ if it is non-monotone and the maximal monotone prefix $(p_0, \ldots, p_i)$ $(i < k)$ is a ϕ-path.

Now for the key definition: the *free ϕ-pyramid* of $\mathcal{B}$ (relative to s) is

$$\Delta_\phi^0(\mathcal{B}; s) = \\ \{t \in FP(\mathcal{B}) \mid \exists \mathcal{B}\text{-geodesic from } s \text{ to } t \text{ belonging to } \phi\}.$$

By an *extended ϕ-path* we mean a path $(p_0, \ldots, p_k, p_{k+1})$ such that $(p_0, \ldots, p_k)$ is a ϕ-path and $\overrightarrow{p_k p_{k+1}}$ has direction ϕ. Note that while a ϕ-path must avoid obstacles, we allow the last segment of an extended ϕ-path to penetrate obstacles. Finally, the ϕ-*pyramid* of $\mathcal{B}$ (relative to s) is

$$\Delta_\phi(\mathcal{B}; s) = \phi\text{-semiaxis} \cup \text{Closure} (\{t \in \mathbb{R}^3 \mid \\ t \text{ lies on some extended } \phi\text{-path to } t' \in \Delta_\phi^0(\mathcal{B})\}).$$

Some subtleties: we first define the "free part" of the ϕ-pyramid because it can be cleanly defined. On the other hand, this set can have rather complicated shape (it can be full of holes caused by the presence of boxes). The "non-free part" of the ϕ-pyramid are then added so as to create a fairly nice shape (as we will see).

EXAMPLE 2: Consider the $(+Z)$-pyramid of just one box B in $\mathbb{R}^3$. If the relative interior of $(+Z)$-face of B does not intersect $(+Z)$-semiaxis then the $(+Z)$-pyramid is just the $(+Z)$-semiaxis. Otherwise it is union of $(+Z)$-semiaxis and the semi-infinite box whose $(+Z)$-face is the $(+Z)$-face of B and extends to infinity in the $(+Z)$-direction.

EXAMPLE 3: To see a more substantial example of a pyramid, see figure 4: the obstacles B are shown on the left side (a), and the corresponding $(+Z)$-pyramid in $(+X)$-semispace (denoted Δ_{+Z}^{+X}) is drawn on the right side (b).

EXAMPLE 4: Figure 4 also shows why the 3-dimensional pyramid is not fully anticipated from our knowledge about 2-dimensional pyramids: it is not true that every point on the boundary of the pyramid is free. That is, some boxes (called "shelf boxes") may protrude through the boundary. One such "shelf box" is shown in dashed outline in figure 4 (b). See figure 6 for more types of shelf boxes.

The following is the main characterization theorem.

Theorem 3 *Let Δ_ϕ be the pyramid of B relative to s for $\phi \in \{\pm X, \pm Y, \pm Z\}$ in $\mathbb{R}^3$.*
(i) There is a monotone path from s to any free point on the boundary of pyramids;
(ii) The six pyramids are disjoint except at s;
(iii) For every free point t lying outside the six pyramids, there is a monotone path from s to t;
(iv) Every geodesic from s to a free point $t \in \text{Int}(\Delta_\phi)$ is a ϕ-monotone path but not a monotone path.

Note that parts (iii) and (iv) implies a result stronger than the monotonicity theorem (theorem 1): it says that for every free point t, there exists a ϕ such that every geodesic from s to t must be ϕ-monotone.

In view of parts (i) and (iii), we define the *monotone region* to be the region outside the interior of all the 6 pyramids.

Outline of proof for theorem 3 (i), (ii), (iii) As usual, B and s are fixed. In order to study the properties of a $(+Z)$-pyramid, we define two different ways to sweep it, although we do not initially know if what we sweep is exactly the $(+Z)$-pyramid. The line of argument proceeds as follows.

1. First we define a *frontal sweep* procedure for the $(+Z)$-pyramid. We have a horizontal plane $[Z = z_0]$ sweeping from $z_0 = 0$ to $z_0 = \infty$. The *events* are the $(+Z)$-faces of the boxes in B, sorted by increasing Z-coordinates. We maintain a *frontal window* $W_{z_0} \subseteq \mathbb{R}^2$ which is a two dimensional set. Intuitively, this window is the Z-projection of the intersection of the $(+Z)$-pyramid with the plane $[Z = z_0]$. Initially, W_{z_0} is a degenerate rectangle $[0:0] \times [0:0]$. The window is monotone non-decreasing: $z < z' \Rightarrow W_z \subseteq W_{z'}$.

2. Consider the set

$$\Delta_{+Z}^{+Z} := \cup_{z_0 \geq 0}(W_{z_0} \times z_0). \quad (1)$$

The superscript of $(+Z)$ indicates that we are sweeping in the $(+Z)$-direction. We initially prove that

$$\Delta_{+Z}^{+Z} \subseteq \Delta_{+Z} \quad (2)$$

although we will eventually prove that they are equal. Furthermore we show that for any free point $t \in \partial(\Delta_{+Z}^{+Z})$ there is a $(+Z)$-path from s, which shows part (i) of theorem 3.

3. Next, we define a *lateral sweep* procedure for the $(+Z)$-pyramid. Actually, there are 4 ways to do this, depending on the choice of $\phi \in \{+X, -X, +Y, -Y\}$. By symmetry, assume $\phi = +X$. We then sweep the plane $[X = x_0]$ from $x_0 = 0$ to $x_0 = +\infty$, while maintaining a *lateral window* $\delta_{x_0} \subseteq \mathbb{R}^2$. The *events* are the $(+X)$-faces and $(-X)$-faces of boxes in B, sorted by increasing X-coordinates.

These lateral windows turn out to be 2-dimensional pyramids of a set of rectangles that can be defined inductively. Unlike frontal sweeping, the initial lateral window δ_{x_0} is monotone non-increasing with x_0, eventually becoming empty.

4. Consider the set

$$\Delta_{+Z}^{+X} := \cup_{x_0 \geq 0}(x_0 \times \delta_{x_0}). \quad (3)$$

The superscript of $(+X)$ indicates that we are sweeping in the $(+X)$-direction. We prove that

$$\Delta_{+Z}^{+Z} = \Delta_{+Z}^{+X} \cup \Delta_{+Z}^{-X}.$$

5. Now perform another sweep in the $+Z$-direction, except that we now simultaneously sweep all pyramids. That is, the sweeping plane $[Z = z_0]$ will *simultaneously* sweep the set Δ_{+Z}^{+Z} frontally, and the sets

$$\Delta_{+X}^{+Z}, \quad \Delta_{-X}^{+Z}, \quad \Delta_{+Y}^{+Z}, \quad \Delta_{-Y}^{+Z} \quad (4)$$

laterally. In this simultaneous sweeping step, we show parts (ii) and (iii) of theorem 3.

6. Using the disjointness of the pyramids, we can indeed prove that no point can belong to more than one pyramid. This immediately implies equality in (2).

4 Frontal and Lateral Sweeping

We give more details of the sweeping of the $(+Z)$-pyramid.

First consider the frontal sweep which constructs a set Δ_{+Z}^{+Z}. Recall that we use a horizontal plane $H_z : [Z = z]$ to sweep the $(+Z)$-semispace. Initially, $W_z = [0:0] \times [0:0]$. Intuitively, W_z is the Z-projection of $H_z \cap \Delta_{+Z}$. The *events* are those z such that the plane H_z contains a $(+Z)$-face R of some box of B; we say R *belongs to* z. Note that the $(-Z)$-faces of boxes do not define events. At an event z, we update W using the face R belonging

to z. There are 2 questions:
1) What does W look like?
2) How do we update W at an event?

The answer to 1) is that W_z is a "rectilinear convex polygon" that is bounded by four stair cases. Furthermore it contains the origin of the coordinate system and the relative interior of its 4 *extreme faces*, corresponding to the 4 coordinate directions, intersects with the four semiaxes.

We now answer question 2). Suppose that we encounter a new horizontal face R. By symmetry, let us focus on the restriction of W to the first quadrant $Q = [X > 0] \cap [Y > 0]$. Let q_a and q_b be (respectively) the north-east and south-west corners of $R \cap Q$. We have several cases:

- If q_b is the origin, then let $W \cap Q \leftarrow (W \cup R) \cap Q$. See figure 3 (a).
- Else, if q_b lies on the Y-axis and southern edge of R has a non-empty intersection I_s with the interior of $W \cap Q$. Then $R_s := R \cap (\text{Closure}(I_s) \times \mathbb{R})$ and $W \cap Q \leftarrow (W \cup R_s) \cap Q$. See figure 3 (b).
- Else, if q_b lies on the X-axis and the western edge of R has a non-empty intersection I_w with the interior of $W \cap Q$. This is treated in analogy with the previous case: $R_w := R \cap (\mathbb{R} \times \text{Closure}(I_w))$ and $W \cap Q \leftarrow (W \cup R_w) \cap Q$. See figure 3 (c).
- Else, if q_b lies in the interior of $W \cap Q$. Then with R_s, R_w as defined above, and $R_{sw} := R_s \cap R_w$, $W \cap Q \leftarrow (W \cup R_{sw}) \cap Q$. See figure 3 (d).

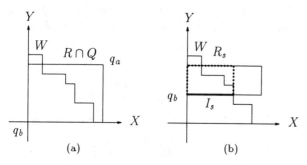
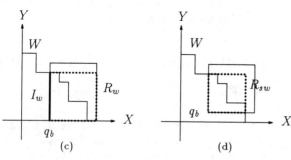

Figure 3:

If q_b lies outside W then none of the 4 cases applies and $W \cap Q$ is unchanged. It should be noted that we cannot simply view (b) and (c) as special cases of (d).

Finally we define the pyramid Δ_{+Z}^{+Z} as in (1).

Lateral sweeping Assume we wish to sweep the $(+Z)$-semispace with a plane $H_x : [X = x]$ in $(+X)$-direction to construct some region Δ_{+Z}^{+X} of $\mathcal{B}$ relative to s. This is illustrated in figure 4.

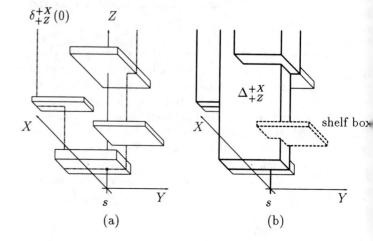

Figure 4: A $(+Z)$-pyramid restricted to the $(+X)$-semispace

The region Δ_{+Z}^{+X} is constructed in the following way: sweep the $(+Z)$-semispace with a sweeping plane H_x with increasing x. The event points are those x to which $(+X)$- or $(-X)$-faces of boxes belong. Let $x_0, \ldots, x_m$ be the sequence of event points in the increasing order, where $x_0 = 0$. At each event point x_i, we compute a set $\mathcal{R}(x_i)$ of rectangles in the sweeping plane. In the sweeping plane H_{x_i} we construct the two-dimensional pyramid $\delta_{+Z}^{+X}(x_i)$ of $\mathcal{R}(x_i)$ relative to the origin of the sweeping plane[1]. At the initial event point $x_0 = 0$, $\mathcal{R}(0)$ comprises the X-projection of $B \cap H_0$ where $B \in \mathcal{B}$. At each event point x_i, $i > 0$, a box may start or end:

- Suppose a new box B starts at x_i. Let R be the X-projection of B. Then $\mathcal{R}(x_i) = \mathcal{R}(x_{i-1}) \cup \{R'\}$, where $R' = R \cap \delta_{+Z}^{+X}(x_{i-1})$. Call R' the *contribution* of B. See figure 5.
- Suppose a box B ends at x_i. Then $\mathcal{R}(x_i) = \mathcal{R}(x_{i-1}) - \{R'\}$, where R' is the contribution of B.

Finally we define the pyramid Δ_{+Z}^{+X} as in (3).

Shelf Boxes Basically, these are boxes that protrude out of the boundary of a pyramid. During the ϕ-sweep (frontal or lateral) of a pyramid, the *shelf boxes* are precisely those boxes whose $|\phi|$-projection are truncated for

---
[1] Here 2-dimensional pyramid $\delta_{+Z}^{+X}(x_i)$ does not have a line segment incident to the origin of sweeping plane, except at $x_i = 0$. $\delta_{+Z}^{+X}(x_i)$ is called δ_{x_i} in the previous informal description.

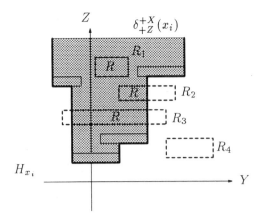

Figure 5: Each R represents the contribution of the respective box whose X-projection is R_1, R_2, R_3. (R_4 has no contribution.)

the purposes of defining our windows (frontal or side). For example, shelf boxes B_1, B_2, B_3 in figure 6 are created by the cases (d), (c), (b) in figure 3, and are called a shelf box of type XY, Y, X, respectively.

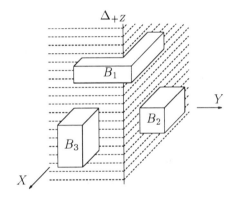

Figure 6: 3 different types of shelf boxes.

5 Monotonicity Property of Shortest Paths

For a direction $(+Z)$, we define a $(+Z)$-*region* of $\mathcal{B}$ relative to s as follows:

$(+Z)$-region := Closure $((+Z)$-semispace $-$
$\{\Delta_{+X} \cup \Delta_{-X} \cup \Delta_{+Y} \cup \Delta_{-Y}\})$.

The boundary of the $(+Z)$-region consists of the following subsets:

1. Boundary of Δ_ϕ in $(+Z)$-semispace, for $\phi \in \{\pm X, \pm Y\}$;
2. A subset of the plane $H = [Z = 0]$ excluding $\text{Int}(\Delta_\phi) \cap H$, for $\phi \in \{\pm X, \pm Y\}$. This subset is the same as the monotone-region in H for the four

2-dimensional pyramids $\delta_{\pm X}, \delta_{\pm Y}$ of the rectangles $\{H \cap B \mid B \in \mathcal{B}\}$ relative to the origin of H.

We will show that all geodesics from s to any free point in $(+Z)$-region are $(+Z)$-monotone. Note that from the definition of pyramids, no geodesic from s to any free point lying inside a pyramid is monotone. For any point t lying inside $\Delta_{+Z}(s)$, let $\Delta_{-Z}(t)$ be a $(-Z)$-pyramid of $\mathcal{B}$ relative to t. Then we have

Lemma 4 *For any point $t \in \text{FP}(\mathcal{B})$, $t \in \text{Int}(\Delta_{+Z}(s))$ if and only if $s \in \text{Int}(\Delta_{-Z}(t))$.*

Proof: Remember that s is the origin of the coordinate system. By symmetry of the claim, it is sufficient to show that if $t \in \text{Int}(\Delta_{+Z}(s))$ then $s \in \text{Int}(\Delta_{-Z}(t))$. Without loss of generality, we assume that t lies in the first orthant of the coordinate system. Since there is no monotone path between s and t, s must lie inside pyramids $\Delta_\phi(t), \phi \in \{\pm X, \pm Y, \pm Z\}$. From the definition of pyramids, $s \notin \Delta_{+X}(t) \cup \Delta_{+Y}(t) \cup \Delta_{+Y}(t)$. Let us assume that $s \notin \Delta_{-Z}(t)$. Then $s \in \Delta_{-X}(t)$ or $s \in \Delta_{-Y}(t)$. Suppose $s \in \Delta_{-X}(t)$. Let $H = [Y = Y(t)]$ and $s_H = (X(s), Y(t), Z(s))$ be the point on H that has the same X-, Z-coordinates as s. Let $\delta(t) = \Delta_{-X}(t) \cap H, \delta(s) = \Delta_{+Z}(s) \cap H$. Then $s_H \in \text{Int}(\delta(t))$ and $t \in \text{Int}(\delta(s))$. Then there is a point $q \in \partial(\delta(s)) \cap \partial(\delta(t))$ such that $s \prec q$ and $q \prec t$. Note that $q \in \text{FP}(\mathcal{B})$ since $\partial(\delta(t)) \subseteq \text{FP}(\mathcal{B})$. Thus there is a monotone path $\mu(s, q)$ from s to q and a monotone path $\mu(q, t)$ from q to t. Therefore concatenating $\mu(s, q)$ and $\mu(q, t)$ gives a monotone path from s to t. This is a contradiction. The same argument applies to the case $s \in \Delta_{-Y}(t)$. Therefore $s \in \text{Int}(\Delta_{-Z}(t))$. $\square$

Lemma 5 *For any point $t \in (+Z)$-region, every geodesic from s to t is a $(+Z)$-monotone path.*

Proof: We prove by induction on the number k of edges of a geodesic. Let μ be a geodesic from s to t.
(1) If $k = 1$ or $k = 2$, then μ is certainly a $(+Z)$-monotone path.
(2) Suppose the claim is true for all paths with less than k edges.
(3) Suppose that μ has k edges. Let $\overrightarrow{qt}$ be the last edge of μ. It is easy to show that every geodesic from s to t must lie in $(+Z)$-region. Therefore if $Z(q) \leq Z(t)$ then μ is a $(+Z)$-monotone path by induction hypothesis. Let us assume that $Z(q) > Z(t)$. For the point t, we construct the set of pyramids $\Delta_\phi(t)$, $\phi \in \{\pm X, \pm Y, \pm Z\}$, of $\mathcal{B}$ relative to t. Let S be the $(-Z)$-region relative to t. Then q must not be in this region since $Z(q) > Z(t)$. Furthermore s must be in this region since (i) if t is in the monotone region relative to s then s is in the monotone region relative to t, and $Z(s) \leq Z(t)$; (ii) if $t \in \text{Int}(\Delta_{+Z}(s))$ then $s \in \text{Int}(\Delta_{-Z}(t))$ by lemma 4. Therefore the path μ from s to t must intersect with the

boundary of S at some point p. Then two cases occur according to the position of p in the boundary of S:

Case (1). If p is on the the plane $H = [Z = Z(t)]$, then p lies in the monotone region relative to t. Thus by part (iii) of theorem 3, there is a monotone path μ_1 on this plane from t to p. Therefore μ_1 is shorter than the sub-path $\mu' = (t, q, \ldots, p)$ of μ since μ' is not monotone.

Case (2). If p is on the boundary of one of the pyramids then by part (i) of theorem 3, there is a monotone path μ_1 from t to p. Therefore μ_1 is shorter than the sub-path $\mu' = (t, q, \ldots, p)$ of μ since μ' is not monotone.

Thus μ cannot be a geodesic from s to t. Therefore $Z(q) \leq Z(t)$ and a geodesic must be $(+Z)$-monotone. $\square$

The part (iv) of theorem 3 follows from this lemma.

6 Query Processing

A face of a pyramid with inward normal direction $\phi \in \{\pm X, \pm Y, \pm Z\}$ is called a ϕ-face. For each pyramid Δ_ϕ, a box $B \in \mathcal{B}$ is called a *supporting* box of Δ_ϕ if the ϕ-face of B intersects a ϕ-face of Δ_ϕ and a box $B \in \mathcal{B}$ is called the *main supporting* box of Δ_ϕ if B is the supporting box of Δ_ϕ closest from s in ϕ-direction.

Constructing pyramids Let us consider the algorithm constructing the pyramid $\Delta_{+Z}(\mathcal{B}; s) \cap [X \geq 0]$ by the lateral sweep along $(+X)$-semiaxis. In the algorithm constructing the pyramid, there are at most $O(n)$ event points. At each event point x_i, we compute a set of rectangles $\mathcal{R}(x_i)$ and construct $\delta_{+Z}^{+X}(x_i)$ of $\mathcal{R}(x_i)$ relative to the origin of the sweeping plane H_{x_i}. It takes $O(n \log n)$ time. Merging two consecutive pyramids $\delta_{+Z}^{+X}(x_i)$ and $\delta_{+Z}^{+X}(x_{i-1})$ to make a $(+X)$-face of Δ_{+Z} on the plane $H_{x_{i-1}}$ and $(+Z)$-, $(+Y)$-, $(+Y)$-face between H_{x_i} and $H_{x_{i-1}}$ takes $O(n)$ time. Similarly we can construct Δ_{+Z}^{-X} and other pyramids in total $O(n^2 \log n)$ time.

After constructing the pyramids, we can classify each box $B \in \mathcal{B}$ as being a shelf box or a box contained in a monotone-region or contained in pyramids. If B is a shelf box, it is divided into subsets that lie inside or outside of pyramids.

Query Processing The main tool for a query processing is the following ray shooting algorithm:

Lemma 6 ([dBvKNO92]) *Given a set of n axis-parallel boxes in $\mathbb{R}^d$, the box hit by any axes-parallel query ray can be computed in $O(\log n)$ time with $O(n^{d-1} \log n)$ preprocessing time.*

Without loss of generality, we assume that a query is t is in the first orthant of the coordinate system. Since pyramids are disjoint by theorem 3, t may lie in $\Delta_{+X}, \Delta_{+Y}, \Delta_{+Z}$ or in the monotone-region. First, we check whether t lies inside Δ_{+X}, Δ_{+Y} or Δ_{+Z} in the following way: shoot $(-X)$-, $(-Y)$- and $(-Z)$-directional rays from t. The following is a characterization of t lying in $\Delta_\phi, \phi \in \{+X, +Y, +Z\}$: let B_ϕ be the box hit by $(-\phi)$-directional ray. Then B_ϕ lies inside Δ_ϕ, or B_ϕ is a shelf box of Δ_ϕ and the hitting point lies inside the pyramid.

If t lies in a monotone-region, shoot a $(-X)$-directional ray ρ in the environment containing $\mathcal{B}$ and $\Delta_\phi, \phi \in \{+X, +Y, +Z\}$, and a plane $H_0 = [X = 0]$. The ray ρ may hit a box in $\mathcal{B}$, a $(-X)$-face of pyramids, or H_0. If ρ hits some face of pyramids, then we can identify the face in the following way: project $(-X)$-faces of Δ_{+Y}, Δ_{+Z} and t to H_0. The projected images define a subdivision of H_0, and they consist of $O(n)$ edges. By the point location algorithm in the plane, we can find the region f of the subdivision containing the projected image of t in $O(\log n)$ time. The original image of f may be H_0 or a $(-X)$-face of Δ_{+Y}, Δ_{+Z}.

We have the following cases.

Case(1). ρ hits H_0: Then the hitting point lies in the monotone region of rectangles $\mathcal{R} := \{B \cap H_0 \mid B \in \mathcal{B}\}$ relative to the origin of H_0. Therefore we need to preprocess $\mathcal{R}$ on H_0, which is a 2-dimensional case.

Case(2). ρ hits a $(-X)$-face of Δ_{+Y} or Δ_{+Z}; We have to preprocess faces of pyramids so that we can find a path from the hitting point to s.

Case(3). ρ hits a $(-X)$-face of a shelf box of Δ_{+Y}, Δ_{+Z} or Δ_{+X}.

Case(4). ρ hits $(-X)$-face of some box in monotone-region.

Next we show how we preprocess to deal with each case. In the preprocessing step, we will construct a shortest path tree SPT(s) (or, briefly SPT) rooted at s with the following properties:
(i) Vertices of SPT represent points lying on edges of boxes;
(ii) An edge (u, v) of SPT represents a "box-path" between u and v. By a *box-path* we mean there is at most one edge parallel to each of the 3 axes.
In the query processing, we find a vertex v of SPT such that a geodesic path from a query point to s passes through the unique sequence of vertices from v to s in SPT.

Let $d_\mathcal{B}(p, q) = d(p, q)$ denote the $\mathcal{B}$-geodesic length from p to q.

Shortest paths from points lying inside pyramids We consider the pyramid Δ_{+Z}. Let $\mathcal{B}(\Delta_{+Z})$ be a set of boxes and subsets of shelf boxes lying inside Δ_{+Z}. Remember that the subset of each shelf box is a box.

If $t \in \text{Int}(\Delta_{+Z}(s))$, then $s \in \text{Int}(\Delta_{-Z}(t))$ by lemma 4 and there is a geodesic $\mu(t,s)$ from t to s belonging to $(-Z)$. Thus $\mu(t,s)$ passes through the boundary of the main supporting box B_t of $\Delta_{-Z}(t)$. Note that B_t is the box hit by a $(-Z)$-directional ray from t. Therefore if we know a geodesic to every point on edges of B_t, then we can get a geodesic from t to s using lemma 6. As we will see later, for a $(-Z)$-face f of a box in $\mathcal{B}(\Delta_{+Z})$, there is a finite number of points on edges of f through which a geodesic from $t \in f$ to s passes. The set of such points is denoted by $C(f)$ and is called *control points* of f. In preprocessing step, we compute a geodesic to every point in $C(f)$.

Therefore if we know the geodesic length to every control point in $C(f)$ of a $(-Z)$-face f of a box in $\mathcal{B}(\Delta_{+Z})$, then the geodesic length $d(t,s)$ from a point $t \in f$ is

$$d(t,s) = \min_{c \in C(f)} \{d(c,s) + \|c-t\|_1\}. \qquad (5)$$

Let us consider the L_1-Voronoi diagram [Lee80] in f of control points $C(f)$ with additive weights $d(c,s), c \in C(f)$. Equation (5) is equal to find the weighted Voronoi vertex $c \in C(f)$ such that t lies in the Voronoi region of c. If t lies in the Voronoi region of c, then we say that c controls t. For example in figure 7 (a), b_k controls t

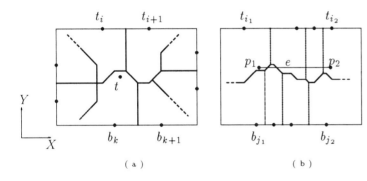

Figure 7: A set of control points that control a point t or an edge e.

To find the point in $C(f)$ that controls t, note that if we project t normally onto an edge of f, it is sufficient to only consider the two nearest control points on that edge. Repeating this for each edge of f, we only need to consider 8 candidate control points. The correct choice among the 8 can be decided in constant time. The projections to each edge takes $O(\log |C(f)|)$ time if we store the control points on an edge appropriately. Thus we do not need to construct explicitly the Voronoi diagram of $C(f)$.

Given a line segment $e = [p_1, p_2] \subset f$ which is parallel to the top edge of f, we want to compute a set $C(e) \subset e$ of control points of e. For example, see figure 7 (b). In

this abstract, we omit details of this computation. It is relatively straight forward: we project the "candidate" control points from each edge of f onto e. Using another pass, we can remove the "redundant candidates". What remains are the control points we seek for e.

Now let us consider how the control points of $(-Z)$-, $(+Z)$-faces of boxes in $\mathcal{B}(\Delta_{+Z})$ are computed. Let $f_1, f_2, \ldots$ be a sequence of $(-Z)$-, $(+Z)$-faces of boxes in $\mathcal{B}(\Delta_{+Z})$ sorted in their Z-coordinates. We will compute the control points of each face by incrementally inserting faces.

1. Initially f_1 is the $(+Z)$-face of the main supporting box of Δ_{+Z}. The control points of f_1 are the four points that are intersection points of $\partial(f_1)$ and the plane $[X = 0]$ and the plane $[Y = 0]$. For each $c \in C(f_1)$, it is connected to s and an edge (c,s) of SPT represent a box-path from c to s. Trivially $d(c,s) = \|c-s\|_1$.

2. After computing the control points for a $(+Z)$-face of a box B we compute the control points for the $(-Z)$-face of B in the following way: for every control point c of the $(+Z)$-face, create a corresponding control point in the boundary of the $(-Z)$-face that has the same X-, Y-coordinates as c. We insert the set of control points of $(+Z)$-, $(-Z)$-face of each box in $\mathcal{B}(\Delta_{+Z})$ to a SPT and each control point c in $(-Z)$-face is connected to its corresponding control point c' in $(+Z)$-face, and $d(c,s) = d(c',s) + \|c-c'\|_1$.

3. Let f_i be the next face to be inserted. If f_i is a $(-X)$-face of some box then we have already created control points of f_i. Otherwise each edge of f_i is divided into a set S of line segments so that each line segment has different bottom face $f_j, 1 \leq j < i$, or has no bottom face. If a line segment $e' \in S$ has such a bottom face $f_k, k < i$ then the control points of e' computed in the following way: let e'' be the projected image of e' on f_k. We compute control points of e'' from $C(f_k)$ and compute control point of e' from $C(e')$. If a line segment has no bottom face, then they are on the boundary of Δ_{+X} and it will be handled in a different way as we will see.

Computing the subdivision of edges of faces can be done by the hidden surface removal algorithm in $O(n^2)$ time [McK87]. Now consider that how many control points lies in each edge of faces. From the rule of inheriting control points, the set of control points is a subset of the intersection points of planes that are affine extension of faces of boxes. For each edge of faces, there are at most $O(n)$ such point. Therefore for each edge of face f, it takes $O(n \log n)$ time to inherit control points from faces below f.

Lemma 7 *For a pyramid Δ_{+Z}, we can compute all control points of faces of boxes lying inside Δ_{+Z} or shelf-*

boxes of Δ_{+Z} in $O(n^2 \log n)$ time.

Shortest paths from points on the boundary of pyramids From theorem 3, there is a monotone path from s to any point on the boundary of pyramids. Here we construct a SPT for points on the boundary of pyramids. Let us consider $(-X)$-faces of pyramid Δ_{+Z} in the first orthant. Let f_i be a $(-X)$-face restricted to the first quadrant swept by the lateral sweeping plane $H_{x_i} = [X = x_i]$. In the sweeping plane, the X-projection of f_i is bounded by the boundary of $\delta_{+Z}^{+X}(x_{i-1})$ and by the boundary of $\delta_{+Z}^{+X}(x_i)$ and possibly by the Z-axis. In Figure 8, the boundary of Δ_{+Z} is projected to the plane $[X = 0]$. In the figure, each unfilled rectangle represents $(-X)$-face of shelf box of type X, hashed rectangle represents $(-X)$-face of shelf boxes of type Y and each double-hashed rectangle represents $(-X)$-face of shelf boxes of type XY. The *base point* of f_i is defined as the point in f_i that has the least Y- and Z-coordinates and it is denoted by $b(f_i)$.

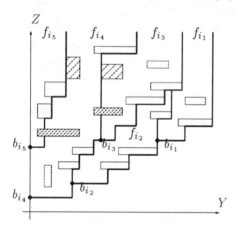

Figure 8: Faces of Δ_{+Z} are projected to the plane $[X = 0]$ and $b_i = b(f_i)$ represents their base points.

After constructing pyramids Δ_{+Z}^{+X} by lateral sweeping along X-axis, we construct a SPT for a set of points on the boundary of the pyramid and in the plane $H_0 = [X = 0]$ in the following way:

1. In plane H_0, let $\mathcal{R} = \{B \cap H_0 \mid B \in \mathcal{B}\}$. We construct four 2-dimensional pyramids of $\mathcal{R}$ relative to s in H_0 and construct $SPT(s)$ for the points in the monotone-region. This preprocessing takes $O(n \log n)$ time.

2. Let f_i be any $(-X)$-face of Δ_{+Z} in the plane H_i. We want to compute a geodesic from any point in f_i to $b(f_i)$ avoiding boxes. For f_i, let $\mathcal{B}(f_i)$ be the set of boxes intersecting f_i and let $\mathcal{R}(f_i) = \{B \cap f_i \mid B \in \mathcal{B}(f_i)\}$. Then the face f_i is a subset of the monotone-region of $\mathcal{R}(f_i)$ relative to $b(f_i)$ in the plane $[X = x_i]$. Therefore we construct an SPT for f_i as in the planar case.

Now we connect the base point b_i to some vertex in SPT. Let p be a point on some face of $f_j, j < i$, such that $[b_i, p]$ lies on $(+Z)$ face of Δ_{+Z}. Then we find a vertex v of SPT constructed for f_j through which a geodesic from p to s passes. We connect b_i to v in SPT and $d(s, b_i) = d(s, v) + \|b_i - v\|_1$. Symmetrically, the above argument applies to $(-Y)$-faces, and the other orthants.

Symmetrically we can construct SPT for the plane $[Y = 0]$, the plane $[Z = 0]$ and the boundary of $\Delta_\phi, \phi = \{\pm X, \pm Y, -Z\}$. Therefore we can find a geodesic from any point lying in those subspace. Let us denote by SPT_1 the SPT constructed for the planes $[X = 0], [Y = 0], [Z = 0]$ and denote by SPT_2 the SPT constructed for the boundary of pyramids.

After constructing the SPT for points on the boundary of pyramids, we construct a SPT for points on the boundary of shelf box. We know that for any point t on the boundary of a shelf box B and outside of Δ_{+Z}, there is a monotone path from t to s. If B is a shelf box of type X, then there exists a point p with the least non-negative Y- and Z-coordinates in the intersection of B and $\partial(\Delta_{+Z})$. We add the point p, which lies some $(+X)$-face of Δ_{+Z}, to the SPT. Note that there is a monotone path μ_1 through the boundary of B from t to p. We can apply the same argument to a shelf box of type Y. For a point t on a shelf box of type XY, there is a monotone path from t to a point p_1 with the least non-negative Y- and Z-coordinates or a point p_2 with the least non-negative X- and Z-coordinates in the intersection of B and $\partial(\Delta_{+Z})$. We add the point p_1 (resp., p_2), which lies some $(+X)$-face (resp., $(+Y)$-face) of Δ_{+Z}, to SPT. Note that there is a monotone path μ_1 through the boundary of B from t to p_1 or p_2.

Lemma 8 *For a pyramid Δ_{+Z}, we can construct a SPT for a set of point on the boundary of Δ_{+Z} and on the boundary of shelf boxes in $O(n^2 \log n)$ time.*

Shortest paths from boxes in monotone-region For a $(-Z)$- or $(+Z)$-face f of a box, let its *minimum point* be the point in f that is closest to the origin. Let $t_1, t_2, \ldots$ be the sequence of minimum points of faces of boxes in monotone-region in the order of their Z-coordinates. In the order of minimum points in the sequence, we add each minimum point to SPT_1 or SPT_2 in the following way. If t_i is a minimum point of $(-Z)$-face of a box B, then it is connected to the minimum point of $(+Z)$-face of B. Otherwise shoot a $(-Z)$-directional ray from t_i. For each point t_i, we need the same case analysis as the query processing. For the case (1), (2) and (3) of query processing, find the vertex v of SPT_1 or SPT_2 through which a geodesic from t_i to s passes and connect v and t_i. The edge (t_i, v) represent the box-path from t to v. If the ray hits some box B in monotone-region, then its minimum point v of B is already inserted to SPT. Therefore we connect t_i to v. Note that $d(t,s) = \|t-s\|_1$.

Lemma 9 *For a set of minimum points of faces of boxes in monotone-region, we can construct the SPT for those points in total $O(n \log n)$ time.*

Theorem 2 follows from lemmas 7, 8 and 9.

7 Conclusion

In this paper we have shown the monotonicity property of rectilinear shortest paths between two points avoiding axis-parallel boxes in $\mathbb{R}^3$. This property has been applied to design an algorithm that finds the rectilinear shortest distance from a query point to the origin of the space in $O(\log n)$ time after $O(n^2 \log n)$ preprocessing time.

An obvious open problem is to prove the analogous monotonicity property of rectilinear shortest paths in $\mathbb{R}^d, d > 3$. One approach would be to generalize the concept of pyramids in this paper. Improving the preprocessing time of our algorithm and designing an algorithm that gives a shortest path between two arbitrary query points will also be interesting.

Another intriguing question is to see what substitutes for the monotonicity property can be found in more general settings.

Acknowledgments

The first author would like to thank J. Sellen and M. Teichman for helpful discussions and M. Bern for bringing his attention to [McK87].

References

[AAG+86] Ta. Asano, Te. Asano, L. J. Guibas, J. Hershberger, and H. Imai. Visibility of disjoint polygons. *Algorithmica*, 1:49–63, 1986.

[CKV87] K. L. Clarkson, S. Kapoor, and P. M. Vaidya. Rectilinear shortest paths through polygonal obstacles in $O(n(\log n)^2)$ time. In *Proc. 3rd Annu. ACM Sympos. Comput. Geom.*, pages 251–257, 1987.

[dBvKNO92] M. de Berg, M. van Kreveld, B. J. Nilsson, and M. H. Overmars. Shortest path queries in rectilinear worlds. *Internat. J. Comput. Geom. Appl.*, 2(3):287–309, 1992.

[dRLW89] P. J. de Rezende, D. T. Lee, and Y. F. Wu. Rectilinear shortest paths in the presence of rectangular barriers. *Discrete Comput. Geom.*, 4:41–53, 1989.

[Ede87] H. Edelsbrunner. *Algorithms in Combinatorial Geometry*, volume 10 of *EATCS Monographs on Theoretical Computer Science*. Springer-Verlag, Heidelberg, West Germany, 1987.

[EM94] H. ElGindy and P. Mitra. Orthogonal shortest route queries among axes parallel rectangular obstacles. *Internat. J. Comput. Geom. Appl.*, 4(1):3–24, 1994.

[GM91] S. K. Ghosh and D. M. Mount. An output-sensitive algorithm for computing visibility graphs. *SIAM J. Comput.*, 20:888–910, 1991.

[HS93] John Hershberger and Subhash Suri. Efficient computation of Euclidean shortest paths in the plane. In *Proc. 34th Annu. IEEE Sympos. Found. Comput. Sci. (FOCS 93)*, pages 508–517, 1993.

[Lee80] D. T. Lee. Two-dimensional Voronoi diagrams in the L_p-metric. *J. ACM*, 27:604–618, 1980.

[McK87] M. McKenna. Worst-case optimal hidden-surface removal. *ACM Trans. Graph.*, 6:19–28, 1987.

[Mit92] J. S. B. Mitchell. L_1 shortest paths among polygonal obstacles in the plane. *Algorithmica*, 8:55–88, 1992.

[Wel85] E. Welzl. Constructing the visibility graph for n line segments in $O(n^2)$ time. *Inform. Process. Lett.*, 20:167–171, 1985.

COMMUNICATIONS

Union of Spheres (UoS) Model for Volumetric Data

Vishwa Ranjan and Alain Fournier
Department of Computer Science, University of British Columbia

Introduction

A *stable* representation of an object means that the representation is unique, is independent of the sampling geometry, resolution, noise, and other small distortions in the data, and is instead linked to the shape of the object. Stable representations help characterize shapes for comparison or recognition; skeletal (or medial axis) and volume primitive models have been popular in vision for the same reason. Piecewise polyhedral representations, e.g., tetrahedra, and voxel representations, e.g., octrees, generally tend to be unstable.

We propose a representation for 3D objects based on the set union of overlapping sphere primitives. This union of spheres (UoS) model has some attractive properties for computer graphics, computational vision, and scientific visualization [1, 2]. This communication briefly explains how we can obtain a

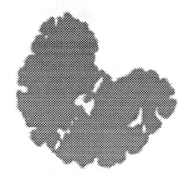

Figure 1: Brain from one slice of MRI head data represented as a union of circles.

stable UoS model for objects from *volumetric* data, i.e., data in which we can test if a given point is inside the defined object or outside. This constraint makes the problem of object reconstruction more manageable (Figure 1). In the absence of such inside-outside information, reconstructing an object is a tough problem [3]. For our method, the input is a discrete set of points on the boundary of an object defined in 3D discrete volumetric data such as a CT scan or MRI. The boundary points may be obtained by choosing a threshold (or isovalue) in the data or by other edge-detection methods.

The Algorithm

Our method has two steps [4]. In the first step, we generate a UoS representation which is truthful to the input volumetric data in the following sense: (a) all inside voxels are included in at least one sphere, (b) no outside voxel is within any sphere, (c) all boundary points are on the surface of at least one sphere, (d) no boundary point is inside any sphere. In the second step, we simplify this representation to make it stable and to reduce the number of primitives. This allows faster display and facilitates comparison of objects by reducing them to nearly the same number of primitives.

Step I: The dense UoS representation is obtained using the Delaunay tetrahedralization (DT) as follows:

1. Calculate the DT of the boundary points.

2. Compute the circumscribing sphere to each tetrahedron.

3. Use the empty sphere property (ESP) of the DT to verify from the original data which spheres are inside the defined object. This makes use of the fact that any circumscribing sphere contains either only outside voxels or only inside voxels. This is true because if any sphere contains both inside and outside voxels, it will contain a boundary point as well; and this is forbidden by the ESP of the DT.

Spheres generated in this fashion satisfy all the truthfulness criteria previously mentioned. The properties of the sphere representation are closely linked to the properties of the underlying tetrahedra representation. Whereas the method is easy to implement, the worst case complexity of the algorithm is $O(n^2)$ in both time and number of primitives, where n is the number of boundary points. However, the highest number of primitives observed in our experiments so far has been about $4n$.

Step II: We simplify the dense sphere representation as generated above by clustering similar or close-by spheres and eliminating redundant or non-significant spheres. Many point clustering algorithms could be used. We defined a new clustering method which uses the fact that the underlying primitives are spheres and not points, and gives a better handle on the error introduced during simplification. It is based on the concept of *sphericity* which is a measure of how well a group of spheres can be represented by a single sphere. Sphericity can be defined as the ratio of the volume of union of spheres in the group to the volume of the smallest bounding sphere. For computational convenience, we define sphericity as the ratio of the radius of the largest sphere in the group to the radius of the smallest enclosing sphere. In either case, sphericity is a number between 0 (very non-spherical) and 1 (perfectly spherical).

The simplification algorithm processes the spheres from the largest to the smallest. In every iteration, using the largest remaining sphere, the algorithm calculates a new sphere such that the sphericity $\geq \sigma$ (a user defined threshold, generally close to 1), and it encloses a maximum number of remaining spheres. All the spheres within the newly created sphere are then deleted and the iterations continue until all the spheres in the original representation have been taken care of. The time complexity of the algorithm is $O(n^2)$, n being the number of spheres in the representation to be simplified.

We have used the algorithm to simplify a variety of objects. It generally yields a stable representation and can be applied with different values of sphericity threshold σ. In the extreme cases, when $\sigma = 0$, we get a single sphere as the output, and when $\sigma = 1$, we get the original representation itself as the output. For any intermediate value of σ, we get some intermediate number of spheres in the output. Since the algorithm ensures that all the generated spheres have a sphericity $\geq \sigma$, it puts a bound on the simplification error. However, like most clustering algorithms, it does not guarantee preservation of topology.

Application to Object Registration

We used the stability property of the UoS representation for the problem of object registration: Given two representations, determine if they represent the same object, and if they do, find the transformation from one to the other. To register objects, we first match a sufficient number of spheres from the UoS representation of one object to the other, and then from the matches find the most likely transformation using the method of least squares. The matches between the spheres of two representations A and B are obtained by formulating the problem as a minimum weight bipartite graph matching (also known as "assignment") problem. First, we assign distances from every sphere $a \in A$ to every sphere $b \in B$ according to a pre-defined metric (which is a function of the size of the sphere, and its location with respect to the coordinate frame determined from principal moments). Next, we do a minimum weight matching on the resulting bipartite graph. This problem has been widely studied and can be solved exactly in $O(n^3)$ time, where n is the number of nodes (in our case, the number of spheres in the simplified representation) in the bipartite graph. The results on the experimental data look promising. To determine matches in a bipartite graph with about 400 nodes in total, an IRIS Crimson takes about 1.5 seconds.

Implementation

The described algorithms have been implemented (for 2D and 3D) in the C language and compiled for SGI and IBM RS/6000 machines. Some of the results are shown in images accessible through the URL address <http://www.cs.ubc.ca/spider/ranjan/>.

Conclusion

Union of spheres (UoS) representation has many desirable properties for graphics, vision, and visualization. In this communication, we described how we can generate and simplify a UoS representation for volumetric data. The derived representation can be used to characterize and compare shapes; we applied it to the problem of object registration and the results are encouraging.

Acknowledgements

We gratefully acknowledge the help of the Geometry Center for the triangulation code, Bernd Gaertner for the code for finding the smallest disk enclosing a set of points, and Robert Kennedy for the code for solving the assignment problem.

References

[1] F. Aurenhammer. "Power diagrams: properties, algorithms and applications", *SIAM Journal of Computing*, Vol. 16, 1987, pp. 78–96.

[2] H. Edelsbrunner. "The union of balls and its dual shape", *Proceedings of the 9th Annual Symposium on Computational Geometry*, 1993, pp. 218–229.

[3] H. Edelsbrunner and E. P.Mücke. "Three-dimensional alpha shapes", *ACM Transactions on Graphics*, Vol. 13, No. 1, January 1994, pp. 43–72.

[4] V. Ranjan and A. Fournier. "Volume models for volumetric data", *IEEE Computer*, Vol. 27, No. 7, July 1994, pp. 28–36.

[5] E. Welzl. "Smallest enclosing disks, balls and ellipsoids", Tech. Report No. B 91-09, Fachbereich Mathematik, Freie Universität Berlin, Berlin, Germany, 1991.

Hexahedral Mesh Generation via the Dual

Steven Benzley Ted D. Blacker Scott A. Mitchell Peter Murdoch Timothy J. Tautges

Abstract

We review the spatial twist continuum (STC), a way of viewing the dual of a hexahedral (cubic) mesh as a simple non-degenerate arrangement of surfaces. Given a quadrilateral mesh of a closed surface, the STC gives insight into how the interior volume can be filled with hexahedra that respect the surface mesh. We review a hexahedral mesh generation heuristic called Whisker Weaving. Whisker Weaving incrementally builds the STC in an advancing front fashion.

Although computational geometry has traditionally focussed on triangular and tetrahedral meshes, quadrilateral and hexahedral meshes are often considered to be more valuable for finite element analysis. The CUBIT environment being developed at Sandia National Laboratories is a suite of quadrilateral and hexahedral meshing tools. Whisker Weaving is one of these tools. CUBIT is used directly by practicing analysts, and is incorporated into a number of commercial mesh generation codes.

1 Spatial Twist Continuum

We first describe the spatial twist continuum (STC) for a two-dimensional mesh and then generalize to three dimensions. It is possible to generalize to arbitrary dimensions.

Consider the dual of a quadrilateral mesh. We use dual in the same sense as the Voronoi diagram is the dual of a Delaunay triangulation, except that here we concentrate on the graph properties of the dual and the geometric embedding is arbitrary. Each vertex of the dual has edge-degree four. As such, each vertex can be considered as the intersection of two curves as in Figure 1 left. Opposite edges at each vertex are identified as belonging to the same curve. In this way, the dual of the mesh can be considered as an arrangement of curves as in Figure 1 right.

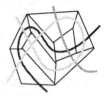

Figure 1: The STC in a single quadrilateral, and in a quadrilateral mesh.

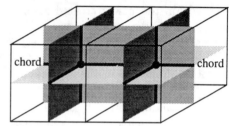

Figure 2: The STC for two cubes that share a quadrilateral.

The points of intersection are simple and non-degenerate, meaning that the curves are nowhere tangent, and exactly two curves intersect at each vertex. Note that a curve may intersect another curve, or itself, an arbitrary number of times. Also, note that an arbitrary simple, non-degenerate arrangement of curves defines a unique quadrilateral mesh (up to embedding).

Now consider three dimensions. The dual of a single cube (hexahedron) can be considered to be the arrangement of three orthogonal planes; see Figure 2. A hexahedral mesh can be considered as a simple, non-degenerate arrangement of surfaces. The surfaces (planes) twist through space, hence the name STC. A column of hexahedra that share opposite faces corresponds to a curve of intersection of two twist planes; see Figure 2. This curve is called a chord.

Suppose we have a quadrilateral mesh on a surface enclosing a volume, and we would like to fill that volume with hexahedra such that the surface mesh is respected. The surface is closed, so the curves of the 2d STC are closed. These curves are called loops. In the STC of any hexahedral mesh respecting the quadrilateral mesh, a loop is a connected component of the intersection of a twist plane with the surface. With this foreknowledge, the loops constrain the possible twist planes, giving valuable insight into how hexahedra must be constructed. For example, suppose twist planes have only one loop, and two loops intersect twice. Then there is a chord between the two points of intersection in any hexahedral mesh! It is unlikely that anyone would guess that two surface quadrilaterals must be joined by a column of hexahedra without considering the STC.

The self-intersections of a loop play a special role in the Whisker Weaving algorithm, and give rise to a number of interesting questions about the interlacement graph of Gauss codes [1]. This will be discussed further in future work.

2 Whisker Weaving

Whisker Weaving is a hexahedral mesh generation heuristic that incrementally builds the STC from a given quadrilateral surface mesh. The algorithm first constructs the loops and assumes that each loop is the only loop of a twist plane. Each twist plane is represented in a two-dimensional logical space, without reference to its three-dimensional embedding.

Each quadrilateral face of the surface mesh corresponds to a chord. A chord appears on the diagram for both of the twist planes that pass through the quadrilateral. It is yet unknown how the chord behaves on the interior. A chord is so far just a short segment called a whisker; see Figure 3 left.

The STC arrangement is incrementally constructed by crossing adjacent chords (i.e. by weaving whiskers). Weaving whiskers creates a vertex which is the dual of a hexahedron. See Figure 3 middle. Weaving whiskers modifies the diagram for each of the twist planes containing the whiskers: the diagrams are interdependent.

Whenever geometrically possible, the dangling ends of two whiskers are joined. This corresponds to having two hexahedra share a quadrilateral face. Usually the chords to be joined have both twist planes in common. Occasionally whiskers that have only one twist plane in common are joined. This merges the two uncommon twist planes into one that has two or more loops.

Once every whisker on every twist plane has been joined to complete a chord, the entire connectivity of the hexahedral mesh has been specified. See Figure 3 right. Twist planes are now embedded in three-dimensional space by smoothing (position averaging). The STC is then dualized back to create hexahedra, and the mesh is smoothed again.

The logical diagrams deserve special note from an implementation perspective. First note that nothing is lost, in that any simple non-degenerate arrangement of surfaces can be represented in this way. For example, if a loop self-intersects, the whisker appears twice on

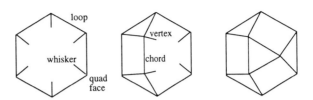

Figure 3: A loop, partially completed twist plane, and completed twist plane.

the diagram in two different locations, rather than on two different diagrams. We have even encountered vertices that are the triple intersection of a single twist plane: the vertex appears in three places on a single diagram. Twist planes that intersect each other only in the interior of the volume are represented by chords that do not intersect the loops, but are closed curves in the interior of the diagrams.

The ability to represent twist planes in this way is a tremendous advantage in developing and debugging the code, in that it allows a complex three dimensional problem to be easily visualized as a collection of interdependent two dimensional problems.

Acknowledgements

This work performed at Sandia National Laboratories and Brigham Young University. This work was supported by the Applied Mathematical Sciences program, U.S. Department of Energy Research and by the U.S. Department of Energy under contract DE-AC04-76DP00789.

References

[1] Pierre Rosenstiehl and Robert E. Tarjan, Gauss codes, planar Hamiltonian graphs, and stack-sortable permutations. *Journal of Algorithms* **5**, 375-390 (1984).

[2] Peter Murdoch, Steven Benzley, Ted D. Blacker and Scott A. Mitchell, A connectivity based method for representing and constructing all-hexahedral finite element meshes, Part I: The spatial twist continuum. Submitted to *Int. J. Numer. Methods Eng.*

[3] Timothy J. Tautges, Scott A. Mitchell and Ted D. Blacker, A connectivity based method for representing and constructing all-hexahedral finite element meshes, Part II: Whisker Weaving. Submitted to *Int. J. Numer. Methods Eng.*

Fast and Robust Computation of Molecular Surfaces

Michel F. Sanner[1], Arthur J. Olson[1], Jean-Claude Spehner[2]

[1]The Scripps Research Institute, 10666 North Torrey Pines Road, La Jolla, California 92037, USA.
[2]Université de Haute Alsace, Laboratoire MAGE, F.S.T., 4 rue des Frères Lumière, Mulhouse 68093, France.

Abstract

In this paper we define the r-reduced surface of a set of n spheres representing a molecule in relation to the r-accessible and r-excluded surfaces. Algorithms are given to compute the outer component of the r-reduced surface in $O[n \log n]$ operations and the analytical description of the corresponding r-excluded surface in time $O[n]$. An algorithm to handle the self-intersecting parts of that surface is described. These algorithms have been implemented in a program called MSMS, which was tested on a set of 709 proteins. The CPU time spent in the different algorithms composing MSMS are given for this set of molecules.

1 Introduction

Molecules are often represented as a set S of overlapping spheres, each having the van der Waals radius of its constituent atom. The solvent accessible surface [1] or r-accessible surface ($Ar(S)$) and the solvent excluded surface [2,3] or r-excluded surface ($Rr(S)$) of a molecule are defined using a spherical probe of radius r representing a solvent molecule. $Ar(S)$ is the topological boundary, denoted $\delta(S')$, of a set of spheres S' obtained by adding r to the radius of each sphere of S. It is also the trajectory of the center of a sphere of radius r rolling over the atoms. $Er(S)$ is made of spherical and toroidal patches lying on the atoms and the probes. For $r = 0$, $Er(S) = \delta(S)$ and for $r = \infty$, $Er(S)$ is the convex hull of S, denoted $conv(S)$. The complexity in size of the set of $Er(S)$ obtained for all possible values of r is in $\Theta[n^2]$ and these surfaces can be found with a worst case complexity in $O[n^2 \log n]$ [4]. $Ar(S)$ and $Er(S)$ can be made of several closed components and the outer one circumscribes all others. Several algorithms to compute analytical models of the $Ar(S)$ [5,6] and triangulations of $Er(S)$ [8,9] have been described, but we are aware of only two other programs which compute explicitly an analytical model of the $Er(S)$: PQMS [7] and AMS [3]. One of the main difficulties in computing this surface is the detection and handling of the self-intersecting parts of the surface called singularities [10,11].

After defining the r-reduced surface of S, denoted $Rr(S)$ we present an algorithm to compute its outer component in $O[n \log n]$ operations and the corresponding $Er(S)$ component in $O[n]$ operations. Then the way singularities are treated is presented and the results of our program MSMS are presented and compared with those obtained with PQMS for our set of 709 proteins.

2 The r-Reduced Surface; $Rr(S)$

$Rr(S)$ has been defined [11] as follows. The center of each sphere of S' holding p ($p>0$) $Ar(S)$ faces defines a vertex of $Rr(S)$ which has multiplicity p. For each circle C, which is the intersection of two spheres of S' and which holds q ($q>0$) edges of $Ar(S)$, the open line segment joining the centers of these spheres is an edge of $Rr(S)$ which has multiplicity q. If the $Ar(S)$ edge c has no vertex, i.e. $c = C$, this edge of the $Rr(S)$ is said to be free. For each vertex v of the $Ar(S)$ there are k spheres $s'_1, s'_2,..., s'_k$ of S' such that v belongs to all these spheres and the circles $s'_k \cap s'_1$ and $s'_i \cap s'_{i+1}$ for i in $\{1, 2,..., k-1\}$ hold an $Ar(S)$ edge ending in v. If $c_1, c_2,..., c_k$ are the centers of the spheres $s'_1, s'_2,..., s'_k$, a surface homeomorphic to an open disk bounded by the polygonal line $c_1 c_2... c_k c_1$ and included in $conv(c_1, c_2,..., c_k)$ is a face of $Rr(S)$.

For any calibrated set S of n spheres, and for any value of r adequate for S, the number of edges and faces of $Rr(S)$ is in $O[n]$ [11, 12]. A calibrated set of spheres is such that the ratio r_{max} / r_{min} of the maximum radius and the minimum radius in S is smaller than a constant T and the distance between the centers of any pair of spheres in S is larger than r_{min} / T. An adequate value of r for S is such that $r_{min} \leq r \leq r_{max}$.

$Rr(S)$ is closely related to alpha-shapes [13]. The set of spheres S representing a molecule is calibrated, and the usual probe radius (1.4-1.5Å) is adequate for S.

3 Computing $Rr(S)$ and $Er(S)$

• $Rr(S)$: The vertices of the first face F of $Rr(S)$ are: the center of the left-most atom s_1 (min{x coordinate - radius}), the center of the atom s_2, such that the circle $s'_1 \cap s'_2$ holds the left-most point, and the center of any atom s_3 such that the probe tangent to s_1, s_2 and s_3 has no overlap with any atom of S. From such a position, the probe can roll over each pair of atoms, whose centers c_i, c_j are connected by an edge of F, until it hits a third atom whose center defines with c_i and c_j a new face of $Rr(S)$. This is done until all edges have been treated. Pairs of $Rr(S)$ vertices which aren't connected by an edge and are closer than $2r + r_1 + r_2$, where r_1 and r_2 are the atom's radii, form free edges of that component of $Rr(S)$. Computational complexity: Because S is calibrated the number of atoms to consider when treating an edge is independent of n [9]. These atoms can be retrieved in $O[\log n]$ operations using a hierarchical subdivision of space [17] obtained by recursively dividing the bounding box of the set of atom's centers into two subsets of comparable sizes. Since $Rr(S)$ has $O[n]$, edges $Rr(S)$ can be computed in $O[n \log n]$.

• $Er(S)$: For every $Rr(S)$ face a spheric reentrant face of $Er(S)$, lying on the probe, is created. For each $Rr(S)$ edge, a toric reentrant face of $Er(S)$ is created. Cycles of convex edges of toric reentrant faces define the contact faces. The complexity to build $Er(S)$ from $Rr(S)$ is a linear function of the number of faces and edges of $Rr(S)$ which is in $O[n]$.

4 Handling singularities

• Radial singularities occur when the radius of the torus, described by a probe σ while rotating around an edge e of $Rr(S)$, is smaller than the radius of the probe. These singularities are detected during the computation of $Rr(S)$. The singular points ($\sigma \cap e$) are stored with the $Rr(S)$ edge. The toric reentrant face created for such an edge is made of two triangular faces connected by a singular vertex or a singular edge.

• Non-radial singularities occur when a part of a spheric reentrant face is inside another (non $Ar(S)$ neighbor) probe. These singularities are handled in three steps:

1 - For each pair of probes tangential to the same three atoms and which both intersect the plane P defined by the centers of the three atoms, clip both spheric reentrant faces with P. The complexity to handle these singularities is a linear function of the number of $Rr(s)$ faces having identical vertices.

2 - For each singular edge e created by a radial singularity, if e, or part of it, is inside a probe, it is entirely or partially "eaten". We add singular vertices at the intersection of the probes eating the biggest part of e, and e. The eaten part of e is replaced by singular edges lying on the circles intersection of the probes on which e was lying and the ones eating e. The created singular edges are added to the list of singular edges to be checked.

3 - For each pair of probes generating the spheric reentrant faces F and F', not tangent to the three same atoms and having an intersection circle C which doesn't intersect the edges of F or F', clip F and F' with the plane holding C.

The complexity of steps 2 and 3 is a function of the number of singular edges and probes generating singularities. In the case of molecules these numbers are small and the experimental results show that the time spent in this part of the program is negligible and behaves as a linear function of the number of atoms (Fig. 1).

5 Results and Discussion

The outer component of $Rr(S)$ and $Er(S)$ of a set of 709 molecules have been computed by MSMS and PQMS for a probe radius of 1.5 Å. These molecules have been selected from the Brookhaven Protein Data Bank to provide a wide range of protein sizes (33 to 43632 atoms) and types. The $Er(S)$ outer components were triangulated with a vertex density of 1.0 vertex/Å^2. Simple tests were performed on the triangulation to verify that every vertex belongs to at least two edges and that each edge belongs to exactly two triangles. All 709 MSMS surfaces successfully passed this test. This compares to 603 surfaces (85%) for PQMS.

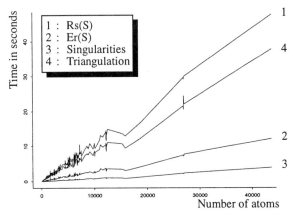

Fig. 1 : Time spend in different parts of MSMS measured for 709 proteins.

Figure 1 shows the CPU time in seconds used by the different algorithms composing MSMS for all molecules in our test set. These times were obtained on an SGI Challenge with 256 Mbytes RAM, running a MIPS R4400 processor at 150 MHz. MSMS computed a triangulation of the outer component of the $Er(S)$ of glutamine synthetase (2gls entry in the Brookhaven Data Base, 46,632 atoms, 494,586 triangles) in 47.7 + 12.14 + 3.87 + 37.7 = 101.48 seconds.

MSMS computes the genus of the analytical $Er(S)$ and of its triangulation. The comparison of these two numbers has proved to be a very inexpensive and powerful test for surface correctness.

For visualization purposes, it is important to avoid very thin triangles which are poorly rendered. The percentage, over all molecules, of such triangles dropped from 0.29% for PQMS to 0.078% for MSMS. Although MSMS produces fairly regular triangulations, the assumption that the spheres of S are in general positions generates a number of very thin toroidal faces which could be avoided by using an $Ar(S)$ multiple vertex approach as proposed by Eisenhaber [6].

6 Conclusion

The r-reduced surface provides a very compact way to store geometric information which can be used to build several molecular surfaces: van der Waals, r-excluded and r-accessible. It enables the checking of atom or residue exposure to the solvent and the computation of accurate molecular surfaces and surface areas. $Rr(S)$ contains all the information needed to test if a given point is inside or outside a specific molecular surface. The implementation of the described algorithms was shown to be very reliable and efficient. The practical performance of MSMS fit the theoretical complexity. The speed achieved allows interactive recomputation of dynamic molecular surfaces for small proteins undergoing molecular dynamics or normal mode motion. Higher speed could probably be achieved by optimizing the code. The program behaves remarkably well with large molecules where the complexity and the implementation are crucial. The surfaces computed with MSMS are widely used at The Scripps Research Institute to illustrate publications, study protein-protein and protein-ligand interactions, describe molecular surfaces by spherical harmonics [15] and study molecular properties.

Acknowledgment

This work was supported, in part by Sandoz Pharmaceutical Ltd., Basel Switzerland and by NIH grant number P01GM38794.

References

[1] Lee B. and Richards F.M. J. Mol. Biol. 55:379-400, 1971.
[2] Richards F.M. Ann. Rev. Biophys. Bioeng. 6:151-176, 1977.
[3] Connolly M.L. J. Appl. Cryst., 16:548-558, 1983.
[4] Boissonnat J.D., et al. J. Mol. Graphics, 12:61-62, 1994.
[5] Perrot G., et al. J. Comp. Chem., 13(1):1-11, 1992.
[6] Eisenhaber F., Argos P. J. Comp. Chem., 14(11):1272-1280, 1993.
[7] Connolly M.L. J. Mol. Graphics, 11:139-141, 1993.
[8] E. Silla et al. J. Mol. Graphics, 8:168-172, Sept. 1990.
[9] Varshney A., Ph D Dissertation. UNC. 1994
[10] Connolly M.L. J. Appl. Cryst., 18:499-505, 1985.
[11] Sanner M.F. PhD thesis. UHA, (France), 1992.
[12] Halperin D. and Overmars M. Proc. 10th ACM Symp. Comp. Geom., pp.113-122, 1994.
[13] Edelsbrunner H. and Mücke E. ACM Transaction on Graphics, 13:1:43-72, 1994.
[14] Barnes J. and Hut P. Nature 324:446-449, 1986.
[15] Duncan B.S., Olson A.J. Biopolymers, 33:219-229, 1993.

Representation and Computation of Boolean Combinations of Sculptured Models [*]

Shankar Krishnan
krishnas@cs.unc.edu

Atul Narkhede
narkhede@cs.unc.edu

Dinesh Manocha
manocha@cs.unc.edu

Department of Computer Science,
University of North Carolina,
Chapel Hill, NC 27599-3175

Abstract

We outline an algorithm and implementation of a system that computes Boolean combinations of *sculptured* solids. We represent the boundary of the solids in terms of trimmed and untrimmed spline surfaces and a connectivity graph. Based on algorithms for trapezoidation of polygons, partitioning of polygons, surface intersection and ray-shooting, we compute the boundaries of the resulting solids after the Boolean operation.

1 Introduction

The field of solid modeling deals with design and representation of physical objects. The two major representation schemata used in solid modeling are constructive solid geometry (CSG) and boundary representations (B-rep). Both these representations have different inherent strengths and weaknesses and for most applications both these representations are desired. At the same time, the field of geometric modeling has been developed to model classes of piecewise surfaces based on particular conditions of shape and smoothness. Such models are called *sculptured* solids. There is considerable interest in building complete solid representations from spline surfaces and their Boolean combinations [Hof89]. However, the major bottlenecks are in performing robust, efficient and accurate Boolean operations on the sculptured models. The topology of the surfaces becomes quite complicated when Boolean operations are performed and finding a convenient representation for these topologies has been a major challenge. As a result, most of the current solid modelers use polyhedral approximations to these surfaces and apply existing algorithms to design and manipulate these polyhedral objects. Not only does it lead to data proliferation, but the resulting algorithms are inefficient and inaccurate.

We outline a system for representation and efficient computation of Boolean operations on CSG models. Every CSG object is built from a set of primitive objects which are of a simpler structure, and we assume that each of these primitives is an *oriented solid* representable as a collection of rational parametric surfaces. This class of models includes all polyhedral models, and primitive solids like cones, spheres, cylinders and tori. Apart from the primitive objects, each intermediate and final solid in the CSG hierarchy is represented as a collection of *trimmed* surfaces. Along with the surface definition, every *trimmed* surface contains a *trimming* curve in its domain which unambiguously determines which part of the surface is to be retained. We also create a *connectivity* graph that maintains the connectivity information between the various surfaces composing the solid. The entire algorithm proceeds in two steps. Initially, the complete intersection curve between the two solids (at each level of the CSG tree) is determined. The intersection curve corresponding to a well-defined set operation partitions the surface of each solid. The second step of the algorithm uses the graph structure of each solid, and depending on the set operation, computes the new solid and its associated graph structure. The overall algorithm makes use of a number of recently developed geometric algorithms for linear programming [Sei90], trapezoidation of polygons [Sei91], partition of polygonal domains and ray-shooting.

2 Geometric Algorithms used in the System

The CSG algorithm is based on a number of geometric operations. These include trapezoidation of simple polygons, partitioning a simple polygon using non-intersecting polygonal chains, computing the intersection of two planar polygons and finding the intersection curve of two parametric surfaces. Some of the algorithms used are not necessarily the most optimal in terms of time complexity. In such cases, we have traded implementation simplicity for efficiency. This section briefly discusses these algorithms.

We use Seidel's algorithm [Sei91] to construct the horizontal and trapezoidal decomposition of a simple polygon. This enables us to answer point location queries in $O(\log n)$ time and find one point inside the polygon in constant time. It is an incremental randomized algorithm whose expected complexity is $O(n \log^* n)$. In practice, it is almost linear time for a simple polygon with n vertices.

A frequently occurring problem in a single CSG operation involves trimming out a parametric surface domain using a set of intersection curves. This reduces to partitioning a simple polygon with a collection of non-intersecting polygonal chains. A simple $O(n \log^2 n)$ algorithm is used for this purpose, where n is the total number of vertices of the polygon and the chains. We compute all the intersections between the chain segments and the polygon edges and label them in the order they occur with respect to the chain and the polygon. This is followed by a stack-based traversal of the polygon boundary and partitions are generated after proper combinations of labeled intersections are found [KNM94].

Computing all components of the intersection curve between two parametrically defined surfaces is an important

---

[*] Supported in part by Sloan Foundation, university research council grant, NSF grant CCR-9319957, ONR contract N00014-94-1-0738, ARPA contract DABT63-93-C-0048, NSF/ARPA Science and Technology Center for Computer Graphics & Scientific Visualization and NSF Prime contract No. 8920219

Permission to copy without fee all or part of this material is granted provided that the copies are not made or distributed for direct commercial advantage, the ACM copyright notice and the title of the publication and its date appear, and notice is given that copying is by permission of the Association of Computing Machinery. To copy otherwise, or to republish, requires a fee and/or specific permission.
11th Computational Geometry, Vancouver, B.C. Canada
© 1995 ACM 0-89791-724-3/95/0006...$3.50

ingredient of the CSG algorithm. We use the intersection algorithm in [KM94], which computes all components of the intersection curve between a pair of surfaces. It represents the intersection curve in terms of matrices and determinants and computes a piecewise linear approximation as well. The accuracy of the linear approximation is user-controlled. Furthermore, we make use of tests based on bounding boxes and linear programming to prune out the number of pairs of intersecting surfaces.

3 Boolean Operations between Solids

In this section, we briefly highlight the important steps in computing the result of Boolean operation of two solids (A and B). Initially, we compute the intersection curve between all pairs of intersecting spline-surfaces. The intersection algorithm finds the intersection curve between a pair of non-trimmed surfaces. However for trimmed surfaces we need to retain only those portions of the intersection curve which are common to trimming regions of both the surfaces. This is accomplished by marching along the piecewise linear approximation in each domain and performing in-out tests all along. The resulting curve partitions the surface of each solid into multiple components. Each such partition is tested for membership in the final solid by performing ray-shooting queries. We make use of connectivity graphs to minimize the number of ray-surface intersection queries [KNM94]. Membership decisions are based on the following criteria:

- *Union:* A component of solid A is part of the new solid if it lies **outside** solid B. A component of solid B is part of the new solid if it lies **outside** solid A.

- *Intersection:* A component of solid A is part of the new solid if it lies **inside** solid B. A component of solid B is part of the new solid if it lies **inside** solid A.

- *Difference:* A component of solid A is part of the new solid if it lies **outside** solid B. A component of solid B is part of the new solid if it lies **inside** solid A.

4 Implementation and Performance

The system has been implemented on a SGI-Onyx workstation with a single processor configuration. The system consists of two main parts: the surface intersection code and the supporting geometric routines for CSG.

The input to the intersection code are two parametric surfaces, and the output is a piecewise linear approximation and analytic representation (as a matrix polynomial) of their intersection curve. The accuracy of the linear approximation directly depends on the *step-size* and *tolerance* used by the numerical tracing algorithm. Our implementation uses EISPACK routines (in Fortran) for various matrix computations. The intersection code is invoked at every level in the CSG tree. We have observed that the choice of step-size was critical in preventing excessive accumulation of floating point error and data proliferation while performing the set operations on multi-level CSG trees. Large errors can result in inconsistent topology of the final solid. In our implementation, we use conservative step-sizes (0.02–0.05) to circumvent this problem.

Geometric algorithms implemented in fixed precision arithmetic introduce various types of computational errors. In general, performing robust and efficient computations on non-linear models is an open problem. Therefore, the comparison tests are based on tolerances. Due to the wide range of input values encountered, setting a fixed tolerance is not sufficient. To reduce this problem, our system

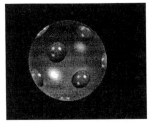

Figure 1: (a) Roller model (b) Cratered sphere

normalizes the surfaces (scaled down) before applying the intersection algorithm and makes use of condition numbers of matrix computations. Degenerate planar cases are handled as special cases by reducing the problem to intersection of polygons in $2D$. The detection of such degenerate overlaps is sensitive to tolerance values and larger tolerance values seem to work better for such cases.

Fig. 1 shows two of the models generated by our system.

- *Roller model*: The complete roller model is generated from a 30 level CSG tree. It consists of 137 trimmed surfaces (some of whose degree is as high as 6×2). The total time for model generation is close to 3.5 minutes.

- *Cratered sphere*: This model is generated from a 15 level CSG tree composed of difference operations between a big sphere and a number of smaller spheres. The resulting solid consists of 96 trimmed surfaces each of degree 2×2 and took 135 seconds to generate.

Our current system has not been optimized and we feel that its overall performance can be improved by a factor of two or more. The complete details of the algorithm and its performance are presented in [KNM94].

Acknowledgements

We thank Greg Angelini, Jom Bourdeaux and Ken Fast at Electric Boat for providing us with a CSG representation of the roller model and the UNC Walkthrough group for their support.

References

[Hof89] C.M. Hoffmann. *Geometric and Solid Modeling*. Morgan Kaufmann, San Mateo, California, 1989.

[KM94] S. Krishnan and D. Manocha. An efficient surface intersection algorithm based on the lower dimensional formulation. Technical Report TR94-062, Department of Computer Science, University of North Carolina, 1994.

[KNM94] S. Krishnan, A. Narkhede, and D. Manocha. Boole: A system to compute boolean combinations of sculptured solids. Technical Report TR95-008, Department of Computer Science, University of North Carolina, 1994.

[Sei90] R. Seidel. Linear programming and convex hulls made easy. In *Proc. 6th Ann. ACM Conf. on Computational Geometry*, pages 211–215, Berkeley, California, 1990.

[Sei91] R. Seidel. A simple and fast randomized algorithm for computing trapezoidal decompositions and for triangulating polygons. *Computational Geometry Theory & Applications*, 1(1):51–64, 1991.

The extensible drawing editor Ipe*

Otfried Schwarzkopf

Department of Computer Science, Utrecht University

A paper in Computational Geometry typically contains quite a number of figures, and many of us spend a significant amount of time on the creation of these figures. Usually, we use some tool such as xfig or idraw to create a Postscript file, which is then included into a LaTeX-document. This works quite well for simple figures, but for more difficult drawings the available tools are just not very satisfying. Ipe is a drawing editor that has specially been written to be useful for the kind of figure that a computational geometry paper might contain. It runs under X11, and currently requires a color screen. In the following, I am trying to list the properties that make Ipe different from other drawing programs.

1 Support for LaTeX-text

If you write your papers using LaTeX or some other TeX-based system, you probably like to be able to put TeX-text into your drawings. This not only means that you can type x_i to get "x_i" instead of having to manually place a little letter "i" next to the "x", but also that you can use the macros from the main paper in the figures as well, and finally the fonts in the figure will nicely match those in the document. However, entering the text in the figure in LaTeX format usually means that you are left to trial and error to figure out whether the text fits in a box or circle, or doesn't overlap other text. Ipe overcomes those problems because it runs TeX in the background. Whenever you enter a piece of text, Ipe simply asks TeX how big the text is, and will show you a bounding box of the proper size. Furthermore, Ipe can LaTeX the whole figure for you, convert it to Postscript, and show it to you in a second window using Ghostscript.

2 Snapping

Ipe is capable of making the objects that you have already drawn "magnetic." This means that once the cursor comes near the object, it automatically "snaps" to the object. This makes it very easy to draw objects that share endpoints, that lie on other object's boundaries, or even to use intersection points between objects. You can easily draw c-oriented polygons, or you can move objects in a direction parallel to another object's edge. A typical example is demonstrated in Figure 1. To draw the two vertical extensions through the endpoints p and q of segment e, you use vertex- and edge-snapping and turn on the 90°-oriented mode. As if by magic, you then only have to click once near p and once near the segments s_1 and s_2 to get the vertical extension of p.

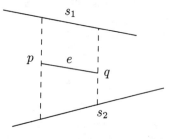

Figure 1:

3 Extensibility

The problem with most drawing programs is that they have a fixed set of commands, whileas in a computational geometry paper you often wish that you had a special function that, say, draws the circle from which a given segment is visible under a certain fixed angle. Ipe allows you to extend its functionality by adding your own code. Your self-defined functions can be used like any other Ipe function, they can appear in menus and can have keyboard shortcuts. The interface is rather simple to use, and allows you to incorporate your own geometric code in Ipe. Among the Ipe user macros that have already been written by me and other Ipe users are the following:

- Importing idraw-files, such that they can be saved and used as Ipe-files.

- Importing Postscript files, turning them in to normal Ipe files. This works for the Postscript output generated by many packages, such as gnuplot, showcase, etc.

- Clipping Ipe objects to a given polygon.

---

*This work was supported by the Netherlands' Organization for Scientific Research (NWO).

- Computing convex hulls, Voronoi diagrams, Delaunay triangulations, higher order Voronoi diagrams and medial axes. (This is based on the `qhull`-code by C. Bradford Barber and Hannu Huhdanpaa.)

- Computing the Minkowski sum and difference of polygons.

Using these functions, making an illustration for Fortune's algorithm such as Figure 2 takes only a few minutes. Figure 3 was created by importing a Postscript file

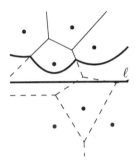

Figure 2:

with a map of the Netherlands, marking the provinces' capital cities, computing their Voronoi diagram and clipping that to some auxiliary polygon, which has been deleted afterwards. Again, this takes only a few minutes. The tiger in Figure 4, finally, has been created by

Figure 3:

importing a Postscript file in the Ghostscript distribution and some slight editing.

4 Making transparancies

Ipe can also be used to prepare transparencies for a presentation. For this purpose, it has a multi-page mode where you can have a template that appears on every page. There is also a special "minipage" text object that can contain `itemize` and other LaTeX-environments. In effect, using Ipe this way is very much like a WYSIWYG version of SLiTeX.

5 Ipe as a visualization tool

It is very easy to create an input file for Ipe. In fact, most of the geometric software currently written at our department has a button to save the current image as an Ipe file. This makes it a matter of seconds to create a figure for a paper, a figure that can still be edited and labeled. In the same spirit, it is possible to use Ipe to create test sets for your implementations.

6 More on Ipe

Ipe is fully documented and comes with a 50-page manual. The manual is also available on-line, and there is a "mouse explainer" that indicates the function of the mouse button at any given moment. Ipe can undo a certain number of operations.

The object types available in Ipe are chosen to be convenient for users in computational geometry: Circles can be entered by specifying center and radius, a diameter, or three points on the boundary. Similarly, there a different input modes for the other objects, and a selection of marks that can be used to indicate point sets, mark vertices differently, and the like.

Ipe stores drawings in a single file. This file is at the same time a valid Encapsulated Postscript file, a TeX source file, and it contains the information that Ipe needs to read it back. You do not need to carry around three different files—a master file, a LaTeX-version and a Postscript version—as some other drawing programs require.

Ipe is highly configurable. Several options can be set in a special configuration window, others are determined by Ipe's X resources.

Figure 4: Find the differences!

More information about Ipe is available on the world wide web.[1] This includes the full Ipe manual, a list of all currently available Ipe user macros, and the Ipe Frequently Asked Questions.

---

[1] The resource locator for the Ipe page is `http://www.cs.ruu.nl/people/otfried/html/ipe.html`. If you have no access to the world wide web, you can ftp to `ftp://ftp.cs.ruu.nl/pub/X11/Ipe/`.

Geomview: a system for geometric visualization *

Nina Amenta, Stuart Levy, Tamara Munzner, Mark Phillips
The Geometry Center

Abstract

Geomview displays objects in three-space and lets you move them around, view them from different angles, and adjust other parameters such as lighting. It is interactive, easy to use, and interfaces well with other software.

1 Introduction.

Geomview is a freely-distributed program that handles the display and interaction aspects of the presentation of 3D geometric data. The idea is that the display and interaction parts of 3D geometric software are generally independent of the content, so that a single display program can be used in various situations. In computational geometry, these situations include experimenting with objects, algorithm animation, making illustrations and producing simple videos.

2 What does Geomview do?

Geomview reads geometric data in a simple file format, displays a window with an image of the object in it, and lets user translate, rotate, and scale the object by moving the mouse in the window. It allows both separate and collective control of the objects. The objects have momentum, and can be set spinning or translating. The user can also control camera position, lighting, and object color and texture interactively.

The input can come from a file - hundreds of input objects are supplied with Geomview - - or from another program that is running simultaneously. In that case, as the other program changes the data, the Geomview image reflects the changes. Geomview itself does not do numeric or symbolic computations. Instead, Geomview is distributed with Unix filters that convert Mathematica or Maple output into the Geomview data format. Of course, users often produce data with their own programs. Geomview does not make photorealistic images as output (although the objects can be Gouraud shaded). Instead, there is a filter to convert Geomview data to Renderman input.

Geomview supports "external modules," which are separate programs that rely on Geomview for their display. Geomview is distributed with a collection of modules. Of particular interest are modules for viewing higher dimensional objects in several projections at once, creating surfaces of revolution, slicing objects, and editing of motion paths for video.

There are commercial products which provide some of these capabilities—most notably AVS, IRIS Inventor, and IRIS Explorer. Geomview is simpler and smaller, and is distributed freely over the Internet.

---

*email: software@geom.umn.edu. The Geometry Center, 1300 South Second Street Minneapolis, MN 55454, USA. The Geometry Center is officially The Center for the Computation and Visualization of Geometric Structures, an NSF Science and Technology Center, supported by grant number NSF/DMS-8920161.

Permission to copy without fee all or part of this material is granted provided that the copies are not made or distributed for direct commercial advantage, the ACM copyright notice and the title of the publication and its date appear, and notice is given that copying is by permission of the Association of Computing Machinery. To copy otherwise, or to republish, requires a fee and/or specific permission.
11th Computational Geometry, Vancouver, B.C. Canada
© 1995 ACM 0-89791-724-3/95/0006...$3.50

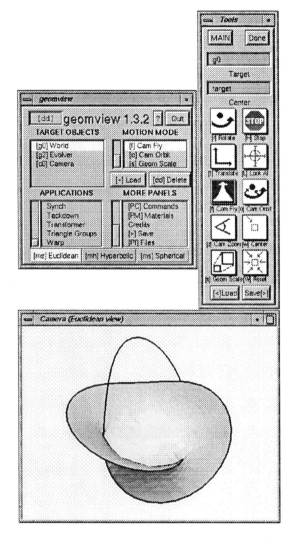

Figure 1: A typical Geomview session. The object in this picture is a soap film whose boundary lies on part of a trefoil knot; it was computed with Ken Brakke's Surface Evolver program.)

3 Visualization

Displaying and interacting with geometric objects is useful in several aspects of a computational geometers work. Examples are essential in doing mathematics. Geomview is useful for developing concrete examples in three and higher dimensions. The polytope in figure 2, for example, is the 4-D cross product of two 2-D heptagons. This is difficult to imagine, but the interactive Geomview display, developed in an afternoon, makes it completely understandable.

Examples are also essential to the comprehensible presentation of mathematics. Complicated

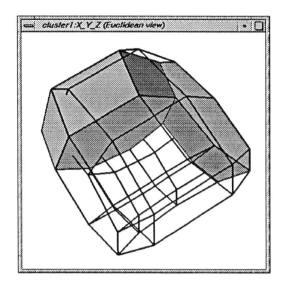

Figure 2: A 4-dimensional cross-polytope. Only two of the 3-dimensional facets are shown as solids.

three- and higher-dimensional examples are better communicated on video, which is easier to produce with Geomview.

Visual feedback is also helpful in developing geometric software, and improves the final product. Geomview vastly reduces the programming overhead for graphical output. Rather than develop independent graphics modules, new computational geometry software packages should use an existing visualization system such as Geomview. The recent "production quality" convex hull programs, for instance, provide Geomview output in low dimensions.

4 Availability

Geomview runs on a number of Unix platforms: Silicon Graphics, NeXT, 486 PCs running NeXTStep, and any graphics workstation running Unix and the X window system. Source code and binaries are available at ftp://geom.umn.edu/software/geomview.

The first release was done in 1991-92 by Geometry Center staff members Stuart Levy, Tamara Munzner, and Mark Phillips. Since that time there have been ten additional public releases, with contributions from Daeron Meyer, Tim Rowley, Nathaniel Thurston, Celeste Fowler, Oliver Goodman, John Sullivan, Daniel Krech and Scott Wisdom.

VideHoc: A Visualizer for Homogeneous Coordinates

Robert R. Lewis
bobl@cs.ubc.ca

University of British Columbia
Department of Computer Science

7 March, 1995

1 Introduction

VideHoc is an interactive graphical program that visualizes two-dimensional homogeneous coordinates. Users manipulate data in one of four views and all views are dynamically updated to reflect the change.

The program is a re-implementation of the program *Etch* developed by Snoeyink [snoe88]. Hanrahan [hanr84] describes a non-graphical tool with a similar purpose. XYZ (Nievergelt, et al. [niev91]) is a similar and in many ways more powerful tool, but without *VideHoc*'s intrinsic support for homogeneous coordinates.

VideHoc is intended to serve two roles: as an instructive tool for classes in computational geometry and as a research framework for testing new algorithms in computational geometry, particularly those exploiting homogeneous duality.

2 A *VideHoc* Overview

Figure 1 shows a typical *VideHoc* session. The canvas, or drawing area, shows the data the user has entered as elements: points, lines, chains (of line segments), and wedges (swept angles). The canvas can show several views: primal flat (Cartesian), primal fisheye (projection of the entire Cartesian plane onto a hemisphere), dual flat, and dual fisheye, either individually or all at the same time. When all views are shown, the same data appears in all of them: any change made to one window is reflected immediately in the other three.

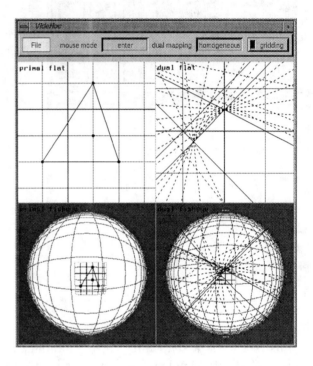

Figure 1: A typical *VideHoc* session

The control bar above the canvas lets the user see and change *VideHoc*'s current state, including:

mouse mode This is what the mouse is currently being used for: **enter** for entering points and lines, **view** for altering the what the canvas is displaying, or **select** for selecting points and lines for use in more complicated commands.

dual mapping This defines the kind of dual mapping that takes place: (classical) homogeneous, Edelsbrunner, or Brown.

gridding Whether or not points and lines (the two points defining them, actually) entered are to be constrained to lie on integer grid points.

All canvas input is done with the mouse. The left button is for single clicks: enter a point, pan to a new center, select a nearby element, etc. The middle button is for "down-drag-up" operations: area-related or two-point commands such as: enter a line, zoom in or out around the center, or select points within a given rectangle. The right button brings up a pop-up menu of commands relevant to the mouse mode.

3 Commands

Non-trivial commands the user can invoke from the **select** mouse mode include:

sort ... brings up a dialog box that allows the user to sort the selected points by x, y, or their selection order. The points are labelled with their resulting sort order, which is maintained until something (like adding or deleting a point) occurs that could possibly change the order.

join if two lines are selected, adds their intersection point or, if two points are selected, adds the line that passes through them. The result is dynamic, so that if either or both of the two lines or points are moved later, the result gets moved accordingly.

chain constructs a line segment going through all selected points (in selection order). This chain is dynamic: it will change if any of the points along it get moved and will vanish if any of the points get deleted.

convex hull constructs the convex hull of the selected points. This hull is dynamic: it may change if any of the points that define it get moved so as to redefine the hull and it will go away if any of the points that it was defined over get deleted.

4 Implementation Lessons

Implementing *VideHoc* has been instructive in itself. The distinction between an algorithms's conceptual form and its robust implementation was abundantly clear on several occasions. It was challenging, for example, to make the Graham convex hull algorithm work with both for points at infinity and for multiple points with the same coordinates[1]!

---
[1] Since *VideHoc* allows gridding, this situation was easy to create.

5 Conclusions

The goals of *VideHoc* are to support computational geometry education and research. The purpose of this communication is, therefore, to release it and elicit a response from the computational geometry community.

VideHoc is far from complete. Possible improvements include on-line help, commands to save and restore a user's configuration, hard copy output (including, optionally, color), support for oriented projective geometry (as specified in Stolfi [stol91]), an "undo" command, an application language for constructing (and debugging) algorithms interactively, support for 3 (or more?) dimensions (possibly making use of stereo viewing), and ports of *VideHoc* to more generally-available platforms.

6 Availability

A Silicon Graphics executable version of *VideHoc* is available via ftp from the node `ftp.cs.ubc.ca` in the directory `/pub/local/bobl/VideHoc`. It is also available as a download via the UBC Imager Web (URL: `http://www.cs.ubc.ca/nest/imager/imager.html`).

References

[hanr84] Pat Hanrahan. "A Homogeneous Geometry Calculator". 3-D Technical Memo 7, NYIT, 1984.

[niev91] Jürg Nievergelt, Peter Schorn, Michele de Lorenzi, Christoph Ammann, and Adrian Brünngger. "XYZ: A project in experimental geometric computation". *Computational Geometry — Methods, Algorithms and Applications: Proc. Internat. Workshop Comput. Geom. CG '91*, Vol. 553 of *Lecture Notes in Computer Science*, pp. 171–186. Springer-Verlag, 1991.

[snoe88] Jack Snoeyink. "Etch: A Drawing Program That Illustrates Geometric Duality". unpublished article, Xerox PARC, Palo Alto, CA, 1988.

[stol91] Jorge Stolfi. *Oriented Projective Geometry: A Framework for Geometric Computations*. Academic Press, 1991.

Evaluation of a new method to compute signs of determinants

F. Avnaim*   J-D. Boissonnat*   O. Devillers*   F. P. Preparata[‡]   M. Yvinec*[°]

*INRIA, BP 93
06902 Sophia Antipolis Cedex,
France

[‡]Dep. of Computer Science
Brown University, Providence
RI 02912-1910, USA

[°]CNRS, URA 1376, I3S
250 rue A. Einstein
06560 Valbonne, France

Introduction

We propose a method to evaluate signs of 2×2 and 3×3 determinants with b-bit integer entries using b and $(b+1)$-bit arithmetic respectively, that is typically half the number of bits usually required. Algorithms of this kind are very relevant to computational geometry, since most of the numerical aspects of geometric applications are reducible to evaluations of determinants. Therefore such algorithms provide a practical approach to robustness. The algorithm has been implemented and experimental results show that it slows down the computing time by only a small factor with respect to (error-prone) floating-point calculation, and compares favorably with other exact methods.

The main practical interest of our method, compared to other exact evaluation methods, is that it allows to use the very fast standard arithmetic of the computer while it does not assume a too small range of the original data.

The complete proofs of the following theorems can be found in the full version[1] of the paper, which contains also some references. The code is currently available.

Theorem 1 *The sign of a 2×2 determinant with b-bit integer entries can be evaluated using b-bit arithmetic in at most b iterations, each one involving $O(1)$ additions, comparisons and Euclidean divisions.*

---

*E-mail : olivier.devillers@sophia.inria.fr. Work of these authors has been supported in part by the ESPRIT Basic Research Action Nr. 7141 (ALCOMII).
[‡]Work of this author has been supported in part by NSF Grant CCR-91-96176.
[1]Same authors and title, INRIA Report 2306.
Code and reports available at http://www.inria.fr:/prisme/personnel/devillers/anglais/determinant.html

Permission to copy without fee all or part of this material is granted provided that the copies are not made or distributed for direct commercial advantage, the ACM copyright notice and the title of the publication and its date appear, and notice is given that copying is by permission of the Association of Computing Machinery. To copy otherwise, or to republish, requires a fee and/or specific permission.
11th Computational Geometry, Vancouver, B.C. Canada
© 1995 ACM 0-89791-724-3/95/0006...$3.50

Theorem 2 *The sign of a 3×3 determinant with b-bit integer entries can be evaluated using $b + 1$-bit arithmetic in at most $3b$ iterations, each one involving $O(b)$ signs of 2×2 determinants.*

Extensive simulations have shown that these worst-case bounds are much too pessimistic in practice even for null determinants.

Two-dimensional case

Let $D = \begin{vmatrix} x_1 & y_1 \\ x_2 & y_2 \end{vmatrix}$ be a 2×2 determinant whose entries are b-bit integers. Without loss of generality, we may assume that $0 < x_1 \leq x_2$ and $0 < y_1 \leq y_2$ otherwise either we can conclude immediately or we perform an elementary manipulation on the matrix to achieve such condition.

Under the above assumptions, we can write $x_2 = x_1 k_1 + x_r$ with $k_1 \in \mathbb{N}$ and $0 \leq x_r < x_1$ and define $y_r = y_2 - k_1 y_1$. Then,

$$D = \begin{vmatrix} x_1 & y_1 \\ x_2 & y_2 \end{vmatrix} = \begin{vmatrix} x_1 & y_1 \\ x_2 - k_1 x_1 & y_2 - k_1 y_1 \end{vmatrix} = \begin{vmatrix} x_1 & y_1 \\ x_r & y_r \end{vmatrix}$$

If $y_1 > \frac{2^b}{k_1}$; then y_r cannot be computed, but in that case y_r is certainly negative and thus $D < 0$. If y_r can be computed but is outside the range $[0, y_1]$ the sign of D can be obtained too: if $y_r < 0$ then $D < 0$ and if $y_r > y_1$ then $D > 0$. Moreover, if $x_r < \frac{x_1}{2}$ and $y_r > \frac{y_1}{2}$ $D > 0$ and if $x_r > \frac{x_1}{2}$ and $y_r < \frac{y_1}{2}$ then $D < 0$. Otherwise we rewrite D as follows:

$$\text{if } \begin{cases} x_r < x_1/2 \\ y_r < y_1/2 \end{cases} \text{ then } D = \begin{vmatrix} x_1 & y_1 \\ x_r & y_r \end{vmatrix} \quad (1)$$

$$\text{if } \begin{cases} x_r > x_1/2 \\ y_r > y_1/2 \end{cases} \text{ then } D = \begin{vmatrix} x_1 & y_1 \\ x_1 - x_r & y_1 - y_r \end{vmatrix} \quad (2)$$

In both cases, we get a new determinant where both entries of the second row have been divided by at least two, and we can iterate the computation.

In conclusion, at each iteration, using only comparisons and Euclidean divisions, either the algorithm stops or it iterates on a reduced problem, where a row is replaced by one whose entries are less than half the size of the original ones. Hence, the number of iterations is bounded from above by the logarithm of the largest representable integer, i.e. the number b of bits in the binary representation of the initial entries. This proves Theorem 1.

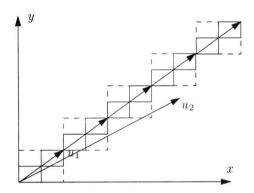

A geometric interpretation of this algorithm is shown on the figure. Point $u_2 = (x_2, y_2)$ is first located in a vertical strip. If u_2 is outside the dashed box in that strip, then the sign of D is immediatly obtained; else if u_2 is outside the solid boxes, then the sign of D is also obtained, otherwise the nearest multiple of u_1 is substracted to u_2 for the next iteration.

Three dimensions

This algorithm can be generalized to compute a three dimensional determinant defined by three vectors U_1, U_2, U_3. The analogous of the dashed boxes of the figure is a parallelepiped whose edges are the projections of U_1 and U_2 in the xy plane and the projection of $U_1 + U_2$ on the z axis. As in the 2D case the dashed boxes approximate the line defined by u_1, in the 3D case the plane spanned by U_1 and U_2 is approximated by parallelepiped boxes. If U_3 is above or below these boxes the sign of the determinant is easy to obtained, otherwise, the dashed box containing U_3 is subdivided into 8 solid boxes. Depending in which of these sub-boxes U_3 lies, the sign of D is obtained or U_3 is replaced by a vector whose z coordinate is divided by 2.

Implementation

Implementation has been done using C++, and the C++ ATT compiler on a Sun SS5-70. Times have been obtained using the `clock` command.

The entries are integers stored in variable of type *double*. This allows to manipulate exact 53 bits integers, to benefit of the fast floating point arithmetic of the processor and to handle easily overflows. In 3D, we do not apply exactly the Theorem above and use some simplification wich need to assume a $(b+2)$-bit arithmetic.

The code had been tested on several kinds of determinants and compared with the direct floating-point computation (denoted *double* in the tables, this give a wrong result in degenerate case), with the quadruple precision (*long double*) provided by the compiler, with the *real* of LEDA (release 3.1.1, see this proceedings) which provides exact computation with floating point filter, and with the *integer* of LEDA (exact computation). It is clear that in practical use, our technique (as any other) must be combined with a floating point filter (which is not done here).

The presented results have been obtained on the following kinds of determinants:
-1- determinants with integer entries uniformly distributed in $[-2^{53}, 2^{53}]$ in 2D and $[-2^{51}, 2^{51}]$ in 2D.

-2- determinants $\begin{vmatrix} a & a \\ a & a \end{vmatrix}$ in 2D and $\begin{vmatrix} a & a & a \\ a & a & a \\ a & a & a \end{vmatrix}$ in 3D with a at random.

-3- a perturbaution of -2- obtained by adding to each entrie a random integer in [-3,3].

-4- $\begin{vmatrix} a_1 & \lfloor \alpha a_1 \rfloor \\ a_2 & \lfloor \alpha a_2 \rfloor \end{vmatrix}$ and $\begin{vmatrix} a_1 & a_2 & \lfloor \alpha a_1 + \beta a_2 \rfloor \\ a_3 & a_4 & \lfloor \alpha a_3 + \beta a_4 \rfloor \\ a_5 & a_6 & \lfloor \alpha a_5 + \beta a_6 \rfloor \end{vmatrix}$ where a_i are random integers and α, β are random real numbers in $[-1/2, 1/2]$. The $\lfloor \rfloor$ means that the computations are done using the rounded floating-point arithmetic.

-5- $\begin{vmatrix} a & b \\ -a & -b \end{vmatrix}$ and $\begin{vmatrix} a & c & e \\ b & d & f \\ -a-b & -c-d & -e-f \end{vmatrix}$ a, b, c, d, e, f at random.

-6- a perturbaution of -5-.

Execution times in micro-seconds.

2D	-1-	-2-	-3-	-4-	-5-	-6-
double	0.5	0.5	0.5	0.5	0.5	0.5
long double	170	160	160	170	160	170
real	50	4800	1600	2500	5000	60
integer	48	48	48	48	48	50
this paper	2.3	4.2	3.0	29	53	27

3D	-1-	-2-	-3-	-4-	-5-	-6-
double	1.9	2.8	2.2	2.0	2.0	2.0
real	180	19000	22000	4500	22000	160
integer	230	190	210	220	220	220
this paper	14	18	160	300	930	310

Applications

Using the 53 bit arithmetic as described above allows the following precision on the entries (number of bits) for the following geometric tests:

function	line side	plane side	in circle	in sphere
det. size	2×2	3×3	2×2	3×3
entries size	52	50	24	23

Exact Geometric Computation in LEDA

C. Burnikel* J. Könemann[†] K. Mehlhorn* S. Näher[‡] S. Schirra* C. Uhrig*

Almost all geometric algorithms are based on the Real-RAM model. Implementors often simply replace the exact real arithmetic of this model by fixed precision arithmetic, thereby making correct algorithms incorrect. Two approaches have been taken to remedy this situation. The first approach is redesigning geometric algorithms for fixed precision arithmetic. The redesign is difficult and can inherently not lead to exact results. Moreover it prevents application areas from making use of the rich literature of geometric algorithms developed in computational geometry. The other approach advocates the use of exact real arithmetic.

In the LEDA platform of combinatorial and geometric computation [4] the new data-type **real** facilitates exact computation for many geometric problems. Every integer is a **real** and **reals** are closed under the operations addition, subtraction, multiplication, division and squareroot. All comparison operators $\{>, \geq, <, \leq, =\}$ are *exact*. In order to determine the sign of a real number x the data type first computes a *separation bound* q such that $|x| \leq q$ implies $x = 0$ and then computes an approximation of x of sufficient precision to decide the sign of x. The user may assist the data type by providing a separation bound q. The data type also allows to evaluate real expressions with arbitrary precision. Figure 1 shows (part of) the LEDA manual page for **reals**.

reals provide exact computation in a convenient way. In an implementation of a geometric algorithm in C++, **reals** can be used like **doubles**. The following example

---

*Max-Planck-Institut für Informatik, Im Stadtwald, 66123 Saarbrücken, Germany. Supported by the ESPRIT Basic Research Actions Program, under contract No. 7141 (project ALCOM II).

[†]Fachbereich 14, Informatik, Universität des Saarlandes, 66041 Saarbrücken, Germany.

[‡]Martin-Luther-Universität Halle, Fachbereich Mathematik und Informatik, 06099 Halle, Germany.

arises in the computation of Voronoi diagrams of line segments [2]. For i, $1 \leq i \leq 3$, let $l_i : a_i x + b_i y + c_i = 0$ be a line in two-dimensional space and let $p = (0, 0)$ be the origin. The circles passing through p and touching l_1 and l_2 have centers $(x_v/z_v, y_v/z_v)$ where

$$x_v = a_1 c_2 + a_2 c_1 \pm \sqrt{2 c_1 c_2 (\sqrt{N} + D)}$$
$$y_v = b_1 c_2 + b_2 c_1 \pm \text{sign}(S)\sqrt{2 c_1 c_2 (\sqrt{N} - D)}$$
$$z_v = \sqrt{N} - a_1 a_2 - b_1 b_2$$

with $S = a_1 b_2 + a_2 b_1$, $N = (a_1^2 + b_1^2)(a_2^2 + b_2^2)$, and $D = a_1 a_2 - b_1 b_2$.

The following C++ procedure tests whether l_3 intersects, touches or misses one of these circles by comparing the distances of $(x_v/z_v, y_v/z_v)$ and p and $(x_v/z_v, y_v/z_v)$ and l_3, see [2].

```
int INCIRCLE(real a_1, real b_1, real c_1, real a_2, real b_2,
             real c_2, real a_3, real b_3, real c_3, int pm)
{
    real RN = sqrt((a_1 * a_1 + b_1 * b_1) * (a_2 * a_2 + b_2 * b_2));
    real A = a_1 * c_2 + a_2 * c_1;
    real B = b_1 * c_2 + b_2 * c_1;
    real C = 2 * c_1 * c_2;
    real D = a_1 * a_2 - b_1 * b_2;
    real S = a_1 * b_2 + a_2 * b_1;
    real x_v = A + pm * sqrt(C * (RN + D));
    real y_v = B + pm * S.sign() * sqrt(C * (RN - D));
    real z_v = RN - (a_1 * a_2 + b_1 * b_2);
    real P = a_3 * x_v + b_3 * y_v + c_3 * z_v;
    real R = P * P - (a_3 * a_3 + b_3 * b_3) * (x_v * x_v + y_v * y_v);
    return R.sign();
}
```

If a_i, b_i, c_i are known to be $2k$-bit integers the sign computation $R.\text{sign}()$ can be replaced by $R.\text{sign}(48 * k)$, see [2]. Without this assistance the sign computation can be extremely slow.

The implementation of **reals** is based on LEDA data types **integer** and **bigfloat** which are arbitrary preci-

\multicolumn{3}{l}{The arithmetic operations $+$, $-$, $*$, $/$, $+=$, $-=$, $*=$, $/=$, $-$(unary), the comparison operations $<$, $<=$, $>$, $>=$, $==$, $!=$ and the stream operations all are available.}
int	x.sign()	returns -1 if (the exact value of) $x < 0$, 1 if $x > 0$, and 0 if $x = 0$.
int	x.sign(integer a)	as above. Precondition: if $|x| \leq 2^{-a}$ then $x = 0$.
void	x.improve(integer a)	(re-)computes the approximation of x such that its final quality is bounded by a, i.e., $x.get\_precision() \geq a$ after the call $x.improve(a)$.
void	x.compute_in(long k)	(re-)computes the approximation of x; each numerical operation is carried out with k binary places.
void	x.compute_up_to(long k)	(re-)computes an approximation of x such that the error of the approximation lies in the k-th binary place.
double	x.todouble()	returns the current double approximation of x.
double	x.get_double_error()	returns the quality of the current double approximation of x, i.e., $|x - x.todouble()| \leq x.get\_double\_error() * |x.todouble()|$.
integer	x.get_precision()	returns the quality of the current internal approximation $x.num$ of x, i.e., $|x - x.num| \leq 2^{-x.get\_precision()}$.
real	sqrt(real x)	squareroot.
real	fabs(real x)	absolute value of x.

Figure 1: (Part of) the LEDA manual page for **reals**

sion integers and floating point numbers with exponents and mantissa of arbitrary length. Similar to [1, 6] a **real** is represented by the expression which defines it. Furthermore it has a **double** approximation $\hat{x}$ together with a relative error bound ϵ_x. In the sign computation a *floating-point filter* is used as suggested in [3]. When the double test is inconclusive the quality of the approximation is repeatedly doubled (using calls of improve) until a decision is possible. A decision is possible if either the absolute value of the approximation is less than its absolute error or the separation bound is reached. The latter bound is either provided by the user or computed as in [5, 6].

The INCIRCLE procedure has been tested with incircle tests (of this special type) that arised in a randomized incremental construction of Voronoi diagrams of line segments. For these realistic data (all were nondegenerate) the exact test with **reals** was about 20 times slower than the equivalent test with **doubles**.

References

[1] M.O. Benouamer, P. Jaillon, D. Michelucci, and J.-M. Moreau. A "lazy" solution to imprecision in computational geometry. 5th CCCG, pp. 73–78, 1993.

[2] C. Burnikel, K. Mehlhorn, and S. Schirra. How to compute the Voronoi diagram of line segments: Theoretical and experimental results. ESA'94, pp. 227–239, 1994.

[3] S. Fortune and C. van Wyk. Efficient exact arithmetic for computational geometry. *Proc. of the 9th Symp. on Computational Geometry*, pages 163–171, 1993.

[4] K. Mehlhorn and S. Näher. LEDA: A platform for combinatorial and geometric computing. CACM, Jan. 1995.

[5] M. Mignotte. *Mathematics for Computer Algebra*. Springer Verlag, 1992.

[6] C.K. Yap and T. Dube. The exact computation paradigm. In *Computing in Euclidean Geometry II*, 1995.

A Global Motion Planner for a Mobile Robot on a Terrain*

Jean-Daniel Boissonnat[1]   Katrin Dobrindt[2]   Bernhard Geiger[1]   Henri Michel[3]

1 Introduction

Moving on a terrain (mountain landscape), a mobile robot must avoid collision with the ground, it must be stable without risk of tumbling and its wheels must touch the ground. The methods for solving the motion planning problem for such an all-terrain mobile robot can be divided into two classes. A *local* method plans the motion having only local information about the configuration space [4] and therefore has the drawback that the robot might get stuck in a deadlock. In contrast, a *global* method performs a preprocessing step to compute a representation of the complete space of the configurations of the robot that are collision free and satisfy the stability constraints. Both methods can be used in a complementary way: the global method that uses a simplified model plans the coarse motion and computes intermediate goal configurations for the local method that does the fine planning and thus tends to avoid deadlocks. There are a few papers presenting local optimization methods for the above motion planning problem. The —to our knowledge— only global method computes an octree representation of the space of collision free and stable configurations [2].

We devise a global method for planning collision free and stable motions of a robot on a terrain that relies on a new model of the robot: it is equipped with active vertical suspensions to keep its body always horizontal—see Figure 1. We believe that this model, although simplified, allows to represent in a satisfactory way a variety of all-terrain mobile robots in motion planning systems. We show that constructing the space of collision free and stable configurations of the robot reduces to computing lower and upper envelopes (the pointwise minimum resp. maximum) of a set of algebraic surfaces patches in four dimensions. We propose an algorithm for computing an approximation of this

---

*This research was partially supported by the ESPRIT II Basic Research Action 6546 (PROMotion) and by the MRE-project n° 91 S 0290, and was mainly done while K.D. and H.M. were employed at INRIA. K.D. is currently supported by the Dutch Organisation for Scientific Research (NWO).
[1] INRIA, B.P. 93, 06902 Sophia-Antipolis Cedex, France.
[2] Vakgroep Informatica, Universiteit Utrecht, Postbus 80.089, 3508 TB Utrecht, the Netherlands.
[3] CRS4, Via Nazario Sauro 10, 09123 Cagliari, Italy.

Figure 1: A robot: a polyhedral body carried by four suspensions one of which is extended to the maximum.

space. Its running-time depends on the complexity of the terrain and the robot as well as on the resolution of discretization used. The motion planner has been implemented and has been tested on various examples.

2 The Method

The terrain is modeled as a *polyhedral terrain* $\mathcal{T}$ with n triangular faces. This is the graph of a continuous function that assigns to every point on the plane an elevation

The robot consists of a polyhedral body with l vertices carried by k vertical active suspensions. The top endpoints of the suspensions lie on the polyhedron, and their bottom endpoints are point wheels.[1] All wheels must be in contact with the terrain. The suspensions can each be extended from zero to a fixed maximal elongation a and ensure that the body of the robot remains always horizontal. Figure 1 shows such a robot. The assumption of horizontality reduces the dimension of the configuration space: though the robot is allowed to translate freely, it can only rotate around the z-axis. Such a configuration in $\mathbb{R}^3 \times [0, 2\pi)$ of this robot is *feasible*,

(1) if it is *free*, that is, at this configuration the body of the robot does not intersect the interior of $\mathcal{T}$, and

(2) if it is *stable*, that is, there is a set of permitted elongations such that at this configuration the body of the robot is horizontal.

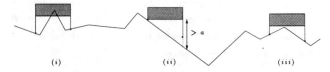

Figure 2: The configuration of the two-dimensional robot in (i) is not free, but satisfies condition (2); although the right suspension in (ii) is extended to the maximum the wheel is not in contact with the ground, the configuration is free, but not stable; the configuration in (iii) is free and stable.

---

[1] The wheels can also be modelled as spheres with a fixed radius. In this case, we can shrink the sphere to a point by replacing the terrain by the Minkowski difference of $\mathcal{T}$ and the sphere.

The *feasible space* is the set of all feasible configurations. The condition (2) models the stability of the robot. Figure 2 illustrates different configurations in $\mathbb{R}^2$.

To solve the motion planning problem we will first construct the feasible space and then generate a path in this space. We first consider the restricted case of a purely translational motion of the robot in $\mathcal{T}$, for a fixed orientation θ. In this case, the configuration space is three-dimensional. The boundary of the set of all free configurations is the upper envelope of $O(ln)$ triangles in $\mathbb{R}^3$, that are the Minkowski differences of the faces of $\mathcal{T}$ and the body of the robot [3]. The complexity of this set is nearly quadratic in ln. The set can be constructed in the same time bound. Consider next the translations of $\mathcal{T}$ that are the Minkowski differences of $\mathcal{T}$ and the top endpoints of each suspension, translated by a in a vertical upwards direction. The configurations below such a translation of $\mathcal{T}$ are the configurations where this suspension is not extended to the maximum. The (upper) boundary of the set of all configurations that satisfy condition (2) is now the lower envelope of these k translations of $\mathcal{T}$ consisting of in total kn triangles. The feasible space (at orientation θ) consists of the configurations that lie above the upper envelope as well as below the lower envelope. The complexity of the feasible space is proportional to the sum of the complexities of these envelopes [3].

Since the terrain is the graph of a function and because of condition (2), the z-value of a feasible configuration belongs to an interval of length at most a. This property allows us to reduce the dimension of the configuration space by one and to consider for a three-dimensional feasible configuration only its two-dimensional projection in direction of the z-axis. We compute the intersection between the upper envelope (that is the boundary of the freespace) and the lower envelope (that is the boundary of the set of all stable configurations) and project this intersection onto $\mathbb{R}^2$. This gives us planar polygonal contours that correspond to the boundary of the projection of the feasible space.

Up to now the robot was only allowed to translate. When it can both translate and rotate, constructing the feasible space is reduced to computing upper and lower envelopes of algebraic surface patches of bounded degree in $\mathbb{R}^3 \times [0, 2\pi)$. Since there is no efficient algorithm known for this problem, we propose an approximate method. We repeat the step of preceding paragraph for a discrete set of orientations. Then we reconstruct the feasible space in $\mathbb{R}^2 \times [0, 2\pi)$ from the planar contours obtained for the different orientations using a method based on the Delaunay triangulation [1]. As a result, we get a tetrahedrization of the feasible space and an adjacency graph of the tetrahedra.

The actual trajectory of the robot is then computed as follows. We find an initial path in the adjacency graph using an $A^\star$ algorithm. This path is smoothed by recursively refining straight line segments inside the feasible space. A local planner calculates then the actual suspension lengths.

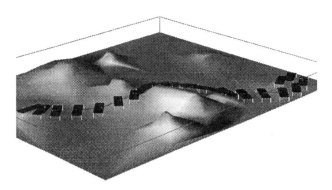

Figure 3: The trajectory of a rectangular robot on a terrain (300 triangles).

Figure 3 shows the trajectory for a rectangular robot moving on a terrain consisting of 300 triangles. The feasible space is discretized in θ in steps of 10 degrees—see Figure 4. Computing the tetrahedrization of the feasible space (150,000 tetrahedra) takes roughly 3 minutes and path finding 16 seconds (SGI Indy R4400SC).

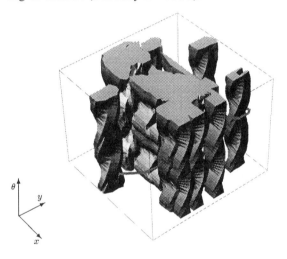

Figure 4: The space of all feasible configurations (complement of the solid) obtained by interpolating planar contours obtained for 36 different orientations.

Acknowledgements. We thank Francesca Fierro, Silvia Bussi and Laurence Foulloy for implementing parts of the motion planner.

References

[1] J-D. Boissonnat and B. Geiger. Three dimensional reconstruction of complex shapes based on the Delaunay triangulation. In *Proc. Biomedical Image Process. Biomedical Visual.*, vol. 1905, pages 964–975. SPIE, 1993.

[2] B. Dacre Wright. *Planification de trajectoires pour un robot mobile sur un terrain accidenté*. Thèse de Doctorat, Ecole Nationale Supérieure des Télécommunications, 1993.

[3] H. Edelsbrunner, L. Guibas, and M. Sharir. The upper envelope of piecewise linear functions: algorithms and applications. *Discrete Comput. Geom.*, 4:311–336, 1989.

[4] J.-C. Latombe. *Robot Motion Planning*. Kluwer Academic Publishers, Boston, 1991.

Complete Algorithms for Reorienting Polyhedral Parts using a Pivoting Gripper

Anil Rao[*]
Utrecht University

David Kriegman[†]
Yale University

Ken Goldberg[‡]
USC

1 Introduction

Achieving a desired spatial configuration of a part is a fundamental issue in robotics. In industrial applications, a familiar task is that of feeding parts: bringing parts into a desired position and orientation (pose). To rapidly feed a stream of industrial parts arriving on a conveyor belt, the vision-based system proposed by Carlisle *et. al.* [1] uses a SCARA-type arm with only 4 DoF (three translatory and one rotatory) due to cost, accuracy, and speed requirements. However, such arms can only reorient parts about the vertical axis due to kinematic limitations (see Fig. 1).

Contact between a part and a supporting plane only occurs along its convex hull. When rotations and translations in the plane are ignored, the part generally assumes one of a finite number of stable configurations [3]. In this communication, we will consider computing a sequence of pivoting actions that will move a *polyhedral* part with n faces from an initial stable configuration $\hat{s}$ to a final stable configuration $\hat{f}$. The decision question is whether or not a *single* pivot action can accomplish this task: we give a $O(n \log n)$ time solution.

It may not be possible to move the part between an arbitrary pair of stable configurations in a single pivot action; in general a sequence of pick-pivot-place operations may be necessary. Therefore, we consider computing the complete graph of possible transitions and give an algorithm that runs in $O(m^2 n \log n)$ time, m being the number of stable configurations. A path through this transition graph represents a plan which moves the part from some initial to final configuration. The algorithm is complete in that whenever a path of pivot

---

[*]Department of Computer Science. Rao is supported by the ESPRIT Basic Research Action No. 6546 (Project PRoMotion). anil@cs.ruu.nl

[†]Department of Electrical Engineering. Kriegman was supported in part by an NSF Young Investigator Award IRI-9257990. kriegman@yale.edu

[‡]Department of Computer Science. Goldberg is supported by NSF Young Investigator Award IRI-9457523 and IRI-9123747 and by Adept Technology, Inc. goldberg@usc.edu

Permission to copy without fee all or part of this material is granted provided that the copies are not made or distributed for direct commercial advantage, the ACM copyright notice and the title of the publication and its date appear, and notice is given that copying is by permission of the Association of Computing Machinery. To copy otherwise, or to republish, requires a fee and/or specific permission.
11th Computational Geometry, Vancouver, B.C. Canada
© 1995 ACM 0-89791-724-3/95/0006...$3.50

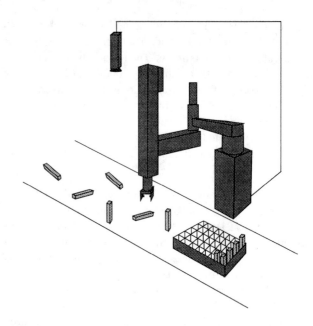

Figure 1: The pivoting gripper mounted on a SCARA arm.

actions exists, each conforming with the gripper accessibility and friction constraints, it will be found.

See the complete paper [4] for the details which also includes a generalization that considers "capture regions" around stable configurations.

2 Problem Statement

We assume *(i)* The worktable is a flat plane orthogonal to gravity at a known height; *(ii)* The parallel-jaw gripper is able to translate with 3 DoF and to rotate about the gravity vector; *(iii)* The gripper has a passive degree of freedom – a pivot axis, parallel to the support plane joining the two points of contact; and *iv* The gripper makes two "hard contact" with the part – point contacts with friction which offers no static resistance to rotation about the pivot axis.

Fig. 1 shows the robot work cell. The **input** to the algorithm is: A polyhedral part $\mathcal{P}$ stored as a boundary representation (B-rep), its center of gravity, c, which is taken to be the origin of the coordinate system used to define the B-rep, and

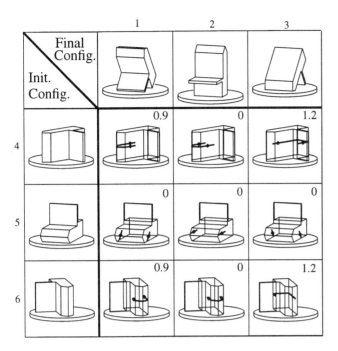

Figure 2: The partial matrix of transitions for the part with six stable configurations. Cell (i, j) indicates the family of accessible pivot grasps that will move configuration i to configuration j; the optimal grasp (requiring minimal μ_{static}) from among this family is shown as a pair of disks. Numbers in the upper right-hand-corner of each cell indicate the minimal required coefficient of friction.

the coefficient of static friction μ_{static}.

The **output** is a transition graph whose nodes are the stable configurations, or faces F_i of the convex hull. The arcs between nodes describe points on the part corresponding to grasp axes that will rotate the part from one stable face to another.

3 Computing the Transition Graph

First compute the convex hull $\mathcal{H}$ for the polyhedron $\mathcal{P}$. A face of $\mathcal{H}$ is stable when the projection of the center of gravity in the normal direction onto the face lies within the face; the stable faces become the nodes of the transition graph. For every ordered pair of stable faces of $\mathcal{H}$, whose normals are given by $\hat{\mathbf{s}}$ and $\hat{\mathbf{f}}$, determine the set of grasp points (if there are any) that will pivot the part to $\hat{\mathbf{f}}$ as described below. The direction of the grasp axis is given by:

$$\hat{\mathbf{a}} = \frac{\hat{\mathbf{s}} \times \hat{\mathbf{f}}}{|\hat{\mathbf{s}} \times \hat{\mathbf{f}}|}. \quad (1)$$

Note: $\hat{\mathbf{a}}$ is undefined when $\hat{\mathbf{s}}$ and $\hat{\mathbf{f}}$ are parallel or anti-parallel. In these cases, precise pivot actions are unnecessary or impossible. The parametric equation of the family of grasp axes indexed by λ is:

$$\mathbf{a}_\lambda(t) = t\hat{\mathbf{a}} - \lambda \hat{\mathbf{f}}, \quad (2)$$

where $\lambda > 0$ can be interpreted as the distance from the center of gravity to the axis. Thus, the grasp axis must lie in the half-plane, $\mathcal{A}$, spanned by $\hat{\mathbf{a}}$ and $-\hat{\mathbf{f}}$.

1. Determine the half-plane $\mathcal{A}$ of grasp axes which will successfully pivot $\mathcal{P}$ according to Eq. (2) and the direction of the grasp axis $\hat{\mathbf{a}}$ from Eq. (1).
2. Compute the intersection of $\mathcal{A}$ with $\mathcal{P}$ which yields a collection of intersection polygons **P** in the grasp plane.
3. In the direction $\hat{\mathbf{f}}$ within the grasp plane, compute the upper $\mathcal{U}$ and lower $\mathcal{L}$ envelope of the polygon(s) **P**. The importance of points on these envelopes is that they are *accessible* to a gripper linearly approaching the part along the grasp axis.
4. For each edge of $\mathcal{U} \cup \mathcal{L}$ whose corresponding face has surface normal $\hat{\mathbf{n}}$, determine if the face can be grasped by a point contact with friction in the direction $\hat{\mathbf{a}}$ according to: $\|\hat{\mathbf{a}} \cdot \hat{\mathbf{n}}\| \leq \cos \alpha$. Here α is the *friction angle* computed from $\tan(\alpha) = \mu_{\text{static}}$.
5. Merge the two sorted envelopes $\mathcal{U}$ and $\mathcal{L}$ into a set $\Lambda = \cup \Lambda_i$ where each Λ_i is a closed interval of λ. Associated with each interval is the pair of functions $u(\lambda)$ and $l(\lambda)$ which return the grasp points.
6. If $\Lambda \neq \emptyset$, create an arc in the transition graph between $\hat{\mathbf{s}}$ and $\hat{\mathbf{f}}$.

The complexity of the algorithm is dominated by the construction of the envelopes which takes $O(n \log n)$ time per iteration [2]; the rest of the steps have linear complexity. Since there are $O(m^2)$ pairs of stable configurations, the complexity of constructing the entire transition graph is $O(m^2 n \log n)$. For star-shaped (wrt the center of gravity) polyhedra, this reduces to $O(m^2 n)$ because the intersections computed in Step 2 will each consist of a single star-shaped polygon.

Implementation: The algorithm for planning pivot actions was implemented in the Symbolic Computing System *Maple V*. The choice of Maple was made because several primitive geometric tests and computations are built in with Maple's *geom3d* package. As an example, consider Fig. 2. The part has $n = 11, m = 6$. The entire transition graph, a fourth of which is shown in the figure, was computed in 28 seconds on a Silicon Graphics workstation (R4400 processor running at 150 MHz, 96.5 SPECfp92, 90.4 SPECint92).

References

[1] B. Carlisle, K. Y. Goldberg, A. S. Rao, and J. Wiegley. A pivoting gripper for feeding industrial parts. In *Intl. Conf. on Robotics and Automation*, pages 1650–1655. IEEE, May 1994.

[2] J. Hershberger. Finding the upper envelope of n line segments in $O(n \log n)$ time. *Inform. Process. Lett.*, 33:169–174, 1989.

[3] D. J. Kriegman. Computing stable poses of piecewise smooth objects. *CVGIP: Image Understanding*, 55(2), March 1992.

[4] A. Rao, D. Kriegman, and K. Goldberg. Complete algorithms for reorienting polyhedral parts using a pivoting gripper. Technical Report UU-CS-1994-49, Utrecht University, 1994. Submitted to the IEEE *Transactions on Robotics and Automation*. Available by anonymous ftp from ftp.cs.ruu.nl /pub/RUU/CS/techreps/CS-1994/1994-49.ps.gz.

Geometry in GIS is not Combinatorial: Segment Intersection for Polygon Overlay

D. S. Andrews   J. Snoeyink*
Dept of Computer Science
University of British Columbia

1 Introduction

Map overlay processing is at the core of most vector-based Geographic Information Systems (GISs). One of the time-consuming steps in overlay processing is detecting intersections from segments in two maps: the *red/blue segment intersection problem*. Computational geometers are prone to say [6] "An easy solution would test all pairs, thus requiring $O(n^2)$ time in the worst case. A *much better* approach exists (Bentley-Ottmann), though, requiring only $O(n \log n)$ time." This is not the end of the story in practice, however; one cannot ignore characteristics of the input data.

In [1], we grouped seven algorithms for segment intersection into two categories: *spatial partitioning algorithms* that divide space into regions and compute intersections among the segments that pass through each region, and *spatial ordering algorithms* that use the structure given by a geometric order to reduce the number of pairs of segments that must be tested for intersection. Although spatial ordering algorithms can be made *output-sensitive*—their running time can be analyzed in terms of the size of the output—our experiments have shown that spatial partitioning algorithms give better performance on GIS data sets, corroborating Pullar's work [8].

While we found this result disappointing, in some ways it is to be expected. In a problem where the input and output are completely unstructured, approaches that use geometric structure internally do not show their full strength. See Andrews et al. [1] for more details.

---

*Supported in part by an NSERC Research Grant and a fellowship from the B.C. Advanced Systems Institute.

Permission to copy without fee all or part of this material is granted provided that the copies are not made or distributed for direct commercial advantage, the ACM copyright notice and the title of the publication and its date appear, and notice is given that copying is by permission of the Association of Computing Machinery.To copy otherwise, or to republish, requires a fee and/or specific permission.
11th Computational Geometry, Vancouver, B.C. Canada
© 1995 ACM 0-89791-724-3/95/0006...$3.50

2 Algorithms & Data Sets

We analyzed brute force (bforce), three spatial partitioning (grid, based on a uniform grid [4]; quad, based on a quadtree [9]; bsp, binary space partition [5]), and three spatial ordering algorithms (BO, Bently-Ottmann [2]; trap, trapezoid sweep [3]; segT, hereditary segment tree [7].)

Data Set	# Segs	Description
Hn	n	Long, near-horizontal segs
Vn	n	Long, near-vertical segs
surv	239	Survey grid, Littleton, CO
rail	447	Railroads. strictly N to S
feat	564	Geographical features. many clusters
bound	1176	Boundaries. Often follow roads
vegit	5562	Vegitation. Small clusters
roads	11074	Roads and trails. clustered in town
hydro	13972	Hydrography. Evenly distrib.
comp	8053	Research, M. Knapp Forest, B.C.
froads	19532	Forest roads. Evenly distrib.
fcover	116359	Forest cover. Evenly distrib.
biogeo	235635	Bio and geo features. Evenly distrib.

Table 1: Characteristics of test data

To test algorithms on data with many intersections we generated a grid pattern. To give more realistic tests of overlay, we used 1:24,000 Digital Line Graph (DLG) data on Littleton, CO, from xerox.spectrum.com and data on the Malcolm Knapp Research Forest from Jerry Maedel in the UBC Department of Forestry. Table 1 describes data characteristics.

3 Results

Running times on the artificial and Littleton data were measured on a Silicon Graphics Personal Iris with a 12 Mhz processor and 16M of memory. The Research Forest data sets were run on a SGI Indigo Elan with a 33 Mhz processor and 32M of memory. The average times for 10 runs are given in seconds. Caveat: the algorithms were implemented by different programmers, which has a significant affect on running times and memory usage.

Fig. 1 displays the cost in milliseconds per *intersection* for the grid of horizontal and vertical segments. The Bentley-Ottmann sweep (BO) is off the graph because it does balanced-tree operations for every intersection. The hered. segment tree runs in subquadratic time—even with a quadratic number of intersections—because it reports intersections in batches.

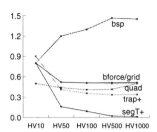

Fig. 1: Cost (μsec) per *intersection*: artificial data

For a more realistic comparisons, we run algorithms on surv and another GIS data set and graph the time taken per line segment in milliseconds. (Our our quad tree algorithm depends on the order in which the files are presented.) Here the spatial partitioning algorithms are the best. They take advantage of the characteristics of GIS data (short segments, evenly distributed) and the fact that the number of intersections per segment is small. They also benefit most from the fact that the surv data set is small; the spatial ordering algorithms have an overhead associated with looking at the structure of the larger data set. The trapezoid sweep is somewhat competitive; it would be more so if initial sorting had been done in preprocessing.

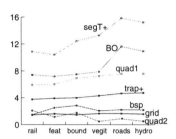

Fig. 2: Cost per *segment*: ∩ with surv

If we concentrate on algorithms that have performed well above, we can attempt larger data sets. We report the times for trap including and excluding preprocessing.

Again, the spatial partitioning algorithms perform the best, but the trapezoid sweep is competitive—especially if the segments are already stored in sorted order. The memory requirements of the quadtree algorithm and the bsp tree algorithm did not allow us to run many tests on the research forest data.

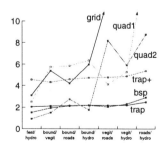

Fig. 3: Large data sets (μsecs/segment)

To obtain problem instances with a single data distribution and number of segments, we intersected two copies of a map—one displaced slightly relative to the other. The trapezoid algorithm without counting preprocessing is best here, followed by the binary space partition, and then the trapezoid algorithm counting preprocessing.

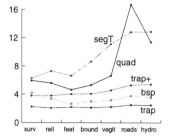

Fig. 4: Bumped data (μsecs/segment)

Acknowledgements

Implementations were by J. Boritz, T. Chan, G. Denham, J. Harrison, C. Zhu, as well as ourselves. This research was supported in part by NSERC and BC ASI. We thank Tom Poiker for encouragement to investigate the GIS overlay problem.

References

[1] D. S. Andrews, et al. Further comparison of algorithms for geometric intersection problems. In *Proc. 6th Intl. Symp. Spatial Data Handling*, pages 709–724, Edinburgh, UK, 1994.

[2] J. L. Bentley and T. A. Ottmann. Algorithms for reporting and counting geometric intersections. *IEEE Trans. Comput.*, C–28(9):643–647, 1979.

[3] T. M. Chan. A simple trapezoid sweep algorithm for reporting red/blue segment intersections. In *Proc. 6CCCG*, 1994.

[4] W. R. Franklin. Efficient intersection calculations in large databases. In *Int'l Cartog Assoc 14th World Conf.*, pages A62 – A63, Budapest, Aug. 1989.

[5] H. Fuchs, Z. Kedem, and B. Naylor. On visible surface generation by a priori tree structures. In *Proc. SIGGRAPH '80*, pages 124–133, 1980.

[6] M. Overmars. Teaching computational geometry. Tech Rep RUU–CS–93–34, Utrecht Univ., 1993.

[7] L. Palazzi and J. Snoeyink. Counting and reporting red/blue segment intersections. *CVGIP: Graph. Mod. Image Proc.*, 56(4):304–311, 1994.

[8] D. Pullar. Comparative study of algorithms for reporting geometrical intersections. *Proc. 4th Int'l Symp Spatial Data Handling*, pages 66–76, 1990.

[9] H. Samet. *The Design and Analysis of Spatial Data Structures*. Addison-Wesley, 1989.

On Levels of Detail in Terrains*

Mark de Berg[1]   Katrin Dobrindt[1]

Abstract. In many applications it is important that one can view a scene at different levels of detail. More precisely, one would like to visualize the part of the scene that is close at a high level of detail, and the part that is far away at a low level of detail. We propose a hierarchy of detail levels for a polyhedral terrain that allows this: given a view point, it is possible to select the appropriate level of detail for each part of the terrain in such a way that the parts still fit together continuously. The new feature of our method is that it allows the use of Delaunay triangulations at every level.

1 Introduction

A *terrain* is the graph of a continuous function that assigns to every point on the plane an elevation. Terrains model mountain landscapes and are thus an important class of scenes in several areas—geographic information systems (GIS) and flight simulation are obvious examples. For rendering purposes it is convenient to model a terrain as a collection of triangles. Such a representation is called a *triangulated irregular network*, or TIN, in geographic information systems, and it is called a *polyhedral terrain* in computational geometry.

For a realistic representation of a terrain millions of triangles are needed. In applications such as flight simulation a terrain should be rendered at real time, but even with modern technology it is impossible to achieve this when the number of triangles is this large. Fortunately, a realistic image of the terrain is only crucial when one is close to the terrain and only a small part of the terrain is visible; when one is flying high above the terrain a coarse representation suffices. What is needed is a *hierarchy* of representations at various levels of detail; this makes it possible for a given view point to render the terrain at an adequate level of detail. Subsequent levels of the hierarchy should not differ too much in appearance: switching to more and more detailed representations should not cause disturbing 'jumps' in the image. On the other hand, the reduction of the number of triangles in subsequent levels should not be too small, otherwise too many levels would be needed, resulting in an unacceptable increase in storage. It is in general insufficient for rendering purposes to use only one level of detail at a time: although some part of the terrain may be close to the view point, another part can still be far away. Hence, one would like to combine parts from different levels into a single representation of the terrain, such that each part of the terrain is rendered with appropriate detail. This means that the levels cannot be completely independent, as it should be possible to glue them together continuously.

We study this problem in the following setting. We are given a collection of data points in the plane, each with its own elevation. We wish to compute a hierarchy of levels, where the most detailed level corresponds to a triangulation of the set of data points. Less detailed levels should correspond to triangulations of subsets of the set of data points.

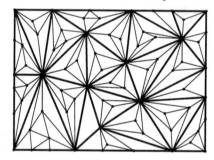

Figure 1: A two-level hierarchy with a tree structure.

There are several papers dealing with this problem—De Floriani et al. [1] give an overview. Previous approaches can be subdivided into two categories. In the first category each triangle at a certain level is replaced by a collection of smaller triangles at the next level. This results in a hierarchy with a tree structure. Hence, it is very easy to combine parts of different levels into one representation of the terrain. The disadvantage of this method is that it produces very skinny triangles at higher detail levels—see Figure 1. (It is also possible to generate extra data points to avoid small angles, but this will result in non-conforming triangulations when detail levels are combined.) The second category uses the Delaunay triangulation [4] of the 2D data points at every level. This way small angles are avoided, which reduces robustness problems and aliasing problems. The resulting hierarchy is not a tree but a directed acyclic graph, which makes it difficult to combine parts from different levels into one representation of the terrain. Whether this could be done was mentioned as an open problem by De Floriani et al. [1].

We show that constructing the hierarchy with Kirkpatrick's

triangulation refinement method [2] makes it possible to use Delaunay triangulations and still be able to combine different levels into one.

2 The method

We start with the finest level, which is the Delaunay triangulation of the given set of n data points. Next we remove an *independent* set of pairwise non-adjacent vertices of bounded degree from the triangulation, and we retriangulate using the Delaunay triangulation. The definition of the Delaunay triangulation ensures that edges whose vertices remain present also remain present themselves. Hence, we only have to retriangulate the polygons surrounding the removed vertices, as in the original triangulation refinement method. Figure 2 illustrates this: the data points indicated by circles are removed and their surrounding polygons now have to be retriangulated using the Delaunay criterion. Thus for each re-

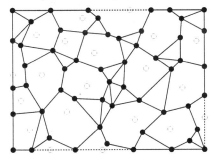

Figure 2: Generating the next level by retriangulation.

moved vertex, there is a group of new triangles that replaces a group of old triangles. We continue the process of deleting an independent set and retriangulating until we have obtained the coarsest level we want. (If we continue the process to a level of constant complexity, we can use the hierarchy for point location and windowing operations as well.) The structure we get is a directed acyclic graph and uses $O(n)$ storage.

Of course we must be careful not to remove vertices that are important for the shape of the terrain, such as peaks, passes, pits, points with high gradient. Therefore we assign every vertex an importance which is computed using existing methods [3]. We try to choose the vertices for the independent set in order of increasing importance. This ensures a small approximation error between two subsequent levels. We also allow the user to *fix* vertices that are very important; fixed vertices are never removed. The vertices of the convex hull of the 2D data points are always fixed, so that the terrain keeps its original size. It can be shown that if we fix k vertices, then it is still possible to find a large enough independent set as long as the number of vertices is some constant factor larger than k.

Because the hierarchy does not have a tree-like shape, the combination of triangles from different levels into one representation seems difficult. But our construction ensures that a group of a constant number of triangles is replaced by a group of triangles at the following level. This is the property that allows us to combine different levels: although we cannot select individual triangles from each level, we *can* select complete groups of triangles. The collection of the so selected triangles always forms the Delaunay triangulation of the vertices in the combination—see Figure 3.

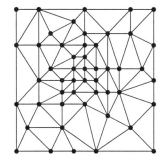

Figure 3: Combination of parts from different levels of detail. The finest level is the Delaunay triangulation of 13x13 grid points.

We have implemented our approach and tested it on data sets that are based on regular grids obtained by anonymous ftp from edcftp.cr.usgs.gov:/pub/data/DEM/250. In our tests, the storage (in bytes) increases by a factor of about 5—see Table 1.

	$k = \sqrt{n}$	k=4
Lake Charles (West)	5.06	5.00
Bangor (East)	4.97	4.94
San Bernardino (East)	5.08	5.03
Mount Marcy	5.01	4.98
Random TIN	5.19	5.10

Table 1: Storage constants (in bytes) for some data sets (based on regular grids and TINS) for different numbers k of fixed points. The constants are computed with respect to the initial triangulation.

The possibility to combine different levels of detail reduces the render time significantly, as compared to the time when the whole terrain is rendered at one (high enough) level of detail. For example, rendering a terrain consisting of 32,258 triangles at the finest level takes 2.55 secs per frame (SGI Indigo2 R4400SC, GU1-Extreme graphics board), while rendering a combination, which consists of 6,107 triangles of the finest up to the coarsest level, takes only 0.460 secs per frame (+ 0.600 secs extraction time).

Acknowledgement. We thank Jack Meijerink for implementing a first version of the algorithm.

References

[1] L. De Floriani, P. Marzano, and E. Puppo. Hierarchical terrain models: Survey and formalization. In *Proc. ACM Sympos. Applied Comput.*, 1994.

[2] D. G. Kirkpatrick. Optimal search in planar subdivisions. *SIAM J. Comput.*, 12:28–35, 1983.

[3] J. Lee. A comparison of existing methods for building irregular networks models of terrain from grid digital elevation models. *Int. J. of GIS*, 5:267–286, 1991.

[4] F. P. Preparata and M. I. Shamos. *Computational Geometry: an Introduction*. Springer-Verlag, New York, NY, 1985.

An Animated Library of Combinatorial VLSI–Routing Algorithms*

Dorothea Wagner[†]   Karsten Weihe[†]

We present an animated library, *CRoP*, of combinatorial algorithms for certain classes of VLSI problems. The aim of the package is to demonstrate and teach those algorithms. For this, *CRoP* comes with a graphical interface for animating the integrated algorithms, and in fact, *CRoP* has already been used for graduate courses on theoretical VLSI design and combinatorial optimization, and for technical presentations of algorithms in workshops. See the full version for details and experiences. Currently, we are going to weave an interface to *CRoP* into the World Wibe Web.[‡]

All algorithms contained in *CRoP* are combinatorial algorithms for various special cases of the *switchbox routing problem*.[§] In its most general version, this means the following. We are given a rectangular grid, and each boundary node is assigned a number $0, \ldots, k$. The problem is to find edge–disjoint Steiner trees (*wires*), $T_1, \ldots, T_k$, in the grid such that T_i connects all boundary nodes with number i (the *terminals* of *net i*) and no boundary edge is occupied. If the size of the grid does not suffice, additional rows and columns may be added. The main objective is to minimize the number of rows and columns added, but other objectives must often be incorporated as well.

During the VLSI design process of integrated circuits, this problem occurs after the *placement phase* and specifies the courses of the wires through the grid (*layout*). More precisely, the layout is only the first step. In order to avoid shortcuts between different wires, a second step, the *wiring*, is necessary. Therein, the Steiner trees are embedded in a three–dimensional grid, which is formed by a number of copies of the switchbox stacked on top of each other (*layers*), and additional connections (*vias*) between corresponding nodes of different layers. This procedure gives rise to two further optimization criteria: the number of layers and the number of vias.

Most combinatorial layout algorithms apply only to special cases of the general switchbox problem. One important special case is the *channel routing problem*, where only terminals on the upper and lower boundary are allowed. Another special case, which appears seldom in practice but is of great theoretical interest, is the case that all nets have exactly two terminals. See [5] for an exhaustive survey and for a discussion of theoretical and practical significance.

Two different types of algorithms are integrated in *CRoP* so far: *layout* algorithms and *wiring* algorithms. The layout algorithms are described in Refs. [2,3,4,6,7,8,10], and the wiring algorithms, in Refs. [1,4,8,9]. Each of the algorithms in Refs. [2,7,9,10] is implemented in two versions as described in the respective reference. The figure shows a channel consisting of two–terminal nets, and the solution constructed by the algorithm in [10].

Moreover, together with two other members of our group, *Jürgen Kaminski* and *Stephan Hartmann*, we are currently developing, implementing, and integrating new heuristics for $\mathcal{NP}$–hard cases. For this, the graphical display provided by *CRoP* has proven to be of great advantage for "fine tuning."

The graphical display not only shows the output of an algorithm. There is also the opportunity to slow down the graphical display until a human eye is able to follow what the algorithm is doing. Besides that, certain auxiliary features, which are used by an algorithm but do not appear in its output, can be displayed in addition. Finally, two algorithms can be compared by displaying the layouts determined by the algorithms both at the same time.

The implementation consists of a callable library of the algorithms and a graphical environment. Therefore, algorithms may be used independently of the graphical

---

[*]This research was supported by the Technische Universität Berlin under grant FIP 1/3. Full version: by *anonymous ftp* as ftp.math.tu-berlin.de/pub/Preprints/combi/Report-354-1993.ps.Z and via *WWW* from *URL*
http://www.informatik.uni-konstanz.de/~weihe/manuscripts.html/#paper10.

[†]Universität Konstanz, Informatik, 78434 Konstanz, Germany, *e–mail*: {dorothea.wagner,karsten.weihe}@uni-konstanz.de

[‡]At a URL accessible via http://winnie.math.tu-berlin.de.

[§]The name *CRoP* is an acronym of **C**hannel **Ro**uting **P**roblems, which is the class of switchbox problems for which the library had originally been designed.

interface. Originally, *CRoP* has been implemented in PASCAL on Apollo workstations, using Apollo's own graphic tools. Recently, *CRoP* has been translated into the C programming language and adapted to the X11 Window system and runs now on Sun Sparc stations. The C sources amount to about 800 Kbytes.

A session with *CRoP* usually consists of several stages: constructing an instance, running a layout algorithm, running a wiring algorithm, and displaying additional information. An instance may be constructed interactively, determined by one of the built-in random generators, or taken from the database of instances.

After generating an instance, one may select one or two layout algorithms. The results are displayed in a separate window, which is popped up right for this aim. The procedure can be shown in slow motion, with an arbitrary delay time between two atomic actions (multiples of 1/10 sec).

Once the layout phase is completed, the user may ask *CRoP* to run one of the wiring algorithms or display additional information. Moreover, while the display window is open, one keystroke suffices to dump the current display to a file in *PostScript* format. (This is the way the figure has come into the paper.)

Our main experience is that a tool like this may provide new, deep, surprising insights into existing theoretical algorithms and, even better, to design *new* algorithms (which have already been integrated or are going to be integrated in *CRoP*, too). See the full version for more details. Currently, we are working on a more extensive library for disjoint paths and Steiner tree algorithms in *general planar graphs*. For this (much more ambitious) system we use the powerful modeling features of C++. Once the latter system is matured, *CRoP* will become an integral part of it.

Acknowledgement We thank *Jürgen Kaminski* for implementing the algorithms in [7] and the second version of [9]. Moreover, we thank *Thomas Hecker, Dietmar Kühl*, and *Jörn Schulze* for translating *CRoP* into the C language and adapting it to the X11 Window System.

References

1 M.L. Brady and D.J. Brown. VLSI routing: Four layers suffice. In F.P. Preparata, editor, *Advances in Computer Research, VOL 2: VLSI Theory*, pages 245–257. JAI Press Inc., 1984.

2 M. Formann, D. Wagner, and F. Wagner. Routing through a dense channel with minimum total wire length. *J. Algorithms*, 15:267–283, 1993.

3 A. Frank. Disjoint paths in a rectilinear grid. *Combinatorica*, 2:361–371, 1982.

4 R. Kuchem, D. Wagner, and F. Wagner. Area-optimal three-layer channel routing. In *Proceedings of the 30th Annual Symposium on Foundations of Computer Science, FOCS'89*, pages 506–511, 1989.

5 T. Lengauer. *Combinatorial Algorithms for Integrated Circuit Layout*. Wiley-Teubner, 1990.

6 K. Mehlhorn and F.P. Preparata. Routing through a rectangle. *J. Assoc. Comp. Mach.*, 33:60–85, 1986.

7 K. Mehlhorn, F.P. Preparata, and M. Sarrafzadeh. Channel routing in knock-knee mode: Simplified algorithms and proofs. *Algorithmica*, 1:213–221, 1986.

8 F.P. Preparata and W. Lipski, Jr. . Optimal three-layer channel routing. *IEEE Trans. Comp.*, C-33:427–437, 1984.

9 M. Sarrafzadeh, D. Wagner, F. Wagner, and K. Weihe. Wiring knock-knee layouts: A global approach. *IEEE Trans. Comp.*, 43:581–589, 1994.

10 D. Wagner. Optimal routing through dense channels. *Int. J. on Comp. Geom. and Appl.*, 3:269–289, 1993.

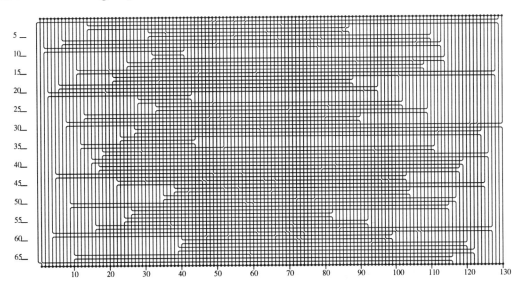

Printed Circuit Board Simplification: Simplifying Subdivisions in Practice*

Berto van de Kraats   Marc van Kreveld   Mark Overmars

Department of Computer Science, Utrecht University

1 Introduction

The line simplification problem—reducing the number of points to represent a polygonal line—has been studied extensively in digital image analysis, cartography, and computational geometry (references can be found in [6, 7]). Motivation includes reducing storage requirements, computation speed, and object recognition. This note discusses the problem of subdivision simplification for printed circuit board (PCB) layout. Due to extra constraints on the simplification, none of the existing methods can be applied directly. When a new circuit board is designed, it is necessary to test if it doesn't emit too much radiation. Simulating the radiation is important in speeding up the design of good printed circuit boards. At Philips Research such simulation tools are being developed [4]. Their approach requires a mesh of quadrilaterals that is built on the space between the components on the circuit board. The simulations are computationally quite expensive (typically measured in hours). The time required is mainly dependent on two parameters: the number of mesh elements, and their shape. The number of mesh elements is for the larger part dependent on the number of corners in the circuit board components. (See [3] for a discussion on mesh generation and [5] for meshes of quadrilaterals.) Laws in physics state that a signal of a certain frequency can only be influenced by an obstacle if the obstacle is larger than the wavelength of the signal. This means that a lot of detail on the circuit board may be eliminated prior to the simulation.

The circuit board is represented by a rectangle with disjoint polygonal holes. The simplification of the holes will be discussed. It is known that some subdivision sim-

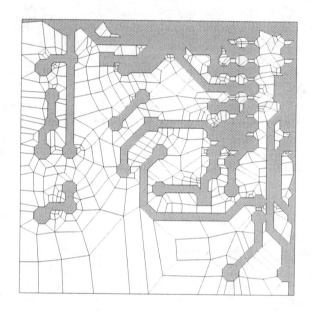

plification problems that minimize the output complexity of the simplification are NP-complete [6] (see also [2] for subdivision simplification). Therefore, a heuristic approach has been taken.

2 PCB constraints

There are several constraints in PCB simplification which are not considered in standard line simplification methods. Besides reducing the complexity of the polygonal holes, we have the following constraints:

1. The approximation of each hole must enclose the hole.
2. The approximation must be topologically equivalent (no holes may join or disappear).
3. The approximation of each hole must remain within distance $\epsilon > 0$ of the hole.
4. No small angles may be introduced (specified by a parameter $\theta > 0$).
5. No narrow passages may be created (a passage may be reduced in narrowness with at most a factor $0 < \delta < 1$).

---

*This research was supported by ESPRIT Basic Research Action 7141 (project ALCOM II: *Algorithms and Complexity*), and by a PIONIER project of the Dutch Organization for Scientific Research N.W.O.

Usually, only the third constraint is considered in line simplification methods. The last constraint is difficult to formulate exactly [8].

3 Simplification ideas

To obtain the simplification, we define three operators, each of which reduces the number of vertices of a hole. Before an operator is applied, we test whether or not its application violates one of the constraints. The operators are:

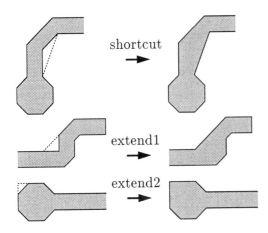

The *shortcut* bypasses a hole vertex by connecting the two neighbors, the *extend1* extends one edge to bypass a hole vertex, and the *extend2* extends two edges to bypass the edge in between.

First the algorithm applies all shortcuts until none is possible, then all possible extend1's are applied, and finally all possible extend2's are applied. It could be that after the extend2's new shortcuts are possible. Hence, we could repeat the process. The implementation doesn't try this. There are two reasons for applying the operators in the given order: complexity reduction and efficiency of computation (see [8] for more details).

Testing the constraints for each operator requires $O(n \log n)$ time in the implementation. The most difficult test is for the third constraint. When a sequence of operators has already been applied, the original and the approximation of a part of a hole may be disjoint over a number of consecutive vertices. To test for the applicability of an operator with respect to the third constraint, the Hausdorff distance between two polygonal chains must be computed [1]. The whole algorithm requires $O(n^2 \log n)$ time. Reduction is expected to be possible but not important in our application due to the complexity of the simulation itself.

4 Results and conclusions

The algorithm has been applied to six quite different circuit boards. Rather than indicating the number of vertices of the PCB we indicate the number of quadrilaterals after the mesh generation, because that is what we were interested in. The table below summarizes the results for reasonable values of the parameters ϵ, θ, δ.

no. of quads	no. after shortcuts	no. after extend1's & extend2's	total reduction
2319	986	925	60%
2146	1244	904	57%
1323	1137	1097	17%
488	412	384	21%
465	304	204	56%
74	52	52	29%

Other experiments showed that in practice, the values of distance parameter ϵ and the corridor parameter δ affect the reduction, but the minimum allowed angle θ hardly affects it. As we hoped the simplification did hardly change the result of the simulation but improved the running time considerably.

References

[1] H. Alt, B. Behrends, and J. Blömer, Approximate matching of polygonal shapes, *Proc. 7th ACM Symp. Comp. Geom.* (1991), pp. 186–193.

[2] M. de Berg, M. van Kreveld, and S. Shirra, A New Approach to Subdivision Simplification, *Proc. Auto-Carto 12* (1995), to appear.

[3] M. Bern and D. Eppstein, Mesh generation and optimal triangulation, in: F.K. Hwang and D.-Z. Du (Eds.), *Euclidean Geometry and the Computer*, World Scientific, 1993.

[4] R. du Cloux, G.P.J.F.M. Maas, A.J.H. Wachters, R.F. Milsom, and K.J. Scott, FASTERIX, an environment for PCB simulation, *Proc. 11th Int. Conf. on EMC* (1993), Zürich, Switzerland.

[5] H. Everett, W. Lenhart, M. Overmars, T. Shermer, and J. Urrutia, Strictly convex quadrilateralizations of polygons, *Proc. 4th Can. Conf. Comp. Geom.* (1992), pp. 77–82.

[6] L.J. Guibas, J. Hershberger, J.S.B. Mitchell, and J.S. Snoeyink, Approximating polygons and subdivisions with minimum link paths, *Int. J. Comp. Geom. & Appl.* **3** (1993), pp. 383–415.

[7] J.D. Hobby, Polygonal approximations that minimize the number of inflections, *Proc. 4th ACM-SIAM Symp. Discr. Alg.* (1993), pp. 93–102.

[8] B. van de Kraats, *Approximation of Polygons with Applications to Circuit Layout*, Report INF/SCR-94-39, Dept. of Comp. Science, Utrecht University, 1994.

An Implementation for Maintaining Arrangements of Polygons

Michael Goldwasser *
Stanford University

Abstract

Constructing arrangements of geometric objects is a basic problem in computational geometry. Applications relying on arrangements arise in such fields as robotics, assembly planning, computer vision, graphics, and computer-assisted surgery. Arrangements are also used as a building block for other theoretical results in computational geometry. Many papers and textbooks have presented algorithms for maintaining arrangements under various conditions.

This paper is a discussion of the practical issues that arose during the development of a software package which constructs an arrangement of polygons and segments using a basic randomized incremental approach. The need to handle polygons in addition to segments, and to deal with arrangements on a sphere as well as a plane, guided many design decisions. Also the need to cope with degeneracies and numerical inaccuracy in an efficient and consistent manner, brought up issues that are often glossed over in theoretical presentations of algorithms.

This software is written in C and is available via ftp at `flamingo.stanford.edu` in the directory `/pub/wass/arrangement/`.

1 Project Background

This software was originally developed for use in the Stanford Assembly Analysis Tool (STAAT), an assembly planner for polyhedral parts [4]. For each pair of parts, analysis is performed to calculate the set of directions in which one part will collide with the other. This region is represented as a polygon on the sphere of directions where vertices are points on the sphere and edges are great circle arcs. The overall arrangement of these polygons is then traversed and analyzed to develop possible assembly plans.

2 Features and Limitations

There are several features of this software which allow extra flexibility that was necessary for the original project.

- The basic objects are polygons rather than isolated segments. The important distinction between the representation of a polygon as an object versus as a set of individual edges is that the resulting arrangement must *guarantee* that common endpoints in the polygon are truly unified in the arrangement.

- A polygon may be degenerate and represent a simple segment, or even an isolated vertex. Therefore this software has the generality to handle arrangements of segments.

- Many algorithms are designed for arrangements of non-intersecting objects; however in this application, polygon segments may intersect segments from other polygons.

- The input polygons may lie on the sphere rather than on the plane. Although this brings up some key programming issues, it turns out to be a minor conceptual issue.

The notable limitation of this software is the fact that it is only semi-dynamic. For this application, polygons could be incrementally inserted however there was no need to delete polygons from the arrangement.

3 Algorithm

We represent the arrangement using a vertical decomposition into trapezoids[1]. To avoid the maintenance of faces with arbitrarily many edges, we decompose each face into a number of trapezoids. This is done by extending an imaginary vertical *thread* from each vertex of the face upwards (downwards) into the face until an edge of the face is reached. See Figure 1.

---

*Department of Computer Science, Stanford University, Stanford, CA, 94305. E-mail: `wass@cs.stanford.edu`. Research supported by a grant from the Stanford Integrated Manufacturing Association (SIMA), by NSF/ARPA Grant IRI-9306544, and by NSF Grant CCR-9215219.

The algorithm used for construction is a basic randomized incremental construction, similar to that of Mulmuley[3] or Seidel[5]. We give a brief explanation here, but we refer the reader to [3], [5] for details.

We maintain two data structures. The primary structure stores the features of the actual arrangement, as well as all the topological information connecting neighboring features. The secondary structure, a *layered dag*[2], allows us to take a new query point, and efficiently determine which cell of the arrangement contains it. Our algorithm for incrementally inserting a new polygon uses the point location query to locate one of the endpoints, and then uses a "travel and split" approach to insert an edge, while updating the data structures. Once we reach the other endpoint, we continue with the next edge of the polygon.

While inserting a new edge, if we intersect an existing vertex, we break the new edge into two pieces at this vertex. If we intersect an existing edge, we insert a new vertex at this intersection point, and we break both edges into two pieces at this new vertex. The vertical threads incident to each edge do not create vertices or break the edge. Instead, each edge maintains an ordered list of incident threads. These lists are used to cross from one side of an edge to the other.

4 Spherical Arrangements

Maintaining arrangements on a sphere versus on a plane seems like a very different task at the implementation level. However, they can both be handled by the identical package with a slight bit of care. The key is that there is an exact correspondence between the spherical coordinate system and the double-sided projective plane by viewing the coordinates (x, y, z) as homogeneous with weight z. A good explanation is given by Stolfi in [6].

To handle the sphere, we break it into two hemispheres using a "great meridian," M. Then we maintain the arrangement on both hemispheres separately, yet in a way so that the features are *glued* properly at M. On each hemisphere, concepts such as "vertical", "left", and "above" can be defined. To define *vertical*, we choose a pair of antipodes lying on the meridian and consider these to be the top and bottom poles. The set of vertical threads corresponds to the pencil of half circles passing through these poles.

5 Experimental Results

This software was developed in C for both a Decstation 3100 as well as a Dec Alpha workstation. It is currently used successfully in the STAAT assembly planning project. The package has been used there to con-

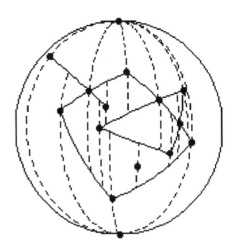

Figure 1: A Spherical Arrangement

struct an arrangement with over 1600 vertices, 3800 edges, and 5300 trapezoids. Such an arrangement was constructed in just under 2 seconds on a Dec Alpha using floating point calculations.

6 Source Code Available

This software is written in C and is available via ftp at flamingo.stanford.edu in the directory /pub/wass/arrangement/.

Acknowledgements: The author wishes to thank Dan Halperin, Leo Guibas, Cyprien Godard, and G. D. Ramkumar for their helpful discussions. Also, extra thanks to Bruce Romney for reviewing thousands of lines of source code.

References

[1] B. Chazelle and J. Incerpi. Triangulation and shape-complexity. *ACM Trans. Graph.*, 3(2):135–152, 1984.

[2] H. Edelsbrunner, L. J. Guibas, and J. Stolfi. Optimal point location in a monotone subdivision. *SIAM J. Comput.*, 15:317–340, 1986.

[3] K. Mulmuley. *Computational Geometry: An Introduction Through Randomized Algorithms*. Prentice Hall, New York, 1993.

[4] B. Romney, C. Godard, M. Goldwasser, and G. Ramkumar. An efficient system for analyzing assembly complexity. Manuscript, Dept. Comput. Sci., Stanford Univ., Stanford, CA, 1994.

[5] R. Seidel. A simple and fast incremental randomized algorithm for computing trapezoidal decompositions and for triangulating polygons. *Comput. Geom. Theory Appl.*, 1:51–64, 1991.

[6] J. Stolfi. *Oriented Projective Geometry: A Framework for Geometric Computations*. Academic Press, 1991.

Minimal enclosing parallelogram with application*

Christian Schwarz[†]   Jürgen Teich[‡]   Alek Vainshtein[§]   Emo Welzl[¶]   Brian L. Evans[‖]

We show how to compute the smallest area parallelogram enclosing a convex n-gon in the plane in linear time, and we describe an application of this result in digital image processing.

Related work has been done on finding a minimal enclosing triangle, see e.g. [OAMB86], a minimal enclosing rectangle [FS75], a minimal enclosing k-gon [ACY85], and a minimal enclosing k-gon that has sides of equal lengths or a fixed-angle sequence [DA84]. Note that whereas e.g. a rectangle would be contained in the latter class of polygons, our problem is different since the angles of the desired enclosing parallelogram are not given in advance. Nevertheless, our method clearly borrows from the techniques developed in the computational geometry literature, and our contribution is to show how these methods can help to obtain the result as requested by the application. In fact, we learned that the linear time algorithm has been previously published in a Russian journal, [Vai90].

There are two key facts which lead to the algorithm. First, let us consider the edges e_1, e_2, e_3 and e_4 of an enclosing parallelogram (in counterclockwise order), and let l_1, l_2, l_3 and l_4, respectively, be their supporting lines. Then there is an optimal enclosing parallelogram which has at least one of the edges e_1 and e_3 flush with an edge of the convex polygon, and also one edge of e_2 and e_4 flush with an edge of the polygon. There are at most n pairs of parallel tangents to an n-gon, where at least one supports an edge. This already leads to an $O(n^2)$ algorithm.

A second stronger condition for any optimal parallelogram reads as follows: There is a line l parallel to l_1 (and so to l_3), which intersects the polygon in two points touched by edges e_2 and e_4. Similarly, a symmetric statement holds for a line parallel to l_2 and l_4. This excludes many pairs of directions for the edges of an optimal parallelogram. There are only a linear number of such combinations possible, and we can scan through those in a "rotating calipers"-fashion in linear time.

The implementation showed that, in fact, the linear-time algorithm is not substantially more difficult to implement than the quadratic algorithm. Furthermore, the linear-time algorithm is faster than the quadratic algorithm for even the smallest problem size of $n = 5$.

The application that motivated this research is compressing two-dimensional signals (e.g. images) based on their frequency content [ETS94]. Specifically, we are interested in designing rational decimation systems that reduce the number of input samples by a rational factor. A rational decimation system extracts a specific portion of the frequency content (the passband) and resamples the resulting signal at its Nyquist rate. Rational decimation systems are realized by a cascade of four linear operators—modulator, upsampler, filter, and downsampler. In the two-dimensional case, the modulation factor n_0, the upsampling matrix L, the filter passband specifications, and the downsampling matrix M can be computed directly from any parallelogram that circumscribes the passband and has vertices which are rational multiples of π. Therefore, in order to optimize the compression ratio of the overall system, we need to find the parallelogram of minimal area that circumscribes the passband.

Our design procedure takes the vertices of the desired passband and returns the decimator system parameters n_0, L, and M. The vertices would be sketched with a mouse, typed in, or defined by mathematical formulas. The design algorithm amounts to (1) snapping the vertices of the desired passband to grid points that are rational multiples of π, (2) finding the convex hull of the rational vertices, (3) computing the minimal enclosing parallelogram, and (4) calculating the design parame-

---

*This research has been carried out while C. Schwarz and E. Welzl were at the International Computer Science Institute (ICSI), Berkeley, CA, USA, and while J. Teich was at the Dept. of Electrical Eng. and Computer Sci., Univ. of California, Berkeley, USA.

[†]Max-Planck-Institut f. Informatik, Im Stadtwald, D-66123 Saarbrücken, Germany. E-mail: schwarz@mpi-sb.mpg.de

[‡]Institut TIK, ETH Zürich, Gloriastr. 35, CH-8092 Zürich, Switzerland. E-mail: teich@tik.ethz.ch

[§]School of Math. Sci., Tel Aviv Univ., Tel Aviv, Israel. E-mail: alek@math.tau.ac.il

[¶]Freie Univ. Berlin, Inst. f. Informatik, Takustr. 9, D-14195 Berlin, Germany. E-mail: emo@inf.fu-berlin.de

[‖]Dept. of Electrical Eng. and Computer Sci., Univ. of California, Berkeley, CA, 94720, USA. E-mail: ble@eecs.berkeley.edu; supported by the Ptolemy Project, which is funded by the ARPA RASSP Program (F33615-93-C-1317), SRC (94-DC-008), NSF (MIP-9201605), ONR, the MICRO Program, and several companies.

Permission to copy without fee all or part of this material is granted provided that the copies are not made or distributed for direct commercial advantage, the ACM copyright notice and the title of the publication and its date appear, and notice is given that copying is by permission of the Association of Computing Machinery. To copy otherwise, or to republish, requires a fee and/or specific permission.
11th Computational Geometry, Vancouver, B.C. Canada
© 1995 ACM 0-89791-724-3/95/0006...$3.50

```
{n0, L, M} =
  DesignDecimationSystem2D[
    sketchedPolyVertexList,
    Justification -> All, Mod -> 10 ];
```

The theoretical upper limit on the compression ratio, computed as the ratio of $4\pi^2$ over the area of the original polygon, is $\frac{50}{9}$-to-1, which is 5.55-to-1.
Packing efficiency by the parallelogram is 80%.
The compression ratio is 40-to-9 (4.44-to-1).

$$n_0 = \begin{bmatrix} \frac{-3\pi}{10} \\ \frac{2\pi}{5} \end{bmatrix} \quad L = \begin{bmatrix} 3 & 0 \\ -48 & 3 \end{bmatrix} \quad M = \begin{bmatrix} 5 & -2 \\ -80 & 40 \end{bmatrix}$$

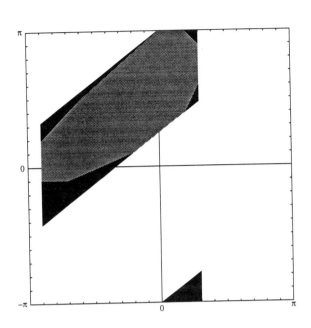

Figure 1: Automatic Design of a Rational Decimation System

ters. Figure 1 shows the automatic design of a rational decimation system from a user's sketch of the desired passband based on our implementation in the Mathematica symbolic mathematics environment. The figure shows one $2\pi \times 2\pi$ period of the frequency domain, which includes most of one circumscribing parallelogram (top) and a small piece of its replica from the period below (bottom).

The full details of the algorithms to find the minimal enclosing parallelogram are available as a technical report [STWE94]. Details of the digital signal application may be found in [ETS94].

References

[ACY85] A. Aggarwal, J. S. Chang, and C. K. Yap. Minimum area circumscribing polygons. *The Visual Computer: International Journal of Graphics*, 1:112–117, 1985.

[DA84] A. DePano and A. Aggarwal. Finding restricted k-envelopes for convex polygons. In *Proc. 22nd Allerton Conf. on Comm. Control and Computing*, pp. 81–90, 1984.

[ETS94] B. L. Evans, J. Teich, and C. Schwarz. Automated design of two-dimensional rational decimation systems. In *IEEE Asilomar Conference on Signals, Systems, and Computers*, Pacific Grove, CA, Oct. 31 - Nov. 2, 1994, pp. 498-502. Available by FTP to ptolemy.eecs.berkeley.edu and by World Wide Web at URL http://ptolemy.eecs.berkeley.edu.

[FS75] H. Freeman and R. Shapira. Determining the minimum-area encasing rectangle for an arbitrary closed curve. *Communications of the ACM*, 18:409–413, 1975.

[OAMB86] J. O'Rourke, A. Aggarwal, S. Maddila, and M. Baldwin. An optimal algorithm for finding enclosing triangles. *J. Algorithms*, 7:258–269, 1986.

[STWE94] C. Schwarz, J. Teich, E. Welzl, and B. Evans. On finding a minimal enclosing parallelogram. Tech. Rep. TR-94-036, International Computer Science Institute, 1947 Center Street, Suite 600, Berkeley, CA, Aug. 1994, available by gopher at gopher.icsi.berkeley.edu and by World Wide Web at URL http://www.icsi.berkeley.edu.

[Vai90] A. Vainshtein. Finding minimal enclosing parallelograms. *Diskretnaya Matematika* 2:72–81, 1990. In Russian.

Topologically sweeping the visibility complex of polygonal scenes

RIVIÈRE Stéphane *

1 Introduction

In order to do efficient visibility computations in the plane, one must deal with *maximal free line segments*, that is line segments of maximal length in free space. To that end, [PV93b] introduce, for scene of convex curved objects, a new data structure, the *visibility complex*, whose elements represent classes of maximal free segments having the same visibilities. In particular the visibility complex contains the visibility graph.

We have studied the visibility complex for general polygonal scenes, and have devised an algorithm for topologically sweeping the visibility complex (and so for computing the visibility graph with the same complexity) in optimal time $O(n \log n + m)$ and optimal space $O(n)$, where n is the total number of polygon vertices, and m the size of the visibility graph. Since it is a sweep algorithm, it improves the space storage of [GM87] for computing the visibility graph of polygonal scenes.

Moreover, we have implemented our algorithm and we have made experimental comparisons with two other sweep algorithms for computing visibility graphs : the algorithm of [OW88] and the algorithm of [PV93a], which both run in time $O(m \log n)$. They were tested on scenes composed of line segments. We see that the gain due to the suppression of the logarithmic cost is significant.

2 Sweeping the visibility complex

We consider a scene of obstacles in the plane. Given a point in free space and a direction, one can extend the point backward and forward along the direction, until the extremities get into an obstacle. We get a maximal free line segment (or ray), and the pair of obstacles delimiting it is its visibility. The visibility complex, is a 2-dimensional cellular complex which componants represent maximal connected sets of rays having the same visibility, such that one can pass continuously from one ray to another while keeping the visibility constant.

[PV93b] (resp. [PV95]) give an incremental (resp. sweep) algorithm for computing the visibility complex of a scene of n curved convex objects in optimal time $O(n \log n + m)$ (where m is the size of the corresponding visibility graph) and in $O(m)$ (resp $O(n)$) space. They use in both their algorithms the concept of *pseudo-triangulation*.

Here we deal with general disjoints polygons, and the elementary objects are the line segments, sides of the polygons. We use duality to study the sweep of the visibility complex. A line $y = ax + b$ in the scene is turned into a point $(a, -b)$ in dual space, and a vertex extremity (v_x, v_y) in the scene is turned into a line $y = -v_x + v_y$ in dual space, this dual line representing all the lines passing through the vertex in the scene.

Since many rays can be contained in one line, the structure of the complex is more involved than the structure of the dual arrangement of lines of the scene (that is the arrangement of dual lines of polygon vertices). For example, the complex has a non planar structure, and dual lines corresponding to polygon vertices hidden from each other in the scene, do not have an intersectioni in the complex.

To handle the sweep, we define a partial order on the vertices of the complex. The we show that a sweep of the complex compatible with this order can be achieved by a succession of discrete events :"passing" one vertex of the complex. Then we extend the concept of *horizon trees*, devised by [EG86] to sweep an arrangement of n lines in optimal time $O(n^2)$, to the complex. These trees allow to find vertices that can be passed at each step of the sweep, according to the order. Finally, using new methods in the complex for finding visibilities between vertices in the scene, we show that our algorithm runs in optimal time $O(n \log n + m)$ and space $O(n)$.

3 Experimental results

The algorithm has been implemented. Extending the notion of *rotation trees*, data structure used in [OW88] to compute the visibility graph of a scene of line segments, we have programmed the sweep in the polygonal scene. The programm also handles degenerate configurations.

Then, we have made experimental comparisons for computing visibility graphs of scenes of line segments between two algorithms and ours. The first one is the algorithm of [OW88]. The second is the more general algorithm of [PV93a] which deals with general curved objects, specially

---

*iMAGIS-IMAG - BP 53
38041 GRENOBLE CEDEX 09 - FRANCE
e-mail : Stephane.Riviere@imag.fr
iMAGIS is a joint project between CNRS/INRIA/INPG/UJF

adapted for the tests to deal with line segments only. Both algorithms make in fact a strait sweep of the visibility complex, and have a time complexity $O(m \log n)$. The strait sweep needs a priority queue that we have implemented here with a heap (for implementation details about these two algorithms, see [Riv93]).

The elementary operation of these algorithms is the comparison of the slope of two line segments, or equivalently testing if a point is above or below a line segment. The complexity measured is the number of these elementary operations, the initialization of the structures not being taken into account.

The algorithms are tested on scenes taken at random. For each line segment $[(x,y),(x+dx,y+dy)]$, the coordinates x and y (resp. dx and dy) are chosen randomly in the interval $[-a,a]$ (resp. $[-b,b]$), using a uniform distribution.

We fix the number n of line segments in the scene, and by tuning a and b we generate scenes such that the sizes of the visibility graphs range from sparse graphs ($m = 4(n-1)$) to dense graphs ($m = 2n(n-1)$). For a given n, the complexity curves are lines of equation $complexity = c(n) * m + m_0$ (see figure 1).

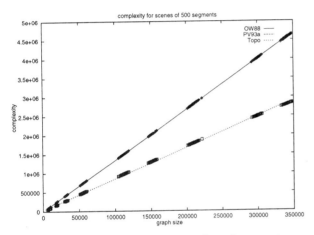

Figure 1: results for a fixed number of segment

By testing the algorithms on 500 scenes and using a linear regression, we compute their constant of complexity $c(n)$. Figure 2 shows the result : we plot the mesured $c(n)$ for $n = 100, 200, \ldots, 1000$.

The algorithms of [OW88] and of [PV93a] are very close, and their logarithmic nature is clearly shown (in fact the curves drawn for these algorithms are a "logarithmic regression"). Finally we see that our algorithm (named Topo in the figures) has better performances (its complexity is about the two thirds of the precedent above) for a same complexity of implementation.

4 Conclusion

We have devised an algorithm for topologically sweeping the visibility complex of a polygonal scene. It runs in optimal time and optimal working space. It is easy to implement : the data structures used are simple and the program is efficient in practice (the constants are small).

This algorithm allows the computation of the the visibility complex. Once built, the complex can be used to solve visibility requests. For example, one can compute a view from a point in time $O(m \log n)$ (n : size of scene, m size of the computed view) using the complex (see [PV93b]).

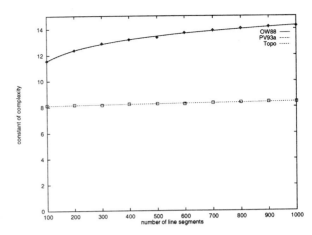

Figure 2: constant of complexity curves

Future research will study the visibility complex for dynamic scenes.

Author/Title Index

[EG86] H. Edelsbrunner and L. J. Guibas. Topologically sweeping an arrangement. In *Proc. 18th Annu. ACM Sympos. Theory Comput.*, pages 389–403, 1986.

[GM87] S. K. Ghosh and D. M. Mount. An output sensitive algorithm for computing visibility graphs. In *Proc. 28th Annu. IEEE Sympos. Found. Comput. Sci.*, pages 11–19, 1987.

[OW88] M. H. Overmars and E. Welzl. New methods for computing visibility graphs. In *Proc. 4th Annu. ACM Sympos. Comput. Geom.*, pages 164–171, 1988.

[PV93a] M. Pocchiola and G. Vegter. Sweep algorithm for visibility graphs of curved obstacles. manuscrit, 1993.

[PV93b] M. Pocchiola and G. Vegter. The visibility complex. In *Proc. 9th Annu. ACM Sympos. Comput. Geom.*, pages 328–337, 1993.

[PV95] M. Pocchiola and G. Vegter. Computing the visibility graph via pseudo-triangulation. In *Proc. 11th Annu. ACM Sympos. Comput. Geom.*, 1995.

[Riv93] S. Rivière. Experimental comparison of two algorithms for computing visibility graphs. manuscrit, 1993.

4th Annual Video Review of Computational Geometry

The following notes give brief descriptions of the video segments that make up the accompanying video proceedings. As in previous years, the videos use images to enhance our understanding of geometric ideas. In some cases this involves the illustration of algorithms in action. In others, we see applications of geometric algorithms in other disciplines. As the field matures, we see the videos serving as valuable adjuncts in each of these functions.

The six video segments were selected by a video program committee meeting in Minneapolis, Minnesota on January 17, 1995. The segments were judged based on their clarity, quality and contribution to the field. Each accepted video has been revised to reflect the comments of the committee. In addition, authors have been encouraged to include screen dumps in the accompanying text presented in this proceedings.

We thank the members of the Video Program Committee for their help in evaluating the entries. The members of the committee were Nina Amenta (Geometry Center), Marc Brown (DEC/SRC), David Dobkin (Princeton University), D.T. Lee (Northwestern), Stephen North (ATT Bell Labs) and Seth Teller (MIT). We also thank the Geometry Center for providing the facilities for our meeting. The conference and the Computer Science Department at Princeton University have provided additional support to assist in the creation of the final video.

David Dobkin
Video Review Chair

The visibility complex made visibly simple
an introduction to 2D structures of visibility

Frédo Durand   Claude Puech
iMAGIS*

Spatial coherence is a very important topic in image Synthesis, particularly in the popular ray-tracing algorithm and all its expensive intersection computations, or the radiosity light modeling which involves huge visibility calculations to determine the *form factors*. Motion planning where objects are seen as obstacles to straight line trajectories deals too with visibility problems. In order to solve them, a data structure called the *visibility graph* has been widely used. In polygonal scene it links the vertices that can see each other, in a scene of convex objects, it is made of the bitangents which don't intersect any object (see [GM91], [OW88]).

But its discrete structure does not allow taking efficiently into account the coherence between the intersections of two "very near" lines, typically like for the computation of a view from a point (visibility polygon). That's why the *visibility complex* of a set of a set of n pairwise disjoint convex objects in the plane was introduced by [PV93]. Working in a dual space that transforms lines into points allows to group the lines sharing the same visibility in faces, and the main computational structure is now the complex, instead of the scene as before with the graph.

The aim of the video is to give a better understanding of various notions involved in the definition and use of the visibility complex. It starts with basic considerations on the duality in the plane, and shows visually the correspondance between the real and the dual space. Then it introduces the dual arrangement of a set of convex objects. It gives a graphical idea of what the visibility complex is by first using lines with origins, which then helps to introduce the notion of *maximal free segments* which are used for the visibility complex. We insist on the structure of a typical face of the complex. Then a sweeping algorithm that can build the complex in $O(mlog(n))$ where n is the size of the visibility graph is shown quickly.

The video illustrates also the use of the complex to compute the view "around" a point. It consists in going through the complex along the curve defined by the set of segments originated at this point, which is a sine curve in the dual space. The object seen changes only where the segment leaves a face ; it is calculated in $O(klog(n))$ where k is the size of the view.

Time coherence is also taken into account, since when an object is moved, the modifications of the complex takes only place along its curve of tangency, and the topology of the complex is modified only when a line becomes tangent to three objects. This is shown in the video.

The complex is build using an algorithm by Michel Pocchiola and Gert Vegter [PV93] implemented by Stéphane Rivière (see [SR93]) and us. The implementation of the interactive representation of the complex was written in C++ using the Inventor library on a Silicon Graphics Indigo2 workstation. The images computed were then transfered on a Macintosh where the editing was realized under Adobe Premiere associated to a Radius VideoVision card. Some schema features were also added with Premiere, such as some arrows, lines or crosses.

Author/Title Index

[PV93] Michel Pocchiola and Gert Vegter. The visibility complex In *Proceedings of the 9th ACM Symposium on Computational Geometry*, San Diego, June 1993.

[RS93] Stéphane Rivière Comparaison d'algorithmes de calcul de graphes de visibilité *Rapport de stage du DEA IMA*

[CG89] B. Chazelle and L. Guibas. Visibility and intersection problems in plane geometry *Discrete and Computational Geometry*, 4:551-581, 1989.

[CGL85] B.M. Chazelle, L.J. Guibas and D.T. Lee. The power of geometric duality *BIT*, 25:76-90, 1985.

[GM91] S.K. Ghosh and D. Mount. An output sensitive algorithm for computing visibility graphs. *Siam J. Comput.*, 20:888-910, 1991.

[OW88] M.H. Overmars and E. Welzl. New methods for computing visibility graphs In *Proceedings of the fourth ACM Symposium on Computational Geometry*, pages 164-171, 1988.

[PO90] Michel Pocchiola. Graphics in flatland revisited. In *Proceedings of the 2nd Scandinavian Workshop on Algorithm Theory, Bergen*, pages 85-96. Springer Lecture Notes in Comp. Sc., vol. 447, July 1990.

[VE90] Gert Vegter. The visibility diagram : a data structure for visibility and motion planning problems in the plane. In *Proceedings of the 2nd Scandinavian Workshop on Algorithm Theory, Bergen*, pages 425-436. Springer Lecture Notes in Comp. Sc., vol 519, July 1990.

[VE91] Gert Vegter. Dynamically maintaining the visibility graph. In *Proceeding of the 2nd Workshop WADS'91, Ottawa, Canada*, pages 425-436. Springer Lecture Notes in Comp. Sc., vol 519, August 1991.

---

*BP 53, 38041 Grenoble Cedex 09, France Frederic.Durand@imag.fr, Claude.Puech@imag.fr iMAGIS is a joint project between CNRS/INRIA/INPG/UJF

An Animation of Euclid's Proposition 47: The Pythagorean Theorem

Steve Glassman
Greg Nelson

DEC Systems Research Center
130 Lytton Ave.
Palo Alto, CA 94301
{gnelson,steveg}@src.dec.com

One of the hardest parts of reading a mathematical proof is keeping track of all the names. We believe that animation can help alleviate this problem. The accompanying videotape shows an animation of the Pythagorean Theorem. One view contains the text of the proof, paragraph by paragraph, while a second view contains graphical reminders of the definitions of the names.

Motivation

"I plan to spend my retirement lying on the beach, drinking beer, and reading mathematics. But I don't suppose I'll be reading from ordinary books, since they will probably be obsolete by then. For better or for worse, I'll be reading from some kind of portable computerized dynabook. So I have a personal interest in what computer technology has to offer to the reader of mathematics.

Now anybody can see that computerized animations will be dynamite for presenting examples. But reading examples is just flirting with mathematics. True lovers read proofs. If examples were proofs, every odd number would be prime. For every minute that students spend studying examples, they spend ten minutes studying proofs. So the question is, can animations can help in proofs as well as examples? I think the answer might be yes.

The hard part of reading a proof is keeping track of all the names. Let N be such-and-such, let θ, ρ', and $\aleph$ be this, that, and the other. *Let, let, let,* and every *let* an imposition on the mind, an interruption to the flow of ideas. Bad writers make the problem worse by introducing unnecessary names (and for a really mind-numbing effect they use unnecessary alphabets and typefaces). But even the best writers have to introduce some names, and I think animation can help the reader keep track of them. For example, as

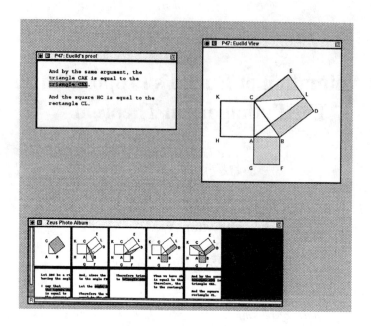

the proof unfolds, the animation could highlight the names and give graphic reminders of their definitions.

To test this theory, Steve Glassman and I animated the proof of Proposition 47 from Book I of Euclid's *Elements* [2], better known as the Pythagorean Theorem. Before showing you the result, I want to emphasize that we are not trying to use animation as a mere attention-getting device, or in a misguided attempt to appeal to your intuition instead of to your reason. This is Euclid's proof in Euclid's words, with computer animation as a mnemonic aid."

— Greg Nelson

Acknowledgments

This animation was prepared as part of the 1992 SRC Algorithm Animation Festival [1].

References

[1] Marc H. Brown, The 1992 SRC Animation Festival, In *Proc. 1993 IEEE Symposium on Visual Languages*, pages 116–123, August 1993.

[2] Euclid *The Elements, Book I*.

HIPAIR: Interactive Mechanism Analysis and Design Using Configuration Spaces

Leo Joskowicz
IBM T.J. Watson Research Center
P.O. Box 704
Yorktown Heights, NY 10598
E-mail: josko@watson.ibm.com

Elisha Sacks*
1398 Computer Science Building
Purdue University
West Lafayette, IN 47907
E-mail: eps@cs.purdue.edu

We present an interactive problem solving environment for reasoning about shape and motion in mechanism design. Reasoning about shape and motion plays a central role in mechanism design because mechanisms perform functions by transforming motions via part interactions. The input motion, the part shapes, and the part contacts determine the output motion. Designers must reason about the interplay between shape and motion at every step of the design cycle.

Reasoning about shape and motion is difficult and time consuming even for experienced designers. The designer must determine which features of which parts interact at each stage of the mechanism work cycle, must compute the effects of the interactions, must identify contact transitions, and must infer the overall behavior from this information. The designer must then infer shape modifications that eliminate design flaws, such as part interference and jamming, and that optimize performance. The difficulty in these tasks lies in the large number of potential contacts, in the complexity of the contact relations, and in the discontinuities induced by contact transitions.

Current computer-aided design programs support only a few aspects of reasoning about shape and motion. Drafting programs provide interactive environments for the design of part shapes, but do not support reasoning about motion. Simulation programs, which compute and animate the motions of the parts of mechanisms, reveal only one of many possible behaviors. Commercial simulators only handle linkages: mechanisms whose parts interact through permanent surface contacts, such as hinges and screws. Other packages handle specialized mechanisms, such as cams and gears. They cannot handle mechanisms whose parts interact intermittently or via point or curve contacts. Yet these *higher pairs* play a central role in mechanism design. Our survey of 2500 mechanisms in an engineering encyclopedia shows that 66% contain higher pairs

*Elisha Sacks is in part supported by the Hypercomputing and Design (HPCD) Project, ARPA contract DABT-63-93-C-0064. The content of the information herein does not necessarily reflect the position of the Government and official endorsement should not be inferred.

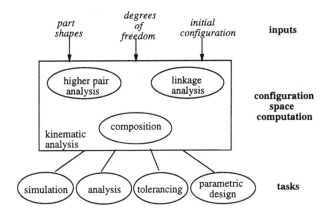

Figure 1: The HIPAIR problem solving environment.

and that 18% involve intermittent contacts.

We have developed a problem solving environment, called HIPAIR, for reasoning about shape and motion in mechanisms (Figure 1). The core of the environment is a module that automates the kinematic analysis of mechanisms composed of linkages and higher pairs. This module provides the computational engine for a range of tasks, including kinematic simulation, analysis, tolerancing, and parametric design. It is comprehensive, robust, and fast. HIPAIR handles higher pairs with two degrees of freedom, including ones with intermittent and simultaneous contacts. This class contains 90% of 2.5D pairs and 80% of all higher pairs according to our survey. It has been tested on over 100 parametric variations of 25 kinematic pairs and on dozen multipart mechanisms, including a Fuji disposable camera with ten moving parts. It analyzes pairs in under one second and mechanisms in under ten seconds.

HIPAIR computes and manipulates configuration spaces. Configuration spaces encode the relations among part shapes and part motions in a uniform geometric format. They are a concise, explicit, and complete representation of the mechanism kinematics. They simplify and systematize reasoning about shape and motion by reducing it to computational geometry.

The configuration space of a mechanism is the space of configurations (positions and orientations) of its parts. The dimension of the configuration space equals the number of degrees of freedom of the parts. For example, a gear pair has a two-dimensional configuration space because each gear has one rotational degree of freedom. The gear orientations provide a natural coordinate system. The configuration space partitions into free space where the parts do not touch and into blocked space where some parts overlap. The common boundary, called contact space, contains the configuration where some parts touch without overlap and the others are free. Only free space and contact space are physically realizable.

HIPAIR computes the configuration space of a mechanism by computing and composing the configuration spaces of all pairs of interacting parts. The correctness of this procedure follows from the compositionality of configuration space. The mechanism free space equals the intersection of the pairwise free spaces because a mechanism configuration is free when every pair of parts is free. The blocked space equals the union of the pairwise blocked spaces because a mechanism configuration is blocked when at least two parts overlap.

HIPAIR computes pairwise configuration spaces by computing and composing the contacts between all pairs of part features. The configuration space is two dimensional because the pair has two degrees of freedom. Each contact occurs along a curve in configuration space. The contact curves of the pair partition configuration space into connected components that form the free and blocked spaces. The component that contains the initial configuration is the realizable space.

HIPAIR enumerates the feature pairs, generates the contact curves, computes the planar partition with a line sweep and retrieves the realizable component. The line sweep algorithm handles all degeneracies, including vertical segments, overlapping segments and multiple segments meeting at a point. The curves come from a table with entries for all combinations of pairwise contact shapes (points, line segments, and arcs) and degrees of freedom (translation along a planar axis or rotation around an orthogonal axis).

HIPAIR composes the pairwise configuration spaces by linearizing the contact zone boundaries and intersecting them with the simplex algorithm. The result is a partition of the mechanism configuration space into free and blocked regions defined by linear inequalities in the motion coordinates. We visualize the configuration space, which can have any dimension, by projecting it onto the pairwise configuration spaces and by animating the part motions under input motions.

The simulator computes the part motions by incrementally tracing their configuration space path under the input motion. It computes the segment of the motion path that lies in the region containing the initial configuration, replaces the initial configuration with the endpoint of the segment, and repeats the process. The analysis module computes a partition of the configuration space and outputs a graph whose nodes are symbolic descriptions of free and contact regions and whose links specify region adjacencies. The parametric design module supports the design of higher pairs by inverting the mapping from parameter values to configuration spaces. It continuously updates the parameter values to track the modified configuration space using differential constraint satisfaction techniques. The kinematic tolerance analysis module computes a generalized configuration space, called kinematic tolerance space, that encodes the variation in kinematic function over the variational class of mechanisms.

The videotape explains configuration spaces and illustrates how HIPAIR supports mechanism design and analysis. The first segment shows the analysis of a cam follower pair. It introduces configuration space, manipulation, and simulation. In the second segment, the user interactively modifies the design to explore the sensitivity to part variations and to improve performance. The third segment shows how HIPAIR reveals subtle failure modes in a half-gear pair. The fourth segment illustrates kinematic simulation of two multi-part mechanisms: a door lock and a Fuji disposable camera.

The videotape consists entirely of HIPAIR output, except for the 3D animations where the snapshots were computed by HIPAIR and the solid models were displayed by the GEOMVIEW graphics package from the University of Minnesota Geometry Center. HIPAIR is written in Commonlisp with a C graphics interface. The video was produced on an SGI Iris Indigo workstation with a low-end graphics board. All the examples were run in real time.

References

1. "Computational Kinematics", L. Joskowicz and E. Sacks, *Artificial Intelligence*, Vol. 51, Nos. 1-3, North-Holland, 1991.

2. "Automated Modeling and Kinematic Simulation of Mechanisms", E. Sacks and L. Joskowicz, *Computer-Aided Design*, Vol. 25, No. 2, 1993.

3. "Configuration Space Computation for Mechanism Design", E. Sacks and L. Joskowicz, *Proc. of the IEEE Int. Conference on Robotics and Automation*, IEEE Computer Society Press, 1994.

4. "Kinematic Tolerance Analysis", L. Joskowicz, E. Sacks, V. Srinivasan, *Proc. of the 3rd ACM Symposium on Solid Modeling and Applications*, ACM Press, 1995.

5. "Computational Kinematic Analysis of Higher Pairs with Multiple Contacts", E. Sacks and L. Joskowicz, to appear, *ASME Journal of Mechanical Design*, 1995.

Incremental collision detection for polygonal models *

Madhav Ponamgi   Ming C. Lin   Dinesh Manocha

Department of Computer Science
University of North Carolina
Chapel Hill, NC 27599
{ponamgi,manocha,lin}@cs.unc.edu

1 Introduction

Fast and accurate collision detection between general polygonal models is a fundamental problem for computational geometry, robotics, and computer-simulated environments. Most earlier algorithms are either restricted to a class of models, such as convex polytopes, or are not fast enough for practical applications. We outline and demonstrate an incremental algorithm for collision detection between general polygonal models in dynamic environments. The algorithm combines hierarchical representation with incremental frame to frame computation to rapidly detect collisions. It makes use of coherence between successive instances to efficiently determine the number of objects interacting. For each pair of objects, it localizes the interference regions on their convex hulls. The features associated with these regions are represented in a precomputed hierarchy. The algorithm uses a coherence based approach to quickly traverse the precomputed hierarchy to check for collisions between the features. More details of the algorithm are given in [5].

The algorithm builds on our previous work for pairs of convex polytopes [6] and multiple moving convex polytopes [4]. For each non-convex polyhedral model, we compute its convex hull and the axis-aligned bounding box enclosing the convex hull. A four level hierarchical and incremental algorithm is used to detect collisions [5]:

1. We determine which pairs of bounding boxes are colliding, using the sweep and prune algorithm described in [3, 4].

2. For each intersecting bounding box pair, we determine if the objects' convex hulls are colliding by walking across the external and pseudo-internal Voronoi regions of the polytopes. When one polytope's feature is within another polytopes pseudo-internal Voronoi regions, interpenetration is detected.

3. For each colliding hull pair, we determine the areas of intersection on the convex hulls. There are several algorithms in the literature for intersection computation [2, 7]. We extend these algorithms using a coherence based scheme.

4. For each intersection region, we examine the vertex, edge, and face features associated with it to determine precise contact points. These features are represented as a bounding box hierarchy, and we apply a hierarchical version of the sweep and prune algorithm to determine collisions.

After the second stage of the algorithm, it is known that for a pair of non-convex objects that their convex hulls are penetrating. We need to identify the regions of contact so that we can examine the interior of the convex hulls for collisions between features of the original polyhedron not lying on its convex hulls. These internal features are organized into pre-processed hierarchies.

We first describe the pre-processing data structures used during the dynamic traversal phase of the algorithm.

2 Pre-Processing

Each non-convex object is described as a set of faces. Each face consists of a list of vertices describing the edges of the face. The faces, edges, and vertices comprising an object are called its *features*. Our first step in pre-processing is to categorize these features. Taking the convex hull of the object separates its features into three distinct regions. The features of the object that are coincident with the faces of the convex hull are identified as the *hull features* of the object. The features that do not lie on the convex hull are identified as the *concavity features*. The vertices and edges bordering the convex and concavity patches are the *boundary features*. These are computed using convex hull algorithms and feature traversal using depth-first search algorithms.

2.1 Computing Caps

We want to compute a "cap" over the concavity features using its boundary features. These caps "hide" the features of the concavity. Multiple caps need to be computed because an object may be composed of disjoint concavity feature subsets.

The caps are the "missing" regions of the object's convex hull. Caps can be computed by subtracting the features of the object coincident with the convex hull from its convex hull. In general, the caps consist of non-convex polygons.

2.2 Constructing the Bounding Box Hierarchy

We construct an octree like hierarchy for the features below the caps. Initially, the vertices of each concavity

---

*Supported in part by a Sloan foundation fellowship, University research council grant, NSF grant CCR-9319957, ONR contract N00014-94-1-0738, ARPA contract DABT63-93-C-0048, NSF/ARPA Science and Technology Center for Computer Graphics & Scientific Visualization and NSF Prime contract No. 8920219

Permission to copy without fee all or part of this material is granted provided that the copies are not made or distributed for direct commercial advantage, the ACM copyright notice and the title of the publication and its date appear, and notice is given that copying is by permission of the Association of Computing Machinery. To copy otherwise, or to republish, requires a fee and/or specific permission.
11th Computational Geometry, Vancouver, B.C. Canada
© 1995 ACM 0-89791-724-3/95/0006...$3.50

are surrounded by an axis-aligned bounding box. This bounding box is split into smaller bounding boxes with cutting planes along the x, y, and z axes. For example, splitting the box in half along the x, y, and z axes results in eight children for the top-level box. The subdivision of child boxes continues recursively until each child box contains a threshold number of vertices. Eventually the faces associated with the vertices are introduced.

3 Determining Regions of Contact between Polytopes

When the convex hulls of two polytopes are determined to be colliding using the Voronoi region based penetration detection algorithm, we enter the third stage of the algorithm. The collision between the objects' the convex hulls has three cases:

1. The collision is between convex feature subsets on both the objects. This is a "real" collision. The application receives the contact status so it can plan an appropriate collision response.

2. The collision is between a cap of one object and a convex feature subset of the other object. We want to determine if the convex feature subset collides with any of the faces underneath the cap belonging to the concavity.

3. The collision is between two caps. We want to determine if any faces underneath both caps collide.

When two convex hulls A and B interpenetrate, then at least one feature of $a \in A$ is contained within the pseudo-internal Voronoi region of B and vice-versa for feature $b \in B$. These two features can be used as the starting basis to construct the complete interpenetrating volume with a depth-first-search technique. Each adjacent feature of a is tested against the pseudo-Voronoi cell containing a and eventually the pseudo-Voronoi cell containing it is determined. This is repeated recursively for adjacent features of a and b. After the initial cost of constructing this interpenetration volume, the update time for each feature, as each of these features move slightly from frame to frame, is expected to be constant due to coherence (for each feature). Moreover, only the penetrating features on both objects are considered so the overall cost is linear in the number of features penetrating. The total expected running time using the internal penetration walk is $O(n)$.

4 Hierarchical Sweep and Prune

Each object is enclosed by an axis-aligned bounding box. The boxes are sorted using a dimension reduction technique. They are projected onto the x, y, and z axes forming three lists of intervals. The intervals are sorted and two bounding boxes overlap whenever their intervals in all three dimensions overlap. Due to coherence, the sorted lists change little from frame to frame so insertion sort is used. Bounding box overlaps trigger the subsequent step of the algorithm.

The hierarchical version of this algorithm sorts the first level of the bounding boxes enclosing the intersecting regions. Then it proceeds to see which of the children at this level are overlapping. The overlapping children are sorted to check for overlaps in a similar way recursively. When both hierarchies are at the leaf level, the faces of

Figure 1: Chain interlocked tori.

the objects are introduced. At this last level, the exact contact points are computed by finding the intersection of the face polygons; these are passed to the application to plan an appropriate response.

The level by level sorting of the hierarchies' bounding boxes is tricky to manage both in terms of efficiency and memory usage for objects composed of thousands of polygons. Multiple lists have to be sorted, and data needs to be managed efficiently. We use a bounding box management scheme that allocates the space for the data once, but can have multiple pointers to it. So instead of allocating this data value at multiple levels in the hierarchy, it efficiently swaps pointers during the sort.

5 Implementation and Performance

These algorithms have been implemented and applied to two non-trivial simulations: screw-threaded insertion and a ring of torii [5]. In the video we demonstrate the partitioning of the features of a polyhedron into hull features, concavity features and boundary features for a polygonal approximation of a torus. These sets of features are used for an interactive simulation of a ring of tori with collision detection and a simple collision response (as shown in Figure 1).

References

[1] B. Chazelle and D. Dobkin. "Intersection of Convex Objects in Two and Three Dimensions" Journal of the ACM (January, 1987) pp.1-27.

[2] B. Chazelle. "An Optimal Algorithm for Intersecting Three-Dimensional Convex Polyhedra" IEEE Symposium on Found. of Comp. Sci 30 (1989) pp.586-591.

[3] J. Cohen, M. Lin, D. Manocha, and M. Ponamgi. "Efficient Collision Detection for Multi-Body Environments" ACM Computational Geometry Video Review, 1994.

[4] J. Cohen, M. Lin, D. Manocha, and M. Ponamgi. "Exact Collision Detection for Multi-Body Environments" Interactive 3-D Symposium April 1995, Monterey, CA.

[5] M. Ponamgi, D. Manocha, and M. Lin "Incremental Algorithms for Collision Detection between Solid Models" ACM Solid Modeling 1995, Salt Lake City, Utah.

[6] M. Lin and J. Canny. "Efficient algorithms for incremental distance computation" IEEE Conf. on Robotics and Automation, Sacramento, CA, 1991.

[7] D.Muller and F.Preparata. Finding the intersection of two convex polyhedra. *Theoretical Computer Science*, 8(2):217-236, 1978.

Convex Surface Decomposition*

BERNARD CHAZELLE[†]   DAVID P. DOBKIN[‡]   NADIA SHOURABOURA[§]   AYELLET TAL[¶]

This video, which is based on [1], illustrates heuristics for decomposing the boundary of a polyhedron into convex-shaped pieces. The problem is NP-complete, so heuristics are the best one can (probably) expect. The problem is motivated by applications in manufacturing, computer graphics, multimedia systems, etc.

Finding meaningful test data was an important component of this work. We collected several hundred real-life polyhedral models from manufacturing companies and industrial AutoCAD users. The first part of the video displays a sample of these objects (e.g., Figure 1).

Figure 1: Glass

The heuristics investigated in this video rely on a flooding strategy, in which facets are collected along the way as long as they form a convex patch. The second part of the video shows an animation of a flooding heuristic based on breadth-first search. It involves decomposing the boundary in a greedy manner, restarting a new patch only when necessary. The examples shown in the video demonstrate cases in which the decomposition which is found in this fashion is optimal.

The next segment exhibits a case where the previous heuristic fails miserably. The decomposition found is off the optimal decomposition by a factor proportional to the input size (Figure 2). Cases such as this one happen in the presence of global failures. Typically this occurs with twisted shapes, e.g., spirals, drills.

Figure 2: BFS on a Bracelet

In the final segment of the video, the "flood-and-retract" heuristic is animated: this is a method specially designed to overcome the "obvious" pitfalls of more naive heuristics. The method works in two steps. First, the surface is flooded by a patch cover: this means that when restarting a new patch, we allow the traversal of old facets as well as new ones. Then, in a second pass, we transform the cover into partition.

One aim of this video was to illustrate the use of an algorithm animation system, in this case GASP [2], [3]. An innovative aspect of this work has been the use of animations to guide our search for good heuristics. We found that nothing was more useful than the ability to look at a decomposition in three dimensions as though we held it in our hands.

The video was prepared in the department of computer science at Princeton University. The programs were written in C to run under UNIX on a Silicon Graphics Iris. Recording was done at the Interactive Computer Graphics Lab at Princeton and editing was done with the assistance of the Princeton Department of Media Services.

Acknowledgments: We thank Kirk Alexander, Mike Mills and Kevin Perry for their help in producing the final video.

---

*Work by Bernard Chazelle, David Dobkin and Ayellet Tal has been supported in part by NSF Grant CCR-93-01254 and The Geometry Center, University of Minnesota, an STC funded by NSF, DOE, and Minnesota Technology, Inc. Work by Nadia Shouraboura has been supported in part by NSF Grant PHY-90-21984.

[†]Department of Computer Science, Princeton University
[‡]Department of Computer Science, Princeton University
[§]Program in Applied and Comput. Math., Princeton University
[¶]Department of Computer Science, Princeton University

References

[1] B. Chazelle, D. Dobkin, N. Shouraboura, A. Tal. *Strategies for Polyhedral Surface Decomposition: An Experimental Study*, Eleventh Annual ACM Symposium on Computational Geometry, June 1995.

[2] D. Dobkin, A. Tal. *GASP - An Animation System for Computational Geometry*, Sixth Canadian Conference on Computational Geometry, August 1994.

[3] A. Tal, D. Dobkin. *GASP - A System for Visualizing Geometric Algorithms*, Visualization '94, October 1994.

3D Modeling using the Delaunay Triangulation

Bernhard Geiger[1]

March 13, 1995

A major problem in medical imaging is the three dimensional reconstruction of human organs from parallel cross sections. Besides visualization, such 3D models are needed for radiation therapy planning, surgical planning and simulation, rapid prototyping and volumetric measurements.

We proposed a solution to the reconstruction problem, which is based on the Delaunay triangulation [1].

The video displays the different steps of this algorithm. As a demonstration object, we use the human pelvis. The pelvis is an ideal structure to demonstrate both the usefulness of shape reconstruction and the algorithmic concept of our method. Firstly, it is a nontrivial exercise to imagine the pelvis from its cross-sections. Even with a radiological training, it is hard to see the shape of the oblique pelvic inlet. Secondly, it is important that the cross-sections show a variety of topologies, like branchings and holes.

The pelvic data is a series of 23 axial MRIs (magnetic resonance images). The bone contours were outlined manually since the image characteristic of MRI is not sufficient to employ automatic segmentation methods.

In each cross-section, we get one or several, possibly nested closed polygons. We calculate the 2D Delaunay triangulation of the polygon vertices. There are several situations where Steiner points are added automatically onto the polygons: when segments are not contained (conforming Delaunay triangulation), when there are obtuse angles opposite of polygon segments (to guarantee that the internal Voronoi skeleton stays inside the polygons). We also add Steiner points inside polygons at the orthogonal projection of external Voronoi vertices of adjacent sections (which improves complex branchings, see [1]).

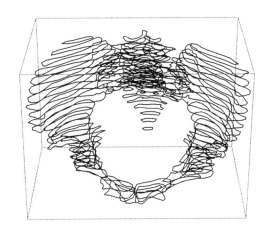

Fig. 1. The contours of the pelvic bones

The space between adjacent cross-sections is then filled with tetrahedra. Each Delaunay triangle of the lower cross-section is connected to the vertex in the upper cross-section that lies closest to its circumcenter. Similarly, each triangle of the upper cross-section is connected to a vertex of the lower cross-section. The missing gaps are filled with tetrahedra having one edge in each cross-section. An edge of the lower cross-section is connected to a edge of the upper cross-section iff their dual Voronoi edges intersect when projecting them orthogonally onto the same plane. It can be shown that this procedure results in the 3D Delaunay triangulation of the vertices in two adjacent planes.

In the next step, we have to eliminate superfluous tetrahedra, mainly those having an edge outside a contour due to concavities or holes. The

---

[1] INRIA BP 93 – 06902 Sophia Antipolis Cedex (France)
E-mail : geiger@sophia.inria.fr

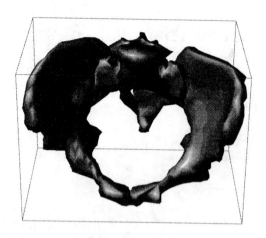

Fig. 2. The reconstruction of the pelvis

surface—the subset of tetrahedral facets without neighbor—is displayed with Gouraud shading.

OTHER RESULTS SHOWN ON THE VIDEO TAPE

The skull is obtained from the Chapel Hill Volume Rendering Test Data Set, about 120 CT slices of a cadaver. The bone was segmented automatically by thresholding.

The brain data, from the Surgical Planning Laboratory at Brigham and Womans Hospital in Boston is automatically segmented. The video shows the outer surface, and the ventricles.

The small bronchi of the lung were manually segmented. We use that data set for the simulation of lung biopsy.

The last example is a simulation of knee kinematics. The model and the description of the movement were obtained from 13 MR acquisitions at a different degree of flexion.

TECHNICAL DETAILS

For the visualization, we used home made software on a SGI Indigo Extreme (75 MHZ IP26 Processor). The reconstruction program called Nuages is available via anonymous ftp on betelgeuse.inria.fr.

ACKNOWLEDGEMENT

Special thanks to: Frank Muennemann, Resonex Inc. (pelvis), UNC Chapel Hill (skull), Olivier Dourthe (trachea), Ron Kikinis, Brigham and womens hospital (segmented brain), Stephane Boisgard, CHU Clermont Ferrand (knee), Sue Whitesides (narrator).

REFERENCES

[1] J-D. Boissonnat and B. Geiger. Three dimensional reconstruction of complex shapes based on the Delaunay triangulation. In R. S. Acharya and D. B. Goldgof, editors, *Biomedical Image Processing and Biomedical Visualization*, pages 964–975, San Jose CA, February 1993. SPIE. vol. 1905, part 2.

Permission to copy without fee all or part of this material is granted provided that the copies are not made or distributed for direct commercial advantage, the ACM copyright notice and the title of the publication and its date appear, and notice is given that copying is by permission of the Association of Computing Machinery. To copy otherwise, or to republish, requires a fee and/or specific permission.
11th Computational Geometry, Vancouver, B.C. Canada
© 1995 ACM 0-89791-724-3/95/0006...$3.50

Index

Abellanas, M. 71
Agarwal, P. 39, 182, 267, 326
Aichholzer, O. 220
Alt, H. 89
Amenta, A. C12
Andrews, D. C24
Aronov, B. 267, 316
Arya, S. 172, 336
Aurenhammer, F. 220
Avis, D. 20
Avnaim, F. C16
Balaban, I. 211
Benzley, S. C4
Bespamyatnikh, S. 152
Blacker, T. C4
Boissonnat, J. 79, C16, C20
Bremner, D. 20
Burnikel, C. C18
Chan, T. 10
Chazelle, B. 297, V9
Chen, D. 370
Choi, J. 350, 380
Dasgupta, B. 162
Davis, A. 316
De Berg, M. C26
Devilliers, O. C16
Dey, T. 316
Dickerson, M. 238
Di Battista, G. 306
Dobkin, D. 297, V9
Dobrindt, K. C20, C26
Drysdale, R. 61
Durand, F. V2
Dyer, M. 345
Edelsbrunner, H. 98
Efrat, A. 39
Erickson, J. 127
Evans, B. C34
Finke, U. 119
Fournier, A. C2
Garg, A. 306
Geiger, B. C20, V11
Glassman, S. V3
Goldberg, K. C22
Goldwasser, M. C32
Gritzmann, P. 1
Guibas, L. 190
Gupta, P. 162
Hernandez, G. 71

Hinrichs, K. 119
Hufnagel, A. 1
Icking, Chr. 258
Janardan, R. 162
Joskowith, L. V5
Klein, R. 71, 258
Klenk, K. 370
Könnemann, J. C18
Kriegman, D. C22
Krishnan, S. C8
Levy, S. C12
Lewis, R. C14
Lin, M. V7
Liotta, G. 306
Lovász, L. 147
Manocha, D. C8, V7
Marimont, D. 190
Mata, C. 360
Matoušek, J. 138
McElfresh, S. 238
Mehlhorn, K. C18
Michel, H. C20
Mitchell, J. 360
Mitchell, S. C4
Montague, M. 238
Mount, D. 172, 336
Munzner, T. C12
Murdoch, P. C4
Näher, S. C18
Narayan, O. 336
Narkhede, A. C8
Nelson, G.
Neumann-Lara, V. 71
Olson, A. C6
Overmars, M. 29, C30
Pach, J. 147
Pal, S. 316
Pellegrini, M. 287
Phillips, M. C12
Pocchiola, M. 248
Ponamgi, M. V7
Prasad, D. 316
Preparata, F. C16
Puech, C. V2
Rao, A. 29, C22
Ranjan, V. C2
Rege, A. 277
Rivière, S. C36
Rote, G. 220
Sacks, E. V5
Sanner, M. C6

Schirra, S. C18
Schömer, E. 51
Schwarz, C. C34
Schwarzkopf, O. 29, 89, 182, C10
Sellen, J. 350
Sharir, M. 39, 79, 182, 326
Shouraboura, N. 297, V9
Smid, M. 162
Snoeyink, J. C24
Spehner, J. C6
Su, P. 61
Suri, S. 267
Szegedy, M. 147
Tagansky, B. 79, 200
Tal, A. 297, V9
Tamaki, H. 230
Tamassia, R. 306
Taschwer, M. 220
Tassinari, E. 306
Tautges, T. C4
Teich, J. C34
Thiel, C 51
Tokuyama, T. 230
Tu, H. 370
Uhrig, C. C18
Urrutia, J. 71
Vainshtein, A. C34
Van de Kraats, B. C30
Van Kreveld, M. C30
Vargiu, F. 306
Vegter, G. 248
Wagner, D. C28
Wagner, F. 109
Waupotitsch, R. 98
Weihe, K. C28
Welzl, E. C34
Wentink, Ch. 29
Wolff, A. 109
Yap, C. 350, 380
Yvinec, M. 79, C16